Complexity Dichotomies for Counting Problems

Volume 1: Boolean Domain

Complexity theory aims to understand and classify computational problems, especially decision problems, according to their inherent complexity. This book uses new techniques to expand the theory for use with counting problems. The authors present dichotomy classifications for broad classes of counting problems in the realm of P and NP. Classifications are proved for partition functions of spin systems, graph homomorphisms, constraint satisfaction problems, and Holant problems. The book assumes minimal prior knowledge of computational complexity theory, developing proof techniques as needed and gradually increasing the generality and abstraction of the theory.

This volume presents the theory on the Boolean domain, and includes a thorough presentation of holographic algorithms, culminating in classifications of computational problems studied in exactly solvable models from statistical mechanics.

JIN-YI CAI is Professor of Computer Science and the Steenbock Professor of Mathematical Sciences at the University of Wisconsin–Madison. He was born in Shanghai, China. He studied at Fudan University (class of 77) and at Cornell University, receiving his Ph.D. in 1986. He held faculty positions at Yale University (1986–1989), Princeton University (1989–1993), and SUNY Buffalo (1993–2000), where he rose from Assistant Professor to Full Professor in 1996. He received a Presidential Young Investigator Award (1990), an Alfred P. Sloan Fellowship (1994), and a John Simon Guggenheim Fellowship (1998). He is a Fellow of Association for Computing Machinery (ACM) and American Association for the Advancement of Science (AAAS).

XI CHEN is Associate Professor of Computer Science at Columbia University. He was born in Suzhou, China. He studied at Tsinghua University and received his Ph.D. in 2007. His research focuses on complexity theory and algorithmic game theory. He is the recipient of a National Science Foundation (NSF) CAREER Award, an Alfred P. Sloan Fellowship (2012), and a European Association of Theoretical Computer Science (EATCS) Presburger Award (2015).

Complexity Dichotomies for Counting Problems

Volume 1: Boolean Domain

JIN-YI CAI
University of Wisconsin, Madison

XI CHEN
Columbia University, New York, New York

CAMBRIDGE
UNIVERSITY PRESS

CAMBRIDGE
UNIVERSITY PRESS

University Printing House, Cambridge CB2 8BS, United Kingdom

One Liberty Plaza, 20th Floor, New York, NY 10006, USA

477 Williamstown Road, Port Melbourne, VIC 3207, Australia

314–321, 3rd Floor, Plot 3, Splendor Forum, Jasola District Centre, New Delhi – 110025, India

79 Anson Road, #06-04/06, Singapore 079906

Cambridge University Press is part of the University of Cambridge.

It furthers the University's mission by disseminating knowledge in the pursuit of education, learning, and research at the highest international levels of excellence.

www.cambridge.org
Information on this title: www.cambridge.org/9781107062375
DOI: 10.1017/9781107477063

First published 2017

Printed in the United States of American by Sheridan Books, Inc.

A catalogue record for this publication is available from the British Library.

Library of Congress Cataloging-in-Publication Data

Names: Cai, Jin-Yi, 1961– | Chen, Xi (Computer science professor)
Title: Complexity dichotomies for counting problems / Jin-Yi Cai, University of Wisconsin, Madison, Xi Chen, Columbia University, New York, New York.
Description: Cambridge : Cambridge University Press, 2017– | Includes bibliographical references and index. Contents: volume 1. Boolean domain
Identifiers: LCCN 2017020174 | ISBN 9781107062375 (hardback : alk. paper)
Subjects: LCSH: Computational complexity. | Combinatorial enumeration problems. | Homomorphisms (Mathematics) | Algebra, Boolean.
Classification: LCC QA267 .C2187 2017 | DDC 511.3/52–dc23 LC record available at https://lccn.loc.gov/2017020174

ISBN 978-1-107-06237-5 Hardback

Contents

Preface

The raison d'être of computational complexity theory is to understand and classify computational problems according to their inherent complexity. In the past fifteen years or so, there have been enormous advances in the study of the complexity of counting problems. A substantial theory has been developed that classifies every problem in a class of Sum-of-Product computations as being either polynomial-time computable or #P-hard. Such theorems are called complexity dichotomies.

This book presents the dichotomy theory for counting problems, expressible as Sum-of-Product computations. More specifically, it presents dichotomy theorems for partition functions of Spin Systems, Graph Homomorphisms, Constraint Satisfaction Problems, and Holant Problems. The theorems have the consequence that, assuming the complexity theory hypothesis P $\neq$ #P, which is implied by P $\neq$ NP, in broad classes of counting problems, up to polynomial-time equivalence, there are only exactly two distinct levels for exact counting complexity.

Volume I presents the theory on the Boolean domain; Volume II presents the theory on general domains. In Volume I a particular feature is a thorough presentation of holographic algorithms, culminating in classifications of computational problems studied in Exactly Solvable Models from statistical mechanics.

Much of the material in the book is presented in book form for the first time. The book assumes minimal prior knowledge of computational complexity theory. The proof techniques are developed as needed. The dichotomy theorems are carefully organized, so that the theory is presented in gradually increasing generality and abstraction.

We want to thank many colleagues who have read parts of the manuscript and shared their thoughts on the subject. Their comments and criticisms have been a source of inspiration, and we wish to express our gratitude for their help. They are Manuel Blum, Andrei Bulatov, Sitan Chen, Martin Dyer, Alan Frieze,

Zhiguo Fu, Leslie Ann Goldberg, Aaron Gorenstein, Heng Guo, Juris Hartmanis, Pavol Hell, Miki Hermann, Steve Homer, Mark Jerrum, Ravi Kannan, Dick Karp, Michael Kowalczyk, JM Landsberg, Liang Li, Dick Lipton, Martin Loebl, Pinyan Lu, Janos Makowsky, Jo Ellis-Monaghan, Jason Morton, Jaroslav Nešetřil, Mike Paterson, David Richerby, Les Valiant, Osamu Watanabe, Tyson Williams, Mingji Xia, Yitong Yin, and Standa Zivny. Two colleagues, Dr. Michael Kowalczyk and Dr. Tyson Williams, deserve our special gratitude. They helped us with early drafts of Chapter 5 and Chapter 8 respectively, which we made use of with their permission. They ought to be coauthors of these chapters. We also thank Mr. Artem Govorov who found a subtle gap in the proof of a lemma in an early draft, which was corrected. Of course, any remaining error is the sole responsibility of the authors.

Jin-Yi Cai
Xi Chen

Madison, Wisconsin
New York, New York

1
Counting Problems

This book is a study of the computational complexity of counting problems, especially those that can be expressed as a sum-of-product computation. Our aim is to give a systematic presentation to a body of work, mostly from the past two decades, that gives some comprehensive classifications to these sum-of-product computations from the perspective of computational complexity. All these sum-of-product problems are computable within the level of the complexity class #P. The classification theorems are stated in the following form, known as a dichotomy: For a class of problems expressible within a framework, every problem in the class is either computable in polynomial time, or it is #P-hard to compute, i.e., it is as hard as any other problem in the class #P.

1.1 Counting Problems and Models of Computation

We assume the readers have some basic and preliminary knowledge about computational complexity such as P and NP, for example, at a level provided by an undergraduate course on the subject of the Theory of Computing. This knowledge is not crucial, however, as we will not use many general results but rather present the framework and develop the necessary proof techniques gradually. Any reader wishing to be more thoroughly acquainted with the full scope of computational complexity theory can consult a standard textbook, e.g., [Sip96, Pap94, AB09]. On the other hand, a reader with no prior exposure to complexity theory but with a strong mathematical background, for example, at a level provided by a solid undergraduate mathematics education, should be able to follow all proofs in the book, provided he or she is willing to accept a handful of results without proof. These results can be found elsewhere, and the particular proofs of these do not impact an understanding of the material in this book. Such a reader, however, would benefit from a wider exposure to complexity theory to gain some insight as to why certain problems are investigated and what questions are asked.

Briefly, in computational complexity theory a *problem* consists of an infinite set of problem instances, rather than one particular instance. For example, the problem DETERMINANT is to compute the determinant of an arbitrarily given matrix, not one particular matrix. Solving a problem is to say that there is an algorithm that solves all problem instances. The formal notion of an algorithm is a Turing machine, an idealized computer that carries out step-by-step operations according to a finitary program, valid for all problem instances. The efficiency of the algorithm is measured in terms of the maximum number of steps $T(n)$ the Turing machine may take for all problem instances of size n. For example, for integer matrices, the problem instance of DETERMINANT is an integer square matrix in which the size is the sum of the bit length of all the matrix entries. Formally the complexity classes P and NP are defined for decision problems, those problems for which the answer to each instance is either Yes or No. For example, the problem VERTEX COVER is the following decision problem: Given a problem instance consisting of a graph $G = (V, E)$ and an integer parameter k, whether there is a subset of vertices $S \subseteq V$ such that $|S| \leq k$ and every edge in E is incident to at least one vertex from S. A decision problem Π is in the complexity class P if there is an algorithm solving the problem, i.e., gives a correct Yes or No answer to every problem instance, such that $T(n)$ is bounded above by a fixed degree polynomial in n. The class NP consists of all decision problems Σ for which there exists some problem Π in P, and some polynomial $p(\cdot)$, such that for all problem instances x of Σ, x is a Yes instance of Σ iff there exists some y with size $|y| \leq p(|x|)$ and the pair $\langle x, y\rangle$ is a Yes instance of Π. Intuitively, y is a certificate of size polynomially bounded in x, such that it can be verified in polynomial time that y certifies that x is a Yes instance of Σ.

There are historical as well as internal logical reasons, mainly from computability theory, why the definitions of P and NP are formulated in terms of decision problems. Even though a problem may be more naturally stated as a search problem or a numerical problem, it can often be restated as a decision problem such that solving the decision problem is equivalent to solving the search or numerical problem within a polynomial factor in efficiency. For example, if there is an algorithm that solves the decision problem VERTEX COVER, then by binary search on the value k one can find the minimum size k_0 of any vertex cover of G. Furthermore, by considering a sequence of at most $O(k_0 n)$ graphs obtained by removing some vertices and their incident edges from G of n vertices, and repeatedly invoking the decision algorithm, one can compute a minimum size vertex cover of G. For the DETERMINANT problem one can formulate a decision problem as whether the determinant of an integer matrix is greater than a threshold value. It is easy to compute an a priori bound for a given determinant; a binary search will then find the exact determinant value in polynomial time from the decision algorithm. Such reductions show that there is a theoretical equivalence between a decision problem and its search or numerical version; however, in reality, a polynomial-time algorithm usually directly computes the output value, without going through the decision version.

In this book we mostly study counting problems. Typically these include counting the number of specific combinatorial configurations, such as vertex covers, matchings, or proper colorings in a graph, or the evaluation of partition functions in statistical physics. For such problems, it is not expected that they are equivalent to their decision problem counterparts in polynomial time [Tod91]. Instead we define #P, a complexity class of functions for counting problems.

For every NP problem Σ one can formulate a corresponding counting problem using the problem Π in P that defines Σ in terms of certificates. For any problem instance x of Σ, we define the function $f(x)$ as the number of certificates y such that the pair $\langle x, y\rangle$ is a Yes instance of Π. The class of all such functions is denoted as #P. This complexity class, defined by Valiant [Val79a, Val79b] in the study of the complexity of the permanent function, will play a central role in this book. Natural counting problems corresponding to decision problems at the level of NP that have a nonnegative integer solution can all be formulated as a problem in #P. To include more problems at this level we also consider problems whose solutions are not necessarily nonnegative integers, such as those from statistical physics. To capture their computational complexity, we consider the closure of #P under polynomial-time Turing reductions, namely, those problems solvable by a polynomial-time algorithm given free access to a hypothetical algorithm to some problem in #P, where each query costs only the time it takes to write the query. This class is formally denoted as $\mathrm{FP}^{\#\mathrm{P}}$, where FP denotes the class of functions computable by a polynomial time Turing machine. Typical problems in #P include counting the number of satisfying assignments to a Boolean formula or the number of vertex covers in a graph. Weighted versions of these problems as well as sum-of-product computations such as partition functions from statistical physics can be easily formulated as problems that are polynomial-time reducible to #P.

The formal Turing machine model is naturally suited to the study of computation over discrete structures such as integers or graphs. However, in this book it is more natural to consider computation over the real or complex numbers. Doing so causes a technical issue of how one may represent exactly the individual real or complex numbers, and how to account for the complexity of various operations on these numbers. This is a question relating to the model of computation. One can, for example, take the computational model on real numbers by Blum–Shub–Smale [BSS89, BCSS98]. This is more intuitive; however, in a strict logical sense this is not equivalent to the classical Turing machine model and the results we obtain would not have the same logical meaning. So, formally we still consider the classical Turing machine model, and restrict the objects of computation to algebraic (complex) numbers. Thus, technically, every number α is specified by a finite irreducible polynomial $f(X) \in \mathbb{Z}[X]$ with integer coefficients, and a small disk of rational radius containing a unique root α of $f(X)$. We discuss models of computation further in Section 1.5. But this is not a central issue for the type of sum-of-product computations we treat in this book; basically the theory can be developed in any reasonable model of computation in which sum and product can be efficiently computed.

Some basic notations. We denote by $\mathbb{N}$ the set of natural numbers $\{0, 1, 2, \ldots\}$, by $\mathbb{Z}$ the set of all integers $\{\ldots, -2, -1, 0, 1, 2, \ldots\}$, and by $\mathbb{Z}_+$ the set of positive integers $\{1, 2, 3, \ldots\}$. We denote by $\mathbb{Q}$, $\mathbb{R}$, and $\mathbb{C}$ the sets of rational (algebraic) real, and complex numbers respectively. We denote $\mathrm{i} = \sqrt{-1}$. All graphs are finite and undirected, unless stated otherwise. Graphs may or may not have loops and parallel edges, which should be clear from the context. If α and β are finite bit strings, then α_i denotes its ith bit, and $\alpha \oplus \beta$ denotes its bitwise XOR string. For a binary string α, its Hamming weight is denoted by $\mathrm{wt}(\alpha)$, i.e., the number of 1's in α. We will generally treat a column vector $v \in \mathbb{C}^n$ and the row vector v^{T}, its transpose, interchangeably. For a matrix $M \in \mathbb{C}^{m \times n}$ and a $v \in \mathbb{C}^n$, we write Mv as its matrix vector product. Sometimes we may write Mv even though we explicitly listed the elements of v as a row vector. The formal meaning is just Mv^{T}, if v is a row vector. Similar comments apply for vM. If $A = (a_{ij}) \in \mathbb{C}^{m \times n}$ and $B = (b_{st}) \in \mathbb{C}^{k \times \ell}$, we denote $A \otimes B$ as the tensor product matrix in $\mathbb{C}^{mk \times n\ell}$, indexed by the pairs $(i, s) \in [m] \times [k]$ for rows and $(j, t) \in [n] \times [\ell]$ for columns, both in lexicographic order.

We use $\leq_{\mathrm{T}}^{\mathrm{P}}$ or simply $\leq_{\mathrm{T}}$ to denote polynomial time Turing reducibility, and $\equiv_{\mathrm{T}}^{\mathrm{P}}$ or simply $\equiv_{\mathrm{T}}$ to denote polynomial time Turing equivalence.

1.2 Three Classes of Counting Problems

We now formally define three frameworks for the types of sum-of-product computations in this book. Fix a finite domain $[q] = \{1, 2, \ldots, q\}$, for a positive integer q. If $q = 2$ it is called the Boolean domain. We consider any set of functions $\mathcal{F}$ on domain $[q]$, where each $f \in \mathcal{F}$ maps from $[q]^k$ to a commutative semiring $\mathbf{R}$, for some $k \geq 0$, called the arity of f. If $k = 0$, then f is a constant. If $k = 1$, 2, or 3, then f is called a unary, binary, or ternary function respectively. In this book, by default we will take the semiring **R** to be the (algebraic) complex numbers, but still denote it as $\mathbb{C}$. Functions in $\mathcal{F}$ are also called *signatures* or local constraint functions. A signature f is symmetric if its value is invariant under permutation of its variables. For the Boolean domain this means that the value of f depends only on the Hamming weight of its input.

1.2.1 Spin Systems or Graph Homomorphism Problems

A spin system on a graph $G = (V, E)$ is the following model and it comes from statistical physics. Let $[q] = \{1, \ldots, q\}$ be a finite domain, where $q \geq 1$ is an integer. We consider all vertex assignments $\sigma : V \to [q]$. There is an edge function $f : [q]^2 \to \mathbb{C}$. For each assignment σ we have an evaluation $\prod_{(u,v)\in E} f(\sigma(u), \sigma(v))$, a product over every edge $(u, v) \in E$. Then we define the *partition function*

$$Z_f(G) = \sum_{\sigma:V\to[q]} \prod_{(u,v)\in E} f(\sigma(u), \sigma(v)). \tag{1.1}$$

In physics, the partition function represents the aggregate of thermodynamic variables or a normalizing factor of a system, as one sums over all possible configurations of the particles.

The value $f(\sigma(u), \sigma(v))$ is the local contribution, the product $\prod_{(u,v)\in E} f(\sigma(u), \sigma(v))$ is the weight of the assignment σ, and the partition function is the sum of weights over all assignments. If G is an undirected graph, as is typically the case, we require f to be a symmetric function, $f(i, j) = f(j, i)$, for all $i, j \in [q]$.

If $q = 2$ this is called a 2-spin system, and for general q it is called a q-spin system.

Well-known examples of a 2-spin system include the Ising model [Isi25], where $f(0, 0) = f(1, 1) = a$ and $f(0, 1) = f(1, 0) = b$ for some two constants a and b. Sometimes there is also a vertex weight function, represented by a unary function normalized to $u(0) = 1$ and $u(1) = \lambda$. Then the partition function of the Ising model is

$$Z_{a,b,\lambda}(G) = \sum_{\sigma:V\to[2]} a^{|\{(u,v)\in E:\sigma(u)=\sigma(v)\}|} b^{|\{(u,v)\in E:\sigma(u)\neq\sigma(v)\}|} \lambda^{|\{v\in V:\sigma(v)=1\}|}. \tag{1.2}$$

Exercise: Fix any constant J. One can define the Hamiltonian function $H(\sigma)$ for any assignment $\sigma : V \to \{-1, +1\}$,

$$H(\sigma) = -J \sum_{(u,v)\in E} \sigma(u)\sigma(v).$$

(It is called ferromagnetic if $J > 0$ and antiferromagnetic if $J < 0$.) Then the partition function of the Ising model can be defined by $Z(G) = \sum_{\sigma:V\to\{-1,+1\}} e^{-\beta H(\sigma)}$, where $\beta \geq 0$ is called the inverse temperature. Show that this function $Z(G)$ can be realized by the sum-of-product expression $Z_{a,b,\lambda}(G)$.

A generalization of the Ising model is called the Potts model [Pot52], where for $q \geq 2$ and $i, j \in [q]$, $f(i,i) = a$ and $f(i,j) = f(j,i) = b$ for $i \neq j$. One can normalize $b = 1$ and write a as $1 + \gamma$ for a parameter γ. In this parameterization the partition function of the Potts model is

$$Z_{\text{Potts}}(G; q, \gamma) = \sum_{\sigma:V\to[q]} \prod_{(u,v)\in E} (1 + \gamma\delta(\sigma(u), \sigma(v))), \tag{1.3}$$

where $\delta(i,j) = 1$ if $i = j$, and 0 otherwise.

The partition function of the Potts model can be linked to the Tutte polynomial $\mathrm{T}(G; x, y)$ by setting $\gamma = y - 1$ and $q = (x-1)(y-1)$. Indeed, one way to define the Tutte polynomial in terms of q and γ is that $(x-1)^{\kappa(V,E)}(y-1)^{|V|}\,\mathrm{T}(G; x, y)$ is equal to (cf. Definition 6.27)

$$Z_{\text{Tutte}}(G; q, \gamma) = \sum_{F\subseteq E} q^{\kappa(V,F)}\gamma^{|F|},$$

where $\kappa(V, F)$ is the number of connected components in the spanning subgraph (V, F).

Although the Tutte polynomial is defined for any q, if we restrict to a positive integer q, then

$$Z_{\text{Tutte}}(G; q, \gamma) = Z_{\text{Potts}}(G; q, \gamma).$$

To prove this equality, we consider expanding the product $\prod_{(u,v)\in E}(1 + \gamma\delta(\sigma(u), \sigma(v)))$ in (1.3) as a sum indexed by $F \subseteq E$, which collects a factor $\gamma^{|F|}$, namely

$$\prod_{(u,v)\in E} (1 + \gamma\delta(\sigma(u), \sigma(v))) = \sum_{F\subseteq E} \gamma^{|F|} \prod_{(u,v)\in F} \delta(\sigma(u), \sigma(v)).$$

The product term indexed by F is 1 iff $\sigma(u) = \sigma(v)$ for all u and v that belong to the same connected component of the subgraph (V, F), and 0 otherwise. In (1.3) the sum $\sum_{\sigma:V\to[q]}$ has exactly $q^{\kappa(V,F)}$ such terms.

Spin systems are also called graph homomorphisms, and they come from a different source. Given two graphs G and H, a graph homomorphism from G to H is a map σ from the vertex set of G to the vertex set of H such that for every edge (u, v) in G, the image is also an edge in H. In general, multigraphs are allowed; thus, for example, an edge in G can be mapped to a loop in H. The

(a) #VERTEXCOVER (b) #ANTICHAIN

Figure 1.1. Target graphs H and the combinatorial counting problems they define as #H-coloring problems.

counting problem is to compute the number of graph homomorphisms from G to H.

We will fix an H, and consider the computational problem where G is the input. A special case is when $H = K_q$, the complete graph on q vertices, without self-loops. In this case, a graph homomorphism from G to H is a valid coloring of the vertices of G, using at most q distinct colors, as any adjacent pair of vertices of G must be mapped to distinct vertices of H. Owing partly to this special case, counting graph homomorphisms to H in general is also called #H-colorings, and H is called the target graph.

Graph homomorphisms can express a variety of combinatorial problems. In fact, their principal purpose is to express and then treat a wide variety of locally defined graph properties in a uniform way. For example, if H is a graph consisting of two vertices $\{\mathtt{T}, \mathtt{F}\}$ and two edges $\{(\mathtt{T}, \mathtt{T}), (\mathtt{T}, \mathtt{F})\}$, a loop and an edge between the two vertices (see Figure 1.1a), then the #H-coloring problem is the counting problem of VERTEX COVER, i.e., to count the number of vertex covers in graph G, denoted as #VC. Indeed a homomorphism from G to this H is a two-coloring of vertices of G such that the subset of vertices of G mapped to T forms a vertex cover. By flipping the intended meaning of T and F, this also counts the number of independent sets in G. Of course this simply reflects the fact that a subset $S \subseteq V(G)$ is a vertex cover iff its complement $S^c \subseteq V(G)$ is an independent set. As another example, suppose H is the two-vertex directed graph connected by a single directed edge and both vertices have one directed self-loop (see Figure 1.1b). Then the #H-coloring problem takes directed graphs as input. If the input is a directed acyclic graph, then it defines a partial order, and the #H-coloring problem is to compute the number of antichains (or equivalently, the number of lower sets, i.e., downward closed sets in the partial order, or equivalently, the number of upper sets) in this partial order. If the input is not acyclic, then every strongly connected component must be mapped to one vertex in H, and the #H-coloring problem is to count the number of antichains in the induced partial order after collapsing each strongly connected component.

Exercise: Show that #VC can be expressed as #H-coloring by defining explicitly the binary constraint function.

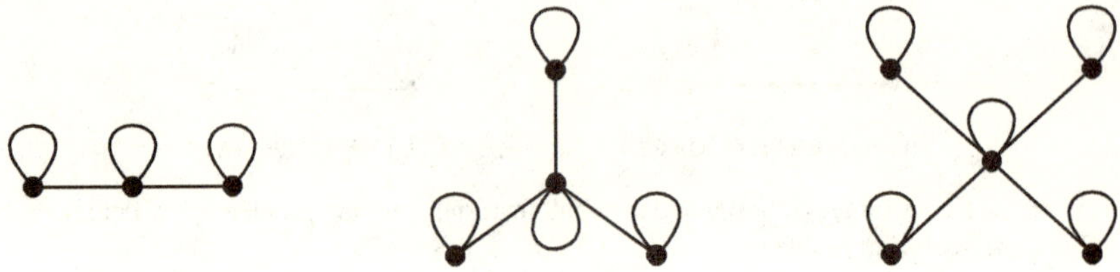

(a) $q=2$ (b) $q=3$ (c) $q=4$

Figure 1.2. Target graph H for counting q-particle Widom–Rowlinson configurations for $q \in \{2, 3, 4\}$ as an #H-coloring problem.

Exercise: Show that there is a bijection between the set of antichains and the set of lower sets. Conclude that the number of antichains is the same as the number of lower sets. By symmetry the same is proved for upper sets.

More generally, we consider weighted graph homomorphisms . Let A be a q-by-q matrix over $\mathbb{C}$. Given a graph or a directed graph $G = (V, E)$, the (weighted) graph homomorphism problem is to compute

$$Z_A(G) = \sum_{\sigma:V \to [q]} \prod_{(u,v) \in E} A_{\sigma(u),\sigma(v)}. \tag{1.4}$$

The target graph H is now defined by the matrix A, which is the weighted adjacency matrix of H. When A is a 0-1 matrix, then this is the unweighted version of graph homomorphism. The matrix A can also be identified as a binary function. In the case of Vertex Cover the function is the binary Boolean Or function. For Independent Set this is the Nand function. For the problem of counting antichains, the function is the binary Implication function. For q-coloring, this is the binary Disequality function on domain $[q]$.

This is exactly the same notion of a partition function in statistical physics. The Ising model corresponds to the partition function with matrix $A = \begin{bmatrix} a & b \\ b & a \end{bmatrix}$. The Potts model corresponds to the partition function with matrix $A = J_q + \gamma I_q$, were J_q is the $q \times q$ matrix of all 1's and I_q is the $q \times q$ identity matrix, and γ is a parameter.

The q-particle Widom–Rowlinson model [WR70] corresponds to the #H-coloring problem in which the domain size is $q+1$, H is the star graph on $q+1$ vertices and all vertices have self-loops (see Figure 1.2). The q-type Beach model [BS94, BS95] corresponds to the #H-coloring problem in which the domain size is $2q$, H is the complete graph on q vertices, each of these q vertices has a pendant vertex, and all $2q$ vertices have a self-loop (see Figure 1.3).

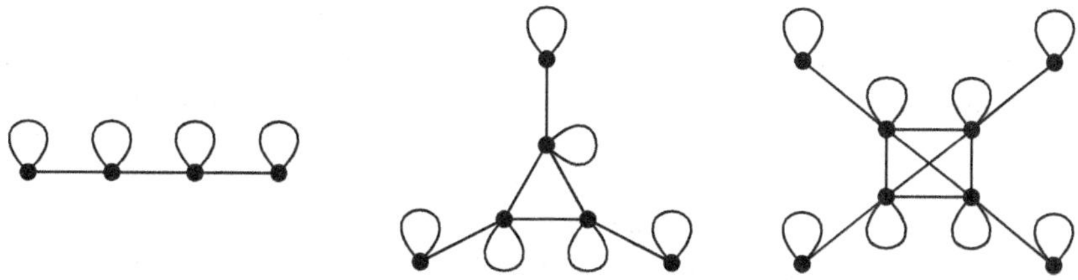

(a) $q = 2$ (b) $q = 3$ (c) $q = 4$

Figure 1.3. Target graph H for counting q-type Beach model configurations for $q \in \{2, 3, 4\}$ as an #H-coloring problem.

For more on this connection with statistical physics, see [Wel93, Chapter 4] or [Thu09, Chapter 2] (as well as [Bax82]).

Graph homomorphism can be viewed as a special case of counting constraint satisfaction problems, where instead of one binary constraint function there can be a set of constraint functions.

1.2.2 Constraint Satisfaction Problems

A counting constraint satisfaction problem (#CSP) [CKS01] is parametrized by a set of local constraint functions $\mathcal{F}$. It is denoted by #CSP$_q(\mathcal{F})$ when the constraint functions in $\mathcal{F}$ are defined over a domain $[q]$ of size q. An instance of #CSP$_q(\mathcal{F})$ is a finite set of variables $x_1, x_2, \ldots, x_n$, and a finite set C of clauses. Each clause is a constraint $f \in \mathcal{F}$ of some arity k depending on f together with a sequence of k (not necessarily distinct) variables $x_{i_1}, \ldots, x_{i_k} \in \{x_1, x_2, \ldots, x_n\}$. The output is

$$\sum_{x_1, \ldots, x_n \in [q]} \prod_{(f, x_{i_1}, \ldots, x_{i_k}) \in C} f(x_{i_1}, \ldots, x_{i_k}). \tag{1.5}$$

In the study of #CSP$_q(\mathcal{F})$, the set $\mathcal{F}$ is usually finite and considered fixed. In particular, there is a maximum arity k among functions in $\mathcal{F}$. An input instance on n variables can be described by n^k bits for a fixed k. Hence we consider the input size is n. When it is the Boolean domain (i.e., $q = 2$), we denote it simply as #CSP($\mathcal{F}$). If $\mathcal{F}$ consists of a single function f we write #CSP(f) for #CSP($\{f\}$). We write similarly #CSP(f, g) if $\mathcal{F} = \{f, g\}$, and #CSP($\mathcal{F}, g$) for #CSP($\mathcal{F} \cup \{g\}$), etc.

The canonical example of a #CSP is #SAT, or counting Boolean Satisfiability, the problem of counting the number of satisfying assignments to a given Boolean formula. As a constraint satisfaction problem, it is #CSP($\mathcal{F}$), with

$\mathcal{F} = \{\text{OR}_k | k \geq 1\} \cup \{\neq_2\}$, where OR_k is the OR function of arity k and $(\neq_2)$ is the binary DISEQUALITY function. Here are several more well-known examples of #CSP($\mathcal{F}$)'s over the Boolean domain and their corresponding constraint sets:

$$\begin{array}{rll}
\text{SAT} & \text{has} & \mathcal{F} = \{\text{OR}_k | k \geq 1\} \cup \{\neq_2\} \\
\text{3SAT} & \text{has} & \mathcal{F} = \{\text{OR}_3, \neq_2\} \\
\text{1-IN-3SAT} & \text{has} & \mathcal{F} = \{\text{EXACT-ONE}_3, \neq_2\} \\
\text{NAE-3SAT} & \text{has} & \mathcal{F} = \{\text{NOT-ALL-EQUAL}_3, \neq_2\}
\end{array}$$

$$\begin{array}{rll}
\text{MON-SAT} & \text{has} & \mathcal{F} = \{\text{OR}_k | k \geq 1\} \\
\text{MON-3SAT} & \text{has} & \mathcal{F} = \{\text{OR}_3\} \\
\text{MON-1-IN-3SAT} & \text{has} & \mathcal{F} = \{\text{EXACT-ONE}_3\} \\
\text{MON-NAE-3SAT} & \text{has} & \mathcal{F} = \{\text{NOT-ALL-EQUAL}_3\}
\end{array}$$

By #CSP^d($\mathcal{F}$), we denote the special case of #CSP($\mathcal{F}$) in which every variable appears a multiple of d times. Note that #CSP($\mathcal{F}$) is the same as #CSP^1($\mathcal{F}$), and #CSP^2($\mathcal{F}$) is the same as every variable appearing an even number of times.

1.2.3 Holant Problems

A Holant problem is parametrized by a set of local constraint functions $\mathcal{F}$, also called signatures . A *signature grid* $\Omega = (G, \pi)$ over $\mathcal{F}$ consists of a graph $G = (V, E)$ and a mapping π that assigns to each vertex $v \in V$ an $f_v \in \mathcal{F}$ and a linear order of the incident edges at v. The arity of f is equal to the degree at v, and the incident edges at v are associated with the input variables of f_v. If all signatures in $\mathcal{F}$ are symmetric then there is no need to assign an order for incident edges at any v.

Definition 1.1. For a set $\mathcal{F}$ of signatures over a domain $[q]$, we define Holant$_q$($\mathcal{F}$) as

Input: A *signature grid* $\Omega = (G, \pi)$ over $\mathcal{F}$
Output:

$$\text{Holant}_q(\Omega; \mathcal{F}) = \sum_{\sigma: E \to [q]} \prod_{v \in V} f_v(\sigma \mid_{E(v)}),$$

where

- $G = (V, E)$, and $E(v)$ denotes the incident edges of v and
- $\sigma \mid_{E(v)}$ denotes the restriction of σ to $E(v)$, and $f_v(\sigma \mid_{E(v)})$ is the evaluation of f_v on the ordered input tuple $\sigma \mid_{E(v)}$.

In Volume I of this book we exclusively present the theory over the Boolean domain $q = 2$. We use Holant($\mathcal{F}$) to denote Holant$_2$($\mathcal{F}$). We also denote

Holant(Ω) or Holant$_\Omega$ for Holant(Ω; $\mathcal{F}$), omitting $\mathcal{F}$ in the expression when $\mathcal{F}$ is clear from the context. We use G in place of Ω when π is clear from context. We write Holant$_q(\mathcal{F}, g)$ for Holant$_q(\mathcal{F} \cup \{g\})$. When $\mathcal{F}$ is a finite set of signatures, we sometimes just list the signatures it contains. For example, if $\mathcal{F} = \{f, g\}$, then instead of writing Holant$_q(\{f, g\})$, we may also write Holant$_q(f, g)$. This is especially true when $\mathcal{F}$ is a singleton set.

A signature f of arity k over the domain $[q]$ can be denoted by $(f_0, \ldots, f_x, \ldots, f_{q^k-1})$, where f_x is the output of f on $x \in [q]^k$, ordered lexicographically as integers from 0 to $q^k - 1$. We also write it as a column vector. This listing of values of f in lexicographical order is the same as in a truth table. It is a vector in $\mathbb{C}^{q^k}$, or a tensor in $(\mathbb{C}^q)^{\otimes k}$ indexed by $[q]^k$. A symmetric signature f of arity k over the Boolean domain can be expressed as $[f_0, f_1, \ldots, f_k]$, where f_w is the value of f on inputs of Hamming weight w. An example is the EQUALITY signature $(=_k) = [1, 0, \ldots, 0, 1]$ of arity k, where there are $k - 1$ entries of 0. Another example is the binary DISEQUALITY function $(\neq_2) = [0, 1, 0]$.

We give some examples of Holant problems using a symmetric signature f. If f is the only available signature and its arity is k, then the graph G in a Holant problem using $\{f\}$ must be k-regular, i.e., every vertex of the graph G has degree k. Here are four examples over the Boolean domain.

$$\text{Holant}(G; f) \text{ counts} \begin{cases} \text{matchings} & \text{in } G \text{ when } f = \text{AT-MOST-ONE}_k; \\ \text{perfect matchings} & \text{in } G \text{ when } f = \text{EXACT-ONE}_k; \\ \text{cycle covers} & \text{in } G \text{ when } f = \text{EXACT-TWO}_k; \\ \text{edge covers} & \text{in } G \text{ when } f = \text{OR}_k. \end{cases}$$

An example problem over domain size q is q-EDGE-COLORING, Holant$_q$(G; ALL-DISTINCT), which counts proper edge colorings in G using at most q colors; i.e., at every vertex of G the incident edges are colored with distinct colors from $[q]$. Each of these five examples is expressed as a Holant problem in a straightforward manner. A less obvious example is that Holant(G; $\frac{1}{2}[3, 0, 1, 0, 3]$) counts the number of Eulerian orientations of a 4-regular graph G; this will become clear when we develop the theory further.

As stated earlier, one can view a signature of arity k as a tensor along with a default choice of basis $\{e_{i_1} \otimes e_{i_2} \otimes \cdots \otimes e_{i_k} \mid i_1 i_2 \ldots i_k \in \{0, 1\}^k\}$, where $e_0 = (1,\ 0)$ and $e_1 = (0,\ 1)$. Readers who are familiar with quantum computation may want to write it as $e_0 = |0\rangle$, $e_1 = |1\rangle$, and $e_{i_1} \otimes e_{i_2} \otimes \cdots \otimes e_{i_k} = |i_1 i_2 \ldots i_k\rangle$. When doing so, a signature grid is equivalent to a tensor network, and the Holant of the signature grid is equal to the scalar that remains after contracting all edges in the corresponding tensor network. (However, in this book we will not assume or use any knowledge of quantum computing, and will not use these quantum computing notations.)

A *planar signature grid* is a signature grid such that its underlying graph is planar and for some planar embedding, for every vertex v, the linear order of the incident edges at v agrees with the clockwise cyclic order of the incident edges at v in the embedding starting with some particular edge. We use Pl-Holant$_q(\mathcal{F})$ to denote the restriction of Holant$_q(\mathcal{F})$ to planar signature grids. For signature sets $\mathcal{F}$ and $\mathcal{G}$, a *bipartite signature grid* over $(\mathcal{F} \mid \mathcal{G})$ is a signature grid $\Omega = (H, \pi)$ over $\mathcal{F} \cup \mathcal{G}$, where $H = (V, E)$ is a bipartite graph with bipartition $V = (V_1, V_2)$ such that $\pi(V_1) \subseteq \mathcal{F}$ and $\pi(V_2) \subseteq \mathcal{G}$. Signatures in $\mathcal{F}$ are considered as row vectors (or covariant tensors); signatures in $\mathcal{G}$ are considered as column vectors (or contravariant tensors) [DP91] (see also chapter 2 of [Rom12]). We use Holant$_q(\mathcal{F} \mid \mathcal{G})$ to denote the restriction of Holant$_q(\mathcal{F} \cup \mathcal{G})$ to bipartite signature grids over $(\mathcal{F} \mid \mathcal{G})$, and Holant$_q(\Omega; \mathcal{F}|\mathcal{G})$ to denote the value on Ω. A *planar bipartite signature grid* is one that is both planar and bipartite. We use Pl-Holant$_q(\mathcal{F} \mid \mathcal{G})$ to denote the restriction to these signature grids, and Pl-Holant$_q(\Omega; \mathcal{F}|\mathcal{G})$ to denote the value on Ω. We use LHS and RHS to stand for the left-hand side and right-hand side respectively.

A signature f of arity k is *degenerate* if there exist unary signatures $u_j \in \mathbb{C}^q$ $(1 \le j \le k)$ such that $f = u_1 \otimes \cdots \otimes u_k$. Replacing such an f by the k unary signatures u_j, one for each associated edge, does not change the Holant value. A symmetric degenerate signature has the form $u^{\otimes k}$, for some unary $u \in \mathbb{C}^q$. Replacing a signature $f \in \mathcal{F}$ by a constant multiple cf, where $c \ne 0$, does not change the complexity of Holant$_q(\mathcal{F})$. It merely introduces a nonzero factor c^n to Holant$_q(\Omega; \mathcal{F})$, where f appears n times in Ω.

Exercise: Prove that the Holant value is unchanged when $f = u_1 \otimes \cdots \otimes u_k$ is replaced by unary signatures u_j, with u_j assigned to the jth incident edge of every vertex assigned f.

Exercise: Prove that if a degenerate signature $f = u_1 \otimes \cdots \otimes u_k$ is symmetric, then it has the form $u^{\otimes k}$, for some unary $u \in \mathbb{C}^q$.

We use the term tractable as a shorthand for polynomial time computable. We allow $\mathcal{F}$ to be an infinite set. For Holant$_q(\mathcal{F})$ to be tractable, the problem must be computable in polynomial time when the input graph G and the description of the signatures appearing in G are included in the input size. In contrast, we say Holant$_q(\mathcal{F})$ is #P-hard if there exists a finite subset of $\mathcal{F}$ for which the problem is #P-hard. We say a signature set $\mathcal{F}$ is tractable (resp. #P-hard) in the context of Holant problems if the corresponding counting problem Holant$_q(\mathcal{F})$ is tractable (resp. #P-hard). Similarly for a signature f, we say f is tractable (resp. #P-hard) if $\{f\}$ is. We also speak of a signature or signature set as being tractable or #P-hard for various restricted classes of Holant problems, such as those defined over planar, bipartite, or planar and bipartite graphs. The meaning of the restriction should be clear from context.

We can express $\#\mathrm{CSP}_q^d(\mathcal{F})$ as a Holant problem. An instance of $\#\mathrm{CSP}_q^d(\mathcal{F})$ has the following bipartite view. Create a node for each variable and each clause. Connect a variable node to a clause node if the variable appears in the clause. This bipartite graph is also known as the *constraint graph* . To each variable vertex, we assign the EQUALITY signature of the appropriate arity. To each clause vertex, we assign the constraint used in that clause. Under this view, we see that

$$\#\mathrm{CSP}_q^d(\mathcal{F}) \equiv_{\mathrm{T}} \mathrm{Holant}_q(\mathcal{EQ}_d \mid \mathcal{F}), \tag{1.6}$$

where $\mathcal{EQ}_d = \{=_{dk} | k \geq 1\}$ is the set of EQUALITY signatures of arities equal to a multiple of d. We denote by $\text{Pl-}\#\mathrm{CSP}_q^d(\mathcal{F})$ the restriction of $\#\mathrm{CSP}_q^d(\mathcal{F})$ to inputs with a planar constraint graph. Again we drop q and write $\#\mathrm{CSP}^d$ or $\text{Pl-}\#\mathrm{CSP}^d$ if $q = 2$, and drop d if $d = 1$. The preceding construction also shows that

$$\text{Pl-}\#\mathrm{CSP}_q^d(\mathcal{F}) \equiv_{\mathrm{T}} \text{Pl-Holant}_q(\mathcal{EQ}_d \mid \mathcal{F}). \tag{1.7}$$

If $d \in \{1, 2\}$, then more is true.

Lemma 1.2. *Let $\mathcal{F}$ be a set of signatures over a domain of size q. If $d \in \{1, 2\}$, then*

$$\#\mathrm{CSP}_q^d(\mathcal{F}) \equiv_{\mathrm{T}} \mathrm{Holant}_q(\mathcal{EQ}_d \cup \mathcal{F}) \quad \text{and}$$
$$\text{Pl-}\#\mathrm{CSP}_q^d(\mathcal{F}) \equiv_{\mathrm{T}} \text{Pl-Holant}_q(\mathcal{EQ}_d \cup \mathcal{F}).$$

Proof. By (1.6) and (1.7), it suffices to show

$$\mathrm{Holant}_q(\mathcal{EQ}_d \mid \mathcal{F}) \equiv_{\mathrm{T}} \mathrm{Holant}_q(\mathcal{EQ}_d \cup \mathcal{F})$$

and correspondingly for $\text{Pl-Holant}_q(\mathcal{EQ}_d \mid \mathcal{F})$. It both cases, the reduction from left to right in the equivalence is trivial; just ignore the bipartite restriction. For the other direction, we take a signature grid for the problem on the right and create a bipartite signature grid for the problem on the left such that both signature grids have the same Holant value up to an easily computable factor. If the initial graph is planar, then the final graph will also be planar, so this will prove both equivalences.

If two signatures in $\mathcal{F}$ are assigned to adjacent vertices, then we subdivide all edges between them and assign the binary EQUALITY signature $(=_2) \in \mathcal{EQ}_d$ to all new vertices. Suppose EQUALITY signatures $(=_n), (=_m) \in \mathcal{EQ}_d$ are assigned to adjacent vertices connected by ℓ edges. If $n = m = \ell$, then these two vertices are disjoint from the rest, and we simply remove these two vertices. The Holant

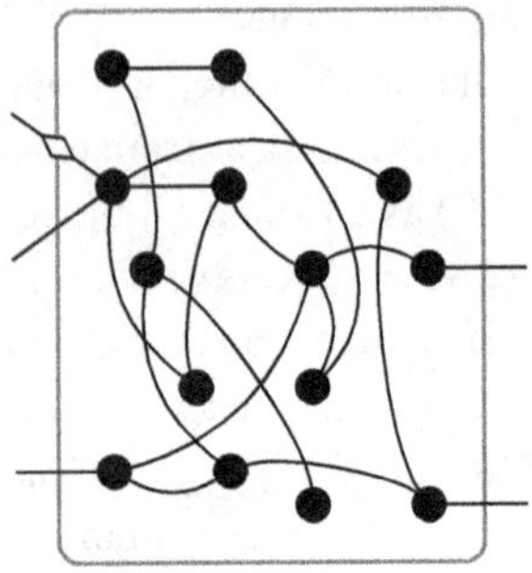

Figure 1.4. An $\mathcal{F}$-gate with five dangling edges.

of the resulting signature grid differs from the original by a factor of q. Otherwise, we contract all ℓ edges, merge the two vertices into one, and assign $(=_{n+m-2\ell}) \in \mathcal{EQ}_d$ to the new vertex. □

1.3 Reductions

We now introduce several types of reductions we will use in this book, including gadget constructions, holographic transformations, and interpolations.

1.3.1 Gadget Construction

Local Gadget Constructions via $\mathcal{F}$-gates

A basic type of reduction is generally known as a gadget construction. In the context of Holant problems, we construct a graph in order to realize a signature. Fix a set $\mathcal{F}$ of signatures over a domain $[q]$ of size q. We say a signature f is *realizable* from $\mathcal{F}$ if there is a graph with some dangling edges where each vertex is assigned a signature from $\mathcal{F}$, and the resulting signature with inputs on the dangling edges is exactly f.

Formally, this is defined in terms of an $\mathcal{F}$-gate. An $\mathcal{F}$-gate F is similar to a signature grid (G, π) for Holant($\mathcal{F}$) except that $G = (V, E, E')$ is a graph with regular edges E and some dangling edges E', and $E \cap E' = \emptyset$. Each $e' \in E'$ has only one end incident to a vertex in V, and the other end dangling. The dangling edges define external variables for the $\mathcal{F}$-gate. They are ordered, and usually presented pictorially by starting at the edge marked with a diamond and proceeding cyclically (see Figure 1.4 for an example.)

An $\mathcal{F}$-gate F with k dangling edges defines the function

$$\Gamma(y_1, \ldots, y_k) = \sum_{\sigma: E \to [q]} \prod_{v \in V} f_v(\widehat{\sigma} \mid_{E(v)}),$$

where $(y_1, \ldots, y_k) \in [q]^k$ is an assignment on the dangling edges, $\widehat{\sigma}$ is the extension of the assignment σ on the internal edges E by the assignment $(y_1, \ldots, y_k)$ on the dangling edges E', and f_v is the signature assigned by π at v. We call this function Γ the signature of the $\mathcal{F}$-gate. We also call an $\mathcal{F}$-gate a gadget. If the signature of an $\mathcal{F}$-gate is invariant under cyclic permutations of inputs, then we omit the diamond because it is unnecessary. We say that such signatures are *rotationally symmetric*. If the signature is invariant under any permutation of inputs, then we say it is *symmetric*. These notions are defined in terms of the signature function Γ. The signature of an $\mathcal{F}$-gate may be rotationally symmetric or symmetric as a function, even though the graph defining it may not be so.

An $\mathcal{F}$-gate is *planar* if the underlying graph can be embedded in the plane without edge crossings and the dangling edges are in the outer face in the cyclic order. Now suppose we have two signature sets $\mathcal{F}$ and $\mathcal{G}$ in the context of a bipartite Holant problem $\text{Holant}_q(\mathcal{F} \mid \mathcal{G})$. Then an $(\mathcal{F} \mid \mathcal{G})$-gate is an $(\mathcal{F} \cup \mathcal{G})$-gate such that the underlying graph is bipartite, the vertices in one part are assigned signatures from $\mathcal{F}$, and the vertices in the other part are assigned signatures from $\mathcal{G}$. Furthermore, we say that an $(\mathcal{F} \mid \mathcal{G})$-gate is *on the left* (resp. *on the right*) if each vertex incident to a dangling edge is assigned a signature from $\mathcal{F}$ (resp. $\mathcal{G}$).

Using $\mathcal{F}$-gates, we can reduce one Holant problem to another.

Lemma 1.3. *Let $\mathcal{F}$ be a set of signatures over a domain of size q. If there exists an $\mathcal{F}$-gate with signature f, then*

$$\text{Holant}_q(\mathcal{F}, f) \leq_T \text{Holant}_q(\mathcal{F}).$$

Similar statements hold for gadgets that are planar, bipartite, or both for Holant problems defined over the same class of graphs.

Proof. Let F be an $\mathcal{F}$-gate with signature f. Given an instance Ω of $\text{Holant}_q(\mathcal{F}, f)$, we replace every occurrence of f by the $\mathcal{F}$-gate F to obtain an instance Ω' of $\text{Holant}_q(\mathcal{F})$. Since f is the signature of the $\mathcal{F}$-gate F, the Holant values for these two signature grids are identical. Furthermore, the size of F is a constant with respect to Ω, so the size of Ω' is only a constant factor larger than that of Ω. □

Even for a very simple signature set $\mathcal{F}$, the signatures for all $\mathcal{F}$-gates can be quite complicated and expressive. In Chapter 4 we will see that for the simple class of weighted EXACT-ONE functions, which define weighted PERFECT MATCHINGS, the resulting signature set can be quite expressive. These are called *matchgates*.

Signature Matrix

It is convenient to write a signature as a matrix. An immediate advantage is that a matrix is more of a pictorial representation than a vector, which aids understanding. However, the more important reason is to simplify the computation of the signature of a gadget.

Definition 1.4. Let f be a signature of arity n over a domain of size q. The *signature matrix of f with parameter ℓ* is an $q^\ell \times q^{n-\ell}$ matrix for some integer $0 \le \ell \le n$ in which the first ℓ inputs (in order) are the row index and the remaining $n - \ell$ inputs (in reverse order) are the column index.

If the arity of f is even, then the *signature matrix of f*, without specifying a parameter, is the signature matrix of f with parameter $\ell = \frac{n}{2}$ and is denoted by M_f.

For example, if $f = (f_{00}, f_{01}, f_{10}, f_{11})$ has arity 2 over the Boolean domain, then $M_f = \begin{bmatrix} f_{00} & f_{01} \\ f_{10} & f_{11} \end{bmatrix}$. If g is a signature of arity 4 over the Boolean domain with $g(i, j, k, \ell) = g_{ijk\ell}$, then

$$M_g = \begin{bmatrix} g_{0000} & g_{0010} & g_{0001} & g_{0011} \\ g_{0100} & g_{0110} & g_{0101} & g_{0111} \\ g_{1000} & g_{1010} & g_{1001} & g_{1011} \\ g_{1100} & g_{1110} & g_{1101} & g_{1111} \end{bmatrix}.$$

Notice the reversal of order in the column index (k, ℓ), i.e., the columns are listed in the order $k\ell = 00, 10, 01, 11$. The purpose of reversing the order of the column index is so that we can use matrix product in gadget computations. Let F be an $\mathcal{F}$-gate with signature f of arity n. We often depict F with ℓ dangling edges protruding to the left and $n - \ell$ dangling edges protruding to the right to aid in the mapping from F to the signature matrix of f with parameter ℓ. If F and F' are two $\mathcal{F}$-gates of arity 4 with signature matrices M and M', then the matrix product MM' is the signature matrix of the $\mathcal{F}$-gate linking the two $\mathcal{F}$-gates, namely by merging the fourth and third edges (in cyclic order) of F with the first and second edges (also in cyclic order) of F' respectively. In particular, if F and F' are planar $\mathcal{F}$-gates, then this linking operation produces a planar $\mathcal{F}$-gate.

1.3.2 Holographic Transformation

Sometimes two counting problems may appear different, but are really the same problem under different guises. Holographic transformation is a primary tool to establish this quantitative connection.

It is a simple fact that in a graph, the number of vertex covers is equal to the number of independent sets. The reason is that the complement of one type is the other. At the vertex assignment level, this is accomplished by an exchange of the assignments 0 and 1. This is a type of combinatorial mapping.

In a holographic transformation the 0-1 assignments are no longer considered purely as discrete objects. Rather, they are embedded in a vector space as basis elements, and then we consider linear transformations of the ambient space. The mapping between vertex covers and independent sets can be described by a linear transformation that is simply a basis exchange $e_0 = (1,\ 0) \leftrightarrow e_1 = (0,\ 1)$. In general, a holographic transformation will be much more algebraically "mixing." Such a transformation puts the assignments of 0 and 1 into a superposition much like the states of a qubit in quantum computing. However, quantum computation is not required – or even any computation at all. Just like the example with counting vertex covers and independent sets, we are merely describing a mathematical proof that two different looking problems are actually the same. By undergoing a change of basis we can connect different problems from this perspective, and prove that they have the same (or closely related) answers.

To formally introduce the idea of holographic transformations, it is convenient to consider bipartite graphs. For a general graph, we can always transform it into a bipartite graph while preserving the Holant value as follows. For each edge in the graph, we replace it by a path of length two. (This operation is called a *2-stretch* of the graph and yields the edge–vertex incidence graph.) Each new vertex is assigned the binary EQUALITY signature $(=_2)$.

For a $q \times q$ matrix T and a signature f of arity n, written as a column vector in $\mathbb{C}^{q^n}$, we write $Tf = T^{\otimes n} f$ as the transformed signature. For a signature set $\mathcal{F}$, define $T\mathcal{F} = \{Tf \mid f \in \mathcal{F}\}$. For a signature f of arity n written as a row vector, we similarly define $fT = fT^{\otimes n}$, and $\mathcal{F}T$. Here the tensor product denoted by $\otimes$ is the Kronecker product, namely, if $X = X_{a \times b}$ and $Y = Y_{c \times d}$ are two matrices, then $X \otimes Y$ is an $ac \times bd$ matrix, with entry $X_{ij}Y_{k\ell}$ at row indexed by $(i, k) \in [a] \times [c]$ and column indexed by $(j, \ell) \in [b] \times [d]$, both ordered lexicographically. Tensor power is defined inductively $X^{\otimes n} = X^{\otimes (n-1)} \otimes X$. Whenever we write $T^{\otimes n} f$ or $T\mathcal{F}$, we view the signatures as column vectors; similarly for $fT^{\otimes n}$ or $\mathcal{F}T$ as row vectors.

Let T be an invertible $q \times q$ matrix. The holographic transformation defined by T is the following operation: given a bipartite signature grid $\Omega = (H, \pi)$ of Holant$(\mathcal{F} \mid \mathcal{G})$, for the same bipartite graph H, we get a new signature grid $\Omega' = (H, \pi')$ of Holant$(\mathcal{F}T \mid T^{-1}\mathcal{G})$ by replacing each signature in $\mathcal{F}$ or $\mathcal{G}$ with the corresponding signature in $\mathcal{F}T$ or $T^{-1}\mathcal{G}$. For this reason, signatures in $\mathcal{F}$ are called convariant and those in $\mathcal{G}$ contravariant. Valiant's Holant Theorem [Val08] , Theorem 1.5, states that the Holant value is unchanged, under a holographic transformation (see also [CC07]).

Theorem 1.5. *Let $\mathcal{F}$ and $\mathcal{G}$ be sets of complex-valued signatures over a domain of size q. Suppose Ω is a bipartite signature grid over $(\mathcal{F} \mid \mathcal{G})$. If $T \in \mathbf{GL}_q(\mathbb{C})$, then*

$$\operatorname{Holant}_q(\Omega; \mathcal{F} \mid \mathcal{G}) = \operatorname{Holant}_q(\Omega'; \mathcal{F}T \mid T^{-1}\mathcal{G}),$$

where Ω' is the corresponding signature grid over $(\mathcal{F}T \mid T^{-1}\mathcal{G})$.

Proof. We transform Ω to Ω' in several steps, preserving the Holant value in each step. The essential idea is that operator application is associative. We illustrate this in Figure 1.5.

Let $\Omega_0 = \Omega$. Let $G = (U, V, E)$ be the graph underlying Ω_0. Vertices in U are assigned signatures in $\mathcal{F}$ by π while vertices in V are assigned signatures in $\mathcal{G}$ by π. Let $e = (u, v) \in E$ be an edge with endpoints $u \in U$ and $v \in V$, assigned signatures $f \in \mathcal{F}$ and $g \in \mathcal{G}$ respectively. Initially, Figure 1.5a depicts the neighborhood of u and v in $\Omega = \Omega_0$. In this example, both u and v are incident to three other edges but the vertices incident to the other ends of these edges are not shown. We do the following operations for each edge $e = (u, v) \in E$: Subdivide e and assign $(=_2)$ to the new vertex w. Let the resulting signature grid be Ω_1 (see Figure 1.5b). It is clear that

$$\operatorname{Holant}_q(\Omega_0; \mathcal{F} \mid \mathcal{G}) = \operatorname{Holant}_q(\Omega_1; \mathcal{F} \cup \mathcal{G} \cup \{=_2\}).$$

Then we subdivide w to get two adjacent vertices u' and v' so that we now have a path (u, u', v', v) for every edge $e = (u, v) \in E$ in G. Assign to u' and v' the binary signature (call them h_u and h_v respectively) whose signature matrix is T and T^{-1} respectively. If T is not a symmetric matrix, then these signatures are not symmetric and it matters which edge corresponds to which input. The first input for h_u, represented by the row index in T, corresponds to the edge $\{u, u'\}$. The first input for h_v, represented by the row index in T^{-1}, corresponds to the edge $\{u', v'\}$, so that it is merged with the second input for h_u, represented by the column index in T. Let the resulting signature grid be Ω_2. (See Figure 1.5c, which indicates the first inputs of h_u and h_v by the rotated placement of T and T^{-1}, with the first input to their left.) If we contract the edge $\{u', v'\}$ within the dashed box, then we get back Ω_1. Thus, the Holant value is unchanged,

$$\operatorname{Holant}_q(\Omega_1; \mathcal{F} \cup \mathcal{G} \cup \{=_2\}) = \operatorname{Holant}_q(\Omega_2; \mathcal{F} \cup \mathcal{G} \cup \{T, T^{-1}\}).$$

Now Ω_3 in Figure 1.5d is really the same as Ω_2, except we think of the order of contraction operations differently by associating the binary T and T^{-1} to the

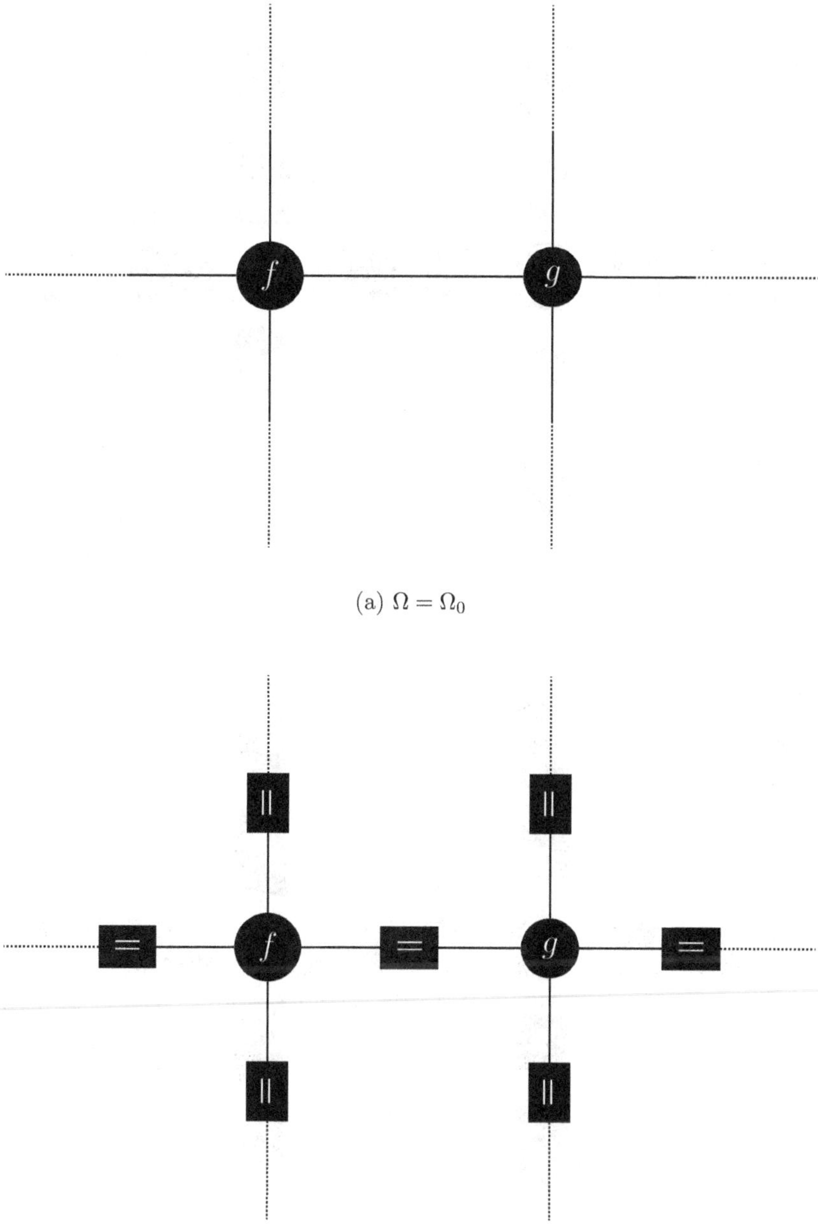

(a) $\Omega = \Omega_0$

(b) Ω_1

Figure 1.5. Neighborhood around two adjacent vertices.

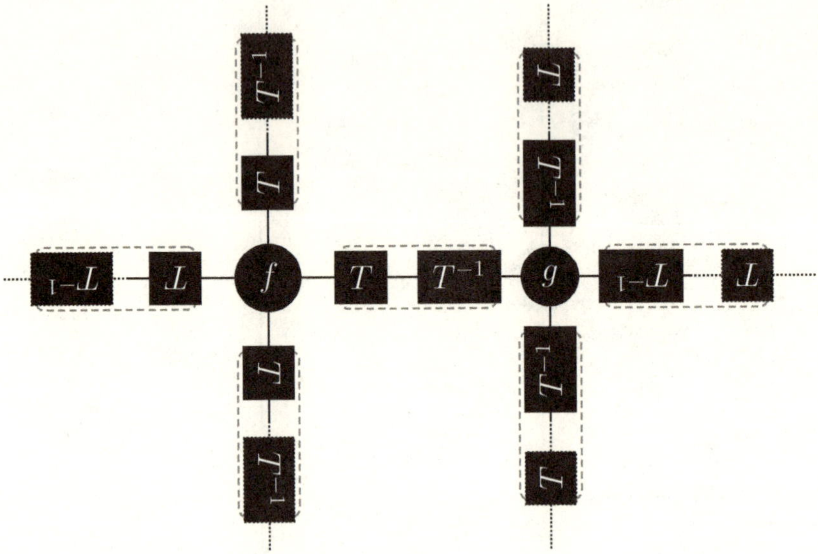

(c) Ω_2

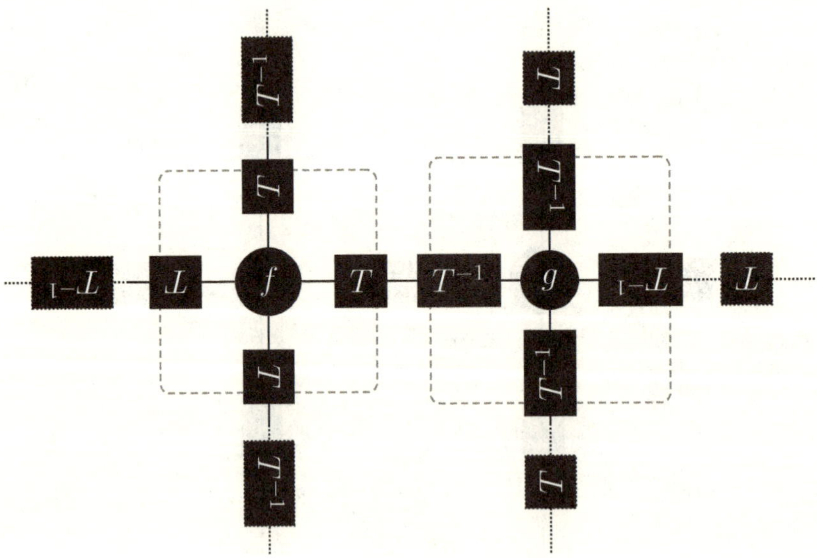

(d) Ω_3

Figure 1.5. (*cont.*)

original signatures f and g first.

$$\text{Holant}_q(\Omega_2; \mathcal{F} \cup \mathcal{G} \cup \{T, T^{-1}\}) = \text{Holant}_q(\Omega_3; \mathcal{F} \cup \mathcal{G} \cup \{T, T^{-1}\})$$

Finally we contract $\{u, u'\}$ and $\{v', v\}$. This defines $\Omega' = \Omega_4$. After doing so, we once again have the bipartite graph G. What has changed is the assignment to each vertex. Any vertex $u \in U$ of degree $\deg(u)$ assigned $f \in \mathcal{F}$ in Ω is now assigned $fT^{\otimes \deg(u)}$. Similarly any vertex $v \in V$ of degree $\deg(v)$ assigned $g \in \mathcal{G}$ in Ω is now assigned $(T^{-1})^{\otimes \deg(v)} g$. In general, the new assignment to each vertex is the transformed signature in $\mathcal{F}T$ or $T^{-1}\mathcal{G}$ respectively, as claimed. Hence,

$$\text{Holant}_q(\Omega_3; \mathcal{F} \cup \mathcal{G} \cup \{T, T^{-1}\}) = \text{Holant}_q(\Omega_4; \mathcal{F}T \mid T^{-1}\mathcal{G}). \qquad \square$$

Therefore, an invertible holographic transformation does not change the complexity of the Holant problem in the bipartite setting. Furthermore, there is a special kind of holographic transformation, the orthogonal transformation, that preserves the binary equality and thus can be used freely in the standard (non-bipartite) setting. We denote by $\mathbf{O}_q(\mathbb{C})$ the group of $q \times q$ orthogonal matrices over $\mathbb{C}$. Note that $T \in \mathbf{O}_q(\mathbb{C})$ iff $T^{-1}T = I_q$ iff the holographic transformation $(=_2) \mapsto (=_2)T^{\otimes 2} = (=_2)$. The latter identity can be checked by its matrix form, because for a binary signature f, if M is its matrix form, then $T^{\mathrm{T}}MT$ is the matrix form of the binary signature $fT^{\otimes 2}$.

Exercise: Verify that for a binary covariant signature (row vector) f with signature matrix M, the signature matrix of $fT^{\otimes 2}$ is $T^{\mathrm{T}}MT$. If f is a binary contravariant signature (column vector) with signature matrix M, then the signature matrix of $(T^{-1})^{\otimes 2} f$ is $T^{-1}M(T^{-1})^{\mathrm{T}}$. Generalize this to signatures of higher arity.

Theorem 1.6. *Let $\mathcal{F}$ be a set of complex-valued signatures over a domain of size q. Suppose Ω is a signature grid over $\mathcal{F}$. If $H \in \mathbf{O}_q(\mathbb{C})$, then*

$$\text{Holant}_q(\Omega; \mathcal{F}) = \text{Holant}_q(\Omega'; H\mathcal{F}),$$

where Ω' is the corresponding signature grid over $H\mathcal{F}$.

Proof. $\text{Holant}_q(\Omega; \mathcal{F}) = \text{Holant}_q(\Omega; (=_2) \mid \mathcal{F})$. $\square$

We use Theorems 1.5 and 1.6 to reduce both between tractable problems and between #P-hard problems. Some of our reductions between tractable problems use the following definition, where $\mathscr{C}$ is usually some known tractable class of signatures.

Definition 1.7. Let $\mathcal{F}$ be any set of complex-valued signatures over a domain of size q. We say $\mathcal{F}$ is $\mathscr{C}$-transformable if there exists a $T \in \mathbf{GL}_q(\mathbb{C})$ such that $(=_2)T^{\otimes 2} \in \mathscr{C}$ and $\mathcal{F} \subseteq T\mathscr{C}$.

Note that if $\text{Holant}_q(\mathscr{C})$ is tractable over any set of graphs, then $\text{Holant}_q(\mathcal{F})$ is tractable over the same set of graphs for any $\mathscr{C}$-transformable set $\mathcal{F}$.

Lemma 1.8. *Let $\mathcal{F}$ be any set of complex-valued signatures over a domain of size q. If $\mathcal{F}$ is $\mathscr{C}$-transformable, then*

$$\text{Holant}_q(\mathcal{F}) \leq_T \text{Holant}_q(\mathscr{C}),$$

where both problems are defined over the same set of graphs.

Another Identity

Holographic transformations between two signature grids give the same Holant value. The following lemma shows a different kind of reason why two Holant problems must always have the same Holant value.

Lemma 1.9. *Let x and y be any numbers in $\mathbb{C}$. Then for every signature grid Ω having a $(2, 4)$-regular bipartite underlying graph,*

$$\text{Holant}(\Omega; [0, 1, 0] \mid [x, y, 1, 0, 0]) = \text{Holant}(\Omega; [0, 1, 0] \mid [0, 0, 1, 0, 0]).$$

Proof. In any signature grid Ω having a $(2, 4)$-regular bipartite underlying graph, because $[0, 1, 0]$ is the only signature on the left, any nonzero term in the Holant sum must assign 1 to exactly half of the edges in G. On the right side, if some copy of $[x, y, 1, 0, 0]$ contributes an x or y in some assignment, then fewer than half of its incident edges are assigned 1. To compensate, some other copy of $[x, y, 1, 0, 0]$ must have more than half of its incident edges assigned 0, so it contributes a factor 0. □

Exercise: Prove that for all $x, y \in \mathbb{C}$ not both 0, there is no holographic transformation between the two Holant problems in Lemma 1.9. Thus the converse of Theorem 1.6 does not hold.

1.3.3 Polynomial Interpolation

Polynomial interpolation is a powerful technique to prove #P-hardness for counting problems. We can introduce this technique with an example from [Val79a] by Valiant.

Lemma 1.10. #PerfectMatching $\leq_T$ #Matching

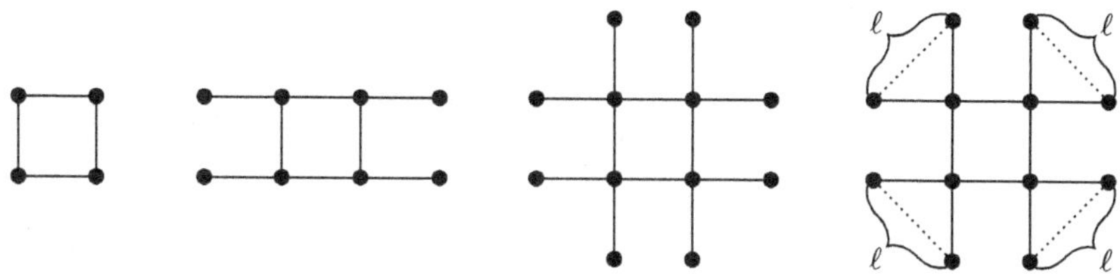

$G = G_0$ G_1 G_2 G_ℓ

Figure 1.6. Some graphs obtained from an initial graph G in the proof of Lemma 1.10.

Proof. Let $G = (V, E)$ be a graph. We want to determine the number of perfect matchings in G assuming that we have an oracle to count (all, not necessarily perfect) matchings.

For every integer $0 \le \ell \le n$, we construct a graph G_ℓ from G in the following way. For each vertex $v \in V$ we will "grow a thistle," namely, we add ℓ new vertices v_k $(1 \le k \le \ell)$, each connected to v by one edge. See Figure 1.6 for some examples of these graphs beginning with a specific graph G. In G_ℓ each v_k has degree 1 and clearly it can only be matched to v. If v is matched to some v_k $(1 \le k \le \ell)$, then v cannot be matched to any other v_ℓ for $\ell \ne k$ nor can it be matched to any other original vertex of G.

Let m_k be the number of matchings in G that omit k vertices. Then we can express the number of matchings in G_ℓ as

$$\sum_{k=0}^{n} (\ell + 1)^k m_k = \#\text{MATCHING}(G_\ell). \tag{1.8}$$

This is because a matching of G that omits exactly k vertices can be extended to exactly $(\ell + 1)^k$ matchings in G_ℓ, and every matching in G_ℓ is obtained uniquely this way from a matching of G. To see this, let M be a matching of G that omits k vertices. Then each of the k vertices in G that is not matched by M has $\ell + 1$ possibilities with respect to a matching M' in G_ℓ extending M, because it can can be matched with one of the ℓ new neighbors in G_ℓ or remain unmatched. Conversely, for every matching M' of G_ℓ, we can consider its subset consisting of edges in G. This defines a matching M in G and M' is one of the extensions from M, and such an M is unique.

We collect these equations to form the linear system

$$\begin{bmatrix} (0+1)^0 & (0+1)^1 & \cdots & (0+1)^n \\ (1+1)^0 & (1+1)^1 & \cdots & (1+1)^n \\ \vdots & \vdots & \ddots & \vdots \\ (n+1)^0 & (n+1)^1 & \cdots & (n+1)^n \end{bmatrix} \begin{bmatrix} m_0 \\ m_1 \\ \vdots \\ m_n \end{bmatrix} = \begin{bmatrix} \#\text{MATCHING}(G_0) \\ \#\text{MATCHING}(G_1) \\ \vdots \\ \#\text{MATCHING}(G_n) \end{bmatrix}. \tag{1.9}$$

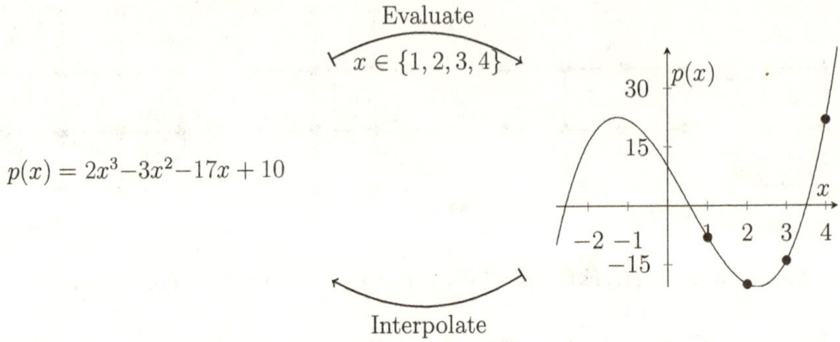

Figure 1.7. Interpolation is the inverse of evaluation.

Using our oracle, we know the right side. On the left, the coefficient matrix is Vandermonde . It is invertible because the values $\ell + 1$, for $0 \leq \ell \leq n$, are all distinct. Therefore, we can invert this matrix and solve for the unknown m_k's in polynomial time. Then m_0, the number of matchings in G that omit no vertices, is the number of perfect matchings G as desired. □

The word "polynomial" did not appear in this proof, so what makes it an example of polynomial interpolation? The polynomial is implicit; it is $p(x) = \sum_{k=0}^{n} m_k x^k$. Asking our oracle for the number of matchings in G_ℓ is equivalent to evaluating $p(x)$ at $\ell + 1$. Of course this is made possible by the design of the gadget of the "thistle," a weed-like structure, which produced such an expression for #MATCHING(G_ℓ). Such designs are where the inventiveness is in an interpolation proof. Polynomial interpolation is the process of recovering the coefficients of a polynomial of known degree from its evaluations at various points (see Figure 1.7), which is what this proof did.

Since our $n + 1$ evaluation points are distinct, we can recover the coefficients of $p(x)$. In this simple interpolation proof, the coefficient matrix is clearly invertible. For more complicated proofs this can be more demanding mathematically, and of course this works hand-in-hand with the inventive process of designing the right gadget. Given these coefficients, our reduction can proceed by computing any polynomial-time computable function from these coefficients. It is also often the case that we are interested in some evaluation of the interpolated polynomial. In the proof of Lemma 1.10, the answer we seek is the evaluation of the polynomial at a particular point 0, obtaining $p(0) = m_0$.

In fact both problems #MATCHING and #PERFECTMATCHING are #P-complete. The reverse reduction, from the problem of counting all matchings #MATCHING to the problem of counting perfect matchings #PERFECTMATCHING, is given in Lemma 1.11. In contrast, this proof is not by interpolation.

Lemma 1.11. #MATCHING $\le_T$ #PERFECTMATCHING

Proof. Let $G = (V, E)$ be a graph. We want to determine the number of all, not necessarily perfect, matchings in G assuming that we have an oracle to count perfect matchings. Suppose G has n vertices $V = \{v_i \mid 1 \le i \le n\}$. For every k such that $0 \le k \le n/2$, consider the graph $G_k = (V', E')$, where $V' = V \cup \{z_j \mid 1 \le j \le n - 2k\}$, a disjoint union of V and $n - 2k$ new vertices, and E' is the disjoint union of E and $n(n - 2k)$ new edges consisting of all pairs $\{(v_i, z_j) \mid 1 \le i \le n, 1 \le j \le n - 2k\}$, i.e., a copy of the complete bipartite graph $K_{n,n-2k}$. Now it is easy to see that every matching in G of size k has exactly $(n - 2k)!$ extensions to a perfect matching in G_k, matching the remaining $n - 2k$ vertices of V to the new vertices in V'. Conversely, every perfect matching in G_k is obtained uniquely in this way. Thus by computing #PERFECTMATCHING on G_k we obtain the number of matchings in G of size k, and summing over all $0 \le k \le n/2$ gives #MATCHING on G. □

We note that the proof in Lemma 1.10 is planar in the sense that if the given G is a planar graph then all graphs G_ℓ constructed from G are also planar graphs. By contrast, the proof in Lemma 1.11 is not planar in the sense that starting from a planar graph G the graphs G_k constructed from G in the proof are in general not planar. There is an intrinsic reason for this discrepancy in the proof. It turns out that #MATCHING is #P-complete even when restricted on planar graphs, while #PERFECTMATCHING is only #P-complete for general graphs; we will see in Chapter 4 that #PERFECTMATCHING can be computed in polynomial time over planar graphs. Consequently, any proof of Lemma 1.11 that is planar in the aforementioned sense would imply a collapse of #P to P. Since collapsing #P to P trivially implies P = NP, this is a more severe collapse than P = NP. This eventuality is considered most unlikely, and ultimately the dichotomy theorems in this book are meaningful only assuming #P does not collapse to P, although the statements and proofs of these theorems do not make this assumption. One of the themes in this book is also to obtain complexity classifications of counting problems, which further delineates the boundary between planar and nonplanar graphs, and the algorithm for #PERFECTMATCHING over planar graphs plays an important role.

The heart of polynomial interpolation as a reduction technique is finding an equation system like (1.9). For the ℓth equation, on the left we have a combination involving some partial sums of the original problem on G. On the right, we have the evaluation of the Holant problem on the constructed graph G_ℓ. The art is to pick G_ℓ so that a relation like (1.8) holds. It is important that the original sum on G is partitioned into at most a polynomial number of partial sums. In this example (1.8) all exponentially many configurations (up to $\binom{n}{k}$)

of matchings in G that omit exactly k vertices are combined, with the common property that each of them has exactly $(\ell + 1)^k$ extensions to a matching in G_ℓ. The construction defines an infinite family $\{G_\ell\}$, and the size of graphs in this family should not grow faster than a polynomial in G and ℓ, so that we can hope to construct polynomially many members in $\{G_\ell\}$ in polynomial time. Finally, they should indeed be constructible individually in polynomial time and some effective relationship such as (1.8) can be established.

Now we consider an example in which the polynomial is quite explicit. For any natural number $k \in \mathbb{N}$ let $\chi(G; k)$ be the number of proper vertex colorings of G using at most k colors, i.e., a labeling of the vertices of G by $[k]$ such that no two neighboring vertices receive the same label. If x, y are two vertices of G, then $G\backslash xy$ (deletion) is the graph obtained from G by deleting all edges between x and y, and G/xy (contraction) is the graph obtained from G by contracting x and y, namely, x and y are merged to form a single vertex and all edges between them are removed. Then

$$\chi(G; k) = \chi(G\backslash xy; k) - \chi(G/xy; k).$$

Hence, by induction, $\chi(G, k)$ is a polynomial in k of degree n, the number of vertices of G. The chromatic polynomial, denoted by $\chi(G; \lambda)$, is the unique polynomial in λ, such that for natural numbers $k \in \mathbb{N}$, $\chi(G; k)$ is the number of proper vertex colorings of G.

The following example of polynomial interpolation is a dichotomy theorem for the chromatic polynomial. The reduction we use comes from Linial [Lin86] and the dichotomy was first explicitly stated in [JVW90]. Let $\chi(\lambda)$ be the problem of evaluating $\chi(G; \lambda)$ on an input graph G.

Lemma 1.12. *Let $\lambda \in \mathbb{C}$. Then $\chi(\lambda)$ is* #P*-hard if $\lambda \notin \{0, 1, 2\}$. The problem is computable in polynomial time if $\lambda \in \{0, 1, 2\}$.*

Proof. If $\lambda = 0$, then $\chi(G, \lambda) = 0^0 = 1$ if G has no vertices and is 0 otherwise. If $\lambda = 1$, then $\chi(G, \lambda) = 1$ if G has no edges and is 0 otherwise. If $\lambda = 2$, then $\chi(G, \lambda) = 2^k$ if G is bipartite with k connected components and is 0 otherwise.

Now suppose $\lambda \notin \{0, 1, 2\}$. We reduce from $\chi(3)$, which is known to be #P-hard (see, e.g., [Lin86, Main Theorem, Case (6)] or [Bar04, Proposition 5]). Let G be a graph with n vertices, and let K_t be the complete graph on t vertices. We use $G + K_t$ to denote the graph obtained from G by adding K_t and all possible edges between the vertices of G and the vertices of K_t. Then clearly

$$\chi(G + K_t; \lambda) = \lambda(\lambda - 1) \cdots (\lambda - t + 1)\chi(G; \lambda - t).$$

when $\lambda \in \mathbb{N}$. Thus, it must also hold as an equation of polynomials when considering λ as an indeterminate. Thus for λ not an integer between 0 and $t - 1$,

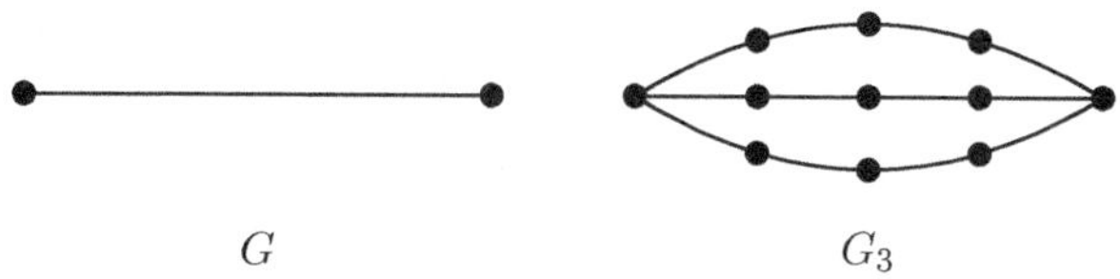

Figure 1.8. Example construction from the proof of Lemma 1.13 with $\ell = 3$.

we have

$$\chi(G; \lambda - t) = \frac{1}{\lambda(\lambda - 1)\cdots(\lambda - t + 1)}\chi(G + K_t; \lambda). \qquad (1.10)$$

If $\lambda \geq 3$ is an integer, then by setting $t = \lambda - 3$, we can directly solve for $\chi(G; 3)$ via (1.10). Otherwise, $\lambda \notin \mathbb{N}$, since $\lambda \neq 0, 1, 2$. Then choosing $0 \leq t \leq n$, using (1.10), we can compute $\chi(G; \lambda - t)$ by evaluating $\chi(G + K_t; \lambda)$ at the given λ for a sequence of graphs $G + K_t$. From these evaluations, we can interpolate the coefficients of $\chi(G; \lambda)$ and evaluate it at $\lambda = 3$. □

The proofs of both Lemmas 1.10 and 1.12 involve interpolation of a single variable polynomial. Some reductions between counting problems are accomplished via interpolation of multivariate polynomials. An early example of this occurs in [PB83, Main Theorem, Case 1], which interpolates a homogeneous polynomial in three variables.

Lemma 1.13. #VERTEXCOVER $\leq_T$ #BIPARTITEVERTEXCOVER

Proof. Given a graph G with n vertices and m edges, we create a graph G_ℓ for every $1 \leq \ell \leq N = \binom{m+2}{2}$ in two steps as follows. First we replace every edge of G with ℓ parallel edges to obtain a graph G'_ℓ; this operation is called an ℓ-thickening on G. Then we replace every edge of G'_ℓ with a path of length 4; this operation is called a 4-stretching on G'_ℓ. Since this stretch is by an even length 4, the resulting graph is bipartite. See Figure 1.8 for an example with $\ell = 3$.

We stratify all subsets S of V according to a detailed account on how many edges in G are not covered by S, how many edges are covered by exactly one vertex in S, and how many are covered by two vertices in S. Specifically, let c_{ijk}, where $i, j, k \geq 0$ and $i + j + k = m$, be the number of $S \subseteq V$ such that exactly

- i edges in G have neither endpoint in S,
- j edges in G have exactly one endpoint in S, and
- k edges in G have both endpoints in S.

Now define

$$p(x, y, z) = \sum_{\substack{i+j+k=m \\ i,j,k \geq 0}} x^i y^j z^k c_{ijk}.$$

Then it can be seen that #VERTEXCOVER$(G) = p(0, 1, 1)$.

Let v_1, v_2, v_3, v_4 and v_5 be the five vertices along a path of length 4. It is a direct verification that there are the following possibilities for vertex covers according to whether exactly zero, one, or two endpoints $\{v_1, v_5\}$ are included in a vertex cover of the path. Indeed

1. There are exactly *two* vertex covers when neither endpoint is in the vertex cover, because in this case both v_2 and v_4 must be in the vertex cover, and v_3 can be arbitrarily placed.
2. There are exactly *three* vertex covers when exactly one endpoint, say v_5, is in the vertex cover, because in this case v_2 and at least one of v_3 and v_4 must be in the vertex cover.
3. There are exactly *five* vertex covers when both endpoints are in the vertex cover, because in this case either v_3 is in the vertex cover and then v_2 and v_4 can be arbitrarily placed, or v_3 is not in the vertex cover and then both v_2 and v_4 must be in the vertex cover.

Thus,

$$\#\text{VERTEXCOVER}(G_\ell) = p(2^\ell, 3^\ell, 5^\ell) = \sum_{\substack{i+j+k=m \\ i,j,k \geq 0}} (2^i 3^j 5^k)^\ell c_{ijk}. \tag{1.11}$$

This defines a Vandermonde system that has full rank iff $2^i 3^j 5^k \neq 2^{i'} 3^{j'} 5^{k'}$ for all distinct pairs $(i, j, k) \neq (i', j', k')$. This is clearly valid by unique prime factorization. Therefore, we can interpolate $p(x, y, z)$ in polynomial time, and then evaluate it at $p(0, 1, 1)$ to obtain the number of vertex covers of G. □

In most interpolation proofs a crucial point is the following. There is a suitable stratification of the Holant sum on the original instance with at most polynomially many different types (in Lemma 1.13 it is $N = \binom{m+2}{2}$). This is usually defined by specifying the number of local configurations induced by a global object (in Lemma 1.13 it is specified by the tuple $\langle i, j, k \rangle$.) Given a type of local configuration, we can find the factor associated with it in the Holant sum on the constructed instance, and this is independent of the exact distribution of these local configurations. Here the fact that we are dealing with sum-of-product computations is the salient reason.

Remark: Let P_n denote the path graph of length n, with vertices $v_0, v_1, \ldots, v_n$. We showed in the proof that the vertex covers of P_4 can be partitioned into three sets represented by the triple $(2, 3, 5)$. A more general relation for P_n is the following recurrence relation

$$\#\text{VERTEXCOVER}(P_n) = \#\text{VERTEXCOVER}(P_{n-1}) + \#\text{VERTEXCOVER}(P_{n-2}).$$

This is the same recurrence relation as the Fibonacci numbers. The recurrence is to partition the vertex covers of P_n according to whether it contains the endpoint v_n or not. The derivation in the proof of Lemma 1.13 corresponds to the following partition: (1) Both v_0 and v_n are not in the vertex cover, in which case v_1 and v_{n-1} must be chosen and then it is the same as vertex covers for P_{n-4}. (2) Exactly one of v_0 and v_n is in the vertex cover, say v_n, in which case v_1 must be chosen and then it is the same as vertex covers for P_{n-3}. However, in the account for #VERTEXCOVER(P_n) this case should be counted twice. (3) Both v_0 and v_n are in the vertex cover in which case it is the same as vertex covers for P_{n-2}. Thus this recurrence is $x_n = x_{n-4} + 2x_{n-3} + x_{n-2}$ which is consistent with the more succinct Fibonacci recurrence relation. However, in the proof of Lemma 1.13 we need to stratify it this way because that is how the common factors $(2^i 3^j 5^k)^\ell$ can be extracted in the expression $p(2^\ell, 3^\ell, 5^\ell)$.

For homogeneous polynomials of degree d in n variables, the number of monomials is $\binom{d+n-1}{n-1}$ and thus the minimum number of points that can interpolate such polynomials is $\binom{d+n-1}{n-1}$, which is polynomial in n for a fixed d. However, often the matrix may not be of full rank. In those case we will have to work harder to prove that interpolation succeeds.

1.4 Further Discussion on Models of Computation

In fact, for the computational problems we will consider in this book, every single problem involves only a fixed number of complex numbers as given parameters $\alpha_1, \alpha_2, \ldots, \alpha_k$, and all other complex numbers involved in the problem belong to the algebraic extension field $\mathbf{F} = \mathbb{Q}(\alpha_1, \alpha_2, \ldots, \alpha_k)$. Thus these parameters $\alpha_1, \alpha_2, \ldots, \alpha_k$ are part of the specification of the problem, and should not be confused with the computational complexity, which is measured by the size of the problem instance.

Because these parameters are part of the problem specification, one may take the following formalistic view. It is known that every finite extension field $\mathbf{F}$ over the rational numbers $\mathbb{Q}$ has the following structural form [Jac85b (Theorem 8.35 on p. 512)]: $\mathbf{F}$ is a finite algebraic extension $\mathbf{E}(\beta_1, \beta_2, \ldots, \beta_\ell)$ of a certain purely transcendental extension field $\mathbf{E}$ over $\mathbb{Q}$, having the form

$\mathbf{E} = \mathbb{Q}(X_1, \ldots, X_m)$ where $X_1, \ldots, X_m$ are algebraically independent. $\mathbf{F}$ is said to have a finite transcendence degree m over $\mathbb{Q}$. It is known that $\mathbf{F}$ has a unique and well-defined transcendence degree over $\mathbb{Q}$. Then the formalistic view demands that $\mathbf{F}$ be specified as such (as part of the task of the one who specifies the problem). The β_j's are specified by irreducible polynomials over $\mathbf{E}$ as usual.

The advantage of this totally general view is that we can then consider the theorems proved in this book logically cover all $\mathbb{C}$, and not restricted to algebraic numbers. Thus, numbers such as e or π need not be excluded on the account of models of computation. However, in realistic terms, this approach is not advisable. There is too much unknown about transcendental numbers; e.g., it is still unknown whether $e + \pi$ or $e\pi$ are rational, algebraic irrational, or transcendental, and it is open whether $\mathbb{Q}(e, \pi)$ has transcendence degree 2 (or 1) over $\mathbb{Q}$, i.e., whether e and π are algebraically independent (although Nesterenko proved that π and e^{π} are algebraically independent [Nes96a, Nes96b]). By placing the foundation of this theory concerning models of computation on the existence of transcendence degree (which is proved nonconstructively), while it may appear to be elegant and sweeping, will obscure the real issue of computational complexity to be addressed in this book. Thus we choose to formally restrict ourselves to algebraic numbers.

However, owing to the mathematical inner connections and the proof techniques to be developed in this book, it is *inadvisable* to restrict ourselves to, say, integers, or $\{0, 1\}$ entries. By allowing complex numbers we discover structures that would be invisible if we were to artificially restrict ourselves to integers, or even real numbers. Moreover, the scope of algebraic numbers is adequate to develop this theory. This is the reason we choose algebraic numbers for our models of computation.

Because our computations mostly concern algebraic operations such as sum and product, the consideration of models of computation appears to be not crucial. For brevity and convenience, we will simply use $\mathbb{C}$ to refer to complex algebraic numbers in $\mathbb{C}$, and simply call them complex numbers.

1.5 An Outline of This Book

Here we give an outline of this book.

The book is divided into two volumes. Volume I of this book is about the dichotomy theory on Boolean domain problems. A characteristic feature for the theory on Boolean domain problems is that we can prove quite explicit and highly effective complexity dichotomies, in all three frameworks of spin systems, #CSP, and Holant problems. Another important feature is that we can give quite a complete classification on problems which are #P-hard in general but polynomial time tractable over planar instances. Volume II of the book will deal with counting problems over general domains. Two main theorems that will be presented are the complexity dichotomy for graph homomorphisms over $\mathbb{C}$ and for #CSP for arbitrary constraint functions taking values in $\mathbb{C}$.

For Volume I, after this introductory Chapter 1, we commence the development of our theory in Chapter 2, where we introduce Holant* problems, a restricted class of counting problems in the Holant framework where all unary (or equivalently all degenerate) functions are assumed to be freely available. We will encounter Fibonacci gates, and show that they constitute a main part of the tractable signatures in Holant* problems. In later chapters it will be seen that Fibonacci gates mainly correspond to what is transformable to the so-called product type constraint functions. The main theorem in Chapter 2 is a complexity dichotomy for Holant* problems for any set of symmetric signatures. One reason we start the presentation of this theory with Holant* problems is that here the theory is relatively simple, very concrete, and yet nontrivial. Also we can illustrate many proof ideas, such as holographic transformations in a simple setting, which will be further developed in later chapters. Another reason is that to be able to obtain all unary functions, either by direct construction or by simulation, is a main theme of proving #P-hardness in later chapters.

In Chapter 3, we consider a second framework, #CSP over Boolean domains, and prove a complexity dichotomy for #CSP for any set of constraint functions, not necessarily symmetric. Here we introduce affine signatures. Together with signatures of product type, these constitute the tractable constraints for Boolean #CSP. The important proof technique – interpolation – is employed here. The dichotomy of Chapter 3 is the special case of a much more general dichotomy of #CSP (in Volume II) restricted to the Boolean domain. Some preparatory theorems can also be derived in a more general setting. However, we choose to give a proof in the Boolean domain independently from the general theory because this proof is elementary, and the resulting dichotomy is explicit. The idea of simulating unary signatures in #P-hardness proofs also appears here, which utilizes the dichotomy from Chapter 2 both explicitly as well as in spirit. It is proved that in Boolean #CSP, one can always obtain two special unary functions, which will define a weaker type of Holant* problems, and will be called Holant^c problems.

Chapter 4 starts with the theory of Pfaffian orientations and Kasteleyn's algorithm, also known as the Fisher–Kasteleyn–Temperley algorithm, or the FKT algorithm, which can count perfect matchings on planar graphs. Then the chapter gives a thorough treatment of the theory of matchgates. It is with matchgates that holographic reductions were first introduced, although in this book we use it already to prove dichotomy theorems in Chapter 2. These matchgates, and together with holographic transformations, are important ingredients in those counting problems that are #P-hard in general but computable in polynomial time for planar graphs. The theory for general matchgates depends on matchgate identities. But for most of Volume I, dichotomy theorems concerning matchgates only use symmetric signatures; for these it is logically possible to skip matchgate identities and only use Theorem 4.11. One can derive the form of symmetric matchgate signatures as given in Theorem 4.11, by the necessary parity constraints and the explicit construction in Section 4.3.3.

Chapter 5 proves a dichotomy for spin systems on regular graphs. A number of proof techniques are introduced in this chapter, which are summarized at the beginning of the chapter. This chapter proves the dichotomy for the case of k-regular graphs, for $k = 3$ and $k = 4$. The more general case for all k is also true. However, for the purpose of the main dichotomy for Holant problems, we will only need the case $k = 3$ and 4. This chapter consciously develops the material in a parallel fashion, by introducing technical tools gradually throughout while moving toward the proof of the desired dichotomy theorem. The statement of this dichotomy for 3-regular and 4-regular graphs will be a key result for later dichotomies.

Chapter 6 is a transitional chapter, serving many purposes. It connects Holant problems and #CSP. It also introduces the transformational perspective of looking at dichotomy theorems so that now not only are holographic transformations used as proof techniques, but also they become the language in which the statement of a dichotomy theorem is expressed. In doing so, naturally we introduce group actions. This chapter also proves a dichotomy for Holant^c problems and for CSP^d; it also proves EULERIAN ORIENTATION is #P-hard. It concludes with a section on Kirchhoff's Matrix Tree theorem and some related results.

Chapter 7 proves a main dichotomy theorem for Volume I, toward which much of the material in previous chapters are preparatory. The dichotomy is proved for any set of symmetric signatures on the Boolean domain without assuming any auxiliary functions. An important new ingredient is the isolation of the vanishing signatures.

Chapter 8 proves a dichotomy for planar #CSP for any set of symmetric signatures on the Boolean domain. It shows that for the #CSP framework, a holographic reduction to Kasteleyn's algorithm, implemented by matchgates, is a universal strategy for those problems that are #P-hard but tractable over planar graphs.

Chapter 9 presents a classification theorem for planar Holant problems for any set of symmetric signatures on the Boolean domain. It turns out that for planar Holant problems there are additional problems that are #P-hard but tractable over planar graphs. These cannot be captured by holographic reductions to Kasteleyn's algorithm. However, together these two strategies are universal.

Chapter 10 presents additional dichotomy theorems for sets of asymmetric signatures. Most of the dichotomy theorems presented so far are for symmetric constraint function sets, except for #CSP in Chapter 3. In this final chapter of Volume I, we first present a classification for planar #CSP problems for all constraint functions that are not necessarily symmetric, over the Boolean domain. Then we revisit Holant* problems of Chapter 2, and prove a Holant* dichotomy for not necessarily symmetric signatures, over the Boolean domain. In doing so, we also gain some new perspective on the nature of the tractable signatures sets.

Volume II presents the theory on the general domain.

1.5.1 Logical Dependence and Omitted Proofs

The logical dependence of the results in Volume I of this book basically follow the order of the chapters. After the preliminary Chapter 1, we present the first nontrivial complexity dichotomy theorem in Chapter 2 for Holant* problems with symmetric constraint functions. Chapter 3 has minimal dependence of Chapter 2 for the proof of the dichotomy theorem for Boolean #CSP. This dependence can be removed with some alternative proof, but we feel that it is a good idea to highlight the general principle early on that the ability to interpolate all unary signatures is a powerful step toward a dichotomy. Then Chapter 4 introduces a new theme, that of counting problems over planar structures. This is the setting where holographic algorithms were originally discovered by Valiant, even though in our presentation we introduce holographic reductions first with Fibonacci gates, not with matchgates. All results in subsequent chapters that concern planar cases depend on the theory presented in Chapter 4. Theorems or parts of theorems that do not have planarity restrictions are independent of Chapter 4. Furthermore, for planar counting problems specified by symmetric local constraint functions, the logical dependence on Chapter 4 can be considerably reduced as indicated above. However, a full appreciation of the theory of what can and cannot be computed efficiently on planar structures cannot be gained with this abridged version.

Chapter 5 presents dichotomy theorems for spin systems on regular graphs. In Chapter 5 we present the proof for k-regular graphs only for $k = 3$ and $k = 4$. The reason is two folded: (1) We have organized the subsequent chapters so that logically the theorem with $k = 3$ and 4 is sufficient for subsequent theorems. (2) The technical details of this chapter are already quite demanding. By restricting to the cases $k = 3$ and $k = 4$ we can avoid some technical complications, while highlighting the essential techniques of this chapter in a more transparent setting. Chapter 6 is a transitional chapter that serves many purposes. In terms of logical dependence, it mainly introduces the transformational perspective, the Eulerian Orientation problem, and the Tutte polynomial, and proves the CSP^d dichotomy for symmetric constraints. After Chapters 5 and 6 that prepared the ground, Chapter 7 presents a dichotomy for Holant problems for symmetric constraints. This proof is quite long; and we have decided to omit the proof of Theorem 7.53, on the complexity of deciding whether a signature set satisfies the dichotomy criteria. This theorem is not needed logically for subsequent chapters. Chapter 8 presents the classification of planar Boolean #CSP for symmetric constraints. The full proof of this classification is included. Chapter 9 presents the corresponding classification of planar Holant problems over the Boolean domain for symmetric constraints. The surprise in Chapter 9 is that unlike planar #CSP problems, planar Holant problems have additional tractable problems besides those that those that can be transformed to Kasteleyn's algorithm. Here, because of the length of the proof of this theorem (Theorem 9.29) we omit most of the proof, but focus on the new tractable problems.

Finally, Chapter 10 presents the few results we currently know about asymmetric constraint functions. For planar Boolean #CSP problems, we present Theorem 10.1. This theorem generalizes the classification in Chapter 8 to all, not necessarily symmetric, constraints. This is also a generalization of the Boolean #CSP dichotomy in Chapter 3. But we omit most of its proof due to its length, merely stating some sketch of proof ideas. The second part of Chapter 10 presents in full proof the Holant* dichotomy that generalizes the main theorem of Chapter 2 to all, not necessarily symmetric constraints. For both theorems, a main proof technique is to reduce to the symmetric case and apply the dichotomy theorem for symmetric constraints. This last chapter serves as a retrospective on the journey we embarked starting from Chapter 2 for Holant* problems with symmetric constraints. Currently our knowledge about the classification program with asymmetric local constraint functions is still very limited. This knowledge is mainly restricted to Holant* problems and planar and nonplanar Boolean #CSP, as presented in Chapters 3 and 10.

2

Fibonacci Gates and Holant* Problems

We start with a special family of signatures called Fibonacci gates. They will be an important building block for tractable Holant problems.

2.1 Fibonacci Gates

Consider the signature $f = [1, 0, 1, 1]$. What is the complexity of Holant(f)? It is plain that f satisfies $f_{k+2} = f_{k+1} + f_k$, for $0 \le k \le 1$, and this turns out to be significant.

Definition 2.1. For any $n \ge 1$, a signature $f = [f_0, f_1, \ldots, f_n]$ is a Fibonacci gate if

$$f_{k+2} = f_{k+1} + f_k, \quad 0 \le k \le n-2.$$

A set of signatures $\mathcal{F}$ is called Fibonacci if every signature in $\mathcal{F}$ is a Fibonacci gate.

Each Fibonacci gate f has two degrees of freedom, as f_0 and f_1 are arbitrary; the rest of the f's the other f_k's are determined by the linear recurrence. In particular, any unary signature is a Fibonacci gate. As with the ordinary Fibonacci sequence, the linear recurrence satisfied by f gives the following formula

$$f_k = c_1\phi^k + c_2\overline{\phi}^k, \quad 0 \le k \le n, \tag{2.1}$$

where ϕ is the golden ratio $\frac{1+\sqrt{5}}{2}$ and $\overline{\phi}$ is its conjugate $\frac{1-\sqrt{5}}{2}$, both roots of the quadratic polynomial $X^2 - X - 1$. To match the initial values f_0 and f_1, we have $n = 3$, $c_1 = (-\overline{\phi}/\sqrt{5})f_0 + (1/\sqrt{5})f_1$ and $c_2 = (\phi/\sqrt{5})f_0 + (-1/\sqrt{5})f_1$. For example, for $f = [1, 0, 1, 1]$, we have $c_1 = -\overline{\phi}/\sqrt{5}$ and $c_2 = \phi/\sqrt{5}$.

For $f = [1, 0, 1, 1]$, Eq. (2.1) can be written as a sum of tensor products:

$$f = c_1 \begin{bmatrix} 1 \\ \phi \end{bmatrix}^{\otimes 3} + c_2 \begin{bmatrix} 1 \\ \overline{\phi} \end{bmatrix}^{\otimes 3}. \tag{2.2}$$

This lists the values of f as a column vector indexed by its input assignments in lexicographic order.

Consider the matrix $M = \begin{bmatrix} 1 & 1 \\ \phi & \bar{\phi} \end{bmatrix}$. We have $M^{\otimes 3}\left(c_1 \begin{bmatrix} 1 \\ 0 \end{bmatrix}^{\otimes 3} + c_2 \begin{bmatrix} 0 \\ 1 \end{bmatrix}^{\otimes 3}\right) = f$. This suggests that we take a *holographic transformation* by the matrix M.

Recall that holographic transformations are defined in Section 1.3.2 of Chapter 1. By Theorem 1.5, for a bipartite Holant problem Holant($\mathcal{F} \mid \mathcal{G}$), the holographic transformation does not change the value

$$\text{Holant}(\Omega; \mathcal{F} \mid \mathcal{G}) = \text{Holant}(\Omega'; \mathcal{F}M \mid M^{-1}\mathcal{G}),$$

where Ω' is the transformed signature grid.

Let's review how we can treat a problem such as Holant(f) as a bipartite Holant problem, and in concrete terms what Theorem 1.5 says in this case. We can view Holant(f) as computing the following tensor *contraction* operation. Given a 3-regular graph Γ where each vertex is labeled f, first we replace each edge e by a path of length 2, and label the new vertex the binary EQUALITY function $(=_2) = [1, 0, 1]$. This transforms the Holant problem Holant(f) to a bipartite Holant problem Holant($=_2 \mid f$) on the Edge-Vertex incidence graph without changing its value on Γ. Next we form the tensor products $\mathcal{R}$ of all left-hand-side copies of $(=_2)$, and $\mathcal{G}$ of all right-hand-side copies of f, respectively. If Γ has m edges, these two tensors, $\mathcal{R}$ and $\mathcal{G}$, both have dimension 2^{2m}, each representing a function of arity $2m$, and there is a one-to-one correspondence between the variables according to the incidence relation in Γ. The contraction operation $\langle \mathcal{R}, \mathcal{G} \rangle$ is $\sum_{x_1, x_2, \ldots, x_{2m}=0,1} \mathcal{R}(x_1, x_2, \ldots, x_{2m})\mathcal{G}(x_1, x_2, \ldots, x_{2m})$. If we view $\mathcal{R}$ as a (row) vector of dimension 2^{2m} and $\mathcal{G}$ as a (column) vector of the same dimension, then the contraction $\langle \mathcal{R}, \mathcal{G} \rangle$ is the dot product of the two vectors.

Now we consider a transformation $(M^{-1})^{\otimes 2m}$ on $\mathcal{G}$ and $M^{\otimes 2m}$ on $\mathcal{R}$, namely

$$\mathcal{G} \mapsto \mathcal{G}' = (M^{-1})^{\otimes 2m}\mathcal{G}, \quad \text{and} \quad \mathcal{R} \mapsto \mathcal{R}' = \mathcal{R}M^{\otimes 2m}. \tag{2.3}$$

It is clear that

$$\langle \mathcal{R}', \mathcal{G}' \rangle = \langle \mathcal{R}, \mathcal{G} \rangle.$$

The transformation in (2.3) can be carried out by distributing M and M^{-1} to each left-hand side copy of $(=_2)$ and right-hand side copy of f. Thus we can consider $(=_2) \mapsto (=_2)M^{\otimes 2}$, where $(=_2)$ is written as a row vector $(1, 0, 0, 1)$, and $f \mapsto (M^{-1})^{\otimes 3} f$, where f is given in (2.2). It follows that

$$(M^{-1})^{\otimes 3} f = c_1 \begin{bmatrix} 1 \\ 0 \end{bmatrix}^{\otimes 3} + c_2 \begin{bmatrix} 0 \\ 1 \end{bmatrix}^{\otimes 3},$$

and in symmetric signature notation this is $(M^{-1})^{\otimes 3} f = [c_1, 0, 0, c_2]$. Note that signatures of the form $[*, 0, \ldots, 0, *]$ force all incident edges to take the same value of 0 or 1. They are called GENERALIZED EQUALITIES, or GEN-EQ for short.

Recall that if we write a binary signature f as a matrix F, with entry $f(i, j)$ at row i and column j (where $i, j = 0, 1$), then MFM^{T} is the matrix form of $M^{\otimes 2} f$, where we write f as a column vector with entries ordered lexicographically. Similarly, $M^{\mathrm{T}} F M$ is the matrix form of $f M^{\otimes 2}$, where we write f as a row vector. It is clear that the identity matrix I is the matrix form of $(=_2)$. It follows that for $M = \begin{bmatrix} 1 & 1 \\ \phi & \overline{\phi} \end{bmatrix}$, $(=_2) M^{\otimes 2}$ takes the form $M^{\mathrm{T}} M = \begin{bmatrix} 1+\phi^2 & 0 \\ 0 & 1+\overline{\phi}^2 \end{bmatrix}$, which is $[1+\phi^2, 0, 1+\overline{\phi}^2]$ in symmetric signature notation. Thus both sides of the bipartite Holant problem $\operatorname{Holant}((=_2)M^{\otimes 2} \mid (M^{-1})^{\otimes 3} f)$ are GEN-EQ.

This means that the Holant value of $\operatorname{Holant}(=_2 \mid f)$, which is the same as the Holant value of $\operatorname{Holant}((=_2)M^{\otimes 2} \mid (M^{-1})^{\otimes 3} f)$, can be easily computed: On each connected component of the given graph Γ, any assignment on a single edge can be uniquely propagated to the entire component. The Holant value on Γ is the product over all connected components.

This can be generalized to any set $\mathcal{F}$ of Fibonacci gates. If $f = [f_0, f_1, \ldots, f_n] \in \mathcal{F}$ is a Fibonacci gate, then (2.1) can be written as

$$f = c_1 \begin{bmatrix} 1 \\ \phi \end{bmatrix}^{\otimes n} + c_2 \begin{bmatrix} 1 \\ \overline{\phi} \end{bmatrix}^{\otimes n}. \tag{2.4}$$

Thus, the same holographic transformation $f \mapsto (M^{-1})^{\otimes n} f$ by M transforms f to a GEN-EQ. This shows that $\operatorname{Holant}(\mathcal{F})$ is tractable.

Exercise: Show that there is an orthogonal matrix $M \in \mathbb{C}^{2\times 2}$ that transforms every signature of the form (2.4) to a GEN-EQ. Use this fact to prove that $\operatorname{Holant}(\mathcal{F})$ is tractable for a set of Fibonacci gates $\mathcal{F}$.

There is an alternative proof of the tractability of $\operatorname{Holant}(\mathcal{F})$ for any finite set of Fibonacci gates $\mathcal{F}$. We turn to this approach next.

Definition 2.2. For any $n \geq 1$, and a parameter $\lambda \in \mathbb{C}$, a signature $f = [f_0, f_1, \ldots, f_n]$ is a generalized Fibonacci gate (with parameter λ) if

$$f_{k+2} = \lambda f_{k+1} + f_k, \quad 0 \leq k \leq n-2. \tag{2.5}$$

A set of signatures $\mathcal{F}$ is called generalized Fibonacci if for some $\lambda \in \mathbb{C}$, every signature in $\mathcal{F}$ is a generalized Fibonacci gate with parameter λ.

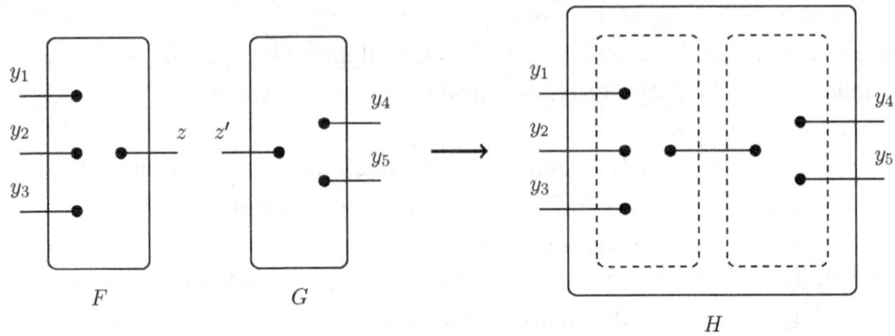

Figure 2.1. First operation.

Theorem 2.3. *For any finite set of generalized Fibonacci gates $\mathcal{F}$, the Holant problem Holant($\mathcal{F}$) is computable in polynomial time.*

Proof. If $\Gamma_1, \Gamma_2, \ldots, \Gamma_\kappa$ are the connected components of a graph Γ, then

$$\text{Holant}_\Gamma = \prod_{j=1}^{\kappa} \text{Holant}_{\Gamma_j}.$$

So we only need to consider connected graphs as inputs.

Suppose Γ has n nodes and m edges. First we cut all the edges in Γ. A node with degree d can be viewed as an $\mathcal{F}$-gate with d dangling edges. Now step by step we merge two dangling edges into one regular edge in the original graph, until we recover Γ after m steps. We prove that all the intermediate $\mathcal{F}$-gates still have generalized Fibonacci signatures with the same parameter λ, and at every step we can compute the intermediate signature in polynomial time. After m steps we get Γ as an $\mathcal{F}$-gate with no dangling edge; the only value of its signature is the Holant value we want. To carry this out, we only need to prove that it is true for one single step. There are two cases, depending on whether the two dangling edges to be merged are in the same component or not. These two operations are illustrated in Figures 2.1 and 2.2.

In the first case, the two dangling edges belong to two components before their merging (Figure 2.1). Let F have dangling edges $y_1, \ldots, y_s, z$ and G have dangling edges $y_{s+1}, \ldots, y_{s+t}, z'$. After merging z with z', we have a new $\mathcal{F}$-gate H with dangling edges $y_1, \ldots, y_{s+t}$. Inductively the signatures of F and G are both generalized Fibonacci gates with the same parameter λ. We show that this remains so for the resulting $\mathcal{F}$-gate H.

We first prove that H is symmetric. We only need to show that the value of H is unchanged if the values of two inputs are exchanged. Because F and G are symmetric, if both inputs are from $\{y_1, \ldots, y_s\}$ or from

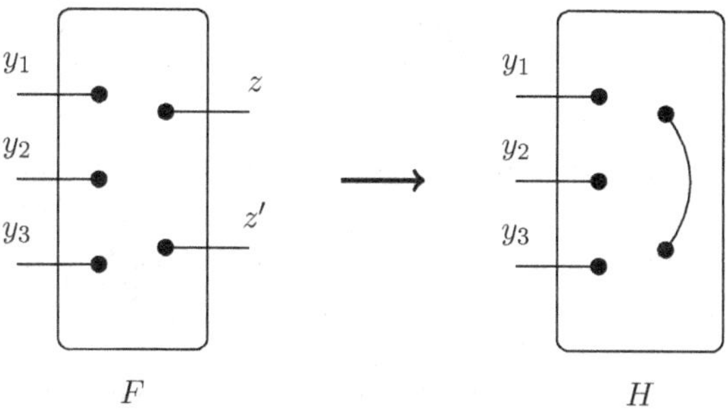

Figure 2.2. Second operation.

$\{y_{s+1}, \ldots, y_{s+t}\}$, the value of H is clearly unchanged. Suppose one input is from $\{y_1, \ldots, y_s\}$ and the other is from $\{y_{s+1}, \ldots, y_{s+t}\}$. By the symmetry of F and G we may assume these two inputs are y_1 and y_{s+1}. Thus we will fix an arbitrary assignment for $y_2, \ldots, y_s, y_{s+2}, \ldots, y_{s+t}$, and we want to show $H(0, y_2, \ldots, y_s, 1, y_{s+2}, \ldots, y_{s+t}) = H(1, y_2, \ldots, y_s, 0, y_{s+2}, \ldots, y_{s+t})$.

We will suppress the fixed values $y_2, \ldots, y_s, y_{s+2}, \ldots, y_{s+t}$ and denote

$$\begin{aligned} F_{xz} &= F(x, y_2, \ldots, y_s, z), \\ G_{yz} &= G(y, y_{s+2}, \ldots, y_{s+t}, z), \quad \text{and} \\ H_{xy} &= H(x, y_2, \ldots, y_s, y, y_{s+2}, \ldots, y_{s+t}). \end{aligned}$$

Then by the definition of Holant, $H_{ab} = F_{a0}G_{b0} + F_{a1}G_{b1}$, for $a, b \in \{0, 1\}$.

Because F and G are generalized Fibonacci gates with parameter λ, we have

$$H_{01} = F_{00}G_{10} + F_{01}(\lambda G_{01} + G_{00}), \quad \text{and} \quad H_{10} = F_{10}G_{00} + (\lambda F_{01} + F_{00})G_{01}.$$

By the symmetry of F and G, we have $H_{01} = H_{10}$.

Now we show that $H(y_1, \ldots, y_{s+t})$ is also a generalized Fibonacci gate with parameter λ. Since we have proved that H is symmetric, we can choose any two input variables to prove it being Fibonacci. Again, we choose y_1 and y_{s+1}. (This assumes that y_1 and y_{s+1} exist, i.e., F and G are not unary functions. If either one of them is unary, the proof is just as easy.) For any fixed values of all other variables, we have $H_{00} = F_{00}G_{00} + F_{01}G_{01}$, $H_{01} = F_{00}G_{10} + F_{01}G_{11}$, and $H_{11} = F_{10}G_{10} + F_{11}G_{11}$. Now using the fact that both F and G are generalized Fibonacci gates with parameter λ, it follows that $H_{11} = \lambda H_{01} + H_{00}$.

If the first two terms of the signatures of F and G are f_0, f_1 and g_0, g_1 respectively, then the first two terms of the signature H are $h_0 = f_0g_0 + f_1g_1$ and $h_1 = f_0g_1 + f_1g_0 + \lambda f_1g_1$.

Next we consider the second case, where the two dangling edges to be merged are in the same component (Figure 2.2). Obviously, the signature for the new gate H is also symmetric. If $F = [f_0, f_1, \ldots, f_n]$ is the signature before the merging operation, then the signature after the merging operation is $H = [f_0 + f_2, f_1 + f_3, \ldots, f_{n-2} + f_n]$. Such an operation preserves the linear recurrence. It follows that H is also a generalized Fibonacci gate with parameter λ, and the signature has already been computed. □

We remark that the proof of tractability in Theorem 2.3 is also valid for an infinite set $\mathcal{F}$ of generalized Fibonacci gates with the same parameter λ; however, for an infinite set $\mathcal{F}$ the input size should include the description of the signatures used in the signature grid. See Remark 1 after Theorem 2.12.

2.2 Orthogonal Transformation of Fibonacci Gates

By Definition 2.1, Fibonacci gates correspond to the case $\lambda = 1$ in Definition 2.2. How much generality does an arbitrary parameter λ provide? Can we recapture the tractability of generalized Fibonacci gates by holographic transformation as we did for $\lambda = 1$?

The recurrence relation $f_{k+2} = \lambda f_{k+1} + f_k$ has the characteristic polynomial $X^2 - \lambda X - 1$. If $\lambda \neq \pm 2\mathrm{i}$ then there are two distinct eigenvalues $\alpha, \beta \neq \pm \mathrm{i}$, and a general solution to the recurrence has the form $f_k = a\alpha^k + b\beta^k$, for some constants a and b determined by f_0 and f_1. Note that $\alpha\beta = -1$. More revealing is the following expression in terms of tensor products:

$$f = a\begin{bmatrix} 1 \\ \alpha \end{bmatrix}^{\otimes n} + b\begin{bmatrix} 1 \\ \beta \end{bmatrix}^{\otimes n}. \tag{2.6}$$

Note that the two vectors $\begin{pmatrix} 1 & \alpha \end{pmatrix}^{\mathrm{T}}$ and $\begin{pmatrix} 1 & \beta \end{pmatrix}^{\mathrm{T}}$ are orthogonal, i.e., their dot product is zero, and $1 + \alpha^2, 1 + \beta^2 \neq 0$. We can scale the entries as follows. Let

$$T = \begin{bmatrix} t_{00} & t_{01} \\ t_{10} & t_{11} \end{bmatrix} = \begin{bmatrix} \frac{1}{\sqrt{1+\alpha^2}} & \frac{1}{\sqrt{1+\beta^2}} \\ \frac{\alpha}{\sqrt{1+\alpha^2}} & \frac{\beta}{\sqrt{1+\beta^2}} \end{bmatrix}.$$

Then T is an orthogonal matrix. Let $a' = a(1 + \alpha^2)^{n/2}$ and $b' = b(1 + \beta^2)^{n/2}$; then we have

$$f = a'\begin{bmatrix} t_{00} \\ t_{10} \end{bmatrix}^{\otimes n} + b'\begin{bmatrix} t_{01} \\ t_{11} \end{bmatrix}^{\otimes n} = T^{\otimes n}\left(a'\begin{bmatrix} 1 \\ 0 \end{bmatrix}^{\otimes n} + b'\begin{bmatrix} 0 \\ 1 \end{bmatrix}^{\otimes n}\right). \tag{2.7}$$

Hence the holographic transformation $(T^{-1})^{\otimes n} f$ is the GEN-EQ signature $a' \begin{bmatrix} 1 \\ 0 \end{bmatrix}^{\otimes n} + b' \begin{bmatrix} 0 \\ 1 \end{bmatrix}^{\otimes n} = [a', 0, \ldots, 0, b']$. Meanwhile the holographic transformation on the left-hand side keeps the binary EQUALITY $(=_2)$ unchanged by T:

$$(=_2) \mapsto (=_2)T^{\otimes 2} = (=_2), \tag{2.8}$$

because T is an orthogonal matrix, and in matrix form $(=_2)T^{\otimes 2}$ is $T^{\mathrm{T}} I T = I$.

Thus for any finite set of generalized Fibonacci gates $\mathcal{F}$ with parameter $\lambda \neq \pm 2\mathrm{i}$, the Holant problem Holant($\mathcal{F}$), being equivalent to Holant($=_2 | \mathcal{F}$), is tractable. This is an alternative proof of Theorem 2.3, when $\lambda \neq \pm 2\mathrm{i}$.

By a limiting argument, one can give a proof of tractability by this *holographic* approach when $\lambda = \pm 2\mathrm{i}$ as well. We will, however, refrain from that, because intrinsically there is a difference for the signatures when $\lambda = \pm 2\mathrm{i}$.

Every generalized Fibonacci gate with $\lambda \neq \pm 2\mathrm{i}$ can be transformed to a GEN-EQ by some orthogonal matrix. By composition, for any λ, $\mu \neq \pm 2\mathrm{i}$, any generalized Fibonacci gate with parameter λ can also be transformed by some orthogonal matrix to another generalized Fibonacci gate with parameter μ. We can take the basic Fibonacci gate with $\lambda = 1$, as well as GEN-EQ, as a normal form for this class. The orthogonal transformation between these two is the normalized version of the matrix M from Section 2.1: $\begin{bmatrix} \frac{1}{\sqrt{1+\phi^2}} & \frac{1}{\sqrt{1+\overline{\phi}^2}} \\ \frac{\phi}{\sqrt{1+\phi^2}} & \frac{\overline{\phi}}{\sqrt{1+\overline{\phi}^2}} \end{bmatrix}$.

An invertible holographic transformation maps nondegenerate signatures to nondegenerate signatures. This is because if $fM^{\otimes n} = (a_1, b_1) \otimes \cdots \otimes (a_n, b_n)$, then $f = (a'_1, b'_1) \otimes \cdots \otimes (a'_n, b'_n)$, where $(a'_k, b'_k) = (a_k, b_k)M^{-1}$. The same argument works for contravariant tensors. Suppose f is a generalized Fibonacci gate satisfying (2.6). The following are equivalent: (1) f is nondegenerate, (2) $ab \neq 0$ in (2.6), (3) its GEN-EQ form $[a', 0, \ldots, 0, b']$ in (2.7) is nondegenerate, and (4) $a'b' \neq 0$.

Define

$$\mathscr{F} = \{f \mid f \text{ satisfies (2.5) for some } \lambda \neq \pm 2\mathrm{i}\} \cup \text{GEN-EQ}. \tag{2.9}$$

Theorem 2.4. *For any $f \in \mathscr{F}$ in (2.9),*

1. *There exists an orthogonal T such that Tf is a* GEN-EQ.
2. *There exists an orthogonal T such that Tf is a Fibonacci gate satisfying Definition 2.1.*
3. *For all orthogonal T, $Tf \in \mathscr{F}$.*

Proof. We only need to prove the last item. By composition and the invertibility of the transformation, we may assume f is a GEN-EQ function $[a', 0, \ldots, 0, b']$.

If $T = \begin{bmatrix} t_{00} & t_{01} \\ t_{10} & t_{11} \end{bmatrix}$, then

$$Tf = a' \begin{bmatrix} t_{00} \\ t_{10} \end{bmatrix}^{\otimes n} + b' \begin{bmatrix} t_{01} \\ t_{11} \end{bmatrix}^{\otimes n}.$$

If any entry $t_{ij} = 0$, then by orthogonality $t_{\bar{i}\bar{j}} = 0$, where $\bar{x} = 1 - x$, and Tf is a GEN-EQ. Suppose T has no zero entries; then Tf satisfies a recurrence relation with the characteristic roots $\lambda_1 = t_{10}/t_{00}$ and $\lambda_2 = t_{11}/t_{01}$. We have $\lambda_1 \neq \lambda_2$ and $\lambda_1\lambda_2 = -1$, by the orthogonality of T. Hence the characteristic polynomial of the recurrence is $X^2 - (\lambda_1 + \lambda_2)X - 1$. Since $\lambda_1 \neq \lambda_2$, we have $\lambda_1 + \lambda_2 \neq \pm 2\mathrm{i}$. Thus $Tf \in \mathscr{F}$. □

We may consider GEN-EQ gates provide a normal form for generalized Fibonacci gates (with $\lambda \neq \pm 2\mathrm{i}$) under orthogonal transformations as in (2.7). Curiously, any GEN-EQ of arity at least 3 does not satisfy any recurrence $f_{k+2} = \lambda f_{k+1} + f_k$. (We can generalize the form of a second-order recurrence to $af_k + bf_{k+1} - af_{k+2} = 0$ for some $(a, b) \neq 0$. Then GEN-EQ does satisfy the recurrence with $(a, b) = (0, 1)$.) The proof of Theorem 2.4 also establishes the following: Other than GEN-EQ itself, any signature f expressible as (2.7), i.e., an orthogonal transformation of GEN-EQ, is in fact a generalized Fibonacci gate.

Exercise: Prove that generalized Fibonacci gates with parameter $\lambda = \pm 2\mathrm{i}$ cannot be transformed to any member of $\mathscr{F}$ in (2.9).

We now ask, what is the right perspective for generalized Fibonacci gates with parameter $\lambda = \pm 2\mathrm{i}$ under holographic transformations?

Let $\mathcal{F}$ be such a set of signatures. To show Holant($\mathcal{F}$) is tractable, we first get rid of degenerate signatures as follows. Each degenerate signature can be replaced by a set of unary signatures. Then combine a unary signature with any $f \in \mathcal{F}$ satisfying (2.5) produces another signature f' with arity$(f') = $ arity$(f) - 1$, which also satisfies (2.5) with the same λ. So we may assume every signature in $\mathcal{F}$ is nondegenerate and has arity ≥ 2.

If every vertex in the given instance graph H has degree ≤ 2, then the Holant value Holant($\mathcal{F}$) can be obtained by computing it as a matrix product and then taking the trace, per each connected component.

Each $f \in \mathcal{F}$ has the form $f_k = ck\mu^{k-1} + d\mu^k$ $(0 \leq k \leq n$ and $\mu = \pm\mathrm{i})$, where $X^2 - 2\mu X - 1$ has double root μ. We will suppose $\mu = \mathrm{i}$. The case

$\mu = -\mathrm{i}$ is similar. Consider

$$f' = [c', d', 0, \ldots, 0] = c' \begin{bmatrix} 1 \\ 0 \end{bmatrix}^{\otimes n} + d' \sum_{j=1}^{n} \left\{ \begin{bmatrix} 1 \\ 0 \end{bmatrix}^{\otimes (j-1)} \otimes \begin{bmatrix} 0 \\ 1 \end{bmatrix} \otimes \begin{bmatrix} 1 \\ 0 \end{bmatrix}^{\otimes (n-j)} \right\}, \tag{2.10}$$

where we set $c' = d - cn\mathrm{i}/2$ and $d' = c\mathrm{i}/2$. Let

$$Z = \begin{bmatrix} 1 & 1 \\ \mathrm{i} & -\mathrm{i} \end{bmatrix}. \tag{2.11}$$

We apply $Z^{\otimes n}$ to f' by distributing each Z among the factors of the tensor product in (2.10):

$$Z^{\otimes n} f' = c' \begin{bmatrix} 1 \\ \mathrm{i} \end{bmatrix}^{\otimes n} + d' \sum_{j=1}^{n} \left\{ \begin{bmatrix} 1 \\ \mathrm{i} \end{bmatrix}^{\otimes (j-1)} \otimes \begin{bmatrix} 1 \\ -\mathrm{i} \end{bmatrix} \otimes \begin{bmatrix} 1 \\ \mathrm{i} \end{bmatrix}^{\otimes (n-j)} \right\}. \tag{2.12}$$

Then $Z^{\otimes n} f' = f$. To wit, any entry of $Z^{\otimes n} f'$ indexed by a bit pattern in $\{0, 1\}^n$ of Hamming weight k is $c'\mathrm{i}^k + d'[k(-\mathrm{i})\mathrm{i}^{k-1} + (n-k)\mathrm{i}^k] = ck\mathrm{i}^{k-1} + d\mathrm{i}^k$. Thus the contravariant transformation $(Z^{-1})^{\otimes n} f = f'$. Let $\mathcal{F}' = Z^{-1}\mathcal{F}$. Under the covariant holographic transformation by Z, $(=_2)$ is transformed to $Z^{\mathrm{T}} I Z = 2\begin{bmatrix} 0 & 1 \\ 1 & 0 \end{bmatrix}$, which is two times the binary DISEQUALITY $(\neq_2)$ function $[0, 1, 0]$ in matrix form. Holant($\mathcal{F}$), which is the same as Holant($=_2 | \, \mathcal{F}$), is equivalent to Holant($\neq_2 | \, \mathcal{F}'$) by Theorem 1.5. The bipartite problem Holant($\neq_2 | \, \mathcal{F}'$), where each $f' \in \mathcal{F}'$ has the form in (2.10), and may have different parameters c' and d', is tractable by the following argument. In any nonzero evaluation, on account of $(\neq_2)$, an assignment must assign exactly half the edges 0 and the other half 1. For any f' in (2.10) with arity greater than 2, to be nonzero it must have strictly more incident edges assigned 0. Any f' with arity 2 requires at least half of the input bits be set to 0. Therefore the Holant value is 0 unless every f' has arity 2. Hence the computation reduces to the case where each function has arity ≤ 2, and the Holant value is computed by matrix product.

What is the class of Holant problems that can be solved by Fibonacci gates, possibly after a suitable holographic transformation? We now characterize this class of problems in terms of signatures that can be transformed to a Fibonacci gate under a holographic transformation.

Theorem 2.5. *A symmetric signature $[f_0, f_1, \ldots, f_n]$ can be transformed by some invertible holographic transformation to a Fibonacci gate according to Definition 2.1 (equivalently to a signature in $\mathscr{F}$ defined in (2.9)) iff there exist three constants a, b and c, such that $b^2 - 4ac \neq 0$, and for all $0 \leq k \leq n-2$,*

$$af_k + bf_{k+1} + cf_{k+2} = 0. \tag{2.13}$$

Proof. For an invertible $M = \begin{bmatrix} \alpha & \gamma \\ \beta & \delta \end{bmatrix}$, we have

$$M^{\otimes n}[u, 0, \ldots, 0, v] = u \begin{bmatrix} \alpha \\ \beta \end{bmatrix}^{\otimes n} + v \begin{bmatrix} \gamma \\ \delta \end{bmatrix}^{\otimes n}.$$

If f has this form, then f satisfies a recurrence relation $af_k + bf_{k+1} + cf_{k+2} = 0$ $(0 \le k \le n-2)$, where a, b, and c satisfy $b^2 - 4ac \neq 0$. In particular, they are not all 0. It can be directly verified that the following choice works: $a = \beta\delta$, $b = -(\alpha\delta + \beta\gamma)$ and $c = \alpha\gamma$.

Conversely, from $af_k + bf_{k+1} + cf_{k+2} = 0$, if $c \neq 0$ we have a second-order linear recurrence with distinct eigenvalues λ and μ, which gives us the expression

$$f = u \begin{bmatrix} 1 \\ \lambda \end{bmatrix}^{\otimes n} + v \begin{bmatrix} 1 \\ \mu \end{bmatrix}^{\otimes n} = M[u, 0, \ldots, 0, v],$$

where $M = \begin{bmatrix} 1 & 1 \\ \lambda & \mu \end{bmatrix}$. If $a \neq 0$, then we have a reverse linear recurrence. If $a = c = 0$, then $b \neq 0$, and we have a GEN-EQ. □

Theorem 2.6. Holant $(\{R_1, R_2, \ldots, R_t\} \mid \{G_1, G_2, \ldots, G_s\})$ *is tractable, where G_i and R_j have arities n_i and m_j respectively, if there exists an invertible M such that all $M^{-1}G_i$ and R_jM are Fibonacci gates. This happens iff there exist a, b, and c, such that $b^2 - 4ac \neq 0$ and the following two conditions are satisfied:*

1. *For any $G_i = [x_1^{(i)}, x_2^{(i)}, \ldots, x_{n_i}^{(i)}]$ and any $0 \le k \le n_i - 2$, $ax_k^{(i)} + bx_{k+1}^{(i)} + cx_{k+2}^{(i)} = 0$.*
2. *For any $R_j = [y_1^{(j)}, y_2^{(j)}, \ldots, y_{m_j}^{(j)}]$ and any $0 \le k \le m_j - 2$, $cy_k^{(j)} - by_{k+1}^{(j)} + ay_{k+2}^{(j)} = 0$.*

Proof. We may take the normal form GEN-EQ, and assume $(M^{-1})^{\otimes n_i} G_i = [*, 0, \ldots, 0, *]$ and $R_j M^{\otimes m_j} = [*, 0, \ldots, 0, *]$, for some $M = \begin{bmatrix} \alpha & \gamma \\ \beta & \delta \end{bmatrix}$. By the proof of Theorem 2.5, there exist $a = \beta\delta$, $b = -(\alpha\delta + \beta\gamma)$ and $c = \alpha\gamma$, such that $b^2 - 4ac \neq 0$, and for all $G_i = [x_1^{(i)}, x_2^{(i)}, \ldots, x_{n_i}^{(i)}]$, $ax_k^{(i)} + bx_{k+1}^{(i)} + cx_{k+2}^{(i)} = 0$ (for $0 \le k \le n_i - 2$). The characteristic polynomial is $cX^2 + bX + a$. If we write R_j in its transpose, then Theorem 2.5 applies to R_j, with $(M^{-1})^{\mathrm{T}} = \frac{1}{\det(M)} \begin{bmatrix} \delta & -\beta \\ -\gamma & \alpha \end{bmatrix}$ replacing M. It follows that each R_j satisfies $cy_k^{(j)} - by_{k+1}^{(j)} + ay_{k+2}^{(j)} = 0$ $(0 \le k \le m_j - 2)$.

Conversely, suppose all G_i and R_j satisfy the linear recurrences. Suppose $c \neq 0$. Then for each G_i the characteristic polynomial of the recurrence is $X^2 + \frac{b}{c}X + \frac{a}{c}$. Let Δ denote the discriminant $b^2 - 4ac$. Let $M = \begin{bmatrix} 2c & 2c \\ -b+\sqrt{\Delta} & -b-\sqrt{\Delta} \end{bmatrix}$. Obviously this is nonsingular, and it can be verified that $G_i = a_i \begin{bmatrix} 2c \\ -b+\sqrt{\Delta} \end{bmatrix}^{\otimes n_i} + b_i \begin{bmatrix} 2c \\ -b-\sqrt{\Delta} \end{bmatrix}^{\otimes n_i}$, for some constants a_i and b_i depending on G_i. Similarly, each R_j satisfies $R_j = c_j \left[b+\sqrt{\Delta} \;\; 2c\right]^{\otimes m_j} + d_j \left[b-\sqrt{\Delta} \;\; 2c\right]^{\otimes m_j}$, for some c_j and d_j depending on R_j. Then both $(M^{-1})^{\otimes n_i} G_i$ and $R_j M^{\otimes m_j}$ are GEN-EQ. The proof is similar if $a \neq 0$. If $a = c = 0$, then $b \neq 0$, and both G_i and R_j are already GEN-EQ. □

2.3 A Dichotomy Theorem for Holant*$(\mathcal{F})$

Every degenerate signature is a tensor product of unary signatures. If $\mathcal{F}$ consists of degenerate signatures only, each signature grid Ω on Holant$(\mathcal{F})$ is decomposed into a disjoint union of components, one for each vertex, and the Holant value is trivially computable. To isolate more interesting tractable families of signatures, we consider the following Holant* problems, where all unary functions $\mathcal{U}$ are assumed to be free.

Definition 2.7. For any signature set $\mathcal{F}$,

$$\text{Holant}^*(\mathcal{F}) = \text{Holant}(\mathcal{F} \cup \mathcal{U}).$$

For bipartite graphs we also denote

$$\text{Holant}^*(\mathcal{F} \mid \mathcal{G}) = \text{Holant}(\mathcal{F} \cup \mathcal{U} \mid \mathcal{G} \cup \mathcal{U}),$$

for any signature sets $\mathcal{F}$ and $\mathcal{G}$.

We begin with a technical proposition.

Proposition 2.8. *Let $n \geq 3$ and let $[x_0, x_1, \ldots, x_n]$ be a nondegenerate symmetric signature. Then for any $2 \leq m \leq n-1$, there exists a nondegenerate subsignature of arity m, unless the signature is of the form $[x_0, 0, \ldots, 0, x_n]$.*

Proof. We assume $(x_1, \ldots, x_{n-1}) \neq 0$, otherwise we are done. We first prove that

$$\text{rank} \begin{bmatrix} x_0 & \cdots & x_{n-2} \\ x_1 & \cdots & x_{n-1} \end{bmatrix} = 2, \quad \text{or} \quad \text{rank} \begin{bmatrix} x_1 & \cdots & x_{n-1} \\ x_2 & \cdots & x_n \end{bmatrix} = 2. \tag{2.14}$$

Suppose otherwise. Then there are λ, μ such that $(x_0, \ldots, x_{n-2}) = \lambda(x_1, \ldots, x_{n-1})$ and $(x_2, \ldots, x_n) = \mu(x_1, \ldots, x_{n-1})$. If $\lambda = 0$ then $x_0 = \ldots = x_{n-2} = 0$ and $x_{n-1} \neq 0$, and $\det \begin{bmatrix} x_{n-2} & x_{n-1} \\ x_{n-1} & x_n \end{bmatrix} \neq 0$. Hence $\lambda \neq 0$. Similarly $\mu \neq 0$. Then any $x_i \neq 0$ $(0 < i < n)$ implies that $x_{i-1} \neq 0$ and $x_{i+1} \neq 0$. Hence $x_i \neq 0$ for all $0 \leq i \leq n$. This implies that $\lambda\mu = 1$, and thus $\operatorname{rank} \begin{bmatrix} x_0 & \ldots & x_{n-1} \\ x_1 & \ldots & x_n \end{bmatrix} = 1$, a contradiction to the hypothesis of the lemma. Hence (2.14) holds. Then an induction completes the proof. Note that if $[x_0, \ldots, x_{n-1}]$ is nondegenerate and has the form $[x_0, 0, \ldots, 0, x_{n-1}]$, then $x_{n-1} \neq 0$. Then for any $2 \leq m \leq n-1$, we have a nondegenerate subsignature $[0, \ldots, 0, x_{n-1}, x_n]$ of arity m. □

The next definition defines some explicit families of signatures. It turns out that they essentially capture the tractable classes of Holant* problems, as will be shown in Theorem 2.12.

Definition 2.9. A signature $[x_0, x_1, \ldots, x_n]$, where $n \geq 2$, has type I, if there exist a and b (not both 0), such that $ax_k + bx_{k+1} - ax_{k+2} = 0$ $(0 \leq k \leq n-2)$. To specify the parameters a and b, we say it is of type I(a, b). We say it is of type II, if $x_k + x_{k+2} = 0$ $(0 \leq k \leq n-2)$.

For arity ≥ 2, type I is the union of generalized Fibonacci gates with GEN-EQ. For any nondegenerate signature of arity ≥ 3, if it is of type I(a, b), then (a, b) is unique up to a scalar multiplier; i.e., this $(a, b) \in \mathbb{PC}^1$ is uniquely determined as a point on the projective line. Similarly, any nondegenerate signature of arity ≥ 3 cannot be both of type I and type II. These can be directly verified.

Proposition 2.10. *$[x_0, x_1, x_2, x_3]$ is of type I iff* $\det \begin{bmatrix} x_0 - x_2 & x_1 \\ x_1 - x_3 & x_2 \end{bmatrix} = 0.$

Let $M = \begin{bmatrix} \alpha & \gamma \\ \beta & \delta \end{bmatrix}$, *and define* $[x_0, x_1, x_2, x_3] = M^{\otimes 3}[c, 0, 0, d] = c\begin{bmatrix} \alpha \\ \beta \end{bmatrix}^{\otimes 3} + d\begin{bmatrix} \gamma \\ \delta \end{bmatrix}^{\otimes 3}$. *Then* $\det \begin{bmatrix} x_0 & x_1 \\ x_1 & x_2 \end{bmatrix} = cd\alpha\gamma(\det M)^2$, $\det \begin{bmatrix} x_1 & x_2 \\ x_2 & x_3 \end{bmatrix} = cd\beta\delta(\det M)^2$, *and* $\det \begin{bmatrix} x_0 - x_2 & x_1 \\ x_1 - x_3 & x_2 \end{bmatrix}$ *is their sum.*

Suppose $[x_0, x_1, x_2, x_3] = M^{\otimes 3}[c, 0, 0, d]$ *is nondegenerate. Then, $[x_0, x_1, x_2, x_3]$ is of type I iff the two columns of M are orthogonal $\alpha\gamma + \beta\delta = 0$. In this case, it belongs to I(a, b), where $a = \alpha\gamma = -\beta\delta$ and $b = \alpha\delta + \beta\gamma$.*

Proof. A straightforward calculation. Note that $[x_0, x_1, x_2]$ is a binary signature having the matrix form $M\begin{bmatrix} c\alpha & 0 \\ 0 & d\gamma \end{bmatrix}M^{\mathrm{T}}$. Similarly, $[x_1, x_2, x_3]$ has the form $M\begin{bmatrix} c\beta & 0 \\ 0 & d\delta \end{bmatrix}M^{\mathrm{T}}$. □

Proposition 2.11. *Let $[x_0, x_1, x_2, x_3] = M^{\otimes 3}[c, 0, 0, d]$ be nondegenerate. Then, $[x_0, x_1, x_2, x_3]$ is of type II iff the columns of M are multiples of $(1, \mathrm{i})^{\mathrm{T}}$ and $(1, -\mathrm{i})^{\mathrm{T}}$ respectively.*

Proof. By nondegeneracy, $cd \neq 0$ and $\det M \neq 0$. The equations $x_0 + x_2 = 0$ and $x_1 + x_3 = 0$ are respectively $c\alpha(\alpha^2 + \beta^2) + d\gamma(\gamma^2 + \delta^2) = 0$ and $c\beta(\alpha^2 + \beta^2) + d\delta(\gamma^2 + \delta^2) = 0$. Viewed as a linear equation system on c and d, there are nonzero solutions iff its determinant is zero:

$$(\alpha^2 + \beta^2)(\gamma^2 + \delta^2)\det M = 0.$$

It follows that at least one of the factors $\alpha^2 + \beta^2 = 0$ or $\gamma^2 + \delta^2 = 0$. If $\alpha^2 + \beta^2 = 0$, by the linear equation system, since (γ, δ) is a nonzero vector, we get the second factor, $\gamma^2 + \delta^2 = 0$, as well. Similarly, starting with $\gamma^2 + \delta^2 = 0$ we also get $\alpha^2 + \beta^2 = 0$.

Conversely, if $\beta = \pm\mathrm{i}\alpha$ and $\delta = \mp\mathrm{i}\gamma$, then $c\begin{bmatrix} \alpha \\ \beta \end{bmatrix}^{\otimes 3} + d\begin{bmatrix} \gamma \\ \delta \end{bmatrix}^{\otimes 3}$ has the form $[x, y, -x, -y]$. □

The essence of Proposition 2.11 is that $c\begin{bmatrix} \alpha \\ \beta \end{bmatrix}^{\otimes 3} + d\begin{bmatrix} \gamma \\ \delta \end{bmatrix}^{\otimes 3}$ satisfies the second-order linear recurrence with eigenvalues β/α and δ/γ, and it is of type II iff these eigenvalues are $\pm\mathrm{i}$.

Our first major dichotomy theorem in this book is for Holant$^*(\mathcal{F})$, where $\mathcal{F}$ is an arbitrary set of symmetric signatures on Boolean variables.

Since all unary signatures can be used for free in Holant$^*(\mathcal{F})$, we assume the arity of every signature in $\mathcal{F}$ is greater than 1. And since any degenerate signature can be decomposed to unary signatures, we also assume that every signature in $\mathcal{F}$ is nondegenerate.

Theorem 2.12. *Let $\mathcal{F}$ be a set of nondegenerate symmetric signatures over $\mathbb{C}$. Then* Holant$^*(\mathcal{F})$ *is computable in polynomial time for the following three classes of $\mathcal{F}$. In all other cases,* Holant$^*(\mathcal{F})$ *is #P-hard.*

𝔄. *Every signature in $\mathcal{F}$ is of arity ≤ 2;*

$\mathfrak{B}$. *There exist a and b (not both 0, depending only on $\mathcal{F}$), such that every signature in $\mathcal{F}$ either (1) has type I(a, b) or (2) has arity 2 and is of the form* $[2a\lambda, b\lambda, -2a\lambda]$.

$\mathfrak{C}$. *Every signature in $\mathcal{F}$ either (1) has type II or (2) has arity 2 and is of the form* $[\lambda, 0, \lambda]$.

The dichotomy is still valid even if the inputs are restricted to planar graphs.

Remark: In order that Holant$^*(\mathcal{F})$ be a finitely specifiable problem with parameter $\mathcal{F}$, we can require that $\mathcal{F}$ be a finite set. However, our dichotomy theorem is stronger, and applies to an infinite set $\mathcal{F}$ in the following sense: When $\mathcal{F}$ falls in one of the tractable classes, the Holant problem is computable in polynomial time even when $\mathcal{F}$ is infinite, but the input size of the signature grid includes a description (in symmetric signature notation) of the functions at each node. On the other hand, when $\mathcal{F}$ does not belong to one of the tractable classes, then there is a finite subset $\mathcal{F}' \subseteq \mathcal{F}$ for which Holant$^*(\mathcal{F}')$ is #P-hard. In more detail, this means that there is a finite set of unary functions $\mathcal{U}'$ such that Holant$(\mathcal{F}' \cup \mathcal{U}')$ is #P-hard.

Remark: For a more conceptual characterization, the reader is encouraged to compare this dichotomy theorem with Theorem 7.19. To formulate and prove Theorem 7.19, however, we need to develop the necessary machinery and preparatory results, of which Theorem 2.12 is the first one. The reader is also encouraged to compare Theorem 2.12 with Theorem 3.5, which is a more abstract statement of Theorem 2.12.

This dichotomy theorem essentially says that for Holant$^*(\mathcal{F})$ the only tractable symmetric signature sets $\mathcal{F}$ are those when, after all degenerate signatures are removed, $\mathcal{F}$ consists of functions of arity ≤ 2, or consists of generalized Fibonacci gates under a holographic transformation and some specific binary functions, or one additional class.

Proof of Tractability: Class $\mathfrak{A}$ is when every signature in $\mathcal{F}$ has arity ≤ 2. Then the graph of the signature grid Ω is a disjoint union of paths and cycles (isolated points contribute a constant; we may assume there are no isolated points). By matrix multiplication, we can compute the Holant value for a path. The Holant value for a cycle is obtained by taking the trace of a path. The value Holant$_\Omega$ is the product over connected components.

Next we consider class $\mathfrak{B}$. If $a = 0$, all functions in $\mathcal{F}$ are GEN-EQ or binary DISEQUALITY, and Holant$^*(\mathcal{F})$ is obviously computable in polynomial time. Suppose $a \neq 0$. All functions in $\mathcal{F}$ having arity > 2 are generalized Fibonacci gates. We first assume $X^2 - (b/a)X - 1$ has two distinct eigenvalues two distinct eigenvalues λ and μ, where $\lambda \neq \mu$. Then $\lambda\mu = -1$ and $\lambda + \mu = b/a$.

All functions in $\mathcal{F}$ have the form $f = u\begin{bmatrix}1\\ \lambda\end{bmatrix}^{\otimes n} + v\begin{bmatrix}1\\ \mu\end{bmatrix}^{\otimes n}$ or multiples of $g = [2, b/a, -2]$. Under the (inverse) holographic transformation by $M = \begin{bmatrix}1 & \lambda\\ 1 & \mu\end{bmatrix}$, $f \mapsto M^{\otimes n} f$ is a GEN-EQ, $g \mapsto M^{\otimes 2} g = [0, 4 + (b/a)^2, 0]$, and $(=_2) = [1, 0, 1]$ is turned into $[1, 0, 1](M^{-1})^{\otimes 2}$, which is a multiple of $[1 + \mu^2, 0, 1 + \lambda^2]$, a GEN-EQ. In the meantime, all unary functions are transformed into unary functions. Connecting any unary function to a GEN-EQ creates another GEN-EQ. Connecting any unary function to any function of arity 2 creates another unary function. Hence Holant*($\mathcal{F}$) is tractable.

Continuing for class $\mathfrak{B}$, if the eigenvalues $\lambda = \mu$, then $b = \pm 2a\mathrm{i}$. Then $a \neq 0$ and we can normalize to $a = 1$, $b = \pm 2\mathrm{i}$, $\lambda = b/2$. The signature $g = [2, b, -2] = [2, \pm 2\mathrm{i}, -2]$ is degenerate. We have proved using the holographic transformation Z that in this case Holant*($\mathcal{F}$) is tractable.

For class $\mathfrak{C}$, we first get rid of all unary functions and binary functions of the form $\lambda[1, 0, 1]$. If there is a function $\lambda[1, 0, 1]$ applied to two variables (edges) e and e', we just merge them into one, and remove this function with a global factor λ. Combining a unary function with a function of the type $[x, y, -x, -y, \ldots]$ does not change its type. Hence, we may assume all functions satisfy $x_k + x_{k+2} = 0$. If the input graph $G(V, E)$ is not bipartite, there is a cycle $v_1, e_1, v_2, \ldots, v_k, e_k, v_1$ of odd length. We partition all 0-1 assignments for E into two parts, with a 1-1 correspondence between them. An assignment σ is mapped to σ', which assigns the same values on $E - \{e_1, \ldots, e_k\}$, but opposite values on $\{e_1, \ldots, e_k\}$. Under σ and σ', all functions on $V - \{v_1, \ldots, v_k\}$ evaluate to the same value, and if $\sigma(e_{j-1}) \neq \sigma(e_j)$ (where the index j is counted mod k), the function at v_j evaluates to the same value under σ and σ', and if $\sigma(e_{j-1}) = \sigma(e_j)$, the function at v_j evaluates to opposite values under σ and σ'. Consider $\sigma(e_1), \sigma(e_2), \ldots, \sigma(e_k), \sigma(e_1)$. There must be an even number of times where the value changes in this sequence. Since k is odd, there are an odd number of v_j's, where the functions at v_j give opposite values under σ and σ'. Hence, in the summation, the contributions of σ and σ' are canceled.

If the input graph is bipartite, the problem is Holant($\mathcal{F} \mid \mathcal{F}$). It is turned into Holant($\mathcal{F}' \mid \mathcal{F}'$), where $\mathcal{F}' \subseteq$ GEN-EQ, by the holographic reduction $Z = \begin{bmatrix}1 & 1\\ \mathrm{i} & -\mathrm{i}\end{bmatrix}$. This is easily seen by writing each contravariant $f \in \mathcal{F}$ as $u\begin{bmatrix}1\\ \mathrm{i}\end{bmatrix}^{\otimes n} + v\begin{bmatrix}1\\ -\mathrm{i}\end{bmatrix}^{\otimes n} = Z[u, 0, \ldots, 0, v]$, and each covariant $g \in \mathcal{F}$ as $u'\begin{bmatrix}1 & \mathrm{i}\end{bmatrix}^{\otimes m} + v'\begin{bmatrix}1 & -\mathrm{i}\end{bmatrix}^{\otimes m}$, so that $gZ = u'\begin{bmatrix}0 & 2\end{bmatrix}^{\otimes m} + v'\begin{bmatrix}2 & 0\end{bmatrix}^{\otimes m} = [2^m v', 0, \ldots, 0, 2^m u']$. Hence Holant($\mathcal{F} \mid \mathcal{F}$) is tractable.

Proof Outline for Hardness: Now for the proof of hardness, we first prove in Lemma 2.13 that the theorem holds if $\mathcal{F}$ contains a single symmetric signature of arity 3. The main technique is holographic reduction. In Lemma 2.14, we prove that if one signature of arity 3 has the form in class $\mathfrak{B}$ of Theorem 2.12, and we combine it with another signature of arity 2 that is *not* in class $\mathfrak{B}$, then the Holant* problem is #P-hard. The idea of the proof of Lemma 2.14 is to reduce it to Lemma 2.13 with holographic reductions. In Lemma 2.15, we prove the same thing for class $\mathfrak{C}$. In Lemma 2.16 we extend Lemma 2.13 to a single signature of arbitrary arity. Finally, we extend the proof to a set of signatures of arbitrary arities.

2.3.1 The First Lemma: A Single Ternary Signature

The following lemma says that Theorem 2.12 holds if $\mathcal{F}$ consists of only one signature of arity 3. It serves as the foundation for subsequent lemmas.

Lemma 2.13. *Let* $f = [x_0, x_1, x_2, x_3]$ *be a nondegenerate symmetric signature with arity three, then* Holant*(f) *is #P-hard unless* f *has type I or type II.*

Proof. We assume f is of neither type I nor type II, and prove that Holant*(f) is #P-hard. Our starting point is that Holant$([0, 1, 1]|[1, 0, 0, 1])$ and Holant$([1, 0, 1]|[1, 1, 0, 0])$ are both #P-complete [XZZ07]. The first problem is to count the number of vertex covers for 3-regular graphs; the second is to count the number of (not necessarily perfect) matchings for 3-regular graphs. Both problems remain #P-complete even for planar graphs [DL92, XZZ07].

Every nondegenerate signature $[x_0, x_1, x_2, x_3]$ belongs to one of the following categories: For $0 \le k \le 3$,

- Category 1. $x_k = \alpha_1^{3-k}\alpha_2^k + \beta_1^{3-k}\beta_2^k$, where $\det\begin{bmatrix}\alpha_1 & \beta_1\\ \alpha_2 & \beta_2\end{bmatrix} \neq 0$; or
- Category 2. $x_k = ck\alpha^{k-1} + d\alpha^k$, where $c \neq 0$; or
- Category 3. $x_k = c(3-k)\alpha^{2-k} + d\alpha^{3-k}$, where $c \neq 0$.

This parametrization is obtained by considering the solutions $(a, b, c)^{\mathrm{T}}$ to $\begin{bmatrix}x_0 & x_1 & x_2\\ x_1 & x_2 & x_3\end{bmatrix}\begin{bmatrix}a\\ b\\ c\end{bmatrix} = 0$, which form a one-dimensional vector space. Suppose $(a, b, c)^{\mathrm{T}}$ is a nonzero solution. If $a = c = 0$, we have $x_1 = x_2 = 0$ and a diagonal matrix $\begin{bmatrix}\sqrt[3]{x_0} & 0\\ 0 & \sqrt[3]{x_3}\end{bmatrix}$ can be used in category 1. If a and c are not both 0, then we may consider $[x_0, x_1, x_2, x_3]$ satisfies a second-order linear recurrence relation (either forward or backward). Depending on whether the characteristic equation has two distinct roots or a double root, we have a case in category 1 or the other two. Category 3 can be viewed as the reversal of category 2, so we will omit the proof for category 3. We take the convention that $k0^{k-1} = 0, 1, 0, 0$ for $k = 0, 1, 2, 3$ respectively.

For category 1, we have

$$f = [x_0, x_1, x_2, x_3] = \begin{bmatrix}\alpha_1\\ \alpha_2\end{bmatrix}^{\otimes 3} + \begin{bmatrix}\beta_1\\ \beta_2\end{bmatrix}^{\otimes 3}.$$

Nondegeneracy of f implies that $\alpha_1\beta_2 - \alpha_2\beta_1 \neq 0$. The fact that f is not of type I or II implies that $\alpha_1\beta_1 + \alpha_2\beta_2 \neq 0$ by Proposition 2.10, and $\alpha_1^2 + \alpha_2^2 \neq 0$ or $\beta_1^2 + \beta_2^2 \neq 0$ by Proposition 2.11. By symmetry, we can assume that $\alpha_1^2 + \alpha_2^2 \neq 0$.

We apply an orthogonal transformation $T = \frac{1}{\sqrt{\alpha_1^2+\alpha_2^2}}\begin{bmatrix}\alpha_1 & \alpha_2\\ -\alpha_2 & \alpha_1\end{bmatrix}$ to map the vector $(\alpha_1, \alpha_2)^{\mathrm{T}}$ to $(\alpha_1', 0)^{\mathrm{T}}$, where $\alpha_1' = \sqrt{\alpha_1^2 + \alpha_2^2} \neq 0$.

$$f' = T^{\otimes 3} f = [x_0', x_1', x_2', x_3'] = \begin{bmatrix}\alpha_1'\\ 0\end{bmatrix}^{\otimes 3} + \begin{bmatrix}\beta_1'\\ \beta_2'\end{bmatrix}^{\otimes 3}.$$

By Theorem 1.6, this transformation does not change the complexity of the Holant problem. So it suffices to prove #P-hardness for f'. By a scalar multiplication we may assume $\alpha_1' = 1$. So, reusing the notation f, we can assume the given signature is

$$f = [x_0, x_1, x_2, x_3] = \begin{bmatrix}1\\ 0\end{bmatrix}^{\otimes 3} + \begin{bmatrix}\alpha\\ \beta\end{bmatrix}^{\otimes 3}. \tag{2.15}$$

The conditions from Lemma 2.13 become simply both $\alpha, \beta \neq 0$.

Consider $M = \begin{bmatrix}1 & \alpha\\ 0 & \beta\end{bmatrix}$. We have

$$f = M^{\otimes 3}[1, 0, 0, 1] = M^{\otimes 3}\left(\begin{bmatrix}1\\ 0\end{bmatrix}^{\otimes 3} + \begin{bmatrix}0\\ 1\end{bmatrix}^{\otimes 3}\right),$$

and

$$(0, 1, 1, 1)(M^{-1})^{\otimes 2} = \frac{1}{\beta^2}(0, \beta, \beta, 1 - 2\alpha).$$

One can check this by computing $M^{\mathrm{T}}\begin{bmatrix}0 & \beta\\ \beta & 1-2\alpha\end{bmatrix}M$. Ignoring $1/\beta^2$, by a holographic reduction the complexity of Holant($[0, \beta, 1-2\alpha] \mid f$) is the same as the #P-complete vertex cover problem Holant($[0, 1, 1] \mid [1, 0, 0, 1]$). In particular, Holant($[0, \beta, 1-2\alpha], f$) is #P-hard. In order to prove the #P-hardness

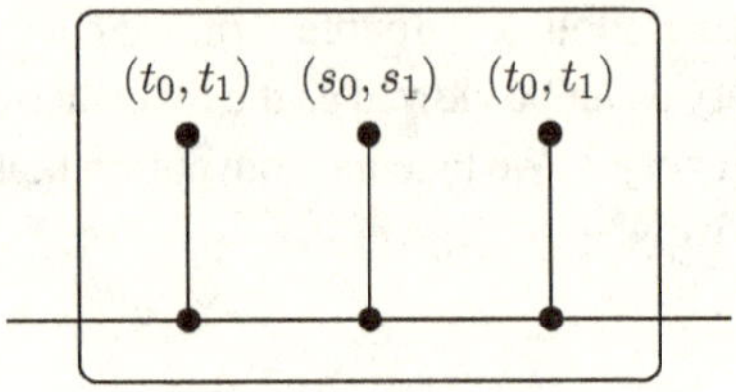

Figure 2.3. We use this gadget to realize the signature $[0, \beta, 1-2\alpha]$. All (three) nodes of degree 3 in this gadget have the signature $f = [x_0, x_1, x_2, x_3]$.

of Holant*(f), we only need to show that $[0, \beta, 1-2\alpha]$ can be realized by f with some unary signatures.

We use the gadget in Figure 2.3 to realize $[0, \beta, 1-2\alpha]$, where the two unary signatures (t_0, t_1) and (s_0, s_1) will be determined later. Let

$$A = \begin{bmatrix}1\\0\end{bmatrix}\begin{bmatrix}1 & 0\end{bmatrix} = \begin{bmatrix}1 & 0\\0 & 0\end{bmatrix}, \quad B = \begin{bmatrix}\alpha\\\beta\end{bmatrix}\begin{bmatrix}\alpha & \beta\end{bmatrix} = \begin{bmatrix}\alpha^2 & \alpha\beta\\\alpha\beta & \beta^2\end{bmatrix}.$$

For f in (2.15), if one input is 0, the induced binary signature has its matrix form $A + \alpha B$. If one input is 1, the induced binary signature has matrix form βB. It follows that the signature in Figure 2.3, as a binary function in matrix form, is

$$\begin{aligned}&(t_0(A+\alpha B)+t_1\beta B)(s_0(A+\alpha B)+s_1\beta B)(t_0(A+\alpha B)+t_1\beta B)\\&\quad=(t_0A+(t_0\alpha+t_1\beta)B)(s_0A+(s_0\alpha+s_1\beta)B)(t_0A+(t_0\alpha+t_1\beta)B).\end{aligned}$$

Exercise: Show that $t_0(A+\alpha B)+t_1\beta B$ is the signature matrix of the binary gadget obtained by connecting the unary signature $[t_0, t_1]$ to f of (2.15). Deduce that the above matrix is the signature matrix of the binary gadget in Figure 2.3.

Now we use a new set of variables:

$$x = t_0, \quad y = t_0\alpha + t_1\beta, \quad z = s_0, \quad w = s_0\alpha + s_1\beta, \tag{2.16}$$

and write the above matrix as $(xA + yB)(zA + wB)(xA + yB)$. We note that, since $\beta \neq 0$, for any given x, y, z, w, we can find t_0, t_1, s_0, s_1 to satisfy the relations (2.16). Then, to realize $[0, \beta, 1-2\alpha]$, we will choose some x, y, z, and w such that

$$(xA+yB)(zA+wB)(xA+yB) = \begin{bmatrix}0 & \beta\\\beta & 1-2\alpha\end{bmatrix}.$$

Substituting A and B, and denoting $\alpha^2+\beta^2$ by γ, we have the following:

$$(xA+yB)(zA+wB)(xA+yB) = w\begin{bmatrix} \alpha^2(x+y\gamma)^2 & y\alpha\beta\gamma(x+y\gamma) \\ y\alpha\beta\gamma(x+y\gamma) & y^2\beta^2\gamma^2 \end{bmatrix}$$
$$+z\begin{bmatrix} (x+y\alpha^2)^2 & y\alpha\beta(x+y\alpha^2) \\ y\alpha\beta(x+y\alpha^2) & y^2\alpha^2\beta^2 \end{bmatrix}.$$

We may choose $w=(x+y\alpha^2)^2$ and $z=-\alpha^2(x+y\gamma)^2$ to make the $(1,1)$ entry zero. The $(1,2)$ (and $(2,1)$) entry is

$$g_1 = xy\alpha\beta^3(x+y\alpha^2)(x+y\gamma);$$

and the $(2,2)$ entry is

$$g_2 = xy^2\beta^4(x(2\alpha^2+\beta^2)+2y\alpha^2\gamma).$$

We want to choose some x and y such that $g_1=\beta$ and $g_2=1-2\alpha$. Recall that $\alpha\beta\neq 0$. We will choose x and $y\neq 0$. As both g_1 and g_2 are homogeneous (of degree 4) in x and y, we can ignore the common factor $xy\beta^3$ of g_1 and g_2. It follows that we only have to satisfy that $g_2/g_1=(1-2\alpha)/\beta$ and $g_1\neq 0$, with $y=1$. The following equation

$$\begin{aligned}\alpha(2\alpha-1)x^2 &+ (2\alpha^2-\alpha+\beta^2)(2\alpha^2+\beta^2)x \\ &+ \alpha^2(\alpha^2+\beta^2)(2\alpha^2-\alpha+2\beta^2)=0 \end{aligned} \tag{2.17}$$

is equivalent to $\beta g_2=(1-2\alpha)g_1$ after removing $xy\beta^3$ and setting $y=1$. What we have to prove is that (2.17) has a root $x\neq 0$ that is not a root of $g_1=g_1(x,1)=0$. The roots of $g_1(x,1)=0$ are $x=0$, $x=-\alpha^2$ and $x=-\gamma$. First, when $x=-\alpha^2$, the left-hand side expression in (2.17) can be simplified to $\alpha^2\beta^4\neq 0$. (Note that when $g_1=0$, the left-hand side expression in (2.17) is βg_2, after removing $xy\beta^3$, and $y=1$.) So $x=-\alpha^2$ is not a root of (2.17). Second, when $x=-\gamma$ the left-hand side expression in (2.17) can be simplified to $-\beta^4\gamma$, and if $x=-\gamma$ is a root, this would force $\gamma=0$. So, assuming the left-hand side expression in (2.17) as a polynomial in x is indeed of degree 2, and it does not have a double root 0, then we can find a root $\xi\neq 0$ of (2.17) that is not a root of g_1. Indeed, This $\xi\neq-\alpha^2$, because $-\alpha^2$ is not a root of (2.17); ξ cannot be $-\gamma$ either, for otherwise $-\gamma$ would be a root of (2.17), which would force $\gamma=0$, and thus $\xi=-\gamma=0$, a contradiction. Thus ξ is a nonzero root of (2.17) but not a root of $g_1(x,1)$, as is needed.

Now let us consider the exceptional cases: either $x = 0$ is a double root of (2.17), or (2.17) has degree < 2. If $x = 0$ is a double root of (2.17), we have

$$(2\alpha^2 - \alpha + \beta^2)(2\alpha^2 + \beta^2) = \alpha^2(\alpha^2 + \beta^2)(2\alpha^2 - \alpha + 2\beta^2) = 0.$$

To satisfy this, since $\alpha\beta \neq 0$, there are only four exceptional cases (A1 to A4): $\alpha = 1, \beta = \pm\mathrm{i}$ or $\alpha = -\frac{1}{2}, \beta = \pm\frac{\mathrm{i}}{\sqrt{2}}$. On the other hand, if the polynomial in (2.17) has degree < 2, by $\alpha \neq 0$, we get $\alpha = \frac{1}{2}$. In this case, the polynomial becomes

$$(1/2 + \beta^2)x + (1/4 + \beta^2)/2 = 0.$$

This gives us four additional exceptional cases (B1 to B4): $\alpha = \frac{1}{2}, \beta = \pm\frac{\mathrm{i}}{2}$, in which case the polynomial is linear with root $x = 0$; or $\alpha = \frac{1}{2}, \beta = \pm\frac{\mathrm{i}}{\sqrt{2}}$, in which case the polynomial degenerates to a (nonzero) constant. In all other cases, there is a nonzero root of (2.17) that is not a root of $g_1(x, 1)$, completing the #P-hardness proof.

For the cases A1 and A2, we use a new starting problem Holant($[1, 1, 0] \mid [1, 0, 0, 1]$), which is the reversal of the previous problem and therefore it is also #P-complete. The problem can also be understood as counting independent sets over 3-regular graphs. Then all previous parts of the proof are still valid, except that the signature of arity 2 to be realized is

$$(1, 1, 1, 0)(T^{-1})^{\otimes 2} = \left(1, \frac{1-\alpha}{\beta}, \frac{1-\alpha}{\beta}, \frac{\alpha^2 - 2\alpha}{\beta^2}\right).$$

Substituting $\alpha = 1, \beta = \pm\mathrm{i}$, the signature is $[1, 0, 1]$ which is trivially realizable by one edge. So we have proved that it is #P-hard in the cases A1 and A2. Now consider the cases A3 and A4: $\alpha = -\frac{1}{2}, \beta = \pm\frac{\mathrm{i}}{\sqrt{2}}$. We will give a different parametrization. For the case A3, we apply an orthogonal transformation $M = \begin{bmatrix} -\mathrm{i} & -\sqrt{2} \\ \sqrt{2} & -\mathrm{i} \end{bmatrix}$ and a scalar multiplier $(2\mathrm{i})^3$ on the signature f in (2.15), to get $f \mapsto (2\mathrm{i})^3 M^{\otimes 3} f = \begin{bmatrix} 1 \\ 0 \end{bmatrix}^{\otimes 3} + \begin{bmatrix} 2 \\ 2\sqrt{2}\mathrm{i} \end{bmatrix}^{\otimes 3}$. This is not one of the exceptional cases and we have proved that it is #P-hard. For the case A4, we apply another orthogonal transformation $M' = \begin{bmatrix} \mathrm{i} & -\sqrt{2} \\ \sqrt{2} & \mathrm{i} \end{bmatrix}$ and a scalar multiplier $(-2\mathrm{i})^3$ on the signature and it becomes $\begin{bmatrix} 1 \\ 0 \end{bmatrix}^{\otimes 3} + \begin{bmatrix} 2 \\ -2\sqrt{2}\mathrm{i} \end{bmatrix}^{\otimes 3}$.

The cases B3 and B4 can be shown by the same method as in A4 and A3, using M' and M respectively. The only remaining cases are B1 and B2. Here we will use another gadget similar to the one in Figure 2.3 except we remove the middle edge (including the node labeled (s_0, s_1) and the middle node of degree 3). For B1, $\alpha = \frac{1}{2}, \beta = \frac{\mathrm{i}}{2}$, the signature of this gadget is

$$(t_0(A + \alpha B) + t_1 \beta B)^2 = (xA + yB)^2,$$

where the matrices A and B are as before, and with the specific values of α, β, $B = \frac{1}{4}\begin{bmatrix} 1 & \mathrm{i} \\ \mathrm{i} & -1 \end{bmatrix}$. By setting $x = \mathrm{i}$ and $y = -2\mathrm{i}$, we have $(xA + yB)^2 = \begin{bmatrix} 0 & \mathrm{i}/2 \\ \mathrm{i}/2 & 0 \end{bmatrix}$, which is the matrix form of the target signature $[0, \beta, 1 - 2\alpha] = [0, \frac{\mathrm{i}}{2}, 0]$. This finishes case B1. Case B2 can be proved in the same way with $x = 1$ and $y = -2$.

Exercise: Prove cases B3 and B4 using the matrices M' and M in the cases A4 and A3, respectively.

Now we prove for category 2. In this case $x_k = ck\alpha^{k-1} + d\alpha^k$, and it satisfies the recurrence $x_{k+2} = 2\alpha x_{k+1} - \alpha^2 x_k$, for $k = 0, 1$. The characteristic polynomial $X^2 - 2\alpha X + \alpha^2$ has a double root α. If $\alpha = \pm\mathrm{i}$, then f has type I$(1, 2\alpha)$. Since we assumed f is not of type I, $\alpha \neq \pm\mathrm{i}$. Then we can choose some orthogonal matrix to transform the signature to the form $[x, y, 0, 0]$ where $y \neq 0$, as follows.

Let $M = \begin{bmatrix} 1 & \frac{d-1}{3} \\ \alpha & c + \frac{d-1}{3}\alpha \end{bmatrix}$ with $\det M = c \neq 0$, then the signature $[x_0, x_1, x_2, x_3]$ can be expressed as

$$(x_0, x_1, x_1, x_2, x_1, x_2, x_2, x_3)^{\mathrm{T}} = M^{\otimes 3}(1, 1, 1, 0, 1, 0, 0, 0)^{\mathrm{T}}.$$

In symmetric signature notation $[x_0, x_1, x_2, x_3] = M^{\otimes 3}[1, 1, 0, 0]$.

Exercise: Verify that $[x_0, x_1, x_2, x_3] = M^{\otimes 3}[1, 1, 0, 0]$. Use M to show that there is another matrix M' such that $[x_0, x_1, x_2, x_3] = M'^{\otimes 3}[0, 1, 0, 0]$.

Let $M = QR$ be its QR-factorization, i.e., where Q is orthogonal and R is upper triangular. In fact, for $M = \begin{bmatrix} 1 & * \\ \alpha & * \end{bmatrix}$, recalling that $1 + \alpha^2 \neq 0$, we can choose our Q as the (complex) orthogonal matrix $Q = Q^{\mathrm{T}} = \frac{1}{\sqrt{1+\alpha^2}}\begin{bmatrix} 1 & \alpha \\ \alpha & -1 \end{bmatrix}$. Then $QM = R = \begin{bmatrix} u & w \\ 0 & v \end{bmatrix}$ is upper triangular, where $u = \sqrt{1 + \alpha^2}$. Because $\det Q = -1$ and $\det R = -\det M = -c \neq 0$, we have $uv \neq 0$. This Q is our choice of the orthogonal transformation. It follows that

$$\begin{aligned}
&Q^{\otimes 3}[x_0, x_1, x_2, x_3]\\
&= (QM)^{\otimes 3}[1, 1, 0, 0]\\
&= R^{\otimes 3}\left\{\begin{bmatrix}1\\0\end{bmatrix}^{\otimes 3} + \begin{bmatrix}1\\0\end{bmatrix}\otimes\begin{bmatrix}1\\0\end{bmatrix}\otimes\begin{bmatrix}0\\1\end{bmatrix} + \begin{bmatrix}1\\0\end{bmatrix}\otimes\begin{bmatrix}0\\1\end{bmatrix}\otimes\begin{bmatrix}1\\0\end{bmatrix} + \begin{bmatrix}0\\1\end{bmatrix}\otimes\begin{bmatrix}1\\0\end{bmatrix}\otimes\begin{bmatrix}1\\0\end{bmatrix}\right\}\\
&= \begin{bmatrix}u\\0\end{bmatrix}^{\otimes 3} + \begin{bmatrix}u\\0\end{bmatrix}\otimes\begin{bmatrix}u\\0\end{bmatrix}\otimes\begin{bmatrix}w\\v\end{bmatrix} + \begin{bmatrix}u\\0\end{bmatrix}\otimes\begin{bmatrix}w\\v\end{bmatrix}\otimes\begin{bmatrix}u\\0\end{bmatrix} + \begin{bmatrix}w\\v\end{bmatrix}\otimes\begin{bmatrix}u\\0\end{bmatrix}\otimes\begin{bmatrix}u\\0\end{bmatrix}.
\end{aligned} \tag{2.18}$$

In the notation for symmetric signatures, this is $[u^3 + 3u^2w, u^2v, 0, 0]$, with $u^2v \neq 0$.

By a scalar multiplication, we can normalize the entry u^2v to 1. So we only have to deal with a signature of the form $[v, 1, 0, 0]$ for an arbitrary given v.

For this signature, we can apply a holographic transformation defined by the matrix $M' = \begin{bmatrix}1 & -\frac{v-1}{3}\\ 0 & 1\end{bmatrix}$ with inverse $M'^{-1} = \begin{bmatrix}1 & \frac{v-1}{3}\\ 0 & 1\end{bmatrix}$. To prove #P-hardness, we will reduce from the MATCHING problem Holant$([1, 0, 1] \mid [1, 1, 0, 0])$. Under a contravariant transformation $(M'^{-1})^{\otimes 3}[1, 1, 0, 0] = [v, 1, 0, 0]$, the signature $[1, 1, 0, 0]$ becomes $[v, 1, 0, 0]$. By the corresponding covariant transformation, $[1, 0, 1]$ becomes $[1, \frac{1-v}{3}, 1 + \frac{(1-v)^2}{9}]$:

$$\begin{aligned}
(1, 0, 0, 1)M'^{\otimes 2} &= ((1, 0)^{\otimes 2} + (0, 1)^{\otimes 2})M'^{\otimes 2}\\
&= \left(1, \frac{1-v}{3}, \frac{1-v}{3}, 1 + \frac{(1-v)^2}{9}\right).
\end{aligned}$$

We complete this proof by using the same gadget in Figure 2.3 to realize this binary signature, using unary signatures and $[v, 1, 0, 0]$.

We will rename the values $x = t_0$, $y = t_1$, $z = s_0$ and $w = s_1$ in Figure 2.3. The signature of this gadget in matrix form is $(xA + yB)(zA + wB)(xA + yB)$, where $A = \begin{bmatrix}v & 1\\ 1 & 0\end{bmatrix}$ and $B = \begin{bmatrix}1 & 0\\ 0 & 0\end{bmatrix}$.

We set $x = 1$. Then $(A + yB)(zA + wB)(A + yB)$ is

$$\begin{bmatrix} w\cdot(v+y)^2 + z\cdot(v(v+y)^2 + 2(v+y) & w\cdot(v+y) + z\cdot(v(v+y)+1)\\ w\cdot(v+y) + z\cdot(v(v+y)+1) & w + z\cdot v\end{bmatrix}.$$

Our goal is to choose y, z and w such that it is equal to $\begin{bmatrix}1 & (1-v)/3\\ (1-v)/3 & 1 + (1-v)^2/9\end{bmatrix}$. We can write this requirement as a system of three linear equations in z and w, whose coefficient matrix depends on y.

Then we can complete the proof, if we can choose y such that the following matrix has determinant 0, yet the first two columns have rank 2.

$$\begin{bmatrix} (v+y)^2 & v(v+y)^2+2(v+y) & 1 \\ (v+y) & v(v+y)+1 & (1-v)/3 \\ 1 & v & 1+(1-v)^2/9 \end{bmatrix}.$$

After some elementary column operations on the first two columns it becomes

$$\begin{bmatrix} -(v+y)^2 & 2(v+y) & 1 \\ 0 & 1 & (1-v)/3 \\ 1 & 0 & 1+(1-v)^2/9 \end{bmatrix},$$

whose first two columns obviously have rank 2. Its determinant is

$$-(v+y)^2(1+(1-v)^2/9)+2(v+y)(1-v)/3-1.$$

Clearly $1+(1-v)^2/9$ and $(1-v)/3$ cannot be both 0. Thus we can set y so that this determinant is 0. This completes the proof of Lemma 2.13. □

2.3.2 A Pair of Signatures

Lemma 2.13 shows what happens when there is a single nondegenerate symmetric signature f of arity 3. It explicitly isolates two tractable cases: If Holant*(f) is *not* #P-hard, then f must be of type I or type II. The next lemma addresses what happens when we have more than one signature, where one signature happens to be of type I but some other signature does not *go along* with it.

Lemma 2.14. *Let $f = [x_0, x_1, x_2, x_3]$ and $g = [y_0, y_1, y_2]$ be nondegenerate symmetric signatures with arity 3 and 2 respectively. Suppose f is of type I(a, b) for some a, b (not both zero), but g is not of type I(a, b), i.e., $ay_0 + by_1 - ay_2 \neq 0$, and g is not of the form $[2a\lambda, b\lambda, -2a\lambda]$. Then* Holant*$(g \mid f)$ *is #P-hard. It follows that* Holant*$(\{f, g\})$ *is also #P-hard.*

Proof. Since f is nondegenerate, the pair (a, b) is unique up to a scalar factor. Our proof plan is as follows. We will show that Holant*$(g \mid f)$ is #P-hard by a holographic reduction where g is transformed to the binary EQUALITY $(=_2)$. Hence we want an invertible matrix M such that $f \mapsto M^{\otimes 3} f = f'$ and $g \mapsto g(M^{-1})^{\otimes 2} = [1, 0, 1]$. Note that $[1, 0, 1]$ can be replaced by an edge, and the unary signatures are transformed to unary signatures. Thus the complexity of the problem Holant*$(g \mid f)$ is the same as Holant*(f'). We then apply Lemma 2.13 to f'. Note that f' is nondegenerate. Otherwise, f' would be a tensor product of unary signatures, and then $f = (M^{-1})^{\otimes 3} f'$ would also be a tensor product of unary signatures, and thus degenerate.

The requirement that $g(M^{-1})^{\otimes 2} = [1, 0, 1]$ is the same as $G = M^{\mathrm{T}} M$ for some matrix $M = \begin{bmatrix} \alpha & \gamma \\ \beta & \delta \end{bmatrix}$, where $G = \begin{bmatrix} y_0 & y_1 \\ y_1 & y_2 \end{bmatrix}$ is the matrix form of g. This is

$$\begin{bmatrix} y_0 & y_1 \\ y_1 & y_2 \end{bmatrix} = M^{\mathrm{T}} M = \begin{bmatrix} \alpha^2 + \beta^2 & \alpha\gamma + \beta\delta \\ \alpha\gamma + \beta\delta & \gamma^2 + \delta^2 \end{bmatrix}. \tag{2.19}$$

Such a factorization for a symmetric complex matrix G exists. If $y_0 \neq 0$, we take $M = \frac{1}{\sqrt{y_0}} \begin{bmatrix} y_0 & y_1 \\ 0 & \Delta \end{bmatrix}$, where $\Delta = \sqrt{\det G}$. Note that $\det G \neq 0$ since g is non-degenerate. If $y_0 = 0$ but $y_2 \neq 0$, we can first exchange y_0 and y_2 by $\begin{bmatrix} 0 & 1 \\ 1 & 0 \end{bmatrix}$, and take $M = \frac{1}{\sqrt{y_2}} \begin{bmatrix} \Delta & 0 \\ y_1 & y_2 \end{bmatrix}$. If $y_0 = y_2 = 0$, then we take $M = \sqrt{y_1} Z$ where $Z = \frac{1}{\sqrt{2}} \begin{bmatrix} 1 & 1 \\ \mathrm{i} & -\mathrm{i} \end{bmatrix}$.

Let $f' = [u_0, u_1, u_2, u_3] = M^{\otimes 3} f$. We want to show that f' is not of type I or II, the two tractable cases of Lemma 2.13.

First we assume for a contradiction that f' is of type II. Then it satisfies the recurrence $u_{k+2} + u_k = 0$, for $k = 0, 1$, with characteristic polynomial $X^2 + 1$. Therefore $f' = c \begin{bmatrix} 1 \\ \mathrm{i} \end{bmatrix}^{\otimes 3} + d \begin{bmatrix} 1 \\ -\mathrm{i} \end{bmatrix}^{\otimes 3}$, for some c and d. We have $M^{-1} = \Delta^{-1} \begin{bmatrix} \delta & -\gamma \\ -\beta & \alpha \end{bmatrix}$, where $\Delta = \det M \neq 0$. Then it follows that

$$f = (M^{-1})^{\otimes 3} f' = c\Delta^{-3} \begin{bmatrix} \delta - \gamma\mathrm{i} \\ -\beta + \alpha\mathrm{i} \end{bmatrix}^{\otimes 3} + d\Delta^{-3} \begin{bmatrix} \delta + \gamma\mathrm{i} \\ -\beta - \alpha\mathrm{i} \end{bmatrix}^{\otimes 3}.$$

Since f is nondegenerate, $cd \neq 0$. By the assumption that f has type I(a, b), and by Proposition 2.10, we have

$$\alpha^2 + \beta^2 + \gamma^2 + \delta^2 = 0,$$

and $a = \gamma^2 + \delta^2 = -\alpha^2 - \beta^2$, and $b = -2(\alpha\gamma + \beta\delta)$ after a scaling. By (2.19), $[y_0, y_1, y_2] = [-a, -b/2, a]$ is of the form $[2a\lambda, b\lambda, -2a\lambda]$, a contradiction.

This completes our first step. Our second step is to show that f' cannot be of type I either. For a contradiction, suppose f' is of type I.

We are given that f is of type I(a, b). If $a = 0$ then $b \neq 0$ and $x_1 = x_2 = 0$. We can write

$$f = x_0 \begin{bmatrix} 1 \\ 0 \end{bmatrix}^{\otimes 3} + x_3 \begin{bmatrix} 0 \\ 1 \end{bmatrix}^{\otimes 3},$$

and

$$f' = [u_0, u_1, u_2, u_3] = M^{\otimes 3} f = x_0 \begin{bmatrix} \alpha \\ \beta \end{bmatrix}^{\otimes 3} + x_3 \begin{bmatrix} \gamma \\ \delta \end{bmatrix}^{\otimes 3}.$$

By Proposition 2.10 the dot product of $(\alpha, \beta)^{\mathrm{T}}$ and $(\gamma, \delta)^{\mathrm{T}}$ is zero, i.e.,

$$y_1 = \alpha\gamma + \beta\delta = 0.$$

This implies that $ay_0 + by_1 - ay_2 = 0$, a contradiction.

Finally we consider the case $a \neq 0$. By a scaling we can assume $a = 1$. Assume temporarily that $b \neq \pm 2\mathrm{i}$, then the recurrence $x_{k+2} = bx_{k+1} + x_k$ has two distinct eigenvalues λ and μ with $\lambda + \mu = b$ and $\lambda\mu = -1$. So we can write $[x_0, x_1, x_2, x_3] = c \begin{bmatrix} 1 \\ \lambda \end{bmatrix}^{\otimes 3} + d \begin{bmatrix} 1 \\ \mu \end{bmatrix}^{\otimes 3} = \begin{bmatrix} 1 & 1 \\ \lambda & \mu \end{bmatrix}^{\otimes 3} [c, 0, 0, d]$. Then $[u_0, u_1, u_2, u_3] = \left(M \begin{bmatrix} 1 & 1 \\ \lambda & \mu \end{bmatrix} \right)^{\otimes 3} [c, 0, 0, d]$. By Proposition 2.10 we have

$$\det \begin{bmatrix} u_0 - u_2 & u_1 \\ u_1 - u_3 & u_2 \end{bmatrix} = \left(\det M \cdot \det \begin{bmatrix} 1 & 1 \\ \lambda & \mu \end{bmatrix} \right)^2 cd \begin{bmatrix} 1 & \lambda \end{bmatrix} M^{\mathrm{T}} M \begin{bmatrix} 1 \\ \mu \end{bmatrix}. \tag{2.20}$$

Note that the dot product of the two columns of $M \begin{bmatrix} 1 & 1 \\ \lambda & \mu \end{bmatrix}$ is $(1\ \lambda) M^{\mathrm{T}} M \begin{pmatrix} 1 \\ \mu \end{pmatrix} = (1\ \lambda) \begin{bmatrix} y_0 & y_1 \\ y_1 & y_2 \end{bmatrix} \begin{pmatrix} 1 \\ \mu \end{pmatrix}$, which is exactly $y_0 + by_1 - y_2 \neq 0$. Also $cd(\lambda - \mu)^2 = \det \begin{bmatrix} x_0 & x_1 \\ x_1 & x_2 \end{bmatrix}$. Indeed the binary signature $[x_0, x_1, x_2] = c \begin{bmatrix} 1 \\ \lambda \end{bmatrix}^{\otimes 2} + d \begin{bmatrix} 1 \\ \mu \end{bmatrix}^{\otimes 2}$ has matrix form $\begin{bmatrix} 1 & 1 \\ \lambda & \mu \end{bmatrix} \begin{bmatrix} c & 0 \\ 0 & d \end{bmatrix} \begin{bmatrix} 1 & 1 \\ \lambda & \mu \end{bmatrix}^{\mathrm{T}}$, with determinant $cd(\lambda - \mu)^2$. Hence (2.20) is equivalent to

$$\det \begin{bmatrix} u_0 - u_2 & u_1 \\ u_1 - u_3 & u_2 \end{bmatrix} = (\det M)^2 \det \begin{bmatrix} x_0 & x_1 \\ x_1 & x_2 \end{bmatrix} (y_0 + by_1 - y_2). \tag{2.21}$$

Equation (2.21) is valid for $b \neq \pm 2\mathrm{i}$. However, we show that it is valid for $b = \pm 2\mathrm{i}$ by a limiting argument. We take a sequence $b^{(n)} \to b$, where $b^{(n)} \neq 0, \pm 2\mathrm{i}$. For the given $g = [y_0, y_1, y_2]$ with $y_0 + by_1 - y_2 \neq 0$ we may define a sequence $g^{(n)} = [y_0, \frac{b}{b^{(n)}} y_1, y_2] \to g$, satisfying $y_0 + b^{(n)}(\frac{b}{b^{(n)}} y_1) - y_2 \neq 0$. From $g^{(n)}$ we may define $M^{(n)}$, with $(M^{(n)})^{\mathrm{T}} M^{(n)}$ being the matrix form of

$g^{(n)}$, and satisfying $M^{(n)} \to M$. Then we can define $x^{(n)} = [x_0, x_1, x_2^{(n)}, x_3^{(n)}]$ by the same recurrence with parameter $b^{(n)}$ and then define $u^{(n)}$ by $M^{(n)}$ from $x^{(n)}$ accordingly. By taking limit in (2.21) for this sequence we get (2.21) valid for $b = \pm 2\mathrm{i}$ as well.

The condition that f' is of type I is equivalent to $\det \begin{bmatrix} u_0 - u_2 & u_1 \\ u_1 - u_3 & u_2 \end{bmatrix} = 0$. We know that $\det M \neq 0$, $y_0 + by_1 - y_2 \neq 0$, hence $\det \begin{bmatrix} x_0 & x_1 \\ x_1 & x_2 \end{bmatrix} = 0$. As the third column in $\begin{bmatrix} x_0 & x_1 & x_2 \\ x_1 & x_2 & x_3 \end{bmatrix}$ is linearly dependent on the first two columns by the recurrence, this says that f is degenerate, a contradiction. □

Lemma 2.15 does the same thing as Lemma 2.14 for tractable cases of type II in Lemma 2.13.

Lemma 2.15. *Let $f = [x, y, -x, -y]$ be of type II and $g = [y_0, y_1, y_2]$, both nondegenerate. If g is not of type II, i.e., $y_0 + y_2 \neq 0$, and g is not of the form $[\lambda, 0, \lambda]$, then* Holant*$(g \mid f)$ *is #P-hard. It follows that* Holant*$(\{f, g\})$ *is also #P-hard.*

Proof. The idea of the proof is the same as for Lemma 2.14.

Let $M = \begin{bmatrix} \alpha & \gamma \\ \beta & \delta \end{bmatrix}$ be the matrix such that $g(M^{-1})^{\otimes 2} = [1, 0, 1]$, then $G = \begin{bmatrix} y_0 & y_1 \\ y_1 & y_2 \end{bmatrix} = M^{\mathrm{T}} M$. Let $f' = M^{\otimes 3} f$. Then f' is nondegenerate since f is. Note that f has the form

$$[x, y, -x, -y] = c \begin{bmatrix} 1 \\ \mathrm{i} \end{bmatrix}^{\otimes 3} + d \begin{bmatrix} 1 \\ -\mathrm{i} \end{bmatrix}^{\otimes 3},$$

where $cd \neq 0$ as f is nondegenerate. We get

$$f' = c \begin{bmatrix} \alpha + \gamma \mathrm{i} \\ \beta + \delta \mathrm{i} \end{bmatrix}^{\otimes 3} + d \begin{bmatrix} \alpha - \gamma \mathrm{i} \\ \beta - \delta \mathrm{i} \end{bmatrix}^{\otimes 3}.$$

We want to show that Holant*(f') is #P-hard. By Lemma 2.13, we have to consider two cases. If f' is of type II, then by Proposition 2.11

$$0 = (\alpha + \gamma \mathrm{i})^2 + (\beta + \delta \mathrm{i})^2 = (\alpha - \gamma \mathrm{i})^2 + (\beta - \delta \mathrm{i})^2.$$

This gives $\alpha\gamma + \beta\delta = 0$ and $\alpha^2 + \beta^2 = \gamma^2 + \delta^2$. By $G = M^{\mathrm{T}} M$, this is precisely $y_1 = 0$ and $y_0 = y_2$. Then $[y_0, y_1, y_2]$ has the form $[\lambda, 0, \lambda]$, a contradiction.

Suppose f' has type I. By Proposition 2.10 we have

$$0 = (\alpha + \gamma i)(\alpha - \gamma i) + (\beta + \delta i)(\beta + \delta i) = \alpha^2 + \beta^2 + \gamma^2 + \delta^2.$$

This is precisely $y_0 + y_2 = 0$, a contradiction. □

2.3.3 A Single Signature of Arity n

The following lemma extends Lemma 2.13 to a signature with an arbitrary arity.

Lemma 2.16. *Let $f = [x_0, x_1, \ldots, x_n]$ be a nondegenerate symmetric signature with arity $n > 3$, then* Holant*(f) *is #P-hard unless f has type I or type II.*

Proof. Assume that Holant*(f) is not #P-hard, we prove that f must be of type I or type II. Using unary signatures $\Delta_0 = [1, 0]$ and $\Delta_1 = [0, 1]$, for any subset S of subsignatures of f, Holant*$(S) \leq_T^P$ Holant*(f). Hence Holant*(S) is not #P-hard.

Case A: For every $0 \leq k \leq n - 2$, the subsignature $[x_k, x_{k+1}, x_{k+2}]$ is nondegenerate.

By Lemma 2.13, there are two tractable types to be considered for $[x_0, x_1, x_2, x_3]$.

(i) $[x_0, x_1, x_2, x_3]$ has type I(a, b) for some nonzero pair (a, b). Such a nonzero pair (a, b) must be unique up to a scalar factor, by the nondegeneracy of $[x_0, x_1, x_2, x_3]$. Then by Lemma 2.14, for every $0 \leq k \leq n - 2$, either $ax_k + bx_{k+1} - ax_{k+2} = 0$ or $[x_k, x_{k+1}, x_{k+2}]$ has the form $[2a\lambda, b\lambda, -2a\lambda]$.

If $a = 0$, then $b \neq 0$, and $x_1 = x_2 = 0$ from the linear equations. Then $[x_0, x_1, x_2]$ is degenerate, a contradiction. So $a \neq 0$ and we can normalize it and assume $a = 1$. Now we show that the form $[2\lambda, b\lambda, -2\lambda]$ cannot appear among $[x_k, x_{k+1}, x_{k+2}]$ $(0 \leq k \leq n - 2)$. This will conclude that f has type I(a, b).

Suppose the form $[2\lambda, b\lambda, -2\lambda]$ does occur. Such a λ certainly is nonzero, otherwise it is degenerate. Also $b \neq \pm 2i$, since $[2, \pm 2i, -2]$ is degenerate. If it occurs as $[x_k, x_{k+1}, x_{k+2}]$ for $k = 0$ or 1, then together with $ax_0 + bx_1 - ax_2 = 0$ and $ax_1 + bx_2 - ax_3 = 0$, and $a = 1$, we get $b = \pm 2i$, a contradiction. Thus let k be the minimum index such that $[x_k, x_{k+1}, x_{k+2}]$ has the form $\lambda[2, b, -2]$; then $k \geq 2$, and there is a subsignature $[x_{k-2}, x_{k-1}, x_k, x_{k+1}, x_{k+2}] = \lambda[x, y, 2, b, -2]$ with arity 4, where $ax + by - 2a = 0$ and $ay + 2b - ab = 0$. So the subsignature is $\lambda[b^2 + 2, -b, 2, b, -2]$, of which there is a subsignature $\lambda[-b, 2, b, -2]$. Applying Lemma 2.15 to the nondegenerate signature pair $[-b, 2, b, -2]$ and $[b^2 + 2, -b, 2]$, we have $b^2 + 2 = -2$ or $b^2 + 2 = 2$. The first case would imply that $b = \pm 2i$. So we have $b^2 + 2 = 2$ and thus $b = 0$. In this case, the arity 4 subsignature is $[2, 0, 2, 0, -2]$. Using a unary signature $[1, 1]$, we can get $[2, 2, 2, -2]$. By Lemma 2.13, Holant*($[2, 2, 2, -2]$) is #P-hard.

(ii) If $[x_0, x_1, x_2, x_3]$ is of the form $[x, y, -x, -y]$, then by Lemma 2.15, for every $0 \le k \le n-2$, we have $x_k + x_{k+2} = 0$ or $[x_k, x_{k+1}, x_{k+2}]$ is of the form $[\lambda, 0, \lambda]$. We prove that $[\lambda, 0, \lambda]$ can not appear. This will conclude that f has type II.

Suppose the form $[\lambda, 0, \lambda]$ does appear among $[x_k, x_{k+1}, x_{k+2}]$. $\lambda \neq 0$ by nondegeneracy. It is easy to see that if it occurred at $k = 0$ or 1, then $[x_0, x_1, x_2, x_3] = [x, y, -x, -y]$ is degenerate. Let k be the minimum index such that $[x_k, x_{k+1}, x_{k+2}] = [\lambda, 0, \lambda]$, then $k \ge 2$. To its left we have $x_\ell + x_{\ell+2} = 0$, for $0 \le \ell < k$. It follows that there must be a subsignature (after a scaling) of the form $[1, 0, -1, 0, -1]$. Again we use a unary signature $[1, 1]$ and get $[1, -1, -1, -1]$. By Lemma 2.13 Holant*($[1, -1, -1, -1]$) is #P-hard.

Case B: There exists some $0 \le k \le n-2$, such that $[x_k, x_{k+1}, x_{k+2}]$ is degenerate.

If f is of the form $[x_0, 0, \ldots, 0, x_n]$, then f has type I(0, 1) and Lemma 2.16 holds. In the following we assume f is not of this form.

By Proposition 2.8, there exists $0 \le s \le n-3$, such that the arity 3 subsignature $f' = [x_s, x_{s+1}, x_{s+2}, x_{s+3}]$ is nondegenerate. We want to find a nondegenerate subsignature of arity 3 that contains a *degenerate* subsignature of arity 2. Starting from a degenerate subsignature $g = [x_k, x_{k+1}, x_{k+2}]$ of arity 2, if $s \le k \le s+1$, then g is already a subsignature of f'. Otherwise, either $k < s$ or $k > s+1$. Suppose $k < s$. Consider the subsignature $[x_{k+1}, x_{k+2}, x_{k+3}]$. If it is degenerate, we can substitute it for g, and continue. If it is nondegenerate, then we can substitute f' by $[x_k, x_{k+1}, x_{k+2}, x_{k+3}]$, and it will also be nondegenerate. The proof for $k > s+1$ is similar.

Thus we can find f' and g as specified. This f' must be of the form $[s^2, sr, r^2, x]$ or $[y, s^2, sr, r^2]$. By symmetry, we consider only $f' = [s^2, sr, r^2, x]$. By Lemma 2.13, we have two cases. For the first case, f' has type I. Then $\det \begin{bmatrix} s^2 - r^2 & sr \\ sr - x & r^2 \end{bmatrix} = 0$, by Proposition 2.10. This implies that $srx = r^4$. If $sr \neq 0$, $srx = r^4$ implies that $[s^2, sr, r^2, x]$ is degenerate, a contradiction. If $sr = 0$, then $r = 0$ by $srx = r^4$. Since $[s^2, sr, r^2, x] = [s^2, 0, 0, x]$ is nondegenerate, we must have $s^2 \neq 0$ and $x \neq 0$. Because $n > 3$, there must be a subsignature of f of the form $[s^2, 0, 0, x, z]$ or $[z', s^2, 0, 0, x]$. Suppose we have $[s^2, 0, 0, x, z]$; the other case is similar. Then consider the pair $[s^2, 0, 0, x]$ and $[0, x, z]$. If $z \neq 0$, then Holant*($\{[s^2, 0, 0, x], [0, x, z]\}$) is #P-hard by Lemma 2.14. If $z = 0$, then we have a subsignature $[0, 0, x, 0]$. Holant*($[0, 0, x, 0]$) is also #P-hard by Lemma 2.13. Finally, for the second case, f' has type II: $s^2 + r^2 = 0$ and

$sr + x = 0$. Then the signature must be $[s^2, s^2\mathfrak{i}, -s^2, -s^2\mathfrak{i}]$ or $[s^2, -s^2\mathfrak{i}, -s^2, s^2\mathfrak{i}]$. Both are degenerate, a contradiction. □

2.3.4 Putting Things Together

Finally we extend these lemmas to handle a set of signatures and finish the proof of Theorem 2.12. The main idea is to find a nondegenerate signature of arity 3, and anchor the proof on this arity 3 signature, namely to show that everything else must conform to it, or else the problem is #P-hard.

Proof. of Theorem 2.12: The tractability part has already been proved. We prove the hardness part. We assume every signature in $\mathcal{F}$ is nondegenerate and has arity ≥ 2.

Suppose Holant*$(\mathcal{F})$ is *not* #P-hard. We want to show that $\mathcal{F}$ must fall into one of the three classes.

If no signature of $\mathcal{F}$ has arity ≥ 3, then this is in class $\mathfrak{A}$ and we are done. Suppose there are some signatures of arity ≥ 3. Each such signature is of type I or II by Lemma 2.13 and Lemma 2.16. First suppose all signatures in $\mathcal{F}$ of arity ≥ 3 are of the same type, namely either for some (a, b) they are all of type I(a, b) or they are all of type II. If they are all of type I(a, b), then we may obtain a nondegenerate signature of arity 3 having the same type by either taking a nondegenerate subsignature (Proposition 2.8), or connecting it with the unary $[1, 1]$ if it is of type I$(0, 1)$. Lemma 2.14 finishes the proof in this case and we are in class $\mathfrak{B}$. If all signatures in $\mathcal{F}$ of arity ≥ 3 are of type II, again we can obtain a nondegenerate subsignature of arity 3. Lemma 2.15 then finishes the proof and we are in class $\mathfrak{C}$.

In the following we assume there is more than one type of signatures in $\mathcal{F}$ having arity ≥ 3. Each such signature declares a type I or II.

- Case 1. Suppose there is a nondegenerate type I$(0, 1)$ signature in $\mathcal{F}$ of arity ≥ 3. By connecting it with the unary $[1, 1]$ we may assume we have a nondegenerate type I$(0, 1)$ signature $f \in \mathcal{F}$ of arity 3. By assumption, there exists another nondegenerate signature g of arity ≥ 3 and of a different type. By Proposition 2.8, there exists a nondegenerate subsignature $[x, y, z]$ of arity two. By being *not* #P-hard, we apply Lemma 2.14 to the pair f and $[x, y, z]$ and conclude that there are two cases: $[x, y, z] = [x, 0, z]$, or $[0, y, 0]$. Being nondegenerate, $xz \neq 0$ in the first case, and $y \neq 0$ in the second case. Within g there is a subsignature containing $[x, y, z]$. In the first case, it has the form $[w, x, 0, z]$ (or $[x, 0, z, w]$); then $[w, x, 0]$ (or $[0, z, w]$) is nondegenerate. We apply Lemma 2.14 once again to f and $[w, x, 0]$ (or $[0, z, w]$), and get $w = 0$. In the second case, it has the form $[u, 0, y, 0]$

(or $[0, y, 0, u]$). If $u = 0$, Holant*($[0, 0, y, 0]$) (or Holant*($[0, y, 0, 0]$)) is #P-hard by Lemma 2.13. Hence in all cases we get a signature of the form $[u, 0, v, 0]$ (or its reversal $[0, v, 0, u]$), where $uv \neq 0$. By symmetry we consider $[u, 0, v, 0]$, as $[0, v, 0, u]$ is changed to $[u, 0, v, 0]$ by a 0-1 exchange, and the type I(0, 1) signature f retains its type. We connect a unary $[s, t]$ to $[u, 0, v, 0]$, and get $[su, tv, sv]$. As long as $st \neq 0$ and $s^2/t^2 \neq v/u$ this is nondegenerate and Lemma 2.14 on the pair f and $[su, tv, sv]$ leads to #P-hardness.

- Case 2. Now we assume case 1 does not apply, but there exists a nondegenerate type I(a, b) signature in $\mathcal{F}$ of arity ≥ 3, where $a \neq 0$ and $b \neq \pm 2a\mathrm{i}$. By Proposition 2.8, we may assume there is a nondegenerate type I(a, b) signature f having arity 3. We may normalize (a, b) to $a = 1$, and $b \neq \pm 2\mathrm{i}$.

  Then the characteristic equation $X^2 - bX - 1$ has two distinct roots $\lambda \neq \mu$. It follows that $\lambda\mu = -1$, and $\lambda, \mu \neq \pm\mathrm{i}$. Define $T = \begin{bmatrix} \frac{1}{\sqrt{1+\lambda^2}} & \frac{1}{\sqrt{1+\mu^2}} \\ \frac{\lambda}{\sqrt{1+\lambda^2}} & \frac{\mu}{\sqrt{1+\mu^2}} \end{bmatrix}$, then T is orthogonal. It is easy to verify that $f = T^{\otimes 3} f'$ for some $f' = [x_0, 0, 0, x_3]$, a nondegenerate signature of type I(0, 1). Performing a holographic transformation by T on all signatures in $\mathcal{F}$, we have reduced case 2 to case 1.
- Case 3. We assume cases 1 and 2 do not apply, but there exists a nondegenerate type I$(1, b)$ signature in $\mathcal{F}$ of arity ≥ 3, where $b = \pm 2\mathrm{i}$. By Proposition 2.8, we may assume such a signature f of arity 3. We still suppose there exists a nondegenerate signature g of arity ≥ 3 and of a different type; for otherwise we are done. g could be of type I$(1, b')$ or II, where $b' = \mp 2\mathrm{i}$ (but $b' \neq b$, and hence $b' = -b$). By Proposition 2.8, there exists a nondegenerate subsignature $[x, y, z]$ of g.

If g has type I$(1, b')$, then $x + b'y - z = 0$. Note that for $b = \pm 2\mathrm{i}$, the binary signature $[2, b, -2]$ is degenerate. Since $[x, y, z]$ is nondegenerate, it cannot be $\lambda[2, b, -2]$. Hence, by applying Lemma 2.14 to f and $[x, y, z]$, we get $x + by - z = 0$. This implies that $y = 0$ since $b' \neq b$, and $x = z \neq 0$ by the nondegeneracy of $[x, y, z]$. But within g there is a subsignature $[w, x, 0, x]$ or $[x, 0, x, w]$. As $x \neq 0$, $[w, x, 0]$ and $[0, x, w]$ are nondegenerate. Applying Lemma 2.14 once again we get a contradiction. If g is of type II, then we still have $x + by - z = 0$ as well as $z = -x$. Since $b = \pm 2\mathrm{i}$, this implies that $[x, y, z]$ is degenerate, also a contradiction.

If cases 1, 2 and 3 all do not apply, then all nondegenerate signatures in $\mathcal{F}$ of arity ≥ 3 are of type II. This case has been proved before; we are in class $\mathfrak{C}$. Theorem 2.12 is proved. □

2.4 The Road Ahead

Theorem 2.12 is our first major dichotomy theorem. Although it covers a fairly broad class of problems, we should also recognize its limitations. There are at least three aspects in which Theorem 2.12 is restricted.

The first is the restriction to a symmetric set of constraint functions $\mathcal{F}$. One can ask what happens when we remove this restriction of symmetry. The answer will be given in Chapter 10, where we present a dichotomy theorem, Theorem 10.14, for Holant* problems applicable to a set of constraint functions $\mathcal{F}$ that are not necessarily symmetric. In so doing we will further reveal an alternative perspective on the classification achieved here. To arrive at Theorem 10.14, we will have to develop the theory much further; in particular, the proof of Theorem 10.14 will use Theorem 2.12, the dichotomy for the symmetric case.

The second is the assumption of the presence of all unary functions $\mathcal{U}$ for free in the definition of Holant* problems. One can ask what happens if this assumption is removed. In Chapter 7 we will present a dichotomy for Holant problems on symmetric signatures without assuming any freely available auxiliary functions, in particular unary functions. The proof will utilize a substantial part of the theory to be developed in the next few chapters. We develop this theory starting with free unary functions. The theory will also consider other freely available auxiliary functions, for example, when EQUALITY functions of all arities are assumed to be free, namely Boolean #CSP problems, which we turn to next in Chapter 3. We will also consider the case where only two special unary functions, Δ_0 and Δ_1, are free. This is called Holant^c problems. It is an interesting open problem when we remove both the symmetry restriction and the presence of unary functions. It is open even for Holant^c problems when the symmetry restriction is removed. However, the study of Holant* problems by assuming unary functions $\mathcal{U}$ are free is more than an arbitrary choice. As we develop the theory, aside from explicitly using Theorem 2.12, the *idea* of simulating the free presence of unary functions is a major theme.

Finally, the third limitation is that we restrict our study to the Boolean domain in Volume I of this book. In Volume II we will develop the theory on a general finite domain. However, the theory on general domains is not as well developed as on the Boolean domain. In particular, the analogue of Theorem 2.12 is still open for the general domain.

3
Boolean #CSP

In this chapter we address the following type of counting problems, called Boolean #CSP. Let $\mathcal{F}$ be a set of functions, where each $f \in \mathcal{F}$ is a function $f : \{0, 1\}^k \to \mathbb{C}$, for some k depending on f, mapping Boolean variables to the complex numbers. In this chapter, the functions in $\mathcal{F}$ are not assumed to be symmetric in general. The Boolean counting constraint satisfaction problem #CSP($\mathcal{F}$) is defined as follows. The input I is a finite sequence of constraints on Boolean variables $x_1, x_2, \ldots, x_n$ of the form $f(x_{i_1}, x_{i_2}, \ldots, x_{i_k})$, where $f \in \mathcal{F}$. The output is called the partition function

$$Z(I) = \sum_{x_1,x_2,\ldots,x_n \in \{0,1\}} \prod f(x_{i_1}, x_{i_2}, \ldots, x_{i_k}),$$

where the product is over all constraints occurring in I. When $\mathcal{F} = \{f\}$, consisting of a single constraint function, we denote #CSP($\{f\}$) simply as #CSP(f).

3.1 The 0-1 Case and Nonnegative Boolean #CSP

In general, functions $f \in \mathcal{F}$ can take arbitrary complex values. If each f takes values 0, 1, then it is identified with a Boolean-valued constraint. In that case, $Z(I)$ counts the number of assignments *satisfying* all the Boolean constraints of I.

Typically $\mathcal{F}$ is a finite set and considered fixed. The complexity is measured in terms of the input size of I. For a finite and fixed $\mathcal{F}$ this input size is equivalent to n, up to a polynomial factor. Occasionally we will also allow $\mathcal{F}$ to be infinite, such as those sets containing all unary functions $\mathcal{U}$, in which case the input size must include the description of the constraints used in the input.

If each $f \in \mathcal{F}$ takes values 0, 1, then we have the following theorem.

Theorem 3.1 (Creignou–Hermann). *For every finite $\mathcal{F}$ consisting of 0-1 valued functions on Boolean variables, #CSP($\mathcal{F}$) is #P-complete, unless each $f \in \mathcal{F}$ is the 0-1 indicator function of some affine linear subspace over $\mathbb{Z}_2$, in which case it is in P.*

In Volume I we present the dichotomy theory only on the Boolean domain. Far reaching generalizations to general domains will be presented in Volume II [DG00, BD07, BG05, Bul13, Thu09, GT11, GGJT10, CCL13, DR13, CCL16, CC12].

A 0-1 indicator function of some affine linear subspace is just a Boolean conjunction of XOR functions on a (possibly empty) subset of variables and possibly a constant $1 \in \mathbb{Z}_2$. This includes the constant 0, 1 functions, the unary functions $\Delta_0 = [1, 0]$ and $\Delta_1 = [0, 1]$, the binary DISEQUALITY $[0, 1, 0]$, and EQUALITY of any arities $(=_k)$. When $\mathcal{F}$ has the form specified in the theorem, one can solve #CSP($\mathcal{F}$) in polynomial time by Gaussian elimination over the field $\mathbb{Z}_2$.

Definition 3.2. A function has *product type* if it can be expressed as a product of unary functions, binary EQUALITY functions ($(=_2) = [1, 0, 1]$), and binary DISEQUALITY functions ($(\neq_2) = [0, 1, 0]$), on not necessarily disjoint subsets of variables. We denote by $\mathscr{P}$ the set of all functions of *product type*.

Exercise: Show that a symmetric signature is in $\mathscr{P}$ iff it is a GEN-EQ, or a (scalar multiple of) binary DISEQUALITY, or a constant function.

Definition 3.3. A function is *pure affine* if it is a constant multiple of a 0-1 indicator function of some affine linear subspace over $\mathbb{Z}_2$.

If all functions in $\mathcal{F}$ take nonnegative values, then Theorem 3.1 can be generalized to the following theorem.

Theorem 3.4 (Dyer–Goldberg–Jerrum). *For every finite $\mathcal{F}$ consisting of nonnegative-valued functions, #CSP($\mathcal{F}$) is solvable in polynomial time in the following two cases, and is #P-hard otherwise:*

1. *Every function in $\mathcal{F}$ is of a* product type*; or*
2. *Every function in $\mathcal{F}$ is a* pure affine *function.*

For either tractable cases, the polynomial time algorithm is quite simple. The main claim is that everything else is #P-hard. Theorem 3.4 is concerned with nonnegative-valued functions. However, the tractability still holds without this restriction. This means in particular that functions in $\mathscr{P}$ can be complex valued, i.e., the unary functions in the definition of a function in $\mathscr{P}$ can be complex valued. Any GEN-EQ is in $\mathscr{P}$. Similarly, the constant multiple in the definition of a pure affine function can be from $\mathbb{C}$.

It is also clear that Theorem 3.1 is a special case of Theorem 3.4.

However, allowing real or complex-valued product type or pure affine functions in Theorem 3.4 turns out to be not sufficient to encompass all tractable cases when functions in $\mathcal{F}$ are no longer restricted to those taking nonnegative values. When the functions $f \in \mathcal{F}$ take possibly negative or complex values, there are more nontrivial tractable #CSP($\mathcal{F}$) problems. This is made possible by algebraic cancelations, and they lead to novel efficient algorithms. The goal of this chapter is to prove an explicit dichotomy theorem for #CSP($\mathcal{F}$), Theorem 3.7, for any complex-valued function set $\mathcal{F}$. Theorem 3.4, and a fortiori Theorem 3.1, are special cases of this dichotomy theorem.

Before we move on to investigate Boolean #CSP in detail, we can restate Theorem 2.12 in terms of what can be transformed to product type $\mathscr{P}$. Note that a symmetric signature is in $\mathscr{P}$ iff it is a GEN-EQ, or a binary DISEQUALITY, or a degenerate function.

Theorem 3.5. *Let $\mathcal{F}$ be a set of nondegenerate symmetric signatures over $\mathbb{C}$. Then* Holant*$(\mathcal{F})$ *is #P-hard, unless $\mathcal{F}$ satisfies the following conditions, in which case it is computable in polynomial time.*

1. *All signatures in $\mathcal{F}$ have arity ≤ 2.*
2. *There exists some $M \in \mathbf{GL}_2(\mathbb{C})$ such that $(=_2)M^{\otimes 2} \in \mathscr{P}$ and $\mathcal{F} \subseteq M\mathscr{P}$.*
3. *There exists $\lambda \in \{2\mathrm{i}, -2\mathrm{i}\}$, such that every signature $f \in \mathcal{F}$ of arity n satisfies the recurrence*

$$f_{k+2} = \lambda f_{k+1} + f_k, \quad \text{for} \quad 0 \leq k \leq n-2.$$

Remark: Case 1 here corresponds to class $\mathfrak{A}$ of Theorem 2.12. Case 2 corresponds to all of class $\mathfrak{B}$ (except type $\mathrm{I}(1, \pm 2\mathrm{i})$) as well as class $\mathfrak{C}$. Case 3 corresponds to type $\mathrm{I}(1, \pm 2\mathrm{i})$ of class $\mathfrak{B}$. As shown in Section 2.2, type $\mathrm{I}(1, \pm 2\mathrm{i})$ is also equivalent to the following. Under the Z transformation, every $f \in \mathcal{F}$ has the form $f = Z[c, d, 0, \ldots, 0]$, or every $f \in \mathcal{F}$ has the form $f = Z[0, \ldots, 0, d, c]$, where $Z = \frac{1}{\sqrt{2}}\begin{bmatrix} 1 & 1 \\ \mathrm{i} & -\mathrm{i} \end{bmatrix}$. Recall also that under the covariant transformation by Z, the binary EQUALITY is transformed to the binary DISEQUALITY: $(=_2)Z^{\otimes 2} = (\neq_2)$.

Exercise: Show that there exists some $M \in \mathbf{GL}_2(\mathbb{C})$ such that $(=_2)M^{\otimes 2}$ is a GEN-EQ and $\mathcal{F} \subseteq M\mathscr{P}$, iff there exists an orthogonal T that does this transformation.

Exercise: Show that there exists some $M \in \mathbf{GL}_2(\mathbb{C})$ such that $(=_2)M^{\otimes 2}$ is $(\neq_2)$ and $\mathcal{F} \subseteq M\mathscr{P}$, iff we may take $M = Z$ to do this transformation.

Exercise: Prove that Theorem 3.5 is an equivalent form of Theorem 2.12.

3.2 Affine Functions $\mathscr{A}$ and $\mathscr{F}_1 \cup \mathscr{F}_2 \cup \mathscr{F}_3$

There is a simple relation between #CSP and Holant problems. We can represent an instance of a #CSP problem by a bipartite graph G where the left-hand side is labeled by variables and the right-hand side is labeled by constraints. We define a signature grid Ω on G by assigning an EQUALITY function to every variable node on the left-hand side, and every constraint node on the right-hand side has the given constraint function. Then Holant_Ω is exactly the same as the #CSP problem. In effect, the EQUALITY function on each variable node forces the incident edges to take the same value; this effectively reduces edge assignments in Holant_Ω to vertex assignments on the left-hand side in the #CSP problem. Thus #CSP problems are precisely the special case of Holant problems on bipartite graphs where every vertex on the left-hand side is assigned an EQUALITY function. Furthermore, the restriction on bipartiteness is not essential; given any constraint function set $\mathcal{F}$, it is easy to show that #CSP($\mathcal{F}$) is exactly the same as Holant($\mathcal{F} \cup \mathscr{E}$), where $\mathscr{E} = \{=_k \mid k \geq 1\}$ is the set of all EQUALITY functions of any arity k. In other words, #CSP problems are Holant problems with the implicit assumption that all EQUALITY functions are freely available. This is similar to Holant* problems where we assume all unary functions are freely available. In fact, we will see that the presence of all EQUALITY functions already implies the presence of two special unary functions Δ_0 and Δ_1. Applying Δ_0 or Δ_1 on a variable is called *pinning*.

On the other hand, Holant problems can be considered as #CSP problems where every variable appears twice. We will show that the following three families of functions are tractable for Holant problems. Below we write a function as a vector, listing its values in lexicographic order of the assignments, which we will also call its truth table. Here $\mathfrak{i} = \sqrt{-1}$.

$$\mathscr{F}_1 = \{\lambda([1, 0]^{\otimes k} + \mathfrak{i}^r[0, \quad 1]^{\otimes k}) \mid \lambda \in \mathbb{C}, k = 1, 2, \ldots, \text{ and } r = 0, 1, 2, 3\};$$

$$\mathscr{F}_2 = \{\lambda([1, 1]^{\otimes k} + \mathfrak{i}^r[1, -1]^{\otimes k}) \mid \lambda \in \mathbb{C}, k = 1, 2, \ldots, \text{ and } r = 0, 1, 2, 3\};$$

$$\mathscr{F}_3 = \{\lambda([1, \ \mathfrak{i}]^{\otimes k} + \mathfrak{i}^r[1, \ -\mathfrak{i}]^{\otimes k}) \mid \lambda \in \mathbb{C}, k = 1, 2, \ldots, \text{ and } r = 0, 1, 2, 3\}.$$

We explicitly list all the signatures in $\mathscr{F}_1 \cup \mathscr{F}_2 \cup \mathscr{F}_3$ up to an arbitrary constant multiple from $\mathbb{C}$:

1. $[1,0,\ldots,0,\pm 1]$; $(\mathscr{F}_1, r = 0, 2)$
2. $[1,0,\ldots,0,\pm \mathfrak{i}]$; $(\mathscr{F}_1, r = 1, 3)$
3. $[1,0,1,0,\ldots,0 \text{ or } 1]$; $(\mathscr{F}_2, r = 0)$
4. $[1,-\mathfrak{i},1,-\mathfrak{i},\ldots,(-\mathfrak{i}) \text{ or } 1]$; $(\mathscr{F}_2, r = 1)$
5. $[0,1,0,1,\ldots,0 \text{ or } 1]$; $(\mathscr{F}_2, r = 2)$
6. $[1,\mathfrak{i},1,\mathfrak{i},\ldots,\mathfrak{i} \text{ or } 1]$; $(\mathscr{F}_2, r = 3)$
7. $[1,0,-1,0,1,0,-1,0,\ldots,0 \text{ or } 1 \text{ or } (-1)]$; $(\mathscr{F}_3, r = 0)$
8. $[1,1,-1,-1,1,1,-1,-1,\ldots,1 \text{ or } (-1)]$; $(\mathscr{F}_3, r = 1)$
9. $[0,1,0,-1,0,1,0,-1,\ldots,0 \text{ or } 1 \text{ or } (-1)]$; $(\mathscr{F}_3, r = 2)$
10. $[1,-1,-1,1,1,-1,-1,1,\ldots,1 \text{ or } (-1)]$. $(\mathscr{F}_3, r = 3)$

We can show that Holant_Ω for any $\Omega = (G, \mathscr{F}_1 \cup \mathscr{F}_2 \cup \mathscr{F}_3)$ is computable in P. These functions are all related to each other by holographic reductions. We will prove this tractability by a more general tractability theorem that covers nonsymmetric functions as well.

We note that expressions in complex numbers appear naturally, even for real-valued functions. The special case where $r = 1$, $k = 2$ and $\lambda = (1 + \mathfrak{i})^{-1}$ in $\mathscr{F}_3$ is noteworthy. In this case we get a real-valued binary symmetric function $H = [1, 1, -1]$. In other words, $H(0, 0) = H(0, 1) = H(1, 0) = 1$ and $H(1, 1) = -1$. The matrix form of this function is the Hadamard matrix $H = \begin{bmatrix} 1 & 1 \\ 1 & -1 \end{bmatrix}$. Furthermore, if we take $r = 0$, any k and $\lambda = 1$ in $\mathscr{F}_1$ we get the EQUALITY function on k variables. If $\Omega = (G, \mathcal{F})$ with $\mathcal{F}$ consisting of exactly the function H and all EQUALITY functions, then Holant_Ω is computing the partition function of #CSP(H). This special case where the constraint function set in #CSP consists of a single binary (symmetric) function is also known as graph homomorphism $Z_H(G)$, or a spin system, on (undirected) graph G. What is the value of $Z_H(G)$? If X (respectively Y) is the number of (vertex subset-) induced subgraphs of G with an even (respectively odd) number of edges, then $Z_H(G) = X - Y$. Since trivially $X + Y = 2^n$, we have $X = \frac{2^n + Z_H(G)}{2}$ and $Y = \frac{2^n - Z_H(G)}{2}$; thus the problem Z_H essentially computes the number of induced subgraphs with an even (respectively odd) number of edges.

A function of arity k can be expressed by its truth table of length 2^k. Define

$$\mathscr{D} = \{f \mid f = [a_1, b_1] \otimes [a_2, b_2] \otimes \cdots \otimes [a_k, b_k], \text{ for some } a_i, b_i \in \mathbb{C}\}$$

to be the set of functions that can be expressed as a tensor product of unary functions; that is, a function in $\mathscr{D}$ on k variables is the product of k unary functions applied to its k variables separately. Functions in $\mathscr{D}$ are called *degenerate*. This notion of degeneracy agrees with the definition of degeneracy for symmetric functions. A binary function is in $\mathscr{D}$ iff its corresponding matrix is singular. Note that $\mathscr{D}$ is a subset of $\mathscr{P}$, the signatures of product type.

The main result in this chapter is a complexity dichotomy theorem, Theorem 3.7, for complex-valued Boolean #CSP. The family $\mathscr{F}_1 \cup \mathscr{F}_2 \cup \mathscr{F}_3$ can be naturally generalized to include unsymmetric functions, called the affine family $\mathscr{A}$. In fact, the family $\mathscr{F}_1 \cup \mathscr{F}_2 \cup \mathscr{F}_3$ consists of precisely the unary functions of $\mathscr{A}$, and (for $\lambda \neq 0$ and $k \geq 2$) symmetric nondegenerate functions of $\mathscr{A}$ of arity ≥ 2. Of course unary functions are also symmetric (but are considered degenerate). The dichotomy theorem says that a Boolean #CSP problem #CSP($\mathcal{F}$) is tractable if *either* all its constraint functions $\mathcal{F}$ are of product type $\mathscr{P}$, *or* all are from this affine family $\mathscr{A}$; otherwise it is #P-hard.

Suppose f is a function on input variables $x_1, x_2, \ldots, x_k$. For $1 \leq s \leq k$ and $c = 0, 1$, $f^{x_s=c}$ denotes the function $f^{x_s=c}(x_1, \ldots, x_{s-1}, x_{s+1}, \ldots, x_k) = f(x_1, \ldots, x_{s-1}, c, x_{s+1}, \ldots, x_k)$, and $f^{x_s=*}$ denotes the function $f^{x_s=*}(x_1, \ldots, x_{s-1}, x_{s+1}, \ldots, x_k) = \sum_{x_s=0,1} f(x_1, \ldots, x_k)$.

For a constraint function f we define its underlying relation, also called its support , by

$$R_f = \{X \in \{0, 1\}^k \mid f(X) \neq 0\}.$$

We say a relation $R \subseteq \{0, 1\}^k$ is affine if it is an affine subspace over $\mathbb{Z}_2$ (including possibly the empty space). It is composed of solutions to a system of linear equations. Equivalently, it satisfies the property that if $\alpha, \beta, \gamma \in R$, then the bitwise XOR string $\alpha \oplus \beta \oplus \gamma \in R$. To see this, first suppose R is the set of solutions to $AX = r$ for some matrix A and vector r over $\mathbb{Z}_2$. Then clearly for any $\alpha, \beta, \gamma \in R$, $A(\alpha \oplus \beta \oplus \gamma) = r \oplus r \oplus r = r$, and $\alpha \oplus \beta \oplus \gamma \in R$. Now suppose R satisfies this property. Suppose $R \neq \emptyset$. Define $L_R = \{\alpha \oplus \beta \mid \alpha, \beta \in R\}$. Then L_R is a linear subspace of $\mathbb{Z}_2^k$. Take any $\alpha_0 \in R$. Then $R \subseteq L_R + \alpha_0 = \{X \oplus \alpha_0 \mid X \in L_R\}$, since for any $\alpha \in R$, $\alpha = \alpha \oplus \alpha_0 \oplus \alpha_0$. Also, $L_R + \alpha_0 \subseteq R$ by the fact that R satisfies the property.

If R_f is affine, we say f has affine support . We also view relations as functions from $\{0, 1\}^k$ to $\{0, 1\}$.

Now we define the affine family of functions $\mathscr{A}$. Let X denote the $k + 1$ dimensional column vector $(x_1, x_2, \ldots, x_k, 1)^{\mathrm{T}}$ over the field $\mathbb{Z}_2$. Suppose A is a matrix over $\mathbb{Z}_2$. χ_{AX} denotes the affine relation on inputs $x_1, x_2, \ldots, x_k$, whose value is 1 if AX is the zero vector, and 0 otherwise.

Definition 3.6. A function is of *affine type* if it can be expressed as

$$\lambda \cdot \chi_{AX} \cdot \mathrm{i}^{L_1(X)+L_2(X)+\cdots+L_n(X)},$$

where $\lambda \in \mathbb{C}$, $\mathrm{i} = \sqrt{-1}$, each L_j is an integer 0-1 indicator function of the form $\langle \alpha_j, X \rangle$, where α_j is a $k + 1$ dimensional vector over $\mathbb{Z}_2$, and the dot product $\langle \cdot, \cdot \rangle$ is computed over $\mathbb{Z}_2$. The set of all functions of *affine type* is denoted by $\mathscr{A}$.

We may compute the dot product as an ordinary integer dot product, and then take the value mod 2, producing an integer value 0 or 1. The additions among $L_j(X)$ are the usual addition in $\mathbb{Z}$. It can be computed mod 4, but not mod 2. If we solve the affine linear system $AX = 0 \bmod 2$ we can replace dependent variables and assume the expressions $L_j(X)$ are over a set of free variables $Y \subseteq X$. We usually will omit any nonzero constant factor λ, since it does not affect complexity and can be ignored.

3.3 A Dichotomy for Boolean #CSP

Theorem 3.7. *Suppose $\mathcal{F}$ is a set of functions mapping Boolean inputs to complex numbers. If $\mathcal{F} \subseteq \mathscr{A}$ or $\mathcal{F} \subseteq \mathscr{P}$, then #CSP($\mathcal{F}$) is computable in polynomial time. Otherwise, #CSP($\mathcal{F}$) is* #P*-hard.*

Proof Outline: The polynomial time algorithm for #CSP($\mathscr{P}$) is easy. Section 3.4 gives a polynomial time algorithm for #CSP($\mathscr{A}$). Cancelations when summing for functions in $\mathscr{A}$ play an essential role. The starting point of the hardness result is Theorem 3.14, which says that if $\mathcal{F}$ contains only one binary symmetric function and is not in $\mathscr{A} \cup \mathscr{P}$, then the #CSP problem is #P-hard. In Lemma 3.18, we prove that #CSP(f) is #P-hard unless f has affine support. This structure is essential in the proof of Lemmas 3.19 and 3.20, the two key lemmas of the hardness reduction. The common strategy of Lemmas 3.19 and 3.20 is to reduce the arity of a given function. In Lemma 3.19, we prove that given a function f not in $\mathscr{A}$, we can simulate (in polynomial time) a unary function $f' \notin \mathscr{A}$; In Lemma 3.20, we prove that given a function g not in $\mathscr{P}$, we can simulate (in polynomial time) a binary or a ternary function $g' \notin \mathscr{P}$. Then we prove that #CSP($\{f', g'\}$) is #P-hard. To prove this, we show that we can always combine functions f' and g' to realize a binary symmetric function that is not in $\mathscr{A} \cup \mathscr{P}$.

Exercise: The tractability criterion in Theorem 3.7 should not be confused with the statement $\mathcal{F} \subseteq \mathscr{A} \cup \mathscr{P}$. Show that there are $\mathcal{F}$ satisfying $\mathcal{F} \subseteq \mathscr{A} \cup \mathscr{P}$ and yet #CSP($\mathcal{F}$) is #P-hard.

There is an alternative expression of a function in $\mathscr{A}$. We claim that the expressibility as

$$\lambda \chi_{AX} i^{L_1(X)+L_2(X)+\cdots+L_n(X)}$$

is equivalent to an expression of the form $\lambda \chi_{AX} i^{Q(X)}$, where Q is a homogeneous quadratic polynomial in $\mathbb{Z}[X]$ with the additional requirement that every cross term $x_s x_t$ has an even coefficient, where $s \neq t$. To see this we observe that each

$L_j(X)$ as an integer sum mod 2 can be replaced by $(L_j(X))^2$ as an integer sum mod 4, since $L_j(X) = 0, 1 \pmod 2$ iff $(L_j(X))^2 = 0, 1 \pmod 4$. After this, all cross terms have an even coefficient, and $\mathfrak{i}^{x_s}$ can be replaced by $\mathfrak{i}^{x_s^2}$, for $x_s = 0, 1$. Conversely, we can express Q mod 4 as a sum of squares of affine forms of X, using the extra condition that all cross terms have even coefficients.

Exercise: Prove that the class of function $\mathscr{A}$ is not changed by the following modifications in its definition:

- Q is a polynomial of degree ≤ 2 (with possible linear and constant terms, but cross terms are still required have even coefficients).
- Q is a multilinear polynomial of degree ≤ 2, where cross terms have even coefficients.

Exercise: Prove that a function $\{0, 1\}^k \to \mathbb{Z}_4$ has a unique multilinear polynomial expression. In particular, suppose $f = \lambda \chi_{AX} \mathfrak{i}^{Q(Y)}$, where $Q(Y)$ is a multilinear polynomial on free variables $Y \subseteq X$ of its support. If Q has degree > 2, or has a cross term with an odd coefficient, then $f \notin \mathscr{A}$.

Exercise: Derive a normal form for any real-valued function $f \in \mathscr{A}$.

The class $\mathscr{F}_1 \cup \mathscr{F}_2 \cup \mathscr{F}_3$ is the restriction of $\mathscr{A}$ to symmetric functions. More precisely, $\mathscr{F}_1 \cup \mathscr{F}_2 \cup \mathscr{F}_3$ consists of scalar multiples of all unary or non-degenerate symmetric functions in $\mathscr{A}$. We first show that $\mathscr{F}_1 \cup \mathscr{F}_2 \cup \mathscr{F}_3 \subset \mathscr{A}$. Omitting the nonzero constants, the functions in $\mathscr{F}_1$ can be expressed as $\chi_{x_1=x_2=\ldots=x_k} \mathfrak{i}^{rx_1^2}$. The functions in $\mathscr{F}_2$ can be expressed as $\chi_{\sum_j x_j=0}$, or $\chi_{\sum_j x_j=1}$, or $\mathfrak{i}^{(\sum_j x_j)^2}$, or $(-\mathfrak{i})^{(\sum_j x_j)^2} = \mathfrak{i}^{3(\sum_j x_j)^2}$, respectively. The functions in $\mathscr{F}_3$ can be expressed as $\chi_{\sum_j x_j=0} \mathfrak{i}^{\sum_j x_j^2}$, or $\chi_{\sum_j x_j=1} \mathfrak{i}^{\sum_j x_j^2 - 1} = -\mathfrak{i} \chi_{\sum_j x_j=1} \mathfrak{i}^{\sum_j x_j^2}$, or $(-1)^{\sum_{s<t} x_s x_t} = \mathfrak{i}^{2\sum_{s<t} x_s x_t}$, or $(-1)^{\sum_{s<t} x_s x_t + \sum_j x_j} = \mathfrak{i}^{2\sum_j x_j^2 + 2\sum_{s<t} x_s x_t}$, respectively.

Conversely suppose $f(x_1, \ldots, x_k) \in \mathscr{A}$ is symmetric. Let $S = \{x \in \mathbb{Z}_2^k \mid f(x) \neq 0\}$ be its affine support. We claim that S must be of the following form: $\emptyset$, $\mathbb{Z}_2^k$, $\{(0, \ldots, 0)\}$, $\{(1, \ldots, 1)\}$, $\{(0, \ldots, 0), (1, \ldots, 1)\}$, $\{(0, 1), (1, 0)\}$, and $\{x \in \mathbb{Z}_2^k \mid \sum_j x_j = \epsilon \bmod 2\}$, where $\epsilon = 0$ or 1. If f is identically zero, then $S = \emptyset$. We assume f is not identically zero.

Consider all affine linear equations over $\mathbb{Z}_2$ that are valid on S, which must be closed under permutations of variables by the symmetry of f. If the set of equations is empty, then $S = \mathbb{Z}_2^k$. So assume the set of equations is nonempty. By symmetry, if there is an equation involving only one variable $x_i = \epsilon$, for some $1 \leq i \leq k$, then for all $1 \leq j \leq k$, $x_j = \epsilon$. Thus $S = \{(0, \ldots, 0)\}$ or $\{(1, \ldots, 1)\}$. Now we assume no equation has only one variable, but there is an equation of two variables $x_i + x_j = \epsilon$, for some $1 \leq i < j \leq k$, then by symmetry, this holds for all $i \neq j$. If $k > 2$ and $\epsilon = 1$, we would have $x_1 + x_2 = 1$ and $x_1 + x_2 = (x_1 + x_3) + (x_2 + x_3) = 0 \bmod 2$, and S would be empty. Thus for $k > 2$, we must have $\epsilon = 0$ and we have $x_1 = x_2 = \cdots = x_k$. Thus $S = \{(0, \ldots, 0), (1, \ldots, 1)\}$. If $k = 2$, additionally we also have possibly $S = \{(0, 1), (1, 0)\}$. Finally, we may assume all equations involve at least three variables. Then $k \geq 3$, and every equation must involve all k variables. Otherwise, we have $x_1 + \cdots + x_j = \epsilon$ and $x_1 + \cdots + x_{j-1} + x_{j+1} = \epsilon$, for some $j < k$, whose sum gives $x_j + x_{j+1} = 0 \bmod 2$ having only two variables. Thus $S = \{x \in \mathbb{Z}_2^k \mid \sum_j x_j = \epsilon \bmod 2\}$.

Let $f = \chi_S \mathfrak{i}^{Q(x_1,\ldots,x_k)} \in \mathscr{A}$ be symmetric, where $Q(x_1, \ldots, x_k) = \sum_j a_j x_j^2 + 2\sum_{s<t} b_{s,t} x_s x_t$. Assume f is either unary or not degenerate if it has arity ≥ 2. The case when the support S has at most two points is easy to prove. Suppose $S = \mathbb{Z}_2^k$. Then it is easy to see that all a_j are the same, and all $b_{s,t}$ are the same, by setting one or two $x_j = 1$ and the other x_j's to 0. Then $Q(x_1, \ldots, x_k)$ takes the form $a\sum_j x_j^2 + 2b\sum_{s<t} x_s x_t = a(\sum_j x_j)^2 + 2(b-a)\sum_{s<t} x_s x_t$. If we denote by $\Sigma = \sum_j x_j$ as an integer sum, and we denote by ($\Sigma \bmod 2$) the integer value 0 or 1, then $\mathfrak{i}^Q = \mathfrak{i}^{a(\Sigma \bmod 2)}(-1)^{(b-a)(\binom{\Sigma}{2} \bmod 2)}$. By listing the values according to Σ, we find that either the function f is degenerate, or $f \in \mathscr{F}_1 \cup \mathscr{F}_2 \cup \mathscr{F}_3$.

Now suppose S is specified by $\sum_j x_j = \epsilon \bmod 2$, where $\epsilon = 0, 1$. We first substitute $x_k = \epsilon + \sum_{j=1}^{k-1} x_j \bmod 2$ in the linear forms L_j in the expression $\mathfrak{i}^{L_1(X)+L_2(X)+\cdots+L_n(X)}$, which can then be turned into the form $\mathfrak{i}^{Q'(x_1,\ldots,x_{k-1})}$, where $Q' = \sum_{j=1}^{k-1} a'_j x_j^2 + 2\sum_{1\leq s<t\leq k-1} b'_{s,t} x_s x_t$ is a quadratic form as above. The remaining proof is similar.

Suppose f is a symmetric degenerate function. Then f has the form $[a, b]^{\otimes n}$ for some unary function $[a, b]$ and $n \geq 1$. If either $a = 0$ or $b = 0$, then clearly $f \in \mathscr{A}$, and also $[a, b] \in \mathscr{F}_1 \cup \mathscr{F}_2 \cup \mathscr{F}_3$. Suppose $ab \neq 0$. Then f is a nonzero scalar multiple of $[1, b/a]^{\otimes n}$. Note that the ratio of any two nonzero entries of a function in $\mathscr{A}$ is a power of $\mathfrak{i}$; it follows that f belongs to $\mathscr{A}$ iff b/a is a power of $\mathfrak{i}$, and also iff $[1, b/a] \in \mathscr{F}_1 \cup \mathscr{F}_2 \cup \mathscr{F}_3$. We conclude that for any $f = [a, b]^{\otimes n}$, $f \in \mathscr{A}$ iff $[a, b] \in \mathscr{F}_1 \cup \mathscr{F}_2 \cup \mathscr{F}_3$.

We summarize the above discussion as the following theorem.

Theorem 3.8. *A function f belongs to $\mathscr{A}$ iff it can be expressed as $\lambda\chi_{AX}\mathfrak{i}^{Q(x_1,\ldots,x_k)}$, where Q is a homogeneous quadratic polynomial over $\mathbb{Z}$ with the additional requirement that every cross term $x_s x_t$ has an even coefficient, where $s \neq t$. We may also use all, not necessarily homogeneous, polynomials over $\mathbb{Z}$ of degree ≤ 2, with the same requirement on cross terms.*

The class $\mathscr{F}_1 \cup \mathscr{F}_2 \cup \mathscr{F}_3$ consists of scalar multiples of all unary and all nondegenerate symmetric functions in $\mathscr{A}$. For any degenerate $f = [a, b]^{\otimes n}$, $f \in \mathscr{A}$ iff $[a, b] \in \mathscr{F}_1 \cup \mathscr{F}_2 \cup \mathscr{F}_3$.

3.4 Tractable Cases

We first show that #CSP($\mathscr{P}$) is tractable. Each constraint function in an instance I of #CSP($\mathscr{P}$) is a product of unary functions, binary EQUALITY functions $(=_2)$, and binary DISEQUALITY functions $(\neq_2)$. We may replace each function by its factors as separate constraints. For the new instance of the #CSP, define a graph G: The vertices are variables, and the edges are $(=_2)$. On each connected component we may assume no $(\neq_2)$ is applied to two variables; otherwise $Z(I) = 0$. Now define another graph G': The vertices are the connected components of G, and there is an edge between two components C_1 and C_2 iff there is a constraint $(\neq_2)(x, y)$ applied on some x in C_1 and y in C_2. Then either G' is bipartite or $Z(I) = 0$. Now suppose G' is bipartite. Take the union of components of G belonging to the same connected component of G'. Then there are exactly two truth assignments to all the variables in this union, propagated by $(=_2)$ or $(\neq_2)$. We can easily calculate these two values by mutliplying the functional evaluations, taking into account the unary functions. From these we can calculate $Z(I)$. Hence, #CSP($\mathscr{P}$) is computable in polynomial time.

Now we analyze #CSP($\mathscr{A}$). We will ignore the global constant factor λ for each $F \in \mathscr{A}$. Firstly, we show how to get rid of the factor χ_{AX}.

Lemma 3.9. *Let* $F(x_1, x_2, \ldots, x_k) = \chi_{AX} i^{L_1(X)+L_2(X)+\cdots+L_n(X)} \in \mathscr{A}$. *If* $AX = 0$ *is infeasible over* $\mathbb{Z}_2$, *then* $\sum_{x_1,x_2,\ldots,x_k} F = 0$. *Suppose* $AX = 0$ *is feasible. Then in polynomial time, we can construct another function* $H(y_1, y_2, \ldots, y_s) = i^{L'_1(Y)+L'_2(Y)+\cdots+L'_n(Y)} \in \mathscr{A}$, *such that* $0 \le s \le k$, *and* $\sum_{x_1,x_2,\ldots,x_k} F = \sum_{y_1,y_2,\ldots,y_s} H$.

Proof. In polynomial time we can solve the linear system $AX = 0$ over $\mathbb{Z}_2$, and decide if it is feasible. Suppose $AX = 0$ is feasible. Without loss of generality, we can assume that $\{y_1, y_2, \ldots, y_s\} \subseteq \{x_1, x_2, \ldots, x_k\}$ is a set of independent variables over $\mathbb{Z}_2$ and the others are dependent variables, where $0 \le s \le k$. Each dependent variable can be expressed by an affine linear sum of $y_1, y_2, \ldots, y_s$. For each $L_j(X)$, we can substitute all the dependent variables and get a 0-1 indicator function of $y_1, y_2, \ldots, y_s$, which we denote by $L'_j(Y)$. So we have

$$\sum_{x_1,x_2,\ldots,x_k} \chi_{AX} i^{L_1(X)+L_2(X)+\cdots+L_n(X)} = \sum_{y_1,y_2,\ldots,y_s} i^{L'_1(Y)+L'_2(Y)+\cdots+L'_n(Y)}.$$

□

The following lemma gives a key property of any function $i^{L_1(X)+L_2(X)+\cdots+L_n(X)} \in \mathscr{A}$. We will also use its alternative form $i^{Q(X)}$. This property plays an important role in both the tractability proof and the hardness proof.

Lemma 3.10. *Let $F = \mathfrak{i}^{Q(X)} \in \mathscr{A}$. Let $G(x_2, \ldots, x_k) = F^{x_1=1}/F^{x_1=0}$. Then exactly one of the following two statements holds:*

1. (Congruity) *There exists a constant $c \in \{1, -1, \mathfrak{i}, -\mathfrak{i}\}$ such that for all $x_2, x_3, \ldots, x_k \in \{0, 1\}$ we have $G(x_2, x_3, \ldots, x_k) = c$;*
2. (Semicongruity) *There exists a constant $c \in \{1, \mathfrak{i}\}$ and an affine subspace S of $\mathbb{Z}_2^{k-1}$ with $\dim S = k - 2$, such that*

$$G(x_2, x_3, \ldots, x_k) = c \text{ on } S, \quad \text{and} \quad G(x_2, x_3, \ldots, x_k) = -c \text{ on } \mathbb{Z}_2^{k-1} - S.$$

Proof. Clearly G has the form

$$G(x_2, \ldots, x_k) = \mathfrak{i}^{a_{11}}(-1)^{a_{12}x_2 + \ldots + a_{1k}x_k}.$$

If $a_{12} = \cdots = a_{1k} = 0 \bmod 2$, then we have congruity. Suppose not all $a_{12} = \cdots = a_{1k} = 0 \bmod 2$; then $a_{12}x_2 + \cdots + a_{1k}x_k = 0 \bmod 2$ defines a linear subspace of $\mathbb{Z}_2^{k-1}$ of dimension $k - 2$, and we have semicongruity. □

Theorem 3.11. *#CSP($\mathscr{A}$) is polynomial time computable.*

Proof. We first observe that $\mathscr{A}$ is closed under multiplication. Therefore, given an instance of #CSP($\mathscr{A}$), the value of the output can be expressed as the summation on a single function $F = \chi_{AX}\mathfrak{i}^{L_1(X)+L_2(X)+\cdots+L_n(X)} \in \mathscr{A}$. We also note that if $F \in \mathscr{A}$, so is $F^{x_s=c}$, for $c = 0, 1$.

In each step of our algorithm, we reduce the number of variables by at least one and still get a summation of this form.

If the linear system $AX = 0$ over $\mathbb{Z}_2$ is infeasible, the function is a totally zero function and we just output 0. Otherwise by Lemma 3.9 we can remove the factor χ_{AX} and possibly decrease the number of variables at the same time, and assume it has the form $F = \mathfrak{i}^{L_1(X)+L_2(X)+\cdots+L_n(X)}$. We apply Lemma 3.10 to remove x_1. There are three cases.

Case 1: We have congruity in Lemma 3.10. Then $F^{x_1=1}/F^{x_1=0}$ is a constant c, and

$$\sum_{x_1, x_2, \ldots, x_k} F = (1 + c) \cdot \sum_{x_2, x_3, \ldots, x_k} F^{x_1=0}.$$

So we get a new summation $\sum_{x_2, x_3, \ldots, x_k} F^{x_1=0}$ and have removed a variable x_1.

Case 2: We have semicongruity in Lemma 3.10, and $c = 1$. Then on the affine subspace S, the ratio $F^{x_1=1}/F^{x_1=0} = 1$, and on the complementary subspace $\mathbb{Z}_2^{k-1} - S$ the ratio $F^{x_1=1}/F^{x_1=0} = -1$. For all $(x_2, x_3, \ldots, x_k) \in \mathbb{Z}_2^{k-1} - S$, the terms cancel, $F^{x_1=1}(x_2, x_3, \ldots, x_k) + F^{x_1=0}(x_2, x_3, \ldots, x_k) = 0$. On S, the terms are equal. It follows that

$$\sum_{x_1,x_2,\ldots,x_k} F = 2 \sum_{x_2,x_3,\ldots,x_k} \chi_S F^{x_1=0}.$$

Note that $\chi_S F^{x_1=0}$ is also a function in $\mathscr{A}$, so we get a new summation of this form and have removed a variable x_1.

Case 3: We have semicongruity in Lemma 3.10, and $c = \mathfrak{i}$. Then for all $(x_2, x_3, \ldots, x_k)$ in the affine subspace S, we have $F^{x_1=1}/F^{x_1=0} = \mathfrak{i}$, and in $\mathbb{Z}_2^{k-1} - S$, we have $F^{x_1=1}/F^{x_1=0} = -\mathfrak{i}$. It follows that

$$\sum_{x_1,x_2,\ldots,x_k} F = \sum_{S}(1+\mathfrak{i})F^{x_1=0} + \sum_{\mathbb{Z}_2^{k-1}-S}(1-\mathfrak{i})F^{x_1=0}.$$

Now we make a simple but crucial observation. The ratio of $1+\mathfrak{i}$ and $1-\mathfrak{i}$ is exactly $\mathfrak{i}$, i.e., $\mathfrak{i} = \frac{1+\mathfrak{i}}{1-\mathfrak{i}}$. As a result we can rewrite the two sums as follows:

$$\sum_{x_1,x_2,\ldots,x_k} F = \sum_{S}(1-i)\cdot F^{x_1=0}\cdot \mathfrak{i}^{L(X')} + \sum_{\mathbb{Z}_2^{k-1}-S}(1-i)\cdot F^{x_1=0}\cdot \mathfrak{i}^{L(X')},$$

where $L(X')$, on $X' = (x_2, x_3, \ldots, x_k, 1)$, is a 0-1 indicator function that takes value 1 on S and 0 on $\mathbb{Z}_2^{k-1} - S$. Thus we can combine the two sums and get

$$\sum_{x_1,x_2,\ldots,x_k} F = (1-i)\cdot \sum_{x_2,x_3,\ldots,x_k}\left(F^{x_1=0}\cdot \mathfrak{i}^{L(X')}\right).$$

Note that $F^{x_1=0}\cdot \mathfrak{i}^{L(X')}$ is also a function in $\mathscr{A}$. So we get a new summation of this form and have removed a variable x_1.

After at most k steps we can eliminate all the variables and obtain the value of the initial summation. Both k and n are bounded by input size. In each iteration, we either resolve a linear system $AX = 0$ or compute a linear equation by Lemma 3.10 representing the affine subspace S, both of which can be done in polynomial time. After each iteration, the formula inside the summation gets at most one more factor $\mathfrak{i}^{L(X')}$ or χ_S, so the whole algorithm is in polynomial time. □

3.5 Hardness

In this section, we prove the hardness part of Theorem 3.7. We start with two useful lemmas that provide tools for constructing reductions.

In #CSP problems, if we have a constraint function F, and we let a variable x_j of F not to occur in any other constraint, then we get $F^{x_j=*}$, which is just $F^{x_j=0} + F^{x_j=1}$. This observation proves the following lemma.

Lemma 3.12. *For any $\mathcal{F}$ and F, and for any variable x_j of F,*

$$\#\text{CSP}(\mathcal{F} \cup \{F^{x_j=*}\}) \leq_T \#\text{CSP}(\mathcal{F} \cup \{F\}).$$

The presence of all EQUALITY gates in #CSP already provides two important unary functions, Δ_0 and Δ_1. Applying Δ_0 or Δ_1 to a variable fixes that variable to the constant 0 or 1 respectively, and thus it is called *pinning*. We can simulate Δ_0 and Δ_1 by the following pinning lemma [DGJ09].

Lemma 3.13 (Dyer–Goldberg–Jerrum). *For any $\mathcal{F}$, $\#\text{CSP}(\mathcal{F} \cup \{\Delta_0, \Delta_1\}) \leq_T \#\text{CSP}(\mathcal{F})$. In particular, for any F and $c \in \{0, 1\}$,*

$$\#\text{CSP}(\mathcal{F} \cup \{F^{x_j=c}\}) \leq_T \#\text{CSP}(\mathcal{F} \cup \{F\}).$$

Proof. Given an instance I of $\#\text{CSP}(\mathcal{F} \cup \{\Delta_0, \Delta_1\})$, let V_0 and V_1 denote the set of variables to which Δ_0 and Δ_1 are applied, respectively. If $V_0 \cap V_1 \neq \emptyset$, then there are no satisfying assignments and the partition function $Z(I) = 0$. We assume $V_0 \cap V_1 = \emptyset$. Let $V_2 = V - (V_0 \cup V_1)$ be the set of other variables. We define an instance I' of $\#\text{CSP}(\mathcal{F})$ that has the variable set $V_2 \cup \{y_0, y_1\}$, where y_0, y_1 are two new variables. The constraints of I' are exactly the same as those of I, except (1) every variable in V_0 (respectively V_1) is replaced by y_0 (respectively y_1), and (2) the constraints of Δ_0 and Δ_1 are not applied. Then $Z(I')$ can be expressed as the sum of four quantities,

$$Z(I') = \sum_{(b_0,b_1)\in\{0,1\}^2} Z(I' \mid b_0 b_1),$$

where $Z(I' \mid b_0 b_1)$ is the sum in $Z(I')$ restricted to assigning (y_0, y_1) to $(b_0, b_1) \in \{0, 1\}^2$.

Now we define a second instance I'' of $\#\text{CSP}(\mathcal{F})$ that has the variable set $V_2 \cup \{y\}$, where y is a new variable. It is the same as I' except we identify y_0 and y_1 by y. Then clearly

$$Z(I'') = \sum_{b_0=b_1} Z(I' \mid b_0 b_1) = Z(I' \mid 00) + Z(I' \mid 11).$$

Hence

$$Z(I') - Z(I'') = Z(I' \mid 01) + Z(I' \mid 10). \tag{3.1}$$

Obviously, $Z(I) = Z(I' \mid 01)$. If every function $F \in \mathcal{F}$ has *complementary invariance*, i.e., if F has arity k, then for every input $(b_1, \ldots, b_k) \in \{0, 1\}^k$, $F(b_1, \ldots, b_k) = F(\overline{b_1}, \ldots, \overline{b_k})$, where $\overline{b_i} = 1 - b_i$, then clearly $Z(I') - Z(I'') = 2Z(I)$, and the lemma is proved.

Suppose for some $F \in \mathcal{F}$ of arity k, and some $(b_1, \ldots, b_k) \in \{0, 1\}^k$, $F(b_1, \ldots, b_k) \neq F(\overline{b_1}, \ldots, \overline{b_k})$, we will define two more instances of

#CSP($\mathcal{F}$). The first is I'_b, which is the same as I', except it has one more constraint F applied to the sequence of variables $(y_{b_1}, \ldots, y_{b_k})$. Note that for any assignment in I'_b, if (y_0, y_1) is assigned to $(0, 1)$, then $F(y_{b_1}, \ldots, y_{b_k})$ will evaluate to $F(b_1, \ldots, b_k)$, and if (y_0, y_1) is assigned to $(1, 0)$, then $F(y_{b_1}, \ldots, y_{b_k})$ will evaluate to $F(\overline{b_1}, \ldots, \overline{b_k})$. Of course if (y_0, y_1) is assigned to $(0, 0)$ or $(1, 1)$, then $F(y_{b_1}, \ldots, y_{b_k})$ will evaluate to $F(0, \ldots, 0)$ and $F(1, \ldots, 1)$ respectively.

The second instance we define is I''_b, which is the same as I'', except it has one more constraint F applied to the sequence of variables $(y, \ldots, y)$. It is now clear that

$$Z(I'_b) - Z(I''_b) = Z(I' \mid 01)F(b_1, \ldots, b_k) + Z(I' \mid 10)F(\overline{b_1}, \ldots, \overline{b_k}). \quad (3.2)$$

From (3.1) and (3.2) we can solve $Z(I) = Z(I' \mid 01)$. □

Remark: Before Lemma 3.13 we have used three reduction methods: gadget construction, holographic transformation, and polynomial interpolation. Suppose f is such a reduction from a problem Π to a problem Π'. When the reduction is applied to an instance Ω of Π, f defines a sequence $\{\Omega'_i\}$ of instances of Π', and outputs $\Omega'_1, \Omega'_2, \ldots, \Omega'_m$, where both m and the sizes of Ω'_i, $(1 \le i \le m)$, are polynomially bounded in the size of Ω. (That there might be multiple output instances, i.e., $m > 1$, is typical for polynomial interpolations.) The three reduction methods all satisfy the following *locality* property: If the reduction is applied to an instance of Π that consists of d identical copies of Ω, denoted by $d\Omega$, it will output a polynomial number of instances of Π', where each instance consists of d identical copies of Ω'_i, namely $d\Omega'_1, d\Omega'_2, \ldots, d\Omega'_{m'}$ (m' is polynomially bounded in the size of $d\Omega$, possibly a larger polynomial than m). Clearly this property is closed under composition.

However, this locality property is not shared by the pinning reduction in Lemma 3.13. For a technical reason in Chapter 6 (Theorem 6.16, specifically Lemma 6.23), we will need this locality property, and thus ascertain that a reduction chain exists from a basic #P-hard problem such as VERTEX COVER that does not use the pinning reduction of Lemma 3.13. For this reason, we will make annotating remarks in the following for #P-hardness results that can be proved without using the pinning reduction. For the convenience of addressing this issue, we denote by #CSP$^c(\mathcal{F})$ the problem #CSP($\mathcal{F} \cup \{\Delta_0, \Delta_1\}$), which is logically equivalent to #CSP($\mathcal{F}$) by Lemma 3.13.

3.5.1 One Symmetric Binary Function

For hardness, we first establish the following theorem, which says that Theorem 3.7 holds for a single symmetric binary function.

Theorem 3.14. *If* $[a, b, c] \notin \mathscr{A} \cup \mathscr{P}$, *then* #CSP($[a, b, c]$) *is #P-hard. Explicitly, for any binary function* $[a, b, c]$, *where* $a, b, c \in \mathbb{C}$, *if* #CSP($[a, b, c]$) *is not #P-hard, then* $[a, b, c]$ *takes the form:* $[x, 0, y]$, $[0, x, 0]$, $[x^2, xy, y^2]$, $x[1, \pm i, 1]$ *or* $x[1, \pm 1, -1]$.

For #CSP^c($[a, b, c]$) $\equiv$ #CSP($[a, b, c], \Delta_0, \Delta_1$), *if* $[a, b, c] \notin \mathscr{A} \cup \mathscr{P}$, *then* #CSP^c($[a, b, c]$) *is #P-hard by a reduction that does not use Lemma 3.13.*

Before proving Theorem 3.14, we first note that every one of the five listed exceptional cases are in $\mathscr{A} \cup \mathscr{P}$, and it can be checked directly that all binary symmetric functions in $\mathscr{A} \cup \mathscr{P}$ take one of these five forms.

In this proof we use an important reduction method called *polynomial interpolation*, first introduced by Valiant [Val79b]. The method involves setting up and then solving a system of linear equations. The solvability of these linear systems is usually based on the fact that it is a Vandermonde system.

Here is a simple example. Suppose $\mathcal{F}$ contains some function $F = [1, a, 1]$, where $a \neq 0$ and is not a root of unity. Suppose we want to simulate a function $H = [1, b, 1]$, that is, we want to reduce #CSP($\mathcal{F} \cup \{H\}$) to #CSP($\mathcal{F}$). Given an instance I of #CSP($\mathcal{F} \cup \{H\}$), where there are n occurrences of the constraint H, we construct instances I_j of #CSP($\mathcal{F}$), for each $j \geq 1$, by replacing each occurrence of the constraint $H(x_{i_1}, x_{i_2})$ in I by j many constraints $F(x_{i_1}, x_{i_2})$. Each occurrence of H takes an input of the form $(0, 0)$, $(0, 1)$, $(1, 0)$, $(1, 1)$. We can write the sum defining $Z(I)$ as a sum over all assignments stratified according to the number of $(1, 0)$ or $(0, 1)$ assigned at the n occurrences of H. If exactly i of n occurrences of H are assigned $(1, 0)$ or $(0, 1)$, then the other $n - i$ are assigned $(0, 0)$ or $(1, 1)$. The value $Z(I)$ can be written as the summation

$$Z(I) = \sum_{i=0}^{n} w_i b^i, \tag{3.3}$$

where w_i is the summation of products of all constraints other than these n occurrences of H, over all assignments such that exactly i of these n occurrences of H receive inputs $(1, 0)$ or $(0, 1)$. Note that for assignments $(1, 0)$ or $(0, 1)$, H evaluates to b; the factor b^i in (3.3) comes from this. The other $n - i$ occurrences of H receiving inputs $(0, 0)$ or $(1, 1)$ contribute a factor 1. Meanwhile, we have

$$Z(I_j) = \sum_{i=0}^{n} w_i a^{ij}. \tag{3.4}$$

Note that the same set of values w_i occur here. We let $j = 1, \ldots, n+1$; then (3.4) becomes a system of linear equations about w_i, whose coefficient matrix is a Vandermonde matrix (a^{ij}). The values $Z(I_j)$ can be obtained by oracle calls to #CSP($\mathcal{F}$). Since a is nonzero and not a root of unity, (a^{ij}) is a nonsingular matrix, and we can solve for all w_i in polynomial time from (3.4), which gives us $Z(I)$ by (3.3). This is essentially how every reduction by polynomial interpolation in this chapter will be done.

To get acquainted with the technique, we start with the following theorem, a special case of Theorem 3.14. We illustrate two features. The first is the use of interpolation in a particularly simple setting, where nonnegativity avoids many complications. The second is to show that the ability to get unary functions is a powerful step to prove #P-hardness. Once we can interpolate all unary functions, we can appeal to the dichotomy theorem for Holant* problems. This theorem is also a very special case of much more general theorems on graph homomorphisms, in particular the Bulatov–Grohe theorem (see Volume II).

Theorem 3.15. *Let $[a, b, c]$ be a symmetric binary function, where a, b, c are nonnegative real numbers. Then #CSP($[a, b, c]$) is #P-hard unless $[a, b, c]$ is of one of the following three forms: $[a, 0, c]$, $[0, b, 0]$ or $[x^2, xy, y^2]$.*

For #CSP^c($[a, b, c]$) the same statement holds, and when #CSP^c($[a, b, c]$) is #P-hard a reduction exists that does not use Lemma 3.13.

Proof. We note that #CSP($[0, 1, 1]$) counts the number of vertex covers in a graph. Given a graph $G = (V, E)$, we may consider it defines a #CSP($[0, 1, 1]$) instance I, where each vertex is a Boolean variable and each edge is given the constraint $[0, 1, 1]$ representing the Boolean OR function. Then the product over all edges evaluates to 1 for an assignment $\sigma : V \to \{0, 1\}$ iff $\sigma^{-1}(1)$ is a vertex cover of G. Hence $Z(I)$ is exactly the number vertex covers in G. This problem is #P-complete.

If we used the constraint $[1, 1, 0]$, then #CSP($[1, 1, 0]$) counts the number of independent sets in a graph, since $\sigma^{-1}(1)$ is an independent set iff the product for σ evaluates to 1. In fact, by exchanging 0 and 1, we see that these two problems always have the same solution.

We will reduce one of these #P-hard problems to #CSP($[a, b, c]$), if $[a, b, c]$ does not have one of the three forms. We assume $b \neq 0$, and not both a and c are 0. Since $b \neq 0$, we may normalize $b = 1$ by replacing $[a, b, c]$ with $[a/b, 1, c/b]$. This only causes $Z(I)$ to change by a factor b^m, where m is the number of occurrences of the constraint $[a, b, c]$ in the input instance I. Thus we may assume the given constraint is $[x, 1, y]$, for some nonnegative x and y, where $(x, y) \neq (0, 0)$, and $xy \neq 1$.

First suppose $y \neq 0, 1$. Using Δ_1 we can get the unary function $[1, y]$, namely

$$\#\text{CSP}([x, 1, y], [1, y]) \leq_T \#\text{CSP}([x, 1, y]).$$

With a positive $y \neq 1$, we can interpolate any unary function using $[1, y]$ as follows. #CSP($\mathcal{F}$) is precisely Holant($\mathcal{EQ} \mid \mathcal{F}$) where $\mathcal{EQ} = \{(=_k) \mid k \geq 1\}$ is the set of all Equalities, and $\mathcal{F}$ is any set of constraints. Given any unary function $[u, v]$ and any $\mathcal{F}$, we reduce #CSP($\mathcal{F} \cup \{[u, v]\}$) to #CSP($\mathcal{F} \cup \{[1, y]\}$). Suppose I is an instance for #CSP($\mathcal{F} \cup \{[u, v]\}$), where $[u, v]$ occurs n times. Then $Z(I) = \sum_{i=0}^{n} w_i u^{n-i} v^i$, where w_i is the summation of products of all constraints other than the n occurrences of $[u, v]$, over all assignments such that exactly i of these n occurrences of $[u, v]$ receive input 1. Define an instance I_j for #CSP($\mathcal{F} \cup \{[1, y]\}$) by replacing each occurrence of $[u, v]$ in I by j occurrences of $[1, y]$. Then $Z(I_j) = \sum_{i=0}^{n} w_i y^{ij}$, for all $j \geq 1$. Taking $1 \leq j \leq n + 1$, we obtain a nonsingular Vandermonde system, as the positive $y \neq 1$. We can solve for all w_i in polynomial time, and compute $Z(I)$.

Similarly, if $x \neq 0, 1$, then using Δ_0 we can get the unary function $[x, 1]$, and then interpolate any other unary function. If $x, y \in \{0, 1\}$, since $(x, y) \neq (0, 0)$ and $xy \neq 1$, the only cases left are $(x, y) = (0, 1)$ or $(1, 0)$. These are the #P-hard problems Vertex Cover and Independent Set respectively.

Now suppose we can interpolate any unary function $[u, v]$. By a finite sequence of reductions, we can interpolate any finite set of unary functions. By the dichotomy theorem for Holant* problems, namely Theorem 2.12 (essentially Lemma 2.14), Holant*$((=_3) \mid [x, 1, y])$ is #P-hard. This means that for some finite subset of unary functions $\mathcal{U}' \subset \mathcal{U}$, Holant($\{(=_3)\} \cup \mathcal{U}' \mid \{[x, 1, y]\} \cup \mathcal{U}'$) is #P-hard. Hence Holant($(=_3) \mid \{[x, 1, y]\} \cup \mathcal{U}''$) is #P-hard for some finite $\mathcal{U}'' \subset \mathcal{U}$. The latter is reducible to #CSP($[x, 1, y]$); hence it is also #P-hard. This argument will be used often, and for simplicity in what follows we will just say

$$\text{Holant}^*((=_3) \mid [x, 1, y]) \leq_T \#\text{CSP}([x, 1, y]),$$

and since Holant*$((=_3) \mid [x, 1, y])$ is #P-hard by Theorem 2.12, we conclude that #CSP($[x, 1, y]$) is also #P-hard. □

Now we consider complex valued functions. First we prove two simple lemmas.

Lemma 3.16. *For any symmetric binary function* $[0, b, c]$, *where* $b, c \in \mathbb{C}$ *and* $bc \neq 0$, *the problem* #CSP($[0, b, c]$) *is #P-hard. For* #CSP^c($[0, b, c]$) *a reduction exists that does not use Lemma 3.13.*

Proof. Since $b \neq 0$, we can normalize it and assume $b = 1$. So we rename the constraint function $[0, 1, c]$. First suppose c is a root of unity. Let $c^k = 1$. We can realize $[0, 1^k, c^k] = [0, 1, 1]$ by k repeated copies of $[0, 1, c]$. The problem #CSP($[0, 1, 1]$) is counting vertex covers, hence #P-hard. Now suppose c is not a root of unity. Also by assumption $c \neq 0$. We can interpolate $[0, 1, x]$, for any x, by polynomial interpolation. In particular, we can interpolate $[0, 1, 1]$ and #CSP($[0, 1, 1]$) is #P-hard. □

Lemma 3.17. *Let* $[1, b, c]$ *be a symmetric binary function, where* $bc \neq 0$ *and* $c \neq b^2$. *Suppose either* b *or* c *is not a root of unity. Then* #CSP($[1, b, c]$) *is #P-hard. For* #CSP^c($[1, b, c]$) *the same statement holds, and when it is #P-hard a reduction exists that does not use Lemma 3.13.*

Proof. Suppose b is not a root of unity. Using Δ_0 we can get the unary function $[1, b]$. Then we can interpolate any finitely many unary functions. Then

$$\text{Holant}^*((=_3) \mid [1, b, c]) \leq_T \#\text{CSP}([1, b, c]).$$

By Theorem 2.12 for Holant* problems, #CSP($[1, b, c]$) is #P-hard.

Next suppose c is not a root of unity. Connect two inputs of $(=_3)$ by $[1, b, c]$, we can get the function $[1, c]$, and interpolate any unary function by polynomial interpolation, as in Theorem 3.15. Hence again by Theorem 2.12, #CSP($[1, b, c]$) is #P-hard. □

Now we prove Theorem 3.14: *If* $[a, b, c] \notin \mathscr{A} \cup \mathscr{P}$, *then* #CSP($[a, b, c]$) *is #P-hard.* The basic idea is to interpolate unary functions and then appeal to the Holant* dichotomy. However, in order to interpolate all unary functions we must handle some special cases. For this purpose we also use a gadget that links two copies of the binary function $F = [a, b, c]$. It is realizable in #CSP as $G(x, y) = \sum_z F(x, z)F(z, y)$, using an extra variable z not appearing elsewhere.

Proof. of Theorem 3.14. If $a = 0$, we know $bc \neq 0$, otherwise it is in the second or the third exceptional case of Theorem 3.14. So by Lemma 3.16, #CSP($[a, b, c]$) is #P-hard. The case $c = 0$ is symmetric. Since $[a, b, c] \notin \mathscr{A} \cup \mathscr{P}$, we know $b \neq 0$. Therefore we will assume in the following that $abc \neq 0$, and by normalizing, we can assume $a = 1$.

Now we consider the complexity of #CSP($[1, b, c]$), with $bc \neq 0$. We assume $c \neq b^2$; otherwise $[1, b, c]$ is degenerate, the third exceptional case of

Theorem 3.14. If either b or c is not a root of unity, then #CSP($[1, b, c]$) is #P-hard by Lemma 3.17.

Suppose both b and c are roots of unity. We can realize $G = \begin{bmatrix} 1 & b_2 \\ b & c \end{bmatrix} = [1 + b^2, b + bc, b^2 + c^2]$, which is the signature of two consecutive copies of $[1, b, c]$. G is also nondegenerate. If #CSP(G) is #P-hard so is #CSP($[1, b, c]$). We consider the following cases.

1. $b = -1$. $G = [2, -1 - c, 1 + c^2]$.
   Since $[1, b, c] = [1, -1, c] \notin \mathscr{A} \cup \mathscr{P}$, we know $c \neq \pm 1$. If $c = \pm \mathrm{i}$, we get $G = [2, -1 \mp \mathrm{i}, 0]$, which is #P-hard by Lemma 3.16 (or rather a version of Lemma 3.16 on a reversed signature, by exchanging the inputs 0 and 1). If $c \notin \{\pm 1 \pm \mathrm{i}\}$, then there are no zero entries in G. Since c is a root of unity, and $c \neq \pm 1$, we have $|1 + c^2| \neq 2$. In particular, $\frac{1+c^2}{2}$ is not a root of unity. Normalizing G, we have $[1, \frac{-1-c}{2}, \frac{1+c^2}{2}]$. Hence #CSP($[1, b, c]$) is #P-hard by Lemma 3.17.
2. $b = -c$. $G = [1 + c^2, -c - c^2, 2c^2]$.
   Since $[1, b, c] = [1, -c, c] \notin \mathscr{A} \cup \mathscr{P}$, we know $c \neq \pm 1$. If $c = \pm \mathrm{i}$, we get $G = [0, 1 \mp \mathrm{i}, -2]$, which is #P-hard by Lemma 3.16. If $c \notin \{\pm 1 \pm \mathrm{i}\}$, then there are no zero entries in G. Normalizing, we get $[1, \frac{-c-c^2}{1+c^2}, \frac{2c^2}{1+c^2}]$. For c a root of unity but $c \neq \pm 1$, we have $|1 + c^2| \neq |2c^2|$. In particular, $\frac{2c^2}{1+c^2}$ is not a root of unity. It follows from Lemma 3.17 that #CSP(G) is #P-hard.
3. $c = 1$. $G = [1 + b^2, 2b, 1 + b^2]$.
   Since $[1, b, c] = [1, b, 1] \notin \mathscr{A} \cup \mathscr{P}$, we know $b \notin \{\pm 1 \pm \mathrm{i}\}$. So $1 + b^2 \neq 0$, and $|\frac{2b}{1+b^2}| = \frac{2}{|1+b^2|} \neq 1$. Hence $\frac{2b}{1+b^2}$ is not a root of unity. Therefore the problem is #P-hard by Lemma 3.17.
4. $b \neq -1, b \neq -c$, and $c \neq 1$. Recall that we are given $c \neq b^2$ since $[1, b, c] \notin \mathscr{A} \cup \mathscr{P}$.
   If $F(x_1, x_2) = [1, b, c]$, then $F^{x_2=*} = [1 + b, b + c]$ with both entries nonzero. By $|b| = |c| = 1$, we claim that $|1 + b| = |b + c|$ if and only if $c = 1$ or $c = b^2$. For now assume the claim holds. Since we assumed $c \neq 1$ and $c \neq b^2$, we have $\frac{|b+c|}{|1+b|} \neq 1$ and thus we can interpolate all unary functions with $[1 + b, b + c]$. Therefore #CSP($[1, b, c]$) is #P-hard by Theorem 2.12 for Holant* problems.
   To prove the claim one direction is obvious. For the reverse direction, suppose $|1 + b| = |b + c|$; then $(1 + b)(1 + \bar{b}) = (b + c)(\bar{b} + \bar{c})$. Expanding this equation, we get $b + \bar{b} = c\bar{b} + \bar{c}b$, i.e., $\mathrm{Re}\, b = \mathrm{Re}\, c/b$. If $b = e^{i\alpha}$ and $c = e^{i\beta}$, then $\cos\alpha = \cos(\beta - \alpha)$, hence $\beta - \alpha = \pm\alpha \bmod 2\pi$. It follows that $c = 1$ or b^2.

□

Theorem 3.14 is the special case of graph homomorphism on the Boolean domain. We will discuss graph homomorphism on a general domain in much more detail in Volume II of the book.

3.5.2 Affine Support

The following lemma gives a necessary condition for #CSP(F) being *not* #P-hard. This affine support criterion will be much generalized later when we discuss dichotomies for graph homomorphisms and #CSP on general, i.e., not necessarily Boolean, domains. The proof given here is elementary.

Lemma 3.18. *If the relation R_F defined by F is not affine, then #CSP(F) is #P-hard. For #CSP^c(F) the same statement holds, and when R_F is not affine a reduction exists that does not use Lemma 3.13.*

Proof. We prove by induction on the arity of the function F.

All functions of arity one have affine support. So we assume the arity of F is at least two. We first consider a function F of arity two. Suppose F does not have affine support. This implies that exactly one of its four values is 0. The matrix form of F is

$$\begin{bmatrix} a & b \\ c & d \end{bmatrix} = \begin{bmatrix} F(0,0) & F(0,1) \\ F(1,0) & F(1,1) \end{bmatrix};$$

then in particular $\det(F) \neq 0$. By taking two copies of F sharing a free variable z in the appropriate order (x, z) and (z, y), we can realize the binary function $H(x, y) = \sum_z F(x, z)F(z, y)$, whose matrix form is $H = FF^{\mathrm{T}} = \begin{bmatrix} a^2 + b^2 & ac + bd \\ ac + bd & c^2 + d^2 \end{bmatrix}$. This H is a symmetric binary function, which can also be denoted by $[a^2 + b^2, ac + bd, c^2 + d^2]$. We can apply Theorem 3.14 to H. Because F is nonsingular, so is the matrix for H. Because exactly one entry of F is 0, $ac + bd \neq 0$ and H is not of the form $[x, 0, y]$. Because either $a^2 + b^2 \neq 0$ or $c^2 + d^2 \neq 0$, H is not of the form $[0, x, 0]$. So the only remaining possibilities for $H \in \mathscr{A} \cup \mathscr{P}$ is that H is of the form $x[1, \pm\mathrm{i}, 1]$ or $x[1, \pm 1, -1]$. By symmetry, we only need to consider the cases $a = 0$ and $bcd \neq 0$, or $b = 0$ and $acd \neq 0$. If $a = 0$, we can assume $b = 1$ by normalization, and then the function H is $[1, d, c^2 + d^2]$. If H is of the form $x[1, \pm\mathrm{i}, 1]$, we have $d = \pm\mathrm{i}$ and $c = \pm\sqrt{2}$. Then we can realize another symmetric binary function by $G(x, y) = F(x, y)F(y, x)$. So $G = [a^2, bc, d^2] = [0, \pm\sqrt{2}, -1]$. #CSP($G$) is #P-hard by Theorem 3.14. If H is of the form $x[1, \pm 1, -1]$, we have $d = \pm 1$ and $c = \pm\sqrt{2}\mathrm{i}$. Then $G = [a^2, bc, d^2] = [0, \pm\sqrt{2}\mathrm{i}, 1]$ and #CSP(G) is #P-hard by Theorem 3.14 again. So we have completed for the $a = 0$ case. If

$b = 0$, we can assume $a = 1$ and the function H is $[1, c, c^2 + d^2]$. If H is of the form $x[1, \pm\mathrm{i}, 1]$, we have $c = \pm\mathrm{i}$ and $d = \pm\sqrt{2}$. Then we can realize another binary function F' by $F'(x, y) = F(x, y)F(x, y)$. In matrix notation $F' = \begin{bmatrix} a^2 & b^2 \\ c^2 & d^2 \end{bmatrix} = \begin{bmatrix} 1 & 0 \\ -1 & 2 \end{bmatrix}$. Next we can simulate $H' = F'F'^{\mathrm{T}} = \begin{bmatrix} 1 & -1 \\ -1 & 5 \end{bmatrix}$ from F'. In symmetric notation $H' = [1, -1, 5]$. By Theorem 3.14 #CSP(H') is #P-hard. Finally, if H is of the form $x[1, \pm 1, -1]$, we have $c = \pm 1$ and $d = \pm\sqrt{2}\mathrm{i}$. Then by the same construction, $F' = \begin{bmatrix} 1 & 0 \\ 1 & -2 \end{bmatrix}$ and $H' = \begin{bmatrix} 1 & 1 \\ 1 & 5 \end{bmatrix}$, which in symmetric notation is $H' = [1, 1, 5]$. So again #CSP(H') is #P-hard. We have completed the proof for the case where the function F is of arity 2.

Inductively we assume the lemma has been proved for functions with arity less than k, for some $k \geq 3$, and now assume the function F has arity k. Since R_F is not affine, there exist $\alpha, \beta, \gamma \in R_F$, where $\alpha = (a_1, \ldots, a_k)$, $\beta = (b_1, \ldots, b_k)$, $\gamma = (c_1, \ldots, c_k)$, such that $\delta = \alpha \oplus \beta \oplus \gamma = (d_1, \ldots, d_k) \notin R_F$. We only need to prove that we can use F to simulate a function of smaller arity that does not have affine support.

Divide the index set $[k]$ of input variables of F into four subsets according to the values of α, β, γ as follows:

$$I = \{j \mid a_j = b_j \neq c_j\}, \quad J = \{j \mid a_j = c_j \neq b_j\}, \quad K = \{j \mid b_j = c_j \neq a_j\},$$
$$L = \{j \mid a_j = b_j = c_j\}.$$

Since a_j, b_j and c_j take Boolean values, $\{I, J, K, L\}$ is a partition of $[k]$. We also remark that, if $j, l \in I$, then either $(a_l, b_l, c_l) = (a_j, b_j, c_j)$ or $(\overline{a_j}, \overline{b_j}, \overline{c_j})$. The same is true for J, K and L. Now we have the following four cases, and for each case, we prove our result.

- L is not empty. There exists j such that $a_j = b_j = c_j$.

  We fix the jth input of F to be a_j, and get a function $F^{x_j = a_j}$. Recall that $\delta = \alpha \oplus \beta \oplus \gamma$; then $d_j = a_j = b_j = c_j$. This function $F^{x_j = a_j}$ does not have affine support.

  Now we may assume $L = \emptyset$ and $[k] = I \cup J \cup K$.
- There are indices $l \neq j$, such that $(a_l, b_l, c_l) = (a_j, b_j, c_j)$.

  Without loss of generality we assume $l = 1$ and $j = 2$. Define a function of arity $k - 1$ by $H(x_1, x_3 \ldots, x_k) = F(x_1, x_1, x_3 \ldots, x_k)$. H can be simulated by F, and by the property that $\alpha, \beta, \gamma \in R_F$ and yet $\delta \notin R_F$, H does not have affine support.
- There are indices $l \neq j$, such that $(a_l, b_l, c_l) = (\overline{a_j}, \overline{b_j}, \overline{c_j})$.

  Clearly both l and j belong to the same set I or J or K. Without loss of generality we assume $l = 1 \in I$ and $j = 2 \in I$. The proofs for J and K are the same. Then we have

$$\alpha = (a, \overline{a}, \alpha'), \quad \beta = (a, \overline{a}, \beta'), \quad \gamma = (\overline{a}, a, \gamma'), \quad \text{and} \quad \delta = (\overline{a}, a, \delta'), \tag{3.5}$$

where $a \in \mathbb{Z}_2$, $\alpha', \beta', \gamma' \in \mathbb{Z}_2^{k-2}$, and $\delta' = \alpha' \oplus \beta' \oplus \gamma' \in \mathbb{Z}_2^{k-2}$. Assume for a contradiction that all functions of the forms $F^{x_i=b}$ and $F^{x_i=*}$ have affine support ($1 \le i \le k$, $b = 0, 1$).

Consider $F^{x_1=a}$, whose underlying relation $R_{F^{x_1=a}}$ is affine. Because $\alpha, \beta \in R_F$, $(\overline{a}, \alpha'), (\overline{a}, \beta') \in R_{F^{x_1=a}}$. The summation of $(\overline{a}, \alpha')$, $(\overline{a}, \beta')$, (a, γ'), (a, δ') is the zero vector in $\mathbb{Z}_2^{k-1}$, so $(a, \gamma') \in R_{F^{x_1=a}}$ iff $(a, \delta') \in R_{F^{x_1=a}}$. This implies that $(a, a, \gamma') \in R_F$ iff $(a, a, \delta') \in R_F$.

Next consider $F^{x_2=a}$. Because $\gamma \in R_F$ and $\delta \notin R_F$, we have $(\overline{a}, \gamma') \in R_{F^{x_2=a}}$, and $(\overline{a}, \delta') \notin R_{F^{x_2=a}}$. As we just proved, $R_F(a, a, \gamma') = R_F(a, a, \delta')$. If this value is 1, then $(a, \gamma'), (a, \delta') \in R_{F^{x_2=a}}$, but this is impossible for the affine relation $R_{F^{x_2=a}}$, in view of the four inputs $(\overline{a}, \gamma')$, $(\overline{a}, \delta')$, (a, γ'), and (a, δ'). So we must have $(a, a, \gamma') \notin R_F$ and $(a, a, \delta') \notin R_F$.

Similarly, we can prove that both $(\overline{a}, \overline{a}, \alpha')$ and $(\overline{a}, \overline{a}, \beta') \notin R_F$. More explicitly, first consider $F^{x_2=\overline{a}}$. Because $\alpha, \beta \in R_F$, $(a, \alpha'), (a, \beta') \in R_{F^{x_2=\overline{a}}}$. Having an affine support, $R_{F^{x_2=\overline{a}}}$ evaluates to the same value on $(\overline{a}, \alpha')$ and $(\overline{a}, \beta')$. Thus R_F evaluates to the same value on $(\overline{a}, \overline{a}, \alpha')$ and $(\overline{a}, \overline{a}, \beta')$. Next consider $F^{x_1=\overline{a}}$. It also has an affine support. Since $\gamma \in R_F$ and $\delta \notin R_F$, we have $(a, \gamma') \in R_{F^{x_1=\overline{a}}}$ and $(a, \delta') \notin R_{F^{x_1=\overline{a}}}$. If $(\overline{a}, \overline{a}, \alpha'), (\overline{a}, \overline{a}, \beta') \in R_F$, then $(\overline{a}, \alpha'), (\overline{a}, \beta') \in R_{F^{x_1=\overline{a}}}$. This is impossible for an affine relation $R_{F^{x_1=\overline{a}}}$, in view of the four inputs (a, γ'), (a, δ'), $(\overline{a}, \alpha')$, and $(\overline{a}, \beta')$. Thus it follows that $(\overline{a}, \overline{a}, \alpha') \notin R_F$ and $(\overline{a}, \overline{a}, \beta') \notin R_F$.

To summarize, we have all $(a, a, \gamma'), (a, a, \delta'), (\overline{a}, \overline{a}, \alpha'), (\overline{a}, \overline{a}, \beta') \notin R_F$.

Finally, we consider $F^{x_1=*}$, and calculate as follows:

$$F^{x_1=*}(\overline{a}, \alpha') = F(\alpha) + F(\overline{a}, \overline{a}, \alpha') = F(\alpha) \neq 0,$$

$$F^{x_1=*}(\overline{a}, \beta') = F(\beta) + F(\overline{a}, \overline{a}, \beta') = F(\beta) \neq 0,$$

$$F^{x_1=*}(a, \gamma') = F(\gamma) + F(a, a, \gamma') = F(\gamma) \neq 0,$$

$$F^{x_1=*}(a, \delta') = F(\delta) + F(a, a, \delta') = F(\delta) = 0.$$

This is a contradiction, since we made the assumption that $R_{F^{x_1=*}}$ is affine.

- If there is more than one element in the set I, or in the set J, or in the set K, it is included in the previous two cases. The remaining case is that the sizes of I, J, K are all at most one and L is empty. Because $k > 2$, the sizes of I, J, K are exactly one, and $k = 3$. Without loss of generality, let $I = \{1\}$, $J = \{2\}$ and $K = \{3\}$.

A moment's reflection shows that we can write α, β, γ and δ as follows:

$$\alpha = (p, q, \overline{r}),$$

$$\beta = (p, \overline{q}, r),$$

$$\gamma = (\overline{p}, q, r),$$

$$\delta = (\overline{p}, \overline{q}, \overline{r}),$$

where $p, q, r \in \mathbb{Z}_2$.

First we consider $F^{x_1=p}$, which has an affine support by our induction hypothesis on arity. Let $u = (p, q, r)$, and suppose $u \in R_F$. Then $(q, r) \in R_{F^{x_1=p}}$. Because $\alpha, \beta \in R_F$, then $(q, \overline{r})$ and $(\overline{q}, r)$ both belong to $R_{F^{x_1=p}}$. Then being affine, $(\overline{q}, \overline{r}) \in R_{F^{x_1=p}}$. Let $v = (p, \overline{q}, \overline{r})$, then $v \in R_F$.

Next we consider $R_{F^{x_2=q}}$. By $\alpha, \gamma \in R_F$, we get $(p, \overline{r}), (\overline{p}, r) \in R_{F^{x_2=q}}$. By assumption $u \in R_F$, then $(p, r) \in R_{F^{x_2=q}}$. By $R_{F^{x_2=q}}$ being affine, we get $(\overline{p}, \overline{r}) \in R_{F^{x_2=q}}$. Let $w = (\overline{p}, q, \overline{r})$; then $w \in R_F$.

Now $\alpha, v, w \in R_F$. This gives us $(p, q), (p, \overline{q}), (\overline{p}, q) \in R_{F^{x_3=\overline{r}}}$. Since $R_{F^{x_3=\overline{r}}}$ is affine, $(\overline{p}, \overline{q}) \in R_{F^{x_3=\overline{r}}}$. This means that $\delta = (\overline{p}, \overline{q}, \overline{r}) \in R_F$, which is a contradiction.

We conclude that in fact $u \notin R_F$.

By tracing the above steps, under the new condition $u \notin R_F$, we get $v \notin R_F$, and also $w \notin R_F$.

Finally we consider $F^{x_3=r}$. By $\beta, \gamma \in R_F$, we get $(p, \overline{q}), (\overline{p}, q) \in R_{F^{x_3=r}}$. By $u \notin R_F$, we have $(p, q) \notin R_{F^{x_3=r}}$. By $R_{F^{x_3=r}}$ being affine, we get $(\overline{p}, \overline{q}) \notin R_{F^{x_3=r}}$. i.e., $(\overline{p}, \overline{q}, r) \notin R_F$.

We have now accounted for all eight points of $\mathbb{Z}_2^3$. Exactly three of them α, β, γ belong to R_F and the other five do not. It can be directly verified that $R_{F^{x_1=*}}$ has exactly three members $(q, r), (q, \overline{r}), (\overline{q}, r)$, but not $(\overline{q}, \overline{r})$, which is a contradiction to $R_{F^{x_1=*}}$ being affine. This contradiction completes our proof. □

3.5.3 Arity Reduction

Now we come to the two key lemmas for the hardness proof. Both proofs inductively reduce the arity of a function. Suppose $\mathcal{F} \not\subseteq \mathscr{A}$ and $\mathcal{F} \not\subseteq \mathscr{P}$. Let $F \notin \mathscr{A}$ and $G \notin \mathscr{P}$, where $F, G \in \mathcal{F}$. (It is possible that $F = G$). From F and from G, we recursively simulate functions with smaller arities, keeping the property of being not in $\mathscr{A}$ and not in $\mathscr{P}$ respectively. After the two lemmas we handle the base case of the induction.

Lemma 3.19. *If $F \notin \mathscr{A}$, then either #CSP(F) is #P-hard, or we can simulate a unary function $H \notin \mathscr{A}$, that is, there is a reduction from #CSP($\{F, H\}$) to #CSP(F).*

The same statement holds for #CSP^c, and a reduction exists that does not use Lemma 3.13.

Proof. We prove by induction on the arity of the function F. If F has arity 1, then we may let $H = F$.

Inductively we assume the lemma has been proved for functions with arity less than k, for some $k \geq 2$. Now let F have arity k. In the following proof, for each case, we construct some functions that can be simulated in #CSP(F), but have arity less than k, and then *assume* they are in $\mathscr{A}$ (otherwise, it is proved by induction). Finally we prove that either the problem is #P-hard, or we get a unary function $H \notin \mathscr{A}$ or reach a contradiction.

Since the constant function 0 is in $\mathscr{A}$, F has a nonempty support R_F. We first assume R_F is not the whole space $\mathbb{Z}_2^k$. By Lemma 3.18, either #CSP(F) is #P-hard, or R_F is affine. Suppose $R_F = \chi_{AX}$, and $x_1, x_2, \ldots, x_s$ $(0 \leq s < k)$ are free variables of $AX = 0$. The function $F^{x_{s+1}=*, x_{s+2}=*, \ldots, x_k=*}$ can be simulated by F and has an arity less than k. Thus by our assumption, $F^{x_{s+1}=*, x_{s+2}=*, \ldots, x_k=*} \in \mathscr{A}$. Then obviously F is equal to $\chi_{AX} F^{x_{s+1}=*, x_{s+2}=*, \ldots, x_k=*}$ and thus $F \in \mathscr{A}$. This is a contradiction to the assumption that $F \notin \mathscr{A}$.

So we may assume $R_F = \mathbb{Z}_2^k$. We normalize F by dividing F by the nonzero constant $F(0, 0, \ldots, 0)$ so that we may assume $F(0, 0, \ldots, 0) = 1$. By our assumption, $F^{x_1=0} \in \mathscr{A}$ so all values of $F(0, x_2, \ldots, x_k)$ are powers of $\mathfrak{i}$. Also, $F^{x_2=0}, F^{x_2=1} \in \mathscr{A}$, and since all values of $F^{x_1=0, x_2=0}$ and $F^{x_1=0, x_2=1}$ are powers of $\mathfrak{i}$, so are all values of $F^{x_2=0}$ and $F^{x_2=1}$. In particular, so are all values of $F^{x_1=1, x_2=1}$.

Now we apply Lemma 3.10 to $F^{x_2=0} \in \mathscr{A}$ and $F^{x_2=1} \in \mathscr{A}$. Accordingly, we have the following three cases.

1. Both $F^{x_2=0}$ and $F^{x_2=1}$ have congruity. We will denote the function $F^{x_1=a, x_2=b}$ by F^{ab}. Let c and $c' \in \{1, -1, \mathfrak{i}, -\mathfrak{i}\}$ be the two constants in the definition of congruity for $F^{x_2=0}$ and $F^{x_2=1}$ respectively, in Lemma 3.10. Thus $F^{10}/F^{00}(x_3, \ldots, x_k) = c$ and $F^{11}/F^{01}(x_3, \ldots, x_k) = c'$.
   (a) $c = c'$.
   This means $F^{x_1=1}/F^{x_1=0}$ is a constant c in $\{1, -1, \mathfrak{i}, -\mathfrak{i}\}$. Suppose $c = \mathfrak{i}^r$. Then $F = (\mathfrak{i}^{x_1})^r F^{x_1=0} = \mathfrak{i}^{r x_1^2} F^{x_1=0}$. Since $F^{x_1=0}$ is in $\mathscr{A}$ by arity, this shows that F is also in $\mathscr{A}$, a contradiction.
   (b) $c = -c'$.
   We will use the notation $[\alpha(X)]$ to denote the 0-1 indicator function for an affine linear function $\alpha(X)$ over $\mathbb{Z}_2$. For any input X, $[\alpha(X)]$ takes value $0 \in \mathbb{Z}$ if $\alpha(X) = 0$ in $\mathbb{Z}_2$, and it takes value $1 \in \mathbb{Z}$ if $\alpha(X) = 1$ in $\mathbb{Z}_2$.

Since $c \in \{1, -1, \mathrm{i}, -\mathrm{i}\}$, there exists an r such that $\mathrm{i}^r = c/\mathrm{i}$. Then we claim that

$$F = \left(\mathrm{i}^{[x_1]}\right)^r \cdot \mathrm{i}^{[x_1 \oplus x_2] + [x_2] + [x_2] + [x_2]} \cdot F^{x_1=0}.$$

To verify this, first suppose $x_1 = 0$, then the right-hand side is $\mathrm{i}^{4[x_2]} \cdot F^{x_1=0} = F^{x_1=0}$. Now let $x_1 = 1$; then the right-hand side is $\mathrm{i}^r \cdot \mathrm{i}^{1-[x_2]+3[x_2]} \cdot F^{x_1=0} = c(-1)^{[x_2]} F^{x_1=0}$. This is $cF^{00} = F^{10}$, if $x_2 = 0$. For $x_2 = 1$, the expression is $-cF^{01} = c'F^{01} = F^{11}$. The claim is proved. Now since $F^{x_1=0}$ has arity less than k, $F^{x_1=0} \in \mathscr{A}$. But then the claim implies that $F \in \mathscr{A}$ as well, a contradiction.

(c) $c = \mathrm{i}c'$ or $c = -\mathrm{i}c'$.

Assign an arbitrary assignment for $x_3, \ldots, x_k$. Let P be the resulting function on x_1, x_2. In matrix form, where the rows are indexed by $x_1 = 0, 1$ and columns are indexed by $x_2 = 0, 1$, we have $P = \begin{bmatrix} u & v \\ \pm\mathrm{i}c'u & c'v \end{bmatrix}$, for some u, v taking values in $\{\pm 1, \pm\mathrm{i}\}$. Let $Q(x_1, x_2) = P^3(0, x_2)P(x_1, x_2)$, a product of functions, realizable by pinning Δ_0. In matrix form, this is $Q = \begin{bmatrix} u^4 & v^4 \\ \pm\mathrm{i}c'u^4 & c'v^4 \end{bmatrix} = \begin{bmatrix} 1 & 1 \\ \pm\mathrm{i}c' & c' \end{bmatrix}$. Here we used the fact that the values of P are powers of i. Now $Q^{x_2=*}$ is a unary function $[2, (1 \pm \mathrm{i})c']$ that has unequal and nonzero norms $2 \neq |(1 \pm \mathrm{i})c'| = \sqrt{2}$ and hence not in $\mathscr{A}$.

2. One of $F^{x_2=0}$ and $F^{x_2=1}$ has congruity and the other has semicongruity. Without loss of generality, assume $F^{x_2=0}$ has congruity and $F^{x_2=1}$ has semicongruity. The other case is similar.

By congruity, there is a constant $c \in \{1, -1, \mathrm{i}, -\mathrm{i}\}$, such that $F^{10}/F^{00}(x_3, \ldots, x_k) = c$ for all $x_3, \ldots, x_k \in \mathbb{Z}_2^{k-2}$. By semicongruity there is a constant $c' \in \{1, -1, \mathrm{i}, -\mathrm{i}\}$, and a $(k-3)$-dimensional affine linear subspace $S \subset \mathbb{Z}_2^{k-2}$, represented by an affine linear function $\alpha(x_3, \ldots, x_k) = 0$, such that on S, $F^{11}/F^{01}(x_3, \ldots, x_k) = c'$ and on $\mathbb{Z}_2^{k-2} - S$, $F^{11}/F^{01}(x_3, \ldots, x_k) = -c'$. We note that to have semicongruity, it must be the case that $k \geq 3$, and one of the coefficients of $x_3, \ldots, x_k$ in $\alpha(x_3, \ldots, x_k)$ must be nonzero. Without loss of generality, let it be the coefficient of x_3.

Fix an arbitrary assignment to $x_4, \ldots, x_k$ (if $k = 3$ this step is vacuous), this gives a function $P(x_1, x_2, x_3)$. By changing the constant term in α and c' to $-c'$ if necessary we may assume $x_3 = 0$ gives a point with $\alpha(x_3, \ldots, x_k) = 0$.

Now we will use a special notation to represent $P(x_1, x_2, x_3)$.

$$P = \begin{array}{cccc} z & & & cz \\ & x & cx & \\ & y & c'y & \\ w & & & -c'w \end{array}.$$

This symbol is intended to suggest a cube and is to be read as follows: The left (right) four entries are function values with $x_1 = 0$ ($x_1 = 1$); the top (bottom) four entries are function values with $x_2 = 0$ ($x_2 = 1$); finally the inner (outer) four entries are values with $x_3 = 0$ ($x_3 = 1$).

Let $Q(x_1, x_2, x_3) = P(x_1, x_2, x_3)(P(0, x_2, x_3))^3$. This corresponds to taking the third power of each of left four nodes (x, y, z, w) and multiplying to itself and the node to its right. We get

$$Q = \begin{array}{cccc} 1 & & & c \\ & 1 & c & \\ & 1 & c' & \\ 1 & & & -c' \end{array},$$

since $x^4 = y^4 = z^4 = w^4 = 1$. Next let $R(x_1, x_2, x_3) = Q(x_1, x_2, x_3)$ $(Q(x_1, x_2, 0))^3$. This gives

$$R = \begin{array}{cccc} 1 & & & 1 \\ & 1 & 1 & \\ & 1 & 1 & \\ 1 & & & -1 \end{array},$$

since $c^4 = c'^4 = 1$. Then $R^{x_1=0} = \begin{bmatrix} 1 & 1 \\ 1 & 1 \end{bmatrix}$, $R^{x_1=1} = \begin{bmatrix} 1 & 1 \\ 1 & -1 \end{bmatrix}$. It follows that $R^{x_1=*} = \begin{bmatrix} 2 & 2 \\ 2 & 0 \end{bmatrix}$, and $R^{x_1=*,x_2=*} = [4, 2]$. It has unequal nonzero norms, hence this unary function $R^{x_1=*,x_2=*} \notin \mathscr{A}$.

3. Both $F^{x_2=0}$ and $F^{x_2=1}$ have semicongruity. Let $F^{10}/F^{00} = c$ on $\alpha(x_3, \ldots, x_k) = 0$ and $-c$ on $\alpha(x_3, \ldots, x_k) = 1$. Similarly $F^{11}/F^{01} = c'$ on $\beta(x_3, \ldots, x_k) = 0$ and $-c'$ on $\beta(x_3, \ldots, x_k) = 1$. Here $c, c' \in \{1, -1, \mathfrak{i}, -\mathfrak{i}\}$, and α, β are two nontrivial affine linear functions.
   (a) $c \neq \pm c'$. Since β is nontrivial, we may assume the coefficient of x_3 in β is nonzero. Fix any assignment to $x_4, \ldots, x_k$, we may assume without loss of generality $x_3 = 0$ satisfies $\beta = 0$. We have the following function

$P(x_1, x_2, x_3)$, which in our symbolic notation is

$$P = \begin{array}{cccc} z & & & \epsilon cz \\ & x & \delta cx & \\ & y & c'y & \\ w & & & -c'w \end{array},$$

where $\epsilon, \delta \in \{\pm 1\}$ depending on the assignment of $x_4, \ldots, x_k$. If $\epsilon = \delta$, the two entries $\pm cz$ and $\pm cx$ both take the same $+c$ or $-c$ multiplier, then we have obtained a ternary function of the same form in case 2 and therefore we can continue in exactly the same way. So assume $\epsilon = -\delta$. By renaming c as $-c$, we may assume the two entries are in fact $-cz$ and $+cx$ respectively. Now we take $Q(x_1, x_2, x_3) = P(0, x_2, x_3)^3 P(x_1, x_2, x_3)$. Finally, let $R(x_1, x_2, x_3) = Q(x_1, 0, x_3)^3 Q(x_1, x_2, x_3)$. Then we have

$$Q = \begin{array}{cccc} 1 & & & -c \\ & 1 & c & \\ & 1 & c' & \\ 1 & & & -c' \end{array} \quad \text{and} \quad R = \begin{array}{cccc} 1 & & & 1 \\ & 1 & 1 & \\ & 1 & c'/c & \\ 1 & & & c'/c \end{array}.$$

Note that $c^3 c' = c'/c$, since $c^4 = 1$. It is easy to see that $R^{x_1=*, x_3=0} = [2, 1 + c'/c]$. Since $c'/c \neq \pm 1$ we have $c'/c = \pm \mathrm{i}$. Then this unary function $[2, 1 \pm \mathrm{i}] \notin \mathscr{A}$ since it has unequal nonzero norms $2 \neq |1 \pm \mathrm{i}|$.

(b) $c = \pm c' \in \{1, -1\}$. In this case $F^{x_1=1}/F^{x_1=0}$ takes only values ± 1. Then $R_{F^{x_1=*}}$ is precisely where $F^{x_1=1}/F^{x_1=0} = +1$. If $R_{F^{x_1=*}}$ is not affine, we have #P-hardness by Lemma 3.18. So let $R_{F^{x_1=*}}$ be defined by an affine linear $\gamma(x_2, \ldots, x_k) = 0$. It can be directly verified that

$$F = F^{x_1=0} \cdot \mathrm{i}^{[x_1]+[x_1]+[x_1]+[x_1 \oplus \gamma]+[\gamma]+[\gamma]+[\gamma]}.$$

We have $F^{x_1=0} \in \mathscr{A}$ by arity, which implies that $F \in \mathscr{A}$, a contradiction.

(c) $c = \pm c' \in \{\mathrm{i}, -\mathrm{i}\}$. In this case $F^{x_1=1}/F^{x_1=0}$ takes only values $\pm \mathrm{i}$. We may assume $c = c' = \mathrm{i}$ by changing α to $\alpha \oplus 1$ and/or β to $\beta \oplus 1$ if necessary. Consider the subset

$$\begin{aligned} S &= \{(x_2, x_3, \ldots, x_k) \mid F^{x_1=1}/F^{x_1=0} = \mathrm{i}\} \\ &= \{(0, x_3, \ldots, x_k) \mid \alpha(x_3, \ldots, x_k) = 0\} \cup \{(1, x_3, \ldots, x_k) \mid \\ &\quad \beta(x_3, \ldots, x_k) = 0\}. \end{aligned}$$

First suppose all coefficients of $x_3, \ldots, x_k$ in α and β are the same. If $\alpha(x_3, \ldots, x_k) = \sum_{i=3}^{k} \alpha_i x_i + a$ and $\beta(x_3, \ldots, x_k) = \sum_{i=3}^{k} \alpha_i x_i + b$ over $\mathbb{Z}_2$, then $(a \oplus b)x_2 + \sum_{i=3}^{k} \alpha_i x_i + a = 0$ over $\mathbb{Z}_2$ defines the set S.

Denote this affine linear function by γ, then it can be verified that

$$F = F^{x_1=0} \cdot \mathfrak{i}^{[x_1 \oplus \gamma]+[\gamma]+[\gamma]+[\gamma]}.$$

We have $F^{x_1=0} \in \mathscr{A}$ by arity, which implies that $F \in \mathscr{A}$, a contradiction.

Now suppose some coefficients of $x_3, \ldots, x_k$ in α and β differ. Without loss of generality, suppose the coefficient of x_3 is 0 in α, and is 1 in β respectively. Fix any assignment to $x_4, \ldots, x_k$; then the value of α is fixed, and yet by setting x_3 to 0 or 1, the value of β flips. Then we get a function

$$P(x_1, x_2, x_3) = \begin{array}{cccc} z & & & \epsilon z \\ & x & \epsilon x & \\ & y & \delta y & \\ w & & & -\delta w \end{array}$$

for some $\epsilon, \delta = \pm\mathfrak{i}$. From here the proof is completed as in case 2. □

We observe that for any $a, b \neq 0$, the ternary functions $H = [a, 0, b, 0]$ and $[0, a, 0, b] \notin \mathscr{P}$. In fact, if H were to be in $\mathscr{P}$, any expression as a member of $\mathscr{P}$ must not use any binary DISEQUALITY ($\neq_2$), since there is an assignment having a nonzero value $H \neq 0$, with $x_i = x_j = 0$, for any $i \neq j \in \{1, 2, 3\}$, and the third variable set accordingly. Similarly it cannot use any binary EQUALITY ($=_2$) since there is also an assignment having a nonzero value $H \neq 0$, with $x_i \neq x_j$. But unary functions alone cannot define these functions either.

Lemma 3.20. *For any function $F \notin \mathscr{P}$, either #CSP(F) is #P-hard, or we can simulate, using F, a function* $[a, 0, 1, 0]$ *(or* $[0, 1, 0, a]$*), where $a \neq 0$, or a binary function $H \notin \mathscr{P}$ having no zero values.*

The same statement holds for #CSP^c, and a reduction exists that does not use Lemma 3.13.

Proof. Suppose F has arity k. Then $k \geq 2$ since $F \notin \mathscr{P}$ and $\mathscr{P}$ contains all unary functions. Define a $\{0, 1\}$-matrix by listing all elements of R_F in some order. The rows are indexed by R_F and the columns are indexed by $1, \ldots, k$ corresponding to $x_1, \ldots, x_k$.

We first remove any column that is all-0 or all-1. If we remove an all-0 column corresponding to x_i, then any $X \in R_F$ has $x_i = 0$. The updated matrix corresponds to $R_{F^{x_i=0}}$. Similarly, if we remove an all-1 column corresponding to x_i, then any $X \in R_F$ has $x_i = 1$. The updated matrix corresponds to $R_{F^{x_i=1}}$. If two columns are identical or are complementary in every bit, we remove one of them. If the columns at x_i and x_j are identical, then any $X \in R_F$ has $x_i = x_j$. Similarly, for a pair of complementary columns at x_i and x_j, any $X \in R_F$

implies that $x_i \oplus x_j = 1$. In both cases, the updated matrix by removing column j corresponds to $R_{F^{x_j=*}}$.

We remove columns as long as possible. We claim that this removal process maintains the property of not belonging to $\mathscr{P}$. Suppose we removed an all-0 column at x_i, to get $G = F^{x_i=0}$. Since any $X \in R_F$ has $x_i = 0$, we have $F = \Delta_0(x_i) \cdot G$, where $\Delta_0(x_i)$ is the unary function $[1, 0]$ applied to x_i. This expression shows that if $G \in \mathscr{P}$ then $F \in \mathscr{P}$, a contradiction. The case with removing an all-1 column is similar, where we use the unary function $\Delta_1(x_i)$ instead. If we removed the column at x_j identical to the column at x_i, then $G = F^{x_j=*}$ and $F = \chi_{x_i=x_j} \cdot G$. Finally, for the removal of a complementary column at x_j we have $G = F^{x_j=*}$ and $F = \chi_{x_i \oplus x_j \oplus 1=0} \cdot G$. In every step, we maintain $G \notin \mathscr{P}$.

Now we suppose there is some $G \notin \mathscr{P}$ where no more columns can be removed by the above process. There must be some columns left in the matrix; otherwise the function just before the last column removal is a unary function, hence in $\mathscr{P}$. In fact since $G \notin \mathscr{P}$, the arity of G is ≥ 2. For simplicity we still denote it by k. Thus $k \geq 2$. We have two cases:

Case 1: $|R_G| < 2^k$**.** By Lemma 3.18, we may assume R_G is affine, given by a linear system on $x_1, \ldots, x_k$ over $\mathbb{Z}_2$. We have that $|R_G| = |R_F| \neq 0$. This is because we never deleted any rows, and distinct rows produce distinct rows in the column removal process; so the number of rows remains the same. Clearly $|R_G| > 1$ since otherwise every column (of length 1) is all-0 or all-1. Without loss of generality, assume $x_1, \ldots, x_s$ are free variables in the linear system defining R_G, and $x_{s+1}, \ldots, x_k$ are dependent variables. $|R_G| = 2^s$ is a power of 2. We have shown that $s \geq 1$. By $|R_G| < 2^k$, we have $s < k$. We claim $s \geq 2$. If instead $s = 1$, then every $x_2, \ldots, x_k$ is dependent on x_1 on R_G, so the column at x_2 must be an all-0 or all-1 column, or be identical or complementary to x_1, and the removal process would have continued. The expression of x_k in terms of $x_1, \ldots, x_s$ must involve at least two nonzero coefficients; otherwise the column at x_k must be an all-0 or all-1 column, or be identical or complementary to another column. Without loss of generality, say the coefficients of x_1, x_2 are nonzero.

Let $P(x_1, x_2, x_k) = G^{x_3=0,\ldots,x_s=0,x_{s+1}=*,\ldots,x_{k-1}=*}$ (the two sets of variables $\{x_3, \ldots, x_s\}$ and $\{x_{s+1}, \ldots, x_{k-1}\}$ could be empty). Note that the sum over $x_{s+1}, \ldots, x_{k-1}$ does not introduce any cancelations because for any assignment $x_1, \ldots, x_s$ there is a unique nonzero value for each variable $x_{s+1}, \ldots, x_k$. It follows that $R_P = \chi_{x_1 \oplus x_2 \oplus x_k = c}$ for some $c \in \mathbb{Z}_2$.

The affine linear equation $x_1 \oplus x_2 \oplus x_k = c$ is symmetric. Now we define a *symmetrized* function $H(x_1, x_2, x_k) = \prod_{\sigma \in S_3} P(x_{\sigma(1)}, x_{\sigma(2)}, x_{\sigma(k)})$, where S_3 is the symmetry group on three letters $\{1, 2, k\}$. This H is a symmetric function on (x_1, x_2, x_k) and has support $R_H = R_P$. Thus, after normalizing, $H = [a, 0, 1, 0]$ or $[0, 1, 0, a]$, where $a \neq 0$. This ternary function $H \notin \mathscr{P}$.

Case 2: $|R_G| = 2^k$. If for all $1 \leq j \leq k$, the ratio $G^{x_j=1}/G^{x_j=0}$ is a constant function c_j, (since $|R_G| = 2^k$ there are no divisions by zeros), then $G = c_0 \cdot [1, c_1] \otimes \cdots \otimes [1, c_k]$, where the constant $c_0 = G^{x_1=0,\ldots,x_k=0}$, and the unary function $[1, c_j]$ is on x_j. This gives $G \in \mathscr{P}$, a contradiction.

Now suppose for some j, $G^{x_j=1}/G^{x_j=0}$ is not a constant function. Without loss of generality, we assume $j = 1$. The Boolean hypercube on $(x_2, \ldots, x_k) \in \{0, 1\}^{k-1}$ is connected by edges that flip just one bit. Without loss of generality, along some flip of x_2 we have two different values of $G^{x_1=1}/G^{x_1=0}$. Suppose $G^{x_1=1}/G^{x_1=0}(0, a_3, \ldots, a_k) \neq G^{x_1=1}/G^{x_1=0}(1, a_3, \ldots, a_k)$, for some $a_3, \ldots, a_k$. Set $x_3 = a_3, \ldots, x_k = a_k$, we get a binary function $H(x_1, x_2) = G(x_1, x_2, a_3, \ldots, a_k)$. We have $H(1, 0)/H(0, 0) \neq H(1, 1)/H(0, 1)$, hence the rank of $H = \begin{bmatrix} H(0,0) & H(0,1) \\ H(1,0) & H(1,1) \end{bmatrix}$ is 2.

Suppose $H \in \mathscr{P}$. If a defining expression of H uses binary EQUALITY or DISEQUALITY functions, then we can set x_1, x_2 so that $H = 0$. But H is never zero. So H is defined by unary functions alone. But such a function has a matrix form of rank ≤ 1. This contradiction proves that $H \notin \mathscr{P}$ and completes our proof. □

3.5.4 Proof of Boolean #CSP Dichotomy

Now we are ready to complete the proof for the main theorem of this chapter: Theorem 3.7.

Proof of Theorem 3.7: In Section 3.4 we have shown that #CSP($\mathscr{A}$) and #CSP($\mathscr{P}$) are computable in polynomial time.

If $\mathcal{F} \nsubseteq \mathscr{A}$, by Lemma 3.19, either #CSP($\mathcal{F}$) is #P-hard, or we can simulate a unary function not in $\mathscr{A}$. After dividing by a nonzero factor, such a function takes the form $F = [1, \lambda]$, where $\lambda \notin \{0, \pm 1, \pm \mathfrak{i}\}$. If $\mathcal{F} \nsubseteq \mathscr{P}$, by Lemma 3.20, either #CSP($\mathcal{F}$) is #P-hard, or we can simulate a function $P = [a, 0, 1, 0]$, or $P' = [0, 1, 0, a]$, where $a \neq 0$, or a binary function $H \notin \mathscr{P}$ having no zero values.

First, we prove #CSP($\{F, P\}$) is #P-hard. Clearly $P^{x_1=*} = [a, 1, 1]$. If $a \notin \{1, -1\}$, #CSP($[a, 1, 1]$) is #P-hard by Theorem 3.14. If $a \in \{1, -1\}$, we can construct $Q(x_1, x_2) = \sum_{x_3} P(x_1, x_2, x_3) F(x_3) = [a, \lambda, 1]$, which is $[\pm 1, \lambda, 1]$. Both problems #CSP($[\pm 1, \lambda, 1]$) are #P-hard by Theorem 3.14. The proof for #CSP($\{F, P'\}$) is the same.

Secondly, we prove #CSP($\{F, H\}$) is #P-hard. After normalizing, we may suppose $H = \begin{bmatrix} 1 & x \\ y & z \end{bmatrix}$, where $xyz \neq 0$. Since $\mathscr{P}$ is a superset of $\mathscr{D}$, H is nondegenerate. Hence $z \neq xy$.

There are two cases, depending on whether $z = -xy$. For the case $z \neq -xy$, we construct a symmetric function $H(x_1, x_2)H(x_2, x_1) = [1, xy, z^2]$. By the conditions $xyz \neq 0, z \neq \pm xy$, it is impossible to be the first three tractable cases in Theorem 3.14. If it is in the last two tractable cases, then xy is a power of i. Now we can form the function $H(x_1, x_2)H(x_2, x_1)F(x_1)F(x_2)$, which is $[1, \lambda xy, \lambda^2 z^2]$. This function has no zero entry and its matrix form has rank 2, so it is not of the first three tractable cases in Theorem 3.14. If it were in the last two tractable cases, then λxy is a power of i, which implies that $\lambda = (\lambda xy)/(xy)$ itself is a power of i. However, since $[1, \lambda] \notin \mathscr{A}$, λ is not a power of i.

For the case $z = -xy$, We construct some binary functions with a nonnegative integer parameter s as follows:

$$\begin{aligned}\sum_{x_3} H(x_1, x_3)H(x_2, x_3)(F(x_3))^s &= [1 + \lambda^s x^2, (y + \lambda^s xz), (y^2 + \lambda^s z^2)] \\ &= [1 + \lambda^s x^2, y(1 - \lambda^s x^2), y^2(1 + \lambda^s x^2)].\end{aligned}$$

As λ is not a power of i, at most one of the two values x^2 and λx^2 can be a power of i. Now we choose $s = 0$ or $s = 1$ so that $\lambda^s x^2 \notin \{\pm 1, \pm \mathrm{i}\}$.

After normalizing, we may write the function $[1 + \lambda^s x^2, y(1 - \lambda^s x^2), y^2(1 + \lambda^s x^2)]$ as $[1, y(1 - \lambda^s x^2)/(1 + \lambda^s x^2), y^2]$, noticing that $1 + \lambda^s x^2 \neq 0$. We claim that this function is not one of the five tractable cases from Theorem 3.14. Since there are no zero entries, clearly it is not the first two cases. Its matrix form has rank 2; therefore it is not the third case. If it were the fourth tractable case $[1, \pm \mathrm{i}, 1]$, then $y = \pm 1$, and $(1 - \lambda^s x^2)/(1 + \lambda^s x^2) = \pm \mathrm{i}$. This implies that $\lambda^s x^2 = \pm \mathrm{i}$, which is impossible. If $[1, y(1 - \lambda^s x^2)/(1 + \lambda^s x^2), y^2] = [1, \pm 1, -1]$, the fifth tractable case, then $y = \pm \mathrm{i}$, and again $(1 - \lambda^s x^2)/(1 + \lambda^s x^2) = \pm \mathrm{i}$, also impossible.

The proof of Theorem 3.7 is complete. □

For #CSP$^c(\mathcal{F})$ we have the following theorem. Notice that #CSP$^c(\mathcal{F})$ and #CSP($\mathcal{F}$) are really equivalent by Lemma 3.13. And since $\Delta_0, \Delta_1 \in \mathscr{A} \cap \mathscr{P}$, the tractability criteria of Theorems 3.7 and 3.21 are the same, as they should be. The point of Theorem 3.21 is that a reduction exists for #CSP$^c(\mathcal{F})$ when it is #P-hard that does not use pinning.

Theorem 3.21. *Suppose $\mathcal{F}$ is a set of functions mapping Boolean inputs to complex numbers. If $\mathcal{F} \subseteq \mathscr{A}$ or $\mathcal{F} \subseteq \mathscr{P}$, then #CSP$^c(\mathcal{F})$ is computable in* *polynomial time. Otherwise,* #CSP^c($\mathcal{F}$) *is* #P-*hard, and a reduction exists that does not use Lemma 3.13.*

If we restrict to real (algebraic)-valued constraint functions in Theorem 3.7, we have the following theorem due to Bulatov, Dyer, Goldberg, Jalsenius, and Richerby.

Theorem 3.22 (Bulatov et al.). *Suppose $\mathcal{F}$ is a set of functions mapping Boolean inputs to (algebraic) real numbers.* #CSP($\mathcal{F}$) *is* #P-*hard, unless*

- *Either every function in $\mathcal{F}$ takes the form $\lambda \cdot \chi_{AX} \cdot (-1)^{Q(X)}$, where $\lambda \in \mathbb{R}$, χ_{AX} is a 0-1 indicator function of an affine linear subspace over $\mathbb{Z}_2$, and $Q(X)$ is a quadratic polynomial on X over $\mathbb{Z}$;*
- *Or every function in $\mathcal{F}$ is a product of $(=_2)$, $(\neq_2)$ and real-valued unary functions.*

In these two cases, #CSP($\mathcal{F}$) *is computable in polynomial time.*

Exercise: Prove that Theorem 3.22 follows from Theorem 3.7. Similarly show that Theorems 3.1 and 3.4 can be derived from Theorem 3.22 or Theorem 3.7.

4

Matchgates and Holographic Algorithms

Some counting problems are #P-hard on general instances, but are tractable over planar structures. In fact, one of the most remarkable algorithms for counting problems is Kasteleyn's algorithm (a.k.a. the FKT algorithm) to count the number of perfect matchings for any planar graph. Starting from this chapter, we will gradually build a theory that aims to show that this is not only a beautiful polynomial time counting algorithm, but also a *universal* strategy for a broad class of counting problems, in a very strong technical sense. This will be one of the highlights in Volume I of this book, and will be built on the theory of matchgates and holographic algorithms. We note that a corresponding theory on problems over general domains has not yet been developed, and will be a challenge for the future. To develop this topic fully we start with Pfaffian orientations on planar graphs.

4.1 Pfaffian Orientations and Kasteleyn's Algorithm

An $n \times n$ matrix A is called skew-symmetric if $A_{i,j} = -A_{j,i}$, for $1 \le i, j \le n$. We assume the entries of A are from a field of characteristic $\neq 2$. Then $A_{i,i} = 0$.

Every permutation $\pi = \begin{pmatrix} 1 & 2 & \dots & 2k \\ i_1 & i_2 & \dots & i_{2k} \end{pmatrix}$ defines a partition of $[2k]$ into disjoint pairs $\tilde{\pi} = \{\{i_1, i_2\}, \{i_3, i_4\}, \dots, \{i_{2k-1}, i_{2k}\}\}$. The Pfaffian of an $n \times n$ skew-symmetric matrix A is defined as follows. Suppose $n = 2k \ge 2$ is even; then

$$\mathrm{Pf}(A) = \sum_{\pi} \epsilon_\pi A_{i_1,i_2} A_{i_3,i_4} \cdots A_{i_{2k-1},i_{2k}}, \tag{4.1}$$

where the sum is over all permutations $\pi = \begin{pmatrix} 1 & 2 & \dots & 2k \\ i_1 & i_2 & \dots & i_{2k} \end{pmatrix}$ such that

$$i_1 < i_2,\ i_3 < i_4,\ \dots,\ i_{2k-1} < i_{2k} \quad \text{and} \quad i_1 < i_3 < \cdots < i_{2k-1}. \tag{4.2}$$

The symbol ϵ_π in (4.1) denotes the sign of the permutation π; it is $+1$ or -1 depending on whether the parity of π is even or odd, respectively. We note that there is a natural 1-1 correspondence between permutations π satisfying the stipulation (4.2) and partitions of $[n]$ into disjoint pairs.

If n is odd, then we define $\text{Pf}(A) = 0$. By convention if $n = 0$ we define $\text{Pf}(A) = 1$.

For any permutation $\pi = \begin{pmatrix} 1 & 2 & \ldots & 2k \\ i_1 & i_2 & \ldots & i_{2k} \end{pmatrix}$, not necessarily satisfying the stipulation (4.2), define

$$a_\pi = \epsilon_\pi A_{i_1,i_2} A_{i_3,i_4} \cdots A_{i_{2k-1},i_{2k}}.$$

We make a simple but important observation: If $\tilde{\pi} = \widetilde{\pi'}$, then $a_\pi = a_{\pi'}$, i.e., the expression a_π has the same value if we list the partition $\tilde{\pi} = \{\{i_1, i_2\}, \{i_3, i_4\}, \ldots, \{i_{2k-1}, i_{2k}\}\}$ in any order of the pairs, as well as in any order of the two labels of each pair. This follows from the fact that A is skew-symmetric. This invariance will be used repeatedly.

We say two pairs of labels $i_{2j-1} < i_{2j}$ and $i_{2\ell-1} < i_{2\ell}$ form a crossover , or an overlapping pair, iff $i_{2j-1} < i_{2\ell-1} < i_{2j} < i_{2\ell}$ or $i_{2\ell-1} < i_{2j-1} < i_{2\ell} < i_{2j}$. Suppose π is a permutation satisfying the stipulation (4.2). Let $c(\pi)$ be the number of crossovers among the pairs in the partition $\tilde{\pi}$. Then the sign ϵ_π is also $(-1)^{c(\pi)}$. Hence it can be computed as the parity of the number of crossovers. To see this, consider the permutation π and consider a sequence of adjacent transpositions that moves the sequence $(1, 2, \ldots, 2k-1, 2k)$ to $(i_1, i_2, \ldots, i_{2k-1}, i_{2k})$. We have $i_1 = 1$ by (4.2). The number of transpositions that will bring i_2 to the position right after 1 is the number of labels strictly between $i_1 = 1$ and the number i_2, and has the same parity as the number of crossovers the pair $\{i_1, i_2\}$ forms with all other pairs of labels $\{\{i_3, i_4\}, \ldots, \{i_{2k-1}, i_{2k}\}\}$ in $\tilde{\pi}$. After i_1, i_2 are placed in the first two positions, if $n > 2$, then i_3 is the minimum among all other labels by (4.2), and is currently located right after the first two elements. Then we move i_4 to the position right after i_3. The proof is completed by induction. We note that the definition of Pfaffian and this notion of crossovers apply to any totally ordered label set, in particular to any index set of a principal submatrix obtained from A by selecting a common subset of rows and columns.

The Pfaffian can be computed in polynomial time [Bar83, GM94]. A key relation to determinant is the following theorem due to Cayley [Hal66] (also see p. 352 of [Jac85a]).

Theorem 4.1. *For any $n \times n$ skew-symmetric matrix A,*

$$\det(A) = [\text{Pf}(A)]^2.$$

Note that for odd n, a skew-symmetric matrix A has $\det(A) = 0$. For even n, an algorithm similar to Gaussian elimination can be derived using the identity $\mathrm{Pf}(BAB^{\mathrm{T}}) = \det(B)\mathrm{Pf}(A)$, for any $n \times n$ matrix B. It can be even computed in NC [MSV04].

Let $G = (V, E)$ be a simple undirected graph without self-loops and parallel edges. Suppose the vertices of G are labeled by a totally ordered set, e.g., $V = \{1, 2, \ldots, n\}$. Assign an indeterminate x_e for every edge $e = \{u, v\} \in E$. Then we define the skew-symmetric adjacency matrix $A = A(G)$ of the graph G to be

$$A_{u,v} = \begin{cases} x_e & \text{if } e = \{u, v\} \in E \text{ and } u < v \\ -x_e & \text{if } e = \{u, v\} \in E \text{ and } u > v \\ 0 & \text{if } \{u, v\} \notin E \end{cases} . \tag{4.3}$$

Note that for any permutation π, the partition $\tilde{\pi}$ is a perfect matching iff all pairs are edges. Denote by $\mathcal{M}(G)$ the set of all perfect matchings of G. There is a 1-1 correspondence between nonzero terms in $\mathrm{Pf}(A)$ and $\mathcal{M}(G)$. For any $M \in \mathcal{M}(G)$ there are $2^k k!$ permutations of the form $\pi = \begin{pmatrix} 1 & 2 & \ldots & 2k \\ i_1 & i_2 & \ldots & i_{2k} \end{pmatrix}$ that can represent M, i.e., $\tilde{\pi} = M$, of which the unique permutation that satisfies the stipulation (4.2) is called a *canonical* expression. All other permutation expressions for M are obtained by permuting the order of the pairs $\{i_{2\ell-1}, i_{2\ell}\}$ and the order within each pair.

We define the weight of a perfect matching M in G to be the product of edge weights x_e for all $e \in M$, namely $\Gamma(M) = \Gamma_G(M) = \prod_{e \in M} x_e$. It follows that $a_\pi = \pm\Gamma(M)$, if $\tilde{\pi} = M$.

An orientation of a graph G assigns one direction to each edge of G. We denote by $u \to v$ if the edge $\{u, v\}$ is oriented from u to v. In this case u is its tail, and v is its head. For an oriented graph $\vec{G}$ we modify the skew-symmetric matrix A in (4.3) to be $B = B(\vec{G})$:

$$B_{u,v} = \begin{cases} x_e & \text{if } e = \{u, v\} \in E \text{ and } u \to v \\ -x_e & \text{if } e = \{u, v\} \in E \text{ and } v \to u \\ 0 & \text{if } \{u, v\} \notin E \end{cases} . \tag{4.4}$$

In other words, we change the sign at both entries $A_{u,v}$ and $A_{v,u}$ if $u < v$ and $\{u, v\}$ is oriented $v \to u$.

Given an orientation, we will consider the Pfaffian of B:

$$\mathrm{Pf}(B) = \sum_{\pi} \epsilon_\pi B_{i_1,i_2} B_{i_3,i_4} \cdots B_{i_{2k-1},i_{2k}}, \tag{4.5}$$

where the sum is over permutations π satisfying (4.2). For a perfect matching M in an oriented graph $\overrightarrow{G}$, suppose $M = \tilde{\pi}$, define the Pfaffian term

$$\mathrm{Pf}_{\overrightarrow{G}}(M) = \epsilon_\pi B_{i_1,i_2} B_{i_3,i_4} \cdots B_{i_{2k-1},i_{2k}}. \tag{4.6}$$

It is a term in $\mathrm{Pf}(B)$ when π is the canonical expression for M and it is equal to either the weight of the perfect matching $\Gamma_G(M)$ or its negation $-\Gamma_G(M)$. However, we can choose any permutation π representing M, and the value $\mathrm{Pf}_{\overrightarrow{G}}(M)$ is defined, which does not depend on the particular partition π corresponding to M, owing to the invariance observed earlier.

Definition 4.2. For any perfect matching M in an oriented graph $\overrightarrow{G}$, the *sign* of the perfect matching M with respect to this orientation is

$$\mathrm{sgn}_{\overrightarrow{G}}(M) = \frac{\mathrm{Pf}_{\overrightarrow{G}}(M)}{\Gamma_G(M)} \in \{-1, 1\}. \tag{4.7}$$

When $\overrightarrow{G}$ is clear from the context, we write $\mathrm{sgn}(M)$ instead.

We note that $\mathrm{sgn}(M)$ can be computed with any partition π corresponding to M, i.e., $\tilde{\pi} = M$, owing to the invariance of $\mathrm{Pf}_{\overrightarrow{G}}(M)$. In particular, $\mathrm{sgn}(M)$ can be computed simply as the sign ϵ_π of the permutation $\pi = \begin{pmatrix} 1 & 2 & \ldots & 2k \\ i_1 & i_2 & \ldots & i_{2k} \end{pmatrix}$, where each matching edge $\{i_{2\ell-1}, i_{2\ell}\} \in M$ is listed by its orientation $i_{2\ell-1} \to i_{2\ell}$. This permutation π is in general not a canonical expression.

If M and M' are two perfect matchings, the symmetric difference $M \oplus M'$ consists of a set of cycles of even length that are alternating between edges from M and M'. For any orientation on a graph G, if C is a cycle of even length, then we say it is evenly oriented if there are an even number of edges oriented in one direction, and oddly oriented otherwise. Clearly this notion does not depend on the direction since C has an even length.

Lemma 4.3. *For any orientation on a graph G, and perfect matchings M and M', if k is the number of evenly oriented cycles in $M \oplus M'$ with respect to the orientation, then*

$$\mathrm{sgn}(M) \cdot \mathrm{sgn}(M') = (-1)^k.$$

Proof. We first show that if this equality holds for one orientation then it holds for all orientations. Let e be an edge in $\overrightarrow{G}$ and suppose we reverse its orientation. If $e \notin M \oplus M'$ then obviously the reversal has no effect on the equality. Suppose $e \in M \oplus M'$. Then k is changed by 1, and exactly one of $\mathrm{sgn}(M)$ and $\mathrm{sgn}(M')$ is changed. Hence we may choose any orientation convenient for M

and M' to prove our equality. So we choose one that gives a cyclic orientation on each cycle of $M \oplus M'$ (either direction of the cyclic orientation for each cycle is acceptable). We may orient all other edges arbitrarily. In this orientation, k is the number of cycles in $M \oplus M'$.

Next we will relabel the vertices of $\overrightarrow{G}$. If we apply a permutation σ on the vertices of $\overrightarrow{G}$, the Pfaffian term for a perfect matching M is changed as follows. Suppose M corresponds to the permutation $\pi = \begin{pmatrix} 1 & 2 & \dots & 2k \\ i_1 & i_2 & \dots & i_{2k} \end{pmatrix}$, where each matching edge $\{i_{2\ell-1}, i_{2\ell}\} \in M$ is listed by its orientation $i_{2\ell-1} \to i_{2\ell}$. We write its corresponding Pfaffian term $b_\pi = \epsilon_\pi B_{i_1,i_2} B_{i_3,i_4} \cdots B_{i_{2k-1},i_{2k}}$. If $\sigma = \begin{pmatrix} i_1 & i_2 & \dots & i_{2k} \\ j_1 & j_2 & \dots & j_{2k} \end{pmatrix}$, then each matching edge is also oriented as $j_{2\ell-1} \to j_{2\ell}$ in the new labeling. The skew-symmetric matrix B' defined by the new labeling has $B'_{j_{2\ell-1},j_{2\ell}} = B_{i_{2\ell-1},i_{2\ell}}$, and thus the corresponding product is unchanged $B'_{j_1,j_2} B'_{j_3,j_4} \cdots B'_{j_{2k-1},j_{2k}} = B_{i_1,i_2} B_{i_3,i_4} \cdots B_{i_{2k-1},i_{2k}}$. But the sign ϵ_π becomes $\epsilon_{\sigma\pi} = \epsilon_\sigma \epsilon_\pi$. Hence the change to $\text{sgn}(M)$ is to multiply it by ϵ_σ. The same is true for the change to $\text{sgn}(M')$. Note that we wrote σ as $\begin{pmatrix} i_1 & i_2 & \dots & i_{2k} \\ j_1 & j_2 & \dots & j_{2k} \end{pmatrix}$ in reference to π that corresponds to M. However, the conclusion that we change $\text{sgn}(M)$ by multiplying it with ϵ_σ is independent of how we express σ. In particular, for M', we can write σ in reference to π' corresponding to M', and the conclusion is the same. Hence the product $\text{sgn}(M) \cdot \text{sgn}(M')$ remains unchanged. Meanwhile the value k is clearly the same. So we may choose to label all vertices as follows. Pick any cycle of $M \oplus M'$ and start at any edge of this cycle in M. Label its tail as 1 and its head as 2. Label the rest of the cycle along the chosen orientation using the next unused integers. Thus if the cycle has length 2ℓ, then its vertices are labeled from 1 to 2ℓ consecutively along the orientation starting at the tail of one edge of M. Then starting with $2\ell + 1$ we label the next cycle of $M \oplus M'$, if there is any, in exactly the same way. After all cycles of $M \oplus M'$ are dealt with, we label all remaining vertices so that each edge in $M \cap M'$ is labeled consecutively with the next unused integers, and increasing from tail to head.

Now the Pfaffian term $\text{Pf}_{\overrightarrow{G}}(M)$ for M is just the product of all edge weights $\Gamma_G(M)$. This is seen easily if we write b_π in the way where the permutation π is the identity, and we list all matched edges in the oriented order. For the Pfaffian term $b_{\pi'}$ corresponding to M', we still list the product part $B_{i_1,i_2} B_{i_3,i_4} \cdots B_{i_{2k-1},i_{2k}}$ in the oriented order for each matched edge. The sign $\epsilon_{\pi'}$ for the permutation is as follows. The part for the first cycle of length 2ℓ in $M \oplus M'$ has the form $\begin{pmatrix} 1 & 2 & \dots & 2\ell-1 & 2\ell \\ 2 & 3 & \dots & 2\ell & 1 \end{pmatrix}$, which is an even cycle as a permutation, and has an odd parity. The permutation π' is simply a product of these even cycles in the permutation group, disjoint and one even cycle for each cycle in the graph $M \oplus M'$. Hence it has parity $\epsilon_{\pi'} = (-1)^k$. □

Definition 4.4. An orientation of a connected plane graph is called a *Pfaffian orientation* if along the boundary of every non-outer face, there are an odd number of clockwise-oriented edges.

Lemma 4.5. *Any Pfaffian orientation in a connected plane graph G satisfies the following property: For every cycle C, the number of clockwise-oriented edges of C is of the opposite parity to the number of vertices contained within C. (This number does not include the vertices on the cycle C).*

Proof. Let V and E be the number of vertices and edges contained within C, respectively, and let ℓ be the number of edges on C, which is also the number of vertices on C. The vertices contained within C are those in the interior of the region bounded by C; they do not include those on the cycle C. Similarly, the edges within C do not include those on C. An edge is a bridge edge iff it has the same face on both sides. Then in a traversal of the boundary of a face Δ, if there is a bridge edge e, it is traversed twice, once in each direction. In particular, in either orientation of e, it contributes one to the count of clockwise-oriented edges along the boundary of Δ. For a nonbridge edge e, depending on its orientation, it contributes one to the count of clockwise-oriented edges to exactly one face it borders on but not the other.

Suppose there are F faces bounded by C, and let c_i be the number of clockwise-oriented edges on the boundary of the ith face ($1 \le i \le \mathsf{F}$). Each c_i is odd by assumption; therefore $\mathsf{F} \equiv \sum_{i=1}^{\mathsf{F}} c_i \pmod 2$. By Euler's formula, counting the face formed by the exterior of C, we have $(\mathsf{V} + \ell) - (\mathsf{E} + \ell) + (\mathsf{F} + 1) = 2$. It follows that $\mathsf{E} = \mathsf{V} + \mathsf{F} - 1$.

If we add up all the clockwise-oriented edges among all boundary edges in F faces, each interior edge within C regardless orientation contributes one and each clockwise-oriented edge on C contributes one. Hence $\sum_{i=1}^{\mathsf{F}} c_i = \mathsf{E} + c$, where c is the number of clockwise-oriented edges on C.

It follows that

$$\mathsf{F} \equiv \sum_{i=1}^{\mathsf{F}} c_i = \mathsf{E} + c = \mathsf{V} + \mathsf{F} - 1 + c \pmod 2,$$

and hence $\mathsf{V} + c \equiv 1 \pmod 2$. □

If G has a Pfaffian orientation, then with respect to this orientation every two perfect matchings M and M' must have the same sign: $\mathrm{sgn}(M) = \mathrm{sgn}(M')$.

In fact, since every cycle of $M \oplus M'$ contains a perfect matching (possibly empty) in its interior, the number of vertices within the cycle must be even. Each cycle of $M \oplus M'$ has an even length, consisting of edges from M and M' alternatingly. By Lemma 4.5 the cycle is oddly oriented. By Lemma 4.3, $\text{sgn}(M) \cdot \text{sgn}(M') = 1$.

Hence for a Pfaffian orientation, every Pfaffian term has the same sign.

Definition 4.6. The perfect matching polynomial, PerfMatch(G), is the following:

$$\text{PerfMatch}(G) = \sum_{M \in \mathcal{M}(G)} \prod_{e \in M} x_e, \tag{4.8}$$

where $\mathcal{M}(G)$ is the set of all perfect matchings in G and x_e corresponds to the edge e in G.

We will usually assign a value $w(e)$ in a field to each x_e as the edge weight for e. The default choice of the field is $\mathbb{C}$; but one can replace $\mathbb{C}$ by any commutative ring. To discuss the computation of the perfect matching polynomial PerfMatch(G), we usually fix an assignment of weights $w(e) \in \mathbb{C}$ for x_e. The case $w(e) = 1$ for all edges e in G is called the unweighted case; in this case PerfMatch(G) is the number of perfect matchings in G. For a graph G with at least one vertex, if it has no perfect matchings, then PerfMatch$(G) = 0$. By convention if G is the empty graph with no vertices, we define PerfMatch$(G) = 1$. Pfaffian orientation can be used to compute PerfMatch(G) for any planar graph G.

Theorem 4.7 (Kasteleyn). *Every connected planar graph has a Pfaffian orientation. Such an orientation can be constructed in polynomial time, leading to a polynomial-time algorithm to compute* PerfMatch(G) *for any weighted planar graph G.*

Proof. A graph can tested for planarity, and if so a planar embedding of the graph can be computed in polynomial time [HT74]. Assume G is a connected planar graph, and given a planar embedding. If G is a tree, then any orientation is acceptable. If G is not a tree, then choose any edge on the boundary of the outer face that belongs to a cycle. (Such an edge exists by a simple induction: Take any edge e on the boundary of the outer face. If e is a bridge edge, then its removal disconnects G into two parts. Since G is not a tree, at least one part is not a tree. The traversal of the boundary of the outer face of G consists of the concatenation of two traversals of the boundaries of the outer faces of the two parts, in sequence. Then by induction, one edge of the traversal belongs to a cycle.) Let F be the non-outer face containing this edge e. By induction we can construct a Pfaffian orientation for $G - \{e\}$, the graph with the same vertex set as G but with edge e removed. By definition, in this Pfaffian orientation every non-outer face of $G - \{e\}$ contains an odd number of clockwise oriented edges. Now add e back, and orient e appropriately we can guarantee that F also has an odd number of clockwise-oriented edges. This completes the proof by induction.

For a Pfaffian orientation, let $B = B(\overrightarrow{G})$ be the skew-symmetric matrix. Then either $\mathrm{Pf}(B) = \mathrm{PerfMatch}(G)$ or $\mathrm{Pf}(B) = -\mathrm{PerfMatch}(G)$. The equality is a polynomial equality: for the given Pfaffian orientation, either $+$ holds for all weight values, or $-$ holds for all weight values. Setting all weight values to 1, we can decide which sign is valid for the particular orientation (unless there is no perfect matching and G is nonempty, in which we can safely output $\mathrm{PerfMatch}(G) = 0$). Then we can compute $\mathrm{PerfMatch}(G)$ for the actual weight values. If G has connected components $G_1, G_2, \ldots, G_m$, then $\mathrm{PerfMatch}(G) = \prod_{i=1}^{m} \mathrm{PerfMatch}(G_i)$. □

Exercise: Show that $\mathrm{PerfMatch}(G)$ can be computed in polynomial time for weighted plannar (multi)graphs G with self-loops and multiple edges.

4.2 Matchgates

Valiant initially introduced *matchgates* as a way to show that a nontrivial, though restricted, fragment of quantum computation can be simulated in classical polynomial time [Val02b]. At the heart of this simulation technique is a way to encode certain quantum states by a classical computation of perfect matchings, and to simulate certain quantum gates by the so-called matchgates. Subsequently he introduced holographic algorithms using matchgates [Val08]. It is in this context that holographic transformations were introduced. This idea turns out to have a wider applicability as a general reduction method. We have already encountered its use for the dichotomy of Holant* problems in Chapter 2. In the context of holographic algorithms based on matchgates, this is essentially a reduction from a problem to the planar PERFECT MATCHING problem. Theorem 4.7 shows that there is a remarkable polynomial time algorithm – Kasteleyn's algorithm – to compute the weighted sum of perfect matchings for a planar graph. We introduce matchgates in this section.

Definition 4.8. A matchgate is an undirected weighted plane graph G with a subset of distinguished nodes on its outer face, called the external nodes, ordered in a clockwise order.

A plane graph is a planar graph given with a particular planar embedding. Without loss of generality, we assume all edge weights are nonzero; zero weighted edges can be deleted.

Let G be a matchgate with k external nodes. For each length-k bitstring α, G defines a subgraph G^α obtained from G by the following operation: For all $1 \le i \le k$, if the ith bit of α is 1, then we remove the ith external node and all its incident edges. Thus, $G^{00\ldots0} = G$, and $G^{11\ldots1}$ is obtained from G by removing *all* external nodes and their incident edges.

Definition 4.9. We define the signature of a matchgate G as the vector $\Gamma_G = (\Gamma_G^\alpha)$, indexed by $\alpha \in \{0, 1\}^k$ in lexicographic order, as follows:

$$\Gamma_G^\alpha = \text{PerfMatch}(G^\alpha) = \sum_{M \in \mathcal{M}(G^\alpha)} \prod_{e \in M} w(e). \tag{4.9}$$

For a perfect matching $M \in \mathcal{M}(G^\alpha)$ we define $\Gamma_G^\alpha(M) = \prod_{e \in M} w(e)$ as the *perfect matching term*, equal to the product of the edge weights for the matching M. Where G is clear, we omit the subscript G, and write Γ for Γ_G, Γ^α for Γ_G^α, and $\Gamma^\alpha(M)$ for $\Gamma_G^\alpha(M)$.

We denote by $\mathscr{M}$ the set of all matchgate signatures .

Definition 4.10. A matchgate signature Γ is *symmetric* if, for all α and β of equal Hamming weight, $\Gamma^\alpha = \Gamma^\beta$.

In other words, the value of a symmetric signature entry is only a function of how many 1's there are in its index, not their particular pattern. These signatures are important because they have a clear combinatorial meaning.

Regarding symmetric signatures that are realizable as matchgate signatures, we will prove

Theorem 4.11. *A symmetric signature is the signature of a matchgate iff it has the following form, for some $a, b \in \mathbb{C}$ and integer k (we take the convention that $0^0 = 1$):*

1. $[a^k b^0, 0, a^{k-1}b, 0, a^{k-2}b^2, 0, \ldots, a^0 b^k]$ *(arity $2k \ge 2$)*
2. $[a^k b^0, 0, a^{k-1}b, 0, a^{k-2}b^2, 0, \ldots, a^0 b^k, 0]$ *(arity $2k + 1 \ge 1$)*
3. $[0, a^k b^0, 0, a^{k-1}b, 0, a^{k-2}b^2, 0, \ldots, a^0 b^k]$ *(arity $2k + 1 \ge 1$)*
4. $[0, a^k b^0, 0, a^{k-1}b, 0, a^{k-2}b^2, 0, \ldots, a^0 b^k, 0]$ *(arity $2k + 2 \ge 2$).*

We will postpone the proof of this theorem until we have properly developed the theory of matchgates.

Some Problems

We list some problems that can be solved in polynomial time using holographic algorithms with matchgates. The first several problems are all from the initial paper by Valiant [Val08].

We start with a problem motivated by statistical physics. An orientation of an undirected graph G is an assignment of a direction to each of its edges. An "ice problem" involves counting the number of orientations of a graph such that certain local constraints are satisfied. Pauling [Pau35] initially proposed such a model for planar square lattices, where the constraint was that an orientation assigned exactly two incoming and two outgoing edges at every node. The question of determining how the number of such orientations grows for various planar repeating structures has been extensively studied [Lie67a, Lie67b, Lie67c, Lie67d, Bax82, Wel93].

#PL-3-NAE-ICE

INPUT: A planar graph $G = (V, E)$ of maximum degree 3

OUTPUT: The number of orientations such that no node has all incident edges directed toward it or all incident edges directed away from it

Hence #PL-3-NAE-ICE counts the number of no-sink-no-source orientations. A node of degree 1 will preclude such an orientation. We assume every node has degree 2 or 3. To solve this problem by a holographic algorithm with matchgates, we design a signature grid based on G as follows. We attach to each node of degree 3 a signature $[0, 1, 1, 0]$ (as a contravariant tensor). This represents a NOT-ALL-EQUAL or NAE gate of arity 3. For any node of degree 2 we use a binary NAE signature, i.e., a binary DISEQUALITY signature $(\neq_2) = [0, 1, 0]$ (also a contravariant tensor). For each edge in E we use a binary DISEQUALITY signature $(\neq_2)$ (as a covariant tensor), which stands for an orientation from one node to the other. (To express such a problem, it is completely arbitrary to label one side as contravariant and the other side as covariant tensors; cf. Section 1.3.2 of Chapter 1. They are called generators and recognizers , respectively.) From the given planar graph G we obtain a signature grid Ω, where the underlying graph G' is the edge–vertex incidence graph of G. By definition, Holant_Ω is an exponential sum where each term is a product of appropriate entries of the signatures. Each term is indexed by a 0-1 assignment on all edges of G'; it has value 0 or 1, and it has value 1 iff it corresponds to an orientation of G such that at every vertex of G the local NAE constraint is satisfied. Therefore Holant_Ω is precisely the number of valid orientations required by #PL-3-NAE-ICE.

Note that the signature $[0, 1, 1, 0]$ is not the signature of any matchgate according to Theorem 4.11. A simple reason for this is that a matchgate signature, being defined in terms of perfect matchings, cannot have nonzero values for inputs of both odd and even Hamming weight.

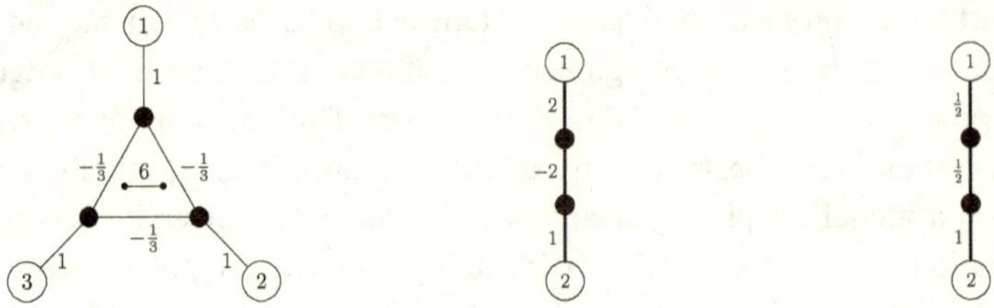

(a) A matchgate with signature $[6,0,-2,0]$, and $H[6,0,-2,0] = [0,1,1,0]$

(b) A matchgate with signature $[2,0,-2]$, and $H[2,0,-2] = [0,1,0]$. Replacing the weights from ± 2 to $\pm\frac{1}{2}$ gives $\frac{1}{2}[1,0,-1]$, and $\frac{1}{2}[1,0,-1]H^{-1} = [0,1,0]$.

(c) A matchgate with signature $\frac{1}{2}[1,0,1]$, and $\frac{1}{2}[1,0,1]H = [1,0,1]$.

Figure 4.1. Some matchgates used in #PL-3-NAE-ICE.

However, under a holographic transformation using $H = \frac{1}{2}\begin{bmatrix} 1 & 1 \\ 1 & -1 \end{bmatrix}$ with $H^{-1} = 2H$,

$$(H^{-1})^{\otimes 3}[0,1,1,0] = (H^{-1})^{\otimes 3}\left\{\begin{bmatrix} 1 \\ 1 \end{bmatrix}^{\otimes 3} - \begin{bmatrix} 1 \\ 0 \end{bmatrix}^{\otimes 3} - \begin{bmatrix} 0 \\ 1 \end{bmatrix}^{\otimes 3}\right\} = [6,0,-2,0],$$

$$(H^{-1})^{\otimes 2}[0,1,0] = H^{\otimes 2}\left\{\begin{bmatrix} 1 \\ 1 \end{bmatrix}^{\otimes 2} - \begin{bmatrix} 1 \\ 0 \end{bmatrix}^{\otimes 2} - \begin{bmatrix} 0 \\ 1 \end{bmatrix}^{\otimes 2}\right\} = [2,0,-2],$$

and

$$[0,1,0]H^{\otimes 2} = \frac{1}{2}[1,0,-1],$$

as the matrix form of this binary signature is $\frac{1}{4}\begin{bmatrix} 1 & 1 \\ 1 & -1 \end{bmatrix}\begin{bmatrix} 0 & 1 \\ 1 & 0 \end{bmatrix}\begin{bmatrix} 1 & 1 \\ 1 & -1 \end{bmatrix} = \frac{1}{2}\begin{bmatrix} 1 & 0 \\ 0 & -1 \end{bmatrix}$. These signatures are all realizable as matchgate signatures, by Theorem 4.11. More concretely, we can exhibit the requisite matchgates in Figures 4.1a and b.

Hence #PL-3-NAE-ICE is precisely the following Holant problem on planar graphs:

$$\text{Holant}([0,1,0] \mid [0,1,0],[0,1,1,0]) \equiv_T$$
$$\text{Holant}\left(\tfrac{1}{2}[1,0,-1] \mid [2,0,-2],[6,0,-2,0]\right).$$

Now we may replace each signature $\frac{1}{2}[1,0,-1]$, $[2,0,-2]$, and $[6,0,-2,0]$ in Ω by its corresponding matchgate; then we can compute Holant_Ω in polynomial time by Kasteleyn's algorithm.

An often useful perspective on holographic transformations is to think of a matrix $M = \begin{bmatrix} a & c \\ b & d \end{bmatrix} \in GL_2(\mathbb{C})$ as representing two generalized truth values of $\left[\mathbf{b}_0 = \begin{pmatrix} a \\ b \end{pmatrix}, \mathbf{b}_1 = \begin{pmatrix} c \\ d \end{pmatrix}\right]$.

We consider a covering problem next. A cycle is a sequence of edges through distinct nodes that starts and ends at the same vertex. A chain is a sequence of edges through distinct nodes that starts and ends at distinct vertices. A cycle-chain cover in $G = (V, E)$ is a set of cycles and chains that are vertex disjoint and the union of their vertex sets is all of V. For real values (x, y), the (x, y) cycle-chain sum of G is the sum of $x^i y^j$ over all cycle-chain covers C, where i and j are the numbers of cycles and chains in C respectively. Note that the $(1, 1)$ cycle-chain sum is the total number of cycle-chain covers G. Note also that the $(2, 0)$ cycle-chain sum is a weighted sum of cycle-chain covers C of G with only cycles, and the weight is 2^i if C has i cycles. In particular, $1/2$ of the $(2, 0)$ cycle-chain sum mod 2 is the parity of the number of Hamiltonian cycles of G. The number of Hamiltonian cycles is #P-hard [LOT03], and the proof can be adapted to show that if nodes of both degrees 2 and 3 are allowed, then the parity problem is $\oplus$P-complete [Val05], which is NP-hard under randomized reduction [VV86]. Similarly, the $(2, 4)$ cycle-chain sum is also $\oplus$P-hard.

#PL-3-CYCLE-CHAIN

INPUT: A planar 3-regular graph $G = (V, E)$

OUTPUT: The $(1, 1)$ cycle-chain sum of G

To solve this problem by a holographic algorithm with matchgates, we design a signature grid as follows. We represent each node of degree 3 in V by a generator with signature $[0, 1, 1, 0]$. For each edge in E we use a recognizer with a binary EQUALITY signature $(=_2) = [1, 0, 1]$, which enforces a 0-1 assignment of the edges, and is interpreted as either taking or not taking the edge in the cycle-chain cover. The signature $[0, 1, 1, 0]$ at each node of degree 3 enforces that the subgraph formed by the edges taken is a spanning subgraph (i.e., a subgraph containing all vertices) and has degree either 1 or 2 at every vertex, hence a cycle-chain cover.

By a holographic transformation using H, both $[6, 0, -2, 0] = (H^{-1})^{\otimes 3}[0, 1, 1, 0]$ and $\frac{1}{2}[1, 0, 1] = [1, 0, 1]H^{\otimes 2}$ are realizable as matchgate signatures (Figures 4.1a and c). Hence #PL-3-CYCLE-CHAIN is computable in polynomial time.

The next problem is a satisfiability problem.

#PL-3-NAE-SAT

INPUT: A planar formula Φ consisting of a conjunction of NAE clauses each of size 3

OUTPUT: The number of satisfying assignments of Φ

This is a variant of 3SAT. Lichtenstein [Lic82] defined the notion of planar formulae. A Boolean formula is planar if it can be represented by a planar graph where vertices represent variables and clauses, and there is an edge iff the variable or its negation appears in that clause. The SAT problem is when the gate for each clause is the Boolean OR. When SAT is restricted to planar formulae it is still NP-complete, and its corresponding counting problem is #P-complete. Moreover, for many connectives other than NAE (e.g., EXACTLY-ONE) the unrestricted or the planar decision problems are still NP-complete, and the corresponding counting problems are #P-complete [IMRS98].

We design a signature grid as follows. To each NAE clause we assign a generator with signature $[0, 1, 1, 0]$. To each Boolean variable we assign a generator with signature $(=_k)$, where k is the number of clauses in which the variable appears, either negated or unnegated. Further, if a variable occurrence is negated we have a recognizer $[0, 1, 0]$ along the edge that joins the variable generator and the NAE generator, and if the variable occurrence is unnegated then we use a recognizer $[1, 0, 1]$ instead. Under a holographic transformation using H, $(=_k)$ is transformed to

$$(H^{-1})^{\otimes k}\left\{\begin{bmatrix}1\\0\end{bmatrix}^{\otimes k}+\begin{bmatrix}0\\1\end{bmatrix}^{\otimes k}\right\}=\begin{bmatrix}1\\1\end{bmatrix}^{\otimes k}+\begin{bmatrix}1\\-1\end{bmatrix}^{\otimes k}=2[1,0,1,0,\ldots].$$

Thus, by Theorem 4.11, all the signatures used can be realized by matchgates under a holographic transformation.

The next problem is related to a class of problems called node deletion problems.

Generally speaking, for a graph G and a graph property Φ, one is interested in subsets of vertices S of G such that $G \setminus S$ has the property Φ, where $G \setminus S$ is the graph obtained from G by the removal of S together with all incident edges of S. Often one can also investigate the problem in a subclass of graphs G, such as planar graphs or graphs with maximum degree at most Δ, for some Δ.

The optimization problem is to find a subset of minimum cardinality that can be removed. With an additional threshold parameter k as input together with G, one can define a decision problem on whether a subset of cardinality at most k exists that can be removed in order to satisfy the property. The problem of finding a minimum cardinality subset can be reduced to the decision problem by a polynomial time Turing reduction (Cook reduction), since for the minimum parameter k the answer to the decision problem with $(G \setminus \{v\}, k-1)$ is Yes iff v belongs to some removal subset of minimum cardinality k. Hence we can find a solution subset S of minimum cardinality by using the oracle for decision problem by an iteration. Note that for any subset S' of cardinality $k-1$ such that $(G \setminus \{v\}) \setminus S'$ has property Φ, the subset $S = S' \cup \{v\}$ is a solution to G of minimum cardinality k.

Furthermore, among all removal subsets of minimum cardinality, one can define a counting problem, or a parity counting problem.

The following problem can be solved in polynomial time by holographic algorithms with matchgates.

PL-NODE-BIPARTITION

INPUT: A planar graph $G = (V, E)$ of maximum degree 3

OUTPUT: The minimum cardinality of a subset $S \subset V$ such that $G \setminus S$ is a bipartite graph

Lewis and Yannakakis [LY80] discuss a general approach to such "node deletion" problems. See also [KD79, Yan81a, Yan81b]. If we consider edge deletion instead of node deletion, then this is just another way of defining the problem MAX-CUT, which is NP-hard (and even NP-hard to approximate [GW95, ALM+98, Hås01]). Hadlock proved that planar MAX-CUT is solvable in polynomial time [Had75]. Choi et al. [CNR89] showed that for maximum degree ≤ 3, the minimization problems for node deletion and for edge deletion are equivalent with respect to bipartiteness. Thus PL-NODE-BIPARTITION is also solvable in polynomial time. Choi et al. also proved that for planar graphs with maximum degree ≤ 4, or for general graphs with maximum degree ≤ 3, the decision problem is NP-complete. This improves an earlier result by Krishnamoorthy and Deo [KD79] that the problem is NP-complete for planar graphs with maximum degree ≤ 6.

We can solve the problem PL-NODE-BIPARTITION using a holographic algorithm with matchgates as follows. We may assume all vertices are of degree 2 or 3, as retaining any vertex of degree 1 does not violate bipartiteness. We will use generator signatures $[x, 1, x]$ and $[x, 1, 1, x]$ for a vertex of degree 2 and 3 respectively. These are realizable by matchgates under H. For each edge we use a recognizer signature $[0, 1, 0]$; this results in an orientation. The recognizer signature $[0, 1, 0]$ is also realizable by matchgates under H^{-1}. For any orientation, each source or sink node contributes a factor x, and the other nodes contribute a factor 1. The Holant sum is a polynomial in x, where the degree, i.e., the largest i for which x^i has a nonzero coefficient, is exactly the maximum number of nodes that a bipartite graph can have that is obtained by deleting nodes and incident edges from G.

The degree of this polynomial is at most $|V|$. Thus by evaluating the Holant at $|V|+1$ distinct values of x, we can interpolate the polynomial and reconstruct it. Consequently we can find its actual degree d, which gives $|V|-d$ as the answer to the instance of PL-NODE-BIPARTITION.

Exercise: Show that, for any x, the signatures $[x, 1, x]$ and $[x, 1, 1, x]$ can be realized by matchgates under H.

Notice that once the minimum cardinality k is obtained to an instance G of PL-NODE-BIPARTITION, we can further compute the parity of the number of subsets S of cardinality k such that S is an independent set and $G \setminus S$ is a connected bipartite graph.

Let S be a subset of cardinality k such that $G \setminus S$ is bipartite. Since k is minimum, any vertex $v \in S$ must have edges connecting to both sides of the bipartite graph $G \setminus S$; otherwise one can keep v and $S - \{v\}$ is a solution to PL-NODE-BIPARTITION with cardinality $k-1$, a contradiction. Let $G[S]$ be the induced subgraph of G on S, then $G[S]$ is a matching, i.e., every vertex in S has degree at most 1 in $G[S]$. Suppose $G[S]$ has m edges. Let κ be the number of connected components of $G \setminus S$. We claim that there are exactly $2^{m+\kappa}$ orientations counted in the coefficient of x^{n-k} of the Holant sum that correspond to S being a solution to PL-NODE-BIPARTITION. Clearly any orientation that corresponds to S being a solution is counted in the $2^{m+\kappa}$ distinct orientations that are obtained by independently flipping the orientations on the m matching edges, and for each connected component C of $G \setminus S$, flipping all edges $\{(u, v) \mid u \in C \text{ or } v \in C\}$. Note that these edges include any edge connecting a sink or a source node of C with a node in S. However, under an arbitrary flip, for any connected component C, the sink nodes and source nodes become source nodes and sink nodes respectively, and the nodes in S must remain nonsink nonsource nodes, because $|S|$ is minimum.

Clearly, S is an independent set iff $m = 0$ in $G[S]$, and $G \setminus S$ is connected iff $\kappa = 1$. Therefore every independent set S of $|S| = k$ such that $G \setminus S$ is connected and bipartite is counted exactly twice in the Holant sum in the coefficient of x^{n-k}, and all other S of $|S| = k$ contribute a number congruent to 0 mod 4. Hence if we divide this coefficient by 2 and then take the number mod 2, we get the parity of the number of independent sets $S \subset V$ such that $|S| = k$ and $G \setminus S$ is a connected bipartite graph. This solves the problem in polynomial time.

In [Val10], Valiant presents several interesting problems of this kind.

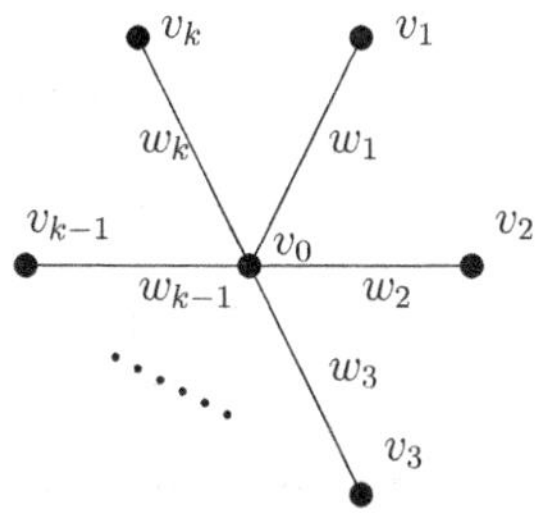

Figure 4.2. A matchgate with an underlying star graph

We now consider a matching problem. Jerrum [Jer87] showed that counting the number of (not necessarily perfect) matchings in a planar graph is #P-complete, and Vadhan [Vad01] subsequently proved that it remains #P-complete even for planar bipartite graphs of degree 6. For degree 2 the problem can be solved easily and one might have conjectured that all other nontrivial cases are #P-complete. However, the following problem [Val08] can be solved in polynomial time.

#X-MATCHINGS

INPUT: A planar weighted bipartite graph $G = (V, E, W)$, where V has bipartition V_1, V_2 and the nodes in V_1 have degree 2

OUTPUT: The sum of the masses of all matchings of all sizes where the mass of a matching is the product of (i) the weights of all the edges present in the matching, and (ii) the quantity $-(w_1 + \cdots + w_k)$ for all the V_2 nodes that are not matched, where $w_1, \ldots, w_k$ are the weights of the edges incident to that (unmatched) node

Note that if every V_2 node has degree 4 and every edge has weight 1, then computing #X-MATCHINGS gives the number of matchings, but each weighted by $(-4)^k$, where k is the number of unmatched V_2 nodes. Computing this mod 5 gives the number of matchings mod 5.

To solve this problem, we use a different matrix $M = \begin{bmatrix} -1 & 1 \\ 1 & 0 \end{bmatrix}$. Consider the matchgate pictured in Figure 4.2. This star graph has a central vertex v_0 and k external nodes labeled $v_1, \ldots, v_k$, and the edge (v_0, v_i) has weight w_i $(1 \leq i \leq k)$. Its signature Γ has nonzero entries only at weight $k - 1$, and the value $\Gamma^\alpha = w_i$, where $\alpha = 1^{i-1}01^{k-i}$ has a single 0 at bit position i. The covariant transformation is $\Gamma M^{\otimes k} = \sum_{i=1}^{k} w_i (1, 0)^{\otimes(i-1)} \otimes (-1, 1) \otimes (1, 0)^{\otimes(k-i)}$. This signature has nonzero values $-(w_1 + \cdots + w_k)$ at 0^k, and w_i at $0^{i-1}10^{k-i}$. All other entries of $\Gamma M^{\otimes k}$ are 0.

To each vertex in V_2 we attach the recognizer signature $\Gamma M^{\otimes k}$ above. To each vertex in V_1 we attach the AT-MOST-ONE signature $[1, 1, 0]$ as a generator. Under a contravariant transformation $(M^{-1})^{\otimes 2}[-1, 0, 1] = [1, 1, 0]$. We verify that $M^{\otimes 2}[1, 1, 0] = [-1, 0, 1]$ by its matrix form

$$\begin{bmatrix} -1 & 1 \\ 1 & 0 \end{bmatrix} \begin{bmatrix} 1 & 1 \\ 1 & 0 \end{bmatrix} \begin{bmatrix} -1 & 1 \\ 1 & 0 \end{bmatrix} = \begin{bmatrix} -1 & 0 \\ 0 & 1 \end{bmatrix},$$

and $[-1, 0, 1]$ can be realized by a matchgate consisting of a path of length 3, where the two end vertices are external nodes, and the weights of the three edges are respectively 1, 1 and -1 along the path.

This problem is motivated by its proximity to counting the number of (not necessarily perfect) matchings in a planar graph, which is #P-complete [Jer87]. Still, the quantity $-(w_1 + \cdots + w_k)$ seems a little artificial. If one were to be able to replace $-(w_1 + \cdots + w_k)$ by 1, then one would be able to count all (not necessarily perfect) matchings in such planar bipartite graphs. Later in this chapter, we will develop the theory of matchgates to address the problem of when a desired set of signatures can be realized by a holographic algorithm with matchgates. This theory will show that one cannot replace $-(w_1 + \cdots + w_k)$ by 1.

Now we consider a curious counting satisfiability problem.

$\#_7$PL-RTW-MON-3CNF

INPUT: A planar 3CNF Boolean formula where each variable appears positively and in exactly two clauses (planar, read-twice, monotone, 3CNF)

OUTPUT: The number of satisfying assignments modulo 7

Let us first consider simply the counting problem for this restricted class of Boolean formulae, which is denoted as #PL-RTW-MON-3CNF. This problem is known to be #P-complete. Its graphic representation is simply a 2-3-regular graph, i.e., a bipartite graph where every vertex on the left has degree 2 representing a Boolean variable, and every vertex on the right has degree 3 representing an OR clause. For a given planar, read-twice, monotone 3CNF formula Φ, we design a signature grid by assigning the binary EQUALITY signature $[1, 0, 1]$ as a generator for each variable, and the Boolean OR signature on three variables $[0, 1, 1, 1]$ as a recognizer on each clause. We will consider these signatures over the finite field $\mathbb{Z}_7$. Let $M = \begin{bmatrix} 4 & 1 \\ 5 & 1 \end{bmatrix}$, with inverse $M^{-1} = \begin{bmatrix} -1 & 1 \\ 5 & -4 \end{bmatrix}$. We verify that, over $\mathbb{Z}_7$,

$$\begin{aligned} [0, 1, 1, 1](M^{-1})^{\otimes 3} &= [(1, 1)^{\otimes 3} - (1, 0)^{\otimes 3}](M^{-1})^{\otimes 3} \\ &= (4, -3)^{\otimes 3} - (-1, 1)^{\otimes 3} = [2, 0, 2, 0]. \end{aligned}$$

Meanwhile $M^{\otimes 2}[1, 0, 1]$ has its matrix form

$$MI_2M^{\mathrm{T}} = \begin{bmatrix} 4 & 1 \\ 5 & 1 \end{bmatrix} \begin{bmatrix} 4 & 5 \\ 1 & 1 \end{bmatrix} = \begin{bmatrix} 3 & 0 \\ 0 & 5 \end{bmatrix} = [3, 0, 5].$$

Both $[2, 0, 2, 0]$ and $[3, 0, 5]$ can be realized as matchgate signatures, by Theorem 4.11.

We remark that $\#_2$PL-RTW-MON-3CNF, the parity version of the counting problem #PL-RTW-MON-3CNF, is $\oplus$P-complete [Val06], thus NP-hard under randomized reductions [VV86]. The polynomial time solvability for mod 7 is a surprise, especially in view of its hardness mod 2. In this chapter we will develop the theory to give an explanation of this mystery.

Finally we consider a geometric problem.

2-COLOR-COUNTING

INPUT: A set S of n points on a plane is given, where no three points are colinear. Also given is a set of line segments between some pairs of points in S. We assume no three line segments intersect at a point not in S, and every point of S is incident to either two or three line segments.

OUTPUT: The number of 2-colorings for the line segments that satisfy the following conditions: (1) for every point in S, the incident line segments are not monochromatic; (2) when two line segments cross each other, they have different colors.

We use the NAE generator signatures $[0, 1, 0]$ or $[0, 1, 1, 0]$ for each point with either two or three incident line segments respectively. For each intersection of two line segments we use the generator signature $G = (H^{-1})^{\otimes 4}\Gamma$, where Γ is the signature of the matchgate given in Figure 4.3, and $H = \begin{bmatrix} 1 & 1 \\ 1 & -1 \end{bmatrix}$. G has exactly two nonzero values $G^{0101} = G^{1010} = 1$. This ensures that the coloring on the line segment is properly propagated past an intersection, and two intersecting line segments are colored differently. For each line segment between two consecutive points which are from S or are intersection points, we add a binary EQUALITY recognizer signature $(=_2)$. Under a holographic transformation by H (which is symmetric and, up to a scalar, orthogonal and thus its own inverse), $[0, 1, 0]$, $[0, 1, 1, 0]$ and $(=_2) = [1, 0, 1]$ are all realizable. It follows that the problem 2-COLOR-COUNTING can be solved in polynomial time.

The fact that the matchgate signature Γ is as stated can be directly verified. However, after Section 4.3.2 it will become clear why this construction works.

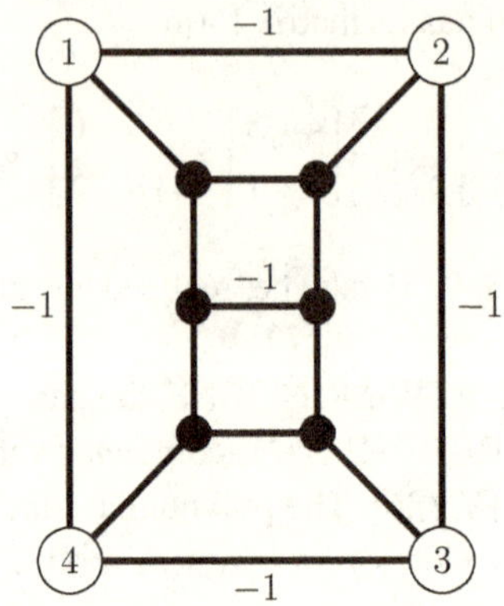

Figure 4.3. A matchgate with signature Γ indexed by $\{0, 1\}^4$. Edges without explicitly labeled weight all have weight 1. Listed lexicographically $\Gamma = (1, 0, 0, -1, 0, 1, -1, 0, 0, -1, 1, 0, -1, 0, 0, 1)^{\mathrm{T}}$. Also see Figure 4.5 in Section 4.3.2. Under the holographic transformation H this matchgate has signature $G = (H^{-1})^{\otimes 4}\Gamma$, up to a nonzero factor, $G^{0101} = G^{1010} = 1$ and $G^{\alpha} = 0$ for other $\alpha \in \{0, 1\}^4$.

Exercise: Show that we may compute the signature $G = (H^{-1})^{\otimes 4}\Gamma$ in problem 2-COLOR-COUNTING as follows. The 4×4 matrix form of G is the product $(H^{-1})^{\otimes 2}M(\Gamma)((H^{-1})^{\otimes 2})^{\mathrm{T}}$, where $M(\Gamma)$ is the 4×4 matrix form of Γ which has rows and columns indexed by $\{0, 1\}^2$, in lexicographic order, and has entry Γ^{ijkl} in entry at row (ij) and column (kl). Conclude that the 4×4 matrix form of G is, up to a nonzero scalar,

$$\begin{bmatrix} 0 & 0 & 0 & 0 \\ 0 & 1 & 0 & 0 \\ 0 & 0 & 1 & 0 \\ 0 & 0 & 0 & 0 \end{bmatrix}.$$

The two nonzero entries are indexed by 0101 and 1010 respectively. Written with the column listed in reverse order it is, up to a nonzero scalar,

$$\begin{bmatrix} G^{0000} & G^{0010} & G^{0001} & G^{0011} \\ G^{0100} & G^{0110} & G^{0101} & G^{0111} \\ G^{1000} & G^{1010} & G^{1001} & G^{1011} \\ G^{1100} & G^{1110} & G^{1101} & G^{1111} \end{bmatrix} = \begin{bmatrix} 0 & 0 & 0 & 0 \\ 0 & 0 & 1 & 0 \\ 0 & 1 & 0 & 0 \\ 0 & 0 & 0 & 0 \end{bmatrix}.$$

For most holographic algorithms that have been studied, the signatures needed are symmetric signatures. For this problem we used an asymmetric signature G. Holographic algorithms that employ asymmetric signatures have not been extensively studied, and will be an interesting topic for the future. Another notable feature about 2-COLOR-COUNTING is that the problem is not a priori stated for a planar graph. It is in the process of forming the signature grid that we obtain a planar graph.

4.3 The Theory of Matchgates

In this section we develop the theory of matchgates. The main result is that there is a set of identities, called matchgate identities, that characterize matchgate signatures.

4.3.1 Matchgate Identities

Without loss of generality we may assume the underlying graph of any matchgate is connected. If it is not connected and there are several connected components G_i, then consider a clockwise traversal of all the external nodes. We may consider the planar embedding is on the sphere with one fixed point in the outer face designated as ∞. We temporarily connect each external node to ∞ by nonintersecting paths. As we clockwise-traverse from one external node u to the next v, if they belong to different components G_i and G_j, we can connect u to v by a path of length 2: $u, e = \{u, w\}, w, e' = \{w, v\}, v$, adding a new node w, and one extra node w' with an edge $\{w, w'\}$. This gadget can be made disjoint from all the temporary paths to ∞, and also disjoint from each other. All new edges (there are three edges on each such gadget) have weight 1. In any perfect matching, w is matched to w' and therefore this gadget has no effect on the signature. Then we remove the temporary paths to ∞. Now the matchgate graph is (1) connected, and (2) its outer face uniquely well-defined for the given planar embedding, with a connected boundary.

By default we assume the edge weights are complex numbers in Definitions 4.8 and 4.9. Thus $\Gamma_G \in \mathbb{C}^{2^k}$, or equivalently we may consider it as a tensor in $(\mathbb{C}^2)^{\otimes k}$. But the theory can be developed in any commutative ring, in particular in a finite field. We will make occasional use of that.

Matchgate signatures satisfy a remarkable set of identities, called the *matchgate identities*. We will show that these identities are a necessary and sufficient condition for a vector of values to be the signature of a matchgate. We state the matchgate identities, or MGI.

Theorem 4.12. *Let Γ be the signature of a matchgate with k external nodes. For any length-k bitstrings $\alpha, \beta \in \{0, 1\}^k$, let $P = \{p_1, \ldots, p_l\}$, where*

$p_1 < \cdots < p_l$, be the subset of $[k]$ such that p_i is the ith bit where α and β differ. Then, the signature Γ satisfies

$$\sum_{i=1}^{l}(-1)^i \Gamma^{\alpha \oplus e_{p_i}} \Gamma^{\beta \oplus e_{p_i}} = 0, \tag{4.10}$$

where e_j denotes a length-k bitstring with a 1 in the jth index, and 0 elsewhere.

We will identify P with its characteristic sequence, which is $\alpha \oplus \beta \in \{0, 1\}^k$, the bitwise XOR of α and β.

Exercise: Show that, for matchgates of arity at most 3, the parity condition is sufficient. For arity 4, there is a single MGI that is sufficient together with the parity condition.

A perfect matching has an even number of vertices. It follows that $\text{PerfMatch}(G^\alpha) = 0$, whenever G^α has an odd number of vertices. Thus, a matchgate signature satisfies the following:

> *Parity Condition*: For any matchgate signature Γ, either for all α of odd Hamming weight, or for all α of even Hamming weight, $\Gamma^\alpha = 0$.

The parity condition is a consequence of MGI.

Theorem 4.13. *If a vector Γ obeys the MGI, then it also obeys the parity condition.*

Proof. For a contradiction assume $\Gamma^\alpha \neq 0$ and $\Gamma^\beta \neq 0$, for some α and β of even and odd Hamming weight respectively. We define $\widetilde{\Gamma}$ by $\widetilde{\Gamma}^\gamma = \Gamma^{\gamma \oplus \alpha}$. Since $\gamma \oplus \gamma' = (\gamma \oplus \alpha) \oplus (\gamma' \oplus \alpha)$, if the vector Γ obeys the MGI then the vector $\widetilde{\Gamma}$ also obeys the MGI. Also $\widetilde{\Gamma}^{00\ldots0} = \Gamma^\alpha \neq 0$ and $\widetilde{\Gamma}^{\alpha \oplus \beta} = \Gamma^\beta \neq 0$. Note that $\alpha \oplus \beta$ has an odd Hamming weight.

Let $\beta' = \{p_1, \ldots, p_l\}$ be of minimum odd Hamming weight such that $\widetilde{\Gamma}^{\beta'} \neq 0$. Now invoke the MGI on the bitstrings $00\ldots0 \oplus e_{p_1}$ and $\beta' \oplus e_{p_1}$. That gives

$$0 = -\widetilde{\Gamma}^{00\ldots0}\widetilde{\Gamma}^{\beta'} + \sum_{i=2}^{l}(-1)^i \widetilde{\Gamma}^{00\ldots0 \oplus e_{p_1} \oplus e_{p_i}} \widetilde{\Gamma}^{\beta' \oplus e_{p_1} \oplus e_{p_i}}. \tag{4.11}$$

We have $\widetilde{\Gamma}^{00\ldots0}\widetilde{\Gamma}^{\beta'} \neq 0$. If $l = 1$ then the sum $\sum_{i=2}^{l}$ is vacuous, and we have a contradiction. So $l \geq 2$ and we consider each term in the sum $\sum_{i=2}^{l}$. Observe that for every $2 \leq i \leq l$, $\beta' \oplus e_{p_1} \oplus e_{p_i}$ has an odd Hamming weight *less* than that of β', hence $\widetilde{\Gamma}^{\beta' \oplus e_{p_1} \oplus e_{p_i}} = 0$. Thus the sum $\sum_{i=2}^{l}$ is 0 but $\widetilde{\Gamma}^{00\ldots0}\widetilde{\Gamma}^{\beta'} \neq 0$, a contradiction. $\square$

Nonetheless, because the parity condition is a set of linear equations, and the MGI are quadratic, in practice the parity condition is a good criterion to apply first.

Exercise: Show that, for matchgates of arity at most 3, the parity condition also implies MGI. For signatures of arity 4 satisfying the parity condition, there is a single MGI that is sufficient to imply all MGI.

MGI were first introduced by Valiant in [Val02a, Val02b] in the context of proving certain 2-input 2-output quantum gates cannot be realized by a matchgate. It was shown that 2-input 2-output matchgates must satisfy certain identities that are named the matchgate identities. These identities are actually concerned with *characters* of matchgates. These so-called characters are defined directly in terms of Pfaffians, and their underlying matchgates need not be planar by definition. In the case of 2-input 2-output matchgates, these character values constitute a 4×4 matrix, called a character matrix. Subsequently [CC06, CCL09, CG14] this theory is generalized to matchgates of an arbitrary number of external nodes. The ultimate result is that there is an equivalence of matchgate characters (of not necessarily planar matchgates) and matchgate signatures (of planar matchgates). Furthermore, matchgate identities characterize matchgate signatures. Theorem 4.12 states that MGI is a necessary condition. Theorem 4.18 states that MGI is also sufficient and will be proved in Section 4.3.2.

The matchgate identities are inspired by a set of identities called Pfaffian signature identities. These identities have the same form as MGI but apply to Pfaffian signatures. The original discovery and proof of MGI are based on the Pfaffian signature identities, but we will present with permission a short and elegant bijective proof of MGI due to Jerrum [Jer15].

If we fix any Pfaffian orientation for G, it defines a directed graph $\overrightarrow{G}$. Then $\overrightarrow{G}^{\alpha}$, which is obtained from $\overrightarrow{G}$ by removing some vertices and their incident edges according to α, is *also* Pfaffian oriented. This is because we remove zero or more vertices only on the outer face, and the removal of these vertices and their incident edges does not create any *non-outer face*. Thus a single fixed Pfaffian orientation for G induces a set of Pfaffian orientations, one for each G^{α}. We assume a Pfaffian orientation for G is fixed, and each G^{α} inherits the induced Pfaffian orientation.

As the orientation in $\overrightarrow{G}$ induces a Pfaffian orientation for all G^{α}, we can naturally refer to the oriented graph $\overrightarrow{G^{\alpha}} = \overrightarrow{G}^{\alpha}$. The skew-symmetric matrix $B(\overrightarrow{G^{\alpha}})$ is obtained from $B(\overrightarrow{G})$ (see (4.4)) by removing the appropriate rows and columns indicated by the 1's in α. We abbreviate $\mathrm{Pf}(B(\overrightarrow{G^{\alpha}}))$ as $\mathrm{Pf}^{\alpha}_{\overrightarrow{G}}$. Where $\overrightarrow{G}$ is clear, we just write Pf^{α}. With a given Pfaffian orientation on the plane graph

G, and a given labeling of its k external nodes in clockwise order, we define the Pfaffian signature of $\overrightarrow{G}$ as follows.

Definition 4.14. The Pfaffian signature of $\overrightarrow{G}$ is the vector (Pf^α) indexed by $\alpha \in \{0,1\}^k$.

Each Pfaffian signature entry Pf^α is a sum of Pfaffian terms (see (4.5) and (4.6)) for $\overrightarrow{G^\alpha}$. Each nonzero Pfaffian term $\mathrm{Pf}_{\overrightarrow{G^\alpha}}(M)$ corresponds to a perfect matching $M \in \mathcal{M}(G^\alpha)$.

By the property of a Pfaffian orientation, $\mathrm{Pf}(B(\overrightarrow{G})) = \pm\mathrm{PerfMatch}(G)$ is a term by term equation. Thus, for every $\alpha \in \{0,1\}^k$, there exists $\epsilon(\alpha) \in \{-1,1\}$, such that for all $M \in \mathcal{M}(G^\alpha)$,

$$\mathrm{Pf}_{\overrightarrow{G^\alpha}}(M) = \epsilon(\alpha)\Gamma_{G^\alpha}(M). \tag{4.12}$$

The sign $\epsilon(\alpha)$ is the same for every $M \in \mathcal{M}(G^\alpha)$.

Theorem 4.15. *Let $\overrightarrow{G}$ be a plane graph with a Pfaffian orientation and k external nodes. For any length-k bitstrings $\alpha, \beta \in \{0,1\}^k$, let $P = \{p_1, \ldots, p_l\}$, where $p_1 < \cdots < p_l$, be the subset of $[k]$ whose characteristic sequence is $\alpha \oplus \beta$. Then,*

$$\sum_{i=1}^{l}(-1)^i \mathrm{Pf}^{\alpha\oplus e_{p_i}}\mathrm{Pf}^{\beta\oplus e_{p_i}} = 0. \tag{4.13}$$

Theorem 4.15 will follow from the Grassmann–Plücker identities over Pfaffian minors of a matrix. We state the following definition of the Grassmann–Plücker identities for a skew-symmetric matrix A. In writing $\mathrm{Pf}(i_1, i_2, \ldots, i_K)$ we mean the Pfaffian of the $K \times K$ matrix whose rows and columns are the $i_1, i_2, \ldots, i_K$th rows and columns of A, in that order. The order matters: $\mathrm{Pf}(i_1, i_2, \ldots) = -\mathrm{Pf}(i_2, i_1, \ldots)$, for instance. In particular, if there are two identical rows and columns, the Pfaffian is 0. When we write $\mathrm{Pf}(i_1, i_2, \ldots, \widehat{i_k}, \ldots, i_K)$, the $\widehat{i_k}$ means that i_k is explicitly excluded from that list.

Theorem 4.16. (The Grassmann–Plücker Identities). *Let $I = \{i_1, i_2, \ldots, i_K\}$, $J = \{j_1, j_2, \ldots, j_L\}$ be subsets of indices of a skew-symmetric A, where $i_1 < i_2 < \cdots < i_K$ and $j_1 < j_2 < \cdots < j_L$. Then*

$$\sum_{\ell=1}^{L}(-1)^{\ell-1}\mathrm{Pf}(j_\ell, i_1, \ldots, i_K)\mathrm{Pf}(j_1, \ldots, \widehat{j_\ell}, \ldots, j_L)$$

$$+\sum_{k=1}^{K}(-1)^{k-1}\mathrm{Pf}(i_1, \ldots, \widehat{i_k}, \ldots, i_K)\mathrm{Pf}(i_k, j_1, \ldots, j_L) = 0. \tag{4.14}$$

Theorem 4.16 has the following short proof [Val02b, Mur00] originally from [Oht92].

Proof of Theorem 4.16. From the definition of Pfaffian:

$$\mathrm{Pf}(j_\ell, i_1, \ldots, i_K) = \sum_{k=1}^{K}(-1)^{k-1}\mathrm{Pf}(j_\ell, i_k)\mathrm{Pf}(i_1, \ldots, \widehat{i_k}, \ldots, i_K)$$

$$\mathrm{Pf}(i_k, j_1, \ldots, j_L) = \sum_{\ell=1}^{L}(-1)^{\ell-1}\mathrm{Pf}(i_k, j_\ell)\mathrm{Pf}(j_1, \ldots, \widehat{j_\ell}, \ldots, j_L)$$

and also $\mathrm{Pf}(j_\ell, i_k) + \mathrm{Pf}(i_k, j_\ell) = 0$. The proof is completed by substituting these into (4.14). □

There is another form of these identities that is more closely related to the Pfaffian signature identities (4.13).

Theorem 4.17. *Let A, I, J be as in Theorem 4.16. For a subset S of indices of A, we write $\mathrm{Pf}(S)$ when S is listed in increasing order. Let $I \Delta J = \{k_1, \ldots, k_m\}$ (listed in increasing order) be the symmetric difference of I and J. Then*

$$\sum_{s=1}^{m}(-1)^{s-1}\mathrm{Pf}(I\Delta\{k_s\})\mathrm{Pf}(J\Delta\{k_s\}) = 0. \tag{4.15}$$

Proof of Theorem 4.17. We prove Theorem 4.17 by Theorem 4.16.

Consider a term in (4.14), and let x be the element that is moved from the index set of one Pfaffian to another. If $x \in I \cap J$, clearly the term is 0. It follows that there is a one-to-one correspondence between the remaining terms in (4.14) and the terms in (4.15). All we need to show is that each such term in (4.14) has the same sign as its counterpart in (4.15).

Suppose $x \in J - I$. In that case, the term in (4.14) is

$$(-1)^z\mathrm{Pf}(x, i_1, \ldots, i_K)\mathrm{Pf}(j_1 \ldots, \hat{x}, \ldots, j_L),$$

where z is the number of elements in J less than x. We write $z = a + b$, where

$$\begin{aligned} a &= |\{y \mid y \in J - I, y < x\}| \\ b &= |\{y \mid y \in J \cap I, y < x\}|, \end{aligned}$$

and we also define

$$c = |\{y \mid y \in I - J, y < x\}|.$$

We can put the indices in $\mathrm{Pf}(x, i_1, \ldots, i_K)$ in increasing order by moving x along until it is in its proper place. To do this we move x exactly $b + c$ entries. Thus

$$\mathrm{Pf}(x, i_1, \ldots i_K) = (-1)^{b+c}\mathrm{Pf}(I \cup \{x\})$$

and so it follows that

$$(-1)^z\mathrm{Pf}(x, i_1, \ldots i_K) = (-1)^{a+c}\mathrm{Pf}(I \cup \{x\}).$$

It is clear that $a + c$ is precisely the number of those in $I \Delta J$ less than x, exactly the sign in front of the corresponding term in (4.15).

The argument for the case $x \in I - J$ is symmetric. □

Now we are ready to prove Theorem 4.15.

Proof of Theorem 4.15. We prove Theorem 4.15 by Theorem 4.17. For a matchgate G, let α and β be two bitstrings of length k, where k is the number of external nodes in G. The ith bit of α, denoted as α_i, corresponds to the ith external node in G in clockwise order.

Let U be the set of all internal (i.e., not external) nodes in G. We define $I = \{v_i \mid \alpha_i = 0\} \cup U$, where v_i is the label of the ith external node in G. Similarly let $J = \{v_i \mid \beta_i = 0\} \cup U$. Observe that $I \Delta J = \{v_i \mid \alpha_i \neq \beta_i\}$. It follows that there is a term-for-term correspondence between (4.15) of Theorem 4.17 and (4.13) of Theorem 4.15. □

Now we present a bijective proof of MGI , Theorem 4.12. This proof is due to Jerrum [Jer15].

We first note that MGI trivially hold if l is odd. In this case, α and β have opposite parity, and so does $\alpha \oplus e_{p_i}$ and $\beta \oplus e_{p_i}$ for every p_i, and thus one of $G^{\alpha \oplus e_{p_i}}$ and $G^{\beta \oplus e_{p_i}}$ has an odd number of nodes, and hence MGI trivially holds by the parity condition. So we may assume l is even.

As before, we denote by $\mathcal{M}(G)$ the set of all perfect matchings of the graph G. We will establish a weight-preserving bijection

$$\bigcup_{i \text{ odd}} \left[\mathcal{M}(G^{\alpha \oplus e_{p_i}}) \times \mathcal{M}(G^{\beta \oplus e_{p_i}})\right] \longleftrightarrow \bigcup_{j \text{ even}} \left[\mathcal{M}(G^{\alpha \oplus e_{p_j}}) \times \mathcal{M}(G^{\beta \oplus e_{p_j}})\right].$$

More precisely, note that the MGI can be stated as

$$\sum_{1 \le i \le l,\ i \text{ odd}} \Gamma^{\alpha \oplus e_{p_i}} \Gamma^{\beta \oplus e_{p_i}} = \sum_{1 \le j \le l,\ j \text{ even}} \Gamma^{\alpha \oplus e_{p_j}} \Gamma^{\beta \oplus e_{p_j}}. \tag{4.16}$$

The product $\Gamma^{\alpha \oplus e_{p_i}} \Gamma^{\beta \oplus e_{p_i}}$ is a sum over all pairs $(M, M') \in \mathcal{M}(G^{\alpha \oplus e_{p_i}}) \times \mathcal{M}(G^{\beta \oplus e_{p_i}})$ of the product of their weight $w(M)w(M')$.

Take any $(M, M') \in \mathcal{M}(G^{\alpha \oplus e_{p_i}}) \times \mathcal{M}(G^{\beta \oplus e_{p_i}})$, consider their XOR $M \oplus M'$. Since M and M' are matchings in G, $M \oplus M'$ consists of a set of alternating paths and alternating cycles of even length. Since $\alpha_{p_i} \neq \beta_{p_i}$, there is an alternating path starting at p_i in $M \oplus M'$. This alternating path must end in some external node p_j. By planarity of G, there must be an even number of external nodes between i and j in cyclic order, among $\{p_1, p_2, \ldots, p_l\} - \{p_i, p_j\}$. This is because each external node in this list is the initial node of an alternating path in $M \oplus M'$, and must end at a node on this list. Alternating paths in $M \oplus M'$ do not cross paths in a plane graph G. An alternating path starting at any p_s where s is between i and j in cyclic order, either $i < s < j$ or $j < s < i$, must end in a node p_t with $i < t < j$ or $j < t < i$ respectively.

If we flip the edges of M and M' along the alternating path from p_i to p_j we obtain another pair $(\widetilde{M}, \widetilde{M'}) \in \mathcal{M}(G^{\alpha \oplus e_{p_j}}) \times \mathcal{M}(G^{\beta \oplus e_{p_j}})$. It is clearly an *involution*, i.e., this mapping is its own inverse. This sets up a bijective map between $\bigcup_{i \text{ odd}} [\mathcal{M}(G^{\alpha \oplus e_{p_i}}) \times \mathcal{M}(G^{\beta \oplus e_{p_i}})]$ and $\bigcup_{j \text{ even}} [\mathcal{M}(G^{\alpha \oplus e_{p_j}}) \times \mathcal{M}(G^{\beta \oplus e_{p_j}})]$.

It is clear that the product of weight $w(M)w(M')$, which is just a product of all edge weights of the alternating paths and cycles of $M \oplus M'$ (together with $w(e)^2$ from length 2 "cycles" for every edge $e \in M \cap M'$), is identical to $w(\widetilde{M})w(\widetilde{M'})$. The bijective map preserves the weight in the two sides of (4.16). This proves Theorem 4.12.

4.3.2 MGI Imply Matchgate-Realizable

Theorem 4.12 showed that the signature of any matchgate must satisfy the MGIs. In this subsection, we show that any $\Gamma \in (\mathbb{C}^2)^{\otimes k} = \mathbb{C}^{2^k}$ satisfying the MGIs can be realized as the signature of a matchgate with k external nodes. Thus MGIs are not only necessary but also sufficient conditions for matchgate signatures.

Consider a length 2^k vector Γ indexed by $\{0, 1\}^k$ satisfying MGI. If it is the all-zeros vector then it is trivially realizable. So assume there is at least one nonzero value.

Preprocessing Assume $\Gamma^\beta \neq 0$, for some $\beta \in \{0, 1\}^k$. Define $\Gamma'^\alpha = \Gamma^{\alpha \oplus \overline{\beta}} / \Gamma^\beta$, where $\overline{\beta} = \beta \oplus 11\ldots1$. Thus, $\Gamma'^{11\ldots1} = 1$, and Γ' also satisfies MGI. In this section we will create a matchgate G' with signature Γ'. Given such a G', we can create a matchgate G with signature Γ as follows. First we add two new nonexternal nodes u, v to G' and an edge $\{u, v\}$ of weight Γ^β. Those two nodes are not connected to any other nodes – in effect they contribute exactly a factor Γ^β to each perfect matching term. Then, if the ith bit of β is 0, we add a new edge $\{v_i, v_i'\}$ of weight one to the ith external node v_i,

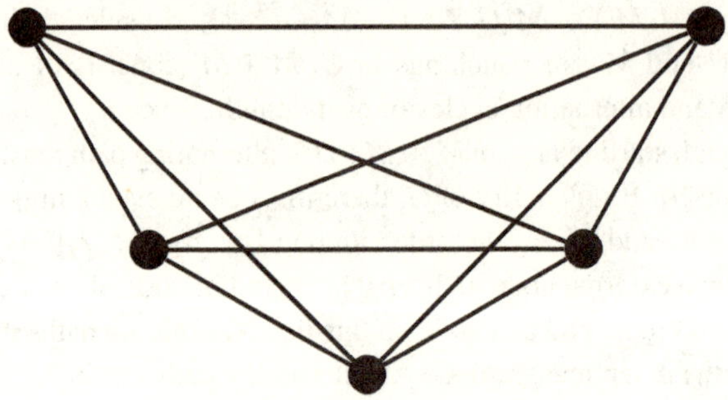

Figure 4.4. The embedding for K_5.

and making v'_i the new ith external node. It follows that the signature of G is exactly Γ.

Construction We now show that we can realize Γ satisfying MGI and $\Gamma^{11\ldots1} = 1$. Let K_k denote the complete graph on k vertices. The labels of K_k are ordered $1 < 2 < \cdots < k$, and correspond to the bit positions in the index for Γ. We place the nodes of K_k on a convex curve, as illustrated in Figure 4.4. The nodes are arranged in clockwise order by their index, and two edges cross each other geometrically in the drawing of the graph iff their labels form an overlapping pair as defined before algebraically. (We assume the k nodes are placed in general position, so that any pair of crossing edges intersect at a unique point. There are exactly $\binom{k}{4}$ such intersection points.) For each α of Hamming weight $k-2$, note that K_k^α has exactly one edge left. For each such α, set the weight of the unique edge in K_k^α to be Γ^α. This defines a weight for every edge of K_k.

Equality with Pfaffian Clearly our embedding of K_k is not planar for a general $k \geq 4$. We first prove the following equality. Let $\mathrm{Pf}(K_k^\alpha)$ be the Pfaffian value of the skew-symmetric matrix representing K_k^α, where the nodes of K_k^α have the induced order $1 < 2 < \cdots < k$. Then for all $\alpha \in \{0,1\}^k$:

$$\mathrm{Pf}\big(K_k^\alpha\big) = \Gamma^\alpha. \qquad (4.17)$$

It follows that the $\binom{k}{2}$ edge weights of K_k determine the 2^k values of any Γ satisfying MGI.

Equation (4.17) holds trivially for any α of Hamming weight $k-l$, for any odd l. The left-hand side is a Pfaffian of a matrix with an odd number of rows and columns, hence 0, and the right-hand side is 0 by the parity condition and Theorem 4.13. Now we consider the case when l is even.

Clearly (4.17) holds for any α of Hamming weight greater than or equal to $k-2$. By assumption Γ satisfies the MGIs (4.10). Inductively, consider α

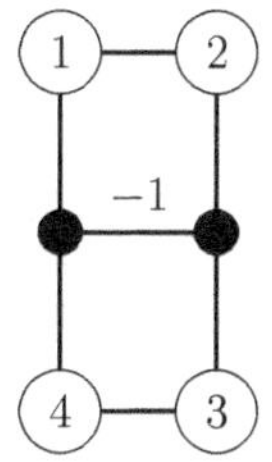

Figure 4.5. The crossover gadget. The external nodes are those labeled, and all edge weights are 1, except the edge labeled −1.

with Hamming weight $k - l$ for some even $l > 2$. Let $\{p_1, \ldots, p_l\}$ be the set of indices listed in increasing order $p_1 < \cdots < p_l$, where α has the bit 0. These are the bit positions where α differs from 1^k. Consider the MGI on $\alpha \oplus e_{p_1}$ and $1^k \oplus e_{p_1}$:

$$\Gamma^{\alpha}\Gamma^{11\ldots1} = \sum_{i=2}^{l}(-1)^i \Gamma^{\alpha\oplus e_{p_1}\oplus e_{p_i}}\Gamma^{1^k\oplus e_{p_1}\oplus e_{p_i}}. \tag{4.18}$$

As $\Gamma^{11\ldots1} = 1$, we see that Γ^{α} is defined by higher Hamming weight terms.

Thus all lower Hamming weight terms of Γ are determined by those of weight $k-2$, or equivalently the $\binom{k}{2}$ edge weights of K_k. However, by Theorem 4.17, the Pfaffian values also satisfy exactly the same identities as MGI. By induction, it follows that $\mathrm{Pf}(K_k^{\alpha}) = \Gamma^{\alpha}$ for all α. We have proved (4.18).

Planarizing K_k We want to show next that there exists a *planar* matchgate G with signature $\Gamma_G = \Gamma$. We construct such a G from K_k. Consider the convex embedding of K_k. For $k \geq 4$ it has some edge crossings, as shown in Figure 4.4. The planar graph G is created by replacing each edge crossing with a *crossover gadget* from Figure 4.5. The crossover gadget is itself a matchgate X with the following signature:

$$X^{0000} = 1, \quad X^{0101} = 1, \quad X^{1010} = 1, \quad X^{1111} = -1$$

and for all other $\beta \in \{0, 1\}^4$, $X^{\beta} = 0$. We note that even though geometrically this gadget is only symmetric under a rotation of π (but not $\pi/2$), its signature is invariant under a cyclic permutation, and thus functionally it is symmetric under a rotation of $\pi/2$. Now we replace every crossing of a pair of edges in the embedded K_k by a copy of X. For example, this replacement by the crossover gadget changes Figure 4.4 to Figure 4.6. If an edge $\{i, j\}$ in K_k crosses some other edges (this happens for every nonadjacent i and j in the cyclic sense), then this replacement breaks the edge $\{i, j\}$ into several parts. If $\{i, j\}$ crosses $t \geq 0$ other edges, then it is replaced by t copies of the crossover gadget, connected by

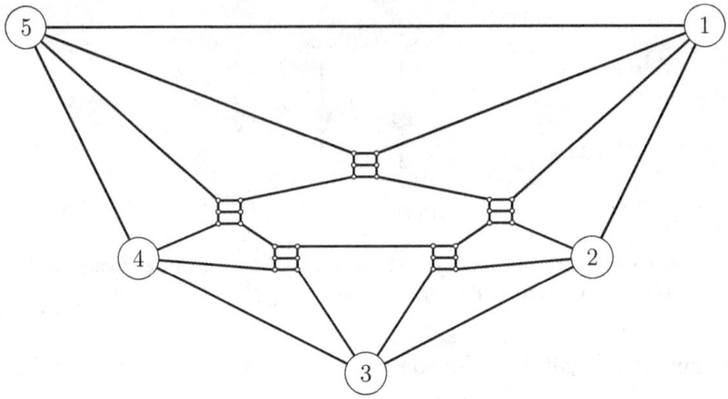

Figure 4.6. The graph from Figure 4.4 with the crossovers replaced by crossover gadgets from Figure 4.5.

$t + 1$ edges outside of crossover gadgets. Of course one copy of the crossover gadget is used for both edges of a pair of crossing edges in this replacement (see Figure 4.7). Define I to be the set of all edges in G that are *not* part of a crossover gadget. Then each edge $\{i, j\}$ in K_k defines a unique subset of edges in I, which we call the *i-j passage*. It is clear that I is a disjoint union of these i-j passages, over all $\binom{k}{2}$ pairs $1 \le i < j \le k$. Finally, we choose one edge in each i-j passage to have the weight $\Gamma^{[k]-\{i,j\}}$, namely the edge weight of $\{i, j\}$

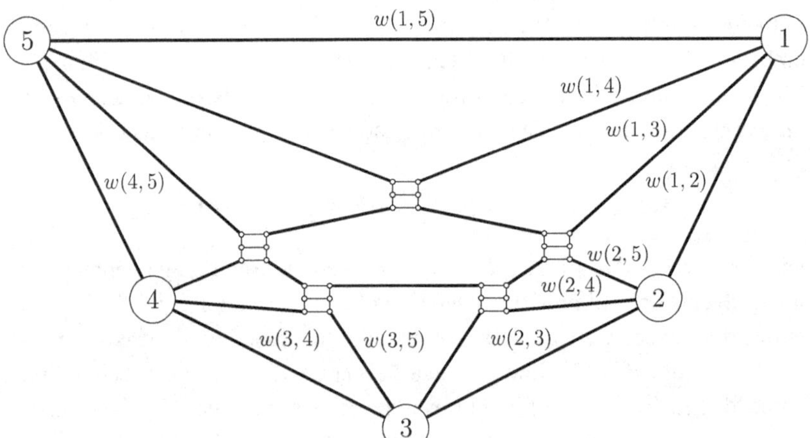

Figure 4.7. The "planarized" K_5 with edge weights. The unlabeled edges have weight 1. For notational simplicity, in the figure we use the notation $w(i, j)$ for $w(\{i, j\})$.

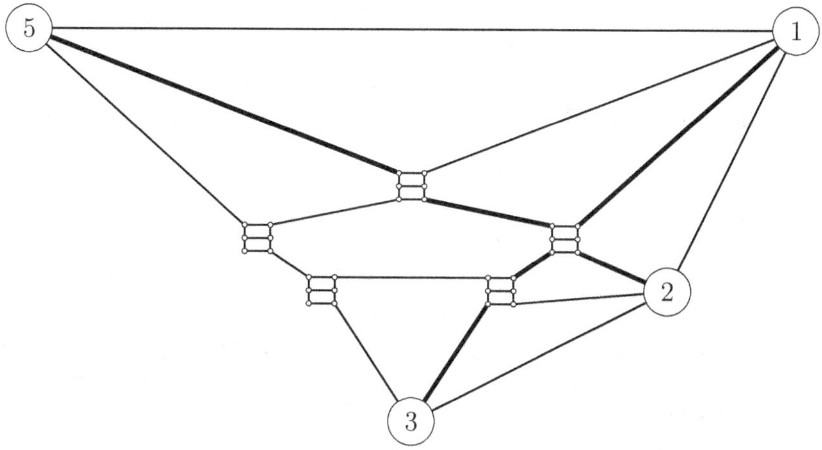

Figure 4.8. The thick edges comprise $S(M)$ for G^{00010}, where $M = \{\{1, 3\}, \{2, 5\}\}$.

in K_k. To be specific, we will choose this edge to be the one adjacent to i, the lower indexed external node of $\{i, j\}$. All other edges of I are assigned weight 1 (see Figure 4.7). This defines our planar matchgate G with external nodes $1 < 2 < \cdots < k$.

We claim that $\Gamma_G = \Gamma$.

Fix any $\alpha \in \{0, 1\}^k$. For any $S \subseteq I$, define $\mathcal{M}_S(G^\alpha)$ to be the subset of all perfect matchings $M' \in \mathcal{M}(G^\alpha)$ such that $M' \cap I = S$. Every perfect matching $M \in \mathcal{M}(K_k^\alpha)$ defines a collection of i-j passages, one for each $\{i, j\} \in M$. Let $S(M)$ be the union of these i-j passages. Clearly the perfect matching $M \in \mathcal{M}(K_k^\alpha)$ can be recovered from $S(M)$, and is unique for the given $S(M)$. There is a 1-1 correspondence between M and $S(M)$. As an example, we consider $M = \{\{1, 3\}, \{2, 5\}\} \in \mathcal{M}(K_5^{00010})$. The set $S(M)$ for G^{00010} are the thick edges in Figure 4.8.

We will show that, for the purpose of computing the signature entry Γ_G^α, we need to consider only those perfect matchings $M' \in \mathcal{M}(G^\alpha)$ that satisfy the following property:

Property: There exists an $M \in \mathcal{M}(K_k^\alpha)$, such that

$$M' \cap I = S(M). \tag{4.19}$$

This is a consequence of the properties of the crossover gadget. If i is an external node in G^α, then any $M' \in \mathcal{M}(G^\alpha)$ must contain a unique edge e' incident to i. There is a unique j, which is another external node in G, such that e' belongs to the i-j passage. Then by the properties of the crossover gadgets along this i-j passage, we may assume M' contains all edges of this i-j passage, saturating j. In particular, j belongs to G^α. All other M' collectively contribute 0, since the evaluation of the crossover gadget X will be 0. More formally, in the computation of $\Gamma_G^\alpha = \sum_{M' \in \mathcal{M}(G^\alpha)} \prod_{e' \in M'} w(e')$, we classify all $M' \in \mathcal{M}(G^\alpha)$ according to $M' \cap I$. If $S \neq S(M)$ for any $M \in \mathcal{M}(K_k^\alpha)$, then

$$\sum_{M' \in \mathcal{M}(G^\alpha):\ M' \cap I = S} \ \prod_{e' \in M'} w(e') = 0. \tag{4.20}$$

In fact, for any $M' \in \mathcal{M}(G^\alpha)$ such that $M' \cap I = S$ that is not $S(M)$ for any $M \in \mathcal{M}(K_k^\alpha)$, it must be the case that at some crossover gadget X, S induces an external removal pattern $\beta \notin \{0000, 0101, 1010, 1111\}$. Then $X^\beta = 0$, and (4.20) follows.

Thus we restrict to those perfect matchings $M' \in \mathcal{M}(G^\alpha)$ that satisfy the property (4.19). For any $M \in \mathcal{M}(K_k^\alpha)$, it is clear that

$$\sum_{M' \in \mathcal{M}_{S(M)}(G^\alpha)} \prod_{e' \in M'} w(e') = (-1)^{c(M)} \prod_{e \in M} w(e), \tag{4.21}$$

where $c(M)$ counts the number of copies of X where the external removal pattern is $\beta = 1111$. Thus $c(M)$ is exactly the number of overlapping pairs in M. It follows that

$$\begin{aligned}
\Gamma_G^\alpha &= \sum_{M' \in \mathcal{M}(G^\alpha)} \prod_{e' \in M'} w(e') \\
&= \sum_{S \subseteq I} \sum_{M' \in \mathcal{M}_S(G^\alpha)} \prod_{e' \in M'} w(e') \\
&= \sum_{M \in \mathcal{M}(K_k^\alpha)} \sum_{M' \in \mathcal{M}_{S(M)}(G^\alpha)} \prod_{e' \in M'} w(e') \\
&= \mathrm{Pf}(K_k^\alpha).
\end{aligned}$$

The last equality is because each Pfaffian term in $\mathrm{Pf}(K_k^\alpha)$ has exactly the same sign as in (4.21). Hence $\Gamma_G = \Gamma$ follows from this and (4.17).

Theorem 4.18. *Any $\Gamma \in (\mathbb{C}^2)^{\otimes k}$ satisfying the MGI is the signature of a matchgate with k external nodes. The matchgate has $O(k^4)$ nodes. If $\Gamma^{11\ldots1} = 1$, which is achievable by a normalization for every nonzero Γ, there exists a skew-symmetric matrix $M \in \mathbb{C}^{k \times k}$ such that $\Gamma^\alpha = \mathrm{Pf}(M^\alpha)$, where M^α is the matrix obtained from M by deleting all rows and columns belonging to the subset denoted by α.*

Thus, after a normalization of $\Gamma^{11\ldots1} = 1$, a matchgate signature of arity k has a general parametrized form as the sub-Pfaffians of a $k \times k$ skew-symmetric matrix.

For example, consider the matchgate in Figure 4.3 for the problem 2-COLOR-COUNTING in Section 4.2. Its signature is $\Gamma = (1, 0, 0, -1, 0, 1, -1, 0, 0, -1, 1, 0, -1, 0, 0, 1)^{\mathrm{T}} \in \{0, 1\}^4$. The signature is given by the sub-Pfaffians of

$$\begin{bmatrix} 0 & -1 & +1 & -1 \\ +1 & 0 & -1 & +1 \\ -1 & +1 & 0 & -1 \\ +1 & -1 & +1 & 0 \end{bmatrix}.$$

After assigning appropriate edge weights to K_4 according to this matrix, we obtain the matchgate by replacing the single crossover of edges $\{1, 3\}$ and $\{2, 4\}$ by the crossover gadget in Figure 4.5. This results in the matchgate depicted in Figure 4.3.

4.3.3 Symmetric Signatures by Matchgates

We say a matchgate signature is *even* if it is the signature of an even matchgate, i.e., a matchgate with an even number of nodes. An even signature has nonzero values only for indices of even Hamming weight. We define an *odd* signature similarly. Recall that a symmetric arity-k signature can be denoted as $[z_0, z_1, \ldots, z_k]$, where z_i is the value of the signature for an index of Hamming weight i. The symmetric signatures that obey the MGI have a very concise description, which we prove next.

Theorem 4.19. *If $[z_0, \ldots, z_k]$ is an even symmetric matchgate signature, then $z_i = 0$ for all odd i, and there exist r_1 and r_2 not both zero such that for all even $i \geq 2$:*

$$r_1 z_{i-2} = r_2 z_i.$$

Conversely, every sequence of values satisfying these conditions is an even symmetric matchgate signature. The statement for odd symmetric signatures is analogous.

Stated equivalently, a sequence is a symmetric matchgate signature iff it takes the following form: Alternate entries of $[z_0, \ldots, z_k]$ are zero and the entries at the other alternate positions form a geometric progression.

Proof. By the parity condition, all odd parity entries of the signature of an even matchgate are 0. Consider any *even* i and j, where $0 \leq i < j \leq k$. We invoke the

MGI for $\alpha = 1^i10^{k-i-1}$, $\beta = 1^i01^{j-i-1}0^{k-j}$, so that $\alpha \oplus \beta = 0^i1^{j-i}0^{k-j}$. We use the exponentiation notation here to denote repetition. The bitstring α has an odd Hamming weight $i+1$ and β has an odd Hamming weight $j-1$. Note that i and j being both even implies that $j-i-1 \geq 1$. Using the fact that Γ is symmetric, the MGI under α, β can be simplified:

$$\sum_{s=1}^{j-i} (-1)^s \Gamma^{\alpha \oplus e_{i+s}} \Gamma^{\beta \oplus e_{i+s}} = -z_i z_j + \sum_{s=2}^{j-i} (-1)^s z_{i+2} z_{j-2} = 0. \tag{4.22}$$

There are an odd number of terms in the sum $\sum_{s=2}^{j-i}(-1)^s z_{i+2} z_{j-2}$ and the terms alternate their signs and begin with a $+$, so we conclude that

$$z_i z_j = z_{i+2} z_{j-2}.$$

In particular, if i is even and $0 \leq i \leq k-4$, then

$$z_i z_{i+4} = z_{i+2}^2.$$

If $z_{i+2} \neq 0$, then both $z_i \neq 0$ and $z_{i+4} \neq 0$. This means that if any even indexed entry that is not the first or the last even indexed entry (call it a nonexternal entry) is nonzero, then all even indexed entries are nonzero. In this case, the geometric progression is established, with common ratio $z_{i+2}/z_i = z_{i+4}/z_{i+2}$, for even $0 \leq i \leq k-4$.

Suppose all nonexternal even indexed entries are 0. If $k \leq 3$ then the theorem is self-evident. Suppose $k \geq 4$. Let $k^* \leq k$ be the maximum even index. Then $k^* \geq 4$ and we have

$$z_0 z_{k^*} = z_2 z_{k^*-2}.$$

Thus $z_2 = 0$. It follows that $z_0 z_{k^*} = 0$ and therefore at most one extremal even indexed entry can be nonzero. It is also easy to verify that a sequence satisfying the conditions of this theorem satisfies MGI, and hence is a matchgate signature. (We will give a direct construction shortly.) This completes the proof for even signatures. The proof for odd signatures is similar. The theorem follows. □

The explicit list of symmetric signatures in Theorem 4.11 follows from Theorem 4.19.

We have already demonstrated how to build a planar matchgate realizing any MGI-satisfying signature, through a planarizing procedure. It is instructive, however, to give a direct construction for the symmetric case. For the symmetric signature EXACT-ONE-1 $= [0, 1, 0, \ldots, 0]$ of arity n that represents a perfect matching, and its reversal EXACT-ONE-0 $= [0, \ldots, 0, 1, 0]$, we can design a matchgate based on a star graph. For EXACT-ONE-0 we take a star graph with

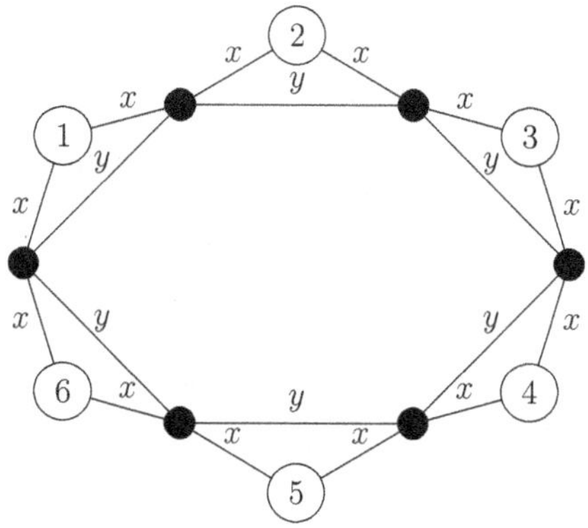

Figure 4.9. A matchgate for an even, symmetric, arity-6 signature.

a single internal vertex and n dangling edges. For EXACT-ONE-1 we can flip the 0's and 1's by extending each dangling edge to be a path of length 2. For weighted versions $[0, \lambda, 0, \ldots, 0]$ and $[0, \ldots, 0, \lambda, 0]$ we can assign the weight λ to all the edges incident to the single internal vertex.

These two cases correspond to some special cases in the list in Theorem 4.11 with $a = 0$ or $b = 0$ ($a = 0$ in item 2, $b = 0$ in item 3, and a or $b = 0$ in item 4.) Now consider a symmetric signature in Theorem 4.11 other than the two cases above. For such a signature, it is difficult to imagine a construction that is planar and also geometrically symmetric for all pairs of external nodes $1 \le i < j \le k$, if $k \ge 4$. In the following construction, the matchgates are not geometrically symmetric for all pairs of external nodes, but functionally they are, in terms of the signatures.

We present two closely related matchgate constructions, one for even symmetric signatures, and the other for odd. The constructions for both these cases work regardless if the signature has odd or even arity.

To construct a symmetric even matchgate G of arity k, we first take k triangles with vertices $\{a_i, b_i, c_i\}$ $(1 \le i \le k)$. The edges $\{a_i, b_i\}$ and $\{a_i, c_i\}$ have weight x, and $\{b_i, c_i\}$ has weight y. We link them in a cycle, identifying c_i with b_{i+1}, where the index is counted modulo k. The matchgate G has k external nodes $\{a_1, \ldots, a_k\}$, and a total of $2k$ nodes. In Figure 4.9 we have an example of a planar matchgate for an even, arity-6 signature. For a symmetric odd signature of arity k, we delete one external node in a matchgate for a symmetric even signature of arity $k + 1$. This is illustrated in Figure 4.10 for arity $k = 5$.

For an even matchgate of $2k$ nodes, we need to consider its values only at entries of even Hamming weight. Consider any $\alpha \in \{0, 1\}^k$ of even Hamming weight. $\alpha_i = 0$ iff a_i remains in G^α. If $\alpha = 1^k$, then G^α is a cycle of length k.

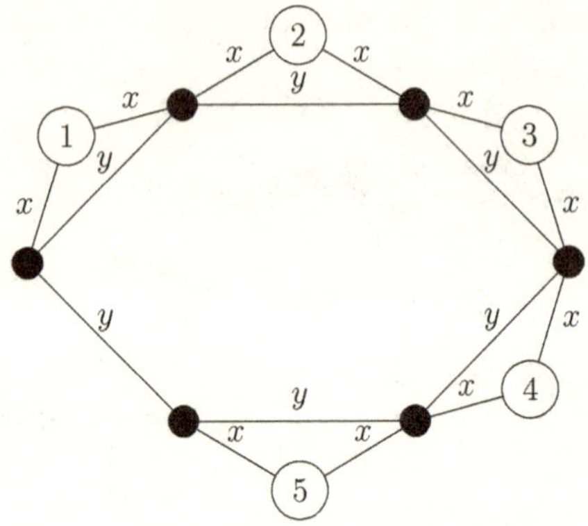

Figure 4.10. A matchgate for an odd, symmetric, arity-5 signature.

Since α also has even Hamming weight, k must be even, and there are exactly two perfect matchings of a cycle of even length, each having weight $y^{k/2}$.

Now assume $\alpha \neq 1^k$. Then α cyclically alternates between consecutive 0's (called a *0-run*) and consecutive 1's (called a *1-run*). Each a_i that remains in G^α must be matched to either b_i (we call it *left-match*) or $c_i = b_{i+1}$ (we call it *right-match*), both with weight x. Consider any 0-run. It is clear that either all a_i within this 0-run left-match or all right-match. Next consider a 1-run of m 1's; it is between two (not necessarily distinct) 0-runs, since $\alpha \neq 1^k$. If m is even, then the path of m edges all with weight y forces the two neighboring 0-runs to take either both left-match or both right-match. Moreover, both possibilities are realizable, and in each case the 1-run contributes a weight $y^{m/2}$. If m is odd, then the path of m edges forces the two neighboring 0-runs to take opposite types of left-match and right-match. Again both possibilities are realizable; in one case the 1-run contributes a weight $y^{(m-1)/2}$, and in another case it contributes a weight $y^{(m+1)/2}$. Furthermore, for two 1-runs 1^m and $1^{m'}$ both of odd length and are consecutive in the sense that the only 1-runs in between are of even length, they contribute a combined weight $y^{(m+m')/2}$. Since α has an even Hamming weight $|\alpha|$, there is an even number of 1-runs of odd length. Hence together the 1-runs contribute a weight $y^{|\alpha|/2}$. There are exactly two perfect matchings in G^α, each uniquely determined by the left-match or right-match choice of any particular a_i in G^α. It follows that the signature value is $\Gamma^\alpha = 2x^{k-|\alpha|}y^{|\alpha|/2}$. Clearly by choosing x and y suitably, we can realize an arbitrary even symmetric signature other than weighted EXACT-ONE-1 and EXACT-ONE-0.

The construction for odd symmetric signatures is to remove one external node in the matchgate for an even symmetric signature of arity one higher. If the even symmetric signature of arity $k+1$ has the form $[z_0, z_1, \ldots, z_{k+1}]$, then the construction gives the signature $[z_1, \ldots, z_{k+1}]$ of arity k. The proof is complete.

4.3.4 Symmetric Signatures Transformable to Matchgates

In Section 4.2 we have seen a number of natural problems that can be solved by holographic reductions to matchgates. For an individual problem, specified by suitable local constraint functions expressed as generator and recognizer signatures, the task is to find suitable matchgates and a holographic transformation such that both the generator and recognizer signatures can be transformed to matchgate signatures. This is called the simultaneous realizability problem. This task for any individual problem is specified by families of algebraic equations. These families of equations are typically exponential in size. Searching for their solutions is what Valiant called "the enumeration" of "freak objects" in his Accidental Algorithms paper [Val06]. Dealing with such algebraic equations can be difficult.

While finding exotic solutions such as the signature for $\#_7$PL-RTW-MON-3CNF can be artistically satisfying, the situation with ever more complicated algebraic constraints on such signatures for any given problem can quickly overwhelm such an artistic approach as well as a computer search. Furthermore, failure to find such solutions to a particular algebraic system yields no proof that such solutions do not exist, and it generally does not give us any insight as to why. The theory of matchgates provides a more systematic understanding. This theory is quite complete for symmetric signatures. We will show that the simultaneous realizability problem can be solved in polynomial time. Using this theory we can see why the modulus 7 is *the* modulus that works for counting PL-RTW-MON-3CNF. Underlying this success is the fact that 7 is $2^3 - 1$, and for any odd prime p, the number 2 is a quadratic residue in $\mathbb{Z}_q$ for any prime factor q of the Mersenne number $2^p - 1$. Generalizing this, we will show that $\#_{2^k-1}$PL-RTW-MON-kCNF is in P for all $k \geq 3$ (the problem is trivial for $k \leq 2$). Furthermore, no suitable signatures exist for any modulus other than factors of $2^k - 1$ for counting PL-RTW-MON-kCNF.

We start with an equivalent set of normal forms from Theorem 4.11 that is more suitable for holographic transformations. If a and b are both nonzero, then all four forms in Theorem 4.11 can be expressed as a sum of two tensor products of the following form: There exist nonzero α and $\beta \in \mathbb{C}$ such that

$$f = [f_0, f_1, \ldots, f_n] = \left[\alpha, \quad \beta\right]^{\otimes n} \pm \left[\alpha, \quad -\beta\right]^{\otimes n}. \tag{4.23}$$

Clearly any signature of the form in (4.23) takes one of the four forms in Theorem 4.11. Conversely, for form 1 (arity $n = 2k$) and form 2 (arity $n = 2k + 1$)

in Theorem 4.11 we can take + for the ± sign in (4.23) and

$$\alpha = \frac{a^{k/n}}{2^{1/n}} \quad \text{and} \quad \beta = \alpha\sqrt{\frac{b}{a}}.$$

We can verify that all entries at odd Hamming weight are 0, and the entry at Hamming weight $2i$ is $2\alpha^{n-2i}\beta^{2i} = a^{k-i}b^i$, for $0 \le i \le k$. For form 3 (arity $n = 2k+1$) and form 4 (arity $n = 2k+2$) in Theorem 4.11 we can take $-$ for the ± sign and

$$\alpha = \frac{a^{(k+\frac{1}{2})/n}}{2^{1/n}b^{1/(2n)}} \quad \text{and} \quad \beta = \alpha\sqrt{\frac{b}{a}}.$$

We can verify that all entries at even Hamming weight are 0, and the entry at Hamming weight $2i+1$ is $2\alpha^{n-2i-1}\beta^{2i+1} = a^{k-i}b^i$, for $0 \le i \le k$.

However, if a or b is 0, but not both 0, there are some additional signatures in Theorem 4.11 that cannot be expressed as (4.23). These cases are as follows:

$$f = [0, \lambda, 0, \ldots, 0] \quad \text{or its reversal} \quad f' = [0, \ldots, 0, \lambda, 0] \tag{4.24}$$

for some $\lambda \neq 0$. These can be expressed as

$$f = \lambda \sum_{k=1}^{n} \left\{ \begin{bmatrix} 1, & 0 \end{bmatrix}^{\otimes(k-1)} \otimes \begin{bmatrix} 0, & 1 \end{bmatrix} \otimes \begin{bmatrix} 1, & 0 \end{bmatrix}^{\otimes(n-k)} \right\} \tag{4.25}$$

and

$$f' = \lambda \sum_{k=1}^{n} \left\{ \begin{bmatrix} 0, & 1 \end{bmatrix}^{\otimes(k-1)} \otimes \begin{bmatrix} 1, & 0 \end{bmatrix} \otimes \begin{bmatrix} 0, & 1 \end{bmatrix}^{\otimes(n-k)} \right\}. \tag{4.26}$$

Of course the case $a = b = 0$ in Theorem 4.11 is the identically 0 signature and corresponds to taking $\alpha = \beta = 0$ in (4.23) and $\lambda = 0$ in (4.25) and (4.26).

Now we can give a closed form expression for symmetric signatures that are transformable to matchgate signatures. We state this for recognizer signatures.

Theorem 4.20. *A symmetric signature $f = [f_0, f_1, \ldots, f_n]$ is transformable to a matchgate signature as a recognizer, i.e., $f = \Gamma M^{\otimes n}$ for some $M = \begin{bmatrix} a & c \\ b & d \end{bmatrix} \in GL_2(\mathbb{C})$ and a matchgate signature Γ iff it takes one of the following forms:*

1. $\exists \lambda, \alpha, \beta \in \mathbb{C}$ and $\epsilon = \pm 1$, such that

$$f = \lambda\left[\left(\begin{bmatrix}\alpha, & \beta\end{bmatrix} M\right)^{\otimes n} + \epsilon\left(\begin{bmatrix}\alpha, & -\beta\end{bmatrix} M\right)^{\otimes n}\right].$$

Thus, for $0 \le k \le n$, the entry of f at weight k is

$$f_k = \lambda\left[(a\alpha + b\beta)^{n-k}(c\alpha + d\beta)^k + \epsilon(a\alpha - b\beta)^{n-k}(c\alpha - d\beta)^k\right].$$

2. *$\exists \lambda \in \mathbb{C}$, such that*

$$f = \lambda \sum_{k=1}^{n} \left\{ \begin{bmatrix} a, & c \end{bmatrix}^{\otimes(k-1)} \otimes \begin{bmatrix} b, & d \end{bmatrix} \otimes \begin{bmatrix} a, & c \end{bmatrix}^{\otimes(n-k)} \right\}.$$

Thus, for $0 \le k \le n$, the entry of f at weight k is

$$f_k = \lambda \left[(n-k)a^{n-k-1}bc^k + ka^{n-k}c^{k-1}d \right].$$

3. *$\exists \lambda \in \mathbb{C}$, such that*

$$f = \lambda \sum_{k=1}^{n} \left\{ \begin{bmatrix} b, & d \end{bmatrix}^{\otimes(k-1)} \otimes \begin{bmatrix} a, & c \end{bmatrix} \otimes \begin{bmatrix} b, & d \end{bmatrix}^{\otimes(n-k)} \right\}.$$

Thus, for $0 \le k \le n$, the entry of f at weight k is

$$f_k = \lambda \left[(n-k)b^{n-k-1}ad^k + kb^{n-k}d^{k-1}c \right].$$

(We take the convention that $x^0 = 1$ and $0x^{0-1} = 0$, even if $x = 0$.)

We have a corresponding statement for a symmetric signature transformable to a matchgate signature as a generator: $g = (M^{-1})^{\otimes n}\Gamma'$ for some $M \in GL_2(\mathbb{C})$ and a matchgate signature Γ'. These formulae can be obtained from those in Theorem 4.20 by substituting $\begin{bmatrix} d & -b \\ -c & a \end{bmatrix}$ for $\begin{bmatrix} a & c \\ b & d \end{bmatrix}$.

From now on, a signature that is transformable to the signature of a matchgate is also called a matchgate signature (or realizable as a matchgate signature). This conforms to the modern concept of tensors, which are given coordinate-free, and not chained to a particular basis. However, to avoid ambiguity, for a signature that is realizable directly as the signature of a matchgate without a transformation, we will call it a *standard* (matchgate) signature.

By the relationship between second order linear recurrences and the expressions obtained here, we have

Theorem 4.21. *A symmetric signature $[x_0, x_1, \ldots, x_n]$ is transformable to a matchgate signature under some holographic transformation iff $\exists (a, b, c) \neq 0$ such that*

$$ax_k + bx_{k+1} + cx_{k+2} = 0$$

for $0 \le k \le n-2$.

The expression of a matchgate realizable signature as the solution to a second-order linear recurrence relation has a uniqueness that we state explicitly below.

Lemma 4.22. *Suppose a sequence* $(x_0, x_1, \ldots, x_n)$*, where* $n \geq 3$*, has the following form:* $x_i = \lambda\alpha^i + \mu\beta^i$*,* $(\lambda\mu \neq 0, \alpha \neq \beta)$*, then the representation is unique. That is, if* $x_i = \lambda'(\alpha')^i + \mu'(\beta')^i$*,* $(0 \leq i \leq n, n \geq 3)$*, then* $\lambda' = \lambda, \mu' = \mu, \alpha' = \alpha, \beta' = \beta$ *or* $\lambda' = \mu, \mu' = \lambda, \alpha' = \beta, \beta' = \alpha$*.*

Lemma 4.23. *Suppose a sequence* $(x_0, x_1, \ldots, x_n)$*, where* $n \geq 3$*, has the following form:* $x_i = \lambda i\alpha^{i-1} + \mu\alpha^i$*,* $(\lambda \neq 0)$*, then the representation is unique. That is, if* $x_i = \lambda' i(\alpha')^{i-1} + \mu'(\alpha')^i$*,* $(0 \leq i \leq n,\ n \geq 3)$*, then* $\lambda' = \lambda, \mu' = \mu, \alpha' = \alpha$*.*

Also, the two cases in Lemma 4.22 and 4.23 are mutually exclusive, depending on whether the eigenvalues of the linear recurrence are distinct or not.

The following observation is due to Valiant [Val08].

Proposition 4.24. *If a generator or recognizer signature is realizable by a matchgate under a holographic transformation M, then it is also realizable by a matchgate under* M'*, where* $M' = \begin{bmatrix} x & 0 \\ 0 & y \end{bmatrix} M$*, for any nonzero* $x, y \in \mathbb{C}$*, or* $M' = \begin{bmatrix} 0 & 1 \\ 1 & 0 \end{bmatrix} M$*.*

Let G be a recognizer signature of arity n realizable under a holographic transformation M by a matchgate with the standard signature Γ, then $G = \Gamma M^{\otimes n}$. If Γ is any standard signature, then so are $\Gamma \begin{bmatrix} x & 0 \\ 0 & y \end{bmatrix}^{\otimes n}$ and $\Gamma \begin{bmatrix} 0 & 1 \\ 1 & 0 \end{bmatrix}^{\otimes n}$. For example, a matchgate with signature $\Gamma \begin{bmatrix} 0 & 1 \\ 1 & 0 \end{bmatrix}^{\otimes n}$ is obtained from the given matchgate for Γ by appending an edge of weight 1 to each external node, which effectively flips the truth values 0 and 1. Proposition 4.24 can also be proved easily using MGI. The proof for generators is similar.

Proposition 4.24 implies that there is an equivalence relation $\sim$ among the holographic transformations, where one can multiply each row by a nonzero constant, or exchange the two rows of the transformation matrix M. Under this equivalence relation, $\mathrm{GL}_2(\mathbb{C})/\sim$ is a two-dimensional manifold. We call this the *basis manifold* $\mathcal{B}$. From now on we identify a basis transformation with its equivalence class containing it. When it is permissible, we use the dehomogenized coordinates $\begin{bmatrix} 1 & \xi \\ 1 & \eta \end{bmatrix}$ to represent a point (i.e., a basis class) in $\mathcal{B}$.

Theorem 4.20 gives a complete characterization of all the symmetric realizable matchgate signatures. These tell us exactly what signatures can be realized over *some* bases. However, to construct a holographic algorithm, one needs to realize some generators and recognizers simultaneously. In terms of $\mathcal{B}$, a given generator (recognizer) defines a (possibly empty) subvariety that consists of all the bases over which it is realizable. The simultaneous realizability is equivalent to a nonempty intersection of these subvarieties. Thus we have to go beyond Theorem 4.20. For every signature that is realizable according to Theorem 4.20, we need to determine the subvariety where it is realizable.

Definition 4.25. Let f be a symmetric recognizer or generator signature of arity n. Then $B_{rec}(f)$ (respectively $B_{gen}(f)$) is the set of all bases in $\mathcal{B}$ for which the symmetric signature f for a recognizer (respectively a generator) is realizable. Formally,

$$B_{rec}(f) = \{M \in \mathcal{B} \mid f = \Gamma M^{\otimes n} \text{ for some matchgate signature } \Gamma\},$$

$$B_{gen}(f) = \{M \in \mathcal{B} \mid f = (M^{-1})^{\otimes n}\Gamma' \text{ for some matchgate signature } \Gamma'\}.$$

Since the identically 0 signature is realizable in every basis, we will assume the signature is not identically 0 in the following discussion.

Lemmas 4.26 to 4.30 give a complete and mutually exclusive list of symmetric matchgate realizable signatures for recognizers that are not identically 0. The corresponding set of results for symmetric matchgate realizable generators can be obtained by substituting $\begin{bmatrix} d & -b \\ -c & a \end{bmatrix}$ for $\begin{bmatrix} a & c \\ b & d \end{bmatrix}$.

Recall that a symmetric signature $[x_0, x_1, \ldots, x_n]$ is nondegenerate iff $\operatorname{rank}\begin{bmatrix} x_0 & \ldots & x_{n-1} \\ x_1 & \ldots & x_n \end{bmatrix} = 2$. The signature is identically 0 iff $\operatorname{rank}\begin{bmatrix} x_0 & \ldots & x_{n-1} \\ x_1 & \ldots & x_n \end{bmatrix} = 0$. It has rank 1 iff it can be expressed as $[a^n, a^{n-1}b, \ldots, b^n]$, for $a, b \in \mathbb{C}$ and not both 0. Lemma 4.26 deals with degenerate signatures.

Lemma 4.26. *For $a, b \in \mathbb{C}$ and not both 0,*

$$B_{rec}([a^n, a^{n-1}b, \ldots, b^n]) = \left\{ \begin{bmatrix} a & b \\ x & y \end{bmatrix} \in \mathcal{B} \,\middle|\, x, y \in \mathbb{C} \right\}.$$

Remark: Every nonzero unary signature is of this form.

In the following we assume the signature is nondegenerate. For the case of arity $n = 2$, we have

Lemma 4.27.

$$B_{rec}([x, y, z]) = \left\{ \begin{bmatrix} a & c \\ b & d \end{bmatrix} \in \mathcal{B} \,\middle|\, \begin{matrix} xd^2 - 2ybd + zb^2 = 0, xc^2 - 2yac + za^2 = 0 \\ \text{or} \;\; xcd - y(ad + bc) + zab = 0 \end{matrix} \right\}.$$

We can prove this lemma by noting that, for arity 2, the only requirement to be a standard signature of a matchgate is the parity constraint. The MGI is satisfied under the parity condition in this case.

In the following we assume the signature has arity $n \geq 3$, and nondegenerate. In this case, we note that the constants a, b, c in Theorem 4.21 are unique up to a scalar factor. In fact, if there are two linearly independent triples (a, b, c), then the following matrix

$$\begin{bmatrix} x_0 & x_1 & \dots & x_{n-2} \\ x_1 & x_2 & \dots & x_{n-1} \\ x_2 & x_3 & \dots & x_n \end{bmatrix}$$

has rank ≤ 1. The first row and the last row are not both 0, otherwise the signature is identically 0 (by $n \geq 3$). It follows that the matrix

$$\begin{bmatrix} x_0 & x_1 & \dots & x_{n-1} \\ x_1 & x_2 & \dots & x_n \end{bmatrix}$$

also has rank 1; hence the signature is degenerate.

The next lemma corresponds to nondegenerate signatures in Theorem 4.21, where $ac = 0$ and $b = 0$. For $b = c = 0$ (and $a \neq 0$) we have $[0, \dots, 0, \lambda, \mu]$, where $\lambda \neq 0$ by nondegeneracy. Its reversal $[\mu, \lambda, 0, \dots, 0]$ corresponds to $a = b = 0$ (and $c \neq 0$).

Lemma 4.28. *Let $\lambda \neq 0$. Then,*

$$B_{rec}([0, \dots, 0, \lambda, \mu]) = \left\{ \begin{bmatrix} 0 & 1 \\ n\lambda & \mu \end{bmatrix} \right\}.$$

Similarly for its reversal,

$$B_{rec}([\mu, \lambda, 0, \dots, 0]) = \left\{ \begin{bmatrix} 1 & 0 \\ \mu & n\lambda \end{bmatrix} \right\}.$$

The reversal signature $[\mu, \lambda, 0, \dots, 0]$ can also be viewed as a special case in Lemma 4.31 *if* we were to allow $\alpha = 0$ there, where for $\alpha = 0$ and $i = 0$, we take the convention that $i\alpha^{i-1} = 0$, and also $\alpha^i = 1$. However, for the listing to be mutually exclusive we disallow $\alpha = 0$ in Lemma 4.31.

Next we consider the case of nondegenerate signatures in Theorem 4.21 where $ac = 0$ and $b \neq 0$.

Lemma 4.29. *For $\lambda\mu \neq 0$,*

$$B_{rec}([\lambda, \lambda\alpha, \dots, \lambda\alpha^{n-1}, \lambda\alpha^n + \mu]) = \left\{ \begin{bmatrix} 1 & \alpha + \omega \\ 1 & \alpha - \omega \end{bmatrix} \middle| \ \omega^n = \pm\frac{\mu}{\lambda} \right\}.$$

Similarly for its reversal,

$$B_{rec}([\lambda\alpha^n + \mu, \lambda\alpha^{n-1}, \ldots, \lambda\alpha, \lambda]) = \left\{ \begin{bmatrix} \alpha + \omega & 1 \\ \alpha - \omega & 1 \end{bmatrix} \middle| \omega^n = \pm\frac{\mu}{\lambda} \right\}.$$

Here $\alpha = 0$ is GEN-EQ, and corresponds to the case $a = c = 0$ and $b \neq 0$. For $\alpha \neq 0$, the reversal signature $[\lambda\alpha^n + \mu, \ldots, \lambda\alpha^2, \lambda\alpha, \lambda]$ can also be written as $\lambda'[1, \alpha', \ldots, \alpha'^n] + [\mu, 0, \ldots, 0]$, where $\lambda' = \lambda\alpha^n$ and $\alpha' = 1/\alpha$, and then can be viewed as a special case in Lemma 4.30 *if* we were to allow $\beta = 0$ there; although for mutual exclusion we disallow $\beta = 0$ in Lemma 4.30.

Now we consider $ac \neq 0$ in Theorem 4.21. We use the fact that the triple (a, b, c) in the statement of Theorem 4.21 is unique up to a scalar factor. We have a unique characteristic equation $cX^2 + bX + a = 0$, which has two roots α and β. In particular the forms 1, 2 and 3 from Theorem 4.20 are mutually exclusive. Suppose $b^2 \neq 4ac$, i.e., $\alpha \neq \beta$, we have the following lemma.

Lemma 4.30. *For* $\lambda\mu \neq 0$, $\alpha\beta \neq 0$, *and* $\alpha \neq \beta$,

$$B_{rec}([\lambda\alpha^i + \mu\beta^i] | i = 0, 1, \ldots, n]) = \left\{ \begin{bmatrix} 1 + \omega & \alpha + \beta\omega \\ 1 - \omega & \alpha - \beta\omega \end{bmatrix} \middle| \omega^n = \pm\frac{\mu}{\lambda} \right\}.$$

If $b^2 = 4ac \neq 0$, i.e., the characteristic roots coincide $\alpha = \beta \neq 0$, we have the following lemma.

Lemma 4.31. *For* $\lambda \neq 0$, *and* $\alpha \neq 0$,

$$B_{rec}([\lambda i\alpha^{i-1} + \mu\alpha^i] | i = 0, 1, \ldots, n]) = \left\{ \begin{bmatrix} 1 & \alpha \\ \mu & n\lambda + \mu\alpha \end{bmatrix} \right\}.$$

Definition 4.32. Simultaneous Realizability Problem (SRP):

Input: A set of symmetric signatures for generators and/or recognizers

Output: A common basis of holographic transformation to matchgate signatures if any exists; "NO" if they are not simultaneously realizable as matchgate signatures by holographic transformations

SRP Algorithm: For every signature $[x_0, x_1, \ldots x_n]$, check if it satisfies Theorem 4.20.

If not, output "NO" and halt.

Otherwise find $B_{gen}([x_0, x_1, \ldots x_n])$ or $B_{rec}([x_0, x_1, \ldots x_n])$ according to one of the lemmas.

Check if these subvarieties have a nonempty intersection.

Theorem 4.33. *The SRP algorithm solves SRP in polynomial time.*

Proof. Checking whether every input signature satisfies Theorem 4.20 can obviously be done in polynomial time. Finding the right form and then the right lemma for a signature that satisfies Theorem 4.21 can also be done in polynomial time, as they are mutually exclusive.

Every subvariety of bases from Lemma 4.26 to 4.31 for recognizers, and similarly for generators, is of one of three kinds: a finite set of points (of linear size), a line, or a quadratic curve. More precisely, expressing things in terms of the manifold $\mathcal{B}$ shows the following. For Lemma 4.26 we get a line with $\xi =$ constant (in the dehomogenized coordinates notation defining $\mathcal{B}$). For Lemma 4.27 we get a union of two sets. The first is finite, where both the dehomogenized coordinates ξ and η satisfy a quadratic polynomial (and by projective closure). The second set is defined by an equation of the form $A\xi\eta + B(\xi + \eta) + C = 0$ (and by projective closure), where A, B, C are known constants. The other cases are all similar. For example, for Lemma 4.29, we get at most n points from the equation $\omega^n =$ constant. To sum up, the SRP algorithm is a polynomial time algorithm for SRP. □

4.3.5 $\#_7$PL-RTW-MON-3CNF Redux

In Section 4.2 we had encountered the problem $\#_7$PL-RTW-MON-3CNF. Valiant [Val06] also introduced another problem $\#_7$Pl-3/2BIP-VC, which counts the number of vertex covers for a planar (3, 2)-regular bipartite graph, modulo 7. They were solved in polynomial time by holographic algorithms based on matchgates. He called these *accidental algorithms*. In this subsection, we apply the theory developed in Section 4.3 to find such algorithms almost *mechanically*. This theory provides more insight and understanding as to what can or cannot be accomplished. With this machinery we are able to generalize to the problems Pl-RTW-MON-kCNF (the same problem for kCNF formulae for a general k) and Pl-k/2BIP-VC (the same problem for planar $(k, 2)$-regular bipartite graphs). We show that there is a unique modulus $2^k - 1$ for which we can design such a holographic algorithm based on matchgates, which counts the number of solutions modulo $2^k - 1$. In the case of $k = 3$, this shows why 7 is special.

$\#_{2^k-1}$PL-RTW-MON-kCNF

INPUT: A planar kCNF Boolean formula where each variable appears positively and in exactly two clauses (planar, read-twice, monotone, kCNF)

OUTPUT: The number of satisfying assignments modulo $2^k - 1$

When $k = 3$ this is $\#_7$PL-RTW-MON-3CNF from Section 4.2. As noted earlier, the counting problem #PL-RTW-MON-3CNF is #P-complete already for $k = 3$.

To solve the problem by a holographic algorithm, we wish to replace each variable by a generator with the signature $[1, 0, 1]$, and each clause by a recognizer with the signature $[0, 1, 1, \ldots, 1]$ (with k 1's). The symmetric signature $[1, 0, 1]$ is the binary EQUALITY function $=_2$ and corresponds to a consistent truth assignment on two edges leading to clauses, and $[0, 1, 1, \ldots, 1]$ (with k 1's) is the Boolean OR function of arity k corresponding to a clause. If we connect the generators and recognizers in a natural way, by Theorems 1.5 and 4.7, this would solve #PL-RTW-MON-kCNF in polynomial time, *if* the signatures are simultaneously realizable by matchgates under a holographic transformation.

Then the question boils down to whether there is a basis in $\mathcal{B}$ where $[1, 0, 1]$ for a generator and $[0, 1, 1, \ldots, 1]$ (with k 1's) for a recognizer can be simultaneously realized. For this, we use our machinery.

From Lemma 4.29, taking $\lambda = 1$, $\mu = -1$ and $\alpha = 1$ for $[\lambda\alpha^k + \mu, \lambda\alpha^{k-1}, \ldots, \lambda\alpha, \lambda]$, we have

$$B_{rec}([0, 1, 1, \ldots, 1]) = \left\{ \begin{bmatrix} 1+\omega & 1 \\ 1-\omega & 1 \end{bmatrix} \middle| \, \omega^k = \pm 1 \right\}.$$

We look for some $\omega^k = \pm 1$, such that $\begin{bmatrix} 1+\omega & 1 \\ 1-\omega & 1 \end{bmatrix} \in B_{gen}([1, 0, 1])$, which is obtained from Lemma 4.27 by substituting $\begin{bmatrix} d & -b \\ -c & a \end{bmatrix}$ for $\begin{bmatrix} a & c \\ b & d \end{bmatrix}$. Explicitly,

$$B_{gen}([x, y, z]) = \left\{ \begin{bmatrix} a & c \\ b & d \end{bmatrix} \in \mathcal{B} \,\middle|\, \begin{matrix} xa^2 + 2yac + zc^2 = 0, \, xb^2 + 2ybd + zd^2 = 0 \\ \text{or} \;\; xab + y(ad + bc) + zcd = 0 \end{matrix} \right\}. \tag{4.27}$$

Thus, we want $(1 + \omega)^2 + 1 = (1 - \omega)^2 + 1 = 0$ or $(1 + \omega)(1 - \omega) + 1 = 0$. The first case is impossible, and in the second case we require $\omega^2 = 2$. Together with the condition $\omega^k = \pm 1$, we have $2^k - 1 = 0$. From this we can already see that for every prime $p | 2^k - 1$, $\#_p$PL-RTW-MON-kCNF is computable in polynomial time. In particular, this is true for every Mersenne prime $2^q - 1$. (Note that $\omega^2 = 2$ means that 2 is a quadratic residue.) If we develop the theory over a finite field $\mathbb{Z}_p$, we can obtain corresponding versions of Lemmas 4.27 and 4.29, and conclude that in order to have these signatures be simultaneously realizable over $\mathbb{Z}_p$, we must have $p | 2^k - 1$.

More generally we have:

Theorem 4.34. *There is a polynomial time algorithm for* $\#_{2^k-1}$*PL-RTW-MON-kCNF. Furthermore, any modulus m for which the appropriate signatures exist must be a divisor of* $2^k - 1$.

Proof. Our discussion above already shows that the modulus $2^k - 1$ is the best we can do. (Formally speaking we should present a generalization such as Theorem 1.5 over a ring such as $\mathbb{Z}_{2^k-1}$, which we will omit here.) We now give the polynomial algorithms in two cases:

Case 1: k is even.
Over the complex numbers $\mathbb{C}$, from (4.27), the generator form of Lemma 4.27, and Lemma 4.29, we can see that a generator for $[1, 0, 1]$ and a recognizer for $[1 + \epsilon 2^{k/2}, 1, 1, \ldots, 1]$ (where there are k 1's, and $\epsilon = \pm 1$) are simultaneously realizable in the basis $\begin{bmatrix} 1+\sqrt{2} & 1 \\ 1-\sqrt{2} & 1 \end{bmatrix}$.

Setting $\epsilon = 1$ and replacing each variable by a generator and each clause by a recognizer with the corresponding signatures, we obtain a matchgrid Ω. If we replace the generators and recognizers by their corresponding matchgates, we get a weighted planar graph G. Then Theorem 1.5 gives

$$\mathrm{Holant}_\Omega = \mathrm{PerfMatch}(G). \tag{4.28}$$

We will denote this value by X.

From the left-hand side of (4.28) we know that X is an integer because every entry in the signatures of generators $[1, 0, 1]$ and recognizers $[1 + 2^{k/2}, 1, 1, \ldots, 1]$ is an integer. Furthermore we have

$$X \equiv \#\text{PL-RTW-MON-}k\text{CNF} \pmod{1 + 2^{k/2}}.$$

From the right-hand side of (4.28) we know that X can be computed in polynomial time using the FKT algorithm for perfect matchings of a planar graph. The planar graph has weights from the subfield $\mathbb{Q}(\sqrt{2}) \subset \mathbb{C}$, which poses no problem to the Pfaffian evaluation of FKT in polynomial time in bit complexity.

Therefore $\#_{2^{k/2}+1}$PL-RTW-MON-kCNF can be computed in polynomial time. Similarly, setting $\epsilon = -1$, we can compute $\#_{2^{k/2}-1}$PL-RTW-MON-kCNF in polynomial time.

Since $\gcd(2^{k/2} + 1, 2^{k/2} - 1) = 1$ and $2^k - 1 = (2^{k/2} + 1)(2^{k/2} - 1)$, we can apply Chinese remaindering to get a polynomial time algorithm for $\#_{2^k-1}$PL-RTW-MON-kCNF.

Case 2: k is odd.
Consider the ring $\mathbb{Z}_{2^k-1}$. (Formally we can develop the theory over such a ring, and consider invertible elements and matrices for the basis manifold, which we will omit.) Let $r = 2^{(k+1)/2} \in \mathbb{Z}_{2^k-1}$. Then r satisfies $r^2 = 2$ in $\mathbb{Z}_{2^k-1}$. We observe that $r^k - 1 = (2^k)^{(k+1)/2} - 1 = 0$ in $\mathbb{Z}_{2^k-1}$.

Therefore over this ring $\mathbb{Z}_{2^k-1}$ and with the basis $\begin{bmatrix} 1+r & 1 \\ 1-r & 1 \end{bmatrix}$, the generator $[1, 0, 1]$ and the recognizer $[0, 1, 1, \ldots, 1]$ (with k 1's) are simultaneously realizable by matchgates according to (4.27) and Lemma 4.29. As a result, we have a polynomial time algorithm for $\#_{2^k-1}$PL-RTW-MON-kCNF. □

Now we consider another problem.

$\#_{2^k-1}$**PL-k/2BIP-VC**

INPUT: A planar regular bipartite graph $G = (U, V, E)$, where every $u \in U$ has degree k and every $v \in V$ has degree 2 (planar $(k, 2)$-regular bipartite graph)

OUTPUT: The number of vertex covers mod $2^k - 1$

Consider an arbitrary subset $S \subseteq V$ of vertices from the right. Every vertex $u \in U$ on the left *either* has all its k adjacent vertices in S, in which case there are exactly two choices to extend S at u to a vertex cover, *or* has some of its k adjacent vertices not in S, in which case there is exactly one choice to extend S at u to a vertex cover. Thus, following the general recipe for holographic algorithms, we want a generator with signature $[1, 0, 1]$ and a recognizer with signature $[1, 1, \ldots, 1, 2]$ (with k 1's), to be simultaneously realized over some basis.

From Lemma 4.29, taking $\lambda = \mu = \alpha = 1$ for $[\lambda, \lambda\alpha, \ldots, \lambda\alpha^{n-1}, \lambda\alpha^n + \mu]$, we have

$$B_{rec}([1, 1, \ldots, 1, 2]) = \left\{ \begin{bmatrix} 1 & 1+\omega \\ 1 & 1-\omega \end{bmatrix} \middle| \, \omega^k = \pm 1 \right\}.$$

Then the same derivation for $\#_{2^k-1}$PL-RTW-MON-kCNF gives a polynomial time algorithm for $\#_{2^k-1}$PL-k/2BIP-VC.

Theorem 4.35. *There is a polynomial time algorithm for $\#_{2^k-1}$PL-k/2BIP-VC. Furthermore, any modulus m for which the appropriate signatures exist must be a divisor of $2^k - 1$.*

There is a 1-1 correspondence between vertex covers and independent sets, by taking complement. Thus we also have a polynomial time algorithm for $\#_{2^k-1}$PL-k/2BIP-IS, which is to count the number of independent sets mod

$2^k - 1$ for planar $(k, 2)$-regular bipartite graph. More directly, we can proceed as before: For any planar $(k, 2)$-regular bipartite graph $G = (U, V, E)$, for any $S \subseteq V$, to extend it to an independent set, every $u \in U$ has either two choices if all k neighbors of u are not in S, or one choice (not to be included) otherwise. It follows that we should simultaneously realize a generator with signature $[1, 0, 1]$ and a recognizer with signature $[2, 1, \ldots, 1, 1]$ (with k 1's), and the rest are the same.

Our general machinery not only can find the required signatures when they exist, but also can prove certain desired signatures cannot be simultaneously realized. As an example, one may wish to extend the previous two problems to allow more than Read-twice as in #Pl-R_ℓ-Mon-kCNF, where $\ell > 2$. This calls for a simultaneous realizability of the Equality function of arity ℓ, $(=_\ell) = [1, 0, \ldots, 0, 1]$ (where there are $(\ell - 1)$ 0's), and the Boolean Or function on k inputs, $[0, 1, 1, \ldots, 1]$ (where there are k 1's). This can be shown to result in an empty intersection on $\mathcal{B}$.

In this book we focus on the exact complexity of counting problems. There is an active research area to study the approximate complexity of counting problems, and to the extent possible to reach classification theorems [JS93, GJP03, LLY12, SST12, MWW09, DFJ02, Wei06, Sly10, GGv+11, DGJ10, DGJR12, SS12, LLZ14, LL15, CGG+16, GGY16, GGJ16, GG16, GL16]. It appears that for approximate counting, there are more than just two levels of complexity, those that are solvable by a polynomial time approximation algorithm and those that are NP-hard to approximate. In particular, the problem to approximately count the number of independent sets for bipartite graphs (#BIS) appears to be a representative problem for a distinct class of intermediate complexity. This is in contrast to $\#_{2^k-1}$PL-k/2BIP-IS, which is solvable in polynomial time exactly.

5

2-Spin Systems on Regular Graphs

In this chapter, we consider the following class of counting problems. An input k-regular graph $G = (V, E)$ is given, where every $e \in E$ is labeled with a symmetric edge function $g : \{0, 1\}^2 \to \mathbb{C}$. We allow self-loops and parallel edges in G, where each self-loop contributes 2 to the degree of the incident vertex. We present a dichotomy theorem for spin systems on k-regular graphs. For simplicity and for what turns out to be sufficient in terms of logical dependence from later chapters, we will restrict our proofs to $k = 3$ and $k = 4$. As a spin system, the vertices of the graph represent Boolean variables. The edge function g takes 0-1 inputs from its incident vertices and outputs arbitrary (algebraic) values in $\mathbb{C}$. The problem is to compute the partition function

$$Z(G) = \sum_{\sigma : V \to \{0,1\}} \prod_{\{u,v\} \in E} g(\sigma(u), \sigma(v)).$$

For example, if g is the Boolean OR function represented by the matrix $\begin{bmatrix} 0 & 1 \\ 1 & 1 \end{bmatrix}$, then $Z(G)$ counts the number of vertex covers on a k-regular graph G.

A spin system can also be defined for a general domain $[q]$, for $q > 2$, where each vertex takes one of q possible values. But in Volume I of this book we focus on the Boolean domain $q = 2$. A 2-spin system is a special case of #CSP where there is a single symmetric binary constraint function. A 2-spin system on a k-regular graph is the further restriction that every variable appears k times. Equivalently, we may view this as a bipartite Holant problem $\text{Holant}(=_k \mid g)$. We treat the binary g as a generator (contravariant) on the right side, and the EQUALITY $(=_k)$ of arity k as a recognizer (covariant) on the left side. The main difference between general Holant problems and #CSP is that EQUALITY functions of arbitrary arity are *presumed* to be present in #CSP. For spin systems on k-regular graphs, we assume the presence of $=_k$, and a single symmetric edge function g. When EQUALITY gates of arbitrary arity are freely available in possible inputs, it is technically easier to prove #P-hardness. Two important techniques, called *stretching* and *thickening*, require the availability of EQUALITY gates of arbitrary arity (equivalently, vertices of arbitrarily high degrees). Stretching replaces an edge by a path, which requires binary EQUALITY gates. Thickening replaces an edge by multiple parallel edges, which requires EQUALITY of arbitrary arity. Proving #P-hardness becomes more challenging in the degree-restricted case. Furthermore, there are indeed cases within this class of counting problems where the problem is #P-hard for general graphs, but is solvable in polynomial time when restricted to k-regular graphs.

We denote the symmetric edge function g by $[x, y, z]$, where $x = g(0, 0)$, $y = g(0, 1) = g(1, 0)$ and $z = g(1, 1)$. Functions will also be called gates or signatures. (For VERTEX COVER , the corresponding function is the OR gate, and is denoted by the signature $[0, 1, 1]$.) The dichotomy theorem presented in this chapter gives a complexity classification of $Z(G)$ on k-regular graphs G in terms of the signature $g = [x, y, z]$, where $x, y, z \in \mathbb{C}$. Note that a self-loop with a binary edge function $g = [x, y, z]$ amounts to a unary vertex weight function $[x, z]$.

First, if $y = 0$, it is easy to compute $Z(G)$ in polynomial time. Assume $y \neq 0$. We may normalize g and set $y = 1$. The main theorem of this chapter is the following.

Theorem 5.1. *Suppose $a, b \in \mathbb{C}$, $k \geq 3$, and let $X = ab$ and $Y = a^k + b^k$. Then $Z(G)$ with $g = [a, 1, b]$ for input k-regular graphs G is* #P-hard *except in the following four cases, for which the problem is in P.*

1. $X = 1$.
2. $X = Y = 0$.
3. $X = -1$ *and* $Y = 0$.
4. $X = -1$ *and* $Y^2 = 4(-1)^k$.

(The same statement holds for Holant$(=_k | \, [a, 1, b], \Delta_0, \Delta_1)$*, and when it is #P-hard a reduction exists that does not use Lemma 3.13.)*

If we restrict the input to planar k-regular graphs, then these four cases are solvable in P, as well as a fifth case, $Y^2 = 4X^k$. The problem remains #P-hard *in all other cases.*

Given a binary symmetric function g, if #CSP(g) is tractable, then of course the 2-spin system $Z(G)$ is also tractable with g on k-regular graphs. Cases 1 to 4 correspond to tractable cases of #CSP by the product type $\mathscr{P}$ (Definition 3.2) and the affine type $\mathscr{A}$ (Definition 3.6), possibly after a holographic transformation. Furthermore, over planar graphs, there is an additional set of tractable cases that are #P-hard over general (nonplanar) k-regular graphs, and these consist entirely of suitable holographic transformations from matchgate computable problems. Finally, everything else is #P-hard, even for planar k-regular graphs. (For #CSP problems, this planar tractable case will be classified in Chapter 8.)

We can parameterize the space by (a, b) and also equivalently by (X, Y). The region defined by cases 1 to 4 in Theorem 5.1 will be called region I. In terms of the categories of tractability for #CSP (cf. Theorem 3.7), case 1 is degenerate, case 2 together with $y = 0$ in $[x, y, z]$ before normalization corresponds to the product type, and cases 3 and 4 correspond to affine binary signatures, including possibly a suitable holographic transformation. Region II is defined to be $Y^2 = 4X^k$ but $X \notin \{0, \pm 1\}$. It can also be specified by $Y^2 = 4X^k$, minus the intersection with region I. It is worth noting that $Y^2 = 4X^k$ specifies precisely those signatures that under a suitable holographic transformation both $=_k$ and g are realizable as standard matchgate signatures. The remaining space is called region III.

Definition 5.2. In terms of (a, b), region I consists of

1. $ab = 1$ (degenerate),
2. $a = b = 0$ (disequality),
3. $ab = -1$ and $a^{2k} = b^{2k}$ (combining cases 3 and 4 in Theorem 5.1).

Region II consists of $a^k = b^k$, minus the intersection with region I. Region III consists of all remaining $(a, b) \in \mathbb{C}^2$.

The union of regions I and II is specified by $X = 1$ or $Y^2 = 4X^k$ or $(X, Y) = (-1, 0)$. In terms of (a, b), this union is $ab = 1$ or $a^k = b^k$ or $[ab = -1$ and $a^{2k} = b^{2k}]$.

We can restate this theorem in terms of a and b. Note that $Y^2 = 4X^k$ is equivalent to $a^k = b^k$.

Theorem 5.3. *Suppose $a, b \in \mathbb{C}$ and $k \geq 3$. Then $Z(G)$ with $g = [a, 1, b]$ for input k-regular graphs G is* #P-hard *except in the following categories, for which the problem is in P.*

1. *$ab = 1$.*
2. *$a = b = 0$.*
3. *$ab = -1$ and $a^{2k} = b^{2k}$.*

(The same statement holds for Holant$(=_k| \ [a, 1, b], \Delta_0, \Delta_1)$*, and when it is #P-hard a reduction exists that does not use Lemma 3.13.)*

If we restrict the input to planar k-regular graphs, then these three categories are solvable in P, as well as a fourth category, $a^k = b^k$. *The problem remains* #P-hard *in all other cases.*

Exercise: Prove that Theorem 5.1 and Theorem 5.3 are equivalent.

Prove that case 3 of Theorem 5.3 is equivalent to $ab = -1$ and $a^{4k} = 1$, as well as $ab = -1$ and $b^{4k} = 1$. Prove that if k and k' are relatively prime, then the tractability conditions of #CSP($[a, 1, b]$) in Theorem 3.7 and that of $Z(G)$ in Theorem 5.3, combined for both k-regular and k'-regular graphs, are equivalent.

We will prove this dichotomy for the cases $k = 3$ and $k = 4$. For technical reasons, the situations for odd and even k are slightly different. But we will proceed in a similar fashion in both cases. We will first establish tractability, in region I for general (not necessarily planar) graphs, and in region II for planar graphs. The main task is to show everything else is #P-hard. For 3-regular graphs, we will first prove that these problems are #P-hard in region II for general graphs (where the problem is tractable when restricted to planar graphs as input). Then we move on to prove #P-hardness within region III, where the problem is #P-hard even for planar graphs. For 4-regular graphs the order is reversed. The main theme in the #P-hardness proof is to show that we can interpolate all unary (or "virtually unary") signatures. After that, in essence, we appeal to ideas and results from the framework of Holant* problems.

To prove this theorem, several new proof techniques are introduced.

1. We present a general approach to interpolate all unary signatures. This involves a recursive iteration gadget, a starter gadget (in some cases), and an analysis of eigenvalues and eigenvectors.
2. We introduce a method to construct gadgets that carry out iterations at a higher dimension, and then collapse it to a lower dimension for the purpose of constructing (virtual) unary signatures. This involves a starter gadget, a recursive iteration gadget, and a finisher gadget. We prove a lemma that guarantees that among polynomially many iterations, some subset of them will be sufficient for interpolation to succeed.
3. Algebraic symmetrization. We derive an expression of $Z(G)$ over k-regular graphs as a graph polynomial, whose degree is reduced by taking symmetry into account. This symmetry enables us to draw connections between different problems, and also reduces the size and complexity of various polynomials that will be analyzed.
4. Eigenvalue shifted pairs. These are coupled pairs of gadgets whose transition matrices differ by δI, where $\delta \neq 0$. They have shifted eigenvalues, and by

analyzing where they may fail individually, we can show that *jointly* they succeed.

5. Anti-gadgets. These are also coupled pairs of gadgets, but in which one gadget has the algebraic effect of canceling out a portion of the other gadget, resulting in a "virtual gadget" with desirable algebraic properties.
6. Gadget syzygy. This is a trio of gadgets that in their alignment we can prove they conjointly succeed.
7. Symmetrizer gadgets. Some of our constructions produce signatures with desirable properties, but they are not necessarily symmetric. If symmetric signatures are required, then gadgets can be used to introduce symmetry in the underlying graph, thereby guaranteeing that the signatures produced from the construction are also symmetric.

When the dichotomy was first proved [CLX11, KC10, CK13, CK12], an additional important idea was to use Tarski's decidability theorem for a real closed fields [Tar51, BOKR86, BPR06] and prove the desired result by a substantial amount of symbolic computation. In the present chapter, we have at least formally eliminated symbolic computation that must require machine verification. However, many constructions in this chapter are found computationally. The fact that they work as stated can be verified by hand with some work, and they have also been verified via computer programs. Symbolic computation remains a valuable tool for exploration and as a proof technique.

Our spin system is the same problem as Holant($=_k$| $[a, 1, b]$), which we will abbreviate as $\text{Hol}_k(a, b)$. In this edge–vertex incidence graph view, the input instances are $(2, k)$-regular bipartite graphs. Throughout this chapter, all $(2, k)$-regular bipartite graphs are arranged with generators (contravariant tensors) on the degree-2 side (right-hand side (RHS)) and recognizers (covariant tensors) on the degree-k side (left-hand side (LHS)).

For convenience, in this chapter we further designate dangling edges as *leading* edges if they are internally connected to a binary generator $g = [a, 1, b]$, or *trailing* edges if they are internally connected to a recognizer $=_k$. We will draw $\mathcal{F}$-gates with leading edges protruding from the left and trailing edges from the right. These $\mathcal{F}$-gates can be composed by merging leading edges of one $\mathcal{F}$-gate with trailing edges of another, respecting the underlying recognizer and generator bipartite structure. Suppose an $\mathcal{F}$-gate has m leading edges and n trailing edges. Then the signature of the $\mathcal{F}$-gate can be organized as a $2^m \times 2^n$ matrix, where the rows and columns are indexed by the 0-1 assignments to the leading and trailing edges respectively. If the number of trailing edges in one $\mathcal{F}$-gate matches the number of leading edges in another, then a new $\mathcal{F}$-gate can be formed by merging these edges, and the associated matrix is obtained

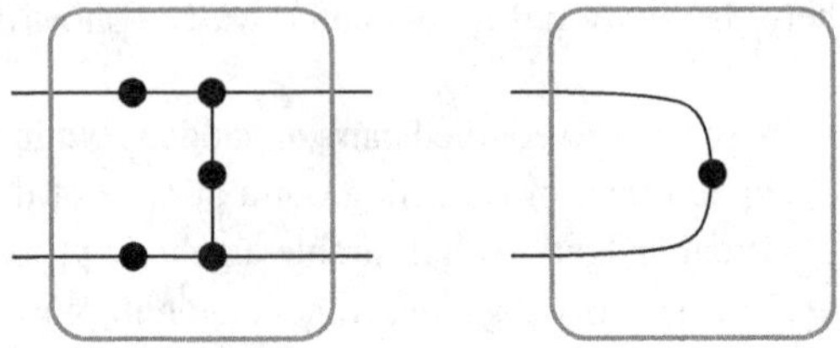

(a) A recursive gadget (b) A starter gadget

Figure 5.1. Examples of binary recursive and starter gadgets.

by multiplying the two original matrices together. In particular, an $\mathcal{F}$-gate with only leading edges would be viewed as a column vector, representing a generator, and then merging with an $\mathcal{F}$-gate having a matching number of trailing edges corresponds to premultiplication by the associated matrix. In this way we can view an $\mathcal{F}$-gate with m leading edges and n trailing edges as transforming $\mathcal{F}$-gates with arity-n generator signatures into $\mathcal{F}$-gates with arity-m generator signatures. Furthermore, the $\mathcal{F}$-gates in this chapter will transform symmetric signatures to signatures that are also symmetric. This implies that there exists an equivalent $(m+1) \times (n+1)$ matrix that operates directly on vectors written in symmetric signature notation. We will henceforth identify the transition matrix with the $\mathcal{F}$-gate itself.

We define some specific types of $\mathcal{F}$-gates. An *arity-r starter gadget* is an $\mathcal{F}$-gate with no trailing edges and r leading edges, representing an arity-r generator. An *arity-r recursive gadget* is an $\mathcal{F}$-gate with r trailing edges and r leading edges. Figure 5.1 gives an example.

When we speak of an $\mathcal{F}$-gate (where $\mathcal{F} = \mathcal{R} \cup \mathcal{G}$) or gadget M being constructible *in the context of* Holant($\mathcal{R} \mid \mathcal{G}$), we mean that the underlying graph of M has a bipartite structure, where vertices in one partition are labeled with signatures from $\mathcal{R}$ and vertices in the other partition from $\mathcal{G}$. In other words, M can appear within an input instance of Holant($\mathcal{R} \mid \mathcal{G}$) without violating any structural bipartite constraints. When we say a generator (or recognizer) is constructible in the context of Holant($\mathcal{R} \mid \mathcal{G}$), we mean that an $\mathcal{F}$-gate is constructible in that context, and that all dangling edges are internally incident to generators (or recognizers). When we say that a sequence of signatures is efficiently constructible in the context of some Holant problem, we mean that a sequence of $\mathcal{F}$-gates having those signatures can be constructed in that context, and that the first n of these can be built in time polynomial in n. When we say that a generator g (or recognizer r) can be efficiently *simulated* in the context of Holant($\mathcal{R} \mid \mathcal{G}$), we mean that Holant($\mathcal{R} \mid \mathcal{G} \cup \{g\}$) $\leq_{\mathrm{T}}^{\mathrm{P}}$ Holant($\mathcal{R} \mid \mathcal{G}$) or Holant($\mathcal{R} \cup \{r\} \mid \mathcal{G}$) $\leq_{\mathrm{T}}^{\mathrm{P}}$ Holant($\mathcal{R} \mid \mathcal{G}$), respectively.

We use Arg to denote the principal value of a complex argument, i.e., $\mathrm{Arg}(z) \in (-\pi, \pi]$ for all nonzero $z \in \mathbb{C}$, such that $z = |z|e^{\mathrm{i}\,\mathrm{Arg}(z)}$. We say that a square nonsingular matrix M has finite projective order if there exists a positive integer j, such that $M^j = \lambda I$ for some (nonzero) $\lambda \in \mathbb{C}$. Otherwise, M has infinite projective order. A nonsingular M has finite projective order iff for some positive integers $i \neq j$, $M^i = \lambda M^j$ for some (nonzero) $\lambda \in \mathbb{C}$.

5.1 3-Regular Graphs

5.1.1 Tractability in Regions I and II

Recall that for 3-regular graphs, region I is specified by $X = 1$ or $(X, Y) \in \{(0,0), (-1,0), (-1, \pm 2\mathrm{i})\}$. Region II is specified by $Y^2 = 4X^3$ but $X \notin \{0, \pm 1\}$. The union of regions I and II is specified by $X = 1$ or $Y^2 = 4X^3$ or $(X, Y) = (-1, 0)$.

The next lemma introduces the technique of algebraic symmetrization. We show that over 3-regular graphs, the Holant value is expressible as a polynomial $P(X, Y)$ with integer coefficients, where $X = ab$ and $Y = a^3 + b^3$. This change of variable, from (a, b) to (X, Y), is important in two ways. First, it allows us to derive tractability and hardness results more easily. It draws a connection between problems that may appear unrelated, but the tractability, and respectively hardness, of one problem implies the other. Second, it facilitates the proof of hardness by reducing the degree of the polynomials involved. Such a reduction in degree also makes symbolic computation more feasible, although we do not use this particular approach here.

Lemma 5.4. *Let G be a 3-regular graph. Then there exists a polynomial $P_G(\cdot, \cdot)$ with two variables and integer coefficients such that for any signature grid Ω having underlying graph G, if the edge function is $[a, 1, b]$, then $Z(G) = P_G(ab, a^3 + b^3)$.*

Proof. Consider any 0-1 vertex assignment σ with a nonzero valuation. If σ' is the complement assignment switching every 0 and 1 in σ, then for σ and σ', we have the sum of valuations $a^i b^j + a^j b^i$ for some i and j. Here i (resp. j) is the number of edges connecting two degree-3 vertices both assigned 0 (resp. 1) by σ. We note that $a^i b^j + a^j b^i = (ab)^{\min(i,j)}(a^{|i-j|} + b^{|i-j|})$.

We prove $i \equiv j \pmod 3$ inductively. For the all-0 assignment, this is clear since in this case i is the number of edges, $j = 0$, and the number of edges is divisible by 3 for a 3-regular graph. Now starting from any assignment σ, if we switch the assignment on one vertex from 0 to 1, it is easy to verify that it changes the valuation from $a^i b^j$ to $a^{i'} b^{j'}$, where $i - j = i' - j' + 3$. As every

0-1 assignment is obtainable from the all-0 assignment by a sequence of switches, the conclusion $i \equiv j \pmod 3$ follows.

Now $a^i b^j + a^j b^i = (ab)^{\min(i,j)}(a^{3\ell} + b^{3\ell})$, for some $\ell \geq 0$ and a simple induction

$$a^{3(\ell+1)} + b^{3(\ell+1)} = (a^{3\ell} + b^{3\ell})(a^3 + b^3) - (ab)^3(a^{3(\ell-1)} + b^{3(\ell-1)})$$

shows that the partition function $Z(G)$ is a polynomial $P_G(ab, a^3 + b^3)$ with integer coefficients. □

Corollary 5.5. *If $X = -1$ and $Y \in \{0, \pm 2\mathrm{i}\}$, then* $\mathrm{Hol}_3(a, b)$ *is in P.*

Proof. The problems $\mathrm{Hol}_3(1, -1)$, $\mathrm{Hol}_3(-\mathrm{i}, -\mathrm{i})$, and $\mathrm{Hol}_3(\mathrm{i}, \mathrm{i})$ are all solvable in P (these fall within the families $\mathscr{F}_1$, $\mathscr{F}_2$, and $\mathscr{F}_3$ from Chapter 3); $X = -1$ for each of these three problems, whereas the value of Y is 0, $2\mathrm{i}$, and $-2\mathrm{i}$ respectively. Since the value $Z(G)$ of any 3-regular signature grid is completely determined by (X, Y) and the polynomial $P_G(\cdot, \cdot)$ (which depends only on the underlying graph G), $\mathrm{Hol}_3(a, b)$ is computable in polynomial time for any a and b such that $ab = -1$ and $a^3 + b^3 \in \{0, \pm 2\mathrm{i}\}$ (i.e., $ab = -1$ and $a^{12} = 1$). □

We now list all the cases where $\mathrm{Hol}_3(a, b)$ is computable in polynomial time.

Theorem 5.6. *If any of the following four conditions is true, then* $\mathrm{Hol}_3(a, b)$ *is solvable in polynomial time:*

1. $X = 1$,
2. $X = Y = 0$,
3. $X = -1$ *and* $Y \in \{0, \pm 2\mathrm{i}\}$
4. $Y^2 = 4X^3$ *and the input is restricted to planar graphs.*

Proof. If $X = 1$ then the signature $[a, 1, b]$ is degenerate and the Holant can be computed in polynomial time. If $X = Y = 0$ then $a = b = 0$, and a 2-coloring algorithm can be employed on the edges. If $X = -1$ and $Y \in \{0, \pm 2\mathrm{i}\}$ then by Corollary 5.5 we are done. If we restrict the input to planar graphs and $4X^3 = Y^2$ (equivalently, $a^3 = b^3$), holographic algorithms with matchgates from Chapter 4 can be applied. Indeed, if $ab = 0$ then $a = b = 0$ and this is case 2, which is tractable even for general graphs. Suppose $ab \neq 0$, and let $\omega = a/b$. We apply a holographic transformation defined by $T = \begin{bmatrix} 1 & 0 \\ 0 & \omega \end{bmatrix}$; then $(=_3)T^{\otimes 3} = (=_3)$, and $(T^{-1})^{\otimes 2}[1, \frac{1}{a}, \frac{b}{a}] = [1, \frac{b}{a^2}, 1]$, using $a^3 = b^3$. Then under a further holographic transformation by $\begin{bmatrix} 1 & 1 \\ 1 & -1 \end{bmatrix}$, both $(=_3)$ and any $[1, c, 1]$ are simultaneously realizable by matchgates. (This can be directly verified, but a more systematic approach is to use the theory developed in Chapter 4, in particular Theorem 4.33.) □

We note that the first three tractable cases also hold for Holant($=_3|$ $[a, 1, b], \Delta_0, \Delta_1$). Also note that diagonal (or antidiagonal) holographic transformations do not change $\{\Delta_0, \Delta_1\}$ up to a nonzero constant.

The main task in Section 5.1 is to prove that all remaining problems are #P-hard for 3-regular graphs.

5.1.2 Planar Tractable Region II Is #P-hard for General Graphs

In this subsection we get acquainted with our technique of proving hardness for Holant problems over regular graphs. Our specific application at the moment is to prove that $\text{Hol}_3(a, b)$ is #P-hard assuming $a^3 = b^3$ and $X \notin \{-1, 0, 1\}$. (By Theorem 5.6, $\text{Hol}_3(a, b)$ is tractable under these assumptions when the input is restricted to planar graphs.) We first make the following simplifying observation: Instead of working under $a^3 = b^3$, we may as well assume $a = b$.

Lemma 5.7. *Suppose $Y^2 = 4X^3$ (equivalently, $a^3 = b^3$). Then there exist $a', b' \in \mathbb{C}$ such that $a' = b'$, $a'b' = X$, and $(a')^3 + (b')^3 = Y$.*

Proof. If $X = 0$ then $Y = 0$ and $a = b = 0$ so we are already done. Otherwise, $X \neq 0$ implies $a \neq 0$ and we may set $a' = b' = b^2a^{-1}$ so that $a'b' = (b^2a^{-1})^2 = b^4a^{-2} = ab = X$ and $(a')^3 + (b')^3 = b^6a^{-3} + b^6a^{-3} = a^3 + b^3 = Y$. □

A recurring theme we will see for proving hardness in this chapter is

1. Obtain an infinite set of pairwise linearly independent signatures.
2. If necessary, project the signatures down to a lower dimension while retaining pairwise linear independence.
3. Use an interpolation argument to simulate *any* signature in that lower dimensional space.

The ability to simulate any signature in the lower dimensional space will lead to a reduction to prove #P-hardness. By "lower dimension" we usually mean unary signatures (i.e., a signature sequence of the form $[x_j, y_j]$ for $j \geq 0$). However, in some circumstances (such as for $\text{Hol}_3(a, a)$), it is impossible to do so without sacrificing pairwise linear independence:

Lemma 5.8. *Let $\mathcal{F}$ be a set of signatures each satisfying the complementary invariance property, namely its value remains unchanged under the complement of all input bits. Then any $\mathcal{F}$-gate has this same property.*

Proof. For any $f \in \mathcal{F}$, if σ and σ' are complementary assignments to its inputs, then they take the same value $f|_{\sigma} = f|_{\sigma'}$. If σ and σ' are complementary assignments to the dangling edges of an $\mathcal{F}$-gate, by complementing the assignments to all internal edges, we have a 1-1 correspondence of all terms defining the signature sum of the $\mathcal{F}$-gate under σ with those under σ', and each pair of corresponding terms has the same value. □

Corollary 5.9. *If $\mathcal{F} = \{[a, 1, a], =_k\}$ for some positive integer k, then any arity 1 or arity 2 $\mathcal{F}$-gate has a signature of the form $[x, x]$ or $[x, y, x]$ respectively.*

Proof. Immediate from Lemma 5.8. □

Given this knowledge, we aim to produce a sequence of "virtual unary" signatures rather than actual unary signatures in such cases. Specifically, we will obtain pairwise linearly independent signatures of the form $[x_j, y_j, x_j]$ instead of $[x_j, y_j]$. Owing to the symmetry exhibited in these binary signatures, we will be able to carry out interpolation without projecting them down to a lower dimension.

Unary Recursive Construction

Given a unary recursive gadget M, we can chain j copies of M together (merging the leading edges of one copy with the trailing edges of the next) to arrive at an arity-2 $\mathcal{F}$-gate with signature M^j. We want such a sequence $\{M^j\}_{j\geq 0}$ of signatures to be pairwise linearly independent.

Lemma 5.10. *Let M be a nonsingular square matrix. Then M has infinite projective order if and only if $\{M^j\}_{j\geq 0}$ are pairwise linearly independent. In particular, if M is the transition matrix for some arity-r recursive gadget, then M has infinite projective order if and only if $\{M^j\}_{j\geq 0}$ is a sequence of pairwise linearly independent signatures.*

Proof. Immediate from definitions. □

The eigenvalues of M give us a convenient way to check that this sequence of $\mathcal{F}$-gates really has the property that the signatures M^j are pairwise linearly independent. In other words, we are reducing the problem of generating a sequence of pairwise linearly independent signatures to a test in terms of eigenvalues.

Lemma 5.11. *Let $M \in \mathbb{C}^{r\times r}$ be a nonsingular matrix, where the ratio of two of its eigenvalues is not a root of unity. Then $\{M^j\}_{j\geq 0}$ is a sequence of pairwise linearly independent matrices.*

Proof. We write $M = TJT^{-1}$ in its Jordan normal form, where T is invertible, and J is a triangular matrix with all eigenvalues of M appearing on its diagonal.

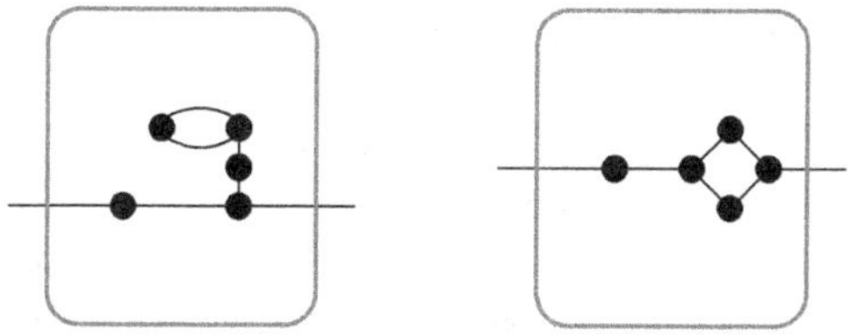

(a) Gadget M_1 (b) Gadget M_2

Figure 5.2. A useful pair of unary recursive gadgets.

Suppose that M has finite projective order. Then for some $m > 0$ we have $M^m = \lambda I$ for some nonzero $\lambda \in \mathbb{C}$. Then $T^{-1}J^mT = \lambda I$ and hence $J^m = \lambda I$. So $\alpha^m = \beta^m$ for all eigenvalues α and β of M. This is a contradiction. We conclude that M has infinite projective order and by Lemma 5.10 $\{M^j\}_{j\geq 0}$ is a sequence of pairwise linearly independent signatures. □

Suppose $a = b$. Depending on the value of a and our gadget M, the sequence $\{M^j\}_{j\geq 0}$ may or may not be pairwise linearly independent, but we will show that for every $[a, 1, a]$ that does not correspond to a tractable case, there exists some unary recursive gadget with this property. There is a technical detail in the interpolation step; we will want these $\mathcal{F}$-gates to take the place of generator signatures in a signature grid to preserve the bipartite structure of the graph. To address this, we tack on an extra degree-2 vertex labeled with the generator $[a, 1, a]$, making the full gadget a binary generator.

Lemma 5.12. *Suppose $a = b$ and $a^4 \notin \{0, 1\}$. Then there exists a unary recursive gadget M for which the generator sequence $\{M^j g\}_{j\geq 0}$ is pairwise linearly independent, where $g = \begin{bmatrix} a & 1 \\ 1 & a \end{bmatrix}$ is a genarator $\mathcal{F}$-gate having a single vertex.*

Proof. Suppose the real part of a is nonzero, and consider gadget M_1 (see Figure 5.2),

$$M_1 = \begin{bmatrix} a^3 + a^2 & a^2 + a \\ a^2 + a & a^3 + a^2 \end{bmatrix}.$$

The eigenvalues of M_1 are $a(a+1)^2$ and $a(a^2 - 1)$, both of which are nonzero. Their ratio is $\frac{a+1}{a-1}$, which is not a root of unity, as $|a + 1| \neq |a - 1|$. Note that $|a + 1| = |a - 1|$ consists of exactly those values a on the imaginary line $i\mathbb{R}$.

Now assume the real part of a is 0, and consider gadget M_2 (see Figure 5.2),

$$M_2 = \begin{bmatrix} a^3 + 1 & a^2 + a \\ a^2 + a & a^3 + 1 \end{bmatrix}.$$

The eigenvalues are $(a+1)(a-1)^2$ and $(a+1)(a^2+1)$, both nonzero. The ratio of these is $\frac{a^2-2a+1}{a^2+1}$. Since a is purely imaginary, a^2+1 is real-valued. Since a is furthermore nonzero and belongs to $\mathrm{i}\mathbb{R}$, $|a^2+1-2a| > |a^2+1|$ and thus $\frac{a^2-2a+1}{a^2+1}$ is not a root of unity.

By Lemma 5.11, we have a unary recursive gadget M for which $\{M^j\}_{j\geq 0}$ is a sequence of pairwise linearly independent signatures. Pairwise linear independence is preserved under invertible linear transformations defined by g on the set $\{M^j\}_{j\geq 0}$. As $M^j \mapsto M^j g$ is an invertible linear transformation, where $g = \begin{bmatrix} a & 1 \\ 1 & a \end{bmatrix}$, we conclude that $\{M^j g\}_{j\geq 0}$ is also pairwise linearly independent. □

Exercise: Compute the signature matrices of gadgets M_1 and M_2 for edge function $[a, 1, b]$. Note that $M_1 - M_2 = (ab-1)I$. Setting $a = b$ as in Lemma 5.12, verify that the signature matrices are as stated.

Interpolation and Reduction

Suppose we have a sequence of pairwise linear independent $\{[x_j, y_j, x_j]\}_{j\geq 0}$ of generators. We show how to simulate *any* generator of the form $[x, y, x]$.

Lemma 5.13. *Let $k \geq 3$ be fixed and suppose that in the context of* $\mathrm{Hol}_k(a, b)$ *we can efficiently construct a sequence of pairwise linearly independent generators of the form $\{[x_j, y_j, x_j]\}_{j\geq 0}$. Then for any $x, y \in \mathbb{C}$ we can efficiently simulate the generator $[x, y, x]$.*

Proof. Let Ω be a bipartite signature grid for the problem $\text{Holant}(\mathcal{R} \mid \mathcal{G} \cup \{[x, y, x]\})$, and suppose the generator $[x, y, x]$ occurs n times in Ω. We stratify the Holant sum according to how many times the generator $[x, y, x]$ evaluates to x. Collecting these terms together, we write $\text{Holant}_\Omega = \sum_{j=0}^{n} c_j x^j y^{n-j}$, where c_j is the sum over all edge assignments of the product of signature evaluations other than the n copies of $[x, y, x]$, subject to the restriction that exactly j of the edge pairs connecting to $[x, y, x]$ are assigned both 0's or both 1's.

We construct a sequence of bipartite signature grids Ω_i for the problem $\text{Holant}(\mathcal{R} \mid \mathcal{G})$, by replacing each occurrence of $[x, y, x]$ with a copy of an $\mathcal{F}$-gate with signature $[x_i, y_i, x_i]$, for $i = 0, 1, \ldots, n$. (By omitting at most one element from our sequence and relabeling, we assume that $y_i \neq 0$ for all i.) Note that the bipartite structure is maintained, and the Holant of Ω_i evaluates to $\sum_{j=0}^{n} c_j x_i^j y_i^{n-j}$, where the same c_j's appear in this sum, for all i. By oracle queries to $\text{Holant}(\mathcal{R} \mid \mathcal{G})$, we obtain a Vandermonde system where c_j are the unknowns.

$$\begin{bmatrix} y_0^{-n} \cdot \text{Holant}_{\Omega_0} \\ y_1^{-n} \cdot \text{Holant}_{\Omega_1} \\ \vdots \\ y_n^{-n} \cdot \text{Holant}_{\Omega_n} \end{bmatrix} = \begin{bmatrix} x_0^0 y_0^0 & x_0^1 y_0^{-1} & \cdots & x_0^n y_0^{-n} \\ x_1^0 y_1^0 & x_1^1 y_1^{-1} & \cdots & x_1^n y_1^{-n} \\ \vdots & \vdots & \ddots & \vdots \\ x_n^0 y_n^0 & x_n^1 y_n^{-1} & \cdots & x_n^n y_n^{-n} \end{bmatrix} \begin{bmatrix} c_0 \\ c_1 \\ \vdots \\ c_n \end{bmatrix}.$$

Note that this Vandermonde system has full rank iff $[x_j, y_j, x_j]$ are pairwise linearly independent, namely x_j/y_j are pairwise distinct. Solving this linear system for c_j $(0 \le j \le n)$ yields the answer to $\text{Holant}_\Omega = \sum_{j=0}^{n} c_j x^j y^{n-j}$. □

Now we will use the interpolatation result to simulate the signature $[1, 0, 1]$, in order to get a reduction from $\text{Holant}(=_1, =_2, =_3, \ldots \mid [a, 1, a])$, namely #CSP($[a, 1, a]$), to prove #P-hardness.

Lemma 5.14. *Suppose the parameters (a, b) are in region II; equivalently, suppose $X, Y \in \mathbb{C}$, $a^3 = b^3$, and $X \notin \{0, 1, -1\}$. Then* $\text{Hol}_3(a, b)$ *is* #P-hard.

The same statement holds for $\text{Holant}(=_3 \mid [a, 1, b], \Delta_0, \Delta_1)$, *and a reduction exists that does not use Lemma 3.13 when it is #P-hard.*

Proof. By Lemmas 5.7 and 5.4 we may assume that $a = b$. Since $X \notin \{0, \pm 1\}$ and $a = b$ we know $a^4 = (ab)^2 \notin \{0, 1\}$. We have a sequence of pairwise linearly independent generator signatures from Lemma 5.12, and by Corollary 5.9 they all have the form $[x, y, x]$. Therefore by Lemma 5.13 we can efficiently simulate generator signature $[1, 0, 1]$. We also know from the dichotomy theorem for Boolean #CSP problems, Theorem 3.7, that #CSP($[a, 1, a]$) is #P-hard. We have a reduction from #CSP($[a, 1, a]$), which is by definition $\text{Holant}(\{=_1, =_2, =_3, \ldots, \} \mid [a, 1, a])$.

$$\begin{aligned} \text{Holant}(\{=_1, =_2, =_3, \ldots, \} \mid [a, 1, a]) &\le_T \text{Holant}(=_3 \mid \{[a, 1, a], [1, 0, 1]\}) \\ &\le_T \text{Holant}(=_3 \mid [a, 1, a]). \end{aligned}$$

In the first step, we can use $(=_2) = [1, 0, 1]$ on the RHS and $(=_3)$ on the LHS to simulate $(=_k)$, for any k. The second step is by interpolation.

For $\text{Holant}(=_3 \mid [a, 1, b], \Delta_0, \Delta_1)$, we cannot apply Lemma 5.7, and we give an alternative proof using a diagonal holographic transformation. Let $\omega = a/b$; then $\omega^3 = 1$. By a holographic transformation $T = \begin{bmatrix} \omega & 0 \\ 0 & \omega^{-1} \end{bmatrix}$ the EQUALITY $(=_3)$ is transformed to $(=_3)T^{\otimes 3} = (=_3)$, we have

$(T^{-1})^{\otimes 2}[a, 1, b] = [a/\omega^2, 1, b\omega^2]$, as seen by its matrix form

$$\begin{bmatrix} \omega^{-1} & 0 \\ 0 & \omega \end{bmatrix} \begin{bmatrix} a & 1 \\ 1 & b \end{bmatrix} \begin{bmatrix} \omega^{-1} & 0 \\ 0 & \omega \end{bmatrix} = \begin{bmatrix} a/\omega^2 & 1 \\ 1 & b\omega^2 \end{bmatrix}.$$

Meanwhile T keeps $\{\Delta_0, \Delta_1\}$ invariant up to a nonzero constant. Note that $a/\omega^2 = b\omega^2$, with the same X and Y. □

Exercise: For any $k \geq 1$, construct a gadget that has signature $(=_k)$ on the LHS, using $(=_2)$ on the RHS and $(=_3)$ on the LHS. For what values of d and d' in place of 2 and 3 can one construct $(=_k)$ for all $k \geq 1$?

5.1.3 Hardness in Region III Where $(X, Y) \notin \mathbb{R} \times \mathbb{R}$

Recall that region III is the complement of the union of regions I and II, which is specified by $X \neq 1$ and $Y^2 \neq 4X^3$ (equivalently $a^3 \neq b^3$) and $(X, Y) \neq (-1, 0)$.

So far, we have seen one complete iteration of the general strategy we use in this chapter. We devised a scheme to produce an infinite set of pairwise linearly independent (virtual) unary signatures, found some gadgets to implement that scheme, used interpolation to simulate *any* such (virtual) unary signature, and applied that result to complete a reduction from a known #P-hard problem to prove #P-hardness. We will continue to develop the theory to make our tools applicable to wider classes of problems. Specifically, in this subsection we will prove #P-hardness for $\text{Hol}_3(a, b)$, subject to the restrictions that $(X, Y) \notin \mathbb{R} \times \mathbb{R}$, $a^3 \neq b^3$, and $X \neq 1$. To prove #P-hardness for region III, we must use planar gadgets and use an initial problem to reduce from that is #P-hard even for planar graphs.

Unary Recursive Construction with Starter Gadgets

As before, we first seek to produce a sequence of pairwise linearly independent signatures. We expand on the unary recursive construction by introducing a *starter gadget*. The following lemma is stated in terms of vectors and matrices so that we can reuse it for region II of $\text{Hol}_4(a, b)$ later on.

Lemma 5.15. *Let $M \in \mathbb{C}^{2\times 2}$ be a nonsingular matrix with infinite projective order, and let $s \in \mathbb{C}^2$ be a vector that is not orthogonal to any row eigenvector of M; i.e., the dot product with any row eigenvector of M is nonzero. Then $\{M^j s\}_{j\geq 0}$ is a sequence of pairwise linearly independent vectors in $\mathbb{C}^2$.*

Proof. Suppose that M is diagonalizable and we write $M = T^{-1}JT$, where $J = \begin{bmatrix} \alpha & 0 \\ 0 & \beta \end{bmatrix}$ and the rows of T are the row eigenvectors of M. As M is nonsingular, we have $\alpha\beta \neq 0$. We know by assumption that $Ts = \begin{bmatrix} x \\ y \end{bmatrix}$ for some $x, y \in \mathbb{C}$, where $xy \neq 0$. Then

$$M^j s = T^{-1}\begin{bmatrix} \alpha^j & 0 \\ 0 & \beta^j \end{bmatrix}\begin{bmatrix} x \\ y \end{bmatrix} = T^{-1}\begin{bmatrix} x & 0 \\ 0 & y \end{bmatrix}\begin{bmatrix} \alpha^j \\ \beta^j \end{bmatrix}.$$

The vectors $\begin{bmatrix} \alpha^j \\ \beta^j \end{bmatrix}$ are pairwise linearly independent since $\frac{\alpha}{\beta}$ is not a root of unity, and this property is preserved by the invertible linear transformation $T^{-1}\begin{bmatrix} x & 0 \\ 0 & y \end{bmatrix}$, so $\{M^j s\}_{j\geq 0}$ is a sequence of pairwise linearly independent vectors.

Now suppose that M is not diagonalizable. Then we write the Jordan normal form of M as $M = T^{-1}JT$, where $J = \begin{bmatrix} \alpha & 0 \\ 1 & \alpha \end{bmatrix}$ and the first row of T is a row eigenvector corresponding to α. We know by assumption that $Ts = \begin{bmatrix} x \\ y \end{bmatrix}$ for some $x, y \in \mathbb{C}$, where $x \neq 0$. Then

$$M^j s = T^{-1}J^j Ts = \alpha^j T^{-1}\begin{bmatrix} 1 & 0 \\ j\alpha^{-1} & 1 \end{bmatrix}\begin{bmatrix} x \\ y \end{bmatrix} = \alpha^j T^{-1} v_j,$$

where the vectors $v_j = \begin{bmatrix} x \\ j\alpha^{-1}x + y \end{bmatrix}$ are pairwise linearly independent. This uses the fact that $x \neq 0$. This pairwise linear independence is preserved by the invertible linear transformation T^{-1}; thus after a scalar multiple α^j, the vectors $\alpha^j T^{-1} v_j$ are also pairwise linearly independent. □

Corollary 5.16. *Let M be a unary recursive gadget with infinite projective order, and let s be a unary starter gadget whose signature is not orthogonal to any row eigenvector of M. Then the sequence $\{M^j s\}_{j\geq 0}$ has pairwise linearly independent generator signatures.*

Proof. Immediate. □

Surprisingly, a set of general-purpose starter gadgets can be made for this construction as long as $ab \neq 1$ and $a^3 \neq b^3$, so we refine this corollary. It is easy to see that $ab = 1$ is a tractable case, as $[a, 1, b] = [a, 1, a^{-1}]$ is degenerate. Recall that $a^3 = b^3$ essentially corresponds to region II.

Theorem 5.17. *In the context of* $\mathrm{Hol}_3(a, b)$ *let $a, b \in \mathbb{C}$ such that $ab \neq 1$ and $a^3 \neq b^3$, and let M be a unary recursive gadget with infinite projective order. Then there exists a starter gadget s (also in the context of* $\mathrm{Hol}_3(a, b)$*) such*

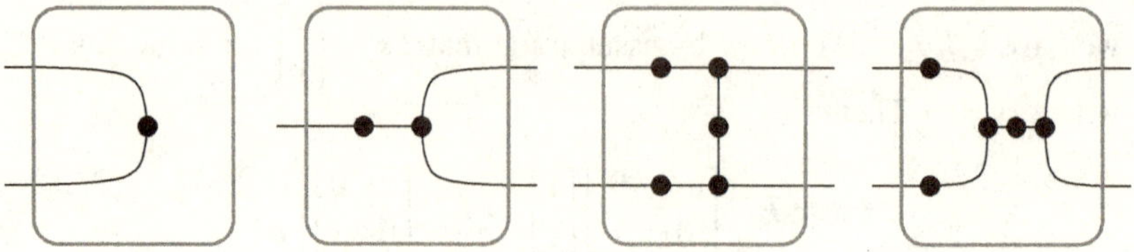

(a) Gadget M_3 (b) Gadget M_4 (c) Gadget M_5 (d) Gadget M_6

Figure 5.3. Gadgets for constructing general-purpose starter gadgets.

that $\{M^j s\}_{j\geq 0}$ is a sequence of pairwise linearly independent unary generator signatures.

Proof. Let M_3, M_4, M_5, and M_6 be defined as in Figure 5.3. We calculate their signature matrices

$$v = M_3 = \begin{bmatrix} a \\ 1 \\ b \end{bmatrix}, \quad M_4 = \begin{bmatrix} a & 0 & 1 \\ 1 & 0 & b \end{bmatrix}, \quad M_5 = \begin{bmatrix} a^3 & 2a & b \\ a^2 & ab+1 & b^2 \\ a & 2b & b^3 \end{bmatrix},$$

$$M_6 = \begin{bmatrix} a^3+1 & 0 & a^2+b \\ a^2+b & 0 & a+b^2 \\ a+b^2 & 0 & b^3+1 \end{bmatrix}.$$

Note that these matrices operate directly on symmetric signatures, e.g., the middle column of M_5 corresponds to the sum of the assignments of 01 and 10 on the trailing edges.

The 2×2 matrices $[M_4 M_5 v \; M_4 v]$ and $[M_4 M_6 v \; M_4 v]$ both have determinant $(a^3 - b^3)(ab-1)^2$, whereas $\det([M_4 M_5 v \; M_4 M_6 v]) = (a^3 - b^3)(ab-1)^3$, so the vectors $M_4 v$, $M_4 M_5 v$, and $M_4 M_6 v$ are pairwise linearly independent provided that $ab \neq 1$ and $a^3 \neq b^3$. The set of vectors that are orthogonal by dot product to at least one row eigenvector of the given matrix $M \in \mathbb{C}^{2\times 2}$ can be expressed as a union of at most two one-dimensional linear subspaces. By the pigeonhole principle, this union cannot contain three pairwise linearly independent vectors. Hence at least one element of $\{M_4 M_5 v, M_4 M_6 v, M_4 v\}$ is not orthogonal to any row eigenvector of M. The corresponding starter gadget can be used with M in a recursive construction and we get a sequence of pairwise linearly independent unary generator signatures by Corollary 5.16. □

Exercise: Verify the signature matrices for gadgets M_3, M_4, M_5, and M_6 are as stated. Notice that these matrices are in a form that can be directly applied to symmetric signatures. Give a simple explanation that the middle columns of M_4 and M_6 are both 0.

We make two remarks:

1. For a nonzero vector $s \in \mathbb{C}^2$, and a nonsingular matrix $M \in \mathbb{C}^{2\times 2}$, the condition that s is not a column eigenvector of M is equivalent to $\det[s\ Ms] \neq 0$, which is also equivalent to the condition that s is not orthogonal by dot product to any row eigenvector of M.
2. A sufficient condition for a nonsingular matrix $M \in \mathbb{C}^{2\times 2}$ to have infinite projective order is that the ratio of the eigenvalues is not a root of unity, which in turn is implied by the two eigenvalues having different norms.

Interpolation with Starter Gadgets

Since Theorem 5.17 already produced a sequence of unary generators (provided a suitable unary recursive gadget is given), there is no need to project it down to a lower dimension. Thus, the plan is to show that we can interpolate any unary generator, along similar lines to Lemma 5.13. However, we will prove it for recognizer signatures of the form $[x, 0, y]$ instead, since this allows us to reuse this lemma later on. Note that we can easily convert from generators of the form $[x, y]$ to recognizers of the form $[x, 0, y]$ by attaching a single vertex with recognizer $=_3$.

Lemma 5.18. *Let $k \geq 3$ be fixed and suppose that in the context of* $\text{Hol}_k(a, b)$, *we can efficiently construct a sequence of pairwise linearly independent recognizers of the form $\{[x_j, 0, y_j]\}_{j\geq 0}$. Then for any $x, y \in \mathbb{C}$ we can efficiently simulate recognizer* $[x, 0, y]$.

Proof. By ignoring at most one signature in the sequence $\{[x_j, 0, y_j]\}_{j\geq 0}$ and relabeling the indices, we assume that each y_j is nonzero. Let Ω be a bipartite signature grid for the problem $\text{Holant}(\mathcal{R} \cup \{[x, 0, y]\} \mid \mathcal{G})$. Suppose recognizer $[x, 0, y]$ occurs n times in Ω. Dropping terms that evaluate to 0, we stratify the Holant sum according to how many times the signature $[x, 0, y]$ evaluates to x. Collecting these terms together, we write $\text{Holant}_\Omega = \sum_{j=0}^{n} c_j x^j y^{n-j}$, where c_j is the sum over all edge assignments of the product of signature evaluations other than the n copies of $[x, 0, y]$, subject to the restriction that exactly j of the edge pairs connecting to $[x, 0, y]$ are both assigned 0, and exactly $n - j$ of the edge pairs connecting to $[x, 0, y]$ are both assigned 1.

We construct a sequence of bipartite signature grids Ω_i for the Holant problem $\text{Holant}(\mathcal{R} \mid \mathcal{G})$, by replacing each occurrence of $[x, 0, y]$ with a copy of an $\mathcal{F}$-gate with signature $[x_i, 0, y_i]$, for $i = 0, 1, \ldots, n$. This has the effect of replacing each occurrence of $[x, 0, y]$ by pairwise linearly independent signatures. Note that the bipartite structure is maintained. By oracle queries to $\text{Holant}(\mathcal{R} \mid \mathcal{G})$, we obtain a nonsingular Vandermonde system where c_j are the unknowns. Solving this linear system for c_j $(0 \le j \le n)$ yields the answer to $\text{Holant}_\Omega = \sum_{j=0}^{n} c_j x^j y^{n-j}$. □

When we apply Theorem 5.17, we will show that the eigenvalues of M have unequal norm. The following lemma gives a concrete way to test this, by examining the characteristic polynomial of M.

Lemma 5.19. *If the roots of a polynomial* $x^2 - Bx + C \in \mathbb{C}[x]$ *have the same norm, then* $B|C| = \overline{B}C$ *and* $B^2\overline{C} = \overline{B}^2 C$. *If further* $B \neq 0$ *and* $C \neq 0$, *then* $\text{Arg}(B^2) = \text{Arg}(C)$.

Proof. If the roots have equal norm, then for some $a, b \in \mathbb{C}$ with $|a| = |b| = 1$, and nonnegative $r \in \mathbb{R}$, we can write $x^2 - Bx + C = (x - ra)(x - rb)$, so $B|C| = r(a + b)r^2 = r(a^{-1} + b^{-1})r^2 ab = \overline{B}C$. Squaring both sides and dividing by C, we have $B^2\overline{C} = \overline{B}^2 C$ (note that this equality still holds when $C = 0$). Multiplying $B|C| = \overline{B}C$ by B we get $B^2|C| = |B^2|C$, and if B and C are both nonzero then $\frac{B^2}{|B^2|} = \frac{C}{|C|}$, that is, $\text{Arg}(B^2) = \text{Arg}(C)$. □

Eigenvalue Shifted Pairs and Other Gadgetry

Given any fixed $a, b \in \mathbb{C}$ for which $\text{Hol}_3(a, b)$ happens to be #P-hard, it is usually a straightforward process of trial and error of using Lemma 5.19 to find a unary recursive gadget with infinite projective order, so as to satisfy the conditions of Theorem 5.17. This would be fine if we were only aiming to prove hardness of a small finite number of problems. However, given some particular unary recursive gadget, there will inevitably be some settings of a and b that do not work. It is natural to add more gadgets to the mix, but the coefficients of the characteristic polynomial of M are themselves polynomials in a and b. Also, the degrees of these polynomials can get rather high and somewhat cumbersome (depending on the size of the gadget M). Algebraic symmetrization can help by reducing the complexity of the polynomials involved, but then we are still left with the question: Given two polynomials $p(X, Y)$ and $q(X, Y)$, under which settings of $X, Y \in \mathbb{C}$ is it the case that $\text{Arg}(p(X, Y)) = \text{Arg}(q(X, Y))$? Moreover, given many such pairs of polynomials, how can we be assured that all relevant X and Y are "covered" by one or more gadgets?

We now introduce a technique called *eigenvalue shifted pairs*, or ESPs for short, to address this problem. Suppose we have unary recursive gadgets M and M', and suppose the eigenvalues of M are those of M' shifted by some complex number δ. Then unless δ "lines up just right," at least one of the two matrices will have eigenvalues with unequal norm, which is just what we need.

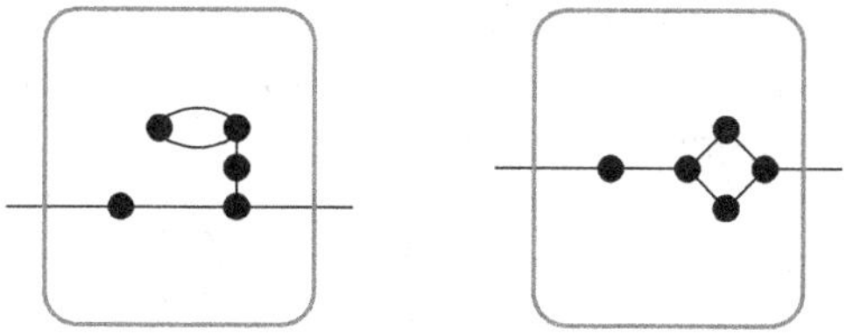

(a) Gadget M_1 (b) Gadget M_2

Figure 5.4. An eigenvalue-shifted pair of gadgets.

Definition 5.20. A pair of nonsingular square matrices M and M' is called an *eigenvalue shifted pair (ESP)* if $M' = M + \delta I$ for some nonzero $\delta \in \mathbb{C}$, and M has distinct eigenvalues.

Clearly for such a pair, M' also has distinct eigenvalues. For example, the transition matrices of gadgets M_1 and M_2 (Figure 5.4) differ only by $ab - 1$ along the main diagonal, and form an ESP for all $a, b \in \mathbb{C}$, except when $ab = 1$ (these are the same gadgets we saw earlier in Figure 5.2). Before we make use of this ESP, we state a technical lemma so we can describe the conditions under which an ESP *doesn't* work.

Lemma 5.21. *Suppose $\alpha, \beta, \delta \in \mathbb{C}$, $|\alpha| = |\beta|$, $\alpha \neq \beta$, $\delta \neq 0$, and $|\alpha + \delta| = |\beta + \delta|$. Then there exist $r, s \in \mathbb{R}$ such that $r\delta = \alpha + \beta$ and $s\delta^2 = \alpha\beta$.*

Proof. After a rotation in the complex plane, namely multiplying α, β, δ by some $e^{i\theta}$, we can assume $\alpha = \overline{\beta}$, and then since $\alpha + \beta, \alpha\beta \in \mathbb{R}$ we just need to prove $\delta \in \mathbb{R}$. Then $(\alpha + \delta)\overline{(\alpha + \delta)} = |\alpha + \delta|^2 = |\beta + \delta|^2 = (\beta + \delta)\overline{(\beta + \delta)} = (\overline{\alpha} + \delta)(\alpha + \overline{\delta})$ and we distribute to get $\alpha\overline{\alpha} + \delta\overline{\delta} + \alpha\overline{\delta} + \overline{\alpha}\delta = \alpha\overline{\alpha} + \delta\overline{\delta} + \overline{\alpha}\overline{\delta} + \alpha\delta$. Canceling repeated terms and factoring, we have $(\overline{\alpha} - \alpha)(\overline{\delta} - \delta) = 0$, and since $\alpha \neq \beta = \overline{\alpha}$ we get $\overline{\delta} = \delta$; therefore $\delta \in \mathbb{R}$. □

Corollary 5.22. *Let M and M' be an eigenvalue shifted pair of 2×2 matrices. If M and M' have the property that both eigenvalues have the same norm, then* $\operatorname{tr}(M)/\delta \in \mathbb{R}$ *and* $\det(M)/\delta^2 \in \mathbb{R}$.

Proof. Let α and β be the eigenvalues of M, so $\alpha + \delta$ and $\beta + \delta$ are the eigenvalues of M'. Suppose that $|\alpha| = |\beta|$ and $|\alpha + \delta| = |\beta + \delta|$. By Definition 5.20, $\alpha \neq \beta$ and $\delta \neq 0$. Then by Lemma 5.21, there exist r and $s \in \mathbb{R}$ such that $\operatorname{tr}(M) = \alpha + \beta = r\delta$ and $\det(M) = \alpha\beta = s\delta^2$, hence $\operatorname{tr}(M)/\delta \in \mathbb{R}$ and $\det(M)/\delta^2 \in \mathbb{R}$. □

We now apply an ESP to prove that for most settings (a, b), the problem $\text{Hol}_3(a, b)$ is #P-hard.

Lemma 5.23. *Suppose X and Y are not both real, $X \neq \pm 1$, $X^2 + X + Y \neq 0$, and $4(X-1)^2(X+1) \neq (Y+2)^2$. Then either gadget M_1 or gadget M_2 has nonzero eigenvalues with distinct norm.*

Proof. The transition matrices are $M_1 = \begin{bmatrix} a^3+ab & a+b^2 \\ a^2+b & ab+b^3 \end{bmatrix}$ and $M_2 = \begin{bmatrix} a^3+1 & a+b^2 \\ a^2+b & b^3+1 \end{bmatrix}$, so $M_1 = M_2 + (X-1)I$, and the eigenvalue shift is nonzero. $\operatorname{tr}(M_2) = Y+2$. Checking the determinants, $\det(M_2) = (X-1)^2(X+1) \neq 0$ and $\det(M_1) = (X-1)(X^2+X+Y) \neq 0$. Also, $\operatorname{tr}(M_2)^2 - 4\det(M_2) = (Y+2)^2 - 4(X-1)^2(X+1) \neq 0$, so the eigenvalues of M_2 are distinct. Therefore by Corollary 5.22, either M_1 or M_2 has nonzero eigenvalues of distinct norm unless

$$Y+2 = \operatorname{tr}(M_2) = r(X-1) \text{ and } (X-1)^2(X+1) = \det(M_2) = s(X-1)^2$$

for some $r, s \in \mathbb{R}$. Then we would have $X = s-1 \in \mathbb{R}$ and $Y = r(X-1) - 2 \in \mathbb{R}$. □

The case where $X \in \mathbb{R}$ and $Y \in \mathbb{R}$ will require different techniques, so we put it aside for now and focus on the remaining three cases in this subsection, where X and Y are not both real and the following conditions hold. Note that $X = 1$ is tractable by Theorem 5.6 and we exclude that.

1. $X = -1$ or
2. $X^2 + X + Y = 0$ or
3. $4(X-1)^2(X+1) = (Y+2)^2$

Having a specific condition such as $X = -1$ or $X^2 + X + Y = 0$ introduces fortuitous cancellations, making it possible to use individual gadgets rather than ESPs. The following gadgets were selected by a trial and error process, using a computer to assist with the search.

Gadgets M_1 and M_7 (see Figure 5.5) can be used to deal with the $X = -1$ case. Recall that any setting of a and b such that $X = -1$ and $Y = \pm 2\mathrm{i}$ is tractable by Theorem 5.6.

Lemma 5.24. *If $X = -1$, $Y \neq \pm 2\mathrm{i}$, and $Y \notin \mathbb{R}$, then either gadget M_1 or gadget M_7 has a transition matrix with nonzero eigenvalues with distinct norm.*

Proof. Suppose $|Y| \neq 2$, $Y \notin \mathbb{R}$. Under the assumption $X = -1$, $\det(M_1) = -2Y \neq 0$ and $\operatorname{tr}(M_1) = Y - 2$, so $\operatorname{tr}(M_1) \cdot |\det(M_1)| - \overline{\operatorname{tr}(M_1)} \cdot \det(M_1) = (Y-2) \cdot |-2Y| + (\overline{Y}-2)(2Y) = 2Y \cdot |Y| - 4|Y| + 2|Y|^2 - 4Y = 2(|Y|-2)(|Y|+Y) \neq 0$. Thus $\overline{\operatorname{tr}(M_1)} \cdot \det(M_1) \neq \operatorname{tr}(M_1) \cdot |\det(M_1)|$ and by Lemma 5.19, M_1 has (nonzero) eigenvalues with distinct norm.

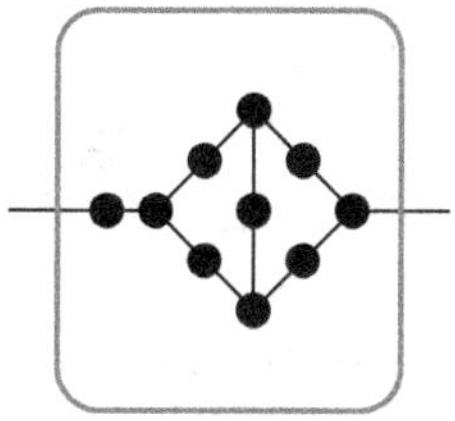

Gadget M_7

Figure 5.5. A unary recursive gadget.

Now suppose $|Y| = 2$, but $Y \neq \pm 2\mathrm{i}$ and $Y \notin \mathbb{R}$.

$$M_7 = \begin{bmatrix} a^6 + 3a^3 + 3ab + b^3 & a^4 + 2a^2b + ab^3 + a + b^5 + 2b^2 \\ a^5 + a^3b + 2a^2 + 2ab^2 + b^4 + b & a^3 + 3ab + b^6 + 3b^3 \end{bmatrix}.$$

Under the assumption $X = -1$, we have $\det(M_7) = -16Y \neq 0$. Substituting $\overline{Y} = 4/Y$,

$$\begin{aligned}
&\mathrm{tr}(M_7)^2\overline{\det(M_7)} - \overline{\mathrm{tr}(M_7)}^2 \det(M_7) \\
&\quad = -16(Y - \overline{Y}) \cdot (-16 + 8Y\overline{Y} + 8Y^2\overline{Y} + Y^3\overline{Y} + 8Y\overline{Y}^2 + Y^2\overline{Y}^2 + Y\overline{Y}^3) \\
&\quad = \frac{-64(Y - \overline{Y})(Y^2 + 4)(Y^2 + 8Y + 4)}{Y^2} \\
&\quad \neq 0.
\end{aligned}$$

Note that $|Y^2 + 4| \leq 8 < |8Y| = 16$. Hence $\mathrm{tr}(M_7)^2\overline{\det(M_7)} \neq \overline{\mathrm{tr}(M_7)}^2 \det(M_7)$ and the eigenvalues of M_7 (which are nonzero) have distinct norm by Lemma 5.19. □

Note that if $X^2 + X + Y = 0$ then $X \in \mathbb{R}$ implies $Y \in \mathbb{R}$. So in the following lemma, the assumption that X and Y are not both real numbers amounts to $X \notin \mathbb{R}$.

Lemma 5.25. *If $X^2 + X + Y = 0$ and $X \notin \mathbb{R}$ then the transition matrix of gadget M_8 has nonzero eigenvalues with distinct norm.*

Proof. Gadget M_8 is defined in Figure 5.6.

$$M_8 = \begin{bmatrix} a^6 + 2a^4b + a^3 + 3a^2b^2 + ab^4 & a^4 + 3a^2b + 2ab^3 + b^5 + b^2 \\ a^5 + 2a^3b + a^2 + 3ab^2 + b^4 & a^4b + 3a^2b^2 + 2ab^4 + b^6 + b^3 \end{bmatrix}.$$

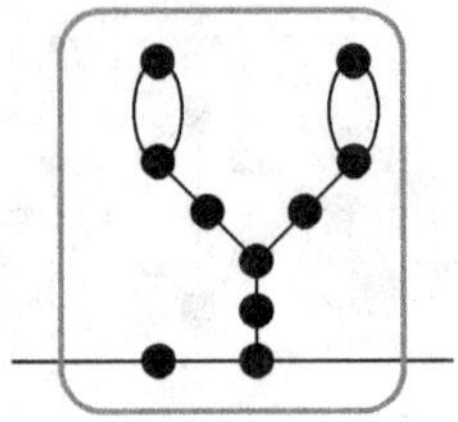

Gadget M_8

Figure 5.6. A unary recursive gadget.

Then the determinant is the polynomial

$$X^6 - 6X^5 - X^4Y + 16X^4 + 11X^3Y - 10X^3 + 5X^2Y^2 \\ - 7X^2Y - X^2 + XY^3 - 4XY^2 - 3XY - Y^3 - Y^2.$$

Amazingly, with the condition $X^2 + X + Y = 0$, this polynomial factors into

$$-X^2(X - 1)^5.$$

Exercise: Verify that the displayed polynomial is congruent to $-X^2(X - 1)^5$ mod $X^2 + X + Y$.

Similarly, the trace, which is $-2X^3 + 6X^2 + 3XY + Y^2 + Y$, also factors into $X(X - 1)^3$. Since $X \notin \mathbb{R}$, we have $\det(M_8) \neq 0$, $\operatorname{tr}(M_8) \neq 0$, and $(1 - X)\det(M_8) = \operatorname{tr}(M_8)^2$. Therefore $\operatorname{Arg}(\det(M_8)) \neq \operatorname{Arg}(\operatorname{tr}(M_8)^2)$ for otherwise X is real, and we conclude by Lemma 5.19 that the eigenvalues of M_8 (which are nonzero) have distinct norm. □

The condition $4(X - 1)^2(X + 1) = (Y + 2)^2$ is somewhat resilient to individual unary recursive gadgets, so we take advantage of another interesting coincidence: two gadgets with transition matrices that have identical trace. In the following lemma, we may assume $X^2 + X + Y \neq 0$ due to Lemma 5.25. We will also assume $X \notin \mathbb{R}$ in this lemma for technical reasons. The case of $X \in \mathbb{R}$ and $Y \notin \mathbb{R}$ will be easy to clean up afterwards.

Lemma 5.26. *Suppose $X \notin \mathbb{R}$, and further assume that $X^2 + X + Y \neq 0$ and $4(X - 1)^2(X + 1) = (Y + 2)^2$. Then the transition matrix of either unary recursive gadget M_9 or unary recursive gadget M_{10} has nonzero eigenvalues with distinct norm.*

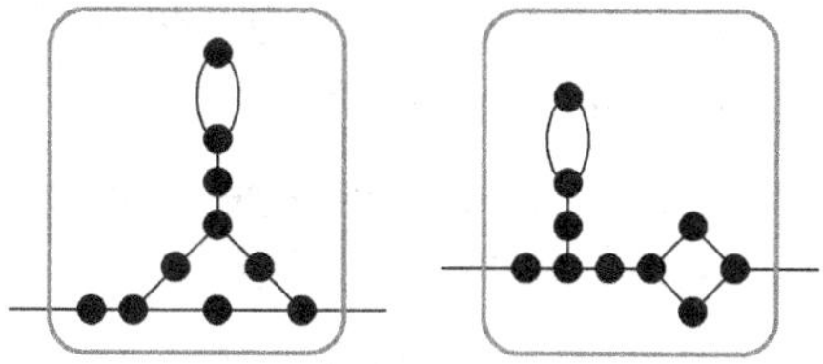

(a) Gadget M_9 (b) Gadget M_{10}

Figure 5.7.

Proof. Gadgets M_9 and M_{10} are defined in Figure 5.7. The transition matrices are

$$M_9 = \begin{bmatrix} a^6 + a^4b + 2a^3 + a^2b^2 + 2ab + b^3 & a^4 + 3a^2b + 2ab^3 + b^5 + b^2 \\ a^5 + 2a^3b + a^2 + 3ab^2 + b^4 & a^3 + a^2b^2 + ab^4 + 2ab + b^6 + 2b^3 \end{bmatrix},$$

$$M_{10} = \begin{bmatrix} a^6 + a^4b + 2a^3 + a^2b^2 + 2ab + b^3 & a^4 + a^3b^2 + a^2b + 2ab^3 + a + b^5 + b^2 \\ a^5 + 2a^3b + a^2b^3 + a^2 + ab^2 + b^4 + b & a^3 + a^2b^2 + ab^4 + 2ab + b^6 + 2b^3 \end{bmatrix}.$$

Let R denote $(Y+2)^2 - 4(X-1)^2(X+1)$. The main diagonals of M_9 and M_{10} are identical, so $\text{tr}(M_9) = \text{tr}(M_{10})$. Furthermore, $\text{tr}(M_9) = \text{tr}(M_9) - R = (X-1)(2X^2+Y)$. Clearly $X - 1 \neq 0$, since $X \notin \mathbb{R}$. If $2X^2 + Y = 0$, then substituting $Y = -2X^2$ into $4(X-1)^2(X+1) = (Y+2)^2$ would yield $X(X-1)^2(X+1) = 0$, again impossible by $X \notin \mathbb{R}$. So $\text{tr}(M_9) \neq 0$. Next, $\det(M_{10}) = (X-1)^3(X+1)(X^2+X+Y)$ and $\det(M_9) = \det(M_9) - R(X-1)^2 = (X-1)^3(X+4)(X^2+X+Y)$, so these are both nonzero. If both M_9 and M_{10} have eigenvalues with equal norm, then applying Lemma 5.19 twice, $\text{Arg}(\det(M_9)) = \text{Arg}(\text{tr}(M_9)^2) = \text{Arg}(\text{tr}(M_{10})^2) = \text{Arg}(\det(M_{10}))$. However, this would imply $\text{Arg}(X+4) = \text{Arg}(X+1)$ and $X \in \mathbb{R}$, so we conclude that either M_9 or M_{10} has nonzero eigenvalues with distinct norm. □

We have already seen a gadget that can handle the remaining case.

Lemma 5.27. *Suppose $X \in \mathbb{R}$ and $Y \notin \mathbb{R}$, and further assume that $4(X-1)^2(X+1) = (Y+2)^2$. Then the transition matrix of unary recursive gadget M_7 has nonzero eigenvalues with distinct norm.*

Proof. Note that $X < -1$, lest Y be real valued. Let R denote $(Y+2)^2 - 4(X-1)^2(X+1)$ as before. We find that $\text{tr}(M_7) = \text{tr}(M_7) - R = 2X(X-1)^2 \neq 0$. We calculate that $\det(M_7) = (X-1)^3(X + 2X^2 + X^3 + 2Y)$, and we claim this is also nonzero. Otherwise, we would have $Y = -(X + 2X^2 + X^3)/2$, and from $X \in \mathbb{R}$ we would have $Y \in \mathbb{R}$, a contradiction. Now $\text{tr}^2(M_7) = 4X^2(X-1)^4 \in \mathbb{R}$, so $\mathrm{tr}^2(M_7)/\det(M_7) \notin \mathbb{R}$ follows from $X \in \mathbb{R}$ and $Y \notin \mathbb{R}$. We conclude by Lemma 5.19 that M_7 has nonzero eigenvalues with distinct norm. □

Now we sum up the results of the previous lemmas. Recall that if $X = 1$ or $(X, Y) = (-1, \pm 2\mathrm{i})$ then $\mathrm{Hol}_3(a, b)$ is tractable.

Lemma 5.28. *Suppose that it is not the case that both X and Y are real valued. Also assume that $X \neq 1$ and $(X, Y) \neq (-1, \pm 2\mathrm{i})$. Then in the context of* $\mathrm{Hol}_3(a, b)$ *there exists a unary recursive gadget M for which the eigenvalues are both nonzero and have distinct norm.*

Proof. We know by Lemma 5.23 that either gadget M_1 or M_2 has a transition matrix with nonzero eigenvalues of distinct norm, except in the following cases, where we will use other gadgets to fill this requirement.

1. $X = -1$,
2. $X^2 + X + Y = 0$, or
3. $4(X - 1)^2(X + 1) = (Y + 2)^2$.

First suppose $X = -1$. Then Y is not real valued and $Y \neq \pm 2\mathrm{i}$, so we get a transition matrix of the required form by Lemma 5.24. If $X^2 + X + Y = 0$ we have $X \notin \mathbb{R}$ (otherwise we would have $X, Y \in \mathbb{R}$), and we have the requirement satisfied by Lemma 5.25. Now we may assume $X^2 + X + Y \neq 0$ and suppose $4(X - 1)^2(X + 1) = (Y + 2)^2$. If $X \notin \mathbb{R}$ then by Lemma 5.26 we are done; otherwise $X \in \mathbb{R}$ but $Y \notin \mathbb{R}$ so we are done by Lemma 5.27. □

Corollary 5.29. *Suppose $a, b \in \mathbb{C}$ are such that it is not the case that both X and Y are real valued. Also assume $X \neq 1$ and $4X^3 \neq Y^2$ (equivalently, $a^3 \neq b^3$). Then in the context of* $\mathrm{Hol}_3(a, b)$ *we can efficiently simulate any recognizer signature of the form* $[x, 0, y]$.

Proof. For any such $a, b \in \mathbb{C}$, we have a unary recursive gadget M by Lemma 5.28 whose transition matrix has nonzero eigenvalues of distinct norm. Then M has infinite projective order, and by Theorem 5.17, we can efficiently construct a sequence of pairwise linearly independent unary generators $[x_j, y_j]$. By attaching an additional vertex labeled with recognizer $=_3$, these become pairwise linearly independent signatures of the form $[x_j, 0, y_j]$. Therefore by Lemma 5.18 we can efficiently simulate any recognizer of the form $[x, 0, y]$. □

Applying Unary Signatures to Prove #P-hardness

The ability to simulate unary signatures (or "virtual unary" signatures) will give us a reduction from Vertex Cover. Note that counting Vertex Cover on

k-regular graphs is just Holant($=_k$| [0, 1, 1]). Counting Vertex Cover is #P-hard even when the input is restricted to k-regular planar graphs. We remark that for $k \geq 6$, any k-regular planar graph must have self-loops or multiple edges. This follows from Euler's formula $V - E + F = 2$.

Lemma 5.30. *Counting* Vertex Covers *on k-regular planar multigraphs (with self-loops and parallel edges) is* #P-hard *for* $k \geq 3$.

Proof. We will give a reduction from counting Vertex Covers on 3-regular planar graphs, which is known to be #P-hard [XZZ07]. If k is odd, we can easily reduce from the given 3-regular planar graph $G = (V, E)$ to a k-regular planar graph by using a gadget consisting of a single vertex incident with $(k - 1)/2$ self-loops and one dangling edge. Connect $k - 3$ copies of this gadget to every vertex $v \in V$. Owing to the self-loops, every copy of this gadget must have its vertex included in the vertex cover, but the graph induced by removing those vertices and their incident edges is identical to G, so the number of vertex covers is the same.

Now we reduce to a k-regular planar graph where k is even. The gadget is similar, consisting of a single vertex incident with $(k - 2)/2$ self-loops and two dangling edges. Being 3-regular, every connected component of G has an even number of vertices, so considering the vertices in pairs and applying $k - 3$ copies of this gadget to each pair of vertices, we are done by the same reasoning as before. However, the resulting graph is not necessarily planar, so we will argue that the vertices in V can be paired in such a way that the gadgets can be introduced while preserving planarity. We do this in two steps. First, we show that there exists a pairing of vertices in which the distance between each pair is ≤ 2. After that we show that planarity is preserved when we add edges of the gadgets according to such a pairing.

We may assume that G is connected, so fix any spanning tree T of G, and note that T has an even number of vertices as G does and the degree of every vertex in T is ≤ 3. It is then sufficient to argue that such a pairing exists in T, because distance measured in G is no more than that measured in T. If there exist adjacent vertices u and v in T such that $\deg(u) = 1$ and $\deg(v) \leq 2$, then we pair u with v. Otherwise, every leaf node is incident with a vertex of degree 3, implying that there exists a vertex t that is adjacent to at least two leaf nodes u and v, and we pair these two leaf nodes together. (To see that such a vertex t exists, fix any node r and make T a rooted tree with root r. Let u be a leaf of maximum depth. As the number of vertices is even, hence more than one, $u \neq r$ and u has a parent t. We have $\deg(t) = 3$, since no neighbor of u has degree ≤ 3. Thus t has another child v. This v must be a leaf by the maximum depth of u.)

In any case, after removing u and v and their incident edges, we are left with an induced subtree T' with an even number of vertices and for which every vertex has degree ≤ 3. This implies a simple recursive algorithm that finds a pairing where paired vertices have distance (on tree T) ≤ 2.

Given such a pairing, we may further assume that there are no pairs (u, v) and (s, t) such that the distance between both pairs is 2 and (u, s, v, t) is a path in G; otherwise we may instead pair u with s and v with t, resulting in two pairs of adjacent vertices. Now for every pair (u, v) of paired vertices, fix a minimal length path in G between them. Draw all gadgets in a neighborhood of the path connecting each pair. Clearly this can be done preserving planarity if (u, v) has distance 1. For (u, v) having distance 2, let the chosen path be (u, s, v). Then s must have degree 3; otherwise s must be paired with some t and (u, s, v, t) or (t, u, s, v) is a path in G. Let the third edge incident to s be (s, w). The gadgets for (u, v) are drawn on the opposite side of (s, w). Note that s is paired with some vertex t through a path (s, w, t) not involving u or v, and causes no edge crossings with the gadgets drawn between u and v. Moreover, gadgets that are drawn along a path involving u or v also introduce no crossings with the gadgets between u and v (they do not connect to s, and the drawing was defined in such a way that they do not cross the path (u, s, v)). Hence planarity is preserved. □

The #P-hardness reduction in Lemma 5.30 is direct from VERTEX COVERS on 3-regular planar graphs; thus clearly it does not use the pinning lemma (Lemma 3.13).

For future reference, we record the planar pairing construction in the proof of Lemma 5.30 separately. The proof also works for multigraphs allowing loops and parallel edges.

Lemma 5.31. *For any 3-regular planar multigraph* $G = (V, E)$, *with* $V = \{v_1, \ldots, v_n\}$, *one can find in polynomial time a pairing* $M = \{\{v_{i_1}, v_{i_2}\}, \ldots, \{v_{i_{n-1}}, v_{i_n}\}\}$, *i.e., a partition of* V *into disjoint pairs, such that* $(V, E \uplus M)$ *is a planar multigraph, where* $E \uplus M$ *is the multiset formed by disjoint union.*

The ability to simulate all unary signatures is what will allow us to prove #P-hardness, although technically we will use "virtual unary" recognizers of the form $[x, 0, y]$. The next lemma says that if we are considering $\text{Hol}_k(a, b)$, then other than on a one-dimensional curve $ab = 1$ and an isolated point $(a, b) = (0, 0)$ we get a reduction from VERTEX COVER by using virtual unary recognizers. As we have seen, the problem $\text{Hol}_k(a, b)$ is tractable on the curve $ab = 1$ and at $(a, b) = (0, 0)$.

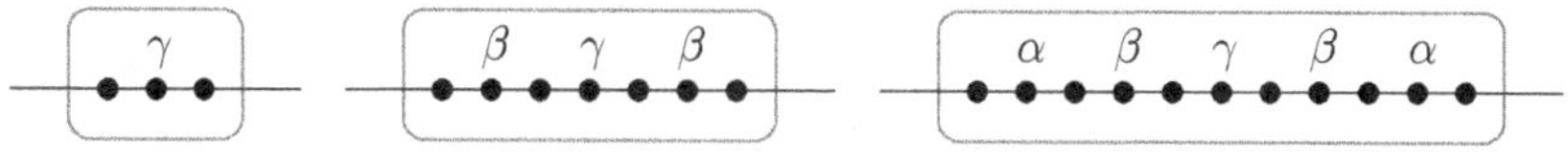

(a) Gadget 1 (b) Gadget 2 (c) Gadget 3

Figure 5.8. Gadgets used to simulate the [0,1,1] signature.

Lemma 5.32. *Suppose $k \geq 3$ and $(a, b) \in \mathbb{C}^2$ satisfies $ab \neq 1$ and $(a, b) \neq (0, 0)$, and assume that for any fixed $m \geq 1$, and any $x_j, y_j \in \mathbb{C}$ $(1 \leq j \leq m)$,*

$$\text{Holant}(\{=_k, [x_1, 0, y_1], \ldots, [x_m, 0, y_m]\} \mid [a, 1, b]) \leq_T \text{Holant}(=_k \mid [a, 1, b]).$$

Then

$$\text{Holant}(=_k \mid [0, 1, 1]) \leq_T \text{Holant}(=_k \mid [a, 1, b])$$

and Holant$(=_k \mid [a, 1, b])$ *is* #P-hard.

The same statement holds for Holant$(=_k \mid [a, 1, b], \Delta_0, \Delta_1)$*, and a #P-hardness reduction exists that does not use Lemma 3.13.*

Proof. Assume $ab \neq 1$ and $(a, b) \neq (0, 0)$. Since $\text{Hol}_k(0, 1)$ is the same problem as Holant$(=_k \mid [0, 1, 1])$, i.e., counting VERTEX COVER on k-regular graphs, which is #P-hard, we only need to show how to simulate the generator signature $[0, 1, 1]$. We split this into three cases, and use a chain of three reductions, each involving a gadget in Figure 5.8. Each gadget will have recognizer signatures of the type $[x, 0, y]$ assigned to the labeled vertices and the generator $[a, 1, b]$ assigned to the unlabeled vertices.

1. $ab \neq 0$ and $ab \neq -1$
2. $ab = 0$
3. $ab = -1$

If $ab \neq 0$ and $ab \neq -1$, then we use gadget 3 in Figure 5.8, and we set its signatures to be

$$\begin{aligned}
\alpha &= [-b/(ab-1), 0, a/(ab-1)] \sim [-b, 0, a], \\
\beta &= [-a^{-2}, 0, b^{-1}(1+ab)^{-1}], \qquad \text{and} \\
\gamma &= [(ab+1)/(1-ab), 0, -a^2(ab+1)/(1-ab)] \sim [1, 0, -a^2].
\end{aligned}$$

Here we use the notation $\sim$ to mean up to a nonzero multiplier. Calculating the resulting signature of gadget 3 using matrix product, we find that it is $[0, 1, 1]$ as desired.

If $ab = 0$ then assume without loss of generality that $a = 0$ and $b \neq 0$ (we can do this because $\text{Hol}_k(0, 1) = \text{Hol}_k(1, 0)$ – also note that $\text{Hol}_k(1, 0)$ is the

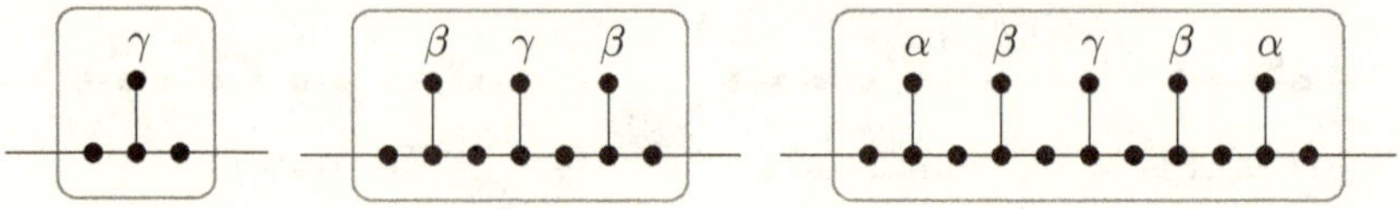

(a) Gadget 1 (b) Gadget 2 (c) Gadget 3

Figure 5.9. Alternate depiction of gadgets 1 through 3.

problem of counting INDEPENDENT SET on k-regular graphs). This time we use gadget 1, setting $\gamma = [b, 0, b^{-1}]$. Then gadget 1 simulates a generator signature $[b^{-1}, 1, 2b]$. But since this signature fits the criteria of case 1 above, we are done by a reduction from case 1.

Similarly, if $ab = -1$, then gadget 2 exhibits a generator signature of the form $[0, 1, 5/(2a)]$ by choosing the signatures $\gamma = [1/(6a), 0, -a/24]$ and $\beta = [-3/a, 0, a]$. Since $5/(2a)$ is nonzero, by a reduction from case 2 we are done. □

Corollary 5.33. *Suppose $a, b \in \mathbb{C}$ are such that X and Y are not both real valued. Also assume $X \neq 1$ and $4X^3 \neq Y^2$ (equivalently, $a^3 \neq b^3$). Then* $\text{Hol}_3(a, b)$ *is* #P-hard, *even when restricted to planar graphs as input.*

The same statement holds for $\text{Holant}(=_3|\ [a, 1, b], \Delta_0, \Delta_1)$*, and a #P-hardness reduction exists that does not use Lemma 3.13.*

Proof. Immediate from Corollary 5.29 and Lemma 5.32. Note that all constructions preserve planarity. □

The proof of Lemma 5.32 is a good demonstration of the power of having all unary signatures. However, the particular choices of the (virtual) unary signatures α, β, γ may appear mysterious. Some explanations will demystify this. As indicated earlier, each signature of the form $[x, 0, y]$ as a recognizer is obtained by connecting a copy of $=_3$ to a unary generator $[x, y]$. The more detailed and direct depiction of gadgets 1 to 3 is in Figure 5.9 with unary signatures. We have seen a similar proof in the dichotomy for Holant* problems, where there were gadgets that resemble "growing weeds." The matrix form of a signature $u = [x, 0, y]$ is diagonal $D_u = \begin{bmatrix} x & 0 \\ 0 & y \end{bmatrix}$. Gadget 3 (in either Figure 5.8c or Figure 5.9c) has the signature $(gD_\alpha g)D_\beta(gD_\gamma g)D_\beta(gD_\alpha g)$, expressed as a matrix product, where $g = \begin{bmatrix} a & 1 \\ 1 & b \end{bmatrix}$. If $D = \begin{bmatrix} x & 0 \\ 0 & y \end{bmatrix}$ then $gDg = \begin{bmatrix} a^2x + y & ax + by \\ ax + by & x + b^2y \end{bmatrix}$, and it can be made a diagonal matrix if we choose $(x, y) \sim (-b, a)$, which is what we did for D_α. On the other hand, if we take $y = -a^2x$

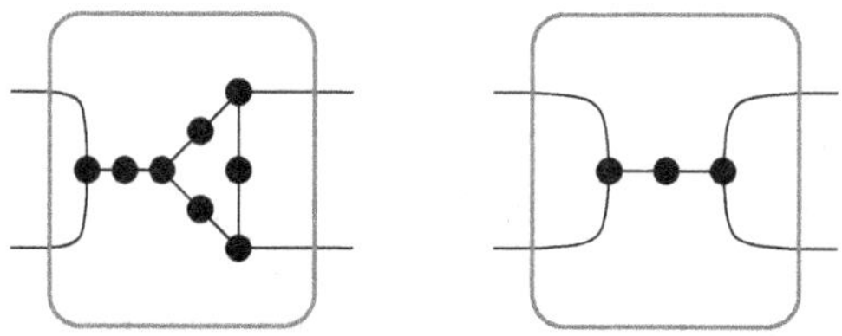

(a) A finisher gadget (b) Another finisher gadget

Figure 5.10. Examples of finisher gadgets.

then gDg has the form $g_1 = x(1-ab)\begin{bmatrix} 0 & a \\ a & 1+ab \end{bmatrix}$, which is what we did for D_γ. Our goal is to realize $\begin{bmatrix} 0 & 1 \\ 1 & 1 \end{bmatrix}$. With a diagonal $D' \sim \begin{bmatrix} 1/a & 0 \\ 0 & 1/(1+ab) \end{bmatrix}$, $D'g_1D' = D'(gD_\gamma g)D'$ takes the form $g_2 = \lambda\begin{bmatrix} 0 & 1 \\ 1 & 1 \end{bmatrix}$. We need the outside factors $gD_\alpha g$ to provide a proper bipartite form, so that gadget 3 constructed in the end is a generator. Thus the strategy is to choose γ such that $gD_\gamma g$ has the form g_1, choose α such that $gD_\alpha g$ takes a diagonal form D''. We assume D'' is invertible and choose $D_\beta \sim (D'')^{-1}D'$. Note that we can choose D_β to be an arbitrary diagonal matrix of the form $\begin{bmatrix} x & 0 \\ 0 & y \end{bmatrix}$, in particular any given $(D'')^{-1}D'$, and because these matrices are diagonal they commute $(D'')^{-1}D' = D'(D'')^{-1}$. Thus $(gD_\alpha g)D_\beta(gD_\gamma g)D_\beta(gD_\alpha g) = D'(gD_\gamma g)D'$ takes the form g_2.

5.1.4 Hardness in Region III Where $(X, Y) \in \mathbb{R} \times \mathbb{R}$

Recall that region III is specified by $X \neq 1$, $Y^2 \neq 4X^3$ (equivalently $a^3 \neq b^3$) and $(X, Y) \neq (-1, 0)$. In this subsection we further assume $(X, Y) \in \mathbb{R} \times \mathbb{R}$.

Interpolation

We now introduce a new type of gadget called a finisher gadget, and adapt our interpolation techniques to the scenario with binary recursive gadgets and finisher gadgets. A finisher gadget has two sets of two dangling edges all connected internally to recognizers. See Figure 5.10 for two examples. It can transform a binary symmetric generator to a binary symmetric recognizer. Symmetry is guaranteed by the symmetry of the underlying graph of the gadget. They have the additional property that one set of two dangling edges (depicted as protruding from the left) are connected to a single $=_k$. These finisher gadgets will be represented by 3×3 matrices with middle row identically 0, and thus rank at most 2. This extension is necessary to treat the case in region III where $(X, Y) \in \mathbb{R} \times \mathbb{R}$, and it is also a natural place to introduce the idea of iterating at a higher dimensional space, which affords more flexibility, and then apply a *projection* down to a lower dimension. The most crucial lemma is Lemma 5.36, which amounts to an exchange of quantifiers. But first we establish tools for binary recursive gadgets as we did for unary recursive gadgets.

Lemma 5.34. *Let $M \in \mathbb{C}^{3\times 3}$ be a nonsingular binary recursive gadget with infinite projective order. Let $s \in \mathbb{C}^3$ be a binary starter gadget such that s is not orthogonal to any row eigenvector of M. Then $\{M^j s\}_{j\geq 0}$ is a sequence of pairwise linearly independent generator signatures.*

Proof. The form of $M^j s$ defines generator signatures, where both s and M are expressed for symmetric signatures.

Let $M = T^{-1}JT$, where J is the Jordan normal form for M, and let $u, v, w \in \mathbb{C}$ be determined by $[u\ v\ w]^{\mathrm{T}} = Ts$. Suppose J is diagonal, let $B = T^{-1}\operatorname{diag}(u, v, w)$, and then

$$M^j s = T^{-1}J^j T s = T^{-1}\begin{bmatrix}\alpha^j & 0 & 0\\ 0 & \beta^j & 0\\ 0 & 0 & \gamma^j\end{bmatrix}\begin{bmatrix}u\\ v\\ w\end{bmatrix}$$

$$= T^{-1}\begin{bmatrix}u & 0 & 0\\ 0 & v & 0\\ 0 & 0 & w\end{bmatrix}\begin{bmatrix}\alpha^j\\ \beta^j\\ \gamma^j\end{bmatrix} = B\begin{bmatrix}\alpha^j\\ \beta^j\\ \gamma^j\end{bmatrix}.$$

There is a pair of eigenvalues of M for which the ratio is not a root of unity (otherwise, we would have some j for which $M^j = \lambda I$ for some nonzero λ). Then the sequence of vectors $[\alpha^j\ \beta^j\ \gamma^j]^{\mathrm{T}}$ are pairwise linearly independent, and since pairwise linear independence is preserved under multiplication by the nonsingular matrix B (note $uvw \neq 0$), we conclude that $\{M^j s\}_{j\geq 0}$ is pairwise linearly independent.

If M has two Jordan blocks, then we have

$$M^j s = T^{-1}J^j T s = T^{-1}\begin{bmatrix}\alpha^j & 0 & 0\\ j\alpha^{j-1} & \alpha^j & 0\\ 0 & 0 & \beta^j\end{bmatrix}\begin{bmatrix}u\\ v\\ w\end{bmatrix},$$

where $uw \neq 0$. If M has one Jordan block, then

$$M^j s = T^{-1}J^j T s = T^{-1}\begin{bmatrix}\alpha^j & 0 & 0\\ j\alpha^{j-1} & \alpha^j & 0\\ \binom{j}{2}\alpha^{j-2} & j\alpha^{j-1} & \alpha^j\end{bmatrix}\begin{bmatrix}u\\ v\\ w\end{bmatrix},$$

where $u \neq 0$. Either way, we have

$$M^j s = T^{-1} J^j T s = T^{-1} \begin{bmatrix} \alpha^j & 0 & 0 \\ j\alpha^{j-1} & \alpha^j & 0 \\ * & * & * \end{bmatrix} \begin{bmatrix} u \\ v \\ w \end{bmatrix}$$

$$= \begin{bmatrix} u\alpha^j \\ uj\alpha^{j-1} + v\alpha^j \\ * \end{bmatrix} = u\alpha^j \begin{bmatrix} 1 \\ j/\alpha + v/u \\ * \end{bmatrix}.$$

The determinant of the 2×2 matrix $\det \begin{bmatrix} 1 & 1 \\ j/\alpha + v/u & i/\alpha^i + v/u \end{bmatrix} = \frac{i-j}{\alpha} \neq 0$. Hence $\{M^j s\}_{j \geq 0}$ is a sequence of pairwise linearly independent signatures. □

We will not construct general-purpose binary starter gadgets (as we did for unary starter gadgets). Instead, we will test starter gadgets individually, using the following lemma.

Lemma 5.35. *Suppose $M \in \mathbb{C}^{m \times m}$ and $s \in \mathbb{C}^m$. If* $\det([s\ Ms\ M^2 s \cdots M^{m-1} s]) \neq 0$ *then s is not orthogonal to any row eigenvector of M.*

Proof. Suppose s is orthogonal to a row eigenvector v of M with eigenvalue λ. Then since $vM^i s = \lambda^i vs = 0$ we have $v[s\ Ms \cdots M^{m-1}s] = 0$, but $v \neq 0$, so this is a contradiction. □

The next lemma is crucial.

Lemma 5.36. *Suppose $\{m_i\}_{i \geq 0}$ is a series of pairwise linearly independent column vectors in $\mathbb{C}^3$. Let F', F'', and $F''' \in \mathbb{C}^{3\times 3}$ be matrices of rank 2 where* $\ker(F')$, $\ker(F'')$, *and* $\ker(F''')$ *are linearly independent. Then for every n, there exists some $S \subseteq \{i \mid 0 \leq i \leq n^3\}$, and some $F \in \{F', F'', F'''\}$, such that $|S| \geq n$ and vectors in $\{Fm_i \mid i \in S\}$ are* pairwise *linearly independent.*

Proof. Let $j > i \geq 0$ be integers, let $N = [m_i\ m_j] \in \mathbb{C}^{3\times 2}$, then $\text{rank}(N) = 2$ and $\dim(\text{Im}(N)) = 2$. Since $\ker(F')$, $\ker(F'')$, and $\ker(F''')$ are linearly independent of $\mathbb{C}^3$, we can let $F \in \{F', F'', F'''\}$ such that $\ker(F) \cap \text{Im}(N) = \{0\}$. Then $FN \in \mathbb{C}^{3\times 2}$ has rank 2, hence Fm_i and Fm_j are linearly independent.

Each $F \in \{F', F'', F'''\}$ defines a coloring of the set $K = \{0, 1, \ldots, n^3\}$ as follows: color $i \in K$ with the linear subspace spanned by Fm_i. Assume for a contradiction that for each $F \in \{F', F'', F'''\}$ there are not n pairwise linearly independent vectors among $\{Fm_i : i \in K\}$. Then, including possibly the 0-dimensional space $\{0\}$, there can be at most n distinct colors assigned by each

$F \in \{F', F'', F'''\}$. By the pigeonhole principle, some i and j with $0 \le i < j \le n^3$ must receive the same color for all $F \in \{F', F'', F'''\}$. This is a contradiction and we are done. □

We remark that Lemma 5.36 amounts to an exchange of quantifiers. As shown in the first paragraph of the proof, the antecedent of the lemma says that for every pair of distinct m_i and m_j from a sequence of pairwise linearly independent $\{m_i\}_{i\ge 0}$, there exists a finisher $F \in \{F', F'', F'''\}$ such that Fm_i and Fm_j are linearly independent. This is a $\forall\exists$ type condition. Then the consequent of the lemma says that there is a single $F \in \{F', F'', F'''\}$ that produces an arbitrarily large set of pairwise linearly independent $\{Fm_i \mid i \in S\}$. This is an $\exists\forall$ type statement (there exist a large S among the first $n^{O(1)}$ many i's and a single F, such that F works for all m_i from S.) The cost of this exchange of quantifiers is that we may have to look at polynomially more vectors in the original list $\{m_i\}_{i\ge 0}$, and we may not know in advance which subset S would work. As we will see, both caveats are harmless in our intended application of this lemma.

In the following lemma we use binary finisher gadgets. Recall that such a gadget has two sets of two dangling edges each, both sets are connected internally to recognizer signatures, and both dangling edges of one set (depicted as protruding from the left) are connected internally to a single EQUALITY signature. The two dangling edges of the other set (depicted as protruding from the right) are to be merged with the two leading edges of a binary recursive gadget. Thus these finisher gadgets have 3×3 transition matrices that have an identically 0 middle row. They transform a generator signature to a recognizer signature.

Lemma 5.37. *Suppose that the following gadgets can be built using complex-valued signatures from a finite generator set $\mathcal{G}$ and a finite recognizer set $\mathcal{R}$.*

1. *A nonsingular binary recursive gadget M with infinite projective order*
2. *A binary starter gadget s where* $\det([s\ Ms\ M^2s]) \neq 0$
3. *Three binary finisher gadgets with rank 2 transition matrices $F_1, F_2, F_3 \in \mathbb{C}^{3\times 3}$, for which* $\ker(F_1)$, $\ker(F_2)$, *and* $\ker(F_3)$ *are linearly independent*

Then for any $x, y \in \mathbb{C}$, $\operatorname{Holant}(\mathcal{R} \cup \{[x, 0, y]\} \mid \mathcal{G}) \le_T^P \operatorname{Holant}(\mathcal{R} \mid \mathcal{G})$. *This reduction does not use Lemma 3.13.*

Proof. By Lemma 5.35 we know that s is not orthogonal to any row eigenvector of M, so $\{M^j s\}_{j\ge 0}$ is a sequence of pairwise linearly independent vectors by Lemma 5.34. We will efficiently find a subsequence of pairwise linearly independent recognizers from the sequence

$$\{F_1 M^j s,\ F_2 M^j s,\ F_3 M^j s \mid j \ge 0\}.$$

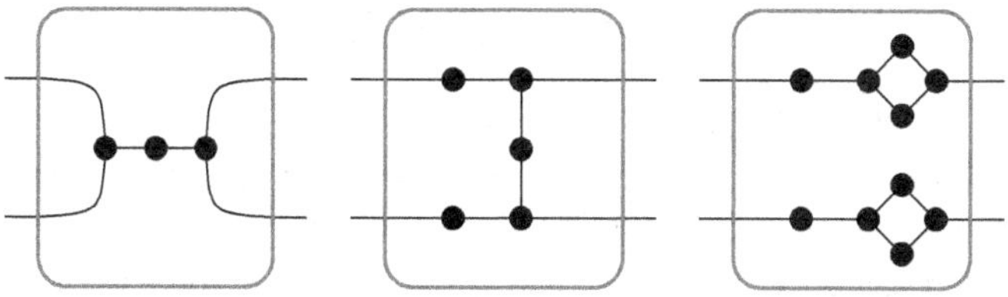

(a) Gadget M_{11} (b) Gadget M_5 (c) Gadget M_{12}

Figure 5.11. Gadgets for constructing finisher gadgets.

By design both sets of dangling edges of each finisher gadget are internally incident to recognizers, making this a list of recognizer signatures of the form $[*, 0, *]$. We can greedily pick a large subset of pairwise linear independent recognizers from the above list in polynomial time. Start with the empty subset, we test each successive signature for pairwise linear independence with all previously chosen signatures, and append it to our subsequence if it passes the test. Note that a signature is eliminated in this process iff it is a multiple of a previously chosen signature. By Lemma 5.36, we will have collected the first n signatures in this process by considering only $0 \le j \le n^3$ among the list. Indeed if at the end of this process we end up having collected fewer than n pairwise linearly independent recognizers, then for all $0 \le j \le n^3$, every $F_1 M^j s$, $F_2 M^j s$ and $F_3 M^j s$ is a multiple of some chosen signature. But then some two distinct members of the set S of size $|S| \ge n$ guaranteed by Lemma 5.36 must be independently dependent, a contradiction. Then by Lemma 5.18, for any $x, y \in \mathbb{C}$ we can efficiently simulate recognizer $[x, 0, y]$. □

Now we explicitly construct some finisher gadgets.

Lemma 5.38. *Suppose $X \neq 1$ and $a^3 \neq b^3$. Then in the context of* $\mathrm{Hol}_3(a, b)$ *there exist three binary finisher gadgets with rank 2 transition matrices $F_1, F_2, F_3 \in \mathbb{C}^{3\times 3}$, for which* $\ker(F_1)$, $\ker(F_2)$, *and* $\ker(F_3)$ *are linearly independent.*

Proof. We start first with the case where $ab \neq 0$. Using the simplest possible choice for a finisher gadget M_{11} (Figure 5.11), we get

$$M_{11} = \begin{bmatrix} a & 0 & 1 \\ 0 & 0 & 0 \\ 1 & 0 & b \end{bmatrix}.$$

We calculate that

$$M_5 = \begin{bmatrix} a^3 & 2a & b \\ a^2 & ab+1 & b^2 \\ a & 2b & b^3 \end{bmatrix}.$$

We build two more finisher gadgets F' and F'' using gadgets M_{11} and M_5 so that $F' = M_{11}M_5$ and $F'' = M_{11}M_5^2$. Since M_{11} has rank 2 and M_5 is nonsingular (note $\det(M_5) = ab(ab-1)^3$), it follows that F' and F'' also have rank 2. Now to show that the kernels of M_{11}, F' and F'' are linearly independent, we check that the cross products of the first and last row vectors of M_{11}, F', and F'' (denoted respectively by v, v', and v'') are linearly independent. Note that M_{11} has an identically 0 middle row, so the cross product of the first row and the third row spans the kernel $\{x \in \mathbb{C}^3 \mid M_{11}x = 0\}$. The same is true for F' and F''. The cross products of the first and last row vectors of M_{11}, F', and F'' are respectively, $[0, 1-ab, 0]$, $(ab-1)^2[2b^2, -ab(ab+1), 2a^2]$, and

$$\begin{aligned}&(ab-1)^3[2b(a^2b^3+a^2+ab^2+b^4),\\ &\quad -ab(a^3b^3+2a^3+2a^2b^2+ab+2b^3), 2a(a^4+a^3b^2+a^2b+b^2)].\end{aligned}$$

As $ab \neq 1$, to see that these three vectors are linearly independent, it suffices to verify that the 2×2 matrix

$$\begin{bmatrix} 2b^2 & 2a^2 \\ 2b(a^2b^3+a^2+ab^2+b^4) & 2a(a^4+a^3b^2+a^2b+b^2) \end{bmatrix}$$

is nonsingular. Since $a \neq 0$, $b \neq 0$, we just need to check

$$\det\begin{bmatrix} b & a \\ a^2b^3+a^2+ab^2+b^4 & a^4+a^3b^2+a^2b+b^2 \end{bmatrix} = (ab-1)(a^3-b^3) \neq 0,$$

so the kernels of M_{11}, F', and F'' are linearly independent when $ab \neq 0$.

Now suppose $ab = 0$. Since $a^3 \neq b^3$, by symmetry, if $ab = 0$ we may assume without loss of generality that $a \neq 0$ and $b = 0$.

$$M_{12} = \begin{bmatrix} a^6+2a^3+1 & 2a^4+2a & a^2 \\ a^5+a^2 & 2a^3+1 & a \\ a^4 & 2a^2 & 1 \end{bmatrix}.$$

Composing M_{11} with M_{12}, we get a finisher gadget with matrix $M_{11}M_{12}$, which has rank 2 since M_{11} has rank 2 and $\det(M_{12}) = 1$. It is also straightforward to see that $F' = M_{11}M_5$ has rank 2, as

$$F' = \begin{bmatrix} a^4+a & 2a^2 & 0 \\ 0 & 0 & 0 \\ a^3 & 2a & 0 \end{bmatrix}$$

using $b = 0$. The cross products of the first and last rows of M_{11}, F', and $M_{11}M_{12}$ are $[0, 1, 0]$, $[0, 0, 2a^2]$, and $[-2a, 2a^3+1, -2a^2(1+a)(a^2-a+1)]$

respectively. Then the matrix of cross products is clearly nonsingular, and we conclude that for any $a, b \in \mathbb{C}$, we have three finisher gadgets satisfying our requirement. □

Corollary 5.39. *Let $a, b \in \mathbb{C}$ such that $X \neq 1$ and $a^3 \neq b^3$. Suppose that in the context of* $\mathrm{Hol}_3(a, b)$*, there exists a nonsingular binary recursive gadget M with infinite projective order and a binary starter gadget s where* $\det([s\ Ms\ M^2 s]) \neq 0$*. Then* $\mathrm{Hol}_3(a, b)$ *is* #P-hard.

The same statement holds for $\mathrm{Holant}(=_3 |\ [a, 1, b], \Delta_0, \Delta_1)$*, and a #P-hardness reduction exists that does not use Lemma 3.13.*

Proof. Immediate from Lemmas 5.38, 5.37, and 5.32. □

Applying Interpolation with Anti-gadgets

Now we will construct specific recursive gadgets to handle the special case of region III when X and Y are real valued: $(X, Y) \in \mathbb{R} \times \mathbb{R}$. We have $X \neq 1$, $Y^2 \neq 4X^3$ (equivalently $a^3 \neq b^3$) and $(X, Y) \neq (-1, 0)$. We will first set $X \neq 0, \pm 1$. There are two cases depending on whether $X + X^2 + Y = 0$. After that we consider $X = -1$ (but $Y \neq 0$) or $X = 0$ (but $Y \neq 0$).

We will introduce the idea of an *anti-gadget* in the next lemma. In a typical reduction from one problem to another in complexity theory, one often designs various gadgets. A reduction starts from an instance of the first problem and constructs an instance of the second problem that may contain a polynomial number of copies of the gadget. But the notion of containing a *negative* number of copies of a gadget seems meaningless. However, the *effect* of including an anti-gadget will indeed be the same as introducing a negative copy of a gadget in a reduction. More precisely, the effect of an anti-gadget, when expressed in algebraic terms, is the same as *erasing* the presence of some other gadget fragment. It is *as if* we managed to include a negative copy of a certain gadget.

We will see that anti-gadget greatly simplifies our work in finding a recursive gadget with infinite projective order. Effectively, we can use one gadget to "cancel out" a portion of another gadget, leaving a "virtual gadget" with a diagonal signature matrix – ideal for our purposes, as it is straightforward to analyze.

Lemma 5.40. *Let $a, b \in \mathbb{C}$ such that $X, Y \in \mathbb{R}$, and assume that $X \notin \{0, 1, -1\}$, $a^3 \neq b^3$, and $X^2 + X + Y \neq 0$. Then* $\mathrm{Hol}_3(a, b)$ *is* #P-hard. *The same statement holds for* $\mathrm{Holant}(=_3 |\ [a, 1, b], \Delta_0, \Delta_1)$*, and a #P-hardness reduction exists that does not use Lemma 3.13.*

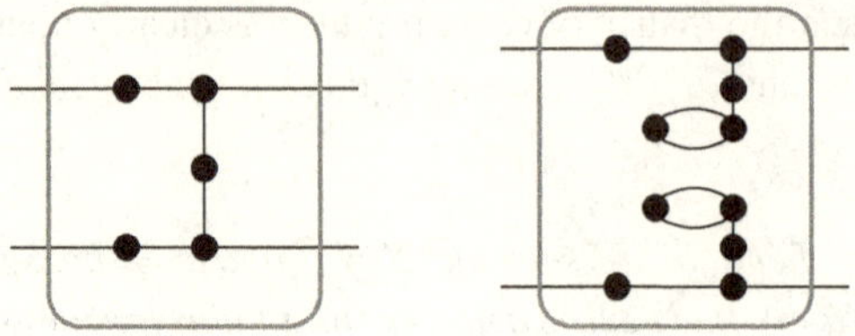

(a) Gadget M_5 (b) Gadget M_{13}

Figure 5.12. An anti-gadget pair.

Proof. We begin this proof with binary recursive gadget

$$M_5 = \begin{bmatrix} a^3 & 2\ a & b \\ a^2 & ab+1 & b^2 \\ a & 2b & b^3 \end{bmatrix}.$$

as depicted in Figure 5.12. Taking determinant, we find $\det(M_5) = X(X-1)^3 \neq 0$. If we let s be the simplest possible binary starter gadget consisting of a single vertex (Figure 5.3(a), Gadget M_3), then $\det([s\ M_5 s\ M_5^2 s]) = -(a^3-b^3)(ab-1)^4 \neq 0$. If M_5 has infinite projective order, then all criteria for Corollary 5.39 are met, and $\mathrm{Hol}_3(a, b)$ is #P-hard.

Now suppose that M_5 has finite projective order, and we let $i \in \mathbb{Z}^+$ and nonzero $\lambda \in \mathbb{C}$ such that $M_5^i = \lambda I$. Then chaining $i-1$ copies of gadget M_5 together, we have an explicit construction for a gadget with signature $M_5^{i-1} = \lambda M_5^{-1}$.

The gadget M_{13} (see Figure 5.12) has a 4×4 signature matrix obtained as a tensor product of the 2×2 matrix of $M_1 = \begin{bmatrix} a(a^2+b) & a+b^2 \\ a^2+b & (a+b^2)b \end{bmatrix}$. Expressed for symmetric signatures, we have

$$M_{13} = \begin{bmatrix} a^2A^2 & 2aAB & B^2 \\ aA^2 & (ab+1)AB & bB^2 \\ A^2 & 2bAB & b^2B^2 \end{bmatrix},$$

where $A = a^2 + b$ and $B = a + b^2$.

Then concatenating the gadget with signature M_5^{i-1} and a copy of M_{13}, we have

$$M = M_5^{i-1} M_{13} = \lambda \begin{bmatrix} \alpha & 0 & 0 \\ 0 & AB & 0 \\ 0 & 0 & \beta \end{bmatrix},$$

where $\alpha = \frac{A^2}{a} = \frac{(a^2+b)^2}{a}$ and $\beta = \frac{B^2}{b} = \frac{(a+b^2)^2}{b}$. Note that $AB = (a^2 + b)(a+b^2) = X^2 + X + Y \neq 0$ and $X \neq 0$ by assumption. It follows that

$\det(M) = \lambda^3\alpha\beta AB = \lambda^3(AB)^3/X \neq 0$, so M is nonsingular. We prove that M has infinite projective order, by showing that some two eigenvalues among $\{\alpha, \beta, AB\}$ have distinct norm. Suppose $|\alpha| = |\beta|$, then $|\alpha\beta X| = |\alpha|^2|X| \neq |\alpha|^2$ since $X \in \mathbb{R}$ and $X \neq \pm 1$. But $|\alpha\beta X| = |AB|^2$ and so $|\alpha| \neq |AB|$. Finally, we check that $\det([s\ Ms\ M^2s]) = \lambda^3(a^3 - b^3)(X-1)^3(X^2+X+Y) \neq 0$, where $s = [a\ 1\ b]^{\mathrm{T}}$. Hence we apply Corollary 5.39 again and conclude that $\mathrm{Hol}_3(a, b)$ is #P-hard. □

The following test is a straightforward way to verify that a matrix has infinite projective order, by looking at the characteristic polynomial.

Lemma 5.41. *If all roots of the complex polynomial $x^3 - Bx^2 + Cx - D$ have the same norm, then $C|C|^2 = \overline{B}|B|^2D$.*

Proof. If the roots have equal norm, then for some $a, b, c \in \mathbb{C}$ with $|a| = |b| = |c| = 1$, and nonnegative $r \in \mathbb{R}$ we can write $x^3 - Bx^2 + Cx - D = (x - ra)(x - rb)(x - rc)$, so $B = r(a+b+c)$, $C = r^2(ab+bc+ca)$, and $D = r^3abc$. Then

$$\begin{aligned} C|C|^2 &= r^2(ab+bc+ca)r^4|ab+bc+ca|^2 \\ &= r(\overline{a+b+c})r^2|a+b+c|^2r^3abc = \overline{B}|B|^2D, \end{aligned}$$

where we used the fact that $|ab+bc+ca| = |ab+bc+ca| \cdot |a^{-1}b^{-1}c^{-1}| = |a^{-1}+b^{-1}+c^{-1}| = |\overline{a+b+c}| = |a+b+c|$. □

Lemma 5.42. *Suppose $X, Y \in \mathbb{R}$, $X^2+X+Y = 0$, $a^3 \neq b^3$, and $X \notin \{0, 1, -1\}$. Then* $\mathrm{Hol}_3(a, b)$ *is* #P-hard. *The same statement holds for* $\mathrm{Holant}(=_3|\ [a, 1, b], \Delta_0, \Delta_1)$*, and a #P-hardness reduction exists that does not use Lemma 3.13.*

Proof. We first verify that $\det(M_5) = X(X-1)^3 \neq 0$. Next we show that M_5 has infinite projective order. The characteristic polynomial of M_5 is $x^3 - Bx^2 + Cx - D$, where $B = X + Y + 1 = 1 - X^2$, $C = (X-1)(X^2+X+Y) = 0$, and $D = X(X-1)^3$. Thus $B, C, D \in \mathbb{R}$ and by Lemma 5.41 it is sufficient to observe $C^3 - B^3D = X(X-1)^6(X+1)^3 \neq 0$. We also have $\det([s\ M_5s\ M_5^2s]) = -(a^3-b^3)(ab-1)^4 \neq 0$ as required by Corollary 5.39, where s is the single-vertex starter gadget. □

We are left with two cases: $X = -1$ (but $Y \neq 0$) or $X = 0$ (but $Y \neq 0$).

Lemma 5.43. *Suppose $X = -1$, $a^3 \neq b^3$, and $Y \in \mathbb{R}$ but $Y \neq 0$. Then* $\mathrm{Hol}_3(a, b)$ *is* #P-hard. *The same statement holds for* $\mathrm{Holant}(=_3|$ $[a, 1, b], \Delta_0, \Delta_1)$*, and a #P-hardness reduction exists that does not use Lemma 3.13.*

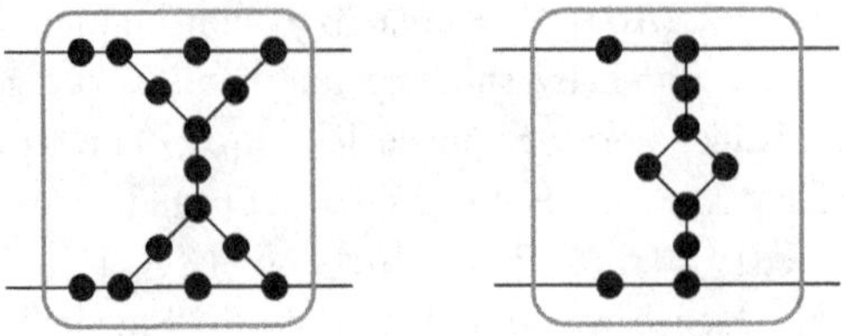

(a) Gadget M_{14} (b) Gadget M_{15}

Figure 5.13. Binary recursive gadgets.

Proof. As before, $\det(M_5) = X(X-1)^3 \neq 0$. Looking at the characteristic polynomial $x^3 - Bx^2 + Cx - D$ of M_5, we have $C^3 - B^3D = -16Y^3 \neq 0$, so by Lemma 5.41 there are eigenvalues for which the ratio is not a root of unity and the matrix has infinite projective order. Once again, $\det([s\ M_5s\ M_5^2s]) = -(a^3 - b^3)(ab-1)^4 \neq 0$, where s is the single-vertex starter gadget, and by Corollary 5.39 we are done. □

Lemma 5.44. *Suppose $X = 0$ and $Y \in \mathbb{R}$ but $Y \neq 0$. Then* $\operatorname{Hol}_3(a,b)$ *is* #P-hard. *The same statement holds for* $\operatorname{Holant}(=_3|\ [a,1,b], \Delta_0, \Delta_1)$*, and a #P-hardness reduction exists that does not use Lemma 3.13.*

Proof. Gadgets M_{14} and M_{15} are depicted in Figure 5.13. Let $M = M_{14}$. Under the condition $X = 0$, we have $\det(M) = Y^3 \neq 0$, and also

$$C^3 - B^3D = 2Y^6(3L^2+1) \neq 0,$$

where we let $L = Y^2 + 6Y + 8$ and the characteristic polynomial of M is $x^3 - Bx^2 + Cx - D$. Also, $\det([s\ Ms\ M^2s]) = 2Y^2(a^3-b^3)(Y+3)$, where s is the single-vertex starter gadget, so by Corollary 5.39 we are done unless $Y = -3$ (note $a^3 \neq b^3$ because either a or b is 0, but not both).

For $X = 0$ and $Y = -3$ we let $M = M_{15}$ and then we have $\det(M) = 6 \neq 0$, $C^3 - B^3D = 21 \neq 0$, and

$$\det([s\ Ms\ M^2s]) = -(X-1)^6(X+1)(a^3-b^3)(Y+2) \neq 0. \qquad \square$$

Finally we are ready to prove the main theorem of Section 5.1: Theorem 5.1 for the case $k = 3$.

Proof of Theorem 5.1 Restricted to $k = 3$

Proof. Tractability (for region I and for region II on planar inputs) is given by Theorem 5.6. Intractability is given by Lemma 5.14 for region II. For region III, where X and Y are not both real valued, we have Corollary 5.33. For region III, where X and Y are both real, we appeal to Lemmas 5.44, 5.43, 5.42, and 5.40 as follows. For $X = 0$ we have Lemma 5.44; for $X = -1$ we have

Lemma 5.43. For the remaining cases we have Lemma 5.40 if $X^2 + X + Y \neq 0$, and Lemma 5.42 if $X^2 + X + Y = 0$. □

5.2 4-Regular Graphs

5.2.1 Tractability in Regions I and II

Recall that for 4-regular graphs, region I is specified by $X = 1$ or $(X, Y) \in \{(0, 0), (-1, 0), (-1, \pm 2)\}$. Region II is specified by $Y^2 = 4X^4$ but $X \notin \{0, \pm 1\}$. The union of regions I and II is specified by $X = 1$ or $Y^2 = 4X^4$ or $(X, Y) = (-1, 0)$.

We will show that X and Y capture the complexity of $\mathrm{Hol}_4(a, b)$, just as we did for $\mathrm{Hol}_3(a, b)$. This will be slightly more involved, since in the case of $\mathrm{Hol}_4(a, b)$ it is not true that the Holant is an integer polynomial in X and Y; there can be an extra factor $a^2 + b^2$. This factor appears precisely when the number of vertices in the graph is odd (which can never happen in a 3-regular graph). We start with a version of Lemma 5.4 adapted to any k.

Lemma 5.45. *Let G be a k-regular graph with n vertices. Then there exists a polynomial $P_G(\cdot, \cdot)$ with two variables and integer coefficients such that, for any signature grid Ω having underlying graph G, $Z(G) = P_G(ab, a^3 + b^3)$ (if n is even) and $Z(G) = (a^{k/2} + b^{k/2})P(ab, a^k + b^k)$ (if n is odd), for any edge function $[a, 1, b]$.*

Proof. In the edge–vertex bipartite view, the partition function $Z(G)$ is the value Holant_Ω in the Holant problem $\mathrm{Holant}(=_k \mid [a, 1, b])$. First note that if k and n are both odd then no k-regular graph exists on n vertices, as kn is twice the number of edges, so k must be even whenever n is odd.

Consider any $\{0, 1\}$-vertex assignment σ. If σ' is the complement assignment switching all 0's and 1's in σ, then the sum of valuations for σ and σ' in Holant_Ω is $a^i b^j + a^j b^i$, where i (resp. j) is the number of edges connecting two degree-k vertices both assigned 0 (resp. 1) by σ. We note that $a^i b^j + a^j b^i = (ab)^{\min(i,j)}(a^{|i-j|} + b^{|i-j|})$.

For the all-0 assignment, $i - j = kn/2$. Now starting from any assignment σ, if we switch the assignment on one vertex v from 0 to 1, it is easy to verify that it changes the valuation from $a^i b^j$ to $a^{i'} b^{j'}$, where $i - j = i' - j' + k$. This takes into account the changes among all edges incident to v, including self loops at v, if any. Every $\{0, 1\}$ assignment σ is obtainable from the all-0 assignment by a sequence of switches, hence $i - j \equiv kn/2 \pmod{k}$.

If n is even, then $kn/2 \equiv 0 \pmod{k}$. Thus, for every assignment σ, we have $i - j \equiv 0 \pmod{k}$. Now $a^i b^j + a^j b^i = (ab)^{\min(i,j)}(a^{k\ell} + b^{k\ell})$, for some integer

$\ell \geq 0$, and a simple induction

$$a^{k(\ell+1)} + b^{k(\ell+1)} = (a^{k\ell} + b^{k\ell})(a^k + b^k) - (ab)^k(a^{k(\ell-1)} + b^{k(\ell-1)})$$

shows that Holant_Ω is a polynomial $P_G(ab, a^k + b^k)$ with integer coefficients.

If n is odd, then in particular k is even. For every assignment σ, we have $i - j \equiv k/2 \pmod{k}$. Now $a^i b^j + a^j b^i = (ab)^{\min(i,j)}(a^{k/2+k\ell} + b^{k/2+k\ell})$, for some integer $\ell \geq 0$. We verify that at $\ell = 0$ and $\ell = 1$, $a^{k/2+k\ell} + b^{k/2+k\ell}$ becomes $a^{k/2} + b^{k/2}$ and $a^{3k/2} + b^{3k/2} = (a^{k/2} + b^{k/2})(a^k + b^k - (ab)^{k/2})$ respectively, both of which are of the form: a product of $a^{k/2} + b^{k/2}$ with an integer polynomial in $(ab, a^k + b^k)$. Then an easy induction

$$\begin{aligned} a^{k/2+k(\ell+1)} + b^{k/2+k(\ell+1)} &= (a^{k/2+k\ell} + b^{k/2+k\ell})(a^k + b^k) \\ &\quad - (ab)^k(a^{k/2+k(\ell-1)} + b^{k/2+k(\ell-1)}) \end{aligned}$$

shows that $a^{k/2+k\ell} + b^{k/2+k\ell}$ is of this form for all $\ell \geq 0$. □

Corollary 5.46. *Let G be any k-regular graph with n vertices, where k is even and n is odd, and let Ω be any signature grid having underlying graph G and every edge labeled $[a, 1, b]$. If $a^{k/2} + b^{k/2} = 0$, then* $\text{Holant}_\Omega = 0$.

Corollary 5.47. *Let $a', b' \in \mathbb{C}$ such that $a'b' = X$ and $(a')^k + (b')^k = Y$. Then* $\text{Hol}_k(a, b)$ *is polynomial time computable (respectively, is* #P-hard*) if and only if* $\text{Hol}_k(a', b')$ *is polynomial time computable (is* #P-hard*).*

Proof. Consider an input instance Ω for $\text{Hol}_k(a, b)$. If n is even then the claim is obvious. Suppose there are an odd number of vertices, and let Ω' be a copy of Ω with an additional vertex having $k/2$ self-loops. Then

$$\text{Holant}_{\Omega'} = (a^{k/2} + b^{k/2})\text{Holant}_\Omega. \tag{5.1}$$

If $a^{k/2} + b^{k/2} \neq 0$ we can easily obtain Holant_Ω from $\text{Holant}_{\Omega'}$ and vice versa. If $a^{k/2} + b^{k/2} = 0$ then both $\text{Holant}_\Omega = 0$ (by Corollary 5.46) and $\text{Holant}_{\Omega'} = 0$ (by (5.1). Thus the complexity of $\text{Hol}_k(a, b)$ is captured by input instances with an even number of vertices, for which the Holant can be expressed as a function of X and Y. □

Now we specialize to $k = 4$. We first list the tractable cases of $\text{Hol}_4(a, b)$.

Theorem 5.48. *For any $a, b \in \mathbb{C}$,* $\text{Hol}_4(a, b)$ *is solvable in polynomial time if any of the following four conditions are true.*

1. $X = 1$.
2. $X = Y = 0$.
3. $X = -1$ *and* $Y \in \{0, \pm 2\}$.
4. $Y^2 = 4X^4$ *and the input is restricted to planar graphs.*

Proof. If $X = 1$ then the signature $[a, 1, b]$ is degenerate and the Holant can be computed in polynomial time. If $X = Y = 0$, then $a = b = 0$ and a 2-coloring argument can be applied to calculate the Holant. If $X = -1$, then applying a holographic transformation under basis $T = \begin{bmatrix} 1 & 0 \\ 0 & a^{-1} \end{bmatrix}$, we get $(T^{-1})^{\otimes 2} g = [a, a, -a]^{\mathrm{T}}$ and $(=_4)T^{\otimes 4} = [1, 0, 0, 0, a^{-4}]$, where $g = [a, 1, -a^{-1}]^{\mathrm{T}}$ (note that the form of this g corresponds to the assumption $X = -1$). Multiplying the signature $[a, a, -a]$ by a^{-1} does not change the complexity of the problem, so Holant$(=_4|\ g)$ is equivalent in complexity to Holant$([1, 0, 0, 0, a^{-4}]\ |\ [1, 1, -1])$, and the problem Holant$(\{[1, 0, 0, 0, a^{-4}], [1, 1, -1]\})$ is tractable if $a^4 \in \{1, -1, \mathfrak{i}, -\mathfrak{i}\}$, by families $\mathscr{F}_1$ and $\mathscr{F}_3$ in Chapter 3. It is easy to see that $a^4 = \pm 1$ or $\pm \mathfrak{i}$ is equivalent to $Y = a^4 + (-a^{-1})^4 = a^4 + a^{-4}$ is equal to ± 2 or 0. Finally, if $Y^2 = 4X^4$, then $a^4 = b^4$ and holographic algorithms using matchgates can be applied when the input graph is planar (see Chapter 4 and the proof of Theorem 5.6). □

We note that the first three tractable cases also hold for Holant$(=_4|\ [a, 1, b], \Delta_0, \Delta_1)$, and diagonal holographic transformations do not change $\{\Delta_0, \Delta_1\}$ up to a nonzero constant.

5.2.2 Hardness in Region III

Recall that for 4-regular graphs, region III is the complement of the union of regions I and II, which is specified by $X \neq 1$ and $Y^2 \neq 4X^4$ (equivalently $a^4 \neq b^4$) and $(X, Y) \neq (-1, 0)$. To prove #P-hardness in region III, we must use planar gadgets, and reduce from a problem that is #P-hard for planar graphs.

For $\mathrm{Hol}_3(a, b)$, we were able to make use of constructions consisting of a unary starter gadget together with iterations of a unary recursive gadget. However, in the context of $\mathrm{Hol}_4(a, b)$ the signature of any $\mathcal{F}$-gate has an even arity. This can be seen by adding up mod 2 all arities of signatures in the $\mathcal{F}$-gate, and noting that every internal edge is counted twice in this sum. So we cannot construct unary starter gadgets. However, we would still like to make use of some nice unary recursive gadgets, such as those we will see in Lemma 5.49. One alternative is to leave off the unary starter gadget, and consider the iterated unary recursive gadgets, which have two dangling edges, as constructions of binary signatures. However, these signatures, like unary recursive gadgets, have one dangling edge e internally incident to a generator and one dangling edge e' internally incident to a recognizer, and are not symmetric. To get around this difficulty, we introduce some gadgets for the sole purpose of transforming the construction into a sequence of *symmetric* binary signatures. Symmetry of the signatures constructed will be guaranteed by the symmetry in the underlying graph fragment ($\mathcal{F}$-gate).

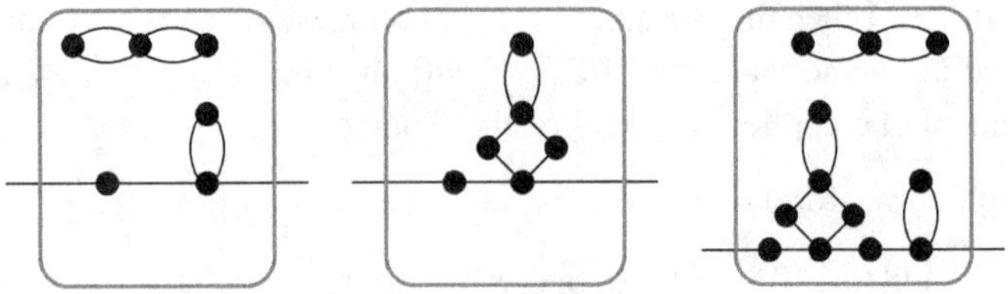

(a) Gadget M_{16} (b) Gadget M_{17} (c) Gadget M_{18}

Figure 5.14. A gadget syzygy.

In the next lemma we introduce a new idea of gadget construction – *syzygy* – a trio of gadgets that work in concert, and their alignment produces their success.

Lemma 5.49. *Suppose either $X \notin \mathbb{R}$, or both $Y \notin \mathbb{R}$ and $X \neq 0, 1$. Then in the context of* $\text{Hol}_4(a, b)$*, there exists a unary recursive gadget M and a gadget A such that $\{M^j A\}_{j \geq 0}$ is a sequence of pairwise linearly independent symmetric binary generator signatures.*

Proof. The gadgets M_{16}, M_{17}, and M_{18} are depicted in Figure 5.14. The matrices for these gadgets are as follows.

$$M_{16} = \begin{bmatrix} a^4 + a^2b^2 & a^2b + b^3 \\ a^3 + ab^2 & a^2b^2 + b^4 \end{bmatrix}, \qquad M_{17} = \begin{bmatrix} a^4 + ab & a + b^3 \\ a^3 + b & ab + b^4 \end{bmatrix},$$

$$M_{18} = \begin{bmatrix} \begin{array}{c} a^8 + a^6b^2 + a^5b + a^4 + 2a^3b^3 \\ + a^2b^2 + ab^5 \end{array} & \begin{array}{c} a^6b + a^4b^3 + 2a^3b^2 \\ + a^2b^5 + 2ab^4 + b^7 \end{array} \\ \\ \begin{array}{c} a^7 + 2a^4b + a^5b^2 + 2a^2b^3 \\ + a^3b^4 + ab^6 \end{array} & \begin{array}{c} a^5b + a^2b^2 + 2a^3b^3 + b^4 \\ + ab^5 + a^2b^6 + b^8 \end{array} \end{bmatrix}.$$

We begin by calculating the trace and determinant of each gadget in the trio of gadgets M_{16}, M_{17}, and M_{18}:

$$\begin{aligned} \text{tr}(M_{16}) &= 2X^2 + Y \\ \text{tr}(M_{17}) &= 2X + Y \\ \text{tr}(M_{18}) &= (2X^2 + Y)(-X^2 + 2X + Y + 1) \\ \det(M_{16}) &= X(X - 1)(2X^2 + Y) \\ \det(M_{17}) &= (X - 1)(X^3 + X + Y) \\ \det(M_{18}) &= X(X - 1)^2(X^3 + X + Y)(2X^2 + Y). \end{aligned}$$

We note that $M_{18} = M_{17}M_{16}$.

We will show that if any one of these traces or determinants is 0, then another one of the three gadgets has nonzero eigenvalues of distinct norms. Since $X \neq 0, 1$ in any case, the only possibilities for any of these traces or determinants to be 0 are

$$2X^2 + Y = 0, \quad \text{or} \quad 2X + Y = 0, \quad \text{or} \quad X^3 + X + Y = 0,$$
$$\text{or} \quad -X^2 + 2X + Y + 1 = 0.$$

First we note that if any of these four equations is satisfied, then $X \notin \mathbb{R}$, for either it is by direct hypothesis of the lemma, or it is by $Y \notin \mathbb{R}$ and then by the equation.

Now we consider the four equations. If $2X^2 + Y = 0$ then $\operatorname{tr}(M_{17}) = -2X(X-1) \neq 0$, $\det(M_{17}) = X(X-1)^3 \neq 0$, and the ratio $\frac{\operatorname{tr}^2(M_{17})}{\det(M_{17})} = \frac{4X}{X-1} \notin \mathbb{R}^+$. If $2X + Y = 0$ then $\operatorname{tr}(M_{18}) = -2X(X-1)^2(X+1) \neq 0$, $\det(M_{18}) = 2X^3(X-1)^4(X+1) \neq 0$, and the ratio $\frac{\operatorname{tr}^2(M_{18})}{\det(M_{18})} = \frac{2(X+1)}{X} \notin \mathbb{R}^+$. If $X^3 + X + Y = 0$ then $\operatorname{tr}(M_{16}) = -X(X-1)^2 \neq 0$, $\det(M_{16}) = -X^2(X-1)^3 \neq 0$, and the ratio $\frac{\operatorname{tr}^2(M_{16})}{\det(M_{16})} = -X + 1 \notin \mathbb{R}^+$. Finally, suppose $-X^2 + 2X + Y + 1 = 0$. Then $\operatorname{tr}(M_{16}) = (X-1)(3X+1) \neq 0$, $\det(M_{16}) = X(X-1)^2(3X+1) \neq 0$, and the ratio $\frac{\operatorname{tr}^2(M_{16})}{\det(M_{16})} = \frac{3X+1}{X} \notin \mathbb{R}^+$. In each case, by Lemma 5.19 one of the gadgets M_{16}, M_{17}, or M_{18} has nonzero eigenvalues of distinct norm.

Now suppose each trace and determinant is nonzero, and suppose all three gadgets fail to have eigenvalues with distinct norm. Then Lemma 5.19 implies that $\frac{\operatorname{tr}^2(M_i)}{\det(M_i)} \in \mathbb{R}^+$ for each M_i. In particular, we claim that in this case $X \neq -1$. Otherwise, by hypothesis $Y \notin \mathbb{R}$ since $X = -1 \in \mathbb{R}$, and yet $\operatorname{tr}(M_{16}) = Y + 2$, $\det(M_{16}) = 2(Y+2)$, and $\frac{\operatorname{tr}^2(M_{16})}{\det(M_{16})} = (Y+2)/2 \in \mathbb{R}^+$; this is a contradiction.

We write $\frac{\operatorname{tr}^2(M_{18})\det(M_{16})\det(M_{17})}{\det(M_{18})\operatorname{tr}^2(M_{16})\operatorname{tr}^2(M_{17})} = \frac{(-X^2+2X+Y+1)^2}{(2X+Y)^2} = \left(1 - \frac{(X-1)(X+1)}{2X+Y}\right)^2$. The cancelation and its factorization as a square all follow from the fact that $\det(M_{18}) = \det(M_{16})\det(M_{17})$. Hence $\frac{(X-1)(X+1)}{2X+Y}$ is real valued. Furthermore, $\frac{(X-1)(X+1)}{2X+Y}$ is nonzero since $X \neq 1$ in either case of the hypothesis and $X \neq -1$ by the proof above. Thus $\frac{2X+Y}{(X-1)(X+1)}$ is well defined and real-valued. We also know that $\frac{2X^2+Y}{X(X-1)} = \frac{\operatorname{tr}^2(M_{16})}{\det(M_{16})} \in \mathbb{R}^+$, so there exist r and $s \in \mathbb{R}$ (both nonzero) such that

$$\begin{aligned} Y &= r(X-1)(X+1) - 2X, \\ Y &= sX(X-1) - 2X^2. \end{aligned} \tag{5.2}$$

Setting these two expressions of Y equal to each other and refactoring, we get

$$(X-1)(r - (s-2-r)X) = 0;$$

therefore $r = (s - 2 - r)X$, but since $r \neq 0$, we get $X \in \mathbb{R}$, and by (5.2), we also get $Y \in \mathbb{R}$. This contradiction establishes that one of the gadgets – M_{16}, M_{17}, or M_{18} – has nonzero eigenvalues of distinct norms. By Lemma 5.11, the powers of one of M_{16}, M_{17}, or M_{18} are pairwise linearly independent. Equivalently, it has infinite projective order by Lemma 5.10.

Now we show that in each case, for the chosen unary recursive gadget M, there exists a gadget A such that $M^j A$ are signature matrices of symmetric binary generator signatures, for all nonnegative integer j. We take g to be a gadget consisting of a single generator vertex with matrix

$$g = \begin{bmatrix} a & 1 \\ 1 & b \end{bmatrix}.$$

If $M = M_{16}$ or $M = M_{17}$, then by the symmetry in the gadget design each function $M^j g$ is a symmetric binary generator signature. Hence we take $A = g$ in these cases. For gadget M_{18}, we set $A = M_{17}g$, and again we get the desired graph symmetry. We know g is nonsingular, as $X \neq 1$. For any case where we need to apply M_{18}, we also have M_{18} is nonsingular. (This includes the case when $2X + Y = 0$, where we explicitly verified that $\det(M_{18}) \neq 0$, and when we supposed all three traces and determinants are nonzero.) However, $M_{18} = M_{17}M_{16}$, so it follows that $M_{17}g$ is also nonsingular in such a case. In any case, applying the nonsingular linear transformation defined by A: $M^j \mapsto M^j A$, does not change the fact that we have a sequence of pairwise linearly independent signatures, and these are symmetric binary generators by design. □

In the next three lemmas we will do the same thing as in Lemma 5.49 to produce a sequence of pairwise linearly independent symmetric binary generator signatures, in the case $X \in \mathbb{R}$. Note that we will avoid tractable cases; thus we will exclude the case $X = 1$ or $(X, Y) = (0, 0), (-1, 0)$ or $(-1, \pm 2)$. In Lemmas 5.50 and 5.51, the generators will be constructed using the technique of a binary recursive gadget and a starter gadget, rather than a unary recursive gadget that is symmetrized as in Lemma 5.49. Thus these generator signatures will be expressed as column vectors in $\mathbb{C}^3$ rather than the matrix form in $\mathbb{C}^{2\times 2}$ as in Lemma 5.49. However, in terms of their functionality as generator signatures there is no difference.

Lemma 5.50. *Suppose $X \in \mathbb{R} - \{-1, 0, 1\}$ and $a^4 \neq b^4$. Then there is a nonsingular binary recursive gadget $M \in \mathbb{C}^{3\times 3}$ and a starter gadget $s \in \mathbb{C}^3$ for which $\{M^j s\}_{j\geq 0}$ is a sequence of pairwise linearly independent symmetric binary generator signatures.*

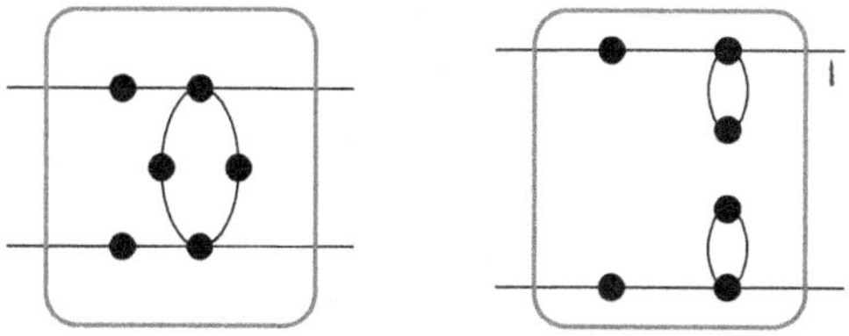

(a) Gadget M_{19} (b) Gadget M_{20}

Figure 5.15. Binary recursive gadgets.

Proof. The gadgets M_{19} and M_{20} are depicted in Figure 5.15.

$$M_{19} = \begin{bmatrix} a^4 & 2a & b^2 \\ a^3 & 1+ab & b^3 \\ a^2 & 2b & b^4 \end{bmatrix}$$

$$M_{20} = \begin{bmatrix} a^4 & 2a^2b & b^2 \\ a^3 & ab+a^2b^2 & b^3 \\ a^2 & 2ab^2 & b^4 \end{bmatrix}.$$

We calculate $\det(M_{19}) = X^2(X-1)^3 \neq 0$ and

$$\det([s\ M_{19}s\ M_{19}^2 s]) = -(a^4 - b^4)(X+1)(X-1)^4 \neq 0,$$

where s is the single vertex starter gadget. If gadget M_{19} also has infinite projective order then by Lemmas 5.35 and 5.34 we are done, so assume otherwise. Then for some integer $i \geq 0$ and nonzero $\lambda \in \mathbb{C}$ we have $M_{19}^i = \lambda M_{19}^{-1}$, and we construct the gadget

$$M = M_{19}^i M_{20} = \lambda M_{19}^{-1} M_{20} = \lambda \begin{bmatrix} 1 & 0 & 0 \\ 0 & X & 0 \\ 0 & 0 & 1 \end{bmatrix}.$$

This is really an application of the idea of an anti-gadget. Then $M^j s = \lambda^j \operatorname{diag}(1, X^j, 1)[a\ 1\ b]^{\mathrm{T}} = \lambda^j [a\ X^j\ b]^{\mathrm{T}}$. Since $X \in \mathbb{R} - \{-1, 0, 1\}$, we have that $\{M^j s\}_{j \geq 0}$ is a sequence of pairwise linearly independent symmetric binary generator signatures. □

Lemma 5.51. *Suppose $X = -1$ and $Y \in \mathbb{R} - \{-2, 0, 2\}$. Then there is a nonsingular binary recursive gadget $M \in \mathbb{C}^{3\times 3}$ and starter gadget $s \in \mathbb{C}^3$ for which $\{M^j s\}_{j \geq 0}$ is a sequence of pairwise linearly independent symmetric binary generator signatures.*

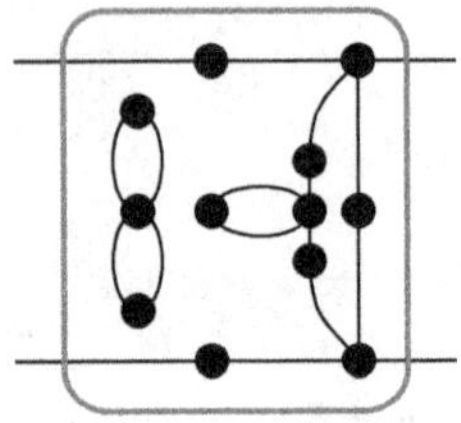

Figure 5.16. A binary recursive gadget.

Proof. The gadget M_{21} is depicted in Figure 5.16.

$$M_{21} = \begin{bmatrix} a^8 + a^5 b + a^6 b^2 & 2a^5 + 4a^3 b^2 & a^3 b + ab^3 \\ + a^3 b^3 & + 2ab^4 & + a^2 b^4 + b^6 \\ & & \\ a^7 + a^4 b + a^5 b^2 & a^4 + a^5 b + 2a^2 b^2 & a^3 b^2 + ab^4 \\ + a^2 b^3 & + 2a^3 b^3 + b^4 + ab^5 & + a^2 b^5 + b^7 \\ & & \\ a^6 + a^3 b & 2a^4 b + 4a^2 b^3 & a^3 b^3 + ab^5 \\ + a^4 b^2 + ab^3 & + 2b^5 & + a^2 b^6 + b^8 \end{bmatrix}.$$

If $X = -1$, then the characteristic polynomial of gadget M_{21} is

$$x^3 - (Y+2)(Y-2)x^2 - 2(Y+2)^2(Y-2)x - 8(Y+2)^2(Y-2),$$

so the determinant is nonzero. By Lemma 5.41, if all roots of the characteristic polynomial have the same norm, then $C^3 = B^3 D$ and this amounts to $(Y+2)^2 = -(Y+2)(Y-2)$, but then $Y(Y+2) = 0$, which is not true. Finally, applying Lemma 5.35 to M_{21} and the single-vertex starter gadget s, we get $\det([s\ M_{21}s\ M_{21}^2 s]) = 16Y^2(a^4 - b^4)(Y+2) \neq 0$. □

Lemma 5.52. *Suppose $X = 0$ and $Y \in \mathbb{C} - \{0\}$. Then there exists a unary recursive gadget M and gadgets A and B such that $\{AM^j B\}_{j \geq 0}$ is a sequence of pairwise linearly independent symmetric binary generator signatures.*

Proof. The gadgets M_{22} and M_{23} are depicted in Figure 5.17.

$$M_{22} = \begin{bmatrix} 1 + a^4 & a + b^3 \\ a^3 + b & 1 + b^4 \end{bmatrix}$$

$$M_{23} = \begin{bmatrix} 2a^4 + a^8 + 2ab + a^5 b + a^3 b^3 + b^4 & a^5 + 2a^2 b + a^3 b^2 + b^3 + 2ab^4 + b^7 \\ a^3 + a^7 + 2a^4 b + 2ab^2 + a^2 b^3 + b^5 & a^4 + 2ab + a^3 b^3 + 2b^4 + ab^5 + b^8 \end{bmatrix}.$$

By $X = 0$ and $Y \neq 0$, without loss of generality we assume $a = 0$, and $b \neq 0$.

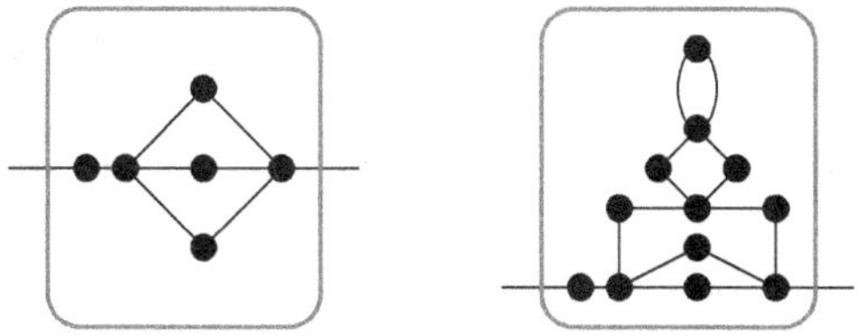

(a) Gadget M_{22} (b) Gadget M_{23}

Figure 5.17. Unary recursive gadgets.

First we show the existence of a unary recursive gadget M that has infinite projective order. Under the condition $X = 0$, $\det(M_{22}) = 1$. If gadget M_{22} has infinite projective order then we define $M = M_{22}$ and we are already done. So suppose otherwise. Then for some integer $i \geq 0$ and nonzero $\lambda \in \mathbb{C}$ we have $M_{22}^i = \lambda M_{22}^{-1}$, and we consider the gadget $M = M_{22}^i M_{23} = \lambda M_{22}^{-1} M_{23} = \lambda b^4 \begin{bmatrix} 1 & b^{-1} \\ 0 & 1 \end{bmatrix}$, which is nonsingular. Then $M^j = \lambda^j b^{4j} \begin{bmatrix} 1 & jb^{-1} \\ 0 & 1 \end{bmatrix}$ for any positive integer j, so M has infinite projective order.

Now we show the existence of appropriate gadgets A and B to symmetrize the construction. If $M = M_{22}$, then we set A to be the identity matrix $I = \begin{bmatrix} 1 & 0 \\ 0 & 1 \end{bmatrix}$ (representing an empty gadget with no vertex and a single straight through edge) and B to be a single generator vertex with one leading edge to be merged with the trailing edge of the recursive gadget, and the other edge becoming a dangling edge. The induced graph symmetry implies that AM^jB is a symmetric binary generator. Furthermore, A and B are nonsingular. If $M = M_{22}^i M_{23}$, then we set A to M_{23} and B to a single generator vertex with one leading edge to be merged with the trailing edge of the recursive gadget, and the other edge becoming a dangling edge. Again, the induced graph symmetry implies that AM^jB is a symmetric binary generator. Furthermore, B is nonsingular, as $X \neq 1$, and A is nonsingular, since if we are using this M as our unary recursive gadget, then we already know that $M_{22}^i M_{23}$ is nonsingular and therefore M_{23} is nonsingular. In any case, the linear transformation $M^j \mapsto AM^jB$ preserves pairwise linear independence, and we have constructed a sequence of pairwise linearly indpendent symmetric binary generator signatures. □

Now we show the existence of a set of general-purpose finisher gadgets. In these finisher gadgets, all dangling edges are internally incident to vertices that are assigned a recognizer signature $(=_4)$. Furthermore, one set of two dangling edges is incident to the same $(=_4)$ vertex. Hence, we are aiming to interpolate recognizer signatures of the form $[*, 0, *]$. These finisher gadgets operate under

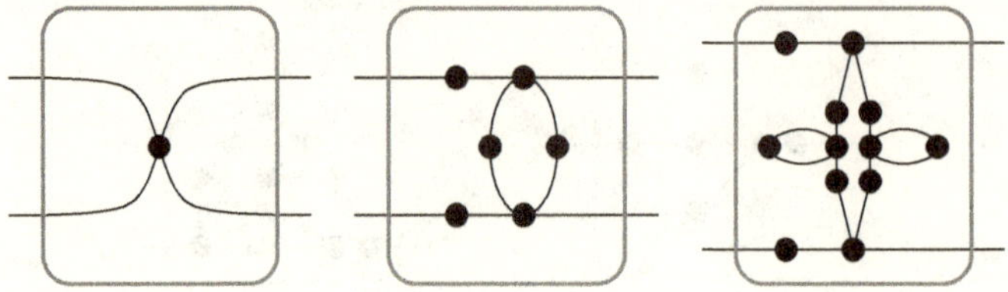

(a) Gadget M_{24} (b) Gadget M_{19} (c) Gadget M_{25}

Figure 5.18. Gadgets for building finisher gadgets.

the assumption that the "incoming" gadget has a binary symmetric generator signature. Our finisher gadgets will be expressed as 3×3 matrices (of rank 2), since the incoming signatures are symmetric. The essential property is that for a trio of finisher gadgets, their one-dimensional kernels are linearly independent.

Lemma 5.53. *Let $a, b \in \mathbb{C}$ such that $ab \neq 1$, and $a^4 \neq b^4$. Then in the context of* $\mathrm{Hol}_4(a, b)$*, there exist three binary finisher gadgets with rank 2 matrices $F, F', F'' \in \mathbb{C}^{3\times 3}$, such that their middle rows are all 0, and the kernels of F, F', and F'' are linearly independent.*

Proof. The gadgets M_{24} and M_{25} are depicted in Figure 5.18.

$$M_{24} = \begin{bmatrix} 1 & 0 & 0 \\ 0 & 0 & 0 \\ 0 & 0 & 1 \end{bmatrix}$$

$$M_{25} = \begin{bmatrix} a^8 + 2a^5b + a^2b^2 & 2a^5 + 4a^3b^2 + 2ab^4 & a^2 + 2ab^3 + b^6 \\ a^7 + 2a^4b + ab^2 & a^4 + a^5b + 2a^2b^2 + 2a^3b^3 + b^4 + ab^5 & a^2b + 2ab^4 + b^7 \\ a^6 + 2a^3b + b^2 & 2a^4b + 4a^2b^3 + 2b^5 & a^2b^2 + 2ab^5 + b^8 \end{bmatrix}.$$

We will break this proof down into two cases: $X \neq 0$ and $X = 0$. For both cases, we will show the existence of three finisher gadgets for which the (one-dimensional) kernels are linearly independent. Every finisher gadget we utilize in this proof has the property that the middle row of the corresponding transition matrix contains only 0's; hence we can get a nonzero vector v that spans its kernel by calculating the cross product of the first and last rows (provided that the matrix has rank 2). Suppose F, F', and F'' are finisher gadgets and let v_1, v_2, and v_3 denote the respective cross products of the first and last rows of each transition matrix. Then we use $\mathrm{cross}(F, F', F'')$ to denote the matrix whose rows are v_1, v_2, and v_3 (in that order). We can test the linear independence of $\{v_1, v_2, v_3\}$ by verifying the determinant $\det(\mathrm{cross}(F, F', F'')) \neq 0$.

Suppose $X \neq 0$. Let $F = M_{24}$, $F' = M_{24}M_{19}$, $F'' = M_{24}M_{19}^2$. Since M_{24} has rank 2 and $\det(M_{19}) = X^2(X-1)^3 \neq 0$, it follows that F, F', and F'' all have rank 2. We calculate that $\det(\text{cross}(F, F', F'')) = 4X^2(a^4 - b^4)(X-1)^4 \neq 0$, so the kernels are linearly independent.

Suppose $X = 0$. Without loss of generality we have $a = 0$ and $b \neq 0$ since $a^4 \neq b^4$. Let $F = M_{24}$, $F' = M_{24}M_{25}$, $F'' = M_{24}M_{25}^2$. Then M_{24} has rank 2,

$$\det(M_{25}) = (ab-1)^3(a^3+b)^2(a^2+b^2)^2(a+b^3)^2, \text{ and}$$
$$\det(\text{cross}(F, F', F'')) = 4(ab-1)^4(a^2-b^2)(a^3+b)^2(a^2+b^2)^7 (a+b^3)^2(a^4+2ab+b^4).$$

Under the condition $a = 0$ these polynomials simplify to $-b^{12} \neq 0$ and $-4b^{28} \neq 0$. Therefore F, F', and F'' all have rank 2 and the kernels of F, F', and F'' are linearly independent. □

Corollary 5.54. *Let $a, b \in \mathbb{C}$ be parameters in region III; namely, (a, b) are such that $X \neq 1$, $Y^2 \neq 4X^4$ (equivalently $a^4 \neq b^4$), and $(X, Y) \neq (-1, 0)$. Then* $\text{Hol}_4(a, b)$ *is* #P-hard *even when restricted to planar inputs.*

The same statement holds for $\text{Holant}(=_4|\ [a, 1, b], \Delta_0, \Delta_1)$*, and a #P-hardness reduction exists that does not use Lemma 3.13.*

Proof. We have produced a sequence of pairwise linearly independent symmetric binary generator signatures by the following lemmas: If $X \notin \mathbb{R}$, then we have Lemma 5.49 that uses the syzygy construction. Next we assume $X \in \mathbb{R}$ but $X \neq 1$. If further $X \notin \{0, -1\}$, then with $a^4 \neq b^4$, we can apply Lemma 5.50. Suppose $X = 0$, then for all $Y \in \mathbb{C} - \{0\}$, we have Lemma 5.52. Finally suppose $X = -1$. In this case if $Y \notin \mathbb{R}$, then we appeal to Lemma 5.49 again. If $X = -1$ and $Y \in \mathbb{R}$, then by the hypothesis of this corollary, $Y \in \mathbb{R} - \{-2, 0, 2\}$. Then we apply Lemma 5.51.

Now we apply Lemma 5.53 and Lemma 5.36 to the sequence of pairwise linearly independent symmetric binary generator signatures just produced, and obtain a sequence of pairwise linearly independent symmetric binary recognizer signatures of the form $\{[x_1, 0, y_1], \ldots, [x_m, 0, y_m]\}$.

Finally we apply Lemma 5.32. The conclusion of Corollary 5.54 follows. We observe that all reductions use only planar gadgets, and Lemma 5.30 is valid for planar inputs; thus the #P-hardness proved holds over planar instances of $\text{Hol}_4(a, b)$. □

5.2.3 Planar Tractable Region II Is #P-hard for General Graphs

Recall that region II is specified by $Y^2 = 4X^4$ (equivalently $a^4 = b^4$) but $X \notin \{0, \pm 1\}$.

It will be convenient to transform the problem to a setting where generator signatures are of the form $[a, 1, a]$. This can be done with a holographic transformation, and we can also reuse some of our work, using the observation that a sequence of pairwise linearly independent signatures remains so after a holographic transformation.

Lemma 5.55. *Let f and f' be $\mathcal{F}$-gates of arity r, where each dangling edge is internally incident to a vertex labeled with a generator signature. Then the property that f and f' are linearly independent is invariant under holographic reductions.*

Proof. Let $T \in \mathbb{C}^{2\times 2}$ be the basis for a holographic reduction. View the signatures f and f' as column vectors in $\mathbb{C}^{2^r}$ in unsymmetric signature notation. After a holographic reduction under basis T, we get the signatures $\hat{f}$ and $\hat{f}'$, where $[\hat{f}\ \hat{f}'] = (T^{-1})^{\otimes r}[f\ f']$. Since T is invertible, the $2^r \times 2$ block matrix $[f\ f']$ has rank 2 if and only if the $2^r \times 2$ block matrix $[\hat{f}\ \hat{f}']$ has rank 2, i.e., f and f' are linearly independent if and only if $\hat{f}$ and $\hat{f}'$ are linearly independent. □

Lemma 5.56. *Suppose $a^4 = b^4$. Then there exists $c \in \mathbb{C}$ such that the problem* $\operatorname{Holant}(=_4|\ [a, 1, b])$ *is equivalent to the problem* $\operatorname{Holant}([1, 0, 0, 0, e]\ |\ [c, 1, c])$ *under a diagonal holographic reduction, where $ab = c^2$ and $e = \pm 1$. Conversely, given a problem of the form* $\operatorname{Holant}([1, 0, 0, 0, e]\ |\ [c, 1, c])$ *where $c \in \mathbb{C}$ and $e = \pm 1$, there is a diagonal holographic reduction to an equivalent problem* $\operatorname{Holant}(=_4|\ [a, 1, b])$ *for which $ab = c^2$.*

Proof. If $ab = 0$, then $a = b = 0$ by $a^4 = b^4$. In this case, the lemma is trivial. Suppose otherwise, then both $a \neq 0$ and $b \neq 0$.

Let ω be a 16th root of unity such that $a\omega^4 = b$. By the diagonal holographic transformation $T = \begin{bmatrix} \omega^{-1} & 0 \\ 0 & \omega \end{bmatrix}$, we have $(T^{-1})^{\otimes 2}[a, 1, b]^{\mathrm{T}} = [a\omega^2, 1, b\omega^{-2}]^{\mathrm{T}} = [a\omega^2, 1, a\omega^2]^{\mathrm{T}}$, and $[1, 0, 0, 0, 1]T^{\otimes 4} = [\omega^{-4}, 0, 0, 0, \omega^4]$, so $\operatorname{Hol}_4(a, b)$ is equivalent to $\operatorname{Holant}([\omega^{-4}, 0, 0, 0, \omega^4]\ |\ [a\omega^2, 1, a\omega^2])$. Multiplying each entry of a signature by a nonzero value does not change the complexity of the problem, and $\omega^8 \in \{\pm 1\}$, so the problem is equivalent to $\operatorname{Holant}([1, 0, 0, 0, \pm 1]\ |\ [a\omega^2, 1, a\omega^2])$.

Conversely, consider a problem of the form $\operatorname{Holant}([1, 0, 0, 0, e]\ |\ [c, 1, c])$ where $c \in \mathbb{C}$ and $e = \pm 1$. If $e = 1$, then we are already done, so assume $e = -1$. Let $\omega = e^{\pi i/8}$, and performing a holographic reduction by $T = \begin{bmatrix} \omega^{-1} & 0 \\ 0 & \omega \end{bmatrix}$, we arrive at $\operatorname{Holant}([-i, 0, 0, 0, -i]\ |\ [\omega^2 c, 1, \omega^{-2} c])$, which is equivalent to Holant($=_4|$ $[a, 1, b]$), where $a = \omega^2 c$, $b = \omega^{-2} c$, $ab = c^2$, and $a^4 = b^4$. $\square$

We observed in Lemma 5.8 that if all signatures in $\mathcal{F}$ are complement-invariant, then any $\mathcal{F}$-gate that we build with them is also complement-invariant. Although we can't get the same conclusion in the setting of Holant($[1, 0, 0, 0, -1] \mid [a, 1, a]$), there is a similar structural property to observe.

Lemma 5.57. *Let $\mathcal{F} = \{[1, 0, 0, 0, -1], [a, 1, a]\}$. Then any $\mathcal{F}$-gate with an even number of degree-4 vertices has the property that its value remains unchanged under the complement of its inputs.*

Proof. Let $f \in \mathcal{F}$. If σ and σ' are complementary assignments to the edges of f, then their values satisfy $f\,|_\sigma = \pm f\,|_{\sigma'}$, where the sign is -1 if f is the function $[1, 0, 0, 0, -1]$, and the sign is $+1$ if $f = [a, 1, a]$. If σ and σ' are complementary assignments to the dangling edges of an $\mathcal{F}$-gate, by complementing the assignments to all internal edges, we have a 1-1 correspondence of all terms defining the signature sum of the $\mathcal{F}$-gate under σ with those under σ', and since the number of occurrences of $[1, 0, 0, 0, -1]$ is even, each pair of complementary assignments has the same value. $\square$

The following lemma is practically identical to Lemma 5.50, except we carry out the proof in region II. Note we do the analysis with a 2×2 iteration matrix owing to the extra degree of symmetry we have in this setting.

Lemma 5.58. *Consider the problem* Holant($[1, 0, 0, 0, e] \mid [a, 1, a]$), *where $a^2 \in \mathbb{R} - \{0, \pm 1\}$ and $e = \pm 1$. Then there exists a binary starter gadget s and a binary recursive gadget M such that $\{M^j s\}_{j \geq 0}$ is a sequence of pairwise linearly independent generator signatures of the form $\{[x_j, y_j, x_j]\}_{j \geq 0}$.*

Proof. In this proof, we will use M_{19}, M_{20}, and the single-vertex starter gadget s.

$$M_{19} = \begin{bmatrix} a^4 & 2ae & a^2 \\ a^3 & e + a^2 e & a^3 \\ a^2 & 2ae & a^4 \end{bmatrix}, \quad M_{20} = \begin{bmatrix} a^4 & 2a^3 e & a^2 \\ a^3 & a^2 e + a^4 e & a^3 \\ a^2 & 2a^3 e & a^4 \end{bmatrix}, \quad s = \begin{bmatrix} a \\ 1 \\ a \end{bmatrix}.$$

Let $C = \{[x\ y\ x]^{\mathrm{T}} : x, y \in \mathbb{C}\}$. Then $s \in C$ and C is closed under M_{19} and M_{20}. For this reason, all $\mathcal{F}$-gate signatures in this proof will be in C and we can restrict our attention to the first two entries of each signature. As such, we form

s', M', and M'' as stand-ins for s, M_{19}, and M_{20} (respectively) as follows:

$$s' = \begin{bmatrix} a \\ 1 \end{bmatrix},$$

$$M' = \begin{bmatrix} 1 & 0 & 0 \\ 0 & 1 & 0 \end{bmatrix} \cdot \begin{bmatrix} a^4 & 2ae & a^2 \\ a^3 & e + a^2 e & a^3 \\ a^2 & 2ae & a^4 \end{bmatrix} \cdot \begin{bmatrix} 1 & 0 \\ 0 & 1 \\ 1 & 0 \end{bmatrix} = \begin{bmatrix} a^4 + a^2 & 2ae \\ 2a^3 & e + a^2 e \end{bmatrix},$$

$$M'' = \begin{bmatrix} 1 & 0 & 0 \\ 0 & 1 & 0 \end{bmatrix} \cdot \begin{bmatrix} a^4 & 2a^3 e & a^2 \\ a^3 & a^2 e + a^4 e & a^3 \\ a^2 & 2a^3 e & a^4 \end{bmatrix} \cdot \begin{bmatrix} 1 & 0 \\ 0 & 1 \\ 1 & 0 \end{bmatrix} = \begin{bmatrix} a^4 + a^2 & 2a^3 e \\ 2a^3 & a^2 e + a^4 e \end{bmatrix}.$$

Note that $M'' = M' \begin{bmatrix} 1 & 0 \\ 0 & a^2 \end{bmatrix}$. First, $\det([s\ M's]) = a(a^2 - 1)(a^2 + e) \neq 0$, so if M' has infinite projective order then we are done by Lemma 5.35 and Lemma 5.15. Now assume otherwise, and since $\det(M') = a^2 e(a^2 - 1)^2 \neq 0$ and M' has finite projective order, there is some integer $i \geq 0$ and nonzero $\lambda \in \mathbb{C}$ for which $M'^i = \lambda M'^{-1}$. Then we construct the gadget $M = M'^i M'' = \lambda M'^{-1} M'' = \lambda \begin{bmatrix} 1 & 0 \\ 0 & a^2 \end{bmatrix}$. Then $M^j s = \lambda^j \operatorname{diag}(1, a^{2j})[a\ 1]^{\mathrm{T}} = \lambda^j [a\ a^{2j}]^{\mathrm{T}}$; hence $\{M^j s\}_{j \geq 0}$ is a sequence of pairwise linearly independent vectors. □

Lemma 5.59. *Consider the problem* Holant$([1, 0, 0, 0, e] \mid [c, 1, c])$, *where* $c^2 \notin \mathbb{R}$ *and* $e = \pm 1$. *Then there exists a unary recursive gadget* M *and gadget* A *such that* $\{M^j A\}_{j \geq 0}$ *is a sequence of pairwise linearly independent generator signatures of the form* $\{[x_j, y_j, x_j]\}_{j \geq 0}$.

Proof. By Lemma 5.56 let $a, b \in \mathbb{C}$ such that Holant$([1, 0, 0, 0, e] \mid [c, 1, c])$ is equivalent to Holant$(=_4 \mid [a, 1, b])$ and $ab = c^2$. Then by Lemma 5.49, since $X = ab = c^2 \notin \mathbb{R}$, we have a recursive unary gadget M and gadget A such that $\{M^j A\}_{j \geq 0}$ is a sequence of pairwise linearly independent symmetric binary generator signatures (in the context of Holant$(=_4 \mid [a, 1, b])$). By Lemmas 5.55 and 5.56, this gadget construction also produces a sequence of pairwise linearly independent binary generator signatures in the context of Holant$([1, 0, 0, 0, e] \mid [c, 1, c])$. All of the gadgets from Lemma 5.49 have an even number of degree-4 vertices, so by Lemma 5.57 (and Lemma 5.8) the signatures of the $\mathcal{F}$-gates we constructed are invariant under the complement of their inputs. □

The following lemma proves the #P-hardness for $\text{Hol}_4(a, b)$ in region II for general graphs.

Lemma 5.60. *Let $a, b \in \mathbb{C}$ such that $a^4 = b^4$ and $ab \notin \{-1, 0, 1\}$. Then* $\text{Hol}_4(a, b)$ *is* #P-hard. *The same statement holds for* $\text{Holant}(=_4 | [a, 1, b], \Delta_0, \Delta_1)$*, and a #P-hardness reduction exists that does not use Lemma 3.13.*

Proof. By Lemma 5.56 there exists $c \in \mathbb{C}$ and $e = \pm 1$ such that $ab = c^2$ and $\text{Hol}_4(a, b)$ is equivalent to $\text{Holant}([1, 0, 0, 0, e] \mid [c, 1, c])$ by a diagonal holographic transformation. Then $c^2 = ab \notin \{0, \pm 1\}$ and by Lemmas 5.59 and 5.58 we have a sequence of pairwise linearly independent generator signatures of the form $\{[x_j, y_j, x_j]\}_{j \ge 0}$. By Lemma 5.13, we can interpolate generator signature $[1, 1, 1]$, which we use in the following chain of reductions.

$$\begin{aligned}
\text{Hol}_3(ec, ec) &\le_T^P \text{Holant}([1, 0, 0, e] \mid [c, 1, c]) \\
&\le_T^P \text{Holant}([1, 0, 0, 0, e] \mid \{[c, 1, c], [1, 1, 1]\}) \\
&\le_T^P \text{Holant}([1, 0, 0, 0, e] \mid [c, 1, c]) \\
&\le_T^P \text{Holant}(=_4 | [a, 1, b]).
\end{aligned}$$

To justify this chain of reductions, observe that any instance of $\text{Hol}_3(ec, ec) = \text{Holant}(=_3 | [ec, 1, ec])$ must have an even number of recognizer vertices $=_3$. The first step follows from a diagonal holographic reduction under basis $\begin{bmatrix} 1 & 0 \\ 0 & e \end{bmatrix}$, so that $\text{Hol}_3(ec, ec)$ is equivalent to $\text{Holant}([1, 0, 0, e] \mid [ec, e, ec])$, as $e^2 = 1$, which has the same complexity as $\text{Holant}([1, 0, 0, e] \mid [c, 1, c])$. Note the number of recognizer vertices is unchanged, thus still even. For the next step of the reduction, any pair of vertices with signature $[1, 0, 0, e]$ can be simulated by a pair of vertices with signature $[1, 0, 0, 0, e]$. Simply introduce a vertex with generator signature $[1, 1, 1] = [1, 1]^{\otimes 2}$ and make it adjacent to both recognizer vertices. (This step of the reduction is in general not planar.) The last two steps are carried out by interpolation and by a diagonal holographic reduction. Since $ec \notin \{0, \pm 1, \pm \mathrm{i}\}$, by Lemma 5.14 (also by Theorem 5.1 for the case $k = 3$ already proved at the end of Section 5.1), we know that $\text{Hol}_3(ec, ec)$ is #P-hard and we are done. □

Proof of Theorem 5.1 Restricted to $k = 4$

Proof. Tractability (for regions I and II) is given by Theorem 5.48. Intractability over region II for nonplanar graphs is given by Lemma 5.60. Intractability over region III even for planar graphs is proved in Corollary 5.54. □

We conclude this chapter by giving an equivalent statement of Theorem 5.1 (hence also equivalent to Theorem 5.3), the dichotomy theorem for spin systems on regular-k graphs. Note that in this book we have only proved the cases when $k = 3$ and $k = 4$, and that will be the only cases we will use in what follows. A full proof of Theorem 5.3 can be found in [CK12, CK13, Kow10]. The following equivalent statements are in a transformational form, which are more conceptual than Theorem 5.3, and will be consistent with the general dichotomy expression in Theorem 7.19 in Chapter 7.

Recall the definition of functions of product type $\mathscr{P}$ (Definition 3.2), and functions of affine type $\mathscr{A}$ (Definition 3.6).

Theorem 5.61. *Suppose $x, y, z \in \mathbb{C}$ and $k \geq 3$. Then $Z(G)$ for a spin system on k-regular graphs G with edge function $g = [x, y, z]$ is* #P-hard *except in the following cases, when the problem is in P: There exists a holographic transformation T such that*

1. *$g \in T\mathscr{P}$ and $(=_k)T^{\otimes k} \in \mathscr{P}$, or*
2. *$g \in T\mathscr{A}$ and $(=_k)T^{\otimes k} \in \mathscr{A}$.*

The transformations in the tractable cases can be chosen to be diagonal. The same statement holds for Holant$(=_4|\ [a, 1, b], \Delta_0, \Delta_1)$*, and a #P-hardness reduction exists that does not use Lemma 3.13.*

If we restrict the input to planar k-regular graphs, then these are solvable in P, as well as when there exists a holographic transformation T such that $g \in T\mathscr{M}$ and $(=_k)T^{\otimes k} \in \mathscr{M}$, where $\mathscr{M}$ denotes all matchgate signatures. The problem remains #P-hard *in all other cases.*

Proof. We first show that condition 1 is equivalent to $g \in \mathscr{P}$. It is obvious that any $g \in \mathscr{P}$ satisfies condition 1 with $T = I$. Suppose g satisfies condition 1 with T. Since $(=_k)T^{\otimes k} \in \mathscr{P}$ is symmetric and $k \geq 3$, it follows that $(=_k)T^{\otimes k}$ is a GEN-EQ, expressible as $[a, 0]^{\otimes k} + [0, b]^{\otimes k}$, where $ab \neq 0$ because $(=_k)T^{\otimes k}$ is nondegenerate. So $(=_k)(TD)^{\otimes k} = (=_k)$ where $D = \begin{bmatrix} a^{-1} & 0 \\ 0 & b^{-1} \end{bmatrix}$. Let $T_1 = TD = \begin{bmatrix} \alpha & \beta \\ \gamma & \delta \end{bmatrix}$. Then

$$[\alpha, \beta]^{\otimes k} + [\gamma, \delta]^{\otimes k} = [1, 0]^{\otimes k} + [0, 1]^{\otimes k}.$$

Dot product with $[-\beta, \alpha]$ and then with $[-\delta, \gamma]$ imply that $\beta\delta[1, 0]^{\otimes(k-2)} + \alpha\gamma[0, 1]^{\otimes(k-2)} = 0$. Since $k \geq 3$, we get $\alpha\gamma = \beta\delta = 0$. So T_1 is either diagonal or antidiagonal since it is nonsingular, thus $T_1\mathscr{P} = \mathscr{P}$. We are also given $g \in T\mathscr{P}$. Since D is diagonal, $D\mathscr{P} = \mathscr{P}$. Thus $g \in T\mathscr{P} = TD\mathscr{P} = T_1\mathscr{P} = \mathscr{P}$.

Clearly, if $y = 0$ in $g = [x, y, z]$ then g is a GEN-EQ, and $g \in \mathscr{P}$. Suppose $y \neq 0$ and we normalize it and write $g = [a, 1, b]$. Then in Theorem 5.3 conditions 1 and 2 correspond to g being degenerate and g being a binary DISEQUALITY. Together these cases are exactly $g \in \mathscr{P}$.

In the following we assume $g \notin \mathscr{P}$. It is easy to verify that condition 3 in Theorem 5.3 is equivalent to $a^{4k} = 1$ and $b = -1/a$. If so, we take $T = \begin{bmatrix} 1 & 0 \\ 0 & a^{-1} \end{bmatrix}$. Then $(=_k)T^{\otimes k} \in \mathscr{A}$, being in the family $\mathscr{F}_1$ of Section 3.2. Furthermore, $(T^{-1})^{\otimes 2}(g) = [a, a, ba^2] = a[1, 1, -1] \in \mathscr{A}$.

Now suppose there exists some T such that $(=_k)T^{\otimes k} \in \mathscr{A}$ and $g \in T\mathscr{A}$. Denote $\mathscr{F}_{123} = \mathscr{F}_1 \cup \mathscr{F}_2 \cup \mathscr{F}_3$ (defined in Section 3.2), then since the signature $(=_k)T^{\otimes k}$ is nondegenerate and symmetric, $(=_k)T^{\otimes k} \in \mathscr{F}_{123}$. By its form, there exists a matrix $M \in \left\{ I, \begin{bmatrix} 1 & 1 \\ 1 & -1 \end{bmatrix}, \begin{bmatrix} 1 & \mathrm{i} \\ 1 & -\mathrm{i} \end{bmatrix} \right\}$ such that $(=_k)(TM^{-1})^{\otimes k}$ has the form $\lambda \left\{ [1, 0]^{\otimes k} + \mathrm{i}^r [0, 1]^{\otimes k} \right\}$ for some $\lambda \neq 0$ and $r \in \{0, 1, 2, 3\}$. Let $\omega = e^{\frac{r\pi \mathrm{i}}{2k}}$, then $\omega^k = \mathrm{i}^r$. Let $T_1 = TM^{-1} \begin{bmatrix} 1 & 0 \\ 0 & \omega \end{bmatrix}^{-1}$, then $(=_k)T_1^{\otimes k} = \lambda \left\{ [1, 0]^{\otimes k} + [0, 1]^{\otimes k} \right\} = \lambda(=_k)$. We write $T_1 = \begin{bmatrix} \alpha & \beta \\ \gamma & \delta \end{bmatrix}$; then

$$[\alpha, \beta]^{\otimes k} + [\gamma, \delta]^{\otimes k} = \lambda \left\{ [1, 0]^{\otimes k} + [0, 1]^{\otimes k} \right\}.$$

Dot product with $[-\beta, \alpha]$ and then with $[-\delta, \gamma]$ imply that $\alpha\gamma = \beta\delta = 0$, as before. So T_1 has the form $\alpha \begin{bmatrix} 1 & 0 \\ 0 & \rho \end{bmatrix}$ or $\alpha X \begin{bmatrix} 1 & 0 \\ 0 & \rho \end{bmatrix}$, where $\alpha \neq 0$, $X = \begin{bmatrix} 0 & 1 \\ 1 & 0 \end{bmatrix}$, and $\rho^k = 1$. It follows that $T = \alpha X' \begin{bmatrix} 1 & 0 \\ 0 & \omega' \end{bmatrix} M$, where $X' = I$ or X, $\omega' = \omega\rho$ and $\omega'^{4k} = 1$.

It is also given that $g \in T\mathscr{A}$. Since $(T^{-1})^{\otimes 2} g$ is a symmetric signature, $(T^{-1})^{\otimes 2} g \in \mathscr{F}_{123}$. By the form of M, $M\mathscr{F}_{123} = \mathscr{F}_{123}$. Hence $\left(\begin{bmatrix} 1 & 0 \\ 0 & \omega' \end{bmatrix}^{-1} X' \right)^{\otimes 2} [a, 1, b] \in \mathscr{F}_{123}$. Then we have $[a\omega', 1, b/\omega'] \in \mathscr{F}_{123}$, or $[b\omega', 1, a/\omega'] \in \mathscr{F}_{123}$. By their forms in Section 3.2, and having assumed $g \notin \mathscr{P}$, we have $ab = -1$ and $a^{4k} = 1$.

Exercise: Suppose $ab = -1$ and $a^{4k} = 1$. Show that there is a diagonal T' of the form $\begin{bmatrix} 1 & 0 \\ 0 & e^{\frac{2\pi \mathrm{i}}{4k} s} \end{bmatrix}$, for some integer s, such that $(=_k)T'^{\otimes k} \in \mathscr{A}$ and $(T'^{-1})^{\otimes 2}[a, 1, b] \in \mathscr{A}$. Thus the transformation in the affine case of Theorem 5.61 can be chosen to be diagonal.

Concerning k-regular planar graphs, we show that there exists a holographic transformation T such that $[a, 1, b] \in T\mathcal{M}$ and $(=_k)T^{\otimes k} \in \mathcal{M}$ iff $a^k = b^k$. Suppose $a^k = b^k$. Let $T = DH_2$, where $D = \begin{bmatrix} 1 & 0 \\ 0 & \omega^{-1} \end{bmatrix}$, $\omega = \sqrt{\frac{a}{b}}$ with $\omega^{2k} = 1$, and $H_2 = \frac{1}{\sqrt{2}}\begin{bmatrix} 1 & 1 \\ 1 & -1 \end{bmatrix}$. Then $(=_k)D^{\otimes k} = [1, 0]^{\otimes k} + \omega^{-k}[0, 1]^{\otimes k}$, and $(=_k)T^{\otimes k}$ is a nonzero multiple of $[1, 1]^{\otimes k} \pm [1, -1]^{\otimes k} \in \mathcal{M}$. Also $(D^{-1})^{\otimes 2}[a, 1, b] = [a, \omega, b\omega^2] = [a, \omega, a]$, and $(T^{-1})^{\otimes 2}[a, 1, b]$ has the matrix form

$$\frac{1}{2}\begin{bmatrix} 1 & 1 \\ 1 & -1 \end{bmatrix}\begin{bmatrix} a & \omega \\ \omega & a \end{bmatrix}\begin{bmatrix} 1 & 1 \\ 1 & -1 \end{bmatrix} = \begin{bmatrix} a+\omega & 0 \\ 0 & a-\omega \end{bmatrix},$$

which is in $\mathcal{M}$.

Now suppose $T = \begin{bmatrix} \alpha & \beta \\ \gamma & \delta \end{bmatrix}$, and both $(=_k)T^{\otimes k} \in \mathcal{M}$ and $[a, 1, b] \in T\mathcal{M}$. By the parity requirements for a matchgate signature $(=_k)T^{\otimes k} = [\alpha, \beta]^{\otimes k} + [\gamma, \delta]^{\otimes k} \in \mathcal{M}$ we have either

$$\alpha^k + \gamma^k = \alpha^{k-2}\beta^2 + \gamma^{k-2}\delta^2 = 0, \tag{5.3}$$

or

$$\alpha^{k-1}\beta + \gamma^{k-1}\delta = \alpha^{k-3}\beta^3 + \gamma^{k-3}\delta^3 = 0. \tag{5.4}$$

First we claim that none of $\alpha, \beta, \gamma, \delta$ can be 0. Suppose $\alpha = 0$, then $\gamma = 0$ or $\gamma^{k-1}\delta = 0$. T is nonsingular, it implies that $\delta = 0$, and $\beta\gamma \neq 0$. But clearly for such a T, $(=_k)T^{\otimes k} \notin \mathcal{M}$ by Theorem 4.11, since $k \geq 3$. Hence $\alpha \neq 0$. By the invariance of $\mathcal{M}$ under $X = \begin{bmatrix} 0 & 1 \\ 1 & 0 \end{bmatrix}$, the proof that $\alpha \neq 0$ also gives $\beta \neq 0$. By the symmetry between $[\alpha, \beta]$ and $[\gamma, \delta]$ we also have $\gamma\delta \neq 0$.

Then dividing in either set of the equations (5.3) and (5.4), we get $\frac{\alpha^2}{\beta^2} = \frac{\gamma^2}{\delta^2}$. By $\det(T) = \alpha\delta - \beta\gamma \neq 0$, we get

$$\frac{\alpha}{\beta} = -\frac{\gamma}{\delta}. \tag{5.5}$$

From (5.3) we get $(\alpha/\gamma)^k = -1$. From (5.4), we get $(\alpha/\gamma)^{k-1} = -\delta/\beta$, and together with (5.5) we get $(\alpha/\gamma)^k = 1$. Thus $(\alpha/\gamma)^{2k} = 1$ from either (5.3) or (5.4).

It follows that

$$T = \begin{bmatrix} \alpha & \beta \\ \alpha\omega & -\beta\omega \end{bmatrix} = \begin{bmatrix} 1 & 0 \\ 0 & \omega \end{bmatrix}\begin{bmatrix} 1 & 1 \\ 1 & -1 \end{bmatrix}\begin{bmatrix} \alpha & 0 \\ 0 & \beta \end{bmatrix}$$

for some $\omega^{2k} = 1$.

Now we consider $(T^{-1})^{\otimes 2}[a, 1, b] \in \mathscr{M}$, where T^{-1} takes the form $\begin{bmatrix} x & 0 \\ 0 & y \end{bmatrix}\begin{bmatrix} 1 & 1 \\ 1 & -1 \end{bmatrix}\begin{bmatrix} 1 & 0 \\ 0 & \rho \end{bmatrix}$, where $xy \neq 0$ and $\rho^{2k} = 1$. Then we can calculate the binary signature $(T^{-1})^{\otimes 2}[a, 1, b]$ by its matrix form to get

$$(T^{-1})^{\otimes 2}[a, 1, b] = [x^2(a + 2\rho + b\rho^2), xy(a - b\rho^2), y^2(a - 2\rho + b\rho^2)].$$

By the parity requirements for a matchgate signature, and $xy \neq 0$, we get either $a - b\rho^2 = 0$ or $a + 2\rho + b\rho^2 = a - 2\rho + b\rho^2 = 0$. The latter leads to $\rho = 0$, a contradiction; thus $a = b\rho^2$. Taking k-th power, we get $a^k = b^k$. □

6

Holant Problems and #CSP

We have by now encountered three representative frameworks for counting problems: Holant problems, #CSP, and spin systems. #CSP can be viewed as the special case of Holant problems where the local constraint function set is assumed to contain EQUALITY of all arities. If we allow other auxiliary functions to be free, such as all unary functions, we get other restricted Holant problems, in this case Holant* problems. Spin systems can be seen as a further specialization of #CSP. At the same time, Holant problems can be viewed as #CSP where each variable appears twice, syntactically.

Their relationship goes deeper than this. We have already seen in Chapters 3 and 5 how results from Holant* problems can be used to derive dichotomy theorems for Boolean #CSP and 2-spin systems. In this chapter we will go the other way. We use results from Boolean #CSP and 2-spin systems to derive several results on Holant problems. We introduce, and prove a dichotomy for, a class of Holant problems called Holant^c problems, an intermediate class between Holant and Holant* problems. We also prove a dichotomy for a version of Boolean #CSP, called #CSP^d where each variable appears a multiple of d times, for some integer $d \geq 1$. Finally we discuss a concrete problem called the EULERIAN ORIENTATION problem, which has a very simple expression as a Holant problem. Although interesting in their own right, the primary purpose of these results is to pave the way for the dichotomy theorem for Holant problems without auxiliary functions in Chapter 7.

In addition to the new theorems, this chapter also introduces a new transformational perspective not only as a proof technique but also how a dichotomy is conceptually understood and even stated. Some group theoretic ideas are naturally introduced in this setting.

6.1 A Dichotomy for a Single Ternary Signature

In this section, we consider the complexity of Holant(f), where $f = [f_0, f_1, f_2, f_3]$ is a symmetric signature of arity 3. We also denote by Pl-Holant(f) the restriction of Holant(f) to planar graphs. If f is degenerate, then Holant(f) is trivially tractable, since a degenerate signature factors as a tensor product of unary signatures, and the signature grid simply decomposes into isolated edges. In the following we assume that f is nondegenerate. In Chapter 2 (Section 2.3.1), we have shown that any nondegenerate signature $f = [f_0, f_1, f_2, f_3]$ can be expressed in one of the following three forms of parameterization (with the convention that $\alpha^0 = 1$, and $k\alpha^{k-1} = 0$ if $k = 0$, even when $\alpha = 0$):

1. $f_k = \alpha^{3-k}\beta^k + \gamma^{3-k}\delta^k$, where $\det \begin{bmatrix} \alpha & \gamma \\ \beta & \delta \end{bmatrix} \neq 0$
2. $f_k = ck\lambda^{k-1} + d\lambda^k$, where $c \neq 0$
3. $f_k = c(3-k)\lambda^{2-k} + d\lambda^{3-k}$, where $c \neq 0$

The first form corresponds to the case when the characteristic equation of a second-order linear recurrence has two distinct roots and we call it the *generic* case. The second form, and its reversal in the third form, correspond to the *double-root* case. We can omit the formal proof for the third form since it is symmetric to the second form.

For the *generic* case, we can apply a holographic transformation using $T = \begin{bmatrix} \alpha & \gamma \\ \beta & \delta \end{bmatrix}$. We have the following reduction:

$$\begin{aligned} \text{Holant}(f) &\equiv_T \text{Holant}(=_2 \mid f) \\ &\equiv_T \text{Holant}([1,0,1]T^{\otimes 2} \mid (T^{-1})^{\otimes 3}[f_0, f_1, f_2, f_3]) \\ &\equiv_T \text{Holant}(g \mid [1,0,0,1]), \end{aligned}$$

where $g = [1,0,1]T^{\otimes 2}$. The crucial step is of course $[f_0, f_1, f_2, f_3] = T^{\otimes 3}[1,0,0,1]$.

Therefore for the *generic* case, the complexity of Holant(f) is equivalent to the spin system on 3-regular graphs Holant($g \mid =_3$), where g is a symmetric binary signature, for which Theorem 5.1 is a dichotomy (the cases $k = 3$ and 4 are proved in Chapter 5). Theorem 5.3 is an equivalent statement of Theorem 5.1. Theorem 5.61 is a more conceptual form of Theorems 5.1 and 5.3. For easy reference we state these together below for $k = 3$.

Theorem 6.1. *The problem* Holant($[x, y, z] \mid =_3$) *is #P-hard for all $x, y, z \in \mathbb{C}$ except in the following cases, where the problem is solvable in polynomial time:*

(1) $xz = y^2$*; (2)* $y = 0$*; (3)* $x = z = 0$*; or (4)* $x^{12} = y^{12}$ *and* $xz = -y^2$*. This is also valid for* Holant$([x, y, z], \Delta_0, \Delta_1 \mid =_3)$*, and when it is #P-hard a reduction exists that does not use Lemma 3.13.*

Furthermore, in addition to these cases Pl-Holant(f) *is solvable in polynomial time if* $x^3 = z^3$*, and remains* #P-hard *in all other cases.*

Equivalently, Holant$([x, y, z] \mid =_3)$ *is in P if there exists* $T \in \mathbf{GL}_2(\mathbb{C})$ *such that*

1. $[x, y, z]T^{\otimes 2} \in \mathscr{P}$ *and* $(=_3) \in T\mathscr{P}$*, or*
2. $[x, y, z]T^{\otimes 2} \in \mathscr{A}$ *and* $(=_3) \in T\mathscr{A}$*.*

Otherwise it is #P-hard*. In addition,* Pl-Holant(f) *is in P if there exists* $T \in \mathbf{GL}_2(\mathbb{C})$ *such that* $[x, y, z]T^{\otimes 2} \in \mathscr{M}$ *and* $(=_3) \in T\mathscr{M}$*. It remains* #P-hard *in all other cases.*

Remark: The signature $[x, y, z]$ in case (1) is degenerate, which is a special case of product type. Cases (2) and (3) are of product type, being binary generalized EQUALITY and DISEQUALITY. These are the only cases for a binary symmetric signature to be of product type. It is easy to show that assuming $y \neq 0$, condition (4) can be equivalently stated as (4′) $x^6 = z^6$ and $xz = -y^2$. Case (4) includes what is transformable to an affine signature by $T = \begin{bmatrix} 1 & 0 \\ 0 & \omega \end{bmatrix}$, where $\omega^3 = 1$ and T keeps $(=_3)$ invariant. One can easily verify that the only matrices T that keep $(=_3)$ invariant are of the form $\begin{bmatrix} \omega & 0 \\ 0 & \omega' \end{bmatrix}$ or $\begin{bmatrix} 0 & \omega \\ \omega' & 0 \end{bmatrix}$, where $\omega^3 = \omega'^3 = 1$, and such matrices T preserve $\mathscr{P}$. The tractable cases of Theorem 6.1 can also be equivalently stated as follows: There exists some T such that $[x, y, z]T^{\otimes 2} \in \mathscr{A} \cup \mathscr{P}$ (additionally in $\mathscr{M}$ for Pl-Holant(f)) and $(T^{-1})^{\otimes 3}(=_3) = (=_3)$.

To get a dichotomy for Holant$([f_0, f_1, f_2, f_3])$, we next deal with the *double-root* case. In this part of the proof we will use nondiagonal holographic transformations. We have the following lemma:

Lemma 6.2. *Let* $f_k = ck\lambda^{k-1} + d\lambda^k$*, where* $c \neq 0$ *and* $0 \leq k \leq 3$*. If* $\lambda = \pm\mathrm{i}$*, then* Holant$([f_0, f_1, f_2, f_3])$ *is in P; in fact, the Holant value is always 0 in this case. If* $\lambda \neq \pm\mathrm{i}$*, then* Holant$([f_0, f_1, f_2, f_3])$ *is #P-hard. Additionally,* Pl-Holant(f) *is in P if there exists orthogonal* $T \in \mathbf{O}_2(\mathbb{C})$ *such that* $(=_2)T^{\otimes 2} = (=_2) \in \mathscr{M}$ *and* $T^{\otimes 3}[f_0, f_1, f_2, f_3] = [0, 1, 0, 0] \in \mathscr{M}$*, and it remains #P-hard otherwise.*

Proof. If $\lambda = \pm\mathrm{i}$, the signature $f = [f_0, f_1, f_2, f_3]$ satisfies the recurrence relation $f_{k+2} = 2\lambda f_{k+1} + f_k$, where $k = 0, 1$. This is a generalized Fibonacci

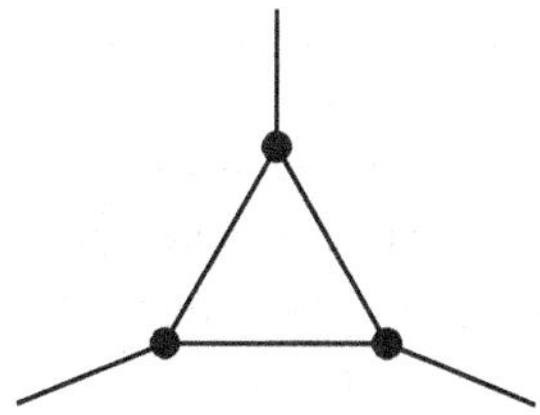

Figure 6.1. All three vertices are assigned the signature $[z, 1, 0, 0]$.

signature. Thus Holant(f) is solvable in P. Note that for $\lambda = \pm\mathrm{i}$, f can also be written as $Z[u, v, 0, 0]$ or $Z[0, 0, v, u]$ where $Z = \begin{bmatrix} 1 & 1 \\ \mathrm{i} & -\mathrm{i} \end{bmatrix}$ is defined in (2.11). In fact, for $\lambda = \mathrm{i}$ we may take $u = -(f_0 + 3\mathrm{i}f_1)/2$ and $v = (f_0 + \mathrm{i}f_1)/2$. Similarly for $\lambda = -\mathrm{i}$. Under a holographic transformation by Z, Holant(f) $\equiv_{\mathrm{T}}$ Holant($\neq_2 | (Z^{-1})^{\otimes 3} f$). As $(Z^{-1})^{\otimes 3} f = [u, v, 0, 0]$ or $[0, 0, v, u]$, in any bipartite signature grid for Holant($\neq_2 | (Z^{-1})^{\otimes 3} f$), the left-hand side (LHS) ($\neq_2$) requires exactly the same number of 0's and 1's, while the right-hand side (RHS) requires strictly more, or strictly less, 0's than 1's. Thus the Holant value is always 0.

In the following we assume $\lambda \neq \pm\mathrm{i}$. We can view Holant($f$) as the bipartite Holant($=_2 | f$). Recall from (2.8) that an orthogonal transformation $T \in \mathbf{O}_2(\mathbb{C})$ keeps the binary EQUALITY ($=_2$) invariant: $(=_2)T^{\otimes 2} = (=_2)$.

Let $M = \begin{bmatrix} 1 & \frac{d-1}{3} \\ \lambda & c + \frac{d-1}{3}\lambda \end{bmatrix}$, then the signature $[f_0, f_1, f_2, f_3]$ can be expressed as

$$(f_0, f_1, f_1, f_2, f_1, f_2, f_2, f_3)^{\mathrm{T}} = M^{\otimes 3}(1, 1, 1, 0, 1, 0, 0, 0)^{\mathrm{T}}.$$

Let $Q = \frac{1}{\sqrt{1+\lambda^2}} \begin{bmatrix} 1 & \lambda \\ \lambda & -1 \end{bmatrix}$, then Q is orthogonal. In exactly the same way as in (2.18) from Chapter 2 (replacing λ for α), the orthogonal transformation Q transforms f to $[z, 1, 0, 0]$ for some $z \in \mathbb{C}$, up to a scalar. Then the complexity of Holant(f) is the same as that of Holant($[z, 1, 0, 0]$).

For $z = 0$ in the signature $[z, 1, 0, 0]$, it is the problem of counting perfect matchings on 3-regular graphs. Because $[0, 1, 0, 0] \in \mathscr{M}$, Pl-Holant($f$) is in P in this case. Over general 3-regular graphs Holant($[0, 1, 0, 0]$) is #P-hard [DL92].

In the following we assume $z \neq 0$. We show that Pl-Holant($[z, 1, 0, 0]$) is #P-hard. We can realize a new ternary signature $f' = [z^3 + 3z, z^2 + 1, z, 1]$ by connecting $[z, 1, 0, 0]$'s in a triangle gadget as illustrated in Figure 6.1. So if Pl-Holant(f') is #P-hard then Pl-Holant($[z, 1, 0, 0]$) is also #P-hard. To see that the gadget in Figure 6.1 has the signature $[z^3 + 3z, z^2 + 1, z, 1]$ note that all three vertices are assigned the weighted matching signature $[z, 1, 0, 0]$ and can have at most one incident edge assigned 1. Thus, for example, on external input 000, there are three ways to assign exactly one internal edge the value 1 giving the value $3z$, and one way to assign all internal edges the value 0 giving the value z^3. The other entries of f' can be computed similarly.

In tensor product notation this signature is

$$f' = [z^3 + 3z, z^2 + 1, z, 1]^{\mathrm{T}} = \frac{1}{2}\left(\begin{bmatrix} z+1 \\ 1 \end{bmatrix}^{\otimes 3} + \begin{bmatrix} z-1 \\ 1 \end{bmatrix}^{\otimes 3}\right).$$

Then the following reduction chain holds:

$$\begin{aligned} \text{Pl-Holant}(f') &\equiv_{\mathrm{T}} \text{Pl-Holant}(=_2 \mid f') \\ &\equiv_{\mathrm{T}} \text{Pl-Holant}([z^2 + 2z + 2, z^2, z^2 - 2z + 2] \mid =_3), \end{aligned} \tag{6.1}$$

where the second step is a holographic reduction using $T = \begin{bmatrix} z+1 & z-1 \\ 1 & 1 \end{bmatrix} \in \mathbf{GL}_2(\mathbb{C})$, and T transforms $(=_2)$ to $(=_2)T^{\otimes 2}$, namely

$$T^{\mathrm{T}} I_2 T = \begin{bmatrix} z^2 + 2z + 2 & z^2 \\ z^2 & z^2 - 2z + 2 \end{bmatrix}.$$

Now we can apply Theorem 6.1 to the Holant problem in (6.1), and check against the tractable cases.

1. $[z^2 + 2z + 2, z^2, z^2 - 2z + 2]$ is nondegenerate. This is because it is the transformed binary signature from the nondegenerate $(=_2)$ under the non-singular matrix $T \in \mathbf{GL}_2(\mathbb{C})$.
2. $z^2 \neq 0$, since by assumption $z \neq 0$.
3. It cannot be the case that $z^2 + 2z + 2 = z^2 - 2z + 2 = 0$, again by $z \neq 0$.
4. We verify the equivalent condition (4′) stated in the Remark after Theorem 6.1, that there is no solution to $(z^2 + 2z + 2)^6 = (z^2 - 2z + 2)^6$ and $(z^2 + 2z + 2)(z^2 - 2z + 2) + z^4 = 0$. This is by a direct computation.

Therefore Holant(f') is #P-hard, and so is Holant($[z, 1, 0, 0]$) for all $z \in \mathbb{C}$.

Next we consider the planar case, with $z \neq 0$. We still have the equivalence Pl-Holant(f) $\equiv_{\mathrm{T}}$ Pl-Holant($[z, 1, 0, 0]$) by the orthogonal transformation Q.

To show that Pl-Holant($[z, 1, 0, 0]$) is also #P-hard for $z \neq 0$, we need to rule out the additional planar tractable case $(z^2 + 2z + 2)^3 = (z^2 - 2z + 2)^3$, if we still use the gadget in Figure 6.1. This condition is

$$3z^4 + 16z^2 + 12 = 0. \tag{6.2}$$

Figure 6.2. Both vertices are assigned the signature $[z, 1, 0, 0]$.

If $3z^4 + 16z^2 + 12 \neq 0$ then Pl-Holant($[z, 1, 0, 0]$) is #P-hard by the gadget in Figure 6.1. There are four exceptional values of z, which we prove separately as follows. Note that for these values, z^2 is real and $z^2 < 0$.

In addition to the gadget in Figure 6.1, we can construct a gadget in Pl-Holant($[z, 1, 0, 0]$) with a binary signature $[z^2 + 2, z, 1]$ in Figure 6.2. Thus the bipartite planar Holant problem Pl-Holant($[z^2 + 2, z, 1] \| f'$) is reducible to Pl-Holant($[z, 1, 0, 0]$). Under the same transformation $T = \begin{bmatrix} z+1 & z-1 \\ 1 & 1 \end{bmatrix} \in$ $\mathbf{GL}_2(\mathbb{C})$, we get an equivalent problem Pl-Holant($[X, Y, Z] \mid [1, 0, 0, 1]$), where $\begin{bmatrix} X & Y \\ Y & Z \end{bmatrix} = T^{\mathrm{T}} \begin{bmatrix} z^2+2 & z \\ z & 1 \end{bmatrix} T$; thus

$$
\begin{aligned}
X &= z^4 + 2z^3 + 5z^2 + 6z + 3, \\
Y &= z^4 + 3z^2 - 1, \\
Z &= z^4 - 2z^3 + 5z^2 - 6z + 3.
\end{aligned}
$$

Now again we can apply Theorem 6.1 to this problem, and verify that no tractable condition is satisfied for any root of (6.2). For the reader's convenience we include this verification as follows.

1. Nondegeneracy of $[X, Y, Z]$ follows from the fact that it is transformed from a nondegenerate binary signature by T.
2. $Y = 0$ is incompatible with (6.2).
3. It cannot be the case that $X = Z = 0$ assuming (6.2). In fact $X = Z$, which is equivalent to $z^3 + 3z = 0$, is already incompatible with (6.2).

4′. There is no solution to $XZ = -Y^2$ and $X^6 = Z^6$. Luckily we don't have to check $X^6 = Z^6$; $XZ = -Y^2$ is already incompatible with (6.2). We can write $X = A + zB$ and $Z = A - zB$, where $A = z^4 + 5z^2 + 3 \in \mathbb{R}$ and $B = 2z^2 + 6 \in \mathbb{R}$ by (6.2). Then $XZ + Y^2 = 0$ is equivalent to $A^2 + Y^2 = z^2B^2$, yet $A^2 + Y^2 > 0$ and $z^2B^2 < 0$ by $z^2 < 0$, which follows from (6.2).

5. Finally, for the planar condition $X^3 = Z^3$, we already know that $X = Z$ is incompatible with (6.2). We can verify that $X^2 + XZ + Z^2 = 0$ is also incompatible. In fact $X^2 + XZ + Z^2 = 3A^2 + z^2B^2$. By (6.2), $3A = -(z^2 + 3)$ and $B = 2(z^2 + 3)$, and therefore $(3A)^2 + 3z^2B^2 = (z^2 + 3)^2(12z^2 + 1)$, which is not zero, again by (6.2).

This completes the proof of Lemma 6.2. □

By Theorem 6.1 and Lemma 6.2, we have a dichotomy theorem for Holant($[f_0, f_1, f_2, f_3]$). From this, we can get a further dichotomy for all bipartite Holant problems on 2-3-regular graphs Holant($[g_0, g_1, g_2] | [f_0, f_1, f_2, f_3]$). The reduction is standard. For any nondegenerate $[g_0, g_1, g_2]$, we can find a transformation M, such that $[g_0, g_1, g_2] = [1, 0, 1]M^{\otimes 2}$, as shown in (2.19) in Chapter 2. Then the bipartite problem Holant($[g_0, g_1, g_2] | [f_0, f_1, f_2, f_3]$) is transformed to the equivalent problem Holant($M^{\otimes 3}[f_0, f_1, f_2, f_3]$), for which we can apply our dichotomy.

We state Theorems 6.3 and 6.4 in a more conceptual form, in terms of what is transformable to various tractable classes $\mathscr{A}$, $\mathscr{P}$ and $\mathscr{M}$.

Theorem 6.3. *Let $f = [f_0, f_1, f_2, f_3]$ be a ternary symmetric signature.* Holant(f) *is #P-hard unless f satisfies one of the following conditions, in which case the problem is in P:*

1. *f is degenerate.*
2. *There exists $T \in \mathbf{GL}_2(\mathbb{C})$ such that $f = T^{\otimes 3}(=_3)$, and the signature $(=_2)T^{\otimes 2} \in \mathscr{A} \cup \mathscr{P}$.*
3. *For $\lambda \in \{2\mathrm{i}, -2\mathrm{i}\}$, $f_2 - \lambda f_1 - f_0 = 0$ and $f_3 - \lambda f_2 - f_1 = 0$.*

Additionally, Pl-Holant(f) *is in P if there exists $T \in \mathbf{GL}_2(\mathbb{C})$ such that $(=_2)T^{\otimes 2} \in \mathscr{M}$ and $f \in T\mathscr{M}$.* Pl-Holant(f) *is #P-hard otherwise.*

As shown in Section 2.2, $f = [f_0, f_1, f_2, f_3]$ satisfies the condition in item (3) in Theorem 6.3 iff f is a generalized Fibonacci gate with parameter $\lambda = \pm 2\mathrm{i}$. This happens iff f has the form $f_k = ck\mu^{k-1} + d\mu^k$ ($0 \le k \le 3$ and $\mu = \pm\mathrm{i}$). Furthermore, this happens iff $(Z^{-1})^{\otimes 3} f$ has the form $[x, y, 0, 0]$ or its reversal $[0, 0, y, x]$, where $Z = \begin{bmatrix} 1 & 1 \\ \mathrm{i} & -\mathrm{i} \end{bmatrix}$. (See also the remark after Theorem 3.5.) Thus, item (3) in Theorem 6.3 can also be stated as $(Z^{-1})^{\otimes 3} f$ has the form $[x, y, 0, 0]$ or its reversal $[0, 0, y, x]$, while $(=_2)Z^{\otimes 2} = 2(\neq_2)$, a constant multiple of the binary Disequality function. Signatures of the form $Z[x, y, 0, 0]$ or $Z[0, 0, y, x]$ are special cases of the so-called *vanishing* signatures; they will be discussed in Chapter 7.

Thus we can state condition (3) of Theorem 6.3 equivalently as (3′): $(Z^{-1})^{\otimes 3} f$ has the form $[*, *, 0, 0]$ or its reversal $[0, 0, *, *]$.

The next theorem is concerned with 2-3-regular graphs.

Theorem 6.4. *Let $f = [f_0, f_1, f_2, f_3]$ and $g = [g_0, g_1, g_2]$ be symmetric signatures of arity 3 and 2 respectively.* Holant($g|f$) *is #P-hard unless f and g satisfy one of the following conditions, in which case the problem is in P:*

1. *f or g is degenerate.*
2. *There exists $T \in \mathbf{GL}_2(\mathbb{C})$ such that $f = T^{\otimes 3}(=_3)$, and the signature $gT^{\otimes 2} \in \mathscr{A} \cup \mathscr{P}$.*
3. *There exists $T \in \mathbf{GL}_2(\mathbb{C})$ such that $f = T^{\otimes 3}[1, 1, 0, 0]$, and $gT^{\otimes 2}$ is of form $[0, *, *]$.*

Additionally, Pl-Holant$(g \mid f)$ *is in P if there exists $T \in \mathbf{GL}_2(\mathbb{C})$ such that $gT^{\otimes 2} \in \mathscr{M}$ and $f \in T\mathscr{M}$.* Pl-Holant$(g \mid f)$ *is #P-hard otherwise.*

Theorem 6.3 is a special case of Theorem 6.4, with $g = (=_2)$. We show that the respective conditions in item (3) are equivalent when $g = (=_2)$. Suppose f is nondegenerate and satisfies the condition (3) of Theorem 6.3. We will write it in the form (3′), namely $(Z^{-1})^{\otimes 3} f = [x, y, 0, 0]$ or its reversal $[0, 0, y, x]$, for some $x, y \in \mathbb{C}$. Suppose $f = Z^{\otimes 3}[x, y, 0, 0]$; the other case is similar. By being nondegenerate, $y \neq 0$. Using this we can find an upper triangular matrix $R = \begin{bmatrix} 1 & \frac{x-1}{3} \\ 0 & y \end{bmatrix} \in \mathbf{GL}_2(\mathbb{C})$ such that $R^{\otimes 3}[1, 1, 0, 0] = [x, y, 0, 0]$. Then we may take $T = ZR$ for the required matrix in item (3) of Theorem 6.4. Note that for an upper triangular matrix R, we have $[0, 1, 0]R^{\otimes 2} = [0, *, *]$. In the other direction, suppose f is nondegenerate and satisfies the condition in item (3) of Theorem 6.4 with $T = \begin{bmatrix} a & c \\ b & d \end{bmatrix}$. The fact that $(=_2)T^{\otimes 2}$ has the form $[0, *, *]$ implies that $a^2 + b^2 = 0$, and so $b = \epsilon a \mathfrak{i}$, where $\epsilon = \pm 1$. Since f is nondegenerate, we have $\det T \neq 0$, thus $a \neq 0$ and $d \neq \epsilon c \mathfrak{i}$. Using this we can find an upper triangular matrix $R = \begin{bmatrix} a & \frac{c-\epsilon d \mathfrak{i}}{2} \\ 0 & \frac{c+\epsilon d \mathfrak{i}}{2} \end{bmatrix}$ such that $TR^{-1} = Z = \begin{bmatrix} 1 & 1 \\ \mathfrak{i} & -\mathfrak{i} \end{bmatrix}$ (for $\epsilon = +1$), or $TR^{-1} = Z' = \begin{bmatrix} 1 & 1 \\ -\mathfrak{i} & \mathfrak{i} \end{bmatrix}$ (for $\epsilon = -1$). Then $(Z^{-1})^{\otimes 3} f$ has the form $[*, *, 0, 0]$ or $[0, 0, *, *]$.

We can also derive Theorem 6.4 from Theorem 6.3. Suppose f and g are nondegenerate. Write $g = (=_2)M$ as in (2.19) from Chapter 2. Then $\operatorname{Holant}(g|f) \equiv_{\mathrm{T}} \operatorname{Holant}(=_2| Mf)$. The condition in item (2) of Theorem 6.3 is the existence of some T such that $Mf = T^{\otimes 3}(=_3)$, and $(=_2)T^{\otimes 2}$ is in $\mathscr{A} \cup \mathscr{P}$. It translates to the condition in item (2) of Theorem 6.4 by taking $M^{-1}T$ as the transformation. For item (3), the condition that $Z^{-1}Mf$ has the form $[c, d, 0, 0]$ (where $d \neq 0$ by nondegeneracy) is equivalent to the condition that for some upper triangular R, we have $R^{-1}Z^{-1}Mf = [1, 1, 0, 0]$. If so we can let $T = M^{-1}ZR$, then $T^{-1}f = [1, 1, 0, 0]$ and $gT = (=_2)MT = (=_2)ZR = 2(\neq_2)R$ has the form $[0, *, *]$. In the other direction, suppose for some T, we have $f = T[1, 1, 0, 0]$ and $gT = [0, a, b]$. By nondegeneracy, $\det T \neq 0$ and $a \neq 0$. Then we may take an upper triangular

$R = \begin{bmatrix} 1 & -b/(2a) \\ 0 & 1 \end{bmatrix}$. Then $[0, a, b]R = a(\neq_2)$, which we can verify by its matrix product form $R^{\mathrm{T}} \begin{bmatrix} 0 & a \\ a & b \end{bmatrix} R = \begin{bmatrix} 0 & a \\ a & 0 \end{bmatrix}$. Then $g = (=_2)M$, with M taken to be a scalar multiple of $ZR^{-1}T^{-1}$, and $\operatorname{Holant}(g|f) \equiv \operatorname{Holant}(=_2|\, Mf)$. With $T^{-1}f = [1, 1, 0, 0]$ and $R^{-1}[1, 1, 0, 0] = [*, *, 0, 0]$, it follows that Mf has the form $Z[*, *, 0, 0]$. We have proved that the condition in item (3) of Theorem 6.4 corresponds precisely to the condition in item (3) of Theorem 6.3 after transforming g to $gM^{-1} = (=_2)$. The same equivalence of Theorem 6.3 and Theorem 6.4 for the planar case can be shown.

The third cases in both Theorems 6.3 and 6.4 are a manifestation of a phenomenon called *vanishing*, which we will discuss in detail in Chapter 7. There is also a symmetric statement to case (3) in Theorem 6.4: $\operatorname{Holant}(g|f)$ is in P, if there is a 2×2 matrix T such that $f = T^{\otimes 3}[0, 0, 1, 1]$ and $gT^{\otimes 2}$ is of form $[*, *, 0]$. This is equivalent to case (3) in Theorem 6.4 if we use $T\begin{bmatrix} 0 & 1 \\ 1 & 0 \end{bmatrix}$ as T.

6.2 Reductions Between Holant and #CSP

In this section, we consider $\operatorname{Holant}(\{g\} \cup \mathcal{G} \mid \{f\} \cup \mathcal{F})$, where $f = [f_0, f_1, f_2, f_3]$ and $g = [g_0, g_1, g_2]$ are nondegenerate, and f is in the generic case of having distinct eigenvalues. We establish an equivalence between this problem and a #CSP problem, under a mild condition, and thus giving a dichotomy via the #CSP dichotomy of Chapter 3. This can be viewed as extending Theorems 6.3 and 6.4 in Section 6.1 for a single ternary signature to a set of signatures containing a nondegenerate ternary signature of the generic case. We will consider the double-root case in Section 6.3.

For f in the generic case, we can apply a holographic reduction to transform f to $(=_3)$. Therefore we need to consider only Holant problems of the form $\operatorname{Holant}(\{g\} \cup \mathcal{G}|\{=_3\} \cup \mathcal{F})$, where g is a nondegenerate symmetric binary signature. Furthermore, we observed after Theorem 6.1 that the ternary EQUALITY signature $(=_3)$ is invariant under any transformation from the following set:

$$\mathscr{T}_3 = \left\{ \begin{bmatrix} 1 & 0 \\ 0 & 1 \end{bmatrix}, \begin{bmatrix} 1 & 0 \\ 0 & \omega \end{bmatrix}, \begin{bmatrix} 1 & 0 \\ 0 & \omega^2 \end{bmatrix} \right\}, \tag{6.3}$$

where $\omega = \omega_3 = e^{2\pi \mathrm{i}/3}$. Thus, for any $T \in \mathscr{T}_3$,

$$\operatorname{Holant}(\{g\} \cup \mathcal{G}|\{=_3\} \cup \mathcal{F}) \equiv_{\mathrm{T}} \operatorname{Holant}(\{gT^{\otimes 2}\} \cup \mathcal{G}T|\{=_3\} \cup T^{-1}\mathcal{F}).$$

As a result, we can normalize g by a holographic reduction $T \in \mathscr{T}_3$. We define a symmetric binary signature $g = [g_0, g_1, g_2]$ to be *normalized* if either $g_0 = 0$ or it is not the case that g_2 is g_0 times a tth primitive root of unity, for some integer $t = 3t'$ where $\gcd(t', 3) = 1$. If g is not normalized, then $g_0 \neq 0$ and $g_2 = g_0\omega_t^s$, where $\omega_t = e^{2\pi \mathrm{i}/t}$, $t = 3t'$, $\gcd(t', 3) = 1$ and $\gcd(s, t) = 1$. Write $1 = 3u + t'v$ for some integers u and v, then $\omega_t = \omega^v \omega_{t'}^u$ and $\omega_t^s = \omega^k \omega_{t'}^l$, for some integers k and l, where $k \equiv sv \bmod 3$, and $\gcd(k, 3) = 1$. Hence we may take $k = 1$ or 2. After applying the transformation $\begin{bmatrix} 1 & 1 \\ 0 & \omega^k \end{bmatrix} \in \mathscr{T}_3$, the signature $[g_0, g_1, g_2]$ becomes $[g_0, g_1\omega^k, g_0\omega_{t'}^l]$, which is normalized. So in the following, we only deal with normalized g. In one case, we also need to normalize a unary signature $[u_0, u_1]$. We say $[u_0, u_1]$ is *normalized* if either $u_0 = 0$ or u_1 is not a multiple of u_0 by a tth primitive root of unity, for some integer $t = 3t'$ where $\gcd(t', 3) = 1$. Again we can normalize the unary signature by a suitable $T \in \mathscr{T}_3$. We note that a normalized signature is still normalized after a nonzero scalar multiple.

Theorem 6.5. *Let $\mathcal{F}$ and $\mathcal{G}$ be arbitrary sets of complex-valued signatures. Let $g = [g_0, g_1, g_2]$ be a normalized and nondegenerate symmetric signature. And in the case of $g_0 = g_2 = 0$, we further assume that $\mathcal{G}$ contains a unary signature $[u_0, u_1]$, which is normalized and $u_0u_1 \neq 0$. Then*

$$\operatorname{Holant}(\{g\} \cup \mathcal{G} | \{=_3\} \cup \mathcal{F}) \equiv_{\mathrm{T}} \#\mathrm{CSP}(\{g\} \cup \mathcal{F} \cup \mathcal{G}).$$

Consequently, by Theorem 3.7, $\operatorname{Holant}(\{g\} \cup \mathcal{G} | \{=_3\} \cup \mathcal{F})$ *is #P-hard unless $\{g\} \cup \mathcal{F} \cup \mathcal{G} \subseteq \mathscr{P}$ or $\{g\} \cup \mathcal{F} \cup \mathcal{G} \subseteq \mathscr{A}$, in which cases the problem is in P. Furthermore, by Theorem 3.21, if $\mathcal{G}$ contains Δ_0 and Δ_1, then a #P-hardness reduction exists for* $\operatorname{Holant}(\{g\} \cup \mathcal{G} | \{=_3\} \cup \mathcal{F})$ *that does not use Lemma 3.13.*

We prove Theorem 6.5 in the rest of this section. We note that we do not require the signatures in $\mathcal{F}$ and $\mathcal{G}$ to be symmetric. This dichotomy establishes an interesting link between Holant and #CSP. The dichotomy on Holant problems established in Theorem 6.5 uses results from #CSP. The applicability of Theorem 6.5 is not limited by the assumption on signature normalization of $[g_0, g_1, g_2]$; for a nonnormalized binary signature, we can first normalize it and then apply the theorem. The additional assumption of the existence of a nonzero unary signature circumvents a technical difficulty.

One direction of the reduction in Theorem 6.5, from Holant to #CSP, is straightforward by definition. Thus the main claim in Theorem 6.5 is a reduction from #CSP to these bipartite Holant problems. The approach is to construct the binary EQUALITY function $(=_2)$ on the LHS in $\operatorname{Holant}(\{g\} \cup \mathcal{G} | \{=_3\} \cup \mathcal{F})$. As soon as we have $(=_2)$ on LHS, together with $(=_3)$ on the RHS, we get EQUALITY gates of all arities $(=_k)$ on RHS. Also with the help of $(=_2)$ on LHS we can transfer $\mathcal{F}$ to LHS. Then we can simulate all of $\#\mathrm{CSP}(\{g\} \cup \mathcal{F} \cup \mathcal{G})$.

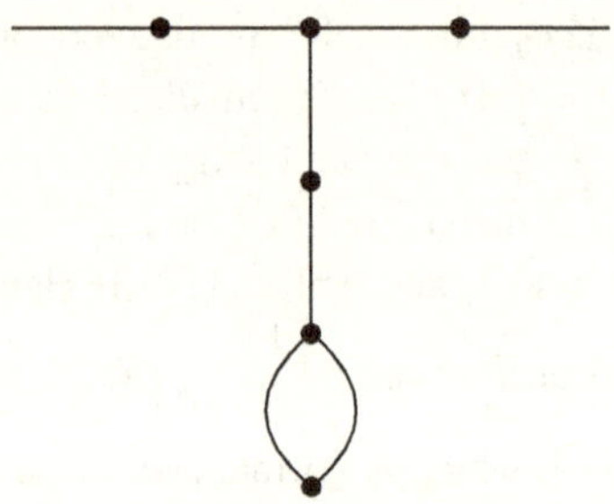

Figure 6.3. A gadget construction for the binary DISEQUALITY function ($\neq_2$) on the LHS. Degree-3 vertices have signature ($=_3$), degree-2 vertices have signature $[1, \pm i, 1]$.

If the problem Holant($\{g\}| =_3$) is already #P-hard, then for any $\mathcal{F}$ and $\mathcal{G}$, the more general problem Holant($\{g\} \cup \mathcal{G} | \{=_3\} \cup \mathcal{F}$) is also #P-hard, and #CSP($\{g\} \cup \mathcal{F} \cup \mathcal{G}$) is reducible to any #P-hard problem. So we only need to consider the case Holant($\{g\}| =_3$) is not #P-hard. For this, we again apply Theorem 6.1 to $g = [g_0, g_1, g_2]$. The first tractable case $g_1^2 = g_0 g_2$ is degenerate, which does not apply here. The following three lemmas deal with the remaining three tractable cases respectively. (Note that Holant($\{g\}| =_3$) is #P-hard (resp. tractable) iff Holant($\{g\}, \Delta_0, \Delta_1 | =_3$) is #P-hard (resp. tractable), by Theorem 5.1. But for the latter problem, when it is #P-hard, there is a reduction that does not use Lemma 3.13.)

Let us consider the case $g_0^{12} = g_1^{12}$ and $g_0 g_2 = -g_1^2$. Assume $g_1 \neq 0$, we can scale it to $[a, 1, b]$, where $a^{12} = 1$ and $ab = -1$. Since $[a, 1, b]$ is assumed to be normalized, it follows that $a^4 = 1$.

Lemma 6.6. *Let $\mathcal{F}$ and $\mathcal{G}$ be two sets of signatures. For all pairs of a and b satisfying $a^4 = 1$ and $ab = -1$,* Holant($[a, 1, b] \cup \mathcal{G} | \{=_3\} \cup \mathcal{F}$) *is #P-hard unless $\mathcal{F} \cup \mathcal{G} \subseteq \mathscr{A}$, in which case it is in P. Furthermore, if $\mathcal{G}$ contains Δ_0 and Δ_1, then a #P-hardness reduction exists for* Holant($[a, 1, b] \cup \mathcal{G} | \{=_3\} \cup \mathcal{F}$) *that does not use Lemma 3.13.*

Proof. We first prove that when $a^4 = 1$ and $ab = -1$,

$$\text{Holant}([a, 1, b] \cup \mathcal{G} | \{=_3\} \cup \mathcal{F}) \equiv_T \#\text{CSP}([a, 1, b] \cup \mathcal{F} \cup \mathcal{G}).$$

To get this, it is sufficient to construct ($=_2$) on the LHS.

Case 1: $a = \pm i$.

It is equivalent to consider Holant($[1, \pm i, 1] \cup \mathcal{G} | \{=_3\} \cup \mathcal{F}$) because they differ only by a constant factor. We can construct $[1, 1]$ on the RHS by connecting the two edges of $[1, \pm i, 1]$ on LHS with two edges of a ($=_3$) on the RHS. With the gadget in Figure 6.3, we can construct the binary DISEQUALITY function ($\neq_2$)

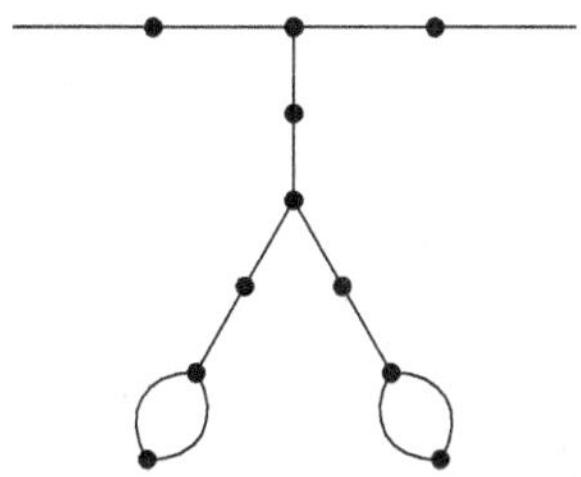

Figure 6.4. Another gadget construction for the binary DISEQUALITY function ($\neq_2$) on the LHS. Degree-3 vertices have signature ($=_3$), degree-2 vertices have signature $[1, \pm 1, -1]$.

on the LHS up to a nonzero scalar. Together with the ($=_3$) on RHS, we can have ($=_3$) on the LHS. Connecting this LHS ($=_3$) with $[1, 1]$ on the RHS, we obtain the binary EQUALITY function ($=_2$) on the LHS.

Case 2: $a = \pm 1$.
It is equivalent to consider Holant($[1, \pm 1, -1] \cup \mathcal{G} | \{=_3\} \cup \mathcal{F}$). With the gadget in Figure 6.4, we can construct ($\neq_2$) on the LHS, and thus ($=_3$) on the LHS, using the given ($=_3$) on the RHS. Furthermore, we can construct $[1, -1]$ and $[1, 0, -1]$ on both sides. For example, $[1, -1]$ is constructed as a fragment of the gadget in Figure 6.4; we first obtain $[1, -1]$ on the RHS by connecting two edges of ($=_3$) with a copy of $[1, \pm 1, -1]$, and later obtain $[1, -1]$ on the LHS as well. Using $[1, -1]$ and ($=_3$) on both sides we get $[1, 0, -1]$ on both sides. By connecting $[1, -1]$ with $[1, 0, -1]$, we can realize $[1, 1]$ on both sides, and consequently ($=_2$) on both sides.

By Theorem 3.7, #CSP($[a, 1, b] \cup \mathcal{F} \cup \mathcal{G}$) is #P-hard unless $[a, 1, b] \cup \mathcal{F} \cup \mathcal{G} \subseteq \mathscr{P}$ or $[a, 1, b] \cup \mathcal{F} \cup \mathcal{G} \subseteq \mathscr{A}$. Since $[a, 1, b] \in \mathscr{A} - \mathscr{P}$, we conclude that the only possible case that is not #P-hard is $\mathcal{F} \cup \mathcal{G} \subseteq \mathscr{A}$. This is also sufficient for tractability. The proof is complete. □

Back to Holant($\{g\} | =_3$) with $g = [g_0, g_1, g_2]$, for the tractable case $g_1 = 0$ in Theorem 6.1, by nondegeneracy, we can scale it to be $[1, 0, a]$, where $a \neq 0$. Then we have the following lemma:

Lemma 6.7. *Let $\mathcal{F}$ and $\mathcal{G}$ be two sets of signatures, and let $a \neq 0$ be a complex number. We assume $[1, 0, a]$ is normalized. Then we have the following dichotomy:*

- *If $a^4 = 1$, then* Holant($[1, 0, a] \cup \mathcal{G} | \{=_3\} \cup \mathcal{F}$) *is #P-hard unless $\mathcal{F} \cup \mathcal{G} \subseteq \mathscr{P}$ or $\mathcal{F} \cup \mathcal{G} \subseteq \mathscr{A}$, in which cases it is in P.*
- *If $a^4 \neq 1$, then* Holant($[1, 0, a] \cup \mathcal{G} | \{=_3\} \cup \mathcal{F}$) *is #P-hard unless $\mathcal{F} \cup \mathcal{G} \subseteq \mathscr{P}$, in which case it is in P.*

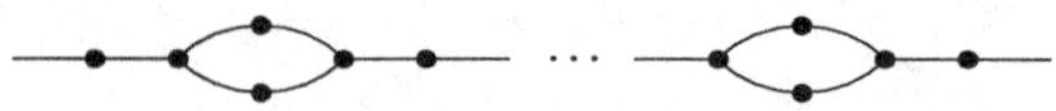

Figure 6.5. A recursive gadget to realize $[1, 0, a^{3k+1}]$ on the LHS. All ternary signatures are $(=_3)$, and all binary signatures are $[1, 0, a]$.

Furthermore, if $\mathcal{G}$ contains Δ_0 and Δ_1, then a #P-hardness reduction exists that does not use Lemma 3.13.

Proof. As above, it is sufficient to construct $(=_2)$ on the LHS, and establish the equivalence $\text{Holant}([1, 0, a] \cup \mathcal{G} | \{=_3\} \cup \mathcal{F}) \equiv_T \#\text{CSP}([1, 0, a] \cup \mathcal{F} \cup \mathcal{G})$. Note that for $a^4 \neq 1$, the signature $[1, 0, a] \notin \mathscr{A}$ and thus $\#\text{CSP}([1, 0, a] \cup \mathcal{F} \cup \mathcal{G})$ being not #P-hard implies that $\mathcal{F} \cup \mathcal{G} \subseteq \mathscr{P}$.

We will use the gadget in Figure 6.5 in our proof. We can use it to realize $[1, 0, a^{3k+1}]$ for any $k \in \mathbb{N}$ on LHS.

If a is not a root of unity, then we will be able to interpolate all signatures of the form $[1, 0, x]$, where $x \in \mathbb{C}$ on LHS. This uses a Vandermonde system as we did in Chapter 3. In particular, we will be able to interpolate $(=_2) = [1, 0, 1]$ on the LHS.

Exercise: If a is not a root of unity, then prove that we can interpolate all signatures of the form $[1, 0, x]$, where $x \in \mathbb{C}$ on the LHS, then complete the proof of Lemma 6.7.

Next we can assume that a is a tth primitive root of unity, that is, $a = \omega_t^b$ for some b relatively prime to t, where $\omega_t = e^{2\pi i/t}$. If t is not a multiple of 3, then we can find an integer k, such that $3k + 1 \equiv 0 \pmod{t}$. Therefore, we can realize $(=_2) = [1, 0, 1]$ on the LHS.

Now we consider the case of $t = 3^l t'$, where $l \geq 1$ and $\gcd(t', 3) = 1$. Since $[1, 0, a]$ is normalized, we have a further condition that $l > 1$. For this case, we cannot construct $(=_2)$ on the LHS directly. Instead we will further apply a diagonal holographic reduction. Also in this case, we have $a^4 \neq 1$, so we want to prove that $\text{Holant}([1, 0, a] \cup \mathcal{F} | \{=_3\} \cup \mathcal{G})$ is #P-hard unless $\mathcal{F} \cup \mathcal{G} \subseteq \mathscr{P}$. The fact that the problem is in P when $\mathcal{F} \cup \mathcal{G} \subseteq \mathscr{P}$ is obvious by Theorem 3.7, since $[1, 0, a] \in \mathscr{P}$.

Since $2t'$ is not a multiple of 3, there exist some integers k and s, such that $3k + 1 = 2st'$. Since

$$a^{3k+1} = a^{2st'} = \omega_{3^l t'}^{2bst'} = \omega_{3^l}^{2bs}$$

we can realize $[1, 0, \omega_{3^l}^{2bs}]$ on the LHS. So

$$\text{Holant}([1, 0, \omega_{3^l}^{2bs}] \cup \mathcal{G} | \{=_3\} \cup \mathcal{F}) \leq_T \text{Holant}([1, 0, a] \cup \mathcal{G} | \{=_3\} \cup \mathcal{F}).$$

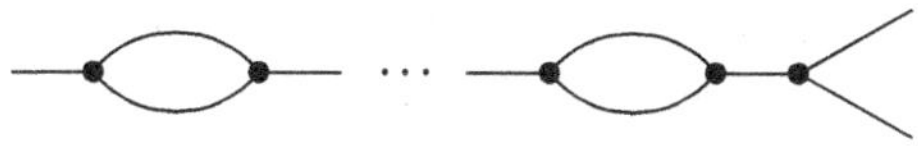

Figure 6.6. A recursive gadget to realize $[1, 0, 0, \omega_{3(l-1)}^{mbs}]$ where m is any odd integer. All ternary signatures are $[1, 0, 0, \omega_{3(l-1)}^{bs}]$. Edges have $(=_2)$.

Therefore it suffices to prove that $\text{Holant}([1, 0, \omega_{3^l}^{2bs}] \cup \mathcal{G}|\{=_3\} \cup \mathcal{F})$ is #P-hard if $\mathcal{F} \cup \mathcal{G} \not\subseteq \mathscr{P}$.

We apply a diagonal holographic reduction $T = \begin{bmatrix} 1 & 0 \\ 0 & \omega_{3^l}^{-bs} \end{bmatrix}$, and get

$$\begin{aligned} &\text{Holant}\big([1, 0, \omega_{3^l}^{2bs}] \cup \mathcal{G}|\{=_3\} \cup \mathcal{F}\big) \\ &\equiv_T \text{Holant}\big(\{=_2\} \cup \mathcal{G}T|[1, 0, 0, \omega_{3(l-1)}^{bs}] \cup T^{-1}\mathcal{F}\big). \end{aligned}$$

We then use the gadget in Figure 6.6 to realize $[1, 0, 0, \omega_{3(l-1)}^{3^{l-1}bs}] = [1, 0, 0, 1] = (=_3)$ on the RHS. Together with $(=_2)$ on the LHS this gives all EQUALITIES on both sides. As a result,

$$\begin{aligned} &\#\text{CSP}\big([1, 0, 0, \omega_{3(l-1)}^{bs}] \cup \mathcal{G}T \cup T^{-1}\mathcal{F}\big) \\ &\leq_T \text{Holant}\big(\{=_2\} \cup \mathcal{G}T|[1, 0, 0, \omega_{3(l-1)}^{bs}] \cup T^{-1}\mathcal{F}\big). \end{aligned}$$

Since $l > 1$, $[1, 0, 0, \omega_{3(l-1)}^{bs}] \notin \mathscr{A}$, as both b and s are relatively prime to 3. So the problem is #P-hard unless $\mathcal{G}T \cup T^{-1}\mathcal{F} \subseteq \mathscr{P}$. Since T and T^{-1} are diagonal matrices, it is equivalent to the statement $\mathcal{F} \cup \mathcal{G} \subseteq \mathscr{P}$. Note that diagonal transformations preserve $\{\Delta_0, \Delta_1\}$. If $\mathcal{G}$ contains Δ_0 and Δ_1, then a #P-hardness reduction exists that does not use Lemma 3.13. This completes the proof of Lemma 6.7. □

For the last tractable case $g_0 = g_2 = 0$ in Theorem 6.1, we can scale it to $(\neq_2) = [0, 1, 0]$.

Lemma 6.8. *Let $\mathcal{F}$ and $\mathcal{G}$ be two sets of signatures, and $a \neq 0$ be a complex number. We assume $[1, a]$ is normalized. Then we have the following dichotomy:*

- *If $a^4 = 1$, then* $\text{Holant}(\{[0, 1, 0], [1, a]\} \cup \mathcal{G}|\{=_3\} \cup \mathcal{F})$ *is #P-hard unless $\mathcal{F} \cup \mathcal{G} \subseteq \mathscr{P}$ or $\mathcal{F} \cup \mathcal{G} \subseteq \mathscr{A}$, in which cases it is in P.*
- *If $a^4 \neq 1$, then* $\text{Holant}(\{[0, 1, 0], [1, a]\} \cup \mathcal{G}|\{=_3\} \cup \mathcal{F})$ *is #P-hard unless $\mathcal{F} \cup \mathcal{G} \subseteq \mathscr{P}$, in which case it is in P.*

Furthermore, if $\mathcal{G}$ contains Δ_0 and Δ_1, then a #P-hardness reduction exists for $\text{Holant}([a, 1, b] \cup \mathcal{G}|\{=_3\} \cup \mathcal{F})$ *that does not use Lemma 3.13.*

Proof. By connecting one copy of $[1, a]$ and two copies of $[0, 1, 0]$ to a $(=_3)$ on the RHS, we can realize $[a, 0, 1]$, or equivalently $[1, 0, 1/a]$, on the LHS. Now Lemma 6.8 follows from Lemma 6.7. □

This completes the proof of Theorem 6.5.

6.3 Holant^c Problems

In Holant* problems, all unary functions are assumed to be freely available. Two special unary functions, IS-ZERO $\Delta_0 = [1, 0]$ and IS-ONE $\Delta_1 = [0, 1]$, are particularly meaningful. They correspond to setting a variable to the constant 0 or 1. This is called pinning, and by Lemma 3.13 these are available for all #CSP problems, namely Δ_0 and Δ_1 are implied by the presence of EQUALITIES of all arities. Note that if we have Δ_0 and Δ_1, then we can construct all subsignatures of any given signature.

We define a class of restricted Holant problems where only the two special unary functions Δ_0 and Δ_1 are assumed to be freely available. This class is called the Holant^c problems.

Definition 6.9. Given a set of signatures $\mathcal{F}$, we use Holant$^c(\mathcal{F})$ to denote Holant$(\mathcal{F} \cup \{\Delta_0, \Delta_1\})$.

In this section, we prove a dichotomy theorem for Holant^c problems with complex-valued symmetric signatures, Theorem 6.12. The proof uses the dichotomies proved in the previous two sections. To use them, we first prove in Lemma 6.10 that we can realize a nondegenerate ternary symmetric signature except in some trivial cases. From this ternary signature we immediately have #P-hardness if it is not in one of the tractable cases in Theorem 6.3. For a tractable ternary signature, we use Theorem 6.5 to extend the dichotomy theorem to a set of signatures. In Theorem 6.5, we considered only the *generic* case of the ternary function. The double-root case is handled here in Lemma 6.11. After that we will be ready to prove Theorem 6.12.

Lemma 6.10. *Given any set of symmetric signatures $\mathcal{F}$ which contains Δ_0 and Δ_1, we can construct a nondegenerate ternary symmetric signature f, except in the following two cases:*

1. *All nondegenerate signatures in $\mathcal{F}$ have arity ≤ 2.*
2. *In $\mathcal{F}$, all unary signatures are of the form $[x, 0]$ or $[0, x]$; all binary signatures are of the form $[x, 0, y]$ or $[0, x, 0]$; and all signatures of arity > 2 are of the form $[x, 0, \ldots, 0, y]$.*

Proof. Suppose case 1 does not hold, and let $f = [f_0, f_1, \ldots, f_n] \in \mathcal{F}$ be a nondegenerate signature of arity ≥ 3. Since we have Δ_0 and Δ_1, we can construct all subsignatures of any signature in $\mathcal{F}$. (A signature is also considered a subsignature of itself.) If there exists a ternary nondegenerate subsignature, we are done. Now suppose all ternary subsignatures of f are degenerate, in particular $n > 3$. By Proposition 2.8, f must be of the form $[f_0, 0, \ldots, 0, f_n]$, where $f_0 f_n \neq 0$. Then, if we have a unary signature or a unary subsignature of the form $[a, b]$ from $\mathcal{F}$ with $ab \neq 0$, we can connect $n-3$ copies of $[a, b]$ to $n-3$ dangling edges of f to get a nondegenerate ternary signature of the form $[x, 0, 0, y]$, and we are done. Otherwise, all unary signatures are of the form $[x, 0]$ or $[0, x]$ and all binary signatures are of the form $[x, 0, y]$ or $[0, x, 0]$. Any nondegenerate signature of arity > 2 must be of the form $[x, 0, \ldots, 0, y]$. If a degenerate signature $[a, b]^{\otimes n}$ has $ab \neq 0$, then we can get a unary subsignature of the form $[a', b']$ with $a'b' \neq 0$. Thus the degenerate signature is also a GEN-EQ of the form $[x, 0, \ldots, 0, y]$, where $x = 0$ or $y = 0$. □

Remark: All the exceptional cases listed in case 2 are special cases of functions of product type.

We next consider Holant$(\{f, g\})$, where f and g are nondegenerate symmetric signatures of arity 3 and 2 respectively, and the characteristic equation for the second-order linear recurrence of $f = [f_0, f_1, f_2, f_3]$ has a double root. By Lemma 6.2, Holant(f) is already #P-hard unless the double eigenvalue is i or $-\mathrm{i}$. Therefore we may assume the double root is $\lambda = \pm\mathrm{i}$, and $f_{k+2} - \lambda f_{k+1} - f_k = 0$ for $k = 0, 1$.

Lemma 6.11. *Let $f = [f_0, f_1, f_2, f_3]$ be a nondegenerate complex signature satisfying $f_{k+2} - \lambda f_{k+1} - f_k = 0$ for $k = 0, 1$, where $\lambda = \pm 2\mathrm{i}$. Let $g = [g_0, g_1, g_2]$ be a nondegenerate binary signature. Then Holant$(g \mid f)$ is #P-hard, and a fortiori Holant$(\{f, g\})$ is also #P-hard, unless $g_2 - \lambda g_1 - g_0 = 0$; if so, the problem Holant$(\{f, g\})$ is in P by generalized Fibonacci gates.*

Proof. We prove this result for $\lambda = 2\mathrm{i}$. The other case is similar.

The sequence $\{f_k\}$ can be written as follows: $f_k = ck\mathrm{i}^{k-1} + d\mathrm{i}^k$, where $c \neq 0$. Thus, we have

$$f = T^{\otimes 3}[1, 1, 0, 0]^{\mathsf{T}}, \quad \text{where} \quad T = \begin{bmatrix} 1 & \frac{d-1}{3} \\ \mathrm{i} & c + \frac{d-1}{3}\mathrm{i} \end{bmatrix}.$$

We can express $\begin{bmatrix} g_0 & g_1 \\ g_1 & g_2 \end{bmatrix} = M^{\mathrm{T}}M$, as shown in (2.19) of Section 2.3.2, for some $M = \begin{bmatrix} \alpha & \gamma \\ \beta & \delta \end{bmatrix}$. M is nonsingular since g is nondegenerate. Then we have

$$g_0 = \alpha^2 + \beta^2, \quad g_1 = \alpha\gamma + \beta\delta, \quad \text{and} \quad g_2 = \gamma^2 + \delta^2.$$

Thus we have the following reduction:

$$\begin{aligned} &\text{Holant}(g \mid f) = \text{Holant}((=_2)M^{\otimes 2} \mid T^{\otimes 3}[1, 1, 0, 0]) \\ &\equiv_{\mathrm{T}} \text{Holant}(=_2 \mid (MT)^{\otimes 3}[1, 1, 0, 0]), \end{aligned}$$

where

$$MT = \begin{bmatrix} \alpha + \gamma\mathrm{i} & * \\ \beta + \delta\mathrm{i} & * \end{bmatrix}, \qquad \text{and we will call it} \quad \begin{bmatrix} p & r \\ q & s \end{bmatrix}.$$

Our next step is to use an orthogonal matrix to transform MT to be upper-triangular, if we can. If $p^2 + q^2 \neq 0$, then we can find a (complex) orthogonal matrix $Q = \frac{1}{\sqrt{p^2+q^2}} \begin{bmatrix} p & q \\ q & -p \end{bmatrix}$ such that QMT is upper-triangular. As shown in Section 6.1, for a nonsingular upper-triangular QMT, $(QMT)^{\otimes 3}[1, 1, 0, 0]$ has the form $[x, y, 0, 0]$ for some $y \neq 0$. We can normalize it to $[z, 1, 0, 0]$.

For any orthogonal matrix Q, the LHS $(=_2)$ is unchanged under the holographic transformation: $(=_2)(Q^{-1})^{\otimes 2} = (=_2)$. This gives us

$$\begin{aligned} &\text{Holant}((=_2) \mid (MT)^{\otimes 3}[1, 1, 0, 0]) \equiv_{\mathrm{T}} \text{Holant}((=_2) \mid [z, 1, 0, 0]) \\ &\equiv_{\mathrm{T}} \text{Holant}([z, 1, 0, 0]), \end{aligned}$$

for some z. This shows the equivalence of $\text{Holant}(g \mid f)$ with $\text{Holant}([z, 1, 0, 0])$. By Lemma 6.2, $\text{Holant}([z, 1, 0, 0])$ is #P-hard. Hence $\text{Holant}(g \mid f)$ is also #P-hard.

Finally, $p^2 + q^2 = 0$ is $(\alpha + \gamma\mathrm{i})^2 + (\beta + \delta\mathrm{i})^2 = (\alpha^2 + \beta^2) + 2\mathrm{i}(\alpha\gamma + \beta\delta) - (\gamma^2 + \delta^2) = 0$. This is exactly $g_2 - \lambda g_1 - g_0 = 0$. □

Exercise: Show that in any instance Ω of $\text{Holant}(\{f, g\})$ if f appears in the signature grid, then the value $\text{Holant}_\Omega = 0$.

A perceptive reader will have noticed that the very explicit dichotomy statement in Lemma 6.11 should be consistent with, and derivable from, the more conceptual dichotomy statement of Theorem 6.4. We now give such a derivation.

As the discussion after Theorems 6.3 and 6.4 shows, the assumption on f here is equivalent to f having the form $Z[c, d, 0, 0]$, or $Z[0, 0, d, c]$, with $d \neq 0$

by nondegeneracy. We will assume the first case. The second case is similar. We have

$$\text{Holant}(g \mid f) = \text{Holant}((=_2)M \mid Z[c,d,0,0]) \equiv_T \text{Holant}(=_2\mid MZ[c,d,0,0]),$$

where $g = (=_2)M$ as shown in (2.19) for some $M = \begin{bmatrix} \alpha & \gamma \\ \beta & \delta \end{bmatrix}$. The property that a ternary symmetric signature is nondegenerate, or is in the generic case (having distinct eigenvalues) is invariant under a nonsingular holographic transformation. This is easily seen in terms of the tensor product expression. Hence the property that it has a double root is also invariant. It follows that, by Theorem 6.4 the only possibility for Holant$(=_2\mid MZ[c,d,0,0])$ to be tractable is when $MZ[c,d,0,0]$ has the form $Z[*,*,0,0,]$ or $Z[0,0,*,*]$.

The following fact can be directly verified: For any nonsingular matrix T and $d \neq 0$, $T[c,d,0,0] = [*,*,0,0,]$ iff T is an upper-triangular matrix. Hence the tractability criterion from Theorem 6.4 is $Z^{-1}MZ$ or $XZ^{-1}MZ$ is upper-triangular, where $X = \begin{bmatrix} 0 & 1 \\ 1 & 0 \end{bmatrix}$. Since

$$\begin{bmatrix} 1 & -\mathrm{i} \\ 1 & \mathrm{i} \end{bmatrix}\begin{bmatrix} \alpha & \gamma \\ \beta & \delta \end{bmatrix}\begin{bmatrix} 1 & 1 \\ \mathrm{i} & -\mathrm{i} \end{bmatrix} = \begin{bmatrix} 1 & -\mathrm{i} \\ 1 & \mathrm{i} \end{bmatrix}\begin{bmatrix} \alpha+\mathrm{i}\gamma & \alpha-\mathrm{i}\gamma \\ \beta+\mathrm{i}\delta & \beta-\mathrm{i}\delta \end{bmatrix}$$
$$= \begin{bmatrix} (\alpha+\mathrm{i}\gamma)-\mathrm{i}(\beta+\mathrm{i}\delta) & * \\ (\alpha+\mathrm{i}\gamma)+\mathrm{i}(\beta+\mathrm{i}\delta) & * \end{bmatrix},$$

the two tractable cases can be expressed as

$$[(\alpha+\mathrm{i}\gamma)+\mathrm{i}(\beta+\mathrm{i}\delta)][(\alpha+\mathrm{i}\gamma)-\mathrm{i}(\beta+\mathrm{i}\delta)] = 0.$$

This is just $(\alpha+\mathrm{i}\gamma)^2 + (\beta+\mathrm{i}\delta)^2 = 0$, and it matches exactly with $g_2 - \lambda g_1 - g_0 = 0$.

The next theorem is the main result of this section. It gives a dichotomy for Holant^c problems. Define

$$\mathscr{T} = \{T \in \mathbf{GL}_2(\mathbb{C}) \mid (=_2)T^{\otimes 2}, \Delta_0 T, \Delta_1 T \in \mathscr{A}\}. \tag{6.4}$$

Theorem 6.12. *Let $\mathcal{F}$ be a set of complex symmetric signatures.* Holant$^c(\mathcal{F})$ *is #P-hard unless $\mathcal{F}$ satisfies one of the following conditions, in which case it is tractable:*

1. Holant$^*(\mathcal{F})$ *is tractable; or*
2. *There exists a $T \in \mathscr{T}$ such that $T^{-1}\mathcal{F} \subseteq \mathscr{A}$.*

Remark: As we have Δ_0 and Δ_1, any degenerate $f = [a,b]^{\otimes n} \in \mathcal{F}$ can be replaced by $[a,b]$, as follows: If $a = b = 0$ then f is identically 0 and applying

Δ_0 to the signature $n-1$ times gives the unary $[0,0]$. If a, b are not both 0, then using Δ_0 or Δ_1 we can obtain a nonzero multiple of $[a, b]$. Thus we may assume all signatures of arity ≥ 2 in $\mathcal{F}$ are nondegenerate. Also, since we deal with symmetric signatures, by Theorem 3.8, the affine class $\mathscr{A}$ can be replaced by $\mathscr{F}_1 \cup \mathscr{F}_2 \cup \mathscr{F}_3$ in the statement of Theorem 6.12.

The dichotomy criterion in Theorem 6.12 can be made explicit. For case 1 we have the very explicit dichotomy for Holant* problems – Theorem 2.12. This also has a more conceptual but less explicit form in Theorem 3.5. For case 2 we will give a finite list of $T \in \mathbf{GL}_2(\mathbb{C})$ in place of $\mathscr{T}$. We will discuss this in more detail after the proof.

Proof. First of all, if $\mathcal{F}$ is in one of the two listed cases of Lemma 6.10, then $\text{Holant}^*(\mathcal{F})$ is tractable and we are done. Now we can assume that we have a nondegenerate symmetric ternary signature $f = [f_0, f_1, f_2, f_3]$ and the problem is $\text{Holant}^c(\mathcal{F} \cup \{f\})$.

As discussed in Section 6.1, there are three categories for f and we only need to consider the first two:

1. $f_k = \alpha^{3-k}\beta^k + \gamma^{3-k}\delta^k$, where $\det \begin{bmatrix} \alpha & \gamma \\ \beta & \delta \end{bmatrix} \neq 0$;
2. $f_k = ck\lambda^{k-1} + d\lambda^k$, where $c \neq 0$.

Case 1: $f_k = \alpha^{3-k}\beta^k + \gamma^{3-k}\delta^k$.

In this case, $f = T^{\otimes 3}(=_3)$, where $T = \begin{bmatrix} \alpha & \gamma \\ \beta & \delta \end{bmatrix}$. Furthermore, we will choose such a T so that $(=_2)T^{\otimes 2}$ is normalized. We may accomplish this by replacing T with TT' for some $T' \in \mathscr{T}_3$ defined in (6.3), as explained in Section 6.2. Then we have the following reduction chain:

$$\begin{aligned} \text{Holant}^c(\mathcal{F}) &\equiv_T \text{Holant}^c(\mathcal{F} \cup \{f\}) \\ &\equiv_T \text{Holant}(\mathcal{F} \cup \{f, \Delta_0, \Delta_1\}) \\ &\equiv_T \text{Holant}((=_2) \mid \mathcal{F} \cup \{f, \Delta_0, \Delta_1\}) \\ &\equiv_T \text{Holant}(\{(=_2), \Delta_0, \Delta_1\} \mid \mathcal{F} \cup \{f\}) \\ &\equiv_T \text{Holant}(\{(=_2)T^{\otimes 2}, \Delta_0 T, \Delta_1 T\} \mid \{=_3\} \cup T^{-1}\mathcal{F}). \end{aligned}$$

Note that $(=_2)$ on the LHS connected with Δ_c on the RHS gives Δ_c on the LHS, for $c = 0$ and 1. Since $(=_2)T^{\otimes 2}$ is a normalized nondegenerate binary signature, we can apply Theorem 6.5. The only thing we need to verify is that in case $(=_2)T^{\otimes 2} = [\alpha^2 + \beta^2, \alpha\gamma + \beta\delta, \gamma^2 + \delta^2] = [0, \alpha\gamma + \beta\delta, 0]$, at least one of $\Delta_0 T = [1, 0]T = [\alpha, \gamma]$ or $\Delta_1 T = [0, 1]T = [\beta, \delta]$ has both entries nonzero. If not, we would have $\alpha\gamma = 0$ and $\beta\delta = 0$, which implies that $(=_2)T^{\otimes 2} = [0, \alpha\gamma + \beta\delta, 0] = [0, 0, 0]$, a contradiction. In the case of $(=_2)T^{\otimes 2} = [0, *, 0]$, we may further normalize the unary signature $\Delta_0 T$ or $\Delta_1 T$ without affecting $(=_2)T^{\otimes 2}$ being normalized. Therefore, by Theorem 6.5, the problem Holant$^c(\mathcal{F})$ is #P-hard unless

$$\{(=_2)T^{\otimes 2}\} \cup T^{-1}\mathcal{F} \subseteq \mathscr{P}$$

(note that unary $\Delta_0 T$, $\Delta_1 T$ are automatically in $\mathscr{P}$), or

$$\{(=_2)T^{\otimes 2}, \Delta_0 T, \Delta_1 T\} \cup T^{-1}\mathcal{F} \subseteq \mathscr{A}.$$

In the first case, Holant$^*(\mathcal{F})$ is tractable, by the dichotomy theorem for Holant* problems, Theorem 2.12 (see also Theorem 3.5). In the second case, this is equivalent to having $T \in \mathscr{T}$ satisfying $T^{-1}\mathcal{F} \subseteq \mathscr{A}$.

Case 2: $f_k = ck\lambda^{k-1} + d\lambda^k$, *where* $c \neq 0$.

In this case, if $\lambda \neq \pm\mathfrak{i}$, the problem is #P-hard by Lemma 6.2 and we are done. Now we consider the case $\lambda = \mathfrak{i}$ (the case $\lambda = -\mathfrak{i}$ is similar). Consider the following equation:

$$z_{k+2} - 2\mathfrak{i}z_{k+1} - z_k = 0. \tag{6.5}$$

We note that $f = [f_0, f_1, f_2, f_3]$ satisfies this equation for $k = 0, 1$. If for all nondegenerate signatures $\Xi = [z_0, z_1, \ldots, z_n]$ in $\mathcal{F}$ with arity $n \geq 2$ we have the following

Condition: Ξ satisfies Equation (6.5) for $0 \leq k \leq n-2$, then, by Theorem 2.12 tractable Class $\mathfrak{B}$ type I$(1, 2\mathfrak{i})$ (equivalently case 3 of Theorem 3.5), Holant$^*(\mathcal{F})$ is tractable, and we are done. So suppose this is not the case, and there is a nondegenerate signature $\Xi = [z_0, z_1, \ldots, z_n] \in \mathcal{F}$, for some $n \geq 2$, that does not satisfy the displayed condition. By Lemma 6.11, if any nondegenerate subsignature $[z_k, z_{k+1}, z_{k+2}]$ does not satisfy (6.5), then, together with f which does satisfy (6.5), we know that the problem is #P-hard and we are done. So we assume every nondegenerate subsignature $[z_k, z_{k+1}, z_{k+2}]$ of Ξ satisfies (6.5). As Ξ is nondegenerate and does not satisfy the displayed condition, we have $n \geq 3$. Observe that Ξ does not satisfy the displayed condition is equivalent to some binary subsignature of Ξ does not satisfy (6.5). Hence there exists some binary subsignature of Ξ that is *degenerate* and does not satisfy (6.5).

Subcase 1: All binary subsignatures of Ξ are degenerate (but Ξ itself is nondegenerate).

By Proposition 2.8, Ξ must be of the form $[z_0, 0, \ldots, 0, z_n]$, where $z_0 z_n \neq 0$.

We claim that there exists a unary subsignature $[f_k, f_{k+1}]$ of f with both entries nonzero. If $f_0 f_1 \neq 0$, then we set $k = 0$ and the claim is proved; if $f_0 = 0, f_1 \neq 0$, then we have $f_2 = 2\mathrm{i} f_1 + f_0 \neq 0$; if $f_0 \neq 0, f_1 = 0$, then we have $f_2 = 2\mathrm{i} f_1 + f_0 \neq 0$ and $f_3 = 2\mathrm{i} f_2 + f_1 \neq 0$. Note that $f_0 = f_1 = 0$ is impossible because $c \neq 0$.

Connect this unary signature to $n - 3$ dangling edges of $[z_0, 0, \ldots, 0, z_n]$, we have a ternary signature $[a, 0, 0, b]$, where $ab \neq 0$. We can use this as the nondegenerate ternary signature f and we have reduced this case to case 1.

Subcase 2: Some binary subsignatures of Ξ are nondegenerate (some others are degenerate).

Then we can find a ternary subsignature $[z_k, z_{k+1}, z_{k+2}, z_{k+3}]$ (or its reversal) where $[z_k, z_{k+1}, z_{k+2}]$ is degenerate and $[z_{k+1}, z_{k+2}, z_{k+3}]$ is nondegenerate and thus satisfies $z_{k+3} - 2\mathrm{i} z_{k+2} - z_{k+1} = 0$. We have

$$z_{k+1}^2 = z_k z_{k+2} \quad \text{and} \quad z_{k+2}^2 \neq z_{k+1} z_{k+3}. \tag{6.6}$$

Thus z_{k+1} and z_{k+2} cannot both be zero. First suppose $z_k \neq 0$. Then $z_{k+1} \neq 0$, for otherwise $z_{k+1} = z_{k+2} = 0$ by $z_{k+1}^2 = z_k z_{k+2}$. Then (z_k, z_{k+1}, z_{k+2}) is a geometric progression with a nonzero ratio $\rho = z_{k+1}/z_k$, and we can assume that $z_k = 1$ after a scaling. Then $[z_k, z_{k+1}, z_{k+2}, z_{k+3}] = [1, \rho, \rho^2, 2\mathrm{i}\rho^2 + \rho]$, and we have

$$\begin{aligned}\begin{bmatrix} 1 & 0 \\ \rho & (\rho + 2\mathrm{i}\rho^2 - \rho^3)^{1/3} \end{bmatrix}^{\otimes 3} [1, 0, 0, 1] &= \begin{bmatrix} 1 \\ \rho \end{bmatrix}^{\otimes 3} + (\rho + 2\mathrm{i}\rho^2 - \rho^3) \begin{bmatrix} 0 \\ 1 \end{bmatrix}^{\otimes 3} \\ &= [1, \rho, \rho^2, 2\mathrm{i}\rho^2 + \rho].\end{aligned}$$

Note that $\rho + 2\mathrm{i}\rho^2 - \rho^3 \neq 0$ so that the 2×2 matrix is nonsingular. We already know that $\rho \neq 0$. If $1 + 2\mathrm{i}\rho - \rho^2 = 0$ then $\rho = \mathrm{i}$, which would imply that $[z_k, z_{k+1}, z_{k+2}, z_{k+3}] = [1, \mathrm{i}, -1, -\mathrm{i}]$ is degenerate, a contradiction. Therefore, the ternary subsignature $[z_k, z_{k+1}, z_{k+2}, z_{k+3}]$ is in the first category and we have reduced the problem to case 1.

Now suppose $z_k = 0$. Then it must be that $z_{k+1} = 0$ and $z_{k+2} \neq 0$ by (6.6). The signature $[z_k, z_{k+1}, z_{k+2}, z_{k+3}]$ becomes $[0, 0, z_{k+2}, z_{k+3}]$, which is equivalent to $[0, 0, 1, z]$ for some $z \in \mathbb{C}$, and as we proved in Lemma 6.2, it is #P-hard. □

There is a different character to this Holant^c dichotomy, as stated, from previous dichotomies we have proved so far. For example, Theorem 2.12 for Holant* problems, or Theorem 3.7 for Boolean #CSP are very *explicit*, so that for any given finite signature set $\mathcal{F}$, it is clear one can decide whether $\mathcal{F}$ satisfies the tractability criteria, in time polynomial in the size of $\mathcal{F}$. In fact, to say that the tractability condition is decidable in polynomial time does not do full justice to the explicitness of these dichotomy theorems; their tractability conditions are expressed essentially in closed form. This level of explicitness is useful, beyond the important issue of decidability in formal logic, or even polynomial time decidability, of the dichotomy criteria. It gives us a most concrete understanding of the dichotomy, as well as it can be used effectively to prove other dichotomy theorems, as shown, for example, when we applied Theorem 2.12 and Theorem 3.7.

Theorem 6.12 for Holant^c problems has a more conceptual form. It has an *existential* statement referring to the set $\mathscr{T}$. The conceptual form provides a better understanding of the reason why the dichotomy works. But it brings up the question of how one may decide whether a given $\mathcal{F}$ satisfies the tractability criteria of Theorem 6.12. One should note that for Holant* problems, Theorem 2.12 has an equivalent form Theorem 3.5, which also has such an existential statement. Our next goal is to show that the Holant^c dichotomy Theorem 6.12 in fact also has a very explicit form. Denote by $\mathscr{F}_{123} = \mathscr{F}_1 \cup \mathscr{F}_2 \cup \mathscr{F}_3$. We will show that the following set

$$\left\{ I, \begin{bmatrix} 1 & 0 \\ 0 & \sqrt{\mathfrak{i}} \end{bmatrix}, \begin{bmatrix} \sqrt{2} & 0 \\ \mathfrak{i} & 1 \end{bmatrix}, \begin{bmatrix} \sqrt{2} & 0 \\ -\mathfrak{i} & 1 \end{bmatrix} \right\} \tag{6.7}$$

can be used in place of the infinite set $\mathscr{T}$ in criterion 2 of Theorem 6.12, namely whether for one of these four matrices T, we have $T^{-1}\mathcal{F} \subseteq \mathscr{F}_{123}$. For convenience, we denote these four matrices by I, A, B, C respectively. We verify that they all belong to $\mathscr{T}$. Trivially $I \in \mathscr{T}$. For A, we have $(=_2)A^{\otimes 2} = [1, 0, \mathfrak{i}] \in \mathscr{F}_1$. Also we have $\Delta_0 A = [1, 0]$ and $\Delta_1 A = \sqrt{\mathfrak{i}}[0, 1]$; thus $A \in \mathscr{T}$. For B, we have $(=_2)B^{\otimes 2} = [1, \mathfrak{i}, 1] \in \mathscr{F}_2$, which has matrix form

$$\begin{bmatrix} \sqrt{2} & \mathfrak{i} \\ 0 & 1 \end{bmatrix} \begin{bmatrix} \sqrt{2} & 0 \\ \mathfrak{i} & 1 \end{bmatrix} = \begin{bmatrix} 1 & \mathfrak{i} \\ \mathfrak{i} & 1 \end{bmatrix}.$$

Also we have $\Delta_0 B = \sqrt{2}[1, 0]$ and $\Delta_1 B = \mathfrak{i}[1, -\mathfrak{i}] \in \mathscr{F}_2$; thus $B \in \mathscr{T}$. For C, we can take conjugation in the above calculations for B.

The key to this simplification of replacing (6.7) for the infinite set $\mathscr{T}$ is a normalizing process afforded by the following group action by $\mathbf{G} = \mathbf{GL}_2(\mathbb{C})$ and a certain stabilizer subgroup. For signatures f written as covaraint tensors, we consider the group action $(f, T) \mapsto fT$ by $\mathbf{G}$, where $T \in \mathbf{G}$. There is a corresponding group action $(T, f) \mapsto T^{-1}f$ by $\mathbf{G}$ on signatures written as contravariant tensors. We wish to investigate those $T \in \mathbf{G}$ such that

$$(=_2)T, \Delta_0 T, \Delta_1 T \in \mathscr{F}_{123}, \quad \text{and} \quad T^{-1}\mathcal{F} \subseteq \mathscr{F}_{123}. \tag{6.8}$$

Let

$$\mathbf{S} = \{S \in \mathbf{G} \mid \mathscr{F}_{123}S \subseteq \mathscr{F}_{123}\} \tag{6.9}$$

be the stabilizer of $\mathscr{F}_{123}$. Clearly $\mathbf{S}$ is a subgroup of $\mathbf{G}$. In fact, by definitions (6.9) and (6.4), it is clear that $\mathbf{S} \subseteq \mathscr{T}$, since $(=_2)$, $\Delta_0, \Delta_1 \in \mathscr{F}_{123}$. The purpose of defining this subgroup $\mathbf{S}$ is that we will use multiplication on the right from $\mathbf{S}$ to normalize the set of transformations.

Let $D = \begin{bmatrix} 1 & 0 \\ 0 & \mathfrak{i} \end{bmatrix}$ and $H = \frac{1}{\sqrt{2}}\begin{bmatrix} 1 & 1 \\ 1 & -1 \end{bmatrix}$. Also let $X = \begin{bmatrix} 0 & 1 \\ 1 & 0 \end{bmatrix}$ and $Z = \frac{1}{\sqrt{2}}\begin{bmatrix} 1 & 1 \\ \mathfrak{i} & -\mathfrak{i} \end{bmatrix}$. Note that $Z = DH$. Also $D^2Z = \frac{1}{\sqrt{2}}\begin{bmatrix} 1 & 1 \\ -i & i \end{bmatrix} = ZX$; hence $X = Z^{-1}D^2Z$. It is easy to verify that $D, H \in \mathbf{S}$, and hence also $X, Z \in \mathbf{S}$, as well as all nonzero scalar multiples of these. We will show that $\mathbf{S}$ consists of exactly the nonzero scalar multiples of members of the group $\langle D, H\rangle$ generated by D and H,

$$\mathbf{S} = \mathbb{C}^* \cdot \langle D, H\rangle. \tag{6.10}$$

Note that X, Z are also in the group $\langle D, H\rangle$ generated by D and H.

To show that $\mathbf{S}$ is contained in the RHS of (6.10), let $T \in \mathbf{S}$. Take $(=_3) \in \mathscr{F}_{123}$, then $(=_3)T^{\otimes 3} \in \mathscr{F}_{123}$. Then by the form of $\mathscr{F}_{123}$, for some $M \in \langle D, H\rangle$, chosen to be either I, or $H^{\mathrm{T}} = H$, or $Z^{\mathrm{T}} = HD$, we have $(=_3)(TM^{-1})^{\otimes 3} \in \mathscr{F}_1$, which is a GEN-EQ. Then either $T' = TM^{-1}$ or $TM^{-1}X$ is a diagonal matrix $\lambda\begin{bmatrix} 1 & 0 \\ 0 & d \end{bmatrix}$. Furthermore, by applying it to $(=_4) \in \mathscr{F}_{123}$ we conclude that $(=_4)T'^{\otimes 4} \in \mathscr{F}_1$, since it is in $\mathscr{F}_{123}$ but not in $\mathscr{F}_2 \cup \mathscr{F}_3$. It follows that d is a power of $\mathfrak{i}$, and hence $\begin{bmatrix} 1 & 0 \\ 0 & d \end{bmatrix}$ is a power of D. Thus $T \in \mathbb{C}^* \cdot \langle D, H\rangle$.

Since both generators D and H are symmetric, $\langle D, H\rangle$ is closed under transpose. Hence we also have $\mathbf{S}$ is the stabilizer of $\mathscr{F}_{123}$ by left action.

$$\mathbf{S} = \{S \in \mathbf{G} \mid S\mathscr{F}_{123} \subseteq \mathscr{F}_{123}\} = \{S \in \mathbf{G} \mid S^{-1}\mathscr{F}_{123} \subseteq \mathscr{F}_{123}\},$$

where equivalently we can replace $\subseteq$ by $=$.

Define the left and right stabilizer groups of $\mathscr{A}$:

$$\mathrm{LStab}(\mathscr{A}) = \{T \in \mathbf{GL}_2(\mathbb{C}) | T\mathscr{A} \subseteq \mathscr{A}\};$$

$$\mathrm{RStab}(\mathscr{A}) = \{T \in \mathbf{GL}_2(\mathbb{C}) | \mathscr{A}T \subseteq \mathscr{A}\}.$$

Then it is easy to verify that both $D, H \in \mathrm{LStab}(\mathscr{A}) \cap \mathrm{RStab}(\mathscr{A})$. It is also clear that both $\mathrm{LStab}(\mathscr{A}) \subseteq \mathbf{S}$ and $\mathrm{RStab}(\mathscr{A}) \subseteq \mathbf{S}$. Hence $\mathbf{S} = \mathbb{C}^* \cdot \langle D, H\rangle = \mathrm{LStab}(\mathscr{A}) = \mathrm{RStab}(\mathscr{A})$. From now on we will write $\mathrm{Stab}(\mathscr{A})$, or $\mathbf{S}$, for this group.

For future reference, we record this as a lemma.

Lemma 6.13. *The left and right stabilizer groups of $\mathscr{A}$ and the left and right stabilizer groups of $\mathscr{F}_{123}$ all coincide, and it is the group $\mathbb{C}^* \cdot \langle D, H\rangle$.*

We remark that $\mathbf{S}$ is not normalized by $\mathscr{T}$. We take $H \in \mathbf{S}$, and calculate that

$$AHA^{-1} = \frac{1}{\sqrt{2}}\begin{bmatrix}1 & 0\\ 0 & \sqrt{\mathfrak{i}}\end{bmatrix}\begin{bmatrix}1 & 1\\ 1 & -1\end{bmatrix}\begin{bmatrix}1 & 0\\ 0 & 1/\sqrt{\mathfrak{i}}\end{bmatrix} = \frac{1}{\sqrt{2}}\begin{bmatrix}1 & 1/\sqrt{\mathfrak{i}}\\ \sqrt{\mathfrak{i}} & -1\end{bmatrix},$$

which maps $[1, 1] \in \mathscr{A}$ to

$$[1, 1]\begin{bmatrix}1 & 1/\sqrt{\mathfrak{i}}\\ \sqrt{\mathfrak{i}} & -1\end{bmatrix} = [1 + \sqrt{\mathfrak{i}}, \frac{1}{\sqrt{\mathfrak{i}}} - 1]$$

(ignoring the scalar $\frac{1}{\sqrt{2}}$), having two nonzero entries of unequal norm, and hence not in $\mathscr{A}$. Therefore $AHA^{-1} \notin \mathbf{S}$.

The following observation is a key point: If T and T' are two elements of $\mathscr{T}$ from the same left coset of $\mathbf{S}$, i.e., $T\mathbf{S} = T'\mathbf{S}$, then T satisfies (6.8) iff T' satisfies (6.8). Thus we may use right multiplication from $\mathbf{S}$ to normalize $\mathscr{T}$ in Theorem 6.12. Formally, we only need to check for $T \in \mathscr{T}$ in the second condition in Theorem 6.12 for a complete set of left coset representatives of $\mathbf{S}$ in $\mathscr{T}$. We claim that the four matrices in (6.7) form such a list.

We show that for every $T \in \mathscr{T}$, T belongs to one of the cosets $\mathbf{S}, A\mathbf{S}, B\mathbf{S}$ and $C\mathbf{S}$. To do that we will first narrow down the list of possible signatures for $[1, 0, 1]T^{\otimes 2}$ and $[1, 0]T$, by right multiplication from $\mathbf{S}$. By assumption that $T \in \mathscr{T}$, both $[1, 0, 1]T^{\otimes 2}$ and $[1, 0]T$ belong to $\mathscr{F}_{123}$. Binary signatures of $\mathscr{F}_{123}$ are nonzero scalar multiples of

$$[1, 0, \pm 1], [1, 0, \pm\mathfrak{i}], [1, \pm 1, -1], [1, \pm\mathfrak{i}, 1], [0, 1, 0].$$

Unary signatures of $\mathscr{F}_{123}$ are nonzero scalar multiples of

$$[1, \pm 1], [1, \pm\mathfrak{i}], [1, 0], [0, 1].$$

Using powers of $D = \begin{bmatrix}1 & 0\\ 0 & \mathfrak{i}\end{bmatrix}$, we may already reduce $[1, 0, -1]$ to $[1, 0, 1]$, reduce $[1, 0, -\mathfrak{i}]$ to $[1, 0, \mathfrak{i}]$, and reduce $[1, 1, -1]$ to $[1, -1, -1]$, then to $[1, -\mathfrak{i}, 1]$, and finally to $[1, \mathfrak{i}, 1]$. We may use Z^{-1} to reduce $[0, 1, 0]$ to $[1, 0, 1]$, as $[1, 0, 1]Z^{\otimes 2} = [0, 1, 0]$. We may further reduce $[1, 0, \mathfrak{i}]$ to $[1, \mathfrak{i}, 1]$ as follows. $[1, 0, \mathfrak{i}]H^{\otimes 2} = \frac{1}{2}\begin{bmatrix}1 & 1\\ 1 & -1\end{bmatrix}\begin{bmatrix}1 & 0\\ 0 & \mathfrak{i}\end{bmatrix}\begin{bmatrix}1 & 1\\ 1 & -1\end{bmatrix} = \frac{1+\mathfrak{i}}{2}[1, -\mathfrak{i}, 1]$, and $[1, -\mathfrak{i}, 1](D^2)^{\otimes 2} = [1, \mathfrak{i}, 1]$.

So we may assume $[1, 0, 1]T^{\otimes 2}$ is a nonzero scalar multiple of $[1, 0, 1]$ or $[1, \mathrm{i}, 1]$. Note that these two signatures are invariant under reversal. Now we further normalize $[1, 0]T$ by the reversal operation $X = \begin{bmatrix} 0 & 1 \\ 1 & 0 \end{bmatrix}$, which does not affect the normalization already done for $[1, 0, 1]T^{\otimes 2}$. Then it is clear that we may assume $[1, 0]T$ is a nonzero scalar multiple of $\{[1, 0], [1, 1], [1, -1], [1, \mathrm{i}]\}$.

In the following, we denote by $\alpha = \frac{1+\mathrm{i}}{\sqrt{2}} = \sqrt{\mathrm{i}}$. We also omit a nonzero scalar factor if it does not cause any confusion.

Suppose $[1, 0, 1]T^{\otimes 2} = \lambda[1, 0, 1]$. If $[1, 0]T = \mu[1, 0]$, then $T = \begin{bmatrix} 1 & 0 \\ 0 & \pm 1 \end{bmatrix} \in \mathbf{S}$, with coset representative I. If $[1, 0]T = \mu[1, 1]$, then $T = \begin{bmatrix} 1 & 1 \\ \pm 1 & \mp 1 \end{bmatrix} \in \mathbf{S}$, with coset representative I. Note that $\begin{bmatrix} 1 & 1 \\ 1 & -1 \end{bmatrix} = \sqrt{2}H$ and $\begin{bmatrix} 1 & 1 \\ -1 & 1 \end{bmatrix} X = \sqrt{2}H$. If $[1, 0]T = \mu[1, -1]$, then $T = \begin{bmatrix} 1 & -1 \\ \pm 1 & \pm 1 \end{bmatrix} \in \mathbf{S}$, with coset representative I. Note that $\begin{bmatrix} 1 & -1 \\ 1 & 1 \end{bmatrix} D^2 = \begin{bmatrix} 1 & 1 \\ 1 & -1 \end{bmatrix} \in \mathbf{S}$ and $\begin{bmatrix} 1 & -1 \\ -1 & -1 \end{bmatrix} D^2 X \in \mathbf{S}$. Finally $[1, 0]T = \mu[1, \mathrm{i}]$ with $[1, 0, 1]T^{\otimes 2} = \lambda[1, 0, 1]$ is impossible, since the latter implies that T is orthogonal up to a nonzero factor, but the dot product $[1, \mathrm{i}] \cdot [1, \mathrm{i}] = 0$.

Next suppose $[1, 0, 1]T^{\otimes 2} = \lambda[1, \mathrm{i}, 1]$. If $[1, 0]T = \mu[1, 0]$, then $T = \begin{bmatrix} \sqrt{2} & 0 \\ \pm \mathrm{i} & \pm 1 \end{bmatrix}$. Clearly with the $+$ sign, $T = B \in B\mathbf{S}$, and with the $-$ sign, $T = CD^2 \in C\mathbf{S}$. If $[1, 0]T = \mu[1, 1]$, then $T = \begin{bmatrix} 1 & 1 \\ \pm\alpha^3 & \mp\alpha^3 \end{bmatrix}$. Note that $A^{-1}\begin{bmatrix} 1 & 1 \\ \alpha^3 & -\alpha^3 \end{bmatrix} = \begin{bmatrix} 1 & 1 \\ \mathrm{i} & -\mathrm{i} \end{bmatrix} \in \mathbf{S}$, being a scalar multiple of Z, and also $A^{-1}\begin{bmatrix} 1 & 1 \\ -\alpha^3 & \alpha^3 \end{bmatrix} = \begin{bmatrix} 1 & 1 \\ -\mathrm{i} & \mathrm{i} \end{bmatrix} \in \mathbf{S}$, being a scalar multiple of ZX. If $[1, 0]T = \mu[1, -1]$, then $T = \begin{bmatrix} 1 & -1 \\ \pm\alpha & \pm\alpha \end{bmatrix}$. Note that $A^{-1}\begin{bmatrix} 1 & -1 \\ \alpha & \alpha \end{bmatrix} = \begin{bmatrix} 1 & -1 \\ 1 & 1 \end{bmatrix} \in \mathbf{S}$, being a multiple of HD^2, and also $A^{-1}\begin{bmatrix} 1 & -1 \\ -\alpha & -\alpha \end{bmatrix} = \begin{bmatrix} 1 & -1 \\ -1 & -1 \end{bmatrix} \in \mathbf{S}$, being a multiple of HXD^2. Finally, if $[1, 0]T = \mu[1, \mathrm{i}]$, then $T = \begin{bmatrix} 1 & \mathrm{i} \\ 0 & \pm\sqrt{2} \end{bmatrix}$. Note that $C^{-1}\begin{bmatrix} 1 & \mathrm{i} \\ 0 & \sqrt{2} \end{bmatrix} = \frac{1}{\sqrt{2}}\begin{bmatrix} 1 & 0 \\ \mathrm{i} & \sqrt{2} \end{bmatrix}\begin{bmatrix} 1 & \mathrm{i} \\ 0 & \sqrt{2} \end{bmatrix} = \frac{1}{\sqrt{2}}\begin{bmatrix} 1 & \mathrm{i} \\ \mathrm{i} & 1 \end{bmatrix} = ZD \in \mathbf{S}$,

and

$$B^{-1}\begin{bmatrix}1 & \mathrm{i}\\ 0 & -\sqrt{2}\end{bmatrix} = \frac{1}{\sqrt{2}}\begin{bmatrix}1 & 0\\ -\mathrm{i} & \sqrt{2}\end{bmatrix}\begin{bmatrix}1 & \mathrm{i}\\ 0 & -\sqrt{2}\end{bmatrix} = \frac{1}{\sqrt{2}}\begin{bmatrix}1 & \mathrm{i}\\ -\mathrm{i} & -1\end{bmatrix} = ZXD \in \mathbf{S}.$$

We have proved that for every $T \in \mathscr{T}$, there exists some $M \in \{I, A, B, C\}$, such that $T \in M\mathbf{S}$. We remark that the four coset representatives $\{I, A, B, C\}$ represent distinct left cosets. This amounts to verifying that $A, B, C, B^{-1}A, C^{-1}A, C^{-1}B$ all do not belong to $\mathbf{S}$. We have $[1, 1]A = [1, \sqrt{\mathrm{i}}]$, $[1, 1]B = [\sqrt{2} + \mathrm{i}, 1]$, and $[1, 1]C = [\sqrt{2} - \mathrm{i}, 1]$, having two nonzero entries with unequal norm or a ratio not a power of i, hence not in $\mathscr{A}$. Similarly $[0, 1]B^{-1}A = \frac{1}{\sqrt{2}}[0, 1]\begin{bmatrix}1 & 0\\ -\mathrm{i} & \sqrt{2}\end{bmatrix}\begin{bmatrix}1 & 0\\ 0 & \sqrt{\mathrm{i}}\end{bmatrix} = \frac{1}{\sqrt{2}}[-\mathrm{i}, \sqrt{2\mathrm{i}}]$, $[0, 1]C^{-1}A = \frac{1}{\sqrt{2}}[0, 1]\begin{bmatrix}1 & 0\\ \mathrm{i} & \sqrt{2}\end{bmatrix}\begin{bmatrix}1 & 0\\ 0 & \sqrt{\mathrm{i}}\end{bmatrix} = \frac{1}{\sqrt{2}}[\mathrm{i}, \sqrt{2\mathrm{i}}]$, and $[0, 1]C^{-1}B = \frac{1}{\sqrt{2}}[0, 1]\begin{bmatrix}1 & 0\\ \mathrm{i} & \sqrt{2}\end{bmatrix}\begin{bmatrix}\sqrt{2} & 0\\ \mathrm{i} & 1\end{bmatrix} = [2\mathrm{i}, 1]$, all have two nonzero entries of unequal norm. Thus these four matrices are distinct coset representatives.

We conclude this discussion by restating Theorem 6.12 in the following effective form:

Theorem 6.14. *Let $\mathcal{F}$ be a set of complex symmetric signatures.* Holant$^c(\mathcal{F})$ *is #P-hard unless $\mathcal{F}$ satisfies one of the following conditions, in which case it is tractable:*

1. Holant$^*(\mathcal{F})$ *is tractable, for which we have an effective dichotomy – Theorem 2.12; or*
2. *There exists some T among the four matrices defined in (6.7), such that $T^{-1}\mathcal{F} \subseteq \mathscr{A}$.*

6.4 A Dichotomy for #CSP^d

In this section we consider a variant of the standard Boolean #CSP problem, where every variable occurs a multiple of d times, for some integer $d \geq 1$. We will prove a complexity dichotomy, due to Huang and Lu [HL16], for this class of problems for any set of symmetric signatures on Boolean variables. This variant is interesting in its own right, but more importantly it is a key step in the proof of the general dichotomy in Chapter 7.

Definition 6.15. Let $d \geq 1$ be an integer. Given a set of signatures $\mathcal{F}$, we use #CSP$^d(\mathcal{F})$ to denote the restriction of Boolean #CSP($\mathcal{F}$) where every variable occurs a multiple of d times.

For example, $\#\mathrm{CSP}^2$ refers to #CSP problems where every variable occurs an even number of times. Define

$$\mathcal{T}_d = \left\{ \begin{bmatrix} 1 & 0 \\ 0 & \omega_d^k \end{bmatrix} \mid 0 \le k < d \right\},$$

where $\omega_d = e^{2\pi \mathrm{i}/d}$. We may apply a holographic transformation by $T \in \mathcal{T}_d$. Since each variable occurs a multiple of d times, this transforms $\#\mathrm{CSP}^d(\mathcal{F})$ to the equivalent $\#\mathrm{CSP}^d(T^{-1}\mathcal{F})$. Consider the problem $\#\mathrm{CSP}^3([1, \omega_3, \omega_6])$, where $\omega_3 = e^{2\pi \mathrm{i}/3} = \frac{-1+\sqrt{3}\mathrm{i}}{2}$ and $\omega_6 = e^{2\pi \mathrm{i}/6} = \frac{1+\sqrt{3}\mathrm{i}}{2}$. Taking $T = \begin{bmatrix} 1 & 0 \\ 0 & \omega_3 \end{bmatrix}$, the problem $\#\mathrm{CSP}^3([1, \omega_3, \omega_6])$ is transformed to the equivalent problem $\#\mathrm{CSP}^3([1, 1, -1])$ (where $-1 = \omega_6/\omega_3^2$), which is computable in polynomial time, by Theorem 3.7. Note that the standard $\#\mathrm{CSP}([1, \omega_3, \omega_6])$ is #P-hard, by the same theorem.

As another example, $\#\mathrm{CSP}^2([1, \mathrm{i}^{3/2}, \mathrm{i}])$ is computable in polynomial time. We can first write $\#\mathrm{CSP}^2([1, \mathrm{i}^{3/2}, \mathrm{i}])$ as an equivalent bipartite Holant problem $\mathrm{Holant}(\{=_2, =_4, \ldots\} \mid [1, \mathrm{i}^{3/2}, \mathrm{i}])$. Then we can perform a holographic transformation by $T = \begin{bmatrix} 1 & 0 \\ 0 & \sqrt{\mathrm{i}} \end{bmatrix} \in \mathcal{T}_8$. The LHS $(=_{2k})$ is transformed to $(=_{2k})T^{\otimes 2k} = [1, 0]^{\otimes 2k} + \mathrm{i}^k[0, 1]^{\otimes 2k} \in \mathscr{A}$. On the RHS, the transformed signature is $T^{-1}[1, \mathrm{i}^{3/2}, \mathrm{i}] = [1, \mathrm{i}, 1] \in \mathscr{A}$ as well. It follows that the problem is tractable in P. Note that the standard $\#\mathrm{CSP}([1, \mathrm{i}^{3/2}, \mathrm{i}])$ is #P-hard by Theorem 3.7.

The dichotomy theorem for $\#\mathrm{CSP}^d$ problems says that these are essentially the only additional tractable cases; all others are #P-hard.

Theorem 6.16 (Huang-Lu). *Let $d \ge 1$ be an integer, and $\mathcal{F}$ be a set of symmetric signatures from Boolean variables to $\mathbb{C}$. Then $\#\mathrm{CSP}^d(\mathcal{F})$ is #P-hard, unless there exists some $T \in \mathcal{T}_{4d}$ such that $\mathcal{F} \subseteq T\mathscr{P}$ or $\mathcal{F} \subseteq T\mathscr{A}$, in which case the problem $\#\mathrm{CSP}^d(\mathcal{F})$ is in P.*

We remark that since T in the theorem is diagonal, the statement $\mathcal{F} \subseteq T\mathscr{P}$ can be replaced by simply $\mathcal{F} \subseteq \mathscr{P}$.

Before we prove this theorem, we recall Theorem 6.5. The condition that the needed binary or unary signatures being normalized can always be achieved by a holographic transformation $T \in \mathcal{T}_3$ defined in (6.3). Thus we can restate Theorem 6.5 as follows:

Theorem 6.17. *Suppose $\mathcal{F}$ and $\mathcal{G}$ are two sets of signatures, where $\mathcal{G}$ contains a nondegenerate binary signature $[g_0, g_1, g_2]$. Furthermore, if $g_0 = g_2 = 0$*

then we assume $\mathcal{G}$ also contains a unary signature $[a, b]$ with $ab \neq 0$. Then

$$\text{Holant}(\mathcal{G} \mid [1, 0, 0, 1] \cup \mathcal{F})$$

is #P-hard unless there exists some $T \in \mathcal{T}_3$ such that $\mathcal{G}T \cup T^{-1}\mathcal{F} \subseteq \mathscr{P}$ or $\mathcal{G}T \cup T^{-1}\mathcal{F} \subseteq \mathscr{A}$, in which case the problem is in P. Furthermore, if $\mathcal{G}$ contains Δ_0 and Δ_1, then a #P-hardness reduction exists that does not use Lemma 3.13.

The tractability proof of Theorem 6.16 is straightforward. The hardness proof of Theorem 6.16 will use Theorem 6.17. For general #CSP problems, there is a pinning lemma (Lemma 3.13) that allows the use of the unary Δ_0 and Δ_1 functions. It turns out that pinning with Δ_0 and Δ_1 is still available in the #CSP^d setting; however, the proof is a bit more subtle. We will show in Lemma 6.22 that a weaker version of pinning holds that lets us pin d occurrences of variables at a time to the same value in the #CSP^d setting. Then we will show that this is sufficient for Theorem 6.16. This weaker version of pinning can be proved constructively. After the proof of Theorem 6.16 we will come back to this issue of pinning in general. The main part of the proof of Theorem 6.16 is to show that the theorem holds when Δ_0 and Δ_1 are present, which becomes Holant^c in essence. The proof uses Theorem 6.17 to repeatedly *regularize* a given signature set, lest the problem be #P-hard. Finally, the signature set is sufficiently regularized that it must satisfy the tractability criteria if it is not #P-hard.

We will postpone the proof of Lemma 6.22 for now, and focus first on the following Theorem 6.18. Note that $T^{-1}\Delta_c$ is a scalar multiple of Δ_c for an invertible diagonal T; thus $T^{-1}\Delta_c \in \mathscr{P} \cap \mathscr{A}$. Hence Theorem 6.18 is also a special case of Theorem 6.16.

Theorem 6.18. *Let $d \geq 1$ be an integer, and $\mathcal{F}$ be a set of symmetric signatures from Boolean variables to $\mathbb{C}$. Then #CSP$^d(\mathcal{F} \cup \{\Delta_0, \Delta_1\})$ is #P-hard, unless there exists some $T \in \mathcal{T}_{4d}$ such that $\mathcal{F} \subseteq T\mathscr{P}$ or $\mathcal{F} \subseteq T\mathscr{A}$, in which case the problem #CSP$^d(\mathcal{F} \cup \{\Delta_0, \Delta_1\})$ is in P. When it is #P-hard, a reduction exists that does not use Lemma 3.13.*

Again we remark that since T in the theorem is diagonal, $\mathcal{F} \subseteq T\mathscr{P}$ is equivalent to $\mathcal{F} \subseteq \mathscr{P}$. We may also remove identically 0 signatures from $\mathcal{F}$ since it does not affect both the complexity of the problem and the tractability criteria.

Before we prove Theorem 6.18 we make some simple observations: Suppose $M = \begin{bmatrix} 1 & 0 \\ 0 & z \end{bmatrix} \in \mathbf{GL}_2(\mathbb{C})$.

1. Signatures of the form $[f_0, 0, f_2, \ldots]$ or $[0, f_1, 0, \ldots]$ maintain their respective form under a holographic transformation by M. In particular, this is true for Δ_0 and Δ_1.
2. $f \in \mathscr{P}$ iff $fM \in \mathscr{P}$. This can be seen by the effect of the transformation M on $(=_2)$, $(\neq_2)$ and the unary functions.
3. For all $d \geq 1$, $\mathcal{T}_d$ is a cyclic group. $M \in \mathcal{T}_d$ iff $M^{-1} \in \mathcal{T}_d$.
4. For all $d \geq 1$, $(=_d)M^{\otimes d}$ is a scalar multiple of $(=_d)$ iff $M \in \mathcal{T}_d$.
5. For all $d \geq 1$, $(=_d)M^{\otimes d} \in \mathscr{A}$ iff $M \in \mathcal{T}_{4d}$.

Proof of Theorem 6.18. We start with the following equivalence:

$$\#\text{CSP}^d(\mathcal{F} \cup \{\Delta_0, \Delta_1\}) \equiv_T \text{Holant}(\mathscr{E}_d \mid \mathcal{F} \cup \{\Delta_0, \Delta_1\}),$$

where we denote $\mathscr{E}_d = \{=_d, =_{2d}, \ldots\}$. We first prove tractability. Suppose there exists some $T \in \mathcal{T}_{4d}$ such that $T^{-1}\mathcal{F} \subseteq \mathscr{P}$, or $T^{-1}\mathcal{F} \subseteq \mathscr{A}$; then we transform the bipartite Holant problem by T:

$$\text{Holant}(\mathscr{E}_d \mid \mathcal{F} \cup \{\Delta_0, \Delta_1\}) \equiv_T \text{Holant}(\mathscr{E}_d T \mid T^{-1}\mathcal{F} \cup \{\Delta_0, \Delta_1\}).$$

Note that a diagonal T preserves $\{\Delta_0, \Delta_1\}$. We have $\mathscr{E}_d T \subset \mathscr{P} \cap \mathscr{A}$, and either all $T^{-1}\mathcal{F} \cup \{\Delta_0, \Delta_1\} \subseteq \mathscr{P}$ or all $T^{-1}\mathcal{F} \cup \{\Delta_0, \Delta_1\} \subseteq \mathscr{A}$. Thus the problem is computable in polynomial time.

We turn to the proof of hardness next. We first prove a general reduction about #CSP^d problems. For a function f, let f^d be the function such that $f^d(\mathbf{x}) = (f(\mathbf{x}))^d$.

Lemma 6.19. *For all $d \geq 1$, #CSP(f^d) $\leq_T$ #CSP^d(f). Similarly, #CSP(f^d, Δ_0, Δ_1) $\leq_T$ #CSP^d(f, Δ_0, Δ_1).*

Proof. Given an instance of #CSP(f^d), we may replace each constraint of f^d by d many copies of f, each applied to the same sequence of occurrences of variables. This gives the same answer to the given instance, and the new instance is a valid instance of #CSP^d(f) since the number of occurrences of every variable is multiplied by d. The lemma is proved. □

By Theorem 3.7, #CSP([0, 1, 0, 0]) is #P-hard, and by Theorem 3.21 a #P-hardness reduction exists for #CSP([0, 1, 0, 0], Δ_0, Δ_1) that does not use Lemma 3.13. The same is true for #CSP([0, 0, 1, 0]) and #CSP([0, 0, 1, 0], Δ_0, Δ_1). Then by Lemma 6.19, we obtain the next lemma.

Lemma 6.20. *For all $d \geq 1$, #CSP^d([0, 1, 0, 0]) and #CSP^d([0, 0, 1, 0]) are #P-hard.*

Reductions for #P-hardness of #CSP^d([0, 1, 0, 0], Δ_0, Δ_1) and #CSP^d([0, 0, 1, 0], Δ_0, Δ_1) exist that do not use Lemma 3.13.

We continue with the proof of Theorem 6.18. Suppose Holant($\mathscr{E}_d \mid \mathcal{F} \cup \{\Delta_0, \Delta_1\}$) is not #P-hard. Connecting $(=_d)$ on the LHS and $d-1$ copies of Δ_c on the RHS we get Δ_c on the LHS, for $c = 0, 1$, with which we can obtain all subsignatures of $\mathcal{F}$ on the RHS. Since we can obtain all subsignatures of $\mathcal{F}$, we can and will replace every degenerate (but not identically 0) $f = [a, b]^{\otimes n} \in \mathcal{F}$ by (a nonzero multiple of) the unary $[a, b]$, without changing the complexity of the problem. If $\mathcal{F}$ consists of only unary functions then $\mathcal{F} \subseteq \mathcal{U} \subset \mathscr{P}$. If all binary subsignatures of $\mathcal{F}$ are degenerate, then by Proposition 2.8, we also have $\mathcal{F} \subseteq \mathscr{P}$. Therefore we may assume $\mathcal{F}$ contains a nondegenerate binary subsignature.

The next claim is crucial: We may assume for any $f = [f_0, f_1, \ldots, f_n] \in \mathcal{F}$ ($n \geq 1$) for any two consecutive terms at least one term must be 0,

$$f_i f_{i+1} = 0 \quad (0 \leq i < n). \tag{6.11}$$

Suppose for some f and some $0 \leq i < n$, $f_i f_{i+1} \neq 0$. Then by pinning we can realize $[f_i, f_{i+1}]$, which is $[1, a]$ up to a scale, for some $a \neq 0$. By connecting $3d-3$ copies of $[1, a]$ to $=_{3d}$ on the LHS we get $[1, 0, 0, a^{3d-3}]$ on the LHS. Now we apply the holographic transformation by $M = \begin{bmatrix} 1 & 0 \\ 0 & a^{-(d-1)} \end{bmatrix}$; the LHS signature $[1, 0, 0, a^{3d-3}]$ then is transformed to $[1, 0, 0, a^{3d-3}]M^{\otimes 3} = [1, 0, 0, 1]$ and the RHS has a nondegenerate binary signature, as well as a unary signature $M^{-1}[1, a] = [1, a^d]$ with both entries nonzero. By Theorem 6.17 Holant($\mathscr{E}_d \mid \mathcal{F} \cup \{\Delta_0, \Delta_1\}$) is #P-hard unless there exists some $M' = MT$ where $T \in \mathcal{T}_3$, such that $\mathscr{E}_d M' \cup M'^{-1}\mathcal{F}$ is a subset of $\mathscr{P}$ or a subset of $\mathscr{A}$. Since M' is diagonal, if $M'^{-1}\mathcal{F} \subseteq \mathscr{P}$ then $\mathcal{F} \subseteq \mathscr{P}$. For the second alternative, $(=_d)M' \in \mathscr{A}$ implies that $M' \in \mathcal{T}_{4d}$. Thus we have established the claim in (6.11), or else we have finished the proof of Theorem 6.18.

Thus, in particular, the only unary signatures in $\mathcal{F}$ are (nonzero multiples of) Δ_0 or Δ_1; the only binary signatures in $\mathcal{F}$ have the form $[0, *, 0]$ (DISEQUALITY) or $[*, 0, *]$ (GEN-EQ). If all signatures in $\mathcal{F}$ have arity ≤ 2, then $\mathcal{F} \subseteq \mathscr{P}$, and Theorem 6.18 is proved.

Therefore we may assume there are signatures in $\mathcal{F}$ of arity ≥ 3. If they are all GEN-EQ, then again $\mathcal{F} \subseteq \mathscr{P}$. So we may assume there is $f = [f_0, f_1, \ldots, f_n] \in \mathcal{F}$ of arity $n \geq 3$ and is not a GEN-EQ. We claim that for any such f, there are no two consecutive zero entries: $f_i = f_{i+1} = 0$. Suppose otherwise. Since it is not a GEN-EQ, there exists some $0 < j < n$ such that $f_j \neq 0$. Either $j < i$, in which case we pick the maximum such $j < i$; or $j > i+1$, in which case we pick the minimum such $j > i+1$. By (6.11), we get a subsignature $[0, f_j, 0, 0]$ or $[0, 0, f_j, 0]$. By Lemma 6.20, we conclude that #CSP^d($\mathcal{F} \cup \{\Delta_0, \Delta_1\}$) is #P-hard.

We can therefore strengthen the claim in (6.11) to say that every signature in $\mathcal{F}$ other than GEN-EQ must have the following strictly alternating form:

$$\forall i, (0 \le i < n), \quad \text{exactly one of } f_i \text{ or } f_{i+1} = 0. \tag{6.12}$$

Thus other than GEN-EQ, every $f \in \mathcal{F}$ has the form $[f_0, 0, f_2, \ldots]$ or $[0, f_1, 0, \ldots]$, where all symbols $f_i \neq 0$, and $\mathcal{F}$ contains such a signature of arity ≥ 3. The strengthened claim (6.12) applies to all signatures in $\mathcal{F}$ of any arity $n \geq 1$, other than GEN-EQ.

In particular there exists some subsignature $[1, 0, a]$ from $\mathcal{F}$ with $a \neq 0$. Suppose there is a nondegenerate GEN-EQ in $\mathcal{F}$ with an odd arity $r \geq 3$. We connect this GEN-EQ on the RHS with $=_{2rd}$ on the LHS having an even arity, to create a nondegenerate GEN-EQ on the LHS of an odd arity $2rd - r \geq 3$. Then we connect zero or more copies of $[1, 0, a]$ on the RHS to it to realize a nondegenerate GEN-EQ on the LHS of arity 3. Now we apply a diagonal holographic transformation M to map it to $[1, 0, 0, 1]$ on the LHS, while $M^{-1}[1, 0, a]$ on the RHS is nondegenerate. It does not have the form $[0, *, 0]$ since M is diagonal. This allows us to apply Theorem 6.17. This completes the proof of Theorem 6.18 in this case.

We summarize what we may assume for $\mathcal{F}$:

> Every unary signature in $\mathcal{F}$ is Δ_0 or Δ_1. Every signature in $\mathcal{F}$ of arity > 1 is nondegenerate, and is either (1) a GEN-EQ of an even arity, or (2) has a strictly alternating form. And there is at least one signature in $\mathcal{F}$ of type (2), with arity ≥ 3.

Without loss of generality, we may assume $\mathcal{F}$ contains a subsignature of the form $[0, 1, 0, a]$, for $a \neq 0$. The other form $[a, 0, 1, 0]$ is symmetric.

By Lemma 6.19, we see that

$$\#\text{CSP}([0, 1, 0, a^d]) \leq_T \#\text{CSP}^d([0, 1, 0, a]).$$

If $\#\text{CSP}^d([0, 1, 0, a])$ is #P-hard, then Theorem 6.18 is proved. Suppose $\#\text{CSP}^d([0, 1, 0, a])$ is not #P-hard, then $\#\text{CSP}([0, 1, 0, a^d])$ is also not #P-hard, and by Theorem 3.7, $[0, 1, 0, a^d] \in \mathscr{P} \cup \mathscr{A}$. However, $[0, 1, 0, a^d] \notin \mathscr{P}$, so we have $[0, 1, 0, a^d] \in \mathscr{A}$. This forces $a^d = \pm 1$ (see the table for $\mathscr{F}_1 \cup \mathscr{F}_2 \cup \mathscr{F}_3$ in Section 3.2.) Now we can apply a holographic transformation with $T = \begin{bmatrix} 1 & 0 \\ 0 & a^{1/2} \end{bmatrix}$ to $\text{Holant}(\mathscr{E}_d \mid \mathcal{F} \cup \{\Delta_0, \Delta_1, [0, 1, 0, a]\})$. Note that $T \in \mathcal{T}_{4d}$, as $a^{2d} = 1$. $(T^{-1})^{\otimes 3}$ maps $[0, 1, 0, a]$ to $[0, 1, 0, 1]$ in the RHS up to a scalar $a^{-1/2}$. Being diagonal T^{-1} maintains Δ_0 and Δ_1. All EQUALITY functions ($=_{4kd}$) in $\mathscr{E}_{4d} \subset \mathscr{E}_d$ on the LHS remain unchanged. Thus we still can obtain Δ_0 and Δ_1 on the LHS and thus obtain all subsignatures of the RHS of $\text{Holant}(\mathscr{E}_d T \mid T^{-1}\mathcal{F} \cup \{\Delta_0, \Delta_1, [0, 1, 0, 1]\})$. In particular, we can realize

$(=_2) = [1, 0, 1]$ on the RHS. Then we can realize all $\mathscr{E}_2 = \{=_2, =_4, \ldots\}$ on the LHS from $\mathscr{E}_{4d}$ on the LHS and $(=_2)$ on the RHS. Since we have $(=_2)$ on both sides now, we can reduce the following nonbipartite Holant problem to the original problem #CSP$^d(\mathcal{F} \cup \{\Delta_0, \Delta_1\})$:

$$\begin{aligned} &\text{Holant}(\mathscr{E}_2 \cup T^{-1}\mathcal{F} \cup \{\Delta_0, \Delta_1, [0, 1, 0, 1]\}) \qquad (6.13) \\ &\leq_T \text{Holant}(\mathscr{E}_d T \mid T^{-1}\mathcal{F} \cup \{\Delta_0, \Delta_1, [0, 1, 0, 1]\}) \\ &\equiv_T \text{Holant}(\mathscr{E}_d \mid \mathcal{F} \cup \{\Delta_0, \Delta_1, [0, 1, 0, a]\}) \\ &\equiv_T \#\text{CSP}^d(\mathcal{F} \cup \{\Delta_0, \Delta_1, [0, 1, 0, a]\}) \\ &\equiv_T \#\text{CSP}^d(\mathcal{F} \cup \{\Delta_0, \Delta_1\}). \end{aligned}$$

Our final step is to prove that the problem in (6.13) is #P-hard, unless $T^{-1}\mathcal{F} \subseteq \mathscr{A}$, which will complete the proof of Theorem 6.18.

Since T is diagonal, the summary on what we may assume for $\mathcal{F}$ is also valid for $T^{-1}\mathcal{F}$. Next we show that if there is a nondegenerate GEN-EQ in $T^{-1}\mathcal{F}$ not in $\mathscr{A}$, then the problem in (6.13) is #P-hard. After a scalar multiplication we may assume it is $[1, 0, \ldots, 0, b] \in T^{-1}\mathcal{F}$ of even arity $2k \geq 2$, with $b \neq 0$. If $2k = 2$, then it is $[1, 0, b]$. If $2k \geq 4$, then by connecting it to $(=_{2k-2}) \in \mathscr{E}_2$ we can realize $[1, 0, b]$. Now we reduce #CSP$([1, 0, b], [0, 1, 0, 1])$ to #CSP$^2([1, 0, b], [1, 0, 1], [0, 1, 0, 1])$, which in turn is reducible to the problem in (6.13): Each variable x that occurs in ℓ constraints in an instance of #CSP$([1, 0, b], [0, 1, 0, 1])$ will be made to occur in 2ℓ constraints. Each constraint $[1, 0, b]$ is replaced by a pair of constraints $[1, 0, b]$ and $(=_2) = [1, 0, 1]$ on the same two occurrences of variables. Each constraint $[0, 1, 0, 1]$ is replaced by two copies of the same constraint $[0, 1, 0, 1]$ on the same three occurrences of variables. This reduction shows that the problem in (6.13) is #P-hard unless $\{[1, 0, b], [0, 1, 0, 1]\} \subset \mathscr{P}$ or $\{[1, 0, b], [0, 1, 0, 1]\} \subset \mathscr{A}$. As $[0, 1, 0, 1] \notin \mathscr{P}$, we get $[1, 0, b] \in \mathscr{A}$, and thus $b^4 = 1$. This shows that all nondegenerate GEN-EQ in $T^{-1}\mathcal{F}$ are in $\mathscr{A}$. In particular we have shown that all signatures in $T^{-1}\mathcal{F}$ of arity ≤ 2 are in $\mathscr{A}$. Note that binary signatures in $T^{-1}\mathcal{F}$ are either GEN-EQ or of the form $[0, *, 0] \in \mathscr{A}$. Also for the problem in (6.13), since Δ_0 and Δ_1 are present, if some nondegenerate GEN-EQ in $T^{-1}\mathcal{F}$ does not belong to $\mathscr{A}$, then there is a #P-hardness reduction that does not use Lemma 3.13.

Now we prove that all signatures in $T^{-1}\mathcal{F}$ of the strictly alternating form are also in $\mathscr{A}$. We may assume it has arity ≥ 3. Suppose $T^{-1}\mathcal{F}$ has a ternary subsignature of the form $[0, 1, 0, b]$ ($b \neq 0$), after a scalar multiple. We can reduce #CSP$([0, 1, 0, b])$ to the problem #CSP$^2([0, 1, 0, b], [0, 1, 0, 1])$, which is in turn reducible to the problem in (6.13). (Again due to the presence of Δ_0 and Δ_1 in (6.13), a #P-hardness proof of this problem exists that does not use Lemma 3.13.) The reduction from #CSP([0, 1, 0, b]) to #CSP^2([0, 1, 0, b], [0, 1, 0, 1]) is as follows: We replace each constraint [0, 1, 0, b] with a copy of [0, 1, 0, b] and [0, 1, 0, 1]. Each variable now occurs an even number of times. Thus the problem #CSP([0, 1, 0, b]) is #P-hard unless $[0, 1, 0, b] \in \mathscr{A}$, which forces $b = \pm 1$. This argument can be extended to ternary subsignatures of the form $[b, 0, 1, 0]$ $(b \neq 0)$. Indeed, within the problem in (6.13) we can construct $(\neq_2) = [0, 1, 0]$ as a subsignature of [0, 1, 0, 1], and then connecting $(\neq_2)$ to all three inputs of $[b, 0, 1, 0]$ converts it to $[0, 1, 0, b]$, and we conclude that $b = \pm 1$. Now extending this pattern to any signature in $T^{-1}\mathcal{F}$ of the strictly alternating form, we see that either it is in $\mathscr{A}$ or there is a subsignature of arity 4 of the form

$$[1, 0, 1, 0, -1] \quad \text{or} \quad [1, 0, -1, 0, -1].$$

By Theorem 6.14, both Holant^c([1, 0, 1, 0, −1]) and Holant^c([1, 0, −1, 0, −1]) are #P-hard. This involves a check with Theorem 2.12 (or more directly Lemma 2.16; see Definition 2.9), and a check with the four matrices in (6.7) with reference to the table for $\mathscr{F}_1 \cup \mathscr{F}_2 \cup \mathscr{F}_3$ in Section 3.2. Hence the problem in (6.13) is also #P-hard.

This essentially completes our proof of Theorem 6.18, except for a technical point of nondependence of Lemma 3.13. Suppose we use the Holant^c dichotomy Theorem 6.14. In the proof of Theorem 6.14 we used nondiagonal holographic transformations, which do not preserve $\{\Delta_0, \Delta_1\}$, and we used, for example, results on Holant(f) for a trnary signature $f = [f_0, f_1, f_2, f_3])$ in the *double-root* case, Lemma 6.2. These proofs ultimately use the pinning reduction (Lemma 3.13). In the following we give an alternative proof that Holant^c([1, 0, 1, 0, −1]) and Holant^c([1, 0, −1, 0, −1]) are #P-hard without using Lemma 3.13.

Lemma 6.21. *If $a \in \mathbb{R}$ and $a \neq \pm 1$, then* Holant^c([0, 1, 0, a]) *is #P-hard. By symmetry,* Holant^c([a, 0, 1, 0]) *is also #P-hard. Furthermore, #P-hardness reductions exist that do not use Lemma 3.13.*

Proof. We start with Holant([0, 1, 0, 0]), which is #P-hard. This is the problem of counting perfect matchings in 3-regular graphs [DL92]. So the problem is #P-hard if $a = 0$.

Now assume that $a \notin \{-1, 0, 1\}$. We will use the signature [0, 1, 0, a] to interpolate all signatures of the form [0, 1, 0, x]; in particular, we can interpolate [0, 1, 0, 0] and finish the hardness reduction.

The recursive construction is depicted in Figure 6.7. We claim that the signature of the jth gadget N_j has the form $[0, x_j, 0, y_j]$. By an easy induction,

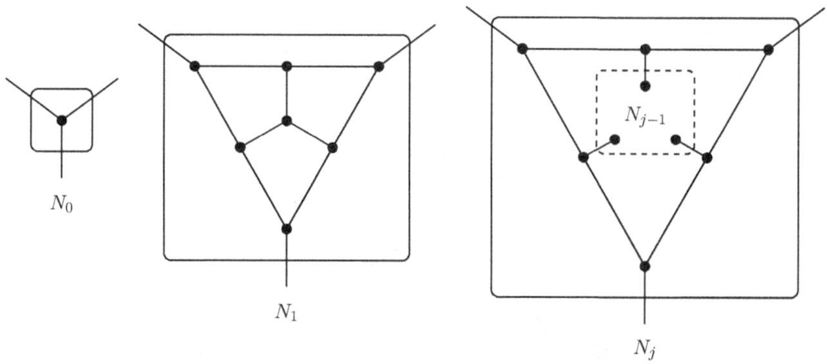

Figure 6.7. The recursive construction. The signature of every vertex in the gadget is $[0, 1, 0, a]$.

there are exactly $6j + 1$ vertices in N_j, which is odd. Let $\sigma \in \{0, 1\}^3$ be any assignment for the external edges of N_j, and let $\hat{\sigma}$ be an extension of σ to all internal edges of N_j. Suppose $\hat{\sigma}$ has a nonzero evaluation. Then every node of N_j must have an input of odd Hamming weight. Let $\mathrm{wt}(\sigma)$ be the Hamming weight of σ. Each internal edge assigned a value 1 by $\hat{\sigma}$ contributes 2 to the sum, over all nodes, of the Hamming weight of inputs, while each external edge assigned a value 1 by σ contributes 1 to the sum. Hence $\mathrm{wt}(\sigma)$ must be odd.

A calculation shows that the sequence (x_j, y_j) satisfies the following recurrence relation:

$$\begin{bmatrix} x_{j+1} \\ y_{j+1} \end{bmatrix} = \begin{bmatrix} 3(a^2+1) & (a^3+a) \\ 3(a^3+a) & a^6+1 \end{bmatrix} \begin{bmatrix} x_j \\ y_j \end{bmatrix},$$

with the initial vector $s = \begin{bmatrix} x_0 \\ y_0 \end{bmatrix} = \begin{bmatrix} 1 \\ a \end{bmatrix}$. Let $A = \begin{bmatrix} 3(a^2+1) & (a^3+a) \\ 3(a^3+a) & a^6+1 \end{bmatrix}$ be the transition matrix. Then $\det A = 3(a^4 - 1)^2 \neq 0$. The characteristic polynomial is $X^2 - (a^6 + 3a^2 + 4)X + 3(a^4 - 1)^2$ with discriminant $(a^6 - 3a^2 - 2)^2 + 12(a^3 + a)^2 > 0$. Therefore A has two distinct real roots, with sum $a^6 + 3a^2 + 4 > 0$. In particular, they are not opposite to each other. Therefore, the ratio of the two eigenvalues is not a root of unity. By Lemma 5.11, $\{A^j \mid j \geq 0\}$ is pairwise linearly independent. By Lemma 5.10, A has infinite projective order.

We claim that $\det[s\ As] \neq 0$. We can calculate that $As = (a^2+1) \begin{bmatrix} a^2+3 \\ a(a^4 - a^2 + 4) \end{bmatrix}$, and therefore $\det[s\ As] = a(a^2+1)(a^2-1)^2 \neq 0$. It follows that for any row eigenvector v of A, the dot product $v \cdot s \neq 0$, for otherwise $v \cdot [s\ As] = 0$ and $v \neq 0$ would imply that $\det[s\ As] = 0$. It follows

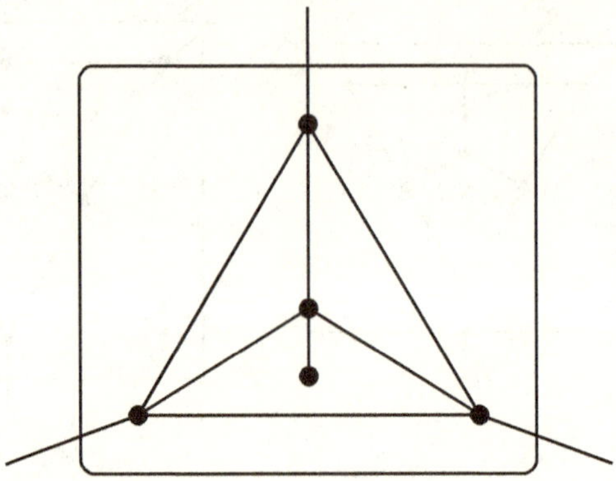

Figure 6.8. The signature of every degree-4 vertex in the gadget is $[1, 0, 1, 0, -1]$. And the signature of the degree-1 vertex in the gadget is $[1, 0]$.

from Lemma 5.15 that $\{A^j s\}$ is pairwise linearly independent. Therefore in $\text{Holant}^c([0, 1, 0, a])$ we can interpolate all signatures of the form $[0, 1, 0, x]$, and thus $\text{Holant}^c([0, 1, 0, a])$ is #P-hard. □

Exercise: Prove that with A and s we can interpolate all signatures of the form $[0, 1, 0, x]$ in $\text{Holant}^c([0, 1, 0, a])$.

Now we complete the proof that $\text{Holant}^c([1, 0, 1, 0, -1])$ is #P-hard. By a reversal $[1, 0, -1, 0, -1]$ becomes $-[1, 0, 1, 0, -1]$ and so $\text{Holant}^c([1, 0, -1, 0, -1])$ is also #P-hard.

Consider the gadget in Figure 6.8, where every vertex of degree 4 is assigned $[1, 0, 1, 0, -1]$, and the degree one vertex is assigned $[1, 0]$. This gadget has $[8, 0, 4, 0]$. So by Lemma 6.21, we know that $\text{Holant}^c([1, 0, 1, 0, -1])$ is #P-hard and the proof of Theorem 6.18 is complete.

Now we give the (weak) pinning lemma for $\#\text{CSP}^d$, to remove the extra assumption that the signature set contains Δ_0 and Δ_1. This will improve Theorem 6.18 to Theorem 6.16.

Lemma 6.22. *For any integer $d \geq 1$ and any $\mathcal{F}$, $\#\text{CSP}^d(\mathcal{F}) \equiv_T \#\text{CSP}^d(\mathcal{F} \cup \{\Delta_0^{\otimes d}, \Delta_1^{\otimes d}\})$.*

Proof. The proof of this lemma is an adaptation of the proof of Lemma 3.13. Given an instance I of $\#\text{CSP}^d(\mathcal{F} \cup \{\Delta_0^{\otimes d}, \Delta_1^{\otimes d}\})$, let V_0 denote the set of variables to which some $\Delta_0^{\otimes d}$ is applied among its d occurrences of variables. Similarly define the set of variables V_1 in terms of $\Delta_1^{\otimes d}$. We may assume $V_0 \cap V_1 = \emptyset$; otherwise the partition function $Z(I) = 0$ trivially. Let $V_2 = V - (V_0 \cup V_1)$ be the set of other variables.

Since I is a #CSP^d instance, each variable occurs a multiple of d times, and therefore the total occurrences of variables in V_0 and in V_1 in all constraints are also multiples of d. If we remove all constraints $\Delta_0^{\otimes d}$ and $\Delta_1^{\otimes d}$, the total occurrences of variables in V_0 and in V_1 in all remaining constraints are still multiples of d.

Now we define an instance I' of #CSP$^d(\mathcal{F})$ that has the variable set $V_2 \cup \{y_0, y_1\}$, where y_0, y_1 are two new variables. The constraints of I' are exactly the same as those of I, except (1) every variable in V_0 (respectively V_1) is replaced by y_0 (respectively y_1), and (2) the constraints $\Delta_0^{\otimes d}$ and $\Delta_1^{\otimes d}$ are not applied. It is clear that I' is indeed a valid instance of #CSP$^d(\mathcal{F})$. The partition function value $Z(I')$ can be expressed as the sum of four quantities,

$$Z(I') = \sum_{(b_0,b_1)\in\{0,1\}^2} Z(I' \mid b_0 b_1),$$

where $Z(I' \mid b_0 b_1)$ is the sum in $Z(I')$ restricted to assigning (y_0, y_1) to $(b_0, b_1) \in \{0, 1\}^2$.

We also define a second instance I'' of #CSP$^d(\mathcal{F})$ that has the variable set $V_2 \cup \{y\}$, where y is a new variable. It is the same as I' except we identify both y_0 and y_1 by y. Then clearly

$$Z(I'') = \sum_{b_0=b_1} Z(I' \mid b_0 b_1) = Z(I' \mid 00) + Z(I' \mid 11).$$

Hence

$$Z(I') - Z(I'') = Z(I' \mid 01) + Z(I' \mid 10). \tag{6.14}$$

Obviously $Z(I) = Z(I' \mid 01)$. If for every possible #CSP$^d(\mathcal{F})$ instance I^* with two distinguished variables z_0 and z_1, it holds that $Z(I^* \mid 01) = Z(I^* \mid 10)$, where $Z(I^* \mid b_0 b_1)$ is the sum in $Z(I^*)$ restricted to assigning (z_0, z_1) to $(b_0, b_1) \in \{0, 1\}^2$, then clearly $Z(I') - Z(I'') = 2Z(I)$, and the lemma is proved.

Suppose there exists some instance I^* of #CSP$^d(\mathcal{F})$ with two distinguished variables z_0 and z_1, such that $Z(I^* \mid 01) \neq Z(I^* \mid 10)$. Then we will define two more instances of #CSP$^d(\mathcal{F})$. The first is J', which is the same as I', except it adds a copy of I^* and identifies the variables z_0 and z_1 with y_0 and y_1, respectively. Then

$$Z(J') = \sum_{(b_0,b_1)\in\{0,1\}^2} Z(I' \mid b_0 b_1) Z(I^* \mid b_0 b_1).$$

The second is J'', which is the same as J', except it further identifies the variables y_0 and y_1. Then

$$Z(J'') = \sum_{b_0=b_1} Z(I' \mid b_0b_1)Z(I^* \mid b_0b_1).$$

Clearly

$$Z(J') - Z(J'') = Z(I' \mid 01)Z(I^* \mid 01) + Z(I' \mid 10)Z(I^* \mid 10). \quad (6.15)$$

Equations (6.14) and (6.15) form a nonsingular linear equation system, from which we can solve $Z(I' \mid 01) = Z(I)$. □

There is a subtlety in the proof of Lemma 6.22. It proves the existence of a polynomial-time Turing reduction from the RHS to the LHS in the statement of the lemma. But the proof given above does not constructively produce such a reduction. Given a finite constraint set $\mathcal{F}$, the proof is nonconstructive in deciding the existence of an instance I^* with $Z(I^* \mid 01) \neq Z(I^* \mid 10)$, because there are an infinite number of potential instances I^* of #CSP$^d(\mathcal{F})$ with two distinguished variables.

Artem Govorov [Gov15] has found an alternative proof that is constructive. With his permission we present it below.

Suppose such an instance I^* exists. It has two distinguished variables, z_0 and z_1, and n other "internal" variables $x_1, x_2, \ldots, x_n$ ($n \geq 0$). The values $Z(I^* \mid 01)$ and $Z(I^* \mid 10)$ are each defined by a sum over 2^n assignments on $x_1, x_2, \ldots, x_n$ over $\{0, 1\}^n$. We define a 1-1 correspondence between $(z_0, z_1, x_1, \ldots, x_n) \in \{0, 1\}^{n+2}$ and $(\overline{z_0}, \overline{z_1}, \overline{x_1}, \ldots, \overline{x_n}) \in \{0, 1\}^{n+2}$, by flipping each bit. Because $Z(I^* \mid 01) \neq Z(I^* \mid 10)$, in the two sums defining $Z(I^* \mid 01)$ and $Z(I^* \mid 10)$, there must be one corresponding pair of terms that are unequal. Let $\sigma \in \{0, 1\}^n$ define such a term in the sum for $Z(I^* \mid 01)$. Let $U_b = \{x_i \mid \sigma(x_i) = b\}$, for $b = 0, 1$. Note that each variable appears a multiple of d times in I^*, and therefore the total number of occurrences in I^* of variables in U_b is also a multiple of d.

Now we modify the instance I^* to I^{**} as follows: We identify every variable $x_i \in U_b$ with z_b, for $b = 0, 1$. This creates a valid instance I^{**} of #CSP$^d(\mathcal{F})$. It has two distinguished variables z_0 and z_1, and no "internal" variables. Clearly $Z(I^{**} \mid 01) \neq Z(I^{**} \mid 10)$. Each application of $f \in \mathcal{F}$ in I^{**} can be only in a finite number of ways. Indeed if f has arity n, then there can be at most 2^n many ways it can be applied to z_0 and z_1 as an ordered sequence of variables. Since $\mathcal{F}$ is finite, there can be at most a finite number of ways all constraints in $\mathcal{F}$ can be applied in I^{**}. Define a type τ as a pair (f, s), where $f \in \mathcal{F}$ has arity n, and $s \in \{0, 1\}^n$ indicating the application $f(z_{s_1}, \ldots, z_{s_n})$. Now let N_τ be the number of times the type τ occurs in I^{**}, and we write $N_\tau = q_\tau d + r_\tau$, with $0 \le r_\tau < d$. Since the total number of occurrences of z_0 and of z_1 are multiples of d, if we remove all $q_\tau d$ applications of constraints of each type τ, we obtain from I^{**} a modified instance of #CSP$^d(\mathcal{F})$ with two distinguished variables, where each type of constraint can occur less than d times each. There is only a finite number of such instances of #CSP$^d(\mathcal{F})$. For each type τ there is also exactly one instance of #CSP$^d(\mathcal{F})$ with two distinguished variables where τ is applied exactly d times. Of these finitely many instances of #CSP$^d(\mathcal{F})$, at least one instance I_* must satisfy

$$Z(I_* \mid 01) \neq Z(I_* \mid 10), \tag{6.16}$$

because $Z(I^{**} \mid 01) \neq Z(I^{**} \mid 10)$, and they are respectively products of the terms from the LHS and RHS in (6.16). This gives us a decidable procedure; consequently the reduction from the RHS to the LHS in Lemma 6.22 is made constructive.

Using Lemma 6.22, we can show the equivalence of #CSP$^d(\mathcal{F})$ and #CSP$^d(\mathcal{F} \cup \{\Delta_0, \Delta_1\})$. *Given* the fact that we have already proved a dichotomy theorem for the latter, this establishes a dichotomy theorem for the former, Theorem 6.16.

Lemma 6.23. *#CSP$^d(\mathcal{F})$ is in P (or is #P-hard, respectively) iff #CSP$^d(\mathcal{F} \cup \{\Delta_0, \Delta_1\})$ is in P (or is #P-hard, respectively).*

Proof. Clearly #CSP$^d(\mathcal{F}) \le_T$ #CSP$^d(\mathcal{F} \cup \{\Delta_0, \Delta_1\})$. We have already proved a dichotomy theorem for the latter problem #CSP$^d(\mathcal{F} \cup \{\Delta_0, \Delta_1\})$. If it is computable in polynomial time, then so is #CSP$^d(\mathcal{F})$. If #CSP$^d(\mathcal{F} \cup \{\Delta_0, \Delta_1\})$ is #P-hard, we show #CSP$^d(\mathcal{F})$ is also #P-hard.

We observe that in all the proofs for #P-hardness so far, which ultimately lead to the #P-hardness of #CSP$^d(\mathcal{F} \cup \{\Delta_0, \Delta_1\})$, we reduce from one of the following three problems to it by a chain of reductions: (a) Holant($=_3$| $[0, 1, 1]$) (Vertex Cover for 3-regular graphs), (b) Holant($[1, 1, 0, 0]$) (Matching), or (c) Holant($[0, 1, 0, 0]$) (Perfect Matching). There are only three reduction methods in this reduction chain, namely, direct gadget construction, polynomial interpolation, and holographic reduction. Importantly, by Theorem 6.18, when #CSP$^d(\mathcal{F} \cup \{\Delta_0, \Delta_1\})$ is #P-hard, a reduction exists that does not use Lemma 3.13.

For any signature grid Ω as an input to one of the three Holant problems listed earlier, the answer Holant$_\Omega$ to the Holant problem is a nonnegative integer. Let $d\Omega$ denote the signature grid that is a disjoint union of d copies of Ω. Then the Holant value Holant$_{d\Omega}$ = (Holant$_\Omega$)^d, the dth power of the Holant value on Ω. Being a nonnegative integer, the value Holant_Ω can be recovered uniquely from $\text{Holant}_{d\Omega}$, and therefore we may start our reduction from $d\Omega$ instead of Ω.

Now we apply the chain of reductions to $d\Omega$. Each step is a reduction f from a problem Π to a problem Π'. When the reduction is applied to an instance I of Π, f defines a sequence $\{I_i'\}$ of instances of Π', and outputs $I_1', I_2', \ldots, I_m'$, where both m and the sizes of I_i', $(1 \le i \le m)$, are polynomially bounded in the size of I. (Typically for polynomial interpolations there will be a polynomial number of output instances.) The three reduction methods all satisfy the following *locality* property: If the reduction is applied to an instance of Π that consists of d identical copies dI, it will output a polynomial number of instances of Π', where each instance consists of d identical copies of I_i', namely $dI_1', dI_2', \ldots, dI_{m'}'$, ($m'$ is polynomially bounded in the size of $d\Omega$, possibly a larger polynomial than m.) Clearly this property is closed under composition. Note that the number of steps in the reduction chain is a constant depending only on the problem, but not the size of the problem instance. (Crucially the reduction method in Lemma 3.13 *does not* satisfy this locality property. This is why we made annotations all along to indicate when a reduction exists without using Lemma 3.13.)

Finally we reach a sequence of polynomially many instances of the form $d\Omega_i$ of polynomially bounded size for the problem $\#\text{CSP}^d(\mathcal{F} \cup \{\Delta_0, \Delta_1\})$, following the #P-hardness reduction of this problem. By the syntactic form of $d\Omega_i$, each Δ_0 and Δ_1 is applied a multiple of d times. Thus we can rewrite this instance as an instance in $\#\text{CSP}^d(\mathcal{F} \cup \{\Delta_0^{\otimes d}, \Delta_1^{\otimes d}\})$. This shows that $\#\text{CSP}^d(\mathcal{F} \cup \{\Delta_0^{\otimes d}, \Delta_1^{\otimes d}\})$ is also #P-hard. By Lemma 6.22, $\#\text{CSP}^d(\mathcal{F})$ is #P-hard. $\square$

This completes the proof of Theorem 6.16. We conclude this section with two remarks. The first is concerned with a subtlety in the proof of Lemma 6.23. The statement that the two problems $\#\text{CSP}^d(\mathcal{F})$ and $\#\text{CSP}^d(\mathcal{F} \cup \{\Delta_0, \Delta_1\})$ are either simultaneously in P or simultaneously #P-hard technically does not depend on the assumption that P is different from #P, although the presumed truth of this hypothesis is what makes the theory of this book interesting. The trivial reduction $\#\text{CSP}^d(\mathcal{F}) \le_T \#\text{CSP}^d(\mathcal{F} \cup \{\Delta_0, \Delta_1\})$ shows that if the former is #P-hard, then so is the latter; and if the latter is in P, then so is the former. The proof of the lemma shows that if the latter is #P-hard, then so is the former. Now suppose the former is in P. We want to claim that the latter is also in P. Suppose not. Then *by the dichotomy theorem* already proved, Theorem 6.18, the latter problem $\#\text{CSP}^d(\mathcal{F} \cup \{\Delta_0, \Delta_1\})$ is #P-hard. The proof of Lemma 6.23 shows that the former problem $\#\text{CSP}^d(\mathcal{F})$ is also #P-hard. Thus #P collapses to P, and therefore $\#\text{CSP}^d(\mathcal{F} \cup \{\Delta_0, \Delta_1\})$ is in P anyhow.

The second remark is that the proof of Lemma 6.23 does not maintain planarity in general. This is because when we substitute $\Delta_0^{\otimes d}$ and $\Delta_1^{\otimes d}$ for d occurrences of Δ_0 and Δ_1 in a signature grid of the form $d\Omega$, there is no guarantee we can do so while maintaining planarity. In the next section we will see that in some circumstances this is still possible.

Very recently Lin and Wang [LW16] showed that there is a general method to extract a constraint function f from its tensor power $f^{\otimes d}$ in the Holant setting. This implies a general pinning lemma in the #CSP^d setting. With their permission we present their result below. Note that this proof is nonconstructive.

Lemma 6.24 (Lin-Wang). *Let f be a signature from Boolean variables to $\mathbb{C}$ and let $d \geq 1$ be an integer. Then we have*

$$\text{Holant}(\mathcal{F} \cup \{f\}) \leq_T \text{Holant}(\mathcal{F} \cup \{f^{\otimes d}\}),$$

for any set $\mathcal{F}$ of signatures from Boolean variables to $\mathbb{C}$.

Proof. We prove this lemma by induction on d. The base case $d = 1$ is trivial. Assume the lemma is true for all integers $1 \leq d' < d$, and we prove it for d.

We ask the question: Is there any signature grid Ω for Holant$(\mathcal{F} \cup \{f\})$ in which the value Holant$_\Omega \neq 0$ and the number of occurrences of f is not divisible by d? If not then the following is a simple reduction proving the lemma: Given any signature grid Ω for Holant$(\mathcal{F} \cup \{f\})$, we return the value Holant$_\Omega = 0$ if the number of occurrences of f is not divisible by d; otherwise, we replace every d occurrences of f in Ω by a single copy of $f^{\otimes d}$, thus creating a signature grid in Holant$(\mathcal{F} \cup \{f^{\otimes d}\})$.

Now suppose the answer to the question is yes, and let Ω_0 be such a fixed signature grid, with the number of occurrences m of f not divisible by d. Write $m = qd + r$, where $1 \leq r < d$. Now replace the qd occurrences of f by q copies of $f^{\otimes d}$. We can use one extra copy of $f^{\otimes d}$, applying its variables in places for the r remaining copies of f in Ω_0; this creates an $(\mathcal{F} \cup \{f^{\otimes d}\})$-gate with dangling edges, defining a signature that is $f^{\otimes(d-r)}$, multiplied by a nonzero constant Holant$_{\Omega_0}$. Hence we have a reduction

$$\text{Holant}(\mathcal{F} \cup \{f^{\otimes(d-r)}\}) \leq_T \text{Holant}(\mathcal{F} \cup \{f^{\otimes d}\}).$$

Note that $1 \leq d - r < d$. Now induction completes the proof. □

Since #CSP$^d(\mathcal{F})$ is nothing but Holant$(\mathcal{F} \cup \mathcal{EQ}_d)$, where $\mathcal{EQ}_d = \{(=_{kd}) \mid k \geq 1\}$ is the set of all Equalities of arity divisible by d, combining Lemma 6.22 obtaining $\Delta_0^{\otimes d}$ and $\Delta_1^{\otimes d}$ for #CSP^d, Lemma 6.24 has the following corollary.

Corollary 6.25. *Let $\mathcal{F}$ be a set of signatures from Boolean variables to $\mathbb{C}$, then*

$$\#\text{CSP}^d(\mathcal{F}) \equiv_T \#\text{CSP}^d(\mathcal{F} \cup \{\Delta_0, \Delta_1\}).$$

Using Corollary 6.25, Theorem 6.16 and Theorem 6.18 are obviously equivalent. In this logical derivation one can bypass the need to carefully annotate along the way when a reduction exists without using Lemma 3.13. The slight drawback is that the proof of Lemma 6.24, hence that of Corollary 6.25, is nonconstructive. It nonconstructively establishes the existence of a polynomial-time reduction. The proof given for Theorem 6.16 is constructive, in that whenever a stated polynomial-time reduction exists, the proof actually constructs one.

6.5 Eulerian Orientations and the Tutte Polynomial

In this section, we prove a specific problem to be #P-hard. This problem will serve as a new basic problem from which we prove #P-hardness in the dichotomy theorems that follow. We first define the problem. Recall that an orientation of an undirected graph is to assign a unique direction, out of two possible directions, to each edge.

Definition 6.26. Given a graph G, an orientation is an *Eulerian orientation* if for each vertex v of G, the number of incoming edges of v equals the number of outgoing edges of v.

We will prove that the problem of counting the number of Eulerian orientations, denoted by #EULERIAN-ORIENTATIONS, or #EO for short, is #P-hard over planar 4-regular graphs. Mihail and Winkler [MW96] proved that #EO is #P-hard over general graphs. Huang and Lu proved it for even-degree regular but not necessarily planar graphs. Guo and Williams improved it to planar 4-regular graphs. There is a related problem counting the number of Eulerian circuits. Given an undirected graph G, an Eulerian circuit is a closed circuit that traverses every edge exactly once. It is well known that an orientation is an Eulerian orientation iff there is an Eulerian circuit that follows the orientation. However, these problems are different. A single Eulerian orientation may correspond to many Eulerian circuits. See Section 6.6.

There is a natural expression of #EO over 4-regular graphs as a bipartite Holant problem. Given a 4-regular undirected graph G, we assign the binary DISEQUALITY signature on each edge, which corresponds to an orientation, then we assign the EXACT-TWO signature $[0, 0, 1, 0, 0]$ of arity 4 for each vertex of

G. This creates a bipartite signature grid out of the edge–vertex incidence graph of G. Then this problem can be expressed as Holant($\neq_2$ | [0, 0, 1, 0, 0]). Over planar 4-regular graphs, we use the notation Pl-Holant($\neq_2$ | [0, 0, 1, 0, 0]) to denote this problem.

The Tutte polynomial of an undirected graph is a polynomial in two variables with integer coefficients. It contains a wealth of information about the graph.

Definition 6.27 (cf. p. 335 of [Bol98]). Let $G = (V, E)$ be an undirected graph; then the Tutte polynomial of G is defined as

$$\mathrm{T}(G; x, y) = \sum_{A \subseteq E} (x-1)^{\kappa(V,A)-\kappa(V,E)} (y-1)^{\kappa(V,A)+|A|-|V|}, \tag{6.17}$$

where $\kappa(V, A)$ denotes the number of connected components of the graph (V, A).

To prove the problem Pl-Holant($\neq_2$ | [0, 0, 1, 0, 0]) #P-hard, we reduce from the problem of evaluating the Tutte polynomial at the point (3, 3), which is #P-hard for planar graphs.

The following theorems give complexity classifications of the Tutte polynomial over general and planar graphs.

Theorem 6.28 (Jaeger, Vertigan, and Welsh [JVW90]). *For $x, y \in \mathbb{C}$, evaluating the Tutte polynomial at (x, y) is #P-hard over graphs unless*

$$(x-1)(y-1) = 1$$

or $(x, y) \in \{(1, 1), (-1, -1), (0, -1), (-1, 0), (\mathrm{i}, -\mathrm{i}), (-\mathrm{i}, \mathrm{i}), (\omega, \omega^2), (\omega^2, \omega)\}$, *where* $\omega = e^{2\pi \mathrm{i}/3}$. *In each exceptional case, the problem is in polynomial time.*

Theorem 6.29 (Theorem 5.1 in [Ver05]). *For $x, y \in \mathbb{C}$, evaluating the Tutte polynomial at (x, y) is #P-hard over planar graphs unless*

$$(x-1)(y-1) \in \{1, 2\} \quad \text{or} \quad (x, y) \in \{(1, 1), (-1, -1), (\omega, \omega^2), (\omega^2, \omega)\},$$

where $\omega = e^{2\pi \mathrm{i}/3}$. *In each exceptional case, the problem is in polynomial time.*

The first step in the reduction from the Tutte polynomial is concerned with a sum of weighted Eulerian orientations on the medial graph of a plane graph. Recall that a plane graph is a planar embedding of a planar graph. We define a medial graph next.

Definition 6.30 (cf. p. 368 of [Bol98]). Given a connected plane graph G, its *medial graph* G_m has a vertex e' for each edge e of G, and vertices e'_1 and e'_2

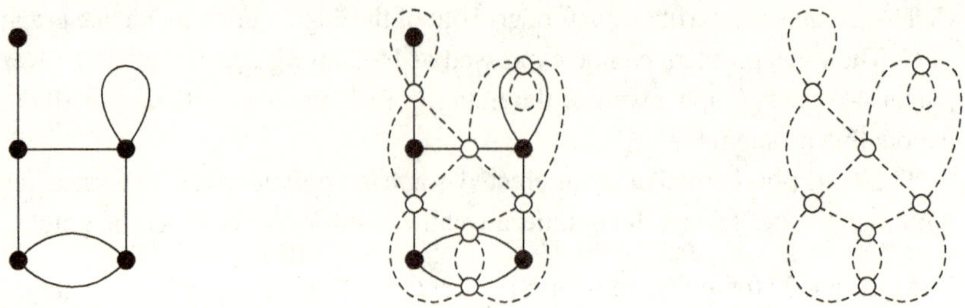

Figure 6.9. A plane graph (a), its medial graph (b), and the two graphs superimposed (c).

in G_m are joined by an edge for each face of G in which their corresponding edges e_1 and e_2 in G occur consecutively.

Note that G may have loops and parallel edges, and the vertices e_1' and e_2' of G_m in Definition 6.30 may not be distinct. An example of a plane graph and its medial graph are given in Figure 6.9. Note that the medial graph of a plane graph is always a planar 4-regular graph. Michel Las Vergnas [Ver88] connected the evaluation of the Tutte polynomial of a planar graph G at the point (3, 3) with a sum of weighted Eulerian orientations on a medial graph of G.

Theorem 6.31 (Theorem 2.1 in [Ver88]). *Let G be a connected plane graph and let $\mathcal{O}(G_m)$ be the set of all Eulerian orientations in the medial graph G_m of G. Then*

$$2 \cdot \mathrm{T}(G; 3, 3) = \sum_{O \in \mathcal{O}(G_m)} 2^{\beta(O)}, \tag{6.18}$$

where $\beta(O)$ is the number of saddle vertices in the orientation O, i.e., the number of vertices in which the edges are oriented "in, out, in, out" in cyclic order.

Although the medial graph depends on a particular embedding of the planar graph G, the RHS of (6.18) is invariant under different embeddings of G. This follows from (6.18) and the fact that the Tutte polynomial does not depend on the embedding of G.

Given a signature g of arity 4, we abbreviate g_{abcd} for the value $g(a, b, c, d)$, for $(a, b, c, d) \in \{0, 1\}^4$. Recall the definition of the *signature matrix* M_g of an arity 4 signature g (Definition 1.4):

$$M_g = \begin{bmatrix} g_{0000} & g_{0010} & g_{0001} & g_{0011} \\ g_{0100} & g_{0110} & g_{0101} & g_{0111} \\ g_{1000} & g_{1010} & g_{1001} & g_{1011} \\ g_{1100} & g_{1110} & g_{1101} & g_{1111} \end{bmatrix}. \tag{6.19}$$

In general when we present g pictorially with four external edges $ABCD$, we may order them counterclockwise. In M_g, the rows are indexed by the bits AB, and the columns are indexed by the bits DC in the *reverse order*, both lexicographically. This reversal is for the convenience that the signature matrix of a construction that links two arity 4 signatures is the matrix product of the signature matrices of the two arity 4 signatures.

In particular, the signature matrix of a symmetric arity 4 signature has identical two middle rows and identical two middle rows. If $f = [f_0, f_1, f_2, f_3, f_4]$ is a symmetric signature of arity 4, then its signature matrix is

$$M_f = \begin{bmatrix} f_0 & f_1 & f_1 & f_2 \\ f_1 & f_2 & f_2 & f_3 \\ f_1 & f_2 & f_2 & f_3 \\ f_2 & f_3 & f_3 & f_4 \end{bmatrix}. \tag{6.20}$$

Now we can prove

Theorem 6.32. #EULERIAN-ORIENTATIONS *is* #P-*hard for planar 4-regular graphs.*

Proof. We reduce the problem of calculating the RHS of (6.18) to the problem Pl-Holant($\neq_2 \mid [0, 0, 1, 0, 0]$). Then by Theorems 6.29 and 6.31, we conclude that #EO is #P-hard for planar 4-regular graphs.

The RHS of (6.18) is the bipartite planar Holant problem Pl-Holant($\neq_2 \mid f$), where the signature matrix of f is

$$M_f = \begin{bmatrix} 0 & 0 & 0 & 1 \\ 0 & 1 & 2 & 0 \\ 0 & 2 & 1 & 0 \\ 1 & 0 & 0 & 0 \end{bmatrix}.$$

Note that $f^{0101} = f^{1010} = 2$ are the two entries respectively listed in the second row indexed by 01 and the third column indexed by 01, and in the third row indexed by 10 and the second column indexed by 10. We perform a holographic transformation by $Z = \frac{1}{\sqrt{2}} \begin{bmatrix} 1 & 1 \\ \mathrm{i} & -\mathrm{i} \end{bmatrix}$ to get

$$\begin{aligned} \text{Pl-Holant}(\neq_2 \mid f) &\equiv_T \text{Pl-Holant}((\neq_2)(Z^{-1})^{\otimes 2} \mid Z^{\otimes 4} f) \\ &\equiv_T \text{Pl-Holant}(=_2 \mid \hat{f}) \\ &\equiv_T \text{Pl-Holant}(\hat{f}), \end{aligned}$$

where crucially $(=_2)Z^{\otimes 2} = (\neq_2)$, and the signature matrix of $\hat{f} = Z^{\otimes 4} f$ is

$$M_{\hat{f}} = \begin{bmatrix} 2 & 0 & 0 & 1 \\ 0 & 1 & 0 & 0 \\ 0 & 0 & 1 & 0 \\ 1 & 0 & 0 & 2 \end{bmatrix}, \tag{6.21}$$

and can be computed as $Z^{\otimes 2} M_f (Z^{\mathrm{T}})^{\otimes 2}$. An alternative way to compute this transformed signature is to write $Z^{\otimes 4} f$ by definition

$$\begin{aligned}
Z^{\otimes 4} \Big\{ & \begin{bmatrix}1\\0\end{bmatrix} \otimes \begin{bmatrix}1\\0\end{bmatrix} \otimes \begin{bmatrix}0\\1\end{bmatrix} \otimes \begin{bmatrix}0\\1\end{bmatrix} + \begin{bmatrix}1\\0\end{bmatrix} \otimes \begin{bmatrix}0\\1\end{bmatrix} \otimes \begin{bmatrix}0\\1\end{bmatrix} \otimes \begin{bmatrix}1\\0\end{bmatrix} \\
& + 2\begin{bmatrix}1\\0\end{bmatrix} \otimes \begin{bmatrix}0\\1\end{bmatrix} \otimes \begin{bmatrix}1\\0\end{bmatrix} \otimes \begin{bmatrix}0\\1\end{bmatrix} + 2\begin{bmatrix}0\\1\end{bmatrix} \otimes \begin{bmatrix}1\\0\end{bmatrix} \otimes \begin{bmatrix}0\\1\end{bmatrix} \otimes \begin{bmatrix}1\\0\end{bmatrix} \\
& + \begin{bmatrix}0\\1\end{bmatrix} \otimes \begin{bmatrix}1\\0\end{bmatrix} \otimes \begin{bmatrix}1\\0\end{bmatrix} \otimes \begin{bmatrix}0\\1\end{bmatrix} + \begin{bmatrix}0\\1\end{bmatrix} \otimes \begin{bmatrix}0\\1\end{bmatrix} \otimes \begin{bmatrix}1\\0\end{bmatrix} \otimes \begin{bmatrix}1\\0\end{bmatrix} \Big\} \\
= \frac{1}{4} \Big\{ & \begin{bmatrix}1\\ \mathrm{i}\end{bmatrix} \otimes \begin{bmatrix}1\\ \mathrm{i}\end{bmatrix} \otimes \begin{bmatrix}1\\ -\mathrm{i}\end{bmatrix} \otimes \begin{bmatrix}1\\ -\mathrm{i}\end{bmatrix} + \begin{bmatrix}1\\ \mathrm{i}\end{bmatrix} \otimes \begin{bmatrix}1\\ -\mathrm{i}\end{bmatrix} \otimes \begin{bmatrix}1\\ -\mathrm{i}\end{bmatrix} \otimes \begin{bmatrix}1\\ \mathrm{i}\end{bmatrix} \\
& + 2\begin{bmatrix}1\\ \mathrm{i}\end{bmatrix} \otimes \begin{bmatrix}1\\ -\mathrm{i}\end{bmatrix} \otimes \begin{bmatrix}1\\ \mathrm{i}\end{bmatrix} \otimes \begin{bmatrix}1\\ -\mathrm{i}\end{bmatrix} + 2\begin{bmatrix}1\\ -\mathrm{i}\end{bmatrix} \otimes \begin{bmatrix}1\\ \mathrm{i}\end{bmatrix} \otimes \begin{bmatrix}1\\ -\mathrm{i}\end{bmatrix} \otimes \begin{bmatrix}1\\ \mathrm{i}\end{bmatrix} \\
& + \begin{bmatrix}1\\ -\mathrm{i}\end{bmatrix} \otimes \begin{bmatrix}1\\ \mathrm{i}\end{bmatrix} \otimes \begin{bmatrix}1\\ \mathrm{i}\end{bmatrix} \otimes \begin{bmatrix}1\\ -\mathrm{i}\end{bmatrix} + \begin{bmatrix}1\\ -\mathrm{i}\end{bmatrix} \otimes \begin{bmatrix}1\\ -\mathrm{i}\end{bmatrix} \otimes \begin{bmatrix}1\\ \mathrm{i}\end{bmatrix} \otimes \begin{bmatrix}1\\ \mathrm{i}\end{bmatrix} \Big\}.
\end{aligned}$$

Then it can be easily calculated that $\hat{f}^{0000} = \hat{f}^{1111} = 2$, $\hat{f}^{0011} = \hat{f}^{0101} = \hat{f}^{1010} = \hat{f}^{1100} = 1$, and all other entries are 0. (A side remark for readers who are familiar with notations from quantum computing: The function f can also be denoted as $|0011\rangle + |0110\rangle + 2|0101\rangle + 2|1010\rangle + |1001\rangle + |1100\rangle$.)

We can perform the same holographic transformation by Z on our target counting problem Pl-Holant$(\neq_2 \mid [0, 0, 1, 0, 0])$ and get

$$\begin{aligned}
&\text{Pl-Holant}(\neq_2 \mid [0, 0, 1, 0, 0]) \\
&\quad \equiv_{\mathrm{T}} \text{Pl-Holant}([0, 1, 0](Z^{-1})^{\otimes 2} \mid Z^{\otimes 4}[0, 0, 1, 0, 0]) \\
&\quad \equiv_{\mathrm{T}} \text{Pl-Holant}([1, 0, 1] \mid \tfrac{1}{2}[3, 0, 1, 0, 3]) \\
&\quad \equiv_{\mathrm{T}} \text{Pl-Holant}([3, 0, 1, 0, 3]).
\end{aligned}$$

We can verify $Z^{\otimes 4}[0, 0, 1, 0, 0] = \frac{1}{2}[3, 0, 1, 0, 3]$ as follows: First note that $Z^{\otimes 4}[0, 0, 1, 0, 0]$ is symmetric, and we may denote it as $[h_0, h_1, h_2, h_3, h_4]$.

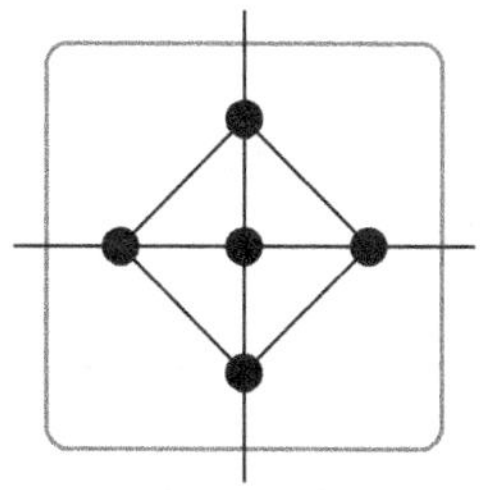

Figure 6.10. The planar tetrahedron gadget. Each vertex is assigned [3, 0, 1, 0, 3].

Writing $Z^{\otimes 4}[0, 0, 1, 0, 0]$, it is by definition $\frac{1}{4}$ of the sum over all six permutations of the tensor product $\begin{bmatrix} 1 \\ \mathrm{i} \end{bmatrix} \otimes \begin{bmatrix} 1 \\ \mathrm{i} \end{bmatrix} \otimes \begin{bmatrix} 1 \\ -\mathrm{i} \end{bmatrix} \otimes \begin{bmatrix} 1 \\ -\mathrm{i} \end{bmatrix}$. Hence it is clear that $h_0 = h_4 = 6/4 = 3/2$ and $h_2 = 1$. The entries of odd weight $h_1 = h_3 = 0$ because there is an involution $\pm\mathrm{i} \mapsto \mp\mathrm{i}$.

Our goal is to reduce Pl-Holant($\hat{f}$) to Pl-Holant([3, 0, 1, 0, 3]).

Consider the planar tetrahedron gadget in Figure 6.10. We assign [3, 0, 1, 0, 3] to every vertex and obtain a signature $32\hat{g}$, where the signature matrix of $\hat{g}$ is

$$M_{\hat{g}} = \frac{1}{2} \begin{bmatrix} 19 & 0 & 0 & 7 \\ 0 & 7 & 5 & 0 \\ 0 & 5 & 7 & 0 \\ 7 & 0 & 0 & 19 \end{bmatrix}.$$

This signature matrix requires some calculation to obtain. One can observe that all entries with odd Hamming weight from $\{0, 1\}^4$ are 0, because [3, 0, 1, 0, 3] has this property. Then, for example, the entry $\hat{g}^{0000}$ can be calculated as follows. Set the external pattern 0000. Then the four outside vertices virtually have the signature [3, 0, 1, 0]. The central vertex v_0 takes nonzero values at eight bit patterns: 0000, 1111, and the six patterns of Hamming weight 2. For 0000 at v_0, the contribution to $\hat{g}^{0000}$ is three times the trace of $\begin{bmatrix} 3 & 0 \\ 0 & 1 \end{bmatrix}^4$, which is 246. Similarly, for 1111 at v_0, the contribution to $\hat{g}^{0000}$ is three times the trace of $\begin{bmatrix} 0 & 1 \\ 1 & 0 \end{bmatrix}^4$, which is 6. For the bit patterns 0101 and 1010 of Hamming weight two at v_0, the contributions to $\hat{g}^{0000}$ is twice the trace of $\begin{bmatrix} 3 & 0 \\ 0 & 1 \end{bmatrix} \begin{bmatrix} 0 & 1 \\ 1 & 0 \end{bmatrix} \begin{bmatrix} 3 & 0 \\ 0 & 1 \end{bmatrix} \begin{bmatrix} 0 & 1 \\ 1 & 0 \end{bmatrix}$, which is 12. Finally, for 0011 at v_0, and its four cyclic permutations, the contributions to $\hat{g}^{0000}$ is four times the trace of $\begin{bmatrix} 0 & 1 \\ 1 & 0 \end{bmatrix} \begin{bmatrix} 0 & 1 \\ 1 & 0 \end{bmatrix} \begin{bmatrix} 3 & 0 \\ 0 & 1 \end{bmatrix} \begin{bmatrix} 3 & 0 \\ 0 & 1 \end{bmatrix}$, which

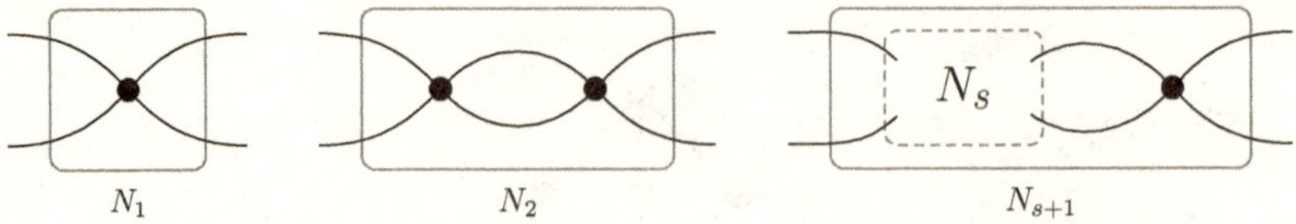

Figure 6.11. Recursive construction to interpolate $\hat{f}$. The vertices are assigned $\hat{g}$.

is 40. This gives the total value $\hat{g}^{0000} = 304$. Moreover, the signature $\hat{g}$ is invariant when we flip 0's with 1's, because $[3, 0, 1, 0, 3]$ has this property. Thus $\hat{g}^{1111} = 304$ as well. Also, $\hat{g}$ is invariant under cyclic permutations owing to the symmetry of the tetrahedron gadget. All nonzero values of $\hat{g}$ can be computed similarly.

Now we show how to reduce Pl-Holant($\hat{f}$) to Pl-Holant($\hat{g}$) by interpolation. Consider an instance Ω of Pl-Holant($\hat{f}$). Suppose that $\hat{f}$ appears n times in Ω. We construct from Ω a sequence of instances Ω_s of Holant($\hat{g}$) indexed by $s \geq 1$. We obtain Ω_s from Ω by replacing each occurrence of $\hat{f}$ with the gadget N_s in Figure 6.11 with $\hat{g}$ assigned to all vertices. Although $\hat{f}$ and $\hat{g}$ are asymmetric signatures, they are invariant under a cyclic permutation of their inputs. Thus, it is unnecessary to specify which edge corresponds to the first input of $\hat{f}$ and $\hat{g}$. We call such signatures *rotationally symmetric*.

To obtain Ω_s from Ω, we effectively replace $M_{\hat{f}}$ with $M_{N_s} = (M_{\hat{g}})^s$, the sth power of the signature matrix $M_{\hat{g}}$. Let

$$T = \begin{bmatrix} 1 & 0 & 0 & 1 \\ 0 & 1 & 1 & 0 \\ 0 & 1 & -1 & 0 \\ 1 & 0 & 0 & -1 \end{bmatrix} \quad \text{and} \quad \Lambda_{\hat{f}} = \begin{bmatrix} 3 & 0 & 0 & 0 \\ 0 & 1 & 0 & 0 \\ 0 & 0 & 1 & 0 \\ 0 & 0 & 0 & 1 \end{bmatrix}, \quad \Lambda_{\hat{g}} = \begin{bmatrix} 13 & 0 & 0 & 0 \\ 0 & 6 & 0 & 0 \\ 0 & 0 & 1 & 0 \\ 0 & 0 & 0 & 6 \end{bmatrix}.$$

Then

$$M_{\hat{f}} = T\Lambda_{\hat{f}}T^{-1} \qquad \text{and} \qquad M_{\hat{g}} = T\Lambda_{\hat{g}}T^{-1}$$

We can view our construction of Ω_s as first replacing each $M_{\hat{f}}$ by $T\Lambda_{\hat{f}}T^{-1}$ to obtain a signature grid Ω', which does not change the Holant value, and then replacing each $\Lambda_{\hat{f}}$ with $\Lambda_{\hat{g}}^s$. We stratify the assignments in Ω' based on the assignment to $\Lambda_{\hat{f}}$. Recall that the rows of $\Lambda_{\hat{f}}$ and $\Lambda_{\hat{g}}$ are indexed by 00, 01, 10, 11 and the columns are indexed by 00, 10, 01, 11, in their respective orders. We only need to consider the assignments to $\Lambda_{\hat{f}}$ that assign

- $(00, 00)$ j many times,
- $(01, 10)$ or $(11, 11)$ k many times, and
- $(10, 01)$ ℓ many times,

where $j + k + \ell = n$, the total number of occurrences of $\Lambda_{\hat{f}}$ in Ω'. Let $c_{jk\ell}$ be the sum over all such assignments of the products of evaluations from T and T^{-1} but excluding $\Lambda_{\hat{f}}$ on Ω'. Then

$$\text{Pl-Holant}_{\Omega} = \sum_{j+k+\ell=n} 3^j c_{jk\ell}$$

and the value of the Holant on Ω_s, for $s \geq 1$, is

$$\text{Pl-Holant}_{\Omega_s} = \sum_{j+k+\ell=n} (13^j 6^k)^s c_{jk\ell}. \tag{6.22}$$

This is a linear equation system with unknowns $c_{jk\ell}$, and a coefficient matrix whose rows are indexed by s and columns are indexed by (j, k), where $0 \leq j, k$ and $j + k \leq n$. We take $1 \leq s \leq \binom{n+2}{2}$. This coefficient matrix in the linear system of (6.22) is Vandermonde and has full rank since for any $j, k, j', k' \geq 0$, if $(j, k) \neq (j', k')$ then $13^j 6^k \neq 13^{j'} 6^{k'}$. Therefore, after obtaining the values of Pl-Holant$_{\Omega_s}$ by oracle calls to #EO, for $1 \leq s \leq \binom{n+2}{2}$, we can solve the linear system for the unknowns $c_{jk\ell}$ (where $j, k, \ell \geq 0$ and $j + k + \ell = n$) and obtain the value of Pl-Holant$_\Omega$. □

6.6 Kirchhoff's Matrix Tree Theorem

In this section we describe a celebrated formula counting the number of spanning trees in an undirected graph due to G. Kirchhoff [Kir47], and an extension to directed graphs due to W. T. Tutte [Tut48]. We then apply this to obtain a counting formula for the number of Eulerian circuits given an Eulerian orientation, known as the *BEST* theorem, originally discovered by de *B*ruijn and van Aardenne-*E*hrenfest [vAEdB51], with a special case due to *S*mith and *T*utte [TS41].

Let $G = (V, E)$ be a simple undirected graph without multiple edges and self loops. A spanning tree of a connected graph G is a subgraph that forms a tree on V. Equivalently, it is a connected subgraph with $|V| - 1$ edges; or minimally connected; or acyclic with $|V| - 1$ edges; or maximally acyclic. The vertex–edge incidence matrix M is defined as follows: The rows and columns of M are indexed by V and E respectively. For every edge $e = \{i, j\}$, we define $M(i, e) = +1$ and $M(j, e) = -1$. (We arbitrarily pick one incident vertex of e to assign $+1$ and the other incident vertex to assign -1.) Then $L = MM^{\mathrm{T}}$ is called a Laplacian matrix of G. L is a symmetric positive semidefinite $|V| \times |V|$ matrix, with diagonal entries $L(i, i) = \deg(i)$, the degree of the vertex i, and $L(i, j) = L(j, i) = -1$ for $i \neq j$, if there is an edge $e = \{i, j\}$, and 0 otherwise. Note that $(1, 1, \dots 1)^{\mathrm{T}}$ is an eigenvector of L with eigenvalue $\lambda_0 = 0$.

Kirchhoff's formula counts the number of labeled spanning trees of G in terms of L.

Theorem 6.33. *(Matrix Tree Theorem) Suppose G is a graph on n vertices with Laplacian matrix L. Then the number of spanning trees of G is the determinant of any $(n-1) \times (n-1)$ submatrix obtained by striking out row i and column i from L (this value is the same for all i).*

Equivalently the number of spanning trees of G is

$$\frac{1}{n}\lambda_1\lambda_2\cdots\lambda_{n-1},$$

where $0 = \lambda_0 \le \lambda_1 \le \lambda_2 \le \cdots \le \lambda_{n-1}$ are the eigenvalues of L.

Proof. For $n = 1$, there is a unique spanning tree with no edges. By convention the determinant of the empty matrix is 1. If $|E| < n - 1$, then there are no spanning trees, and the rank of L is at most $|E|$ by the definition $L = MM^{\mathrm{T}}$. So we may assume $n \ge 2$ and $|E| \ge n - 1$.

Consider the submatrix M' obtained from M by striking out row i. We are interested in the determinant of $M'M'^{\mathrm{T}}$. By the Cauchy–Binet formula,

$$\det(M'M'^{\mathrm{T}}) = \sum_{S \subseteq E:\ |S|=n-1} \det(M'_S)\det((M'_S)^{\mathrm{T}}) = \sum_{S \subseteq E:\ |S|=n-1} [\det(M'_S)]^2,$$

where the sum is over all subsets S of E of cardinality $n-1$, and M'_S is the submatrix of M' by choosing the columns indexed by S.

Suppose (V, S) is disconnected, i.e., the subset of edges S does not correspond to a spanning tree of G. Then there is a proper subset of vertices not containing i that is disconnected from the rest, i.e., a subset $V' \subseteq V - \{i\}$ such that V' is disconnected from $V - V'$ in (V, S). Then, in M'_S, the sum of all rows indexed by V' is 0, because for every edge $e = \{j, k\} \in S$, either both j and k are in V' or both are not. Hence for such S, $\det(M'_S) = 0$.

Suppose (V, S) is connected, i.e., the subset of edges S corresponds to a spanning tree of G. Every spanning tree on $n \ge 2$ vertices has at least two leaves. If $n = 2$ then M'_S has a single ± 1 entry, indicating an edge incident to the remaining vertex and the removed vertex with index i. Then clearly $[\det(M'_S)]^2 = 1$. Suppose $n \ge 3$. There is at least one leaf $j \in [n] - \{i\}$. We can expand $\det(M'_S)$ by row j, which has a single ± 1 entry corresponding to an edge e incident to j, and the rest are all 0. By striking out row j and column e, we effectively remove the leaf j. By induction we get $[\det(M'_S)]^2 = 1$.

It follows that $\det(M'M'^{\mathrm{T}})$ counts the number of spanning trees of G.

To prove the second claim, observe that L is symmetric with row/column sum 0. Thus 0 is an eigenvalue, with eigenvector $v_0 = \frac{1}{\sqrt{n}}(1, 1, \ldots 1)^{\mathrm{T}}$. Being symmetric, the orthogonal complement $\langle v_0 \rangle^\perp$ of v_0 is an invariant subspace of L. The restriction of L to $\langle v_0 \rangle^\perp$ is also symmetric. Thus we have a set of orthonormal eigenvectors $v_0, v_1, \ldots, v_{n-1}$. Let T be an $n \times n$ matrix with column vectors $v_0, v_1, \ldots, v_{n-1}$, then

$$L = T \begin{bmatrix} 0 & 0 & 0 & \ldots & 0 \\ 0 & \lambda_1 & 0 & \ldots & 0 \\ 0 & 0 & \lambda_2 & \ldots & 0 \\ \vdots & \vdots & \vdots & \ddots & \vdots \\ 0 & 0 & 0 & \ldots & \lambda_{n-1} \end{bmatrix} T^{\mathrm{T}}. \tag{6.23}$$

We are interested in the determinant of the $(n-1) \times (n-1)$ submatrix L_{11} obtained by striking out the first row and first column of L, which is the same as striking out the first row of T and first column of T^{T} in the RHS of (6.23).

Notice that the first row and the first column of the diagonal matrix are both 0. Thus

$$L_{11} = T_{11} \begin{bmatrix} \lambda_1 & 0 & \ldots & 0 \\ 0 & \lambda_2 & \ldots & 0 \\ \vdots & \vdots & \ddots & \vdots \\ 0 & 0 & \ldots & \lambda_{n-1} \end{bmatrix} (T_{11})^{\mathrm{T}},$$

where T_{11} is obtained by striking out the first row and first column of T. Hence $\det(L_{11}) = [\det(T_{11})]^2 \lambda_1 \lambda_2 \cdots \lambda_{n-1}$.

If we add rows 2 to n of T to the first row, the $(1, 1)$ entry of T becomes $\sqrt{n}$, and all other entries of the first row become 0, because $v_1, \ldots, v_{n-1}$ are all orthogonal to $(1, 1, \ldots 1)^{\mathrm{T}}$. Thus $\det(T) = \sqrt{n} \det(T_{11})$. It follows that $[\det(T_{11})]^2 = \frac{1}{n}$, since for an orthogonal T, $\det(T) = \pm 1$. □

Corollary 6.34. *(Cayley) There are exactly n^{n-2} labeled spanning trees in the complete graph K_n of n vertices.*

Next we extend Theorem 6.33 to directed graphs. Suppose G is a directed graph. We assume G has no loops and no multiple edges. Every edge $e = (i, j)$ has a direction from i to j. The reverse ordered pair (j, i), if it is present, is considered a distinct edge.

Let $r \in V$ be any vertex. An *arborescence* rooted at r is a set of $|V| - 1$ directed edges, such that the underlying undirected graph is a spanning tree rooted at r, and all directed edges point from child to parent. Thus from any vertex $v \in V$ there is a unique directed path from v to r in an arborescence rooted at r. We want to count the number of arborescences rooted at any vertex in a directed graph.

We define two $|V| \times |E|$ matrices A and B. The rows and columns are indexed by V and E respectively. Suppose $e = (i, j)$ is a directed edge from i to j. Then $A(i, e) = +1$, and $A(i', e) = 0$ for all $i' \neq i$ (including $i' = j$). For B, we define $B(i, e) = +1$, $B(j, e) = -1$, and $B(k, e) = 0$ for all $k \neq i, j$. Then we define the modified Laplacian matrix $L^+ = AB^{\mathrm{T}}$. It is easy to see that $L^+(i, i)$ is the outdegree of i. For $i \neq j$, $L^+(i, j) = -1$ if there is an edge $e = (i, j)$ from i to j, and $L^+(i, j) = 0$ otherwise.

Theorem 6.35. *Suppose G is a directed graph with an underlying connected undirected graph with n vertices and the modified Laplacian matrix L^+. Then the number of arborescences rooted at i is the determinant of the $(n-1) \times (n-1)$ submatrix obtained by striking out row i and column i from L^+.*

Proof. If we denote by A' and B' the submatrices of A and B obtained by deleting the ith rows respectively, then we are interested in $\det(A'B'^{\mathrm{T}})$.

By the Cauchy–Binet formula,

$$\det(A'B'^{\mathrm{T}}) = \sum_{S \subseteq E:\ |S| = n-1} \det(A'_S) \det(B'_S),$$

where the sum is over all subsets S of E of cardinality $n-1$, and A'_S and B'_S are submatrices of A' and B' by choosing the columns indexed by S.

Again, as in the proof of Theorem 6.33, if the underlying undirected graph defined by S on V disregarding directions is disconnected, then $\det(B'_S) = 0$.

Suppose S is such that $\det(B'_S) \neq 0$. This happens iff the underlying undirected graph defined by S on V disregarding directions is a spanning tree. Then by picking off leaves recursively to expand the determinant $\det(B'_S)$, it defines an order (corresponding to a permutation π) of $V - \{i\}$, and all the entries from B'_S used in the expansion are ± 1. According to this order, at each step we remove a leaf of the spanning tree. If we consider expanding $\det(A'_S)$ in this same order, we find that at each step, inductively the row corresponding to the next leaf to be removed has at most one nonzero entry, and that entry is $+1$ iff the directed edge is pointing from the child (current leaf) to the parent in the spanning tree. And if these entries are all nonzero in $n-1$ steps, they are all $+1$, and the same $+1$ entries are encountered in the expansion of $\det(B'_S)$, corresponding to the same permutation π. The two factors of the same sign of the permutation π in $\det(A'_S)$ and $\det(B'_S)$ cancel. Finally, $\det(A'_S)\det(B'_S) = 1$ for each arborescence rooted at i. □

An Eulerian circuit in an undirected graph G is a traversal of all its edges exactly once, i.e., a cyclically ordered sequence of edges of G where consecutive edges are adjacent (they share a common vertex) and every edge appears exactly once. An Eulerian circuit in a directed graph G has the requirement that every edge must be traversed in the direction of the edge. An Eulerian circuit has a direction but no distinguished starting point.

Perhaps the earliest result in graph theory is Euler's observation [Eul36] that an undirected graph G has an Eulerian circuit if and only if G is connected and every vertex has an even degree. Euler proved that this is a necessary condition, and stated without proof that connected graphs with all vertices of even degree have an Eulerian circuit. The first complete proof of the sufficiency of the condition was published in 1873 by Hierholzer [Hie73]. There is a simple generalization to directed graphs, namely the underlying undirected graph is connected, and for every vertex its indegree is equal to its outdegree.

Any Eulerian circuit defines an Eulerian orientation, as there is an equal number of times an Eulerian circuit enters and exits each vertex. Using the method of Theorem 6.35 to count the number of arborescences in a directed graph, we have the following remarkable *BEST* theorem to count the number of Eulerian circuits in a graph given an Eulerian orientation.

Theorem 6.36. *(BEST) Let G be connected and given an Eulerian orientation π. Let r be any vertex of G and $t_r(G)$ be the number of arborescences rooted at r in G. Then the number of Eulerian circuits consistent with π is*

$$t_r(G) \prod_{v \in V(G)} (d_v - 1)!$$

where d_v is both the outdegree as well as the indegree of v under π.

Notice that this theorem shows that in particular $t_r(G)$ does not depend on r for any G given an Eulerian orientation.

Proof. Fix any vertex r of G. Consider any Eulerian circuit consistent with π starting at r. It defines an arborescence T rooted at r as follows: For any vertex $v \neq r$, let exit(v) be the last directed edge leaving v on the Eulerian circuit before it completes the entire traversal and returns to r. Since the undirected underlying graph G is connected, every vertex has $d_v > 0$ many incoming and outgoing edges. Since the Eulerian circuit traverses every edge, every vertex is visited d_v times, and exit(v) is well defined. It has the form (v, w), for some w. If exit$(v) = (v, w)$ then the last visit of w by the Eulerian circuit is after the last visit of v. Hence starting at any $v \neq r$, successively follow the directed edges from $\{\text{exit}(u) \mid u \neq r\}$, it cannot repeat any previously visited vertex, and thus must end in r. Hence these $|V| - 1$ edges define an arborescence T rooted at r.

Now, given π and an arborescence T rooted at r consistent with π, we claim there are exactly $\prod_{v \in V(G)}(d_v - 1)!$ many Eulerian circuits that define T. Given π and T, we define a set of Eulerian circuits as follows: Starting at r, at any vertex follow any unused outgoing edge, with the only rule that at $v \neq r$, use $\text{exit}(v)$ only when there are no other choices. Once we prove these are indeed Eulerian circuits, it will be clear that they all define T and that any Eulerian circuit that defines T is among these.

This walk ends at some vertex because there are only finitely many edges. Since at any $v \neq r$, the walk has used one more incoming edge than outgoing edge, the walk cannot end there. Hence the walk ends at r. When it ends at r, it must have used up all outgoing edges at r. Since π is an Eulerian orientation, there are an equal number of incoming and outgoing edges at r. Every time it is at r, it has traversed the same number of incoming and outgoing edges at r. So it must have traversed all incoming edges at r. These incoming edges include at least one tree edge belonging to T. Hence this walk traversed at least one tree edge. Consider the first time this walk used an edge from T, say $\text{exit}(v)$. We claim that this v is a leaf in T. Otherwise, there is at least one tree edge ending in v, and it has not been traversed because $\text{exit}(v)$ is the first. But then v has at least one outgoing edge other than $\text{exit}(v)$, since when visiting any $v \neq r$ the walk has used one more incoming edge than outgoing edge at v, thus one more unused outgoing edge than incoming edge at v. This is a contradiction by our rule of using the tree edge $\text{exit}(v)$ last at v. Hence, v is a leaf, and when $\text{exit}(v)$ is used all incident edges of the leaf v have been used.

Now we keep track of the progress of this walk by removing this leaf in T. Inductively we maintain an arborescence T' obtained by successively removing leaves, starting from T. By the same argument, the walk ends at r, it must use some tree edge of T' before it ends since it must use some tree edge of T' pointing to r. The vertex u where the first time (after forming T') a tree edge $\text{exit}(u)$ is used must be a leaf in the current arborescence T', at which point all incident edges of u must have been used, and we may remove u, updating T'. It follows that when the walk ends at r, the walk has traversed every edge exactly once, in the direction consistent with the given Eulerian orientation π. Clearly this walk is an Eulerian circuit that defines the given arborescence T rooted at r.

There are exactly $(d_v - 1)!$ choices at every $v \neq r$ and $d_r!$ choices at r, giving rise to $d_r! \prod_{v \neq r}(d_v - 1)!$ many Eulerian circuits consistent with π and T.

However, each such Eulerian circuit is counted exactly d_r times since it passes through r exactly d_r times by a cyclic shift. The theorem follows. □

Regarding the problem of counting Eulerian circuits on undirected graphs, Brightwell and Winkler [BW05] proved that this problem is #P-Complete. Ge and Štefankovič [GŠ12] improved that to show that the problem remains #P-Complete for 4-regular graphs.

7

Holant Dichotomy for Symmetric Constraints

Most of the results in previous chapters can be viewed as preparatory for the dichotomy theorems to be achieved in this and the next two chapters. In this chapter we prove a dichotomy theorem valid for all Holant problems on an arbitrary set of complex-valued *symmetric* constraint functions on Boolean variables, Theorem 7.19. This includes as special cases the dichotomy theorems for Holant* in Chapter 2 (when all unary functions are free), Holant^c in Chapter 6 (when the two unary functions Is-ZERO $[1, 0]$ and Is-ONE $[0, 1]$ are free), and #CSP in Chapter 3 (when all EQUALITIES are free) *restricted* to symmetric constraints.

For Holant problems without freely available auxiliary functions such as unary signatures, the construction of gadgets becomes more constrained, and thus more difficult. This difficulty has shown itself in Chapter 5. In particular, the only natural local operations available are to connect two edges, or to make a loop. If we want to prove hardness with some inductive argument based on the arity of constraint functions, there will be two base cases of arity 3 and 4. Note that signatures of arity ≤ 2 define tractable Holant problems. The arity 3 case is understood from Theorem 6.3 (and also Theorem 6.4), with the main work done in Chapter 5. Therefore it will be natural to study carefully arity 4 signatures.

However, before we can develop the theory satisfactorily for arity 4 signatures we must first deal with a phenomenon called *vanishing*.

7.1 Vanishing Signatures

Definition 7.1. A set of signatures $\mathcal{F}$ is called *vanishing* if the value $\text{Holant}(\Omega; \mathcal{F})$ is zero for every signature grid Ω. A signature f is called *vanishing* if the singleton set $\{f\}$ is vanishing.

In this section, we characterize all sets of symmetric vanishing signatures. The following lemma makes the simple observation that vanishing signatures are closed under gadget constructions, and under linear combinations. Let $af + bg$ denote the entry-wise linear combination of two signatures f and g with the same arity n ($a, b \in \mathbb{C}$), i.e., $(af + bg)(x) = af(x) + bg(x)$ for all $x \in \{0, 1\}^n$.

Lemma 7.2. *Let $\mathcal{F}$ be a vanishing signature set. If a signature f can be realized by a gadget using signatures in $\mathcal{F}$, then $\mathcal{F} \cup \{f\}$ is also vanishing. If f and g are two signatures in $\mathcal{F}$ of the same arity, and $a, b \in \mathbb{C}$, then $\mathcal{F} \cup \{af + bg\}$ is vanishing as well.*

Proof. That $\mathcal{F} \cup \{f\}$ is vanishing is obvious, since any vertex in a signature grid assigned the function f can be replaced by the $\mathcal{F}$-gate defining it. If $f, g \in \mathcal{F}$, then the Holant value on any signature grid using $\mathcal{F} \cup \{af + bg\}$ can be expressed as a linear combination of Holant values using $\mathcal{F}$, one term for a choice of f or g at every vertex labeled $af + bg$. □

Obviously, the identically 0 signature, in which all entries are 0, is vanishing. This is trivial. However, there are less trivial examples of vanishing signatures. Notice that the unary signature $[1, \mathrm{i}]$ when connected to another $[1, \mathrm{i}]$ has a Holant value 0. Consider a signature set $\mathcal{F}$ where every signature is degenerate, and moreover, as a tensor product of unary signatures, more than half of them are $[1, \mathrm{i}]$. For any signature grid Ω with signatures from $\mathcal{F}$, each signature is decomposed into unary signatures, and Ω is decomposed into pairs of vertices, one for each edge. The total Holant value is the product of the Holant on each pair. Since more than half of the unaries in each signature are $[1, \mathrm{i}]$, more than half of the unaries in Ω are $[1, \mathrm{i}]$. Then two $[1, \mathrm{i}]$'s must be paired up and hence $\mathrm{Holant}_\Omega = 0$. Thus, all such signatures form a vanishing set. We also observe that this argument holds when $[1, \mathrm{i}]$ is replaced by $[1, -\mathrm{i}]$.

These signatures described above are generally not symmetric and our present aim is to characterize symmetric vanishing signatures. To this end, we define the following symmetrization operation.

Definition 7.3. Let S_n be the symmetric group of degree $n \geq 1$. Then for any integer t ($0 \leq t \leq n$) and unary signatures $v, v_1, \ldots, v_{n-t}$, we define

$$\mathrm{Sym}_n^t(v; v_1, \ldots, v_{n-t}) = \sum_{\pi \in S_n} \bigotimes_{k=1}^{n} u_{\pi(k)}, \tag{7.1}$$

where the ordered sequence $(u_1, u_2, \ldots, u_n) = (\underbrace{v, \ldots, v}_{t \text{ copies}}, v_1, \ldots, v_{n-t})$.

Note that we include redundant permutations of v in the definition. Identical v_i's also introduce redundant permutations. These redundant permutations simply introduce a nonzero constant factor. However, the allowance of redundant permutations may simplify calculations. If $t = 1$, then $\mathrm{Sym}_n^1(u_1; u_2, \ldots, u_n)$ is a sum of $n!$ tensor products by permuting $u_1, \ldots, u_n \in \mathbb{C}^2$ in all possible ways, and the n unary signatures are treated completely symmetrically.

An illustrative example of Definition 7.3 is

$$\begin{aligned}\mathrm{Sym}_3^2([1, \mathfrak{i}]; [a, b]) &= 2[a, b] \otimes [1, \mathfrak{i}] \otimes [1, \mathfrak{i}] + 2[1, \mathfrak{i}] \otimes [a, b] \otimes [1, \mathfrak{i}] \\ &\quad + 2[1, \mathfrak{i}] \otimes [1, \mathfrak{i}] \otimes [a, b] \\ &= 2[3a, 2\mathfrak{i}a + b, -a + 2\mathfrak{i}b, -3b].\end{aligned}$$

Definition 7.4. A nonzero symmetric signature f of arity n has *positive* $(+)$ *vanishing degree* $k \geq 0$, denoted by $\mathrm{vd}^+(f) = k$, if $k \leq n$ is the largest integer such that there exist $n - k$ unary signatures $v_1, \ldots, v_{n-k}$ satisfying

$$f = \mathrm{Sym}_n^k([1, \mathfrak{i}]; v_1, \ldots, v_{n-k}).$$

If f is the all-zero signature, we define $\mathrm{vd}^+(f) = n + 1$.

We define *negative* $(-)$ *vanishing degree* vd^- similarly, using $-\mathfrak{i}$ instead of $\mathfrak{i}$.

We will prove in Section 7.1.1 (Lemma 7.8) that any nonzero symmetric signature has an expression as a symmetrization of unary signatures as in Definition 7.4. Thus $\mathrm{vd}^\sigma(f)$ is well defined.

Note that $\mathrm{vd}^+(f) = 0$ iff f cannot be expressed as a symmetrization of n unary signatures consisting of at least one $[1, \mathfrak{i}]$ (but an expression as in Definition 7.4 for n unary signatures $v_1, \ldots, v_n$ still exists). Also $\mathrm{vd}^+(f) = n$ iff f is the degenerate signature $[1, \mathfrak{i}]^{\otimes n}$ (recall that we treat two signatures as the same if their ratio is a nonzero constant).

In Chapter 2, we showed in (2.10) to (2.12) that any generalized Fibonacci gate f with parameter $\lambda = 2\mathfrak{i}$, i.e., $f_k = ck\mathfrak{i}^{k-1} + d\mathfrak{i}^k$ (for some $c, d \in \mathbb{C}$, and for all $0 \leq k \leq n$), has the form $f = Z^{\otimes n}[a, b, 0, \ldots, 0]$, for some $a, b \in \mathbb{C}$, where $Z = \begin{bmatrix} 1 & 1 \\ \mathfrak{i} & -\mathfrak{i} \end{bmatrix}$. A similar expression is true if the parameter $\lambda = -2\mathfrak{i}$; in that case $f = Z^{\otimes n}[0, \ldots, 0, b, a]$. The expression in (2.12) shows that $\mathrm{vd}^+(Z^{\otimes n}[1, 0, 0, \ldots, 0]) = n$ and $\mathrm{vd}^+(Z^{\otimes n}[0, 1, 0, \ldots, 0]) = n - 1$. For the linear combination $f = Z^{\otimes n}[a, b, 0, \ldots, 0] = aZ^{\otimes n}[1, 0, 0, \ldots, 0] + bZ^{\otimes n}[0, 1, 0, \ldots, 0]$, Lemma 7.8 in Section 7.1.1 will show that $\mathrm{vd}^+(f) = n - 1$ if $b \neq 0$. Similarly for $f = Z^{\otimes n}[0, \ldots, 0, b, a]$, $\mathrm{vd}^-(f) = n - 1$ if $b \neq 0$.

Definition 7.5. If f is a symmetric signature and $\sigma \in \{+, -\}$, then we define its recurrence degree $\text{rd}^\sigma(f)$ to be

$$\text{rd}^\sigma(f) = \text{arity}(f) - \text{vd}^\sigma(f).$$

For generalized Fibonacci gates f of arity n and parameter $2\mathrm{i}$, $\text{vd}^+(f) \geq n-1$ and $\text{rd}^\sigma(f) \leq 1$, and equalities hold iff f is nondegenerate. All nonzero entries in the expression $(Z^{-1})^{\otimes n} f = [a, b, 0, \ldots, 0]$ have indices of Hamming weight ≤ 1, and there are $\text{vd}^+(f) \geq n-1$ many trailing zeros.

Notice that it is possible for a signature f to have both $\text{vd}^+(f)$ and $\text{vd}^-(f)$ nonzero. For example, $f = [1, 0, 1]$ has $\text{vd}^+(f) = \text{vd}^-(f) = 1$.

By Lemma 7.2 and the discussion after Lemma 7.2, we know that for a signature f of arity n, if $\text{vd}^\sigma(f) > \frac{n}{2}$ for some $\sigma \in \{+, -\}$, then f is a vanishing signature. In Section 7.1.1, Lemma 7.8 will give another simple explanation for this fact. In particular, all generalized Fibonacci gates of arity $n \geq 3$ with parameter $\lambda = \pm 2\mathrm{i}$ are vanishing signatures.

This is easily generalized to a set of signatures.

Definition 7.6. For $\sigma \in \{+, -\}$, we define $\mathscr{V}^\sigma = \{f | 2\,\text{vd}^\sigma(f) > \text{arity}(f)\}$.

Equivalently,

$$\mathscr{V}^\sigma = \{f | 2\,\text{rd}^\sigma(f) < \text{arity}(f)\}.$$

Lemma 7.7. *For a set of symmetric signatures $\mathcal{F}$, if $\mathcal{F} \subseteq \mathscr{V}^+$ or $\mathcal{F} \subseteq \mathscr{V}^-$, then $\mathcal{F}$ is vanishing.*

In Theorem 7.11, we show that these two sets capture all symmetric vanishing signature sets.

7.1.1 Characterization of Vanishing Signatures via Z

There is a simple characterization of the set of vanishing signatures in terms of the holographic transformation by $Z = \frac{1}{\sqrt{2}} \begin{bmatrix} 1 & 1 \\ \mathrm{i} & -\mathrm{i} \end{bmatrix}$. Suppose f has $\text{arity}(f) = n$ and $\text{rd}^+(f) = d$. Then $f \in \mathscr{V}^+$ iff $d < n/2$. We perform a holographic transformation with Z,

$$\begin{aligned} \text{Holant}(=_2 \mid f) &\equiv_T \text{Holant}([1, 0, 1] Z^{\otimes 2} \mid (Z^{-1})^{\otimes n} f) \\ &\equiv_T \text{Holant}([0, 1, 0] \mid \hat{f}), \end{aligned}$$

where $\hat{f} = (Z^{-1})^{\otimes n} f$. We claim that $\hat{f}$ has the form $[\hat{f}_0, \hat{f}_1, \ldots, \hat{f}_d, 0, \ldots, 0]$, and $\hat{f}_d \neq 0$.

Lemma 7.8. *For any symmetric signature f, its vanishing degree $\mathrm{vd}^\sigma(f)$ and recurrence degree $\mathrm{rd}^\sigma(f)$ are well defined, for $\sigma \in \{+,-\}$.*

Let $\hat{f} = (Z^{-1})^{\otimes n} f$. If $\mathrm{rd}^+(f) = d$, then $\hat{f} = [\hat{f}_0, \ldots, \hat{f}_d, 0, \ldots, 0]$ and $\hat{f}_d \neq 0$. The number of trailing 0 entries after $\hat{f}_d$ is $\mathrm{vd}^+(f)$. In particular, $f \in \mathscr{V}^+$ iff all nonzero entries $\hat{f}_i$ occur in strictly the first half of $\hat{f}$, i.e., all have index $i < \frac{n}{2}$.

Similarly, if $\mathrm{rd}^-(f) = d$, then $\hat{f} = [0, \ldots, 0, \hat{f}_{n-d}, \ldots, \hat{f}_n]$ and $\hat{f}_{n-d} \neq 0$. The number of leading 0 entries before $\hat{f}_{n-d}$ is $\mathrm{vd}^-(f)$. In particular, $f \in \mathscr{V}^-$ iff all nonzero entries $\hat{f}_i$ occur in strictly the second half of $\hat{f}$, i.e., all have index $i > \frac{n}{2}$.

Proof. Suppose a nonzero symmetric signature $f = Z^{\otimes n}\hat{f}$ has an expression

$$f = \mathrm{Sym}_n^{n-t}([1, \mathfrak{i}]; v_1, \ldots, v_t), \tag{7.2}$$

for some $0 \le t \le n$ and $v_1, \ldots, v_t \in \mathbb{C}^2$. Since

$$Z = \frac{1}{\sqrt{2}}\begin{bmatrix} 1 & 1 \\ \mathfrak{i} & -\mathfrak{i} \end{bmatrix}, Z^{-1} = \tfrac{1}{\sqrt{2}}\begin{bmatrix} 1 & -\mathfrak{i} \\ 1 & \mathfrak{i} \end{bmatrix},$$

we have $Z^{-1}[1, \mathfrak{i}] = \sqrt{2}[1, 0]$. Thus, f having an expression of the form (7.2) is equivalent to $\hat{f}$ having an expression of the form

$$\hat{f} = \mathrm{Sym}_n^{n-t}([1, 0]; u_1, \ldots, u_t), \tag{7.3}$$

up to a nonzero scalar factor, where $u_i = Z^{-1}v_i \in \mathbb{C}^2$ for $1 \le i \le t$.

From the expression (7.3) for $\hat{f} = [\hat{f}_0, \hat{f}_1, \ldots, \hat{f}_n]$, it is clear that all entries of Hamming weight greater than t in $\hat{f}$ are 0. Moreover, if t is the minimum integer ($0 \le t \le n$) such that (7.2) holds for some $v_1, \ldots, v_t \in \mathbb{C}^2$, then $\hat{f}_t \neq 0$. This is clear if $t = 0$. Suppose $t > 0$ and for a contradiction $\hat{f}_t = 0$. Then one of the u_i must be a multiple of $[1, 0]$. This would imply that one of the v_i is a multiple of $[1, \mathfrak{i}]$ in (7.2), contradicting the minimality of t.

Now suppose

$$\hat{f} = [\hat{f}_0, \ldots, \hat{f}_t, 0, \ldots, 0],$$

where $0 \le t \le n$ and $\hat{f}_t \neq 0$. Such a t is uniquely defined for any nonzero symmetric signature $\hat{f}$, hence also for $f = Z^{\otimes n}\hat{f}$. We can normalize $\hat{f}_t = t!(n-t)!$ by a nonzero scalar multiple. We prove that there are $u_1 = [a_1, 1], \ldots, u_t = [a_t, 1] \in \mathbb{C}^2$ such that

$$\hat{f} = \mathrm{Sym}_n^{n-t}([1, 0]; u_1, \ldots, u_t).$$

Let F denote the signature defined by the right-hand side (RHS), for some $a_1, \ldots, a_t \in \mathbb{C}$ to be determined. Then clearly $F_k = 0$ for $k > t$. Define

$$c_k = \frac{\hat{f}_k}{k!(n-k)!}, \quad \text{for } 0 \le k \le t.$$

Then $c_t = 1$, and over the algebraically closed field $\mathbb{C}$ we can factor the polynomial

$$\sum_{k=0}^{t} c_k X^k = (X + a_1)(X + a_2) \cdots (X + a_t)$$

for some $a_1, a_2, \ldots, a_t \in \mathbb{C}$. It follows that $c_{t-k} = p_k(a_1, a_2, \ldots, a_t)$, where p_k is the elementary symmetric polynomial of degree k on t variables.

For $0 \le k \le t$,

$$\begin{aligned} F_k &= \binom{n-k}{n-t}(n-t)! \sum_{S \subseteq [t]: |S| = t-k} (t-k)!k! \prod_{i \in S} a_i \\ &= (n-k)!k! p_{t-k}(b_1, a_2, \ldots, a_t) = \hat{f}_k, \end{aligned}$$

because on input $(\underbrace{0, \ldots, 0}_{n-k}, \underbrace{1, \ldots, 1}_{k})$ of Hamming weight k, a term with a nonzero contribution in the sum $\text{Sym}_n^{n-t}([1, 0]; u_1, \ldots, u_t)$ defining F must have all $n - t$ copies of $[1, 0]$ occurring in the first $n - k$ positions, and there is a subset of $[a_i, 1]$'s to be in the remaining $(n - k) - (n - t) = t - k$ positions.

The proof for $\text{vd}^-(f)$ and $\text{rd}^-(f)$ is symmetric by replacing $\mathfrak{i}$ by $-\mathfrak{i}$. □

For future reference, we also note the following. If $f = g + h$ is of arity n, where $\text{rd}^+(g) = d$, $\text{rd}^-(h) = d'$, and $d + d' < n$, then after a holographic transformation by Z, $\hat{f} = (Z^{-1})^{\otimes n} f$ takes the form $[\widehat{g}_0, \ldots, \widehat{g}_d, \underbrace{0, \ldots, 0}_{n-d-d'-1}, \widehat{h}_{n-d'}, \ldots, \widehat{h}_n]$, where $n - d - d' - 1 \ge 0$ zeros are present.

Suppose $f \in \mathscr{V}^+$. The expression $f = Z^{\otimes n}[\hat{f}_0, \ldots, \hat{f}_d, 0, \ldots, 0]$ with $d < n/2$ also offers an alternative explanation for why f is vanishing. In any bipartite graph for Holant$([0, 1, 0] \mid \hat{f})$, the binary DISEQUALITY $(\neq_2) = [0, 1, 0]$ on the left imposes the condition that half of the edges must take the value 0 and the other half must take the value 1. On the right side we have $d < n/2$; thus $\hat{f}$ requires that less than half of the edges are assigned the value 1. Therefore the Holant is always 0.

Under the Z transformation, one can observe another interesting phenomenon. For any $a, b \in \mathbb{C}$,

$$\text{Holant}([0, 1, 0] \mid [a, b, 1, 0, 0]) \quad \text{and} \quad \text{Holant}([0, 1, 0] \mid [0, 0, 1, 0, 0])$$

take exactly the same value on every signature grid. This is because, to contribute a nonzero term in the Holant, exactly half of the edges must be assigned 1. Then for the first problem, the signature on the right can never contribute a nonzero value involving a or b. Thus the Holant values of these two problems on any signature grid are always the same. Nevertheless, for every $a, b \in \mathbb{C}$ not both 0, there is no holographic transformation between these two problems. This shows that the converse of the Holant Theorem (Theorem 1.5) does not hold.

We know that Holant($[0, 1, 0] \mid [0, 0, 1, 0, 0]$) counts the number of Eulerian orientations in a 4-regular graph. This problem is #P-hard (Theorem 6.32) and in fact its planar version Pl-Holant($\neq_2 \mid [0, 0, 1, 0, 0]$) is also #P-hard for 4-regular graph. This will play an important role in proofs later. Translating back to the standard setting, the problem of counting Eulerian orientations in a 4-regular graph is Holant($[3, 0, 1, 0, 3]$). The problem Holant($[0, 1, 0] \mid [a, b, 1, 0, 0]$) corresponds to a certain signature $f = Z^{\otimes 4}[a, b, 1, 0, 0]$ of arity 4 with recurrence degree 2. It has a different appearance but induces exactly the same Holant value as the signature for counting Eulerian orientations. Therefore, all such signatures are #P-hard as well. We use this fact later.

We remark that for $f = Z^{\otimes n}[\hat{f}_0, \ldots, \hat{f}_n]$, the signature obtained by a self-loop on f is simply $f' = 2Z^{\otimes n}[\hat{f}_1, \ldots, \hat{f}_{n-1}]$, by removing the extremal entries $\hat{f}_0$ and $\hat{f}_n$. This is easily seen by considering it as an $\mathcal{F}$-gate construction in Holant($\neq_2 \mid \hat{f}$), where two input bits of $\hat{f}$ on the loop are each forced to take exactly 0 and 1, or 1 and 0 respectively. This operation is very handy, and can be repeated.

7.1.2 Characterization of Vanishing Signature Sets

Now we show that $\mathscr{V}^+$ and $\mathscr{V}^-$ capture all symmetric vanishing signature sets. To begin, we show that a vanishing signature set cannot contain both types of nonzero vanishing signatures.

Lemma 7.9. *Let $f \in \mathscr{V}^+$ and $g \in \mathscr{V}^-$. If neither f nor g is the zero signature, then the signature set $\{f, g\}$ is not vanishing.*

Proof. Let arity$(f) = n$ and $\text{rd}^+(f) = d$, so $2d < n$. Since f is not identically 0, we have $d \geq 0$. By Lemma 7.8, $\hat{f} = (Z^{-1})^{\otimes n} f = [\hat{f}_0, \ldots, \hat{f}_d, 0, \ldots, 0]$, where $\hat{f}_d \neq 0$. By forming d self-loops on f we get a nonzero multiple of $Z^{\otimes(n-2d)}[\hat{f}_d, 0, \ldots, 0]$, which is a nonzero multiple of $h = Z^{\otimes(n-2d)}[1, 0]^{\otimes(n-2d)}$. Similarly, from g of arity n' and $\text{rd}^-(g) = d'$, where $2d' < n'$, we can get a nonzero multiple of $h' = Z^{\otimes(n'-2d')}[0, 1]^{\otimes(n'-2d')}$.

Then form a bipartite signature grid with $n' - 2d'$ vertices on one side, each assigned h and $n - 2d$ vertices on the other side, each assigned h'. Connect edges between the two sides arbitrarily as long as it is a 1-1 correspondence. Under the Z-transformation this is equivalent to connecting $n' - 2d'$ copies of $[1, 0]^{\otimes(2(n-2d))}$ with $n - 2d$ copies of $[0, 1]^{\otimes(2(n'-2d'))}$, via $(\neq_2)$. The resulting Holant is clearly nonzero. □

Lemma 7.10. *Every symmetric vanishing signature is in* $\mathscr{V}^+ \cup \mathscr{V}^-$.

Proof. Let f be a symmetric vanishing signature. We prove the lemma by induction on n, the arity of f. For $n = 1$, by connecting $f = [f_0, f_1]$ to itself, we have $f_0^2 + f_1^2 = 0$. Then up to a constant factor, we have either $f = [1, \mathrm{i}]$ or $f = [1, -\mathrm{i}]$. The lemma holds.

For $n = 2$, first we do a self-loop. The Holant value is $f_0 + f_2$. Also, we can connect two copies of f, in which case the Holant value is $f_0^2 + 2f_1^2 + f_2^2$. Since f is vanishing, both $f_0 + f_2 = 0$ and $f_0^2 + 2f_1^2 + f_2^2 = 0$. Solving them, we get $f = [1, \mathrm{i}, -1] = [1, \mathrm{i}]^{\otimes 2}$ or $[1, -\mathrm{i}, -1] = [1, -\mathrm{i}]^{\otimes 2}$ up to a constant factor.

Now assume $n > 2$ and the lemma holds for any signature of arity $k < n$. Let $f = Z^{\otimes n}\hat{f} = Z^{\otimes n}[\hat{f}_0, \hat{f}_1, \ldots, \hat{f}_n]$ be a vanishing signature. Let f' be the signature of arity $n - 2$ by taking a self-loop on f. This is equivalent to forming a self-loop via $(\neq_2)$ in the Z basis, namely $f' = Z^{\otimes(n-2)}[\hat{f}_1, \ldots, \hat{f}_{n-1}]$ up to a nonzero scalar.

If f' is identically 0, then $f = Z^{\otimes n}[\hat{f}_0, 0, \ldots, 0, \hat{f}_n]$. We form a graph with two vertices, both assigned f, and n edges between them. The value of this Holant problem can be calculated in Holant$(\neq_2 | \hat{f})$ and is clearly $2\hat{f}_0\hat{f}_n$. As f is vanishing, it must be 0. Hence either $\hat{f}_0 = 0$ in which case $f \in \mathscr{V}^-$, or $\hat{f}_n = 0$ in which case $f \in \mathscr{V}^+$.

Now suppose f' is not identically 0. Since f is vanishing, f' is vanishing as well. By induction, $f' \in \mathscr{V}^+ \cup \mathscr{V}^-$, and without loss of generality we assume $f' \in \mathscr{V}^+$. By Lemma 7.8, $f' = Z^{\otimes(n-2)}[\hat{f}_1, \ldots, \hat{f}_d, 0, \ldots, 0]$, where $0 \le d - 1 = \mathrm{rd}^+(f') < (n-2)/2$, and $\hat{f}_d \neq 0$. Then $f = Z^{\otimes n}[\hat{f}_0, \hat{f}_1, \ldots, \hat{f}_d, 0, \ldots, 0, \hat{f}_n]$, and we only need to prove $\hat{f}_n = 0$. Suppose otherwise.

By forming $d - 1$ self-loops on f', we get $Z^{\otimes(n-2d)}[1, 0]^{\otimes(n-2d)}$, or $[1, \mathrm{i}]^{\otimes t}$, up to a nonzero constant, where $t = n - 2d$. This construction is valid because $n > 2d$ and $d - 1 \ge 0$.

Define $g = Z^{\otimes(n-1)}[\hat{f}_1, \ldots, \hat{f}_n]$, which is the signature of connecting f with a single unary signature $[1, \mathrm{i}]$. Note that $[1, \mathrm{i}] = Z[1, 0]$ connected via $(\neq_2)$ to $f = Z^{\otimes n}\hat{f}$ simply removes the $\hat{f}_0$ term in the Z-form. We do not claim that g is constructible from f; however, we show that $\{f, g\}$ is vanishing, in particular $\{f', g\}$ is also vanishing. Let Ω be any signature grid on a graph $G = (V, E)$

using $\{f, g\}$. We show that the Holant value $\text{Holant}_\Omega = 0$. Suppose g occurs m times in Ω. Define a new signature grid Ω' on a graph G' having $t|V| + m$ vertices as follows. For each i $(1 \le i \le t)$ we make a copy G_i of G on $V_i = \{v_i \mid v \in V\}$, i.e., for each edge $(u, v) \in E$, include the edge (u_i, v_i) in the new graph. Every vertex in $\bigcup_{i=1}^t V_i$ is assigned the same signature f in Ω'. Next, for each vertex $v \in V$ assigned g, we introduce an additional vertex v', and assign to v' the degenerate signature $[1, \mathfrak{i}]^{\otimes t}$ that we just constructed. Then connect v' to each of $v_1, \ldots, v_t$. This effectively changes the signature at each of these v_i to g. Thus $\text{Holant}_{\Omega'} = (\text{Holant}_\Omega)^t$. Since Ω' is a signature grid on $\{f\}$ and f is vanishing, we have $\text{Holant}_{\Omega'} = 0$. Hence $\text{Holant}_\Omega = 0$.

We have shown that $\{f', g\}$ is vanishing; in particular g is vanishing. By induction $g \in \mathscr{V}^+ \cup \mathscr{V}^-$. We had assumed $\hat{f}_n \ne 0$, and so both f' and g are not identically 0. Lemma 7.9 applies, and we get $g \in \mathscr{V}^+$, since $f' \in \mathscr{V}^+$. Hence $\hat{f}_n = 0$ by the Z-form of g, and thus $f \in \mathscr{V}^+$ Lemma 7.8. □

Combining Lemmas 7.7, 7.9, and 7.10, we obtain the following theorem that characterizes all symmetric vanishing signature sets.

Theorem 7.11. *Let $\mathcal{F}$ be a set of symmetric signatures. Then $\mathcal{F}$ is vanishing if and only if $\mathcal{F} \subseteq \mathscr{V}^+$ or $\mathcal{F} \subseteq \mathscr{V}^-$.*

To finish this section, we prove some useful properties regarding vanishing degrees and recurrence degrees. For two symmetric signatures f and g such that $\text{arity}(f) \ge \text{arity}(g)$, let

$$\langle f, g \rangle = \langle g, f \rangle$$

denote the signature that results after connecting all edges of g to f. (If $\text{arity}(f) = \text{arity}(g)$, then $\langle f, g \rangle$ is a constant, which is considered as a signature of arity 0.)

Lemma 7.12. *For $\sigma \in \{+, -\}$, suppose symmetric signatures f and g satisfy* $\text{vd}^\sigma(g) = 0$ *and* $\text{arity}(f) - \text{arity}(g) \ge \text{rd}^\sigma(f)$. *Then* $\text{rd}^\sigma(\langle f, g \rangle) = \text{rd}^\sigma(f)$.

Proof. We consider $\sigma = +$ since the case $\sigma = -$ is similar. Let $\text{arity}(f) = n$, $\text{arity}(g) = m$, and $\text{rd}^+(f) = d$. Denote the signature $\langle f, g \rangle$ by h.

If $d = -1$, then f is identically 0 and so is h. Hence $\text{rd}^+(h) = -1$.

Suppose $d \ge 0$. By Lemma 7.8, $\text{vd}^\sigma(g) = 0$ is equivalent to $\hat{g}_m \ne 0$ in $g = \mathrm{Z}^{\otimes m}\hat{g} = \mathrm{Z}^{\otimes m}[\hat{g}_0, \ldots, \hat{g}_m]$. We also have $f = \mathrm{Z}^{\otimes n}[\hat{f}_0, \ldots, \hat{f}_d, 0, \ldots, 0]$, with $\hat{f}_d \ne 0$. Then h has the form $\mathrm{Z}^{\otimes (n-m)}[\hat{h}_0, \ldots, \hat{h}_{n-m}]$ of arity $n - m \ge d$, where

$$\hat{h}_i = \sum_{j=0}^{m} \binom{m}{j} \hat{f}_{i+j}\hat{g}_{m-j}, \quad (0 \le i \le n - m).$$

For $i > d$, all terms in the sum are 0 because $\hat{f}_{i+j} = 0$. For $i = d$, this is a nonzero sum because for $j = 0$, $f_{i+j} = f_d \neq 0$ exists, and so does $\hat{g}_m \neq 0$. However, all other terms for $j > 0$ in the sum for $\hat{h}_d$ are 0. Hence $\mathrm{rd}^+(h) = d$, by Lemma 7.8. □

Lemma 7.13. *For $\sigma \in \{+, -\}$, let f be a nonzero symmetric signature and suppose that f' is obtained from f by a self-loop. If $\mathrm{vd}^\sigma(f) > 0$, then $\mathrm{vd}^\sigma(f) - \mathrm{vd}^\sigma(f') = \mathrm{rd}^\sigma(f) - \mathrm{rd}^\sigma(f') = 1$.*

Proof. By Lemma 7.8, $\mathrm{vd}^+(f) > 0$ iff we have $\hat{f}_n = 0$ in the expression $f = Z^{\otimes n}[\hat{f}_0, \ldots, \hat{f}_n]$. The corresponding expression for f' is $Z^{\otimes n}[\hat{f}_1, \ldots, \hat{f}_{n-1}]$, which has one fewer trailing 0 after the last nonzero entry $\hat{f}_d$, where $d = \mathrm{rd}^+(f)$. Hence $\mathrm{vd}^+(f) - \mathrm{vd}^+(f') = 1$. Since $\mathrm{rd}^+(f) = n - \mathrm{vd}^+(f)$ and $\mathrm{rd}^+(f') = (n-2) - \mathrm{vd}^+(f')$, it follows that $\mathrm{rd}^+(f) - \mathrm{rd}^+(f') = 1$. □

Moreover, the set of vanishing signatures is closed under orthogonal transformations. This is because under any orthogonal transformation, the unary signatures $[1, \mathrm{i}]$ and $[1, -\mathrm{i}]$ are either invariant or transformed into each other, up to a nonzero multiple. Then considering the symmetrized form of any signature, we have the following lemma.

Lemma 7.14. *For a symmetric signature f of arity n, $\sigma \in \{+, -\}$, and an orthogonal matrix $T \in \mathbb{C}^{2\times 2}$, either $\mathrm{vd}^\sigma(f) = \mathrm{vd}^\sigma(T^{\otimes n} f)$ or $\mathrm{vd}^\sigma(f) = \mathrm{vd}^{-\sigma}(T^{\otimes n} f)$.*

7.1.3 Characterization by Recurrence Relations

Now we give another characterization of vanishing signatures.

Definition 7.15. An arity n symmetric signature of the form $f = [f_0, f_1, \ldots, f_n]$ is in $\mathscr{R}_t^+$ for a nonnegative integer $t \geq 0$ if

1. $t > n$; or
2. For any $0 \leq k \leq n - t$, $f_k, \ldots, f_{k+t}$ satisfy the recurrence relation of order t

$$\binom{t}{t}\mathrm{i}^t f_{k+t} + \binom{t}{t-1}\mathrm{i}^{t-1} f_{k+t-1} + \cdots + \binom{t}{0}\mathrm{i}^0 f_k = 0. \qquad (7.4)$$

We define $\mathscr{R}_t^-$ similarly but with $-\mathrm{i}$ in place of i in (7.4).

We note that some particular categories of tractable cases in previous dichotomies are in $\mathscr{R}_2^\pm$. This includes case 3 of Theorem 6.3, the case of $I(1, \pm 2\mathrm{i})$ in Class $\mathfrak{B}$ of Theorem 2.12, and the subcase of type $I(1, \pm 2\mathrm{i})$ in case 1 of Theorem 6.12.

It is easy to see that $\mathscr{R}_0^+ = \mathscr{R}_0^-$ is the set of all zero signatures (of any arity). Also, for $\sigma \in \{+, -\}$, we have $\mathscr{R}_t^\sigma \subseteq \mathscr{R}_{t'}^\sigma$ when $t \le t'$. By definition, if arity$(f) = n$ then $f \in \mathscr{R}_{n+1}^\sigma$.

Let $f = [f_0, f_1, \ldots, f_n] \in \mathscr{R}_t^+$ with $0 < t \le n$. Then the characteristic polynomial of its recurrence relation is $(1 + x\mathrm{i})^t$. Thus there exists a polynomial $p(x)$ of degree $\le t - 1$ such that $f_k = \mathrm{i}^k p(k)$, for $0 \le k \le n$. This statement extends to $\mathscr{R}_{n+1}^+$ since a polynomial of degree n can interpolate any set of $n + 1$ values. Furthermore, such an expression is unique. If there are two polynomials $p(x)$ and $q(x)$, both of degree $\le n$, such that $f_k = \mathrm{i}^k p(k) = \mathrm{i}^k q(k)$ for $0 \le k \le n$, then $p(x)$ and $q(x)$ must be the same polynomial. Now suppose $f_k = \mathrm{i}^k p(k)$ $(0 \le k \le n)$ for some polynomial p of degree $\le t - 1$, where $0 < t \le n$. Then f satisfies the recurrence (7.4) of order t. Hence $f \in \mathscr{R}_t^+$.

Thus $f \in \mathscr{R}_t^+$ iff there exists a polynomial $p(x)$ of degree $\le t - 1$ such that $f_k = \mathrm{i}^k p(k)$ $(0 \le k \le n)$, for all $1 \le t \le n + 1$. For $\mathscr{R}_t^-$, just replace i by $-\mathrm{i}$.

It follows that the set of all symmetric signatures of arity n in $\mathscr{R}_t^\sigma$ forms a vector space over $\mathbb{C}$ of dimension t, where $\sigma \in \{+, -\}$.

Lemma 7.16. *Let $f = [f_0, \ldots, f_n]$ be a symmetric signature of arity n, not identically 0. Then for any nonnegative integer $0 \le d < n$ and $\sigma \in \{+, -\}$, the following are equivalent:*

1. $\mathrm{rd}^\sigma(f) \le d$
2. $f \in \mathscr{R}_{d+1}^\sigma$.

Proof. We prove the case $\sigma = +$; the case $\sigma = -$ is symmetric.

By Lemma 7.8, the set V_n^d of symmetric signatures of arity n with $\mathrm{rd}^+(f) \le d$ consists of all expressions of the form $f = Z^{\otimes n}[\hat{f}_0, \ldots, \hat{f}_d, 0, \ldots, 0]$. Hence it forms a vector space over $\mathbb{C}$.

Let $\mathrm{Ex}_n^t = [0, \ldots, 0, 1, 0, \ldots, 0]$ be the Exact-t function of arity n, where $0 \le t \le n$, and the unique nonzero entry 1 appears at weight t. For $0 \le t \le d$, these are linearly independent, and since the linear transformation on the space of symmetric signatures of arity n induced by $Z^{\otimes n}$ is invertible, the image $\{Z^{\otimes n}\mathrm{Ex}_n^t \mid 0 \le t \le d\}$ forms a basis of V_n^d. Thus V_n^d has dimension $d + 1$, the same as that of $\mathscr{R}_{d+1}^+$. We only need to prove that each $Z^{\otimes n}\mathrm{Ex}_n^t \in \mathscr{R}_{d+1}^+$, for $0 \le t \le d$.

We prove this by showing that $Z^{\otimes n}\mathrm{Ex}_n^t$ can be expressed as $p(k)\mathrm{i}^k$, where p is a polynomial of degree $\le t$.

We can express Ex_n^t as a sum of $\binom{n}{t}$ terms of the form $v_1 \otimes v_2 \otimes \ldots \otimes v_n$ where $v_i \in \mathbb{C}^2$ $(1 \le i \le n)$ and exactly t of them are $[0, 1]$ and $n - t$ of them are $[1, 0]$. Then up to a nonzero scalar, $Z^{\otimes n}\mathrm{Ex}_n^t$ is a sum of $\binom{n}{t}$ terms of the form $u_1 \otimes u_2 \otimes \ldots \otimes u_n$ where $u_i \in \mathbb{C}^2$ and exactly t of them are $[1, -\mathrm{i}]$ and $n - t$

of them are $[1, \mathfrak{i}]$. The signature entry of $Z^{\otimes n}\mathrm{Ex}_n^t$ at weight k $(0 \le k \le n)$ is

$$\sum_s \binom{k}{s}\binom{n-k}{t-s}(-\mathfrak{i})^s \mathfrak{i}^{k-s}. \tag{7.5}$$

This is because the $\binom{n}{t}$ terms of $Z^{\otimes n}\mathrm{Ex}_n^t$ are each uniquely determined by the locations of the t factors $[1, -\mathfrak{i}]$. On input $(\underbrace{1, \ldots, 1}_{k}, \underbrace{0, \ldots, 0}_{n-k})$, a term with some s of the first k bit positions and $t-s$ of the last $n-k$ bit positions occupied by $[1, -\mathfrak{i}]$ contributes $(-\mathfrak{i})^s\mathfrak{i}^{k-s}$. The sum in (7.5) is over the range of s satisfying both $0 \le s \le k$ and $0 \le t-s \le n-k$, i.e., $\max\{0, k+t-n\} \le s \le \min\{k, t\}$. This is justified by the combinatorial argument above. However, we will rewrite the sum (7.5) as over the range $0 \le s \le t$, and note that if $s > k$ then the first factor $\binom{k}{s} = 0$, and if $s < k+t-n$ then the second factor $\binom{n-k}{t-s} = 0$. Hence (7.5) can be written as

$$\mathfrak{i}^k \sum_{0 \le s \le t} \binom{k}{s}\binom{n-k}{t-s}(-1)^s. \tag{7.6}$$

It is easy to see that, for $0 \le s \le t$, each $\binom{k}{s}\binom{n-k}{t-s}$ is a polynomial in k of degree t. The conclusion follows. □

Note a slight subtlety in the proof of Lemma 7.16. The combinatorial justification for the sum over s results in a sum whose range depends on k. In order to be rigorous in arguing that the sum over s is a polynomial in k of degree $\le t$, we must replace the range of this sum to be independent of k.

Corollary 7.17. *For a nonzero symmetric signature f of arity n, $\mathrm{rd}^+(f) = t$, if and only if $f \in \mathscr{R}_{t+1}^+ - \mathscr{R}_t^+$. Similarly $\mathrm{rd}^-(f) = t$ if and only if $f \in \mathscr{R}_{t+1}^- - \mathscr{R}_t^-$.*

Remark: Membership in $\mathscr{R}_t^\pm$ is easily decidable by (7.4). Therefore by Lemma 7.16, for any symmetric signature f, its recurrence degree $\mathrm{rd}^\pm(f)$ can be computed in polynomial time in the size of its symmetric signature. This is one advantage of using $\mathscr{R}_t^\pm$ compared to the concept of vanishing degree $\mathrm{vd}^\pm$, while conceptually the latter relates more directly to the notion of vanishing. Lemmas 7.8 and 7.16 show that the two concepts are equivalent. By Lemma 7.16 membership in $\mathscr{V}^\pm$ is decidable in polynomial time in the size of a symmetric signature. This also follows from Lemma 7.8 because the Z-transformation is also computable for a symmetric signature in polynomial time in the size of a symmetric signature. However, checking for concrete signatures by the recurrence (7.4) in Definition 7.15 is more straightforward.

When the results in this chapter were first obtained, we relied heavily in the concept of recurrence degree, which was defined not by Definition 7.5, but in terms of (7.4) and Corollary 7.17 (made into a definition). The reorganization of the proofs in this chapter emphasizes the Z-form as the primary tool. This in hindsight simplifies many proof details. We thank Artem Govorov and Heng Guo especially for discussions that led to some new and simplified proofs; in particular the simplified proof of Lemma 7.16 is due to Govorov, which we present here with his permission.

Note that although we call it the recurrence degree, it refers to a particular recurrence relation. For any nonzero symmetric signature f, by the uniqueness of the representing polynomial $p(x)$, it follows that $\mathrm{rd}^\sigma(f) = t$ iff $\deg(p) = t$, where $0 \le t \le n$. We remark that $\mathrm{rd}^\sigma(f)$ is the maximum integer t such that f does *not* belong to $\mathscr{R}_t^\sigma$. Also, for an arity n signature f, $\mathrm{rd}^\sigma(f) = n$ if and only if f does not satisfy any recurrence relation (7.4) of order $t \le n$ for $\sigma \in \{+, -\}$.

7.2 Theorem Statement and Proof of Tractability

For a class $\mathscr{C}$ of signatures we define the notion of $\mathscr{C}$-transformable in the context of Holant problems. We will particularly use this notion for $\mathscr{C} = \mathscr{A}$ or $\mathscr{C} = \mathscr{P}$.

Definition 7.18. A signature f (respectively a signature set $\mathcal{F}$) is $\mathscr{C}$-transformable if there exists a holographic transformation T such that $f \in T\mathscr{C}$ (respectively the set $\mathcal{F} \subseteq T\mathscr{C}$) and $[1, 0, 1]T^{\otimes 2} \in \mathscr{C}$.

Now we can formally state the dichotomy theorem for Holant problems on an arbitrary set of symmetric signatures.

Theorem 7.19. *Let $\mathcal{F}$ be any set of symmetric, complex-valued signatures in Boolean variables. Then* Holant($\mathcal{F}$) *is* #P*-hard unless $\mathcal{F}$ satisfies one of the following conditions, in which case the problem is in P:*

1. *All nondegenerate signatures in $\mathcal{F}$ are of arity ≤ 2.*
2. *$\mathcal{F}$ is $\mathscr{A}$-transformable.*
3. *$\mathcal{F}$ is $\mathscr{P}$-transformable.*
4. $\mathcal{F} \subseteq \mathscr{V}^\sigma \cup \{f \in \mathscr{R}_2^\sigma \mid \mathrm{arity}(f) = 2\}$ *for some $\sigma \in \{+, -\}$.*
5. *All nondegenerate signatures in $\mathcal{F}$ are in $\mathscr{R}_2^\sigma$ for some $\sigma \in \{+, -\}$.*

Note that any signature in $\mathscr{R}_2^\sigma$ having arity ≥ 3 is a vanishing signature. Thus all nondegenerate signatures of arity ≥ 3 in case 5 are vanishing. Although both cases 4 and 5 are largely concerned with vanishing signatures, these two cases differ. In case 4, all signatures in $\mathcal{F}$ including unary signatures, but excluding binary signatures, must be vanishing of a single type σ; the binary signatures are only required to be in $\mathscr{R}_2^\sigma$. In contrast, case 5 has no requirement placed on degenerate signatures which include all unary signatures. Instead case 5 requires that all nondegenerate signatures of arity ≥ 2 belong to $\mathscr{R}_2^\sigma$. Note that $f \in \mathscr{R}_2^\sigma$ is equivalent to $\mathrm{rd}^\sigma \leq 1$, and therefore for any signature f of arity ≥ 3 this implies that $f \in \mathscr{V}^\sigma$. For large arity $n \geq 3$, this is a strong form of vanishing, meaning that it has a large vanishing degree $\mathrm{vd}^\sigma \geq n-1$.

Case 5 is tractable due to Fibonacci gates. In fact every nondegenerate signature in $\mathscr{R}_2^\sigma$ is a generalized Fibonacci signature with the parameter $\lambda = \sigma 2\mathfrak{i}$ in equation (2.5), and after replacing all degenerate signatures with the corresponding unary signatures we can invoke Holant* tractability. However, in the following we present a unified proof of tractability based on vanishing signatures, which leads to an alternative algorithm for this case.

Proof of Tractability of Theorem 7.19. For any signature grid Ω, Holant_Ω is the product of the Holant on each connected component, so we only need to compute over connected components.

For case 1, after decomposing all degenerate signatures into unary ones, a connected component of the graph is either a path or a cycle and the Holant can be computed using matrix product and trace. Cases 2 and 3 are tractable because, after a particular holographic transformation, their instances are tractable instances of #CSP($\mathcal{F}$), by the dichotomy theorem for #CSP, Theorem 3.7. For case 4, any binary signature $g \in \mathscr{R}_2^\sigma$ has $\mathrm{rd}^\sigma(g) \leq 1$, and thus $\mathrm{vd}^\sigma(g) \geq 1 = \mathrm{arity}(g)/2$. Any signature $f \in \mathscr{V}^\sigma$ has $\mathrm{vd}^\sigma(f) > \mathrm{arity}(f)/2$. If $\mathcal{F}$ contains a signature f of arity ≥ 3, then it must belong to $\mathscr{V}^\sigma$. Then by the combinatorial view, more than half of the unary signatures are $[1, \sigma\mathfrak{i}]$, so Holant_Ω vanishes. On the other hand, if the arity of every signature in $\mathcal{F}$ is ≤ 2, then we have reduced to tractable case 1.

Now consider case 5. First, replace each degenerate signature by its unary signature tensor factors. Any remaining signature of arity ≥ 2 is nondegenerate and thus belongs to $\mathscr{R}_2^\sigma$. Then recursively absorb any unary signature into its neighboring signature. If it is connected to another unary signature, then this produces a global constant factor. If it is connected to a binary signature, then this creates another unary signature. We observe that if $f \in \mathscr{R}_2^\sigma$ has $\mathrm{arity}(f) \geq 2$, then for any unary signature u, after connecting f to u, the signature $\langle f, u\rangle$ still belongs to $\mathscr{R}_2^\sigma$. Hence after recursively absorbing all unary signatures in the above process, we have a signature grid where all signatures belong to $\mathscr{R}_2^\sigma$. Any signature f that remains and has arity ≥ 3 belongs to $\mathscr{V}^\sigma$ since $\mathrm{rd}^\sigma(f) \leq 1$ and thus $\mathrm{vd}^\sigma(f) \geq \mathrm{arity}(f) - 1 > \mathrm{arity}(f)/2$. Thus we have reduced to tractable case 4. □

7.3 A Sample of Problems

We illustrate the scope of Theorem 7.19 by several concrete problems. Note that some problems are naturally expressed with real weights, but they are linked to other problems that use complex weights. Sometimes the inherent link between two real-weighted problems is provided by a transformation through $\mathbb{C}$.

Problem: #VERTEXCOVER

Input: An undirected graph G

Output: The number of vertex covers in G

This classic problem is most naturally expressed as a spin system with edge function being the Boolean OR function, or as a counting CSP problem #CSP([0, 1, 1]). As a real-weighted bipartite Holant problem it is Holant([0, 1, 1] | $\mathcal{EQ}$), where $\mathcal{EQ} = \{(=_k) \mid k \geq 1\}$ is the set of all EQUALITIES. A vertex assigned an equality signature forces all its incident edges to be assigned the same value; this is equivalent to assigning a value to the vertex. The degree 2 vertices assigned the binary OR = [0, 1, 1] should be thought of as an edge between its neighboring vertices. These edge-like vertices force at least one of its neighbors to be selected. The number of assignments satisfying these requirements is exactly the number of vertex covers.

To apply Theorem 7.19, we perform a holographic transformation by $T = \begin{bmatrix} 0 & -\mathrm{i} \\ 1 & \mathrm{i} \end{bmatrix}$. Note that $T^{-1} = \begin{bmatrix} 1 & 1 \\ \mathrm{i} & 0 \end{bmatrix}$, and

$$\begin{aligned}[0, 1, 1] = (0\ 1\ 1\ 1) = [1, 1]^{\otimes 2} + [\mathrm{i}, 0]^{\otimes 2} &= \left\{[1, 0]^{\otimes 2} + [0, 1]^{\otimes 2}\right\}(T^{-1})^{\otimes 2} \\ &= (1\ 0\ 0\ 1)(T^{-1})^{\otimes 2} = (=_2)(T^{-1})^{\otimes 2}.\end{aligned}$$

Thus, a holographic transformation by T yields

$$\begin{aligned}\mathrm{Holant}([0, 1, 1] \mid \mathcal{EQ}) &\equiv_{\mathrm{T}} \mathrm{Holant}([0, 1, 1]T^{\otimes 2} \mid T^{-1}\mathcal{EQ}) \\ &\equiv_{\mathrm{T}} \mathrm{Holant}(=_2 \mid T^{-1}\mathcal{EQ}) \\ &\equiv_{\mathrm{T}} \mathrm{Holant}(T^{-1}\mathcal{EQ}).\end{aligned}$$

The EQUALITY signature $(=_k)$ of arity k in $\mathcal{EQ}$ is transformed by T^{-1} to

$$\begin{aligned} f_{(k)} &= (T^{-1})^{\otimes k}(=_k) \\ &= \begin{bmatrix} 1 & 1 \\ \mathrm{i} & 0 \end{bmatrix}^{\otimes k} \left\{ \begin{bmatrix} 1 \\ 0 \end{bmatrix}^{\otimes k} + \begin{bmatrix} 0 \\ 1 \end{bmatrix}^{\otimes k} \right\} \\ &= \begin{bmatrix} 1 \\ \mathrm{i} \end{bmatrix}^{\otimes k} + \begin{bmatrix} 1 \\ 0 \end{bmatrix}^{\otimes k} \\ &= [2, \mathrm{i}, -1, -\mathrm{i}, 1, \mathrm{i}, -1, -\mathrm{i}, 1, \mathrm{i}, \dots] \end{aligned}$$

of arity k. By Theorem 7.19, Holant($T^{-1}\mathcal{EQ}$) is #P-hard. Indeed, even Holant($f_{(k)}$), the restriction of this problem to k-regular graphs is #P-hard, for $k \geq 3$. This can also be obtained more directly by Theorem 7.52, a dichotomy for a single signature, which is a special case of Theorem 7.19.

Problem: #λ-VERTEXCOVER

Input: An undirected graph G

Output: $\sum_{C \in \mathcal{C}(G)} \lambda^{e(C)}$, where $\mathcal{C}(G)$ denotes the set of all vertex covers of G, and $e(C)$ is the number of edges with both endpoints in the vertex cover C

Our dichotomy also easily handles this edge-weighted vertex cover problem that is denoted by Holant($[0, 1, \lambda] \mid \mathcal{EQ}$). Suppose $\lambda \neq 0$. On regular graphs, this problem is equivalent to the so-called *hardcore gas model*, which is the vertex-weighted problem denoted by Holant($[1, 1, 0] \mid \mathcal{F}$), where $\mathcal{F}$ consists of signatures of the form $[1, 0, \dots, 0, \mu]$. By flipping 0 and 1, this is the same as Holant($[0, 1, 1] \mid \mathcal{F}'$) with $\mathcal{F}'$ containing $[\mu, 0, \dots, 0, 1]$. For k-regular graphs, the following diagonal transformation $T = \begin{bmatrix} 1 & 0 \\ 0 & \frac{1}{\lambda} \end{bmatrix}$ shows this equivalence, where $\lambda = 1/\mu^{1/k}$;

$$\begin{aligned} \text{Holant}([0, 1, \lambda] \mid =_k) &\equiv_T \text{Holant}([0, 1, \lambda]T^{\otimes 2} \mid (T^{-1})^{\otimes k}(=_k)) \\ &\equiv_T \text{Holant}(\tfrac{1}{\lambda}[0, 1, 1] \mid [1, 0, \dots, 0, \lambda^k]) \\ &\equiv_T \text{Holant}([0, 1, 1] \mid [\mu, 0, \dots, 0, 1]). \end{aligned}$$

This problem, denoted by #k-λ-VERTEXCOVER, is #P-hard for $k \geq 3$. To see this, apply the holographic transformation $T = \begin{bmatrix} 0 & -\mathrm{i}\lambda \\ 1 & \mathrm{i} \end{bmatrix}$ to the edge-weighted form of the problem Holant($[0, 1, \lambda] \mid =_k$). Then $[0, 1, \lambda]$ is transformed to $[0, 1, \lambda]T^{\otimes 2} = \lambda(=_2)$ and $(=_k)$ is transformed to

$(T^{-1})^{\otimes k}(=_k) = g_{(\lambda,k)} = \frac{1}{\lambda^k}[\lambda^k + 1, \mathrm{i}, -1, -\mathrm{i}, 1, \dots]$, where $T^{-1} = \frac{1}{\lambda}\begin{bmatrix} 1 & \lambda \\ \mathrm{i} & 0 \end{bmatrix}$. Since Holant($g_{(\lambda,k)}$) is #P-hard by Theorem 7.19, or more directly by Theorem 7.52, we conclude that #k-λ-VERTEXCOVER is also #P-hard.

If $\lambda = 0$, then the above problem is Holant($[0, 1, 0] \mid \mathcal{EQ}$), which is tractable. However, the transformation T above is singular in this case. We can in fact apply another transformation $T' = \begin{bmatrix} 1 - \frac{\lambda}{2} & -\left(1 + \frac{\lambda}{2}\right)\mathrm{i} \\ 1 & \mathrm{i} \end{bmatrix}$ such that it transforms the problem Holant($[0, 1, \lambda] \mid =_k$) into Holant($h_{(\lambda,k)}$) for some $h_{(\lambda,k)}$ regardless of whether $\lambda = 0$ or not. Then by applying Theorem 7.52, we reach the same conclusion that #λ-VERTEXCOVER is #P-hard on k-regular graphs when $\lambda \neq 0$. We note that when $\lambda = 0$, $T' = \begin{bmatrix} 1 & -\mathrm{i} \\ 1 & \mathrm{i} \end{bmatrix} = \sqrt{2}Z^{-1}$, where $Z = \frac{1}{\sqrt{2}}\begin{bmatrix} 1 & 1 \\ \mathrm{i} & -\mathrm{i} \end{bmatrix}$.

We now consider some orientation problems.

Problem: #NOSINKORIENTATION

Input: An undirected graph G

Output: The number of orientations of G such that each vertex has at least one outgoing edge.

This problem is denoted by Holant($[0, 1, 0] \mid \mathcal{F}$), where $\mathcal{F}$ consists of $f_{(k)} = [0, 1, \dots, 1, 1]$ for any arity k. Each degree 2 vertex on the left side of the bipartite graph must have different values assigned to its incident edges. We associate an oriented edge between the neighbors of such vertices with the head on the side assigned 0 and the tail on the side assigned 1. This problem is #P-hard even over k-regular graphs provided $k \geq 3$. Just as with the bipartite form of the vertex cover problem, we do a holographic transformation to apply Theorem 7.19. This time, we pick $T = \frac{1}{2}\begin{bmatrix} 1 & -\mathrm{i} \\ 1 & \mathrm{i} \end{bmatrix} = \frac{1}{\sqrt{2}}Z^{-1}$, with $T^{-1} = \sqrt{2}Z = \begin{bmatrix} 1 & 1 \\ \mathrm{i} & -\mathrm{i} \end{bmatrix}$ and get

$$\begin{aligned} \text{Holant}([0, 1, 0] \mid f_{(k)}) &\equiv_T \text{Holant}([0, 1, 0]T^{\otimes 2} \mid (T^{-1})^{\otimes k} f_{(k)}) \\ &\equiv_T \text{Holant}(\tfrac{1}{2}[1, 0, 1] \mid \hat{f}_{(k)}) \\ &\equiv_T \text{Holant}(\hat{f}_{(k)}), \end{aligned}$$

where $\hat{f}_{(k)} = [2^k - 1, -\mathrm{i}, 1, \mathrm{i}, -1, \dots]$. The signature $-\hat{f}_{(k)} = [1 - 2^k, \mathrm{i}, -1, -\mathrm{i}, 1, \dots]$ is actually a special case of the #k-λ-VERTEXCOVER problem with $\lambda = 2e^{\pi \mathrm{i}/k}$. Therefore, this problem is #P-hard. However, if we consider this problem modulo 2^k, $\hat{f}_{(k)}$ becomes $[-1, -\mathrm{i}, 1, \mathrm{i}, -1, \dots]$, and it belongs to one of the tractable cases in Theorem 7.19. Thus, #NoSinkOrientation is tractable modulo 2^t, where t is the minimum degree of the input graph.

Problem: #NoSinkNoSourceOrientation

Input: An undirected graph G

Output: The number of orientations of G such that each vertex has at least one incoming and one outgoing edge

This problem is denoted by Holant($[0, 1, 0] \mid \mathcal{F}$), where $\mathcal{F}$ consists of $f_{(k)} = [0, 1, \dots, 1, 0]$ for any arity k. This problem is also #P-hard on k-regular graphs for $k \geq 3$. We pick the same T as in the previous problem and get

$$\begin{aligned} \text{Holant}([0,1,0] \mid f_{(k)}) &\equiv_{\mathrm{T}} \text{Holant}([0,1,0]T^{\otimes 2} \mid (T^{-1})^{\otimes k} f_{(k)}) \\ &\equiv_{\mathrm{T}} \text{Holant}(\tfrac{1}{2}[1,0,1] \mid \hat{f}_{(k)}) \\ &\equiv_{\mathrm{T}} \text{Holant}(\hat{f}_{(k)}), \end{aligned}$$

where $\hat{f}_{(k)} = [2^k - 2, 0, 2, 0, -2, \dots]$. Here we transform from one real-weighted Holant problem to another real-weighted Holant problem via a complex-weighted transformation. The #P-hardness follows from Theorem 7.52. Like the previous problem, this problem is tractable modulo 2^t, where t is the minimum degree of the input graph.

Theorem 7.19 is applicable to a set of signatures when different vertices may have different constraints.

Problem: #One-In-Or-One-Out-Orientation

Input: An undirected graph G with each vertex labeled "One-In" or "One-Out"

Output: The number of orientations of G such that each vertex has exactly one incoming or exactly one outgoing edge as specified by its label

This problem is denoted by Holant($[0, 1, 0] \mid \mathcal{F}$), where the set $\mathcal{F}$ consists of signatures of the form $f = [0, 1, 0, \dots, 0]$ and $g = [0, \dots, 0, 1, 0]$. Once again, it is #P-hard on k-regular graphs for $k \geq 3$. We apply the same transformation as in the above two orientation problems. The result is Holant($\{\hat{f}, \hat{g}\}$), where $\hat{f} = [k, (k-2)\mathrm{i}, -(k-4), \dots]$ and $\hat{g} = [k, -(k-2)\mathrm{i}, -(k-4), \dots]$ of arity k. In fact, the entries of $\hat{f}$ satisfy a second-order recurrence relation with characteristic polynomial $(x - \mathrm{i})^2$ while the entries of $\hat{g}$ satisfy one with characteristic polynomial $(x + \mathrm{i})^2$. Thus $\hat{f} \in \mathscr{R}_2^+$ and $\hat{g} \in \mathscr{R}_2^-$. The hardness follows from Theorem 7.19. See also Lemma 7.30. However, the restriction of this problem to planar graphs is tractable since these signatures are all matchgate signatures, by Theorem 4.11. If we consider only one signature, then both Holant($\hat{f}$) and Holant($\hat{g}$) are tractable. This is also true for a set of signatures with different arities but of one type, either of the type $\hat{f}$ or of the type $\hat{g}$. This is case 5 of Theorem 7.19. One can also observe that the problem Holant($\hat{f}$) is equivalent to the problem Holant($[0, 1, 0] \mid [0, 1, 0, \ldots, 0]$), which is always 0 provided $k \geq 3$ by a simple counting argument. Similarly for Holant($\hat{g}$). For $k \geq 3$, $\hat{f} \in \mathscr{V}^+$ and $\hat{g} \in \mathscr{V}^-$ are vanishing signatures. However, combining two such vanishing signatures of the opposite type $\mathscr{V}^+$ and $\mathscr{V}^-$, the problem Holant($\{\hat{f}, \hat{g}\}$) is #P-hard.

One sufficient condition for a signature of arity ≥ 3 to be vanishing is that its entries satisfy a second-order recurrence relation with characteristic polynomial $(x \pm \mathrm{i})^2$. If the entries of a nondegenerate signature f satisfy a second-order recurrence relation with characteristic polynomial $(x - \mu)^2$ for $\mu \neq \pm\mathrm{i}$, then there exists an orthogonal holographic transformation such that f is transformed into a weighted matching signature.

Problem: #λ-WEIGHTEDMATCHING

Input: An undirected graph G

Output: $\sum_{M \in \mathcal{M}(G)} \lambda^{v(M)}$, where $\mathcal{M}(G)$ is the set of all matchings in G and $v(M)$ is the number of unmatched vertices in the matching M

The Holant expression of this problem is Holant($\mathcal{F}$), where $\mathcal{F}$ consists of signatures of the form $[\lambda, 1, 0, \ldots, 0]$. When $\lambda = 0$, this problem counts perfect matchings, which is #P-hard even for bipartite graphs [Val79a] but tractable over planar graphs by Kasteleyn's algorithms (Theorem 4.7). When $\lambda = 1$, this problem counts general matchings. Vadhan [Vad01] proved that counting general matchings is #P-hard over k-regular graphs for $k \geq 5$, but left open the question for $k = 4$. Theorem 7.19, or more directly Theorem 7.52, shows that #λ-WEIGHTEDMATCHING is #P-hard, for any weight λ and on any k-regular graphs for $k \geq 3$.

7.4 Outline of Hardness Proof for Theorem 7.19

The hardness proof of Theorem 7.19 will take up the remainder of this chapter. Before proving Theorem 7.19 which deals with a set of signatures $\mathcal{F}$, we first prove a special case Theorem 7.52, where $\mathcal{F}$ consists of a single signature. The proof is by induction on the arity of the signature. The induction is done by taking a self-loop, which causes the arity to go down by 2. Thus, we need two base cases, a dichotomy for an arity 3 signature, Theorem 6.3, and a dichotomy

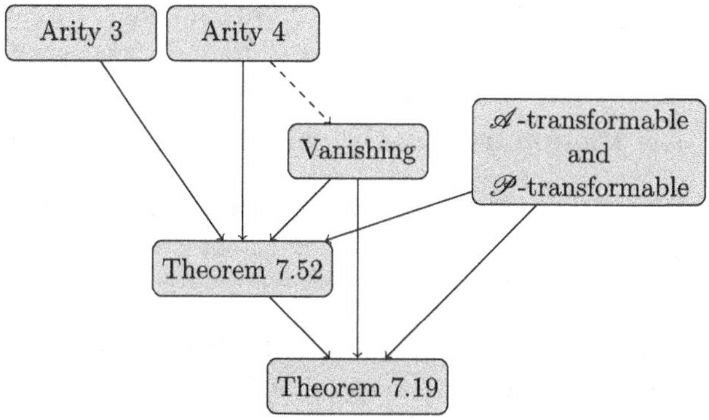

Figure 7.1. Dependency graph of key hardness results for our main dichotomy, Theorem 7.19. The dashed edge indicates a dependence in terms of techniques rather than the result itself. "Arity 3(4)" stands for the arity 3(4) single signature dichotomy. "Vanishing" ("𝒜-transformable and 𝒫-transformable") stands for the lemmas regarding vanishing (𝒜-transformable and 𝒫-transformable) signatures. Dependencies on previous dichotomy theorems are not shown.

for an arity 4 signature, Theorem 7.29, which is a crucial ingredient in our proof of Theorem 7.19. The arity 4 dichotomy not only is a base case of the single signature dichotomy Theorem 7.52, but also is utilized several times in the inductive step.

After obtaining the dichotomy for an arity 4 signature, the proof continues by revisiting the vanishing signatures to determine what signatures combine with them to give #P-hardness. When adding unary or binary signatures, the only possible combinations that maintain the tractability of the vanishing signatures are as described in cases 4 and 5 in Theorem 7.19. Moreover, combining two vanishing signatures of the opposite type of arity ≥ 3 implies #P-hardness. The proof of this last statement uses techniques that are similar to those in the proof of the arity 4 dichotomy.

Another ingredient of the proof is to understand the signatures that are $\mathscr{A}$-transformable or $\mathscr{P}$-transformable. We obtain explicit characterizations of these signatures. We use these characterizations to prove dichotomy theorems for any signature set containing an $\mathscr{A}$- or $\mathscr{P}$-transformable signature. Unless every signature in the set is $\mathscr{A}$- or $\mathscr{P}$-transformable, the problem is #P-hard. The proofs of these dichotomy theorems utilize the #CSP^d dichotomy, Theorem 6.16.

The main dichotomy, Theorem 7.19, depends on Theorem 7.52 and the results regarding vanishing signatures as well as $\mathscr{A}$- and $\mathscr{P}$-transformable signatures. Figure 7.1 summarizes the dependencies among these results. The material in the next four sections is technical. It may be helpful to take a glance now of the *Proof of hardness in Theorem 7.19* at the end of Section 7.8. This may provide some understanding for the organization of the material.

7.5 Dichotomy for One Signature of Arity 4

Definition 7.20. A 4×4 matrix is *redundant* if its middle two rows and middle two columns are the same. Denote the set of all redundant 4×4 matrices over $\mathbb{C}$ by $\mathrm{RM}_4(\mathbb{C})$.

Recall the notion of a signature matrix in (6.19) of a signature of arity 4. As an example, the signature matrix in (6.20) of a symmetric arity 4 signature is redundant.

Define a function $\varphi : \mathbb{C}^{4\times 4} \to \mathbb{C}^{3\times 3}$ by

$$\varphi(M) = AMB,$$

where

$$A = \begin{bmatrix} 1 & 0 & 0 & 0 \\ 0 & \frac{1}{2} & \frac{1}{2} & 0 \\ 0 & 0 & 0 & 1 \end{bmatrix} \quad \text{and} \quad B = \begin{bmatrix} 1 & 0 & 0 \\ 0 & 1 & 0 \\ 0 & 1 & 0 \\ 0 & 0 & 1 \end{bmatrix}.$$

Intuitively, the operation φ replaces the middle two columns of M with their sum and then the middle two rows of M with their average. (These two steps commute.) In the other direction, we define the following function $\psi : \mathbb{C}^{3\times 3} \to \mathrm{RM}_4(\mathbb{C})$ by

$$\psi(N) = BNA.$$

Intuitively, the operation ψ duplicates the middle row of N and then splits the middle column evenly into two columns. Notice that $\varphi(\psi(N)) = N$. When restricted to $\mathrm{RM}_4(\mathbb{C})$, φ is an isomorphism between the semigroup of 4×4 redundant matrices and the semigroup of 3×3 matrices, under matrix multiplication, and ψ is its inverse. This can be seen by noticing that

$$AB = \begin{bmatrix} 1 & 0 & 0 \\ 0 & 1 & 0 \\ 0 & 0 & 1 \end{bmatrix} \quad \text{and} \quad BA = \begin{bmatrix} 1 & 0 & 0 & 0 \\ 0 & \frac{1}{2} & \frac{1}{2} & 0 \\ 0 & \frac{1}{2} & \frac{1}{2} & 0 \\ 0 & 0 & 0 & 1 \end{bmatrix}$$

are the identity elements of their respective semigroups.

Definition 7.21. If g is a signature of arity 4 with a redundant signature matrix M_g, then we define the *compressed signature matrix* of g as $\widetilde{M_g} = \varphi(M_g)$.

If all signatures in an $\mathcal{F}$-gate have even arity, then the $\mathcal{F}$-gate also has even arity. Knowing that binary signatures alone in Holant problems do not produce #P-hardness (see tractable case 1 of Theorem 7.19), with the above constraint in mind, we would like to interpolate other arity 4 signatures using a given arity 4 signature. We are particularly interested in the following signature g with signature matrix

$$M_g = \begin{bmatrix} 1 & 0 & 0 & 0 \\ 0 & \frac{1}{2} & \frac{1}{2} & 0 \\ 0 & \frac{1}{2} & \frac{1}{2} & 0 \\ 0 & 0 & 0 & 1 \end{bmatrix}. \tag{7.7}$$

This is the identity element in the semigroup $\text{RM}_4(\mathbb{C})$. Thus, $\widetilde{M_g} = I_3$. Lemma 7.24 will show that the Holant problem with this signature is #P-hard. In Lemma 7.23, we consider when we can interpolate it. This is a key lemma. There are three cases in Lemma 7.23 and one of them requires the following technical lemma.

Lemma 7.22. *Let $M = [B_0\, B_1\, \cdots\, B_t]$ be an $n \times n$ block matrix such that there exists $\lambda \in \mathbb{C}$, for all integers $0 \le k \le t$, block B_k is an $n \times c_k$ matrix for some integer $c_k \ge 0$, and the entry of a nonempty block B_k (for $c_k \ge 1$) at row r and column c is $(B_k)_{rc} = r^{c-1}\lambda^{kr}$, where $1 \le r \le n$, $1 \le c \le c_k$. If λ is nonzero and is not a root of unity, then M is nonsingular.*

Proof. We prove by induction on n. If $n = 1$, then the sole entry is λ^k for some nonnegative integer k. This is nonzero since $\lambda \neq 0$. Assume $n > 1$ and let the leftmost nonempty block be B_j. We divide row r by λ^{jr}, which is allowed since $\lambda \neq 0$. This effectively changes block B_ℓ into a block of the form $B_{\ell-j}$. Thus, we have another matrix of the same form as M but with a nonempty block B_0. To simplify notation, we denote this matrix again by M. The first column of B_0 is all 1's. We subtract row $r-1$ from row r, for r from n down to 2. This gives us a new matrix $M' = [B'_0\, B'_1\, \cdots\, B'_t]$, and $\det M = \det M'$. Then $\det M'$ is the determinant of the $(n-1) \times (n-1)$ submatrix M'' obtained from M' by removing the first row and column. Now we do column operations on M'' to return the blocks to the proper form so that we can apply induction.

For any nonempty block B'_k other than B'_0, we prove by induction on the number of columns in B'_k that B'_k can be repaired. In the base case, the rth element of the first column is $(B'_k)_{r1} = \lambda^{kr} - \lambda^{k(r-1)} = \lambda^{k(r-1)}(\lambda^k - 1)$ for $r \ge 2$. We divide this column by $\lambda^k - 1$ to obtain $\lambda^{k(r-1)}$, which is allowed since λ is not a root of unity and $k \neq 0$. This is now the correct form for the r'-th element of the first column of a block in M'', for the row index $r' = r - 1 \ge 1$.

Now for the inductive step of B'_k, assume that the first $d-1$ columns of the block B'_k have been repaired to the correct form to be a block in M''. That is, for row index $r \geq 2$, which denotes the $(r-1)$-th row of M'', the rth element in the first $d-1$ columns of B'_k has the form $(B'_k)_{rc} = (r-1)^{c-1}\lambda^{k(r-1)}$. The rth element in column d of B'_k currently has the form $(B'_k)_{rd} = r^{d-1}\lambda^{kr} - (r-1)^{d-1}\lambda^{k(r-1)}$. Then we do column operations

$$\begin{aligned}(B'_k)_{rd} - \sum_{c=1}^{d-1}\binom{d-1}{c-1}(B'_k)_{rc} &= r^{d-1}\lambda^{kr} - (r-1)^{d-1}\lambda^{k(r-1)} \\ &\quad - \sum_{c=1}^{d-1}\binom{d-1}{c-1}(r-1)^{c-1}\lambda^{k(r-1)} \\ &= r^{d-1}\lambda^{kr} - r^{d-1}\lambda^{k(r-1)} \\ &= r^{d-1}\lambda^{k(r-1)}(\lambda^k - 1)\end{aligned}$$

and divide by $(\lambda^k - 1)$ to get $r^{d-1}\lambda^{k(r-1)}$. Once again, this is allowed since λ is not a root of unity and $k \neq 0$. Then more column operations of the same kind yield

$$\begin{aligned}&r^{d-1}\lambda^{k(r-1)} - \sum_{c=1}^{d-1}\binom{d-1}{c-1}(r-1)^{c-1}\lambda^{k(r-1)} \\ &= \lambda^{k(r-1)}\left(r^{d-1} + (r-1)^{d-1} - \sum_{c=1}^{d}\binom{d-1}{c-1}(r-1)^{c-1}\right)\end{aligned}$$

and the expression inside the large parentheses on the RHS is precisely $(r-1)^{d-1}$. This gives the correct form for the rth element in column d of B'_k in M''.

Finally we repair the columns in B'_0, also by induction on the number of columns. In the base case, if B'_0 has only one column, then there is nothing to prove, since this block has disappeared in M''. Otherwise, $(B'_0)_{r2} = r - (r-1) = 1$, so the second column is already in the correct form to be the first column in M'', and there is still nothing to prove. For the inductive step, assume that columns 2 to $d-1$ are in the correct form to be the first block in M'' for $d \geq 3$. That is, the entry at row $r \geq 2$ and column c from 2 through $d-1$ has the form $(B'_0)_{rc} = (r-1)^{c-2}$. The rth element in column d currently has the form $(B'_0)_{rd} = r^{d-1} - (r-1)^{d-1}$. Then we do the column operations

$$\begin{aligned}(B'_0)_{rd} - \sum_{c=2}^{d-1}\binom{d-1}{c-2}(B'_0)_{rc} &= r^{d-1} - (r-1)^{d-1} - \sum_{c=2}^{d-1}\binom{d-1}{c-2}(r-1)^{c-2} \\ &= (d-1)(r-1)^{d-2}\end{aligned}$$

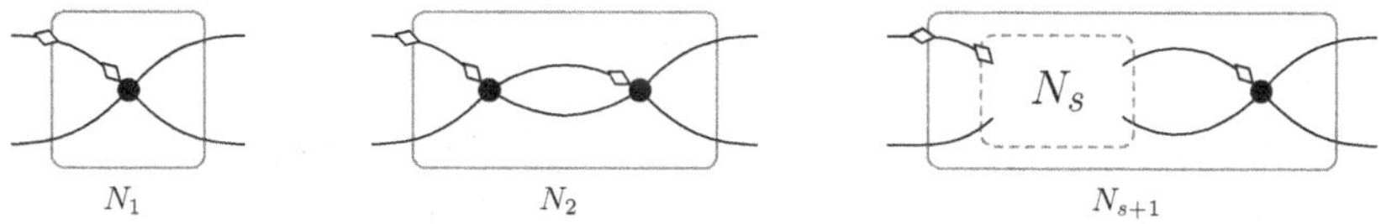

Figure 7.2. Recursive construction to interpolate g. The vertices are assigned f. For not necessarily symmetric signatures we place a diamond on the edge corresponding to the most significant index bit. The remaining index bits are in order of decreasing significance as one travels counterclockwise around the vertex.

and divide by $d-1$, which is nonzero, to get $(r-1)^{d-2}$. Recall that the block B_0' in M'' is indexed by r and d starting at 2. This is the correct form for the rth element in column d of B_0' in M''. Therefore, we invoke our original induction hypothesis that the $(n-1) \times (n-1)$ matrix M'' has a nonzero determinant, which completes the proof. □

Lemma 7.23. *Let g be the arity 4 signature with M_g given in Equation (7.7) and let f be an arity 4 signature with complex weights. If M_f is redundant and $\widetilde{M_f}$ is nonsingular, then for any set $\mathcal{F}$ containing f, we have*

$$\text{Holant}(\mathcal{F} \cup \{g\}) \leq_T \text{Holant}(\mathcal{F}).$$

The proof also works for planar graphs, thus,

$$\text{Pl-Holant}(\mathcal{F} \cup \{g\}) \leq_T \text{Pl-Holant}(\mathcal{F}).$$

Proof. Consider an instance Ω of Holant$(\mathcal{F} \cup \{g\})$. Suppose that g appears n times in Ω. We construct from Ω a sequence of instances Ω_s of Holant$(\mathcal{F})$ indexed by $s \geq 1$. We obtain Ω_s from Ω by replacing each occurrence of g with the gadget N_s in Figure 7.2 with f assigned to all vertices. In Ω_s, the edge corresponding to the ith significant index bit of N_s connects to the same location as the edge corresponding to the ith significant index bit of g in Ω. Since f may not be symmetric, in Figure 7.2 we place a diamond on the edge corresponding to the most significant index bit. The remaining index bits are in order of decreasing significance as one travels counterclockwise around the vertex.

Now to determine the relationship between Holant$_\Omega$ and Holant$_{\Omega_s}$, we use the isomorphism between RM$_4(\mathbb{C})$ and $\mathbb{C}^{3\times 3}$. To compute Holant$_{\Omega_s}$, we effectively replace M_g in Holant$_\Omega$ with $M_{N_s} = (M_f)^s$, the sth power of the signature matrix M_f. By the Jordan normal form of $\widetilde{M_f}$, there exists T and $\Lambda \in \mathbb{C}^{3\times 3}$ such that

$$\widetilde{M_f} = T\Lambda T^{-1} = T \begin{bmatrix} \lambda_1 & b_1 & 0 \\ 0 & \lambda_2 & b_2 \\ 0 & 0 & \lambda_3 \end{bmatrix} T^{-1},$$

where $b_1, b_2 \in \{0, 1\}$. By assumption $\lambda_1\lambda_2\lambda_3 = \det(\widetilde{M_f}) \neq 0$. Also since $\widetilde{M_g} = \varphi(M_g) = I_3$, and $TI_3T^{-1} = I_3$, we have $\psi(T)M_g\psi(T^{-1}) = M_g$. We can view our construction of Ω_s as first replacing each M_g by $\psi(T)M_g\psi(T^{-1})$, which does not change the Holant value, and then replacing each new M_g, sandwiched between $\psi(T)$ and $\psi(T^{-1})$, by $\psi(\Lambda^s) = \psi(\Lambda)^s$ to obtain Ω_s. Observe that

$$\begin{aligned}\varphi(\psi(T)\psi(\Lambda^s)\psi(T^{-1})) &= T\Lambda^sT^{-1} = (\widetilde{M_f})^s = (\varphi(M_f))^s \\ &= \varphi((M_f)^s) = \varphi(M_{N_s}),\end{aligned}$$

and both $\psi(T)\psi(\Lambda^s)\psi(T^{-1})$ and M_{N_s} are redundant, hence $\psi(T)\psi(\Lambda^s)\psi(T^{-1}) = M_{N_s}$. Since $M_g = \psi(T)M_g\psi(T^{-1})$ and $M_{N_s} = \psi(T)\psi(\Lambda^s)\psi(T^{-1})$, replacing each M_g, sandwiched between $\psi(T)$ and $\psi(T^{-1})$, by $\psi(\Lambda^s)$ indeed transforms Ω to Ω_s. (We note that, by the isomorphism, $\psi(T^{-1})$ *is* the multiplicative inverse of $\psi(T)$ within the semigroup $\text{RM}_4(\mathbb{C})$; but we prefer not to write it as $\psi(T)^{-1}$ since it is not the usual matrix inverse as a 4×4 matrix. Indeed, $\psi(T)$ is not invertible as a 4×4 matrix.)

In the analysis that follows, we stratify assignments in Ω_s based on assignment values to $\psi(\Lambda^s)$. The inputs to $\psi(\Lambda^s)$ are from $\{0, 1\}^2 \times \{0, 1\}^2$. However, we can combine the inputs 01 and 10, since $\psi(\Lambda^s)$ is redundant. Thus we actually stratify assignments in Ω_s based on assignment values to Λ^s, which are from $\{0, 1, 2\} \times \{0, 1, 2\}$. In this compressed form, the row and column assignments to Λ^s are the Hamming weight of the two actual inputs from $\{0, 1\}^2$ to the uncompressed form $\psi(\Lambda^s)$.

Now we begin the analysis based on the values of b_1 and b_2, i.e., the shape Λ takes as the Jordan normal form of $\widetilde{M_f}$.

1. $b_1 = b_2 = 0$. In this case $\widetilde{M_f}$ is diagonalizable, we only need to consider the assignments to Λ^s that assign
   - $(0, 0)$ i many times,
   - $(1, 1)$ j many times, and
   - $(2, 2)$ k many times,

   such that $i + j + k = n$, the number of times g occurs in Ω, which is also the number of times the gadget N_s occurs in Ω_s, for all $s \geq 1$. Note that any other assignment contributes 0 to the Holant sum. Also note that

$$\psi(\Lambda^s) = \begin{bmatrix} \lambda_1^s & 0 & 0 & 0 \\ 0 & \lambda_2^s/2 & \lambda_2^s/2 & 0 \\ 0 & \lambda_2^s/2 & \lambda_2^s/2 & 0 \\ 0 & 0 & 0 & \lambda_3^s \end{bmatrix}.$$

Let c_{ijk} be the sum over all such assignments of the products of evaluations (excluding Λ^s but including the contributions from T and T^{-1}) on Ω_s. Note that c_{ijk} is independent of s. Then by the form of M_g in (7.7), we have

$$\text{Holant}_\Omega = \sum_{i+j+k=n} \frac{c_{ijk}}{2^j}$$

and the value of the Holant on Ω_s, for $s \geq 1$, is

$$\text{Holant}_{\Omega_s} = \sum_{i+j+k=n} \left(\lambda_1^i \lambda_2^j \lambda_3^k\right)^s \left(\frac{c_{ijk}}{2^j}\right),$$

where the factor $\frac{1}{2}$ comes from the middle four entries in M_g in (7.7) and in $\psi(\Lambda^s)$. The coefficient matrix is Vandermonde, but it may not have full rank because it may be that $\lambda_1^i \lambda_2^j \lambda_3^k = \lambda_1^{i'} \lambda_2^{j'} \lambda_3^{k'}$ for some $(i, j, k) \neq (i', j', k')$. However, this is not a problem since we are interested only in the sum $\sum c_{ijk}/2^j$. If two coefficients are the same, we replace their corresponding unknowns $\frac{c_{ijk}}{2^j}$ and $\frac{c_{i'j'k'}}{2^{j'}}$ with their sum as a new variable. After all such combinations, we have a Vandermonde system of full rank. In particular, none of the entries are zero since $\lambda_1\lambda_2\lambda_3 = \det(\widetilde{M_f}) \neq 0$. Therefore, we can solve the linear system in polynomial time and obtain the value of Holant_Ω.

2. $b_1 \neq b_2$. We can assume that the Jordan blocks in Λ have been permuted so that $b_1 = 1$ and $b_2 = 0$. Then $\lambda_1 = \lambda_2$, and we denote it by λ. Denote λ_3 by μ. We have

$$\Lambda^s = \begin{bmatrix} \lambda^s & s\lambda^{s-1} & 0 \\ 0 & \lambda^s & 0 \\ 0 & 0 & \mu^s \end{bmatrix}, \text{ and } \psi(\Lambda^s) = \begin{bmatrix} \lambda^s & s\lambda^{s-1}/2 & s\lambda^{s-1}/2 & 0 \\ 0 & \lambda^s/2 & \lambda^s/2 & 0 \\ 0 & \lambda^s/2 & \lambda^s/2 & 0 \\ 0 & 0 & 0 & \mu^s \end{bmatrix}.$$

We only need to consider the assignments to Λ^s that assign
– $(0, 0)$ i many times,
– $(1, 1)$ j many times,
– $(2, 2)$ k many times, and
– $(0, 1)$ ℓ many times,
with $i + j + k + \ell = n$, since any other assignment contributes 0 to the Holant sum. Let $c_{ijk\ell}$ be the sum over all such assignments of the products of evaluations (excluding Λ^s but including the contributions from T and T^{-1}) on Ω_s; $c_{ijk\ell}$ is independent of s. Then

$$\text{Holant}_\Omega = \sum_{i+j+k=n} \frac{c_{ijk0}}{2^j} \tag{7.8}$$

and the value of the Holant on Ω_s, for $s \geq 1$, is

$$\begin{aligned}\text{Holant}_{\Omega_s} &= \sum_{i+j+k+\ell=n} \lambda^{(i+j)s} \mu^{ks} \left(s\lambda^{s-1}\right)^{\ell} \left(\frac{c_{ijk\ell}}{2^{j+\ell}}\right) \\ &= \lambda^{ns} \sum_{i+j+k+\ell=n} \left(\frac{\mu}{\lambda}\right)^{ks} s^{\ell} \left(\frac{c_{ijk\ell}}{\lambda^{\ell} 2^{j+\ell}}\right). \end{aligned} \tag{7.9}$$

If μ/λ is a root of unity, then take a t such that $(\mu/\lambda)^t = 1$. Then

$$\text{Holant}_{\Omega_{st}} = \lambda^{nst} \sum_{i+j+k+\ell=n} s^{\ell} \left(\frac{t^{\ell} c_{ijk\ell}}{\lambda^{\ell} 2^{j+\ell}}\right).$$

For $s \geq 1$, this gives a coefficient matrix $\left(\left(\frac{st}{2\lambda}\right)^{\ell}\right)$ (with rows indexed by s and columns indexed by the tuple (i, j, k, ℓ)) that is Vandermonde. Although this system is not full rank, we can replace all the unknowns $c_{ijk\ell}/2^j$ having $i + j + k = n - \ell$ by their sum to form a new unknown $c'_{\ell} = \sum_{i+j+k=n-\ell} c_{ijk\ell}/2^j$, where $0 \leq \ell \leq n$. The unknown c'_0 is the Holant of Ω that we seek. The resulting Vandermonde system

$$\text{Holant}_{\Omega_{st}} = \lambda^{nst} \sum_{\ell=0}^{n} \left(\frac{st}{2\lambda}\right)^{\ell} c'_{\ell}$$

has full rank, so we can solve for the unknowns in polynomial time and obtain the value of $c'_0 = \mathrm{e} \sum_{i+j+k=n} \frac{c_{ijk0}}{2^j}$.

If μ/λ is not a root of unity, then we replace all the unknowns $c_{ijk\ell}/(\lambda^{\ell} 2^{j+\ell})$ having $i + j = m$ with their sum to form new unknowns $c'_{mk\ell}$, for any $0 \leq m, k, \ell$ and $m + k + \ell = n$. Then by (7.8) and (7.9),

$$\text{Holant}_{\Omega} = \sum_{m+k=n} c'_{mk0} \quad \text{and} \quad \text{Holant}_{\Omega_s} = \lambda^{ns} \sum_{m+k+\ell=n} \left(\frac{\mu}{\lambda}\right)^{ks} s^{\ell} c'_{mk\ell}.$$

After a suitable ordering of the columns, the matrix of coefficients whose rows are indexed by s satisfies the hypothesis of Lemma 7.22. Therefore, the linear system has full rank. We can solve for the unknowns in polynomial time and obtain the value of Holant_{Ω}.

3. $b_1 = b_2 = 1$. In this case, we have $\lambda_1 = \lambda_2 = \lambda_3$, denoted by λ. We have

$$\Lambda^s = \begin{bmatrix} \lambda^s & s\lambda^{s-1} & \frac{s(s-1)}{2}\lambda^{s-2} \\ 0 & \lambda^s & s\lambda^{s-1} \\ 0 & 0 & \lambda^s \end{bmatrix},$$

and

$$\psi(\Lambda^s) = \begin{bmatrix} \lambda^s & \frac{s}{2}\lambda^{s-1} & \frac{s}{2}\lambda^{s-1} & \frac{s(s-1)}{2}\lambda^{s-2} \\ 0 & \frac{\lambda^s}{2} & \frac{\lambda^s}{2} & s\lambda^{s-1} \\ 0 & \frac{\lambda^s}{2} & \frac{\lambda^s}{2} & s\lambda^{s-1} \\ 0 & 0 & 0 & \lambda^s \end{bmatrix}.$$

We only need to consider the assignments to Λ^s that assign

- $(0, 0)$ or $(2, 2)$ i many times,
- $(1, 1)$ j many times,
- $(0, 1)$ k many times,
- $(1, 2)$ ℓ many times, and
- $(0, 2)$ m many times,

with $i + j + k + \ell + m = n$, since any other assignment contributes 0 to the Holant sum. Let $c_{ijk\ell m}$ be the sum over all such assignments of the products of evaluations (excluding Λ^s but including the contributions from T and T^{-1}) on Ω_s; $c_{ijk\ell m}$ is independent of s. Then

$$\text{Holant}_\Omega = \sum_{i+j=n} \frac{c_{ij000}}{2^j}$$

and the value of the Holant on Ω_s, for $s \geq 1$, is

$$\begin{aligned} \text{Holant}_{\Omega_s} &= \sum_{i+j+k+\ell+m=n} \lambda^{(i+j)s} \left(s\lambda^{s-1}\right)^{k+\ell} \left(s(s-1)\lambda^{s-2}\right)^m \left(\frac{c_{ijk\ell m}}{2^{j+k+m}}\right) \\ &= \lambda^{ns} \sum_{i+j+k+\ell+m=n} s^{k+\ell+m}(s-1)^m \left(\frac{c_{ijk\ell m}}{\lambda^{k+\ell+2m}2^{j+k+m}}\right). \end{aligned}$$

We replace all the unknowns $c_{ijk\ell m}/(\lambda^{k+\ell+2m}2^{j+k+m})$ having $i + j = p$ and $k + \ell = q$ with their sum to form new unknowns c'_{pqm}, for any $0 \leq p, q, m$ and $p + q + m = n$. The Holant of Ω is now c'_{n00}. This new linear system is

$$\text{Holant}_{\Omega_s} = \lambda^{ns} \sum_{p+q+m=n} s^{q+m}(s-1)^m c'_{pqm}.$$

But this linear system is still rank deficient. We now index the columns by the tuple (q, m), where $q \geq 0$, $m \geq 0$, and $q + m \leq n$. Correspondingly, we rename the variables $x_{q,m} = c'_{pqm}$. Note that $p = n - q - m$ is determined by (q, m). Observe that the column indexed by (q, m) is the sum of the columns indexed by $(q - 1, m)$ and $(q - 2, m + 1)$ provided $q - 2 \geq 0$. Namely, $s^{q+m}(s-1)^m = s^{q-1+m}(s-1)^m + s^{q-2+m+1}(s-1)^{m+1}$. Of course this is meaningful only if $q \geq 2, m \geq 0$ and $q + m \leq n$. Let

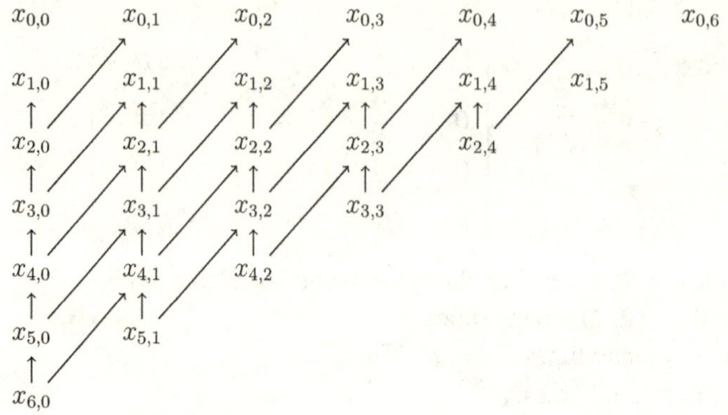

Figure 7.3. A triangular table of variables and their update patterns, for $n = 6$. Rows are indexed by q and columns are indexed by m.

$a_{s,(q,m)} = s^{q+m}(s-1)^m$. Then the linear system is

$$\sum_{q \geq 0,\ m \geq 0,\ q+m \leq n} a_{s,(q,m)} x_{q,m} = \frac{\text{Holant}_{\Omega_s}}{\lambda^{ns}}. \tag{7.10}$$

For any fixed $s \geq 1$, we write $\alpha_{q,m} = a_{s,(q,m)}$, then $\alpha_{q,m} x_{q,m} = (\alpha_{q-1,m} + \alpha_{q-2,m+1}) x_{q,m}$. Therefore we can eliminate variable $x_{q,m}$ by adding its value to $x_{q-1,m}$ and to $x_{q-2,m+1}$. More precisely we define new variables

$$x_{q-1,m} \leftarrow x_{q-1,m} + x_{q,m}$$

$$x_{q-2,m+1} \leftarrow x_{q-2,m+1} + x_{q,m}$$

for every $0 \leq m \leq n-2$ and for q from $n-m$ down to 2. The order of these updates is important. Once a variable $x_{q,m}$ has been added to $x_{q-1,m}$ and $x_{q-2,m+1}$, there should not be any update involving $x_{q,m}$, which has been eliminated. One valid sequence is a double nested loop, with the outer loop being "For $m = 0$ to $n-2$" and for each m in the range there is the inner loop "For $q = n-m$ down to 2". This is illustrated schematically for $n = 6$ in Figure 7.3.

Observe that in each update, some variable $x_{q,m}$ is added to a variable $x_{q',m'}$, where $q' < q$ (a strictly lower index value of q), and $m' \geq m$ (an index value that is $\geq m$). Thus in the order by the double nested loop, once $x_{q,m}$ has been added to $x_{q-1,m}$ and $x_{q-2,m+1}$, there will be no more updates involving $x_{q,m}$.

A more crucial observation is that the column in the linear system (7.10) indexed by the tuple $(q, m) = (0, 0)$ is never changed. This is because, in order to be an updated entry, there must be some $q \geq 2$ and $m \geq 0$ such that

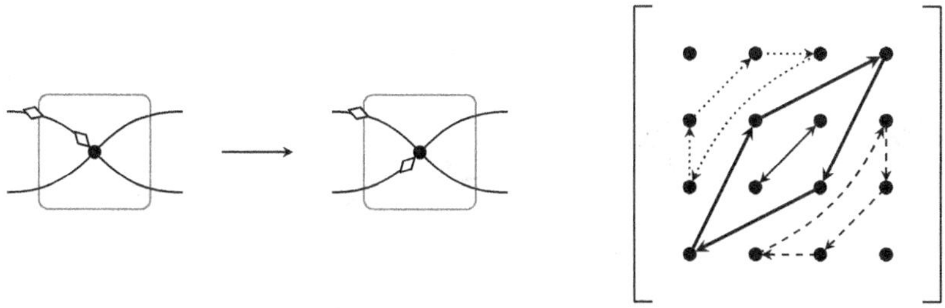

(a) A counterclockwise rotation.

(b) Movement of signature matrix entries under a counterclockwise rotation.

Figure 7.4. The movement of the entries in the signature matrix of an arity 4 signature under a counterclockwise rotation of the input edges. The Hamming weight 1 entries are in the dotted cycle, the Hamming weight 2 entries are in the two solid cycles (one has length 4 and the other one is a swap), and the entries of Hamming weight 3 are in the dashed cycle.

$(q-1, m) = (0, 0)$ or $(q-2, m+1) = (0, 0)$, which is clearly impossible. Hence $x_{0,0} = c'_{n00}$ is still the Holant value on Ω. The $2n+1$ unknowns that remain are

$$x_{0,0},\ x_{1,0},\ x_{0,1},\ x_{1,1},\ x_{0,2},\ x_{1,2},\ \ldots,\ x_{0,n-1},\ x_{1,n-1},\ x_{0,n}$$

and their coefficients in row s are

$$1, s, s(s-1), s^2(s-1), s^2(s-1)^2, \ldots, s^{n-1}(s-1)^{n-1},$$
$$s^n(s-1)^{n-1}, s^n(s-1)^n.$$

For $0 \le \delta \le 2n$, it is clear that the δth entry in this row is a monic polynomial in s of degree δ, and thus s^δ is a linear combination of the first $\delta + 1$ entries indexed by 0 to δ. It follows that the coefficient matrix is a product of the standard Vandermonde matrix multiplied to its right by an upper triangular matrix with all 1's on the diagonal. Hence the matrix is nonsingular, and we can solve the linear system in polynomial time and, in particular, compute c'_{n00}. □

For an asymmetric signature, we often want to reorder the input bits under a circular permutation. For a single counterclockwise rotation of 90°, the effect on the entries of the signature matrix of an arity 4 signature is given in Figure 7.4.

We will derive most of our #P-hardness results through Lemma 7.23 and the following Lemma 7.24. This is done by a reduction from the problem of

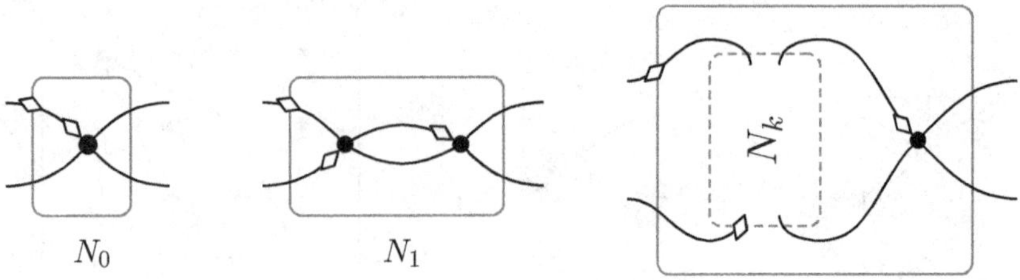

Figure 7.5. Recursive construction to approximate $[1, 0, \frac{1}{3}, 0, 1]$. The vertices are assigned g.

counting Eulerian orientations on 4-regular graphs, which is the Holant problem $\text{Holant}((\neq_2) \mid [0, 0, 1, 0, 0])$. In Theorem 6.32 we have shown that this problem #EO is #P-hard even for planar graphs. Recall that under a holographic transformation by $\begin{bmatrix} 1 & 1 \\ \mathfrak{i} & -\mathfrak{i} \end{bmatrix}$, this bipartite Holant problem becomes the Holant problem $\text{Holant}([1, 0, 1/3, 0, 1])$, up to a nonzero factor.

Lemma 7.24. *Let g be the arity 4 signature with M_g given in Equation (7.7) so that $\widetilde{M_g} = I_3$. Then* Holant(g) *is* #P*-hard. In fact* Pl-Holant(g) *is* #P*-hard.*

Proof. We reduce from the #EO problem $\text{Holant}(\mathscr{O})$, where $\mathscr{O} = [1, 0, 1/3, 0, 1]$, which is #P-hard by Theorem 6.32 even for 4-regular planar graphs. We achieve this via an arbitrarily close approximation using the recursive construction in Figure 7.5 with g assigned to every vertex.

We claim that the signature matrix M_{N_k} of gadget N_k is

$$M_{N_k} = \begin{bmatrix} 1 & 0 & 0 & a_k \\ 0 & a_{k+1} & a_{k+1} & 0 \\ 0 & a_{k+1} & a_{k+1} & 0 \\ a_k & 0 & 0 & 1 \end{bmatrix},$$

where $a_k = \frac{1}{3} - \frac{1}{3}\left(-\frac{1}{2}\right)^k$. This is true for N_0. Inductively assume M_{N_k} has this form. Then the rotated form of the signature matrix for N_k, as described in Figure 7.4, is

$$\begin{bmatrix} 1 & 0 & 0 & a_{k+1} \\ 0 & a_k & a_{k+1} & 0 \\ 0 & a_{k+1} & a_k & 0 \\ a_{k+1} & 0 & 0 & 1 \end{bmatrix}. \tag{7.11}$$

The action of g on the far right side of N_{k+1} in Figure 7.5 is to replace each of the middle two entries in the middle two rows of the matrix in (7.11) with their average, $(a_k + a_{k+1})/2 = a_{k+2}$. This gives $M_{N_{k+1}}$.

Let G be a graph with n vertices and $H_{\mathcal{O}}$ (resp. H_{N_k}) be the Holant value on G with all vertices assigned $\mathcal{O}$ (resp. N_k). Since each signature entry in $\mathcal{O}$ can be expressed as a rational number with denominator 3, each term in the sum of $H_{\mathcal{O}}$ can be expressed as a rational number with denominator 3^n, and $H_{\mathcal{O}}$ itself is a sum of 2^{2n} such terms, where $2n$ is the number of edges in G. If the error $|H_{N_k} - H_{\mathcal{O}}|$ is $\leq 1/3^{n+1}$, then we can recover $H_{\mathcal{O}}$ from H_{N_k} by selecting the nearest rational number to H_{N_k} with denominator 3^n.

For each signature entry x in $M_{\mathcal{O}}$, its corresponding entry $\tilde{x}$ in M_{N_k} satisfies $|\tilde{x} - x| \leq x/2^k$. Then for each term t in the Holant sum $H_{\mathcal{O}}$, its corresponding term $\tilde{t}$ in the sum H_{N_k} satisfies $t(1 - 1/2^k)^n \leq \tilde{t} \leq t(1 + 1/2^k)^n$, thus $-t(1 - (1 - 1/2^k)^n) \leq \tilde{t} - t \leq t((1 + 1/2^k)^n - 1)$. Since $1 - (1 - 1/2^k)^n \leq (1 + 1/2^k)^n - 1$, which is easy to see by their binomial expansions, we get $|\tilde{t} - t| \leq t((1 + 1/2^k)^n - 1)$. Also each term $t \leq 1$. Hence if we take $k = 4n + 4$, then

$$|H_{N_k} - H_{\mathcal{O}}| \leq 2^{2n}((1 + 1/2^k)^n - 1) < 1/3^{n+1}.$$

Note that $(1 + x)^n - 1 < xn(1 + x)^n < exn$, for $x = 1/2^{4n+4}$. □

We summarize our progress with the following corollary of Lemmas 7.23 and 7.24.

Corollary 7.25. *Let f be an arity 4 signature with complex weights. If M_f is redundant and $\widetilde{M_f}$ is nonsingular, then* Holant(f) *is* #P*-hard. In fact* Pl-Holant(f) *is* #P*-hard.*

Exercise: By Lemma 7.24, Pl-Holant(g) is #P-hard, where g is defined in Equation (7.7). The reduction chain is from the #P-hardness of T$(G; 3, 3)$ of the Tutte polynomial to #EO (by Theorem 6.32), then to Pl-Holant(g) (by Lemma 7.24).

Prove this directly from the #P-hardness of T$(G; 3, 3)$, skipping #EO. (Hint: First apply a rotation (Figure 7.4) to $2g$, then interpolate Pl-Holant$(\hat{f})$, where $\hat{f}$ is defined in (6.21).)

In order to make Corollary 7.25 more applicable, we show that for an arity 4 signature f, the redundancy of M_f and the nonsingularity of $\widetilde{M_f}$ are invariant under an invertible holographic transformation.

Lemma 7.26. *Let f be an arity 4 signature with complex weights, $T \in \mathbb{C}^{2\times 2}$ a matrix, and $\hat{f} = T^{\otimes 4} f$. If M_f is redundant, then $M_{\hat{f}}$ is also redundant and*

$$\det(\varphi(M_{\hat{f}})) = \det(\varphi(M_f))\det(T)^6.$$

Proof. Since $\hat{f} = T^{\otimes 4} f$, we can express $M_{\hat{f}}$ in terms of M_f and T as

$$M_{\hat{f}} = T^{\otimes 2} M_f \left(T^{\mathrm{T}}\right)^{\otimes 2}. \tag{7.12}$$

This can be directly checked. Let $T = (t^i_j)$, where row index i and column index j are from $\{0, 1\}$. The entry $f(a, b, c, d)$ appears in M_f at row $(a, b) \in \{0, 1\}^2$ and column $(d, c) \in \{0, 1\}^2$ in lexicographic order. The matrix $T^{\otimes 2}$ has the form $(t^a_{a'} t^b_{b'})$ at row (a, b) and column (a', b') in lexicographic order. Similarly $(T^{\mathrm{T}})^{\otimes 2}$ has the form $(t^{c'}_c t^{d'}_d)$ at row (c, d) and column (c', d'). Thus row (a, b) and column (d, c) of $T^{\otimes 2} M_f \left(T^{\mathrm{T}}\right)^{\otimes 2}$ has entry $\sum_{a',b',c',d' \in \{0,1\}} t^a_{a'} t^b_{b'} f(a', b', c', d') t^d_{d'} t^c_{c'}$, which is the entry $\hat{f}(a, b, c, d)$ placed at row (a, b) and column (d, c) of $M_{\hat{f}}$.

An alternative, but essentially the same, proof of (7.12) is by linearity, and observing that it is valid for the basis vectors $f = e_a \otimes e_b \otimes e_c \otimes e_d$, for $a, b, c, d \in \{0, 1\}$.

Let

$$\mathcal{E} = \begin{bmatrix} 1 & 0 & 0 & 0 \\ 0 & 0 & 1 & 0 \\ 0 & 1 & 0 & 0 \\ 0 & 0 & 0 & 1 \end{bmatrix}.$$

Then $X \in \mathrm{RM}_4(\mathbb{C})$ iff $\mathcal{E}X = X = X\mathcal{E}$. Then it follows that $M_{\hat{f}} \in \mathrm{RM}_4(\mathbb{C})$ if $M_f \in \mathrm{RM}_4(\mathbb{C})$, since $\mathcal{E}T^{\otimes 2}\mathcal{E} = T^{\otimes 2}$, i.e., a simultaneous row flip $ab \leftrightarrow ba$ and column flip $a'b' \leftrightarrow b'a'$ keep $T^{\otimes 2}$ unchanged. For the two matrices A and B in the definition of φ, we note that $BA = M_g$, where M_g given in Equation (7.7) is the identity element of the semigroup $\mathrm{RM}_4(\mathbb{C})$. Since $M_f \in \mathrm{RM}_4(\mathbb{C})$, we have $BAM_f = M_f = M_f BA$. Then we have

$$\begin{aligned} \varphi(M_{\hat{f}}) &= AM_{\hat{f}}B = A\left(T^{\otimes 2} M_f \left(T^{\mathrm{T}}\right)^{\otimes 2}\right) B \\ &= (AT^{\otimes 2}B)(AM_f B)(A\left(T^{\mathrm{T}}\right)^{\otimes 2} B) \\ &= \varphi(T^{\otimes 2})\varphi(M_f)\varphi(\left(T^{\mathrm{T}}\right)^{\otimes 2}). \end{aligned} \tag{7.13}$$

Another direct calculation shows that

$$\det(\varphi(T^{\otimes 2})) = \det(T)^3 = \det(\varphi(\left(T^{\mathrm{T}}\right)^{\otimes 2})). \tag{7.14}$$

To verify (7.14) one can save some calculation by noticing that for any T_1 and T_2, $\varphi((T_1 T_2)^{\otimes 2}) = AT_1^{\otimes 2} T_2^{\otimes 2} B$ by definition, and $AT_1^{\otimes 2}$ has identical two middle columns, thus $AT_1^{\otimes 2} = AT_1^{\otimes 2}BA$, thus φ is multiplicative on $T^{\otimes 2}$, i.e., $\varphi((T_1T_2)^{\otimes 2}) = \varphi(T_1^{\otimes 2})\varphi(T_2^{\otimes 2})$. Note that (7.14) is easy to check for triangular matrices.

Thus, by applying determinant to both sides of Equation (7.13), we have

$$\det(\varphi(M_{\hat{f}})) = \det(\varphi(M_f))\det(T)^6,$$

as claimed. □

In particular, for a nonsingular matrix $T \in \mathbb{C}^{2\times 2}$, M_f is redundant and $\widetilde{M_f}$ is nonsingular iff $M_{\hat{f}}$ is redundant and $\widetilde{M_{\hat{f}}}$ is nonsingular. From Corollary 7.25 and Lemma 7.26 we have the following corollary.

Corollary 7.27. *Let f be an arity 4 signature with complex weights. If there exists a nonsingular matrix $T \in \mathbb{C}^{2\times 2}$ such that $\hat{f} = T^{\otimes 4} f$, where $M_{\hat{f}}$ is redundant and $\widetilde{M_{\hat{f}}}$ is nonsingular, then* Holant(f) *is* #P*-hard. In fact* Pl-Holant(f) *is* #P*-hard.*

The following lemma applies Corollary 7.25.

Lemma 7.28. *Let $f_k = ck\lambda^{k-1} + d\lambda^k$, where $c \neq 0$ and $0 \leq k \leq 4$. Then the problem* Holant$([f_0, f_1, f_2, f_3, f_4])$ *is* #P*-hard unless $\lambda = \pm i$, in which case $[f_0, f_1, f_2, f_3, f_4]$ is a vanishing signature and the Holant vanishes.*

Proof. If $\lambda = \pm i$, then $\text{rd}^{\pm}(f) = 1$, $\text{vd}^{\pm}(f) = 3$, and so $f = [f_0, f_1, f_2, f_3, f_4]$ is vanishing by Theorem 7.11. Otherwise, a holographic transformation with the orthogonal matrix $T = \frac{1}{\sqrt{1+\lambda^2}}\begin{bmatrix} 1 & \lambda \\ \lambda & -1 \end{bmatrix}$ transforms f to $\hat{f} = [t, 1, 0, 0, 0]$ for some $t \in \mathbb{C}$ after normalizing the second entry.

This orthogonal transformation can be directly verified; but a more informative derivation is as follows: Consider a symmetric signature $f = [f_0, \ldots, f_n]$ of general arity n, such that $f_k = ck\lambda^{k-1} + d\lambda^k$, where $c \neq 0$, and $\lambda \neq \pm i$. Let $S = \begin{bmatrix} 1 & \frac{d-1}{n} \\ \lambda & c + \frac{d-1}{n}\lambda \end{bmatrix}$. Note that $\det S = c \neq 0$. Then the signature f can be expressed as

$$f = S^{\otimes n}[1, 1, 0, \ldots, 0],$$

where $[1, 1, 0, \ldots, 0]$ should be understood as a column vector of dimension 2^n, which has a 1 in entries with index weight ≤ 1 and 0 elsewhere. This identity can be verified by observing that

$$[1, 1, 0, \ldots, 0] = [1, 0]^{\otimes n} + \frac{1}{(n-1)!}\,\text{Sym}_n^{n-1}([1, 0]; [0, 1]).$$

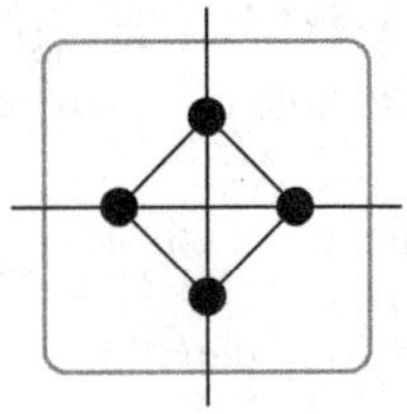

Figure 7.6. The tetrahedron gadget. Each vertex is assigned $\hat{f} = [t, 1, 0, 0, 0]$.

Let $T = \frac{1}{\sqrt{1+\lambda^2}} \begin{bmatrix} 1 & \lambda \\ \lambda & -1 \end{bmatrix}$, then $T = T^{\mathrm{T}} = T^{-1} \in \mathbf{O}_2(\mathbb{C})$ is orthogonal, and $R = TS = \begin{bmatrix} u & w \\ 0 & v \end{bmatrix}$ is upper triangular, where $v, w \in \mathbb{C}$ and $u = \sqrt{1+\lambda^2} \neq 0$. However, $\det R = \det T \det S = (-1)c \neq 0$, so we also have $v \neq 0$. It follows that

$$\begin{aligned} T^{\otimes n} f &= (TS)^{\otimes n}[1, 1, 0, \ldots, 0] \\ &= R^{\otimes n}[1, 1, 0, \ldots, 0] \\ &= R^{\otimes n}\left([1, 0]^{\otimes n} + \frac{1}{(n-1)!}\operatorname{Sym}_n^{n-1}([1, 0]; [0, 1])\right) \\ &= [u, 0]^{\otimes n} + \frac{1}{(n-1)!}\operatorname{Sym}_n^{n-1}([u, 0]; [w, v]) \\ &= [u^n + nu^{n-1}w, u^{n-1}v, 0, \ldots, 0]. \end{aligned}$$

Since $u^{n-1}v \neq 0$, we can normalize the entry of Hamming weight 1 to the value 1 by a scalar multiplication. Thus, we have $[t, 1, 0, \ldots, 0]$ for some $t \in \mathbb{C}$.

Using the tetrahedron gadget in Figure 7.6 with $\hat{f}$ assigned to each vertex, we have a gadget with signature

$$h = [t^4 + 6t^2 + 3, t^3 + 3t, t^2 + 1, t, 1].$$

One way to verify this signature is to treat $[t, 1, 0, 0, 0]$ as a weighted matching signature, and compute the signature of the gadget similarly to a matchgate. Thus, for example, the Hamming weight 0 entry has a contribution t^4 from the empty matching of the six internal edges, a contribution $6t^2$ from the six matchings of a single edge, and a contribution 3 from the three perfect matchings. The other entries can be computed similarly.

Since the determinant of $\widetilde{M_h}$ is 4, the compressed signature matrix of this gadget is nonsingular, so we are done by Corollary 7.25. □

Notice that Lemma 7.28 does not address the complexity of Pl-Holant($[f_0, f_1, f_2, f_3, f_4]$). In particular Pl-Holant($[0, 1, 0, 0, 0]$) is solvable in polynomial time, since $[0, 1, 0, 0, 0]$ is the PERFECT MATCHING signature. Even though Corollary 7.25 applies to Pl-Holant problems as well, the reduction in Lemma 7.28 is nonplanar, as the tetrahedron gadget in Figure 7.6 is not planar. We will show that Pl-Holant($[t, 1, 0, 0, 0]$) remains #P-hard for all $t \neq 0$ in Chapter 8 (Corollary 8.11).

Now we are ready to prove a dichotomy for a single arity 4 signature. For future convenience we state the following theorem for both the case of an arity 3 and arity 4 signature. The arity 3 case is essentially a restatement of Theorem 6.3. Theorem 7.29 does not address the complexity of Pl-Holant problems; Theorem 8.12 in Chapter 8 will be the version of Theorem 7.29 for Pl-Holant problems for arity 4 signatures.

Theorem 7.29. *If f is a nondegenerate, symmetric, complex-valued signature of arity 3 or arity 4 in Boolean variables, then* Holant(f) *is* #P-*hard unless f is $\mathscr{A}$-transformable, or $\mathscr{P}$-transformable, or vanishing, in which case the problem is in P.*

Proof. We first consider f of arity 3. We show that the tractability conditions in Theorem 6.3 are equivalent to the statement that f is $\mathscr{A}$- or $\mathscr{P}$-transformable, or vanishing, for any symmetric signature of arity 3. Clearly conditions 1 and 2 in Theorem 6.3 imply that f is $\mathscr{A}$- or $\mathscr{P}$-transformable. Condition 3 is $f \in \mathscr{R}_2^\sigma$ for $\sigma \in \{+, -\}$, which is the same as $\operatorname{rd}^\sigma(f) \leq 1$ by Lemma 7.16. This is equivalent to f is vanishing for arity 3.

Suppose f is $\mathscr{P}$-transformable (this includes the case when f is degenerate). Then for some T, $(=_2)T^{\otimes 2} \in \mathscr{P}$, and $(T^{-1})^{\otimes 3} f \in \mathscr{P}$. Since $(T^{-1})^{\otimes 3} f$ is symmetric and has arity 3, either it is degenerate, which would imply f is degenerate, or

$$(T^{-1})^{\otimes 3} f = \begin{bmatrix} a \\ 0 \end{bmatrix}^{\otimes 3} + \begin{bmatrix} 0 \\ b \end{bmatrix}^{\otimes 3},$$

for some $ab \neq 0$. Let $D = \begin{bmatrix} a & 0 \\ 0 & b \end{bmatrix}$, then $D \in \operatorname{Stab}(\mathscr{P})$. Then $((TD)^{-1})^{\otimes 3} f = (=_3)$ and $(=_2)(TD)^{\otimes 2} \in \mathscr{P}$.

Suppose f is $\mathscr{A}$-transformable and nondegenerate. Then for some T, $(=_2)$ $T^{\otimes 2} \in \mathscr{A}$, and $(T^{-1})^{\otimes 3} f \in \mathscr{A}$. Since $(T^{-1})^{\otimes 3} f$ is symmetric, it is in $\mathscr{F}_{123} = \mathscr{F}_1 \cup \mathscr{F}_2 \cup \mathscr{F}_3$ defined in Chapter 3. By the form of $\mathscr{F}_{123}$, there exists a matrix

$$M \in \left\{ I, \begin{bmatrix} 1 & 1 \\ 1 & -1 \end{bmatrix}, \begin{bmatrix} 1 & 1 \\ i & -i \end{bmatrix} \right\}$$

such that

$$((TM)^{-1})^{\otimes 3} f = \lambda \left\{ \begin{bmatrix} 1 \\ 0 \end{bmatrix}^{\otimes 3} + \mathfrak{i}^r \begin{bmatrix} 0 \\ 1 \end{bmatrix}^{\otimes 3} \right\},$$

for some $\lambda \neq 0$ and $r \in \{0, 1, 2, 3\}$, and $(=_2)(TM)^{\otimes 2} \in \mathscr{A}$ since $M \in \text{Stab}(\mathscr{A})$. We take $\rho = \mathfrak{i}^{3r}$, and $D = \lambda^{\frac{1}{3}} \begin{bmatrix} 1 & 0 \\ 0 & \rho \end{bmatrix} \in \text{Stab}(\mathscr{A})$. Note that $\rho^3 = \mathfrak{i}^r$. We conclude that $(=_2)(TMD)^{\otimes 2} \in \mathscr{A}$ and $((TMD)^{-1})^{\otimes 3} f = (=_3)$. Hence f satisfies condition 2 of Theorem 6.3.

Now we prove the dichotomy for a single arity 4 signature. Let $f = [f_0, f_1, f_2, f_3, f_4]$. If the compressed signature matrix $\widetilde{M_f}$ is nonsingular, then Holant(f) is #P-hard by Corollary 7.25, so assume that the rank of $\widetilde{M_f}$ is ≤ 2. Then we have

$$a \begin{pmatrix} f_0 \\ f_1 \\ f_2 \end{pmatrix} + 2b \begin{pmatrix} f_1 \\ f_2 \\ f_3 \end{pmatrix} + c \begin{pmatrix} f_2 \\ f_3 \\ f_4 \end{pmatrix} = \begin{pmatrix} 0 \\ 0 \\ 0 \end{pmatrix}.$$

for some $a, b, c \in \mathbb{C}$ that are not all zero. If $a = c = 0$, then $b \neq 0$, so $f_1 = f_2 = f_3 = 0$. In this case, f is a generalized equality GEN-EQ, then $f \in \mathscr{P}$ and in particular f is $\mathscr{P}$-transformable. Now suppose a and c are not both 0. Then f satisfies a second-order recurrence relation. If the roots of the characteristic polynomial of the recurrence relation are distinct, then $f_k = \alpha^{4-k}\beta^k + \gamma^{4-k}\delta^k$ $(0 \leq k \leq 4)$, for some α, β, γ and δ, where $\alpha\delta - \beta\gamma \neq 0$. Let $T_1 = \begin{bmatrix} \alpha & \beta \\ \gamma & \delta \end{bmatrix}$, then $f = (=_4)T_1^{\otimes 4}$, and

$$\text{Holant}(f) \equiv_T \text{Holant}((=_4)T_1^{\otimes 4} \mid (=_2)) \equiv_T \text{Holant}((=_4) \mid T_1^{\otimes 2}(=_2)).$$

By Theorem 5.61 this problem is #P-hard except when there exists some T_2 such that for $\mathscr{C} \in \{\mathscr{P}, \mathscr{A}\}$, $(=_4)(T_2^{-1})^{\otimes 4} \in \mathscr{C}$ and $(T_2T_1)^{\otimes 2}(=_2) \in \mathscr{C}$. Let $T = (T_2T_1)^{-1}$, then $fT^{\otimes 4} \in \mathscr{C}$ and $(T^{-1})^{\otimes 2}(=_2) \in \mathscr{C}$. Hence f is either $\mathscr{A}$- or $\mathscr{P}$-transformable. Note that Theorem 5.61 is the transformational form of Theorem 5.3, which is equivalent to Theorem 5.1. The case $k = 4$ of Theorem 5.1 is proved in Chapter 5.

Now suppose the characteristic polynomial has a double root λ and there are two cases. In the first case, for any $0 \leq k \leq 4$, $f_k = ck\lambda^{k-1} + d\lambda^k$, where $c \neq 0$. In the second case, for any $0 \leq k \leq 4$, $f_k = c(4-k)\lambda^{3-k} + d\lambda^{4-k}$, where $c \neq 0$. This case is mapped to the first case by a holographic transformation by $\begin{bmatrix} 0 & 1 \\ 1 & 0 \end{bmatrix}$. Then we are done by Lemma 7.28. □

(a) The circle is assigned f, the triangle is assigned g, and the squares are assigned $(\neq_2)$.

(b) The circle is assigned h', the triangle is assigned h'', and the squares are given $(\neq_2)$.

Figure 7.7. Gadget constructions used in Lemma 7.30. Part (a) constructs h. Its rotated versions h' and h'' are used in (b) to construct r.

7.6 Vanishing Signatures Revisited

7.6.1 Orientation with a Pair from $\mathscr{V}^+$ and $\mathscr{V}^-$

We prove a lemma on an orientation problem with a pair of vanishing signatures of the opposite type. The proof uses techniques similar to those in Section 7.5. This problem is also called #One-In-Or-One-Out-Orientation in Section 7.3, where a set of such signatures is allowed.

Lemma 7.30. *If $f = [0, 1, 0, \ldots, 0]$ and $g = [0, \ldots, 0, 1, 0]$ are both of the same arity n, for some $n \geq 3$, then the problem* $\operatorname{Holant}(\neq_2 | \{f, g\})$ *is* #P-*hard.*

Proof. Our goal is to obtain a signature that satisfies the hypothesis of Corollary 7.27.

The gadget in Figure 7.7a, with f assigned to the circle vertex, g assigned to the triangle vertex, and $(\neq_2)$ assigned to the square vertices, has signature h with signature matrix

$$M_h = \begin{bmatrix} 0 & 0 & 0 & m \\ 0 & 1 & 1 & 0 \\ 0 & 1 & 1 & 0 \\ 0 & 0 & 0 & 0 \end{bmatrix},$$

where $m = n - 2$ is positive since $n \geq 3$. Although this signature matrix is redundant, its compressed form is singular. Rotating this gadget 90° clockwise and 90° counterclockwise yield signatures h' and h'' respectively, with signature matrices

$$M_{h'} = \begin{bmatrix} 0 & 0 & 0 & 1 \\ 0 & m & 1 & 0 \\ 0 & 1 & 0 & 0 \\ 1 & 0 & 0 & 0 \end{bmatrix} \quad \text{and} \quad M_{h''} = \begin{bmatrix} 0 & 0 & 0 & 1 \\ 0 & 0 & 1 & 0 \\ 0 & 1 & m & 0 \\ 1 & 0 & 0 & 0 \end{bmatrix}.$$

The gadget in Figure 7.7b, with h' assigned to the circle vertex, h'' assigned to the triangle vertex, and $(\neq_2)$ assigned to the square vertices, has a signature r

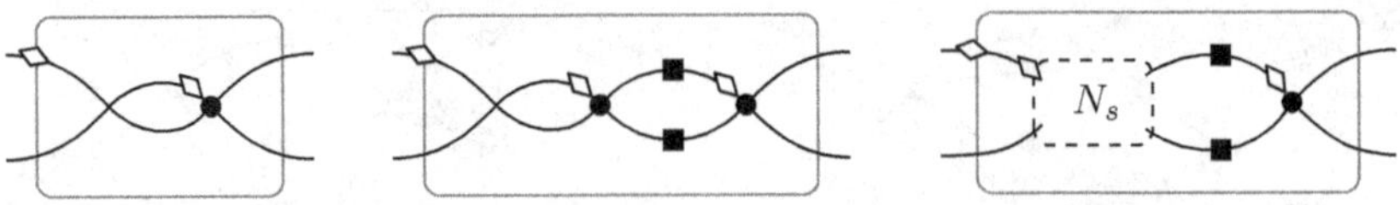

Figure 7.8. Recursive construction to interpolate a signature R that is only a rotation away from having a redundant signature matrix and nonsingular compressed matrix. The circles are assigned r and the squares are assigned $(\neq_2)$.

with signature matrix

$$M_r = M_{h'} \begin{bmatrix} 0 & 0 & 0 & 1 \\ 0 & 0 & 1 & 0 \\ 0 & 1 & 0 & 0 \\ 1 & 0 & 0 & 0 \end{bmatrix} M_{h''} = \begin{bmatrix} 0 & 0 & 0 & 1 \\ 0 & m & m^2+1 & 0 \\ 0 & 1 & m & 0 \\ 1 & 0 & 0 & 0 \end{bmatrix}.$$

Note that the effect of the $(\neq_2)$ signatures is to reverse all four rows of $M_{h''}$ before multiplying it to the right of $M_{h'}$. Although this signature matrix is not redundant, every entry of Hamming weight 2 is nonzero since m is positive.

Now we claim that we can use r to interpolate the following signature R, for any nonzero value $t \in \mathbb{C}$, via the construction in Figure 7.8. Define $p^{\pm} = (m \pm \sqrt{m^2+4})/2$, $P = \begin{bmatrix} 1 & 1 \\ p^+ & p^- \end{bmatrix}$, and $T = P \begin{bmatrix} t & 0 \\ 0 & t^{-1} \end{bmatrix} P^{-1}$, where $t \in \mathbb{C}$ is any nonzero value. We define the signature R by its signature matrix

$$M_R = \begin{bmatrix} 0 & 0 & 0 & 1 \\ 0 & & & 0 \\ 0 & \multicolumn{2}{c}{T} & 0 \\ 1 & 0 & 0 & 0 \end{bmatrix}. \tag{7.15}$$

Consider an instance Ω of Holant$(\neq_2 \mid \mathcal{F} \cup \{R\})$ where $\mathcal{F} = \{f, g\}$. Suppose that R appears n times in Ω. We construct from Ω a sequence of instances Ω_s of Holant$(\neq_2 \mid \mathcal{F})$ indexed by $s \geq 1$. We obtain Ω_s from Ω by replacing each occurrence of R with the gadget N_s in Figure 7.8 with r assigned to the circle vertices and $(\neq_2)$ assigned to the square vertices. In Ω_s, the edge corresponding to the ith significant index bit of N_s connects to the same location as the edge corresponding to the ith significant index bit of R in Ω.

The signature matrix of N_s is the sth power of the matrix obtained from M_r after reversing all rows, and then switching the first and last rows of the final product, namely

$$\begin{bmatrix} 0 & 0 & 0 & 1 \\ 0 & 1 & 0 & 0 \\ 0 & 0 & 1 & 0 \\ 1 & 0 & 0 & 0 \end{bmatrix} \begin{bmatrix} 1 & 0 & 0 & 0 \\ 0 & 1 & m & 0 \\ 0 & m & m^2+1 & 0 \\ 0 & 0 & 0 & 1 \end{bmatrix}^s$$

$$= \begin{bmatrix} 0 & 0 & 0 & 1 \\ 0 & 1 & m & 0 \\ 0 & m & m^2+1 & 0 \\ 1 & 0 & 0 & 0 \end{bmatrix} \begin{bmatrix} 1 & 0 & 0 & 0 \\ 0 & 1 & m & 0 \\ 0 & m & m^2+1 & 0 \\ 0 & 0 & 0 & 1 \end{bmatrix}^{s-1}.$$

The twist of the two input edges on the left side for the first copy of M_r switches the middle two rows, which is equivalent to a total reversal of all rows, followed by the switching of the first and last rows. The total reversals of rows for all subsequent $s-1$ copies of M_r are due to the presence of $(\neq_2)$ signatures.

After such reversals of rows, it is clear that the matrix is a direct sum of block matrices indexed by $\{00, 11\} \times \{00, 11\}$ and $\{01, 10\} \times \{10, 01\}$. Furthermore, in the final product, the block indexed by $\{00, 11\} \times \{00, 11\}$ is $\begin{bmatrix} 0 & 1 \\ 1 & 0 \end{bmatrix}$. Thus in the gadget N_s, the only entries of M_{N_s} that vary with s are the four entries in the middle. These middle four entries of M_{N_s} form the 2×2 matrix $\begin{bmatrix} 1 & m \\ m & m^2+1 \end{bmatrix}^s$. Since $\begin{bmatrix} 1 & m \\ m & m^2+1 \end{bmatrix} = P \begin{bmatrix} \lambda_+ & 0 \\ 0 & \lambda_- \end{bmatrix} P^{-1}$, where $\lambda_\pm = (m^2 + 2 \pm m\sqrt{m^2+4})/2$ are the eigenvalues, we have

$$\begin{bmatrix} 1 & m \\ m & m^2+1 \end{bmatrix}^s = P \begin{bmatrix} \lambda_+^s & 0 \\ 0 & \lambda_-^s \end{bmatrix} P^{-1}.$$

As $\det \begin{bmatrix} 1 & m \\ m & m^2+1 \end{bmatrix} = 1$, $\lambda_+\lambda_- = 1$, so the eigenvalues are nonzero. Since m is positive, $\lambda_+ > \lambda_-$ and so neither λ_+ nor λ_- is a root of unity.

Now we determine the relationship between Holant_Ω and Holant_{Ω_s}. We can view our construction of Ω_s as first replacing M_R with the RHS of

$$M_R = \begin{bmatrix} 1 & 0 \quad 0 & 0 \\ 0 & \multicolumn{2}{c}{\multirow{2}{*}{P}} & 0 \\ 0 & & & 0 \\ 0 & 0 \quad 0 & 1 \end{bmatrix} \begin{bmatrix} 0 & 0 & 0 & 1 \\ 0 & t & 0 & 0 \\ 0 & 0 & t^{-1} & 0 \\ 1 & 0 & 0 & 0 \end{bmatrix} \begin{bmatrix} 1 & 0 \quad 0 & 0 \\ 0 & \multicolumn{2}{c}{\multirow{2}{*}{P^{-1}}} & 0 \\ 0 & & & 0 \\ 0 & 0 \quad 0 & 1 \end{bmatrix},$$

which does not change the Holant value, and then replacing the 4×4 matrix in the middle with the matrix

$$\begin{bmatrix} 0 & 0 & 0 & 1 \\ 0 & \lambda_+^s & 0 & 0 \\ 0 & 0 & \lambda_-^s & 0 \\ 1 & 0 & 0 & 0 \end{bmatrix}.$$

We stratify the assignments in Ω based on the assignments to the n occurrences of the signature matrix

$$\begin{bmatrix} 0 & 0 & 0 & 1 \\ 0 & t & 0 & 0 \\ 0 & 0 & t^{-1} & 0 \\ 1 & 0 & 0 & 0 \end{bmatrix}. \tag{7.16}$$

The inputs to this matrix are from $\{0, 1\}^2 \times \{0, 1\}^2$, which correspond to the four input bits. Recall the way rows and columns of a signature matrix are ordered from Definition 1.4. Thus, e.g., the entry t corresponds to the cyclic input bit pattern 0110 in counterclockwise order. We only need to consider the assignments that assign

- i many times the bit pattern 0110,
- j many times the bit pattern 1001, and
- k many times the bit patterns 0011 or 1100,

such that $i + j + k = n$, since any other assignment contributes 0 to the Holant sum. Let c_{ijk} be the sum over all such assignments of the product of evaluations (excluding (7.16), but including the contributions from the block matrices containing P and P^{-1}) on Ω. Then

$$\text{Holant}_\Omega = \sum_{i+j+k=n} t^{i-j} c_{ijk}$$

and the value of the Holant on Ω_s, for $s \geq 1$, is

$$\text{Holant}_{\Omega_s} = \sum_{i+j+k=n} \lambda_+^{si} \lambda_-^{sj} c_{ijk} = \sum_{i+j+k=n} \lambda_+^{s(i-j)} c_{ijk}.$$

This Vandermonde system does not have full rank. However, we can define for $-n \leq \ell \leq n$,

$$c'_\ell = \sum_{\substack{i-j=\ell \\ i+j+k=n}} c_{ijk}.$$

Then the Holant of Ω is

$$\text{Holant}_{\Omega} = \sum_{-n \le \ell \le n} t^{\ell} c'_{\ell}$$

and the Holant of Ω_s is

$$\text{Holant}_{\Omega_s} = \sum_{-n \le \ell \le n} \lambda_{+}^{s\ell} c'_{\ell}.$$

Now this Vandermonde has full rank because λ_+ is neither zero nor a root of unity. Therefore, we can solve for the unknowns $\{c'_\ell\}$ in polynomial time and obtain the value of Holant_{Ω}. This proves the claim that we can interpolate the signature R in (7.15), for any nonzero $t \in \mathbb{C}$.

Let $t = (\sqrt{m^2+8} + \sqrt{m^2+4})/2$ so $t^{-1} = (\sqrt{m^2+8} - \sqrt{m^2+4})/2$. Let $a = (\sqrt{m^2+8} - m)/2$ and $b = (\sqrt{m^2+8} + m)/2$; then $ab = 2$ and both a and b are nonzero. One can verify that

$$P \begin{bmatrix} t & 0 \\ 0 & t^{-1} \end{bmatrix} P^{-1} = \begin{bmatrix} a & 1 \\ 1 & b \end{bmatrix}.$$

Thus, the signature matrix for R for this particular choice of t is

$$M_R = \begin{bmatrix} 0 & 0 & 0 & 1 \\ 0 & a & 1 & 0 \\ 0 & 1 & b & 0 \\ 1 & 0 & 0 & 0 \end{bmatrix}.$$

After a counterclockwise rotation of 90° on the edges of R, we have a signature R' with a redundant signature matrix

$$M_{R'} = \begin{bmatrix} 0 & 0 & 0 & a \\ 0 & 1 & 1 & 0 \\ 0 & 1 & 1 & 0 \\ b & 0 & 0 & 0 \end{bmatrix}.$$

Its compressed signature matrix

$$\widetilde{M_{R'}} = \begin{bmatrix} 0 & 0 & a \\ 0 & 2 & 0 \\ b & 0 & 0 \end{bmatrix}$$

is nonsingular. After a holographic transformation by Z^{-1}, where $Z = \frac{1}{\sqrt{2}} \begin{bmatrix} 1 & 1 \\ \mathfrak{i} & -\mathfrak{i} \end{bmatrix}$, the binary DISEQUALITY ($\neq_2$) is transformed to the binary

EQUALITY ($=_2$). Thus Holant($\neq_2 | R'$) is transformed to Holant($=_2 \mid Z^{\otimes 4}R'$), which is the same as Holant($Z^{\otimes 4}R'$). We conclude that this Holant problem Holant($Z^{\otimes 4}R'$) is #P-hard by Corollary 7.27, where we choose $T = Z^{-1}$.

To summarize, Holant($Z^{\otimes 4}R'$) is #P-hard implies that Holant($\neq_2 | R'$) is #P-hard, which implies that Holant($\neq_2 | R$) is #P-hard as R' is a rotated form of R, which implies that Holant($\neq_2 | r$) is #P-hard since it can interpolate Holant($\neq_2 |$ R) for all choices of t, which implies that Holant($\neq_2 | \{f, g\}$) is #P-hard since it can directly realize Holant($\neq_2 | r$). □

We note that the planar problem Pl-Holant($\neq_2 | \{f, g\}$) is computable in P by matchgates. The twist introduced in Figure 7.8 is unavoidable; otherwise we would have collapsed #P and P, as all other steps of the reduction chain are planar, all the way back to the #P-hard planar #EULERIAN-ORIENTATIONS problem Pl-Holant($\neq_2 | \mathscr{O}$).

7.6.2 Combining Vanishing Signatures with Others

With Corollary 7.25, Corollary 7.27, and Lemma 7.30 in hand, we consider Holant problems when vanishing signatures are combined with other signatures. We begin with degenerate symmetric signatures, which include unary signatures. Note that in the following lemma proving #P-hardness, to avoid the tractable case 5 in Theorem 7.19, it is necessary to require $\mathrm{rd}^\sigma(f) \geq 2$, for otherwise $f \in \mathscr{R}_2^\sigma$. We prove that in this case, combining a vanishing signature with any degenerate or unary signature of not exactly the same type produces #P-hardness. For future references we prove the stronger statement that it is #P-hard even restricted to planar graphs.

Lemma 7.31. *Let $f \in \mathscr{V}^\sigma$ be a symmetric signature of arity n with $\mathrm{rd}^\sigma(f) = d \geq 2$, where $\sigma \in \{+, -\}$. Suppose $v = u^{\otimes m}$ is a symmetric degenerate signature for some unary signature u and some integer $m \geq 1$. If u is not a multiple of $[1, \sigma \mathrm{i}]$, then* Pl-Holant(f, v) *is #P-hard.*

Proof. We consider $\sigma = +$ since the other case is similar. Since $f \in \mathscr{V}^+$, we have $n > 2d \geq 4$. Under a holographic transformation by $Z = \begin{bmatrix} 1 & 1 \\ \mathrm{i} & -\mathrm{i} \end{bmatrix}$, we have

$$\text{Pl-Holant}(f, v) \equiv \text{Pl-Holant}(\neq_2 | \hat{f}, [a, b]^{\otimes m}),$$

where $\hat{f} = \left(Z^{-1}\right)^{\otimes n} f$ and $[a, b]^{\otimes m} = \left(Z^{-1}\right)^{\otimes m} v = \left(Z^{-1}u\right)^{\otimes m}$. Since u is not a multiple of $[1, \mathrm{i}]$, we have $b \neq 0$. Moreover, $\hat{f} = [\hat{f}_0, \hat{f}_1, \ldots, \hat{f}_d, 0, \ldots, 0]$ with $\hat{f}_d \neq 0$ by Lemma 7.8.

By $d-2$ self-loops on $\hat{f}$ via $\neq_2$, we get $\widehat{f'} = [\hat{f}_{d-2}, \hat{f}_{d-1}, \hat{f}_d, 0, \ldots, 0]$ of arity $n-2d+4$. This $\widehat{f'}$ is on the RHS in Pl-Holant($\neq_2 | \hat{f}, [a,b]^{\otimes m}$). With two more self-loops, we get $[1,0]^{\otimes(n-2d)}$, also on the RHS.

We claim that we can use $[1,0]^{\otimes(n-2d)}$ and $[a,b]^{\otimes m}$ to create $[a,b]^{\otimes(n-2d)}$ on the RHS. If $n-2d > m$, then we connect $[a,b]^{\otimes m}$ to $[1,0]^{\otimes(n-2d)}$ via $\neq_2$ to get $[1,0]^{\otimes(n-2d-m)}$ up to a nonzero factor $b^m \neq 0$. We repeat this process until we get a tensor power $[1,0]^{\otimes \ell}$ for some $\ell \leq m$. If $m > n-2d$, we can get a tensor power $[a,b]^{\otimes \ell}$ for some $\ell \leq n-2d$ by a similar process. Repeat this process, which is a subtractive Euclidean algorithm, and we end up getting both $[1,0]^{\otimes t}$ and $[a,b]^{\otimes t}$, where $t = \gcd(m, n-2d)$. Then we combine $\frac{n-2d}{t}$ copies of $[a,b]^{\otimes t}$ to get $[a,b]^{\otimes(n-2d)}$.

Now connecting $[a,b]^{\otimes(n-2d)}$ back to $\widehat{f'}$ via $\neq_2$, gives $\widehat{f''} = b^{n-2d}[\widehat{f}_{d-2}, \widehat{f}_{d-1}, \widehat{f}_d, 0, 0]$ of arity 4. Notice that Pl-Holant($\neq_2 \mid [\widehat{f}_{d-2}, \widehat{f}_{d-1}, \widehat{f}_d, 0, 0]$) $\equiv_{\mathrm{T}}$ Pl-Holant($\neq_2 \mid [0,0,1,0,0]$). The latter problem is the Eulerian orientation problem over planar 4-regular graphs, which is #P-hard by Theorem 6.32. Thus, Pl-Holant(f, v) is #P-hard. □

Next we consider what binary signatures can be combined with a vanishing signature. Before we prove Lemma 7.33, we prove a simple interpolation lemma.

Lemma 7.32. *Let $x \in \mathbb{C}$. If $x \neq 0$, then for any set $\mathcal{F}$ containing $[x,1,0]$, we have*

$$\text{Holant}(\neq_2 \mid \mathcal{F} \cup \{[v,1,0]\}) \leq_{\mathrm{T}} \text{Holant}(\neq_2 \mid \mathcal{F})$$

for any $v \in \mathbb{C}$.

Proof. Consider an instance Ω of Holant($\neq_2 \mid \mathcal{F} \cup \{[v,1,0]\}$). Suppose that $[v,1,0]$ appears n times in Ω. We stratify the assignments in Ω based on the assignments to $[v,1,0]$. We only need to consider assignments of Hamming weight ≤ 1 since an assignment of Hamming weight 2 contributes a factor of 0. We have

$$\text{Holant}_\Omega = \sum_{i=0}^{n} v^i c_i,$$

where c_i is the sum, over all assignments with exactly i many $[v,1,0]$'s in Ω assigned with a local assignment 00, and $n-i$ many $[v,1,0]$'s in Ω assigned 01 or 10, of the product of evaluations of all other signatures on Ω.

We construct from Ω a sequence of instances Ω_s of Holant($\neq_2 \mid \mathcal{F}$) indexed by $s \geq 1$, by replacing each occurrence of $[v,1,0]$ with a gadget g_s created from s copies of $[x,1,0]$, connected sequentially but with $(\neq_2) = [0,1,0]$ between each sequential pair. The signature of g_s is $[sx, 1, 0]$, which can be verified by the matrix product

$$\left(\begin{bmatrix} x & 1 \\ 1 & 0 \end{bmatrix}\begin{bmatrix} 0 & 1 \\ 1 & 0 \end{bmatrix}\right)^{s-1}\begin{bmatrix} x & 1 \\ 1 & 0 \end{bmatrix} = \begin{bmatrix} 1 & x \\ 0 & 1 \end{bmatrix}^{s-1}\begin{bmatrix} x & 1 \\ 1 & 0 \end{bmatrix}$$
$$= \begin{bmatrix} 1 & (s-1)x \\ 0 & 1 \end{bmatrix}\begin{bmatrix} x & 1 \\ 1 & 0 \end{bmatrix} = \begin{bmatrix} sx & 1 \\ 1 & 0 \end{bmatrix}.$$

The Holant on Ω_s is

$$\text{Holant}_{\Omega_s} = \sum_{i=0}^{n} (sx)^i c_i.$$

For $s \geq 1$, this gives a coefficient matrix that is Vandermonde. Since x is nonzero, sx is distinct for each s. Therefore, the Vandermonde system has full rank. We can solve for the unknowns c_i in polynomial time and obtain the value of Holant_Ω. □

For binary signatures we have the following lemma. Owing to the tractable case 4 in Theorem 7.19, the condition $g \notin \mathscr{R}_2^\sigma$ in Lemma 7.33 is needed.

Lemma 7.33. *Let $f \in \mathscr{V}^\sigma$ be a symmetric nondegenerate signature where $\sigma \in \{+, -\}$. If $g \notin \mathscr{R}_2^\sigma$ is a symmetric nondegenerate binary signature, then* Holant$(\{f, g\})$ *is* #P-*hard.*

Proof. We consider $\sigma = +$ since the other case is similar. A unary signature is degenerate. If a binary symmetric signature f is vanishing, then $\text{vd}^+(f) > 1$, and so $\text{vd}^+(f) \geq 2$; therefore f is also degenerate. Since we assume f is nondegenerate, $\text{arity}(f) \geq 3$.

We prove the lemma by induction on the arity of f. There are two base cases, $\text{arity}(f) = 3$ and $\text{arity}(f) = 4$. However, the arity 3 case is easily reduced to the arity 4 case. We show this first, and then show that the lemma holds for the arity 4 case.

Assume $\text{arity}(f) = 3$. Since $f \in \mathscr{V}^+$, we have $\text{rd}^+(f) < 3/2$, thus $f \in \mathscr{R}_2^+$. From $\text{rd}^+(f) \leq 1$ we have $\text{vd}^+(f) \geq 2$. On the other hand, f is nondegenerate, and so $\text{vd}^+(f) < 3$, thus $\text{vd}^+(f) = 2$ and $\text{rd}^+(f) = 1$.

We connect two copies of f together by one edge to get an arity 4 signature f'. By construction, it may not appear that f' is a symmetric signature; but we show that f' is in fact *symmetric*, nondegenerate and vanishing. It is clearly a vanishing signature, since f is vanishing. Consider the Z transformation, under which f is transformed into $\hat{f} = (Z^{-1})^{\otimes 3} f = [t, 1, 0, 0]$ for some t, up to a nonzero constant. The $(=_2)$ on the connecting edge between the two copies of f

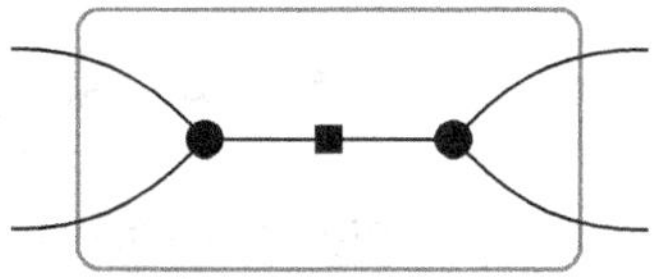

Figure 7.9. The circles are assigned $[t, 1, 0, 0]$ and the square is assigned $(\neq_2)$.

is transformed into $(=_2)Z^{\otimes 2} = (\neq_2)$. In the bipartite setting, our construction is the same as the gadget in Figure 7.9. One can verify that the resulting signature is $\widehat{f'} = [2t, 1, 0, 0, 0]$. The crucial observation is that it takes the same value 0 on inputs 1010 and 1100, where the left two bits are input to one copy of f and the right two bits are for another. The corresponding signature $f' = Z^{\otimes 4}\widehat{f'}$ is symmetric, nondegenerate, with $\mathrm{rd}^+(f') = 1$ and vanishing.

Next we consider the base case of $\mathrm{arity}(f) = 4$. Since $f \in \mathscr{V}^+$, we have $\mathrm{vd}^+(f) > 2$ and $\mathrm{rd}^+(f) < 2$. Since f is nondegenerate we have $\mathrm{rd}^+(f) \neq -1, 0$, hence $\mathrm{rd}^+(f) = 1$ and $\mathrm{vd}^+(f) = 3$. Also by assumption, the given binary $g \notin \mathscr{R}_2^+$, we have $\mathrm{rd}^+(g) = 2$. Once again, consider the holographic transformation by Z. By $\mathrm{rd}^+(g) = 2$, the transformation $(Z^{-1})^{\otimes 2} g$ has the form $[a, b, c]$ with $c \neq 0$ by Lemma 7.8. This gives

$$\begin{aligned}\mathrm{Holant}(=_2 \mid \{f, g\}) &\equiv_T \mathrm{Holant}((=_2)Z^{\otimes 2} \mid \{(Z^{-1})^{\otimes 4} f, (Z^{-1})^{\otimes 2} g\}) \\ &\equiv_T \mathrm{Holant}(\neq_2 \mid \{\hat{f}, \hat{g}\}),\end{aligned}$$

where up to a nonzero constant, $\hat{f} = [t, 1, 0, 0, 0]$ and $\hat{g} = [a, b, 1]$, for some $t, a, b \in \mathbb{C}$. We have $a - b^2 \neq 0$ since g is nondegenerate.

Our next goal is to show that we can realize a signature of the form $[c, 0, 1]$ where $c \neq 0$. If $b = 0$, then $\hat{g}$ is what we want since in this case $a = a - b^2 \neq 0$. Now we assume $b \neq 0$.

Connecting both edges of $\hat{g}$ to $\hat{f}$ via $(\neq_2)$, we get $[t + 2b, 1, 0]$. If $t \neq -2b$, then by Lemma 7.32, we can interpolate any binary signature of the form $[v, 1, 0]$. If $t = -2b$, then we connect two copies of $\hat{g}$ via $(\neq_2)$, and get $\widehat{g'} = [2ab, a + b^2, 2b]$. Connecting this $\widehat{g'}$ to $\hat{f}$ via $(\neq_2)$, we get $[2(a - b^2), 2b, 0]$, using $t = -2b$. Since $a \neq b^2$ and $b \neq 0$, we can interpolate any $[v, 1, 0]$ again by Lemma 7.32.

Hence, we can have any signature $[v, 1, 0]$, where $v \in \mathbb{C}$ is for us to choose. We construct the gadget in Figure 7.10 with the circles assigned $[v, 1, 0]$, the squares assigned $(\neq_2)$, and the triangle assigned $[a, b, 1]$. The resulting gadget has signature $[a + 2bv + v^2, b + v, 1]$, which can be verified by the

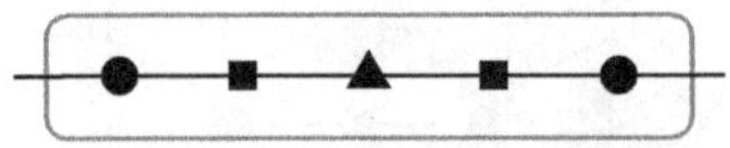

Figure 7.10. A sequence of binary gadgets that forms another binary gadget. The circles are assigned $[v, 1, 0]$, the squares are assigned $\neq_2$, and the triangle is assigned $[a, b, 1]$.

matrix product

$$\begin{bmatrix} v & 1 \\ 1 & 0 \end{bmatrix}\begin{bmatrix} 0 & 1 \\ 1 & 0 \end{bmatrix}\begin{bmatrix} a & b \\ b & 1 \end{bmatrix}\begin{bmatrix} 0 & 1 \\ 1 & 0 \end{bmatrix}\begin{bmatrix} v & 1 \\ 1 & 0 \end{bmatrix} = \begin{bmatrix} a + 2bv + v^2 & b + v \\ b + v & 1 \end{bmatrix}.$$

By setting $v = -b$, we get $[c, 0, 1]$, where $c = a - b^2 \neq 0$.

Now we construct a signature $\hat{h}$ by the tetrahedron gadget given in Figure 7.11, using the signature $[c, 0, 1]$, in two steps. First in Figure 7.11b we assign $[c, 0, 1]$ to the circle vertex of arity two and $(\neq_2)$ to the two square vertices. The resulting signature $[1, 0, c]$ is then used in Figure 7.11a. We assign $[1, 0, c]$ to the six internal edges labeled with triangles, and assign $\hat{f} = [t, 1, 0, 0, 0]$ to the four circle vertices of arity four in Figure 7.11a. By construction, $\hat{h}$ is a symmetric signature, and in fact

$$\hat{h} = [t^4 + 6ct^2 + 3c^2, t^3 + 3ct, t^2 + c, t, 1].$$

In fact the special case with $c = 1$ has already been used in the proof of Lemma 7.28. We can compute these signature entries similarly to a matchgate. The internal edge function $[1, 0, c]$ is a GEN-EQ signature, and the signature $\hat{f}$

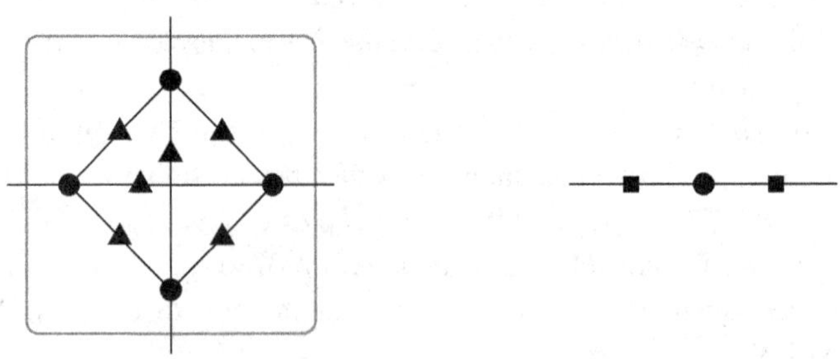

(a) The tetrahedron gadget with edge signatures given in (b).

(b) The gadget representing an edge labeled by a triangle in (a).

Figure 7.11. The tetrahedron gadget with each triangle replaced by the edge in (b), where the circle is assigned $[c, 0, 1]$ and the squares are assigned $\neq_2$. The four circles in (a) are assigned $[t, 1, 0, 0, 0]$.

on the four circle vertices is a weighted version of the matching function AT-MOST-ONE. For example, we may verify that $\hat{h}_0 = t^4 + 6ct^2 + 3c^2$ as follows: For the external input pattern 0000, the term t^4 comes from the empty matching among six internal edges, the term $6ct^2$ comes from the six matchings of size one, each matching edge having weight c, and the term $3c^2$ comes from the three perfect matchings of size two. The other entries can be computed similarly.

The compressed signature matrix of $\hat{h}$ is

$$\widetilde{M_{\hat{h}}} = \begin{bmatrix} 3c^2 + 6ct^2 + t^4 & 2(3ct + t^3) & c + t^2 \\ 3ct + t^3 & 2(c + t^2) & t \\ c + t^2 & 2t & 1 \end{bmatrix}$$

and quite amazingly its determinant is $4c^3 \neq 0$ (check it!). Thus $\widetilde{M_{\hat{h}}}$ is nonsingular. After a holographic transformation by Z^{-1}, where $Z = \frac{1}{\sqrt{2}} \begin{bmatrix} 1 & 1 \\ \mathfrak{i} & -\mathfrak{i} \end{bmatrix}$, the binary DISEQUALITY function ($\neq_2$) is transformed to the binary EQUALITY ($=_2$). Thus Holant($\neq_2 | \hat{h}$) is transformed to Holant($=_2 \mid Z^{\otimes 4}\hat{h}$), which is the same as Holant($Z^{\otimes 4}\hat{h}$). Then, by Corollary 7.27, we are done.

Now we do the induction step. The idea is to produce from f and g another symmetric nondegenerate signature $f' \in \mathscr{V}^+$ that has a smaller arity than f. Assume f has arity $n \geq 5$. Since f is nondegenerate, $\text{rd}^+(f) \neq -1, 0$. Under the Z transformation, $f = Z^{\otimes n}\hat{f}$, $g = Z^{\otimes 2}\hat{g}$, where $\hat{f} = [\hat{f}_0, \ldots, \hat{f}_d, 0, \ldots, 0]$, $d = \text{rd}^+(f) < n/2$, and $\hat{g} = [a, b, 1]$ up to a nonzero scalar.

First suppose $\text{rd}^+(f) = 1$. By Lemma 7.12 we can get $f' = \langle f, g \rangle$ by connecting both edges of g to f. This is a symmetric signature of arity $n - 2 \geq 3$ with $\text{rd}^+(f') = \text{rd}^+(f) = 1$. In particular f' is nondegenerate. Since $\text{rd}^+(f') < \text{arity}(f')/2$, $f' \in \mathscr{V}^+$ by definition. So we are done by induction.

We now suppose $\text{rd}^+(f) = d \geq 2$. Since f is nondegenerate it is certainly nonzero. Since it is vanishing, certainly $\text{vd}^+(f) > 0$. Hence we may apply Lemma 7.13. Let f' be obtained from f by a self-loop, then $\text{rd}^+(f') = d - 1 \geq 1$ and $\text{arity}(f') = n - 2$. Clearly f' is still vanishing. We have $f' = Z^{\otimes(n-2)}[\widehat{f'}_0, \ldots, \widehat{f'}_{d-1}, 0, \ldots, 0]$ where $\widehat{f'}_{d-1} \neq 0$. Since $d - 1 \geq 1$, f' is nondegenerate. The proof is complete by induction. □

Finally, we consider a pair of vanishing signatures of the opposite type, both of arity ≥ 3. We show that opposite types of vanishing signatures cannot mix. More formally, vanishing signatures of opposite types, when put together, lead to #P-hardness.

Lemma 7.34. *Let $f \in \mathscr{V}^+$ and $g \in \mathscr{V}^-$ be nondegenerate symmetric signatures both of arity ≥ 3. Then* Holant($\{f, g\}$) *is #P-hard.*

Proof. Let $\mathrm{rd}^+(f) = d, \mathrm{rd}^-(g) = d'$, $\mathrm{arity}(f) = n$ and $\mathrm{arity}(g) = n'$, then $2d < n$ and $2d' < n'$. We perform a holographic transformation by $Z = \frac{1}{\sqrt{2}}\begin{bmatrix} 1 & 1 \\ \mathrm{i} & -\mathrm{i} \end{bmatrix}$,

$$\begin{aligned}\mathrm{Holant}(=_2 \mid \{f, g\}) &\equiv_{\mathrm{T}} \mathrm{Holant}((=_2)Z^{\otimes 2} \mid \{(Z^{-1})^{\otimes n} f, (Z^{-1})^{\otimes n'} g\}) \\ &\equiv_{\mathrm{T}} \mathrm{Holant}((\neq_2) \mid \{\hat{f}, \hat{g}\}),\end{aligned}$$

where $\hat{f} = (Z^{-1})^{\otimes n} f$ and $\hat{g} = (Z^{-1})^{\otimes n'} g$. By Lemma 7.8 $\hat{f}$ and $\hat{g}$ take the form $\hat{f} = [\hat{f}_0, \hat{f}_1, \ldots, \hat{f}_d, 0, \ldots, 0]$ and $\hat{g} = [0, \ldots, 0, \hat{g}_{n'-d'}, \ldots, \hat{g}_{n'}]$, where $\hat{f}_d \neq 0$ and $\hat{g}_{n'-d'} \neq 0$.

If $d \geq 2$, we will use Lemma 7.31. We can do d' many self-loops on $\hat{g}$ via $(\neq_2)$, getting $\hat{g}' = [0, \ldots, 0, \hat{g}_{n'-d'}]$ of arity $n' - 2d' \geq 1$. If we define $g' = Z^{\otimes(n'-2d')}\hat{g}'$, then $g' = [1, -\mathrm{i}]^{\otimes(n'-2d')}$ up to a nonzero constant. This g' is constructible in the RHS of $\mathrm{Holant}(=_2 \mid \{f, g\})$. Then applying Lemma 7.31 on the pair $\{f, g'\}$, we get $\mathrm{Holant}(f, g)$ is #P-hard. (In fact, Pl-$\mathrm{Holant}(f, g)$ is also #P-hard in this case.) If $d' \geq 2$, we can similarly get $[1, \mathrm{i}]^{\otimes(n-2d)}$ and apply Lemma 7.31. Thus we may assume that $d = d' = 1$.

Therefore, up to a nonzero constant, $\hat{f} = [a, 1, 0, \ldots, 0]$ and $\hat{g} = [0, \ldots, 0, 1, b]$, for some $a, b \in \mathbb{C}$. We show that it is always possible to get two such signatures of the same arity $n^* = \min\{n, n'\}$. Suppose $n > n'$. We form a loop from $\hat{f}$, where the loop is really a path consisting of one vertex and two edges, with the vertex assigned the signature $(\neq_2)$. It is easy to see that this signature is the degenerate signature $2[1, 0]^{\otimes(n-2)}$. Similarly, we can form a loop from $\hat{g}$ and can get $2[0, 1]^{\otimes(n'-2)}$. Thus we have both $[1, 0]^{\otimes(n-2)}$ and $[0, 1]^{\otimes(n'-2)}$. We can connect all $n' - 2$ edges of the second to the first, connected by $(\neq_2)$. This gives $[1, 0]^{\otimes(n-n')}$. We can continue subtracting the smaller arity from the larger one. We continue this process in a subtractive version of the Euclidean algorithm, and end up with both $[1, 0]^{\otimes t}$ and $[0, 1]^{\otimes t}$, where $t = \gcd(n - 2, n' - 2) = \gcd(n - n', n' - 2)$. In particular, $t \mid n - n'$ and by taking $(n - n')/t$ many copies of $[0, 1]^{\otimes t}$, we can get $[0, 1]^{\otimes(n-n')}$. Connecting this back to $\hat{f}$ via $(\neq_2)$, we get a symmetric signature of arity n' consisting of the first $n' + 1$ entries of $\hat{f}$. A similar proof works when $n' > n$.

Thus without loss of generality, we may assume we have two signatures $[a, 1, 0, \ldots, 0]$ and $[0, \ldots, 0, 1, b]$, both of arity $n^* \geq 3$. We also have $[1, 0]^{\otimes t}$ and $[0, 1]^{\otimes t}$ thus both $[1, 0]^{\otimes(n^*-2)}$ and $[0, 1]^{\otimes(n^*-2)}$. Connect $[0, 1]^{\otimes(n^*-2)}$ to $[a, 1, 0, \ldots, 0]$ of arity n^* via $(\neq_2)$ we get $\hat{h} = [a, 1, 0]$. For $a \neq 0$, translating this back by Z, we have a binary signature h together with the given $g \in \mathscr{V}^-$. We claim that $h \notin \mathscr{R}_2^-$. Indeed if it were the case that $h \in \mathscr{R}_2^-$, then $\mathrm{rd}^-(h) \leq 1$ and thus $\hat{h} = [0, *, *]$, a contradiction. By Lemma 7.33, $\mathrm{Holant}(\{g, h\})$ is

#P-hard, and thus Holant($\{f, g\}$) is #P-hard. A similar proof works for the case $b \neq 0$.

The only case left is when $\hat{f} = [0, 1, 0, \ldots, 0]$, and $\hat{g} = [0, \ldots, 0, 1, 0]$, both of arity $n^* \geq 3$. This is #P-hard by Lemma 7.30. □

Remark: Notice that in the last case of the proof of Lemma 7.34, $\hat{f} = [0, 1, 0, \ldots, 0]$, and $\hat{g} = [0, \ldots, 0, 1, 0]$. The planar Holant problem Pl-Holant($(\neq_2) \mid \{\hat{f}, \hat{g}\}$) in this case is computable in polynomial time, and therefore so is Pl-Holant(f, g), because $(\neq_2), \hat{f}, \hat{g}$ are all matchgate signatures. One can prove that, excluding this case, the planar Holant problem Pl-Holant(f, g) in Lemma 7.34 is also #P-hard. The additional condition is that either f is not the Z-transform of the matchgate signature $[0, 1, 0, \ldots, 0]$ or g is not the Z-transform of the matchgate signature $[0, \ldots, 0, 1, 0]$, up to a constant. In Chapter 9 these will be called $\mathscr{M}_4$ signatures.

7.7 $\mathscr{A}$-transformable and $\mathscr{P}$-transformable Signatures

In this section, we study $\mathscr{A}$-transformable signatures and $\mathscr{P}$-transformable signatures. We denote by $\alpha = \frac{1+\mathrm{i}}{\sqrt{2}} = \sqrt{\mathrm{i}} = e^{\frac{\pi \mathrm{i}}{4}}$ and use $\mathbf{O}_2(\mathbb{C})$ to denote the group of 2×2 orthogonal matrices over $\mathbb{C}$. Recall that $\mathscr{F}_{123} = \mathscr{F}_1 \cup \mathscr{F}_2 \cup \mathscr{F}_3$, where $\mathscr{F}_1$, $\mathscr{F}_2$, and $\mathscr{F}_3$ are defined in Chapter 3, Section 3.2.

7.7.1 Characterization of $\mathscr{A}$- and $\mathscr{P}$-transformable Signatures

Recall that by definition, if a set of signatures $\mathcal{F}$ is $\mathscr{A}$-transformable (resp. $\mathscr{P}$-transformable), then the binary Equality $(=_2)$ must be simultaneously transformed into $\mathscr{A}$ (resp. $\mathscr{P}$) along with $\mathcal{F}$. We first characterize all possible matrices for such transformations by considering only the transformation of $(=_2)$. Although there are many binary signatures in $\mathscr{A} \cup \mathscr{P}$, it turns out that it is sufficient to consider only three signatures. The following proposition is best understood in terms of a group action, its stabilizer and orbits.

Proposition 7.35. *Let $T \in \mathbb{C}^{2\times 2}$ be a matrix. Then the following hold:*

1. $[1, 0, 1]T^{\otimes 2} = [1, 0, 1]$ *iff* $T \in \mathbf{O}_2(\mathbb{C})$;
2. $[1, 0, 1]T^{\otimes 2} = [1, 0, \mathrm{i}]$ *iff there exists an* $H \in \mathbf{O}_2(\mathbb{C})$ *such that* $T = H\begin{bmatrix} 1 & 0 \\ 0 & \alpha \end{bmatrix}$;
3. $[1, 0, 1]T^{\otimes 2} = [0, 1, 0]$ *iff there exists an* $H \in \mathbf{O}_2(\mathbb{C})$ *such that* $T = \frac{1}{\sqrt{2}} H \begin{bmatrix} 1 & 1 \\ \mathrm{i} & -\mathrm{i} \end{bmatrix}$.

Proof. The first item is clear since

$$[1, 0, 1]T^{\otimes 2} = [1, 0, 1] \iff T^{\mathsf{T}} I_2 T = I_2 \iff T^{\mathsf{T}} T = I_2,$$

the definition of a 2×2 orthogonal matrix T.

We let $M_1 = I_2$, $M_2 = \begin{bmatrix} 1 & 0 \\ 0 & \alpha \end{bmatrix}$ and $M_3 = Z = \frac{1}{\sqrt{2}} \begin{bmatrix} 1 & 1 \\ \mathrm{i} & -\mathrm{i} \end{bmatrix}$, let $T_j = HM_j$ (for $j = 1, 2, 3$), where $H \in \mathbf{O}_2(\mathbb{C})$. Then

$$[1, 0, 1]T_j^{\otimes 2} = [1, 0, 1](HM_j)^{\otimes 2} = [1, 0, 1]M_j^{\otimes 2} = f_j,$$

where $f_1 = [1, 0, 1]$, $f_2 = [1, 0, \mathrm{i}]$ and $f_3 = [0, 1, 0]$.

Conversely, suppose that $[1, 0, 1](T_j)^{\otimes 2} = f_j$. Then we have

$$[1, 0, 1]\left(T_j M_j^{-1}\right)^{\otimes 2} = f_j \left(M_j^{-1}\right)^{\otimes 2} = [1, 0, 1],$$

so $H = T_j M_j^{-1} \in \mathbf{O}_2(\mathbb{C})$ by case 1. Thus $T_j = HM_j$ as desired. □

We also need the following proposition; the proof is direct.

Proposition 7.36. *If a symmetric signature $f = [f_0, f_1, \ldots, f_n]$ can be expressed in the form $f = a[1, \lambda]^{\otimes n} + b[1, \mu]^{\otimes n}$, for some $a, b, \lambda, \mu \in \mathbb{C}$, then the f_k's satisfy the recurrence relation $f_{k+2} = (\lambda + \mu) f_{k+1} - \lambda\mu f_k$ for $0 \le k \le n - 2$.*

Recall the stabilizer group $\mathrm{Stab}(\mathscr{A}) = \mathbb{C}^* \cdot \langle D, H_2 \rangle$ that was determined in Lemma 6.13, where $D = \begin{bmatrix} 1 & 0 \\ 0 & \mathrm{i} \end{bmatrix}$ and $H_2 = \frac{1}{\sqrt{2}} \begin{bmatrix} 1 & 1 \\ 1 & -1 \end{bmatrix}$.

Lemma 7.37. *Let $\mathcal{F}$ be a set of signatures. Then $\mathcal{F}$ is $\mathscr{A}$-transformable under T iff $\mathcal{F}$ is $\mathscr{A}$-transformable under any $T' \in T\,\mathrm{Stab}(\mathscr{A})$.*

Proof. Sufficiency is trivial since $I_2 \in \mathrm{Stab}(\mathscr{A})$. If $\mathcal{F}$ is $\mathscr{A}$-transformable under T, then by definition, we have $(=_2)T^{\otimes 2} \in \mathscr{A}$ and $\mathcal{F}' = T^{-1}\mathcal{F} \subseteq \mathscr{A}$. Let $T' = TM \in T\,\mathrm{Stab}(\mathscr{A})$ for any $M \in \mathrm{Stab}(\mathscr{A})$. It then follows that $(=_2)T'^{\otimes 2} = (=_2)T^{\otimes 2}M^{\otimes 2} \in \mathscr{A}M = \mathscr{A}$ and $T'^{-1}\mathcal{F} = M^{-1}\mathcal{F}' \subseteq M^{-1}\mathscr{A} = \mathscr{A}$. Therefore $\mathcal{F}$ is $\mathscr{A}$-transformable under any $T' \in T\,\mathrm{Stab}(\mathscr{A})$. □

After restricting by Proposition 7.35 and normalizing by Lemma 7.37, one only needs to check a small subset of $\mathbf{GL}_2(\mathbb{C})$ to determine if $\mathcal{F}$ is $\mathscr{A}$-transformable.

Lemma 7.38. *Let $\mathcal{F}$ be a set of signatures. Then $\mathcal{F}$ is $\mathscr{A}$-transformable iff there exists an $H \in \mathbf{O}_2(\mathbb{C})$ such that $\mathcal{F} \subseteq H\mathscr{A}$ or $\mathcal{F} \subseteq H \begin{bmatrix} 1 & 0 \\ 0 & \alpha \end{bmatrix} \mathscr{A}$.*

Proof. Sufficiency is easily verified by checking that $(=_2)$ is transformed into $\mathscr{A}$ by H and by $H\begin{bmatrix}1 & 0\\ 0 & \alpha\end{bmatrix}$ in respective cases.

If $\mathcal{F}$ is $\mathscr{A}$-transformable, then by definition, there exists a matrix T such that $(=_2)T^{\otimes 2} \in \mathscr{A}$ and $T^{-1}\mathcal{F} \subseteq \mathscr{A}$. Since $(=_2)$ is nondegenerate and symmetric, $(=_2)T^{\otimes 2} \in \mathscr{A}$ is equivalent to $(=_2)T^{\otimes 2} \in \mathscr{F}_{123}$.

Any signature in $\mathscr{F}_{123}$ is expressible as $c(v_1^{\otimes n} + i^t v_2^{\otimes n})$, where $t \in \{0, 1, 2, 3\}$ and (v_1, v_2) is a pair of vectors in the set

$$\{([1,\ \ 0], [0,\ \ 1]), ([1,\ \ 1], [1,\ \ -1]), ([1,\ \ i], [1,\ \ -i])\}.$$

We use D and $H_2 \in \text{Stab}(\mathscr{A})$ to normalize $\mathscr{F}_{123}$ by Lemma 7.37. In particular, $\mathscr{F}_1 = \mathscr{F}_2 H_2$ and $\mathscr{F}_1 = \mathscr{F}_3(D^{-1}H_2)$. Furthermore, the binary signatures in $\mathscr{F}_1$ are just the four signatures $[1, 0, 1]$, $[1, 0, i]$, $[1, 0, -1]$, and $[1, 0, -i]$ up to a scalar. We also normalize these four by D, as $[1, 0, 1] = [1, 0, -1]D^{\otimes 2}$ and $[1, 0, i] = [1, 0, -i]D^{\otimes 2}$. Hence $\mathcal{F}$ being $\mathscr{A}$-transformable implies that there exists a matrix T such that $(=_2)T^{\otimes 2} \in \{[1, 0, 1], [1, 0, i]\}$ and $T^{-1}\mathcal{F} \subseteq \mathscr{A}$. Now we apply Proposition 7.35.

1. If $(=_2)T^{\otimes 2} = [1, 0, 1]$, then by case 1 of Proposition 7.35, we have $T \in \mathbf{O}_2(\mathbb{C})$. Therefore $\mathcal{F} \subseteq H\mathscr{A}$, where $H = T \in \mathbf{O}_2(\mathbb{C})$.
2. If $(=_2)T^{\otimes 2} = [1, 0, i]$, then by case 2 of Proposition 7.35, there exists an $H \in \mathbf{O}_2(\mathbb{C})$ such that $T = H\begin{bmatrix}1 & 0\\ 0 & \alpha\end{bmatrix}$. Therefore $\mathcal{F} \subseteq T\mathscr{A} = H\begin{bmatrix}1 & 0\\ 0 & \alpha\end{bmatrix}\mathscr{A}$ where $H \in \mathbf{O}_2(\mathbb{C})$.

This completes the proof. $\square$

Using these two lemmas, we can characterize all $\mathscr{A}$-transformable signatures. We first define the three sets $\mathscr{A}_1$, $\mathscr{A}_2$, and $\mathscr{A}_3$.

Definition 7.39. A symmetric signature f of arity n is in $\mathscr{A}_1$ if there exists an $H \in \mathbf{O}_2(\mathbb{C})$ and a nonzero constant c such that

$$f = cH^{\otimes n}\left(\begin{bmatrix}1\\1\end{bmatrix}^{\otimes n} + \beta\begin{bmatrix}1\\-1\end{bmatrix}^{\otimes n}\right),$$

where $\beta = \alpha^{tn+2r}$ for some $r \in \{0, 1, 2, 3\}$ and $t \in \{0, 1\}$.

When such an H exists, we say that $f \in \mathscr{A}_1$ with transformation H. If $f \in \mathscr{A}_1$ with I_2, then we say f is in the canonical form of $\mathscr{A}_1$. If f is in the canonical form of $\mathscr{A}_1$, then by Proposition 7.36, for any $0 \le k \le n-2$, we have $f_{k+2} = f_k$ and one of the following holds:

- $f_0 = 0$, or
- $f_1 = 0$, or
- $f_1 = \pm \mathfrak{i} f_0 \neq 0$, or
- n is odd and $f_1 = \pm(1 \pm \sqrt{2})\mathfrak{i} f_0 \neq 0$ (all four sign choices are permissible).

Notice that when n is odd and $t = 1$ in Definition 7.39, it has some complication as described by the factor α^{n+2r} ($r \in \{0, 1, 2, 3\}$), which runs through $\alpha \mathfrak{i}^s$ ($s \in \{0, 1, 2, 3\}$).

Definition 7.40. A symmetric signature f of arity n is in $\mathscr{A}_2$ if there exists an $H \in \mathbf{O}_2(\mathbb{C})$ and a nonzero constant c such that

$$f = cH^{\otimes n}\left(\begin{bmatrix}1\\ \mathfrak{i}\end{bmatrix}^{\otimes n} + \begin{bmatrix}1\\ -\mathfrak{i}\end{bmatrix}^{\otimes n}\right).$$

Similarly, when such an H exists, we say that $f \in \mathscr{A}_2$ with transformation H. If $f \in \mathscr{A}_2$ with I_2, then we say f is in the canonical form of $\mathscr{A}_2$. If f is in the canonical form of $\mathscr{A}_2$, then by Proposition 7.36, for any $0 \le k \le n-2$, we have $f_{k+2} = -f_k$. Since f is nondegenerate, $f_1 \neq \pm \mathfrak{i} f_0$ is implied.

It is worth noting that $H\begin{bmatrix}1\\ \mathfrak{i}\end{bmatrix}$ is a nonzero multiple of either $\begin{bmatrix}1\\ \mathfrak{i}\end{bmatrix}$ or $\begin{bmatrix}1\\ -\mathfrak{i}\end{bmatrix}$, for any orthogonal $H \in \mathbf{O}_2(\mathbb{C})$. The same is true for $H\begin{bmatrix}1\\ -\mathfrak{i}\end{bmatrix}$. Using this fact, the following lemma gives a characterization of $\mathscr{A}_2$. It says that any signature in $\mathscr{A}_2$ is essentially in canonical form.

Lemma 7.41. *Let f be a nondegenerate symmetric signature. Then $f \in \mathscr{A}_2$ iff f is of the form*

$$c\left(\begin{bmatrix}1\\ \mathfrak{i}\end{bmatrix}^{\otimes n} + \beta\begin{bmatrix}1\\ -\mathfrak{i}\end{bmatrix}^{\otimes n}\right)$$

for some $c, \beta \neq 0$.

Proof. Assume that

$$f = c\left(\begin{bmatrix}1\\ \mathfrak{i}\end{bmatrix}^{\otimes n} + \beta\begin{bmatrix}1\\ -\mathfrak{i}\end{bmatrix}^{\otimes n}\right)$$

for some $c, \beta \neq 0$. Consider the orthogonal transformation $H = \begin{bmatrix}a & b\\ b & -a\end{bmatrix}$, where $a = \frac{1}{2}(\beta^{\frac{1}{2n}} + \beta^{-\frac{1}{2n}})$ and $b = \frac{1}{2\mathfrak{i}}(\beta^{\frac{1}{2n}} - \beta^{-\frac{1}{2n}})$. We pick a and b in this way so that $a + bi = \beta^{\frac{1}{2n}}$, $a - bi = \beta^{-\frac{1}{2n}}$, and $(a + bi)(a - bi) = a^2 + b^2 = 1$. Also $\left(\frac{a+bi}{a-bi}\right)^n = \beta$. Then

$$\begin{aligned}
H^{\otimes n} f &= c\left(\begin{bmatrix} a + bi \\ -ai + b \end{bmatrix}^{\otimes n} + \beta \begin{bmatrix} a - bi \\ ai + b \end{bmatrix}^{\otimes n}\right) \\
&= c\left((a + bi)^n \begin{bmatrix} 1 \\ -i \end{bmatrix}^{\otimes n} + (a - bi)^n \beta \begin{bmatrix} 1 \\ i \end{bmatrix}^{\otimes n}\right) \\
&= c\sqrt{\beta}\left(\begin{bmatrix} 1 \\ -i \end{bmatrix}^{\otimes n} + \begin{bmatrix} 1 \\ i \end{bmatrix}^{\otimes n}\right),
\end{aligned}$$

so f can be written as

$$f = c\sqrt{\beta}(H^{-1})^{\otimes n}\left(\begin{bmatrix} 1 \\ i \end{bmatrix}^{\otimes n} + \begin{bmatrix} 1 \\ -i \end{bmatrix}^{\otimes n}\right).$$

Therefore $f \in \mathscr{A}_2$.

On the other hand, the desired form

$$f = c\left(\begin{bmatrix} 1 \\ i \end{bmatrix}^{\otimes n} + \beta \begin{bmatrix} 1 \\ -i \end{bmatrix}^{\otimes n}\right)$$

follows from the fact that

$$\left\{\begin{bmatrix} 1 \\ i \end{bmatrix}, \begin{bmatrix} 1 \\ -i \end{bmatrix}\right\}$$

is fixed setwise under any orthogonal transformation up to nonzero constants. □

Definition 7.42. A symmetric signature f of arity n is in $\mathscr{A}_3$ if there exists an $H \in \mathbf{O}_2(\mathbb{C})$ and a nonzero constant c such that

$$f = cH^{\otimes n}\left(\begin{bmatrix} 1 \\ \alpha \end{bmatrix}^{\otimes n} + i^r \begin{bmatrix} 1 \\ -\alpha \end{bmatrix}^{\otimes n}\right)$$

for some $r \in \{0, 1, 2, 3\}$.

Again, when such an H exists, we say that $f \in \mathscr{A}_3$ with transformation H. If $f \in \mathscr{A}_3$ with I_2, then we say f is in the canonical form of $\mathscr{A}_3$. If f is in the canonical form of $\mathscr{A}_3$, then by Proposition 7.36, for any $0 \le k \le n - 2$, we have $f_{k+2} = i f_k$ and one of the following holds:

- $f_0 = 0$, or
- $f_1 = 0$, or
- $f_1 = \pm\alpha i f_0 \neq 0$.

Now we characterize the $\mathscr{A}$-transformable signatures.

Lemma 7.43. *Let f be a nondegenerate symmetric signature. Then f is $\mathscr{A}$-transformable iff $f \in \mathscr{A}_1 \cup \mathscr{A}_2 \cup \mathscr{A}_3$.*

Proof. Assume that f is $\mathscr{A}$-transformable of arity n. By applying Lemma 7.38 to $\{f\}$, there exists an $H \in \mathbf{O}_2(\mathbb{C})$ such that $f \in H\mathscr{A}$ or $f \in H\begin{bmatrix}1 & 0\\ 0 & \alpha\end{bmatrix}\mathscr{A}$. This is equivalent to $(H^{-1})^{\otimes n} f \in \mathscr{A}$ or $(H^{-1})^{\otimes n} f \in \begin{bmatrix}1 & 0\\ 0 & \alpha\end{bmatrix}\mathscr{A}$. Since f is nondegenerate and symmetric, we can replace $\mathscr{A}$ in the previous expressions with $\mathscr{F}_{123}$. Now we consider the possible cases.

1. If $(H^{-1})^{\otimes n} f \in \mathscr{F}_1$, then a further transformation by $H_2 = \frac{1}{\sqrt{2}}\begin{bmatrix}1 & 1\\ 1 & -1\end{bmatrix} \in \mathbf{O}_2(\mathbb{C})$ puts it into the canonical form of $\mathscr{A}_1$, i.e., $f \in \mathscr{A}_1$ with transformation HH_2.
2. If $(H^{-1})^{\otimes n} f \in \mathscr{F}_2$, then it is already in the canonical form of $\mathscr{A}_1$.
3. If $(H^{-1})^{\otimes n} f \in \mathscr{F}_3$, then it is already of the equivalent form of $\mathscr{A}_2$ given by Lemma 7.41.
4. If $(H^{-1})^{\otimes n} f \in \begin{bmatrix}1 & 0\\ 0 & \alpha\end{bmatrix}\mathscr{F}_1$, then a further transformation by $H_2 = \frac{1}{\sqrt{2}}\begin{bmatrix}1 & 1\\ 1 & -1\end{bmatrix} \in \mathbf{O}_2(\mathbb{C})$ puts it into the canonical form of $\mathscr{A}_1$, i.e., $f \in \mathscr{A}_1$ with transformation HH_2.
5. If $(H^{-1})^{\otimes n} f \in \begin{bmatrix}1 & 0\\ 0 & \alpha\end{bmatrix}\mathscr{F}_2$, then it is already in the canonical form of $\mathscr{A}_3$.
6. If $(H^{-1})^{\otimes n} f \in \begin{bmatrix}1 & 0\\ 0 & \alpha\end{bmatrix}\mathscr{F}_3$, then it is of the form $\begin{bmatrix}1\\ \alpha^3\end{bmatrix}^{\otimes n} + i^r \begin{bmatrix}1\\ -\alpha^3\end{bmatrix}^{\otimes n}$ and a further transformation by $\begin{bmatrix}0 & -1\\ 1 & 0\end{bmatrix} \in \mathbf{O}_2(\mathbb{C})$ puts it into the canonical form of $\mathscr{A}_3$. To see this,

$$\begin{aligned}
&\begin{bmatrix}0 & -1\\ 1 & 0\end{bmatrix}^{\otimes n}\left(\begin{bmatrix}1\\ \alpha^3\end{bmatrix}^{\otimes n} + i^r \begin{bmatrix}1\\ -\alpha^3\end{bmatrix}^{\otimes n}\right)\\
&= \begin{bmatrix}-\alpha^3\\ 1\end{bmatrix}^{\otimes n} + i^r \begin{bmatrix}\alpha^3\\ 1\end{bmatrix}^{\otimes n} = (-\alpha^3)^n \left(\begin{bmatrix}1\\ -\frac{1}{\alpha^3}\end{bmatrix}^{\otimes n} + (-1)^n i^r \begin{bmatrix}1\\ \frac{1}{\alpha^3}\end{bmatrix}^{\otimes n}\right)\\
&= (-\alpha^3)^n \left(\begin{bmatrix}1\\ \alpha\end{bmatrix}^{\otimes n} + i^{2n+r} \begin{bmatrix}1\\ -\alpha\end{bmatrix}^{\otimes n}\right).
\end{aligned}$$

So $f \in \mathscr{A}_3$ with transformation $H\begin{bmatrix} 0 & 1 \\ -1 & 0 \end{bmatrix}$. We have shown that if f is $\mathscr{A}$-transformable then $f \in \mathscr{A}_1 \cup \mathscr{A}_2 \cup \mathscr{A}_3$.

Conversely, if there exists a matrix $H \in \mathbf{O}_2(\mathbb{C})$ such that $(H^{-1})^{\otimes n} f$ is in one of the canonical forms of $\mathscr{A}_1$, $\mathscr{A}_2$, or $\mathscr{A}_3$, then f is $\mathscr{A}$-transformable by applying Lemma 7.38. □

We also have a similar characterization for $\mathscr{P}$-transformable signatures. We define the stabilizer group of $\mathscr{P}$ similar to Stab($\mathscr{A}$). It is easy to see the left and right stabilizers of $\mathscr{P}$ coincide, which we denote by Stab($\mathscr{P}$). Furthermore, Stab($\mathscr{P}$) is generated by nonzero scalar multiples of matrices of the form $\begin{bmatrix} 1 & 0 \\ 0 & \nu \end{bmatrix}$ for any nonzero $\nu \in \mathbb{C}$ and $R = \begin{bmatrix} 0 & 1 \\ 1 & 0 \end{bmatrix}$.

Lemma 7.44. *Let $\mathcal{F}$ be a set of signatures. Then $\mathcal{F}$ is $\mathscr{P}$-transformable iff there exists an $H \in \mathbf{O}_2(\mathbb{C})$ such that $\mathcal{F} \subseteq H\mathscr{P}$ or $\mathcal{F} \subseteq H\begin{bmatrix} 1 & 1 \\ i & -i \end{bmatrix}\mathscr{P}$.*

Proof. Sufficiency is easily verified by checking that $=_2$ is transformed into $\mathscr{P}$ in both cases. In particular, H leaves $=_2$ unchanged.

If $\mathcal{F}$ is $\mathscr{P}$-transformable, then by definition, there exists a matrix T such that $(=_2)T^{\otimes 2} \in \mathscr{P}$ and $T^{-1}\mathcal{F} \subseteq \mathscr{P}$. The nondegenerate binary signatures in $\mathscr{P}$ are either $[0, 1, 0]$ or of the form $[1, 0, \nu]$, up to a scalar. However, notice that $[1, 0, 1] = [1, 0, \nu]\begin{bmatrix} 1 & 0 \\ 0 & \nu^{-\frac{1}{2}} \end{bmatrix}^{\otimes 2}$ and $\begin{bmatrix} 1 & 0 \\ 0 & \nu^{-\frac{1}{2}} \end{bmatrix} \in$ Stab($\mathscr{P}$). Thus, we only need to consider $[1, 0, 1]$ and $[0, 1, 0]$. Now we apply Proposition 7.35.

1. If $(=_2)T^{\otimes 2} = [1, 0, 1]$, then by case 1 of Proposition 7.35, we have $T \in \mathbf{O}_2(\mathbb{C})$. Therefore $\mathcal{F} \subseteq H\mathscr{P}$ where $H = T \in \mathbf{O}_2(\mathbb{C})$.
2. If $(=_2)T^{\otimes 2} = [0, 1, 0]$, then by case 3 of Proposition 7.35, there exists an $H \in \mathbf{O}_2(\mathbb{C})$ such that $T = \frac{1}{\sqrt{2}}H\begin{bmatrix} 1 & 1 \\ i & -i \end{bmatrix}$. Therefore $\mathcal{F} \subseteq H\begin{bmatrix} 1 & 1 \\ i & -i \end{bmatrix}\mathscr{P}$ where $H \in \mathbf{O}_2(\mathbb{C})$.

□

We also have similar definitions of the sets $\mathscr{P}_1$ and $\mathscr{P}_2$.

Definition 7.45. A symmetric signature f of arity n is in $\mathscr{P}_1$ if there exists $H \in \mathbf{O}_2(\mathbb{C})$ and a nonzero constant c such that $f = cH^{\otimes n}\left(\begin{bmatrix} 1 \\ 1 \end{bmatrix}^{\otimes n} + \beta\begin{bmatrix} 1 \\ -1 \end{bmatrix}^{\otimes n}\right)$, where $\beta \neq 0$.

We also define $\mathscr{P}_2 = \mathscr{A}_2$.

When such an H exists, we say that $f \in \mathscr{P}_1$ with transformation H. If $f \in \mathscr{P}_1$ with I_2, then we say f is in the canonical form of $\mathscr{P}_1$. If f is in the canonical form of $\mathscr{P}_1$, then by Proposition 7.36, for any $0 \le k \le n-2$, we have $f_{k+2} = f_k$. Since f is nondegenerate, $f_1 \neq \pm f_0$ is implied.

It is easy to check that $\mathscr{A}_1 \subset \mathscr{P}_1$. Now we characterize the $\mathscr{P}$-transformable signatures as we did for the $\mathscr{A}$-transformable signatures in Lemma 7.43.

Lemma 7.46. *Let f be a nondegenerate symmetric signature. Then f is $\mathscr{P}$-transformable iff $f \in \mathscr{P}_1 \cup \mathscr{P}_2$.*

Proof. Assume that f is $\mathscr{P}$-transformable of arity n. By applying Lemma 7.44 to $\{f\}$, there exists an $H \in \mathbf{O}_2(\mathbb{C})$ such that $f \in H\mathscr{P}$ or $f \in H\begin{bmatrix} 1 & 1 \\ \mathrm{i} & -\mathrm{i} \end{bmatrix}\mathscr{P}$. This is equivalent to $(H^{-1})^{\otimes n} f \in \mathscr{P}$ or $(H^{-1})^{\otimes n} f \in \begin{bmatrix} 1 & 1 \\ \mathrm{i} & -\mathrm{i} \end{bmatrix}\mathscr{P}$.

The nondegenerate symmetric signatures in $\mathscr{P}$ take the form $[0,1,0]$, or $[a,0,\ldots,0,b] = a[1,0]^{\otimes n} + b[0,1]^{\otimes n}$, where $ab \neq 0$. Now we consider the possible cases.

1. If $(H^{-1})^{\otimes n} f = [0,1,0]^{\mathrm{T}} = \frac{1}{2\mathrm{i}}\left(\begin{bmatrix} 1 \\ \mathrm{i} \end{bmatrix}^{\otimes 2} - \begin{bmatrix} 1 \\ -\mathrm{i} \end{bmatrix}^{\otimes 2}\right)$, then it is already of the equivalent form of $\mathscr{P}_2 = \mathscr{A}_2$ given by Lemma 7.41.
2. If $(H^{-1})^{\otimes n} f = a\begin{bmatrix} 1 \\ 0 \end{bmatrix}^{\otimes n} + b\begin{bmatrix} 0 \\ 1 \end{bmatrix}^{\otimes n}$, then a further transformation by $H_2 = \frac{1}{\sqrt{2}}\begin{bmatrix} 1 & 1 \\ 1 & -1 \end{bmatrix} \in \mathbf{O}_2(\mathbb{C})$ puts it into the canonical form of $\mathscr{P}_1$, i.e., $f \in \mathscr{P}_1$ with transformation HH_2.
3. If $(H^{-1})^{\otimes n} f = \begin{bmatrix} 1 & 1 \\ \mathrm{i} & -\mathrm{i} \end{bmatrix}^{\otimes 2} [0,1,0]^{\mathrm{T}} = 2[1,0,1]^{\mathrm{T}} = \begin{bmatrix} 1 \\ 1 \end{bmatrix}^{\otimes 2} + \begin{bmatrix} 1 \\ -1 \end{bmatrix}^{\otimes 2}$, then it is already in the canonical form of $\mathscr{P}_1$.
4. If $(H^{-1})^{\otimes n} f = \begin{bmatrix} 1 & 1 \\ \mathrm{i} & -\mathrm{i} \end{bmatrix}^{\otimes n}\left(a\begin{bmatrix} 1 \\ 0 \end{bmatrix}^{\otimes n} + b\begin{bmatrix} 0 \\ 1 \end{bmatrix}^{\otimes n}\right)$, then it is already of the equivalent form of $\mathscr{P}_2 = \mathscr{A}_2$ given by Lemma 7.41.

Conversely, if there exists a matrix $H \in \mathbf{O}_2(\mathbb{C})$ such that $(H^{-1})^{\otimes n} f$ is in one of the canonical forms of $\mathscr{P}_1$ or $\mathscr{P}_2$, then one can directly check that f is $\mathscr{P}$-transformable by applying Lemma 7.44. □

Combining Lemmas 7.43 and 7.46, we have a necessary and sufficient condition for a single nondegenerate symmetric signature to be $\mathscr{A}$- or $\mathscr{P}$-transformable.

Corollary 7.47. *Let f be a nondegenerate symmetric signature. Then f is $\mathscr{A}$- or $\mathscr{P}$-transformable iff $f \in \mathscr{P}_1 \cup \mathscr{P}_2 \cup \mathscr{A}_3$.*

Notice that our definitions of $\mathscr{P}_1$, $\mathscr{P}_2$, and $\mathscr{A}_3$ each involve an orthogonal transformation. For any single signature $f \in \mathscr{P}_1 \cup \mathscr{P}_2 \cup \mathscr{A}_3$, Holant($f$) is tractable. However, this does not imply that Holant($\mathscr{P}_1$), Holant($\mathscr{P}_2$), or Holant($\mathscr{A}_3$) is tractable. One can check, using Theorem 7.19, that Holant($\mathscr{P}_2$) is tractable while Holant($\mathscr{P}_1$) and Holant($\mathscr{A}_3$) are #P-hard.

Given a symmetric signature f, whether f is $\mathscr{A}$- or $\mathscr{P}$-transformable is decidable in polynomial time in the size of the symmetric signature f given as $n+1$ numbers. We will discuss some necessary conditions in Section 7.9.

7.7.2 Dichotomies When $\mathscr{A}$- or $\mathscr{P}$-transformable Signatures Appear

Our characterizations of $\mathscr{A}$-transformable signatures in Lemma 7.43 and $\mathscr{P}$-transformable signatures in Lemma 7.46 are up to transformations in $\mathbf{O}_2(\mathbb{C})$. Since an orthogonal transformation never changes the complexity of the problem, in the proofs of the following lemmas, we assume any signature in $\mathscr{A}_i$ for $i = 1, 2, 3$, or $\mathscr{P}_j$ for $j = 1, 2$, is already in the canonical form. These lemmas state the general principle that if a set $\mathcal{F}$ of symmetric signatures contains a nondegenerate $\mathscr{A}$-transformable or a $\mathscr{P}$-transformable signature of arity $n \geq 3$, then the only way Holant($\mathcal{F}$) is tractable is when the entire set is $\mathscr{P}$-transformable or $\mathscr{A}$-transformable.

We first give a simple lemma to interpolate $(=_4)$.

Lemma 7.48. *Let $a, b \in \mathbb{C}$. If $ab \neq 0$, then for any set $\mathcal{F}$ of complex-weighted signatures containing $[a, 0, \ldots, 0, b]$ of arity $m \geq 3$,*

$$\text{Holant}(\mathcal{F} \cup \mathcal{EQ}_2) \leq_T \text{Holant}(\mathcal{F}),$$

where $\mathcal{EQ}_2 = \{=_2, =_4, \ldots, =_{2k}, \ldots\}$ is the set of all Equalities *with an even arity.*

The same reduction also works for planar graphs, thus Pl-Holant($\mathcal{F} \cup \mathcal{EQ}_2$) $\leq_T$ Pl-Holant($\mathcal{F}$).

Proof. We only need to show how to get $(=_4)$. Note that a self-loop on $(=_4)$ gives $(=_2)$, and a chain of m copies of $(=_4)$ gives $(=_{2m+2})$, where we link one edge of the ith copy with one edge of the $(i+1)$st copy $(1 \leq i \leq m-1)$.

Since $a \neq 0$, we can normalize the first entry to get $[1, 0, \ldots, 0, x]$, where $x \neq 0$. First, we show how to obtain an arity 4 Gen-Eq signature. If $m = 3$, then we connect two copies together by a single edge to get an arity 4 signature. For larger arities, we form self-loops until realizing a signature of arity 3 or 4. By this process, we have a signature $g = [1, 0, 0, 0, y]$, where $y \neq 0$. If y is a pth root of unity, then we can directly realize $(=_4)$ by connecting p copies of g together, two edges at a time as in Figure 7.2. Otherwise, y is not a root of unity and we can interpolate $(=_4)$ as follows.

Consider an instance Ω of $\operatorname{Holant}(\mathcal{F} \cup \{=_4\})$. Suppose that $(=_4)$ appears n times in Ω. We stratify the assignments in Ω based on the assignments to $(=_4)$. We only need to consider the all-zero and all-one assignments since any other assignment contributes a factor of 0. We have

$$\operatorname{Holant}_\Omega = \sum_{i=0}^{n} c_i,$$

where c_i is the sum, over all assignments with exactly i many $(=_4)$'s in Ω assigned 1111, and $n - i$ many $(=_4)$'s in Ω assigned 0000, of the product of evaluations of all other signatures on Ω.

We construct from Ω a sequence of instances Ω_s of $\operatorname{Holant}(\mathcal{F})$ indexed by $s \geq 1$. We obtain Ω_s from Ω by replacing each occurrence of $(=_4)$ with a gadget g_s created from s copies of $[1, 0, 0, 0, y]$, connecting two edges together at a time as in Figure 7.2. Then

$$\operatorname{Holant}_{\Omega_s} = \sum_{i=0}^{n} (y^s)^i c_i.$$

For $s \geq 1$, this gives a coefficient matrix that is Vandermonde. Since y is neither 0 nor a root of unity, y^s is distinct for each s. Therefore, the Vandermonde system $(1 \leq s \leq n + 1)$ has full rank. We can solve for the unknowns c_i in polynomial time and obtain the value of $\operatorname{Holant}_\Omega$. □

Lemma 7.49. *Let $\mathcal{F}$ be a set of symmetric signatures. Suppose $\mathcal{F}$ contains a nondegenerate signature $f \in \mathscr{P}_1$ of arity $n \geq 3$. Then* $\operatorname{Holant}(\mathcal{F})$ *is* #P-*hard unless $\mathcal{F}$ is $\mathscr{P}$-transformable or $\mathscr{A}$-transformable.*

Proof. Without loss of generality we assume f is in canonical form of $\mathscr{P}_1$ after an orthogonal transformation. The conclusion that $\mathcal{F}$ is $\mathscr{P}$-transformable or $\mathscr{A}$-transformable is invariant under this transformation. So by assumption, for any $0 \leq k \leq n - 2$, $f_{k+2} = f_k$ and $f_1 \neq \pm f_0$ since f is not degenerate. We can express f (after an orthogonal transformation) as

$$f = a_0 \begin{bmatrix} 1 \\ 1 \end{bmatrix}^{\otimes n} + a_1 \begin{bmatrix} 1 \\ -1 \end{bmatrix}^{\otimes n},$$

where $a_0 = (f_0 + f_1)/2$, $a_1 = (f_0 - f_1)/2$, and $a_0 a_1 \neq 0$. For this f, we can further perform an orthogonal transformation by $H_2 = \frac{1}{\sqrt{2}}\begin{bmatrix} 1 & 1 \\ 1 & -1 \end{bmatrix}$ so that f is transformed into the GEN-EQ signature $[a_0, 0, \ldots, 0, a_1]$ of arity n, up to a nonzero constant. By Lemma 7.48, we can obtain $(=_4)$. With $(=_4)$, we can realize any equality signature of even arity. Thus, $\#\text{CSP}^2(H_2\mathcal{F}) \leq_{\text{T}} \text{Holant}(\mathcal{F})$.

Now we apply Theorem 6.16, the $\#\text{CSP}^d$ dichotomy for $d = 2$, to the set $H_2\mathcal{F}$. If $\#\text{CSP}^2(H_2\mathcal{F})$ is #P-hard, then Holant($\mathcal{F}$) is #P-hard as well. Otherwise, this problem is tractable. Therefore, there exists some T of the form $\begin{bmatrix} 1 & 0 \\ 0 & \alpha^k \end{bmatrix}$, where the integer $k \in \{0, 1, \ldots, 7\}$, such that $TH_2\mathcal{F}$ is a subset of $\mathscr{A}$ or $\mathscr{P}$.

If $TH_2\mathcal{F} \subseteq \mathscr{P}$, then we have $H_2\mathcal{F} \subseteq T^{-1}\mathscr{P}$. Notice that $T \in \text{Stab}(\mathscr{P})$, so $T^{-1}\mathscr{P} = \mathscr{P}$. Thus, $\mathcal{F}$ is $\mathscr{P}$-transformable under this H_2 transformation. Otherwise, $TH_2\mathcal{F} \subseteq \mathscr{A}$. It is easy to verify that $(=_2)((TH_2)^{-1})^{\otimes 2}$ is $[1, 0, \mathfrak{i}^{-k}] \in \mathscr{A}$. Thus, $\mathcal{F}$ is $\mathscr{A}$-transformable under this TH_2 transformation. □

Note that the hypothesis $f \in \mathscr{P}_1$ in Lemma 7.49 may seem to demand that all $\mathcal{F}$ is $\mathscr{P}$-transformable in order that the problem Holant($\mathcal{F}$) remain not #P-hard. However, this is not so. Since $\mathscr{A}_1$ is a proper subset of $\mathscr{P}_1$, the condition $f \in \mathscr{P}_1$ is consistent with possibly $f \in \mathscr{A}_1$. In that case $\mathcal{F}$ might be $\mathscr{A}$-transformable and not $\mathscr{P}$-transformable, and then Holant($\mathcal{F}$) is tractable (thus not #P-hard if #P does not collapse to P).

Lemma 7.50. *Let $\mathcal{F}$ be a set of symmetric signatures. Suppose $\mathcal{F}$ contains a nondegenerate signature $f \in \mathscr{P}_2$ of arity $n \geq 3$. Then* Holant($\mathcal{F}$) *is* #P-*hard unless $\mathcal{F}$ is $\mathscr{P}$-transformable or $\mathscr{A}$-transformable.*

Proof. By assumption, for any $0 \leq k \leq n-2$, $f_{k+2} = -f_k$ and $f_1 \neq \pm\mathfrak{i} f_0$ since f is not degenerate. By Lemma 7.41, we have

$$f = a_0 \begin{bmatrix} 1 \\ \mathfrak{i} \end{bmatrix}^{\otimes n} + a_1 \begin{bmatrix} 1 \\ -\mathfrak{i} \end{bmatrix}^{\otimes n},$$

where $a_0 = (f_0 - \mathfrak{i} f_1)/2$ and $a_1 = (f_0 + \mathfrak{i} f_1)/2$, and $a_0 a_1 \neq 0$. Then under the holographic transformation

$$Z' = \begin{bmatrix} \sqrt[n]{a_0} & \sqrt[n]{a_1} \\ \mathfrak{i}\sqrt[n]{a_0} & -\mathfrak{i}\sqrt[n]{a_1} \end{bmatrix}^{-1},$$

we have

$$Z'^{\otimes n} f = (=_n) = \begin{bmatrix} 1 \\ 0 \end{bmatrix}^{\otimes n} + \begin{bmatrix} 0 \\ 1 \end{bmatrix}^{\otimes n}$$

and

$$\begin{aligned}\operatorname{Holant}(=_2 \mid \mathcal{F} \cup \{f\}) &\equiv_{\mathrm{T}} \operatorname{Holant}([1,0,1](Z'^{-1})^{\otimes 2} \mid Z'\mathcal{F} \cup \{Z'^{\otimes n} f\}) \\ &\equiv_{\mathrm{T}} \operatorname{Holant}([0,1,0] \mid Z'\mathcal{F} \cup \{=_n\}).\end{aligned}$$

Thus, we have a bipartite graph having $(=_n)$ together with $Z'\mathcal{F}$ on the right and $(\neq_2) = [0, 1, 0]$ on the left up to a nonzero scalar. Inductively we can realize $(=_{kn})$ on the right for every integer $k > 0$. Suppose we have already constructed $(=_{(k-1)n})$. Place one copy of $(=_{(k-1)n})$ and three copies of $(=_n)$ on the right, and call them A, B, and C. Connect $(n-2)$ edges of A to $(=_{(k-1)n})$, one edge of A to one edge of B, and one edge of A to one edge of C, with all connections via $(\neq_2)$ on the left. This creates a $(=_{kn})$ on the right. Moreover, we can move these EQUALITY signatures to the left side by connecting each edge of $(=_{kn})$ by a copy of $(\neq_2)$. Thus, $\#\mathrm{CSP}^n(Z'\mathcal{F}) \leq_{\mathrm{T}} \operatorname{Holant}(\mathcal{F})$.

Now we apply Theorem 6.16, the $\#\mathrm{CSP}^d$ dichotomy with $d = n$, to the set $Z'\mathcal{F}$. Note that this n is a constant independent of the input size of any signature grid Ω. If $\#\mathrm{CSP}^n(Z'\mathcal{F})$ is #P-hard, then $\operatorname{Holant}(\mathcal{F})$ is #P-hard as well. Otherwise, $\#\mathrm{CSP}^n(Z'\mathcal{F})$ is tractable. Let ω be a primitive $4n$th root of unity. Then under the holographic transformation $T = \begin{bmatrix} 1 & 0 \\ 0 & \omega^k \end{bmatrix}$ for some integer k, $TZ'\mathcal{F}$ is a subset of $\mathscr{A}$ or $\mathscr{P}$. If $TZ'\mathcal{F} \subseteq \mathscr{P}$, then we have $Z'\mathcal{F} \subseteq T^{-1}\mathscr{P}$. Notice that $T \in \operatorname{Stab}(\mathscr{P})$, so $T^{-1}\mathscr{P} = \mathscr{P}$. Also recall that $(=_2)(Z'^{-1})^{\otimes 2}$ is a multiple of $(\neq_2)$. Thus, $\mathcal{F}$ is $\mathscr{P}$-transformable under this Z' transformation.

Otherwise, $TZ'\mathcal{F} \subseteq \mathscr{A}$. It is easy to verify that $(=_2)((TZ')^{-1})^{\otimes 2}$ is $[0, 1, 0] \in \mathscr{A}$ up to a scalar. Thus, $\mathcal{F}$ is $\mathscr{A}$-transformable under this TZ' transformation. □

Lemma 7.51. *Let $\mathcal{F}$ be a set of symmetric signatures. Suppose $\mathcal{F}$ contains a nondegenerate signature $f \in \mathscr{A}_3$ of arity $n \geq 3$. Then* Holant($\mathcal{F}$) *is* #P-*hard unless $\mathcal{F}$ is $\mathscr{A}$-transformable.*

Proof. After an orthogonal transformation we can express f as

$$f = \lambda \left(\begin{bmatrix} 1 \\ \alpha \end{bmatrix}^{\otimes n} + \mathfrak{i}^r \begin{bmatrix} 1 \\ -\alpha \end{bmatrix}^{\otimes n} \right),$$

for some $\lambda \neq 0$ and integer r. In this canonical form, we have $f_{k+2} = \mathfrak{i} f_k$ for any $0 \leq k \leq n-2$.

A self-loop on f yields f', where $f'_k = f_k + f_{k+2} = (1+\mathfrak{i}) f_k$. Thus up to the constant $(1+\mathfrak{i})$, f' is just the first $n-2$ entries of f. By doing zero or more self-loops, we eventually obtain an arity 4 signature when n is even or a ternary one when n is odd. There are eight cases depending on the first two entries of f and the parity of n. However, in each case, we can realize the signature $[1, 0, \mathfrak{i}]$.

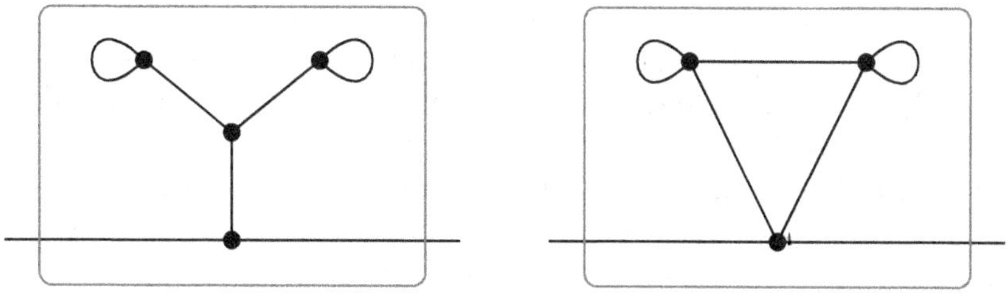

(a) Vertices assigned $[1, \alpha i, i, -\alpha]$. (b) Vertices assigned $[1, \alpha i, i, -\alpha, -1]$.

Figure 7.12. Constructions to realize $[1, 0, i]$.

We list them here. (In the calculations below, we omit certain nonzero constant factors without explanation.)

- $[0, 1, 0, i]$: Another self-loop gives $[0, 1]$. Connect it back to the ternary, we get $[1, 0, i]$.
- $[1, 0, i, 0]$: Another self-loop gives $[1, 0]$. Connect it back to the ternary, we get $[1, 0, i]$.
- $[1, \alpha i, i, -\alpha]$: Another self-loop gives $[1, \alpha i]$. Connect two copies of it to the ternary, we get $[1, -\alpha]$. Then connect this back to the ternary, we finally get $[1, 0, i]$. See Figure 7.12a.
- $[1, -\alpha i, i, \alpha]$: Same construction as in the previous case.
- $[0, 1, 0, i, 0]$: Another self-loop gives $[0, 1, 0]$. Connect it back to the arity 4 signature, we get $[1, 0, i]$.
- $[1, 0, i, 0, -1]$: Another self-loop gives $[1, 0, i]$ directly.
- $[1, \alpha i, i, -\alpha, -1]$: Another self-loop gives $[1, \alpha i, i]$. Connect two copies of it together, we get $[1, -\alpha, -i]$. Connect this back to the arity 4 signature, we get $[1, 0, i]$. See Figure 7.12b.
- $[1, -\alpha i, i, \alpha, -1]$: Same construction as in the previous case.

With $[1, 0, i]$ in hand, we can connect three copies to get $[1, 0, -i]$. Now we construct a bipartite graph, with $\mathcal{F} \cup \{=_2\}$ on the right side and $[1, 0, -i]$ on the left. This problem is reducible to Holant($\mathcal{F}$). We perform a holographic transformation by $M^{-1} = \frac{1}{2\alpha}\begin{bmatrix} 1 & 1 \\ \alpha & -\alpha \end{bmatrix}$, with $M = \begin{bmatrix} \alpha & 1 \\ \alpha & -1 \end{bmatrix}$, and get

$$\begin{aligned}
&\operatorname{Holant}([1, 0, -i] \mid \mathcal{F} \cup \{f, =_2\}) \\
&\quad \equiv_T \operatorname{Holant}([1, 0, -i](M^{-1})^{\otimes 2} \mid M\mathcal{F} \cup \{M^{\otimes n} f, M^{\otimes 2}(=_2)\}) \\
&\quad \equiv_T \operatorname{Holant}([1, 0, 1] \mid M\mathcal{F} \cup \{[1, 0, \ldots, 0, i^r], [1, i, 1]\}) \\
&\quad \equiv_T \operatorname{Holant}\left(M\mathcal{F} \cup \{[1, 0, \ldots, 0, i^r], [1, i, 1]\}\right).
\end{aligned}$$

Note that $M^{\otimes n} f$ is $[1, 0, \ldots, 0, \mathrm{i}^r]$ (after normalizing the first entry); similarly $M^{\otimes 2}(=_2)$ is $[1, \mathrm{i}, 1]$. On the left side, $[1, 0, -\mathrm{i}](M^{-1})^{\otimes 2}$ is a nonzero multiple of $[1, 0, 1] = (=_2)$. Therefore, we can construct all EQUALITY signatures of even arity as follows: Use four copies of $[1, 0, \ldots, 0, \mathrm{i}^r]$ of arity n linked as a chain by three edges we get $(=_{4n-6})$ with even arity $4n - 6 \geq 6$. Then we can get $(=_2)$ and $(=_4)$ by self-loops. Finally, connecting k copies of $(=_4)$ as a chain by $k-1$ edges gives $(=_{4k-2(k-1)}) = (=_{2k+2})$ for all $k \geq 2$. Thus,

$$\#\mathrm{CSP}^2(M\mathcal{F} \cup \{[1, \mathrm{i}, 1]\}) \leq_{\mathrm{T}} \mathrm{Holant}(\mathcal{F}).$$

Now we apply Theorem 6.16, the $\#\mathrm{CSP}^d$ dichotomy for $d = 2$, to the set $M\mathcal{F} \cup \{[1, \mathrm{i}, 1]\}$. If $\#\mathrm{CSP}^2(M\mathcal{F} \cup \{[1, \mathrm{i}, 1]\})$ is #P-hard, then $\mathrm{Holant}(\mathcal{F})$ is #P-hard as well. Otherwise, $\#\mathrm{CSP}^2(M\mathcal{F} \cup \{[1, \mathrm{i}, 1]\})$ is tractable, and there exists some T of the form $\begin{bmatrix} 1 & 0 \\ 0 & \alpha^d \end{bmatrix}$, where the integer $d \in \{0, 1, \ldots, 7\}$, such that $TM\mathcal{F} \cup \{T^{\otimes 2}[1, \mathrm{i}, 1]\}$ is a subset of $\mathscr{A}$ or $\mathscr{P}$.

However, $T^{\otimes 2}[1, \mathrm{i}, 1]$ cannot be in $\mathscr{P}$ for a diagonal T. Thus $TM\mathcal{F} \cup \{T^{\otimes 2}[1, \mathrm{i}, 1]\} \subseteq \mathscr{A}$. Further notice that if $d \in \{1, 3, 5, 7\}$ in the expression of T, then $T^{\otimes 2}[1, \mathrm{i}, 1]$ is not in $\mathscr{A}$. Hence, T must be of the form $\begin{bmatrix} 1 & 0 \\ 0 & \mathrm{i}^d \end{bmatrix}$, where the integer $d \in \{0, 1, 2, 3\}$. For such T, $T^{\otimes 2}[1, \mathrm{i}, 1] \in \mathscr{A}$, and $T^{-1}\mathscr{A} = \mathscr{A}$ as $T \in \mathrm{Stab}(\mathscr{A})$. Thus, $TM\mathcal{F} \cup \{T^{\otimes 2}[1, \mathrm{i}, 1]\} \subseteq \mathscr{A}$ simply becomes $M\mathcal{F} \subseteq \mathscr{A}$. Moreover, $(=_2)(M^{-1})^{\otimes 2}$ is a nonzero multiple of $[1, -\mathrm{i}, 1] \in \mathscr{A}$. Therefore, $\mathcal{F}$ is $\mathscr{A}$-transformable under the transformation by $M = \begin{bmatrix} \alpha & 1 \\ \alpha & -1 \end{bmatrix}$. □

7.8 Proof of Theorem 7.19

In this section, we finish the proof of the main theorem of this chapter, Theorem 7.19. We begin with a dichotomy for a single signature, which we prove by induction on its arity.

Theorem 7.52. *If f is a nondegenerate symmetric signature of arity ≥ 3 with complex weights in Boolean variables, then* Holant(f) *is* #P*-hard unless $f \in \mathscr{P}_1 \cup \mathscr{P}_2 \cup \mathscr{A}_3$ or f is vanishing, in which case the problem is in* P.

Recall that $\mathscr{A}_1 \subset \mathscr{P}_1$ and $\mathscr{A}_2 = \mathscr{P}_2$. Thus $f \in \mathscr{P}_1 \cup \mathscr{P}_2 \cup \mathscr{A}_3$ iff f is $\mathscr{A}$-transformable or $\mathscr{P}$-transformable by Corollary 7.47.

Proof. Let the arity of f be n. The base cases of $n = 3$ and $n = 4$ are proved in Theorem 7.29. Now assume $n \geq 5$.

Let f' be the signature of arity ≥ 3 obtained from f by forming a self-loop. We consider the cases separately whether f' is degenerate or not.

- Suppose $f' = [a, b]^{\otimes(n-2)}$ is degenerate. There are three cases to consider.

  1. If $a = b = 0$, then f' is the all-zero signature. For f, this means $f_{k+2} = -f_k$ for $0 \leq k \leq n-2$, so $f \in \mathscr{A}_2 = \mathscr{P}_2$ by Lemma 7.41, and therefore Holant(f) is tractable.
  2. If $a^2 + b^2 \neq 0$, then f' is nonzero and $[a, b]$ is not a constant multiple of either $[1, \mathrm{i}]$ or $[1, -\mathrm{i}]$. We may normalize so that $a^2 + b^2 = 1$. Then the orthogonal transformation $\begin{bmatrix} a & b \\ -b & a \end{bmatrix}$ transforms the column vector $[a, b]$ to $[1, 0]$. Let $\hat{f}$ be the transformed signature from f, and $\widehat{f'} = [1, 0]^{\otimes(n-2)}$ the transformed signature from f'.

     Since an orthogonal transformation keeps $(=_2)$ invariant, this transformation commutes with the operation of taking a self-loop, i.e., $\widehat{f'} = (\hat{f})'$. Here $(\hat{f})'$ is the function obtained from $\hat{f}$ by taking a self-loop. So $\hat{f}_0 + \hat{f}_2 = 1$ and for every integer $1 \leq k \leq n-2$, we have $\hat{f}_k = -\hat{f}_{k+2}$. With one or more self-loops, we eventually obtain either $[1, 0]$ when n is odd or $[1, 0, 0]$ when n is even. In either case, we connect an appropriate number of copies of this signature to $\hat{f}$ to get an arity 4 signature $\hat{g} = [\hat{f}_0, \hat{f}_1, \hat{f}_2, -\hat{f}_1, -\hat{f}_2]$. We show that Holant($\hat{g}$) is #P-hard. To see this, we first compute $\det(\widetilde{M_{\hat{g}}}) = -2(\hat{f}_0 + \hat{f}_2)(\hat{f}_1^2 + \hat{f}_2^2) = -2(\hat{f}_1^2 + \hat{f}_2^2)$, since $\hat{f}_0 + \hat{f}_2 = 1$. Therefore if $\hat{f}_1^2 + \hat{f}_2^2 \neq 0$, Holant($\hat{g}$) is #P-hard by Corollary 7.25. Otherwise $\hat{f}_1^2 + \hat{f}_2^2 = 0$, and we consider $\hat{f}_2 = \mathrm{i}\hat{f}_1$ since the other case $\hat{f}_2 = -\mathrm{i}\hat{f}_1$ is similar. Since f is nondegenerate, $\hat{f}$ is nondegenerate, which implies $\hat{f}_2 \neq 0$. We can express $\hat{g}$ as $[1, 0]^{\otimes 4} - \hat{f}_2[1, \mathrm{i}]^{\otimes 4}$. Under the holographic transformation by $T = \begin{bmatrix} 1 & (-\hat{f}_2)^{1/4} \\ 0 & \mathrm{i}(-\hat{f}_2)^{1/4} \end{bmatrix}$, we have

$$\begin{aligned} \operatorname{Holant}(=_2 \mid \hat{g}) &\equiv_T \operatorname{Holant}([1, 0, 1]T^{\otimes 2} \mid (T^{-1})^{\otimes 4}\hat{g}) \\ &\equiv_T \operatorname{Holant}(\hat{h} \mid =_4), \end{aligned}$$

     where

$$\hat{h} = [1, 0, 1]T^{\otimes 2} = [1, (-\hat{f}_2)^{1/4}, 0]$$

     and $\hat{g}$ is transformed by T^{-1} into the arity 4 equality $(=_4)$, which can be verified by (also the reason for the choice of T)

$$T^{\otimes 4}\left(\begin{bmatrix} 1 \\ 0 \end{bmatrix}^{\otimes 4} + \begin{bmatrix} 0 \\ 1 \end{bmatrix}^{\otimes 4}\right) = \begin{bmatrix} 1 \\ 0 \end{bmatrix}^{\otimes 4} - \hat{f}_2 \begin{bmatrix} 1 \\ \mathrm{i} \end{bmatrix}^{\otimes 4} = \hat{g}.$$

(a) The circles are assigned $\hat{f}$ and the squares are assigned $(\neq_2)$.

(b) The circles are assigned f.

Figure 7.13. Two gadgets used when $f' = [1, \pm i]^{\otimes(n-2)}$.

By Theorem 5.3 for the case $k = 4$ (which is proved in Chapter 5), Holant$(\hat{h} \mid =_4)$ is #P-hard as $\hat{f}_2 \neq 0$.

3. If $a^2 + b^2 = 0$ but $(a, b) \neq (0, 0)$, then $[a, b]$ is a nonzero multiple of $[1, \pm i]$. We have $f' = [1, i]^{\otimes(n-2)}$ or $[1, -i]^{\otimes(n-2)}$ up to a nonzero constant. We consider the first case since the other case is similar.

Under the holographic transformation $Z = \frac{1}{\sqrt{2}}\begin{bmatrix} 1 & 1 \\ i & -i \end{bmatrix}$, $f' = Z^{\otimes(n-2)}[1, 0]^{\otimes(n-2)}$, up to a nonzero constant, and we have

$$\begin{aligned} \text{Holant}(=_2 \mid f) &\equiv_T \text{Holant}(=_2 \mid f, f') \\ &\equiv_T \text{Holant}([1, 0, 1]Z^{\otimes 2} \mid (Z^{-1})^{\otimes n} f, (Z^{-1})^{\otimes(n-2)} f') \\ &\equiv_T \text{Holant}([0, 1, 0] \mid \hat{f}, [1, 0]^{\otimes(n-2)}), \end{aligned}$$

where $\hat{f} = (Z^{-1})^{\otimes n} f$ takes the form $[\hat{f}_0, \hat{f}_1, 0, \ldots, 0, \hat{f}_n]$ with $\hat{f}_1 \neq 0$. We can normalize $\hat{f}_1 = 1$. If $\hat{f}_n = 0$ then $f \in \mathscr{V}^+$ and we are done. Suppose $\hat{f}_n \neq 0$.

Now consider the gadget in Figure 7.13a with $\hat{f}$ assigned to both vertices, which clearly has a symmetric signature. The Hamming weight 0 entry of this binary signature is 0. This is because the arity of $\hat{f}$ is $n \geq 5$ and so either the left or the right side copy of $\hat{f}$ has at least two incident edges assigned 1 as they are linked by $(\neq_2)$ signatures. But $\hat{f}$ takes value 0 for input of Hamming weight w if $2 \leq w < n$. Similarly one can see that for input 01, the only nonzero configuration is all incident edges to the left side copy of $\hat{f}$ take 0, and those of the right take 1, which has the value $\hat{f}_0\hat{f}_n$. Finally, for input 11, we have two nonzero configurations and get $2\hat{f}_n$ as $\hat{f}_1 = 1$. Thus the binary signature of this gadget is $[0, \hat{f}_0\hat{f}_n, 2\hat{f}_n]$, which is equivalent to $[0, \hat{f}_0, 2]$ since $\hat{f}_n \neq 0$.

By Lemma 7.8, $g = Z^{\otimes 2}[0, \hat{f}_0, 2]$ is a binary symmetric signature with $\text{vd}^+(g) = 2$, and so $g \notin \mathscr{R}_2^+$.

If $\hat{f}_0 \neq 0$, then $g = Z^{\otimes 2}[0, \hat{f}_0, 2]$ is nondegenerate. In this case we construct some function in $\mathscr{V}^+$. First we connect f' back to f, getting a binary signature $Z^{\otimes 2}[0, 0, \hat{f}_n]$. This is best seen in the Z basis, where we connect all $n-2$ edges of $[1, 0]^{\otimes(n-2)}$ via $(\neq_2)$ to $\hat{f}$. So up to a nonzero factor we have $Z^{\otimes 2}[0, 0, 1]$. Then we connect both edges of $Z^{\otimes 2}[0, 0, 1]$ to f, the resulting signature is $h = Z^{\otimes(n-2)}[\hat{f}_0, 1, 0, \ldots, 0]$ of arity $n-2 \geq 3$, up to a nonzero scalar. This is again best seen in the Z basis. Notice that h is nondegenerate and $h \in \mathscr{V}^{+}$. By Lemma 7.33, Holant($\{h, g\}$) is #P-hard, and hence Holant(f) is also #P-hard.

Otherwise suppose $\hat{f}_0 = 0$. Then we have $g = Z^{\otimes 2}[0, 0, 2] = 2Z^{\otimes 2}[0, 1]^{\otimes 2} = [1, -i]^{\otimes 2}$. Connecting this degenerate signature to f, we get a signature $h = \langle f, g\rangle$. Under the Z basis, $(Z^{-1})^{\otimes(n-2)}h = [\hat{f}_0, \hat{f}_1, 0, \ldots, 0] = [0, 1, 0, \ldots, 0]$ of arity $n-2 \geq 3$, up to a nonzero scalar. Hence h is nondegenerate and $\mathrm{rd}^{+}(h) = 1$. Therefore $h \in \mathscr{V}^{+}$.

Moreover, assigning f to both vertices in the gadget of Figure 7.13b, we get a nondegenerate signature $h' \in \mathscr{V}^{-}$ of arity 4. To see this, consider this gadget after a holographic transformation by Z. In this bipartite setting, it is the same as assigning $\hat{f} = [0, 1, 0, \ldots, 0, \hat{f}_n]$ to both the circle and triangle vertices in the gadget of Figure 7.7a. The square vertices there are still assigned $(\neq_2) = [0, 1, 0]$. Although it is not apparent from the gadget's geometry, this signature is in fact symmetric. In particular, its values on inputs 1010 and 1100 are both zero. Note that as $n \geq 5$, the number of parallel paths with $(\neq_2)$ in the middle between the two copies of $\hat{f}$ is $n-2 \geq 3$, and thus at least one copy of $\hat{f}$ has two internal incident edges set to 1 due to the $(\neq_2)$ in the middle. For Hamming weight ≥ 2, the only nonzero value of $\hat{f}$ is at weight n. From this, one can see that the resulting signature is $\widehat{h'} = (Z^{-1})^{\otimes 4}h' = [0, 0, 0, \hat{f}_n, 0]$. Hence $\mathrm{rd}^{-}(h') = 1$ by Lemma 7.8. Therefore h' is nondegenerate and $h' \in \mathscr{V}^{-}$.

By Lemma 7.34, Holant($\{h, h'\}$) is #P-hard, hence Holant(f) is also #P-hard.

- Suppose f' is nondegenerate. If Holant(f') is #P-hard then so is Holant(f). We now assume Holant(f') is not #P-hard. Then, by the inductive hypothesis, $f' \in \mathscr{P}_1 \cup \mathscr{P}_2 \cup \mathscr{A}_3$ or f' is vanishing. If $f' \in \mathscr{P}_1 \cup \mathscr{P}_2 \cup \mathscr{A}_3$, then applying Lemma 7.49, Lemma 7.50, or Lemma 7.51 to f' and the set $\{f, f'\}$, we either have that Holant($\{f, f'\}$) is #P-hard, so Holant(f) is #P-hard as well, or that f is $\mathscr{A}$- or $\mathscr{P}$-transformable, so by Corollary 7.47, $f \in \mathscr{P}_1 \cup \mathscr{P}_2 \cup \mathscr{A}_3$.

  Otherwise, f' is vanishing, so $f' \in \mathscr{V}^{\sigma}$ for $\sigma \in \{+, -\}$ by Theorem 7.11. We may assume that $f' \in \mathscr{V}^{+}$; the other case is similar. Let $\mathrm{rd}^{+}(f') = d - 1$. Then $f' \in \mathscr{V}^{+}$ implies that $2d < n$ and f' is nondegenerate implies that $d \geq 2$.

Under the transformation $Z = \frac{1}{\sqrt{2}}\begin{bmatrix} 1 & 1 \\ i & -i \end{bmatrix}$, we have

$$\begin{aligned}\operatorname{Holant}(=_2 \mid f) &\equiv_T \operatorname{Holant}((=_2)Z^{\otimes 2} \mid (Z^{-1})^{\otimes n} f) \\ &\equiv_T \operatorname{Holant}(\neq_2 \mid \hat{f}),\end{aligned}$$

where $\hat{f} = (Z^{-1})^{\otimes n} f$. We have $\hat{f} = [\hat{f}_0, \hat{f}_1, \ldots, \hat{f}_d, 0, \ldots, 0, \hat{f}_n]$, with $\hat{f}_d \neq 0$, because $\operatorname{rd}^+(f') = d - 1$. If $\hat{f}_n = 0$ then f is vanishing, and we are done. We may assume $\hat{f}_n \neq 0$.

Note that $d < n/2$ and $\hat{f}_d$ occurs strictly in the first half, and there are $n - d - 1$ many 0's between $\hat{f}_d$ and $\hat{f}_n$. Taking a self-loop in the original setting is equivalent to connecting both edges of $(\neq_2)$ to two inputs of a signature after this transformation. Thus, doing this once on $\hat{f}$, we can get $\widehat{f'} = 2[\hat{f}_1, \ldots, \hat{f}_d, 0, \ldots, 0]$ corresponding to f', namely $(Z^{-1})^{\otimes(n-2)} f'$. Doing this $d - 2$ times on $\hat{f}$, up to a nonzero factor, we get a signature $\hat{h} = [\hat{f}_{d-2}, \hat{f}_{d-1}, \hat{f}_d, 0, \ldots, 0, 0/\hat{f}_n]$ of arity $n - 2(d - 2) = n - 2d + 4$. The last entry is $\hat{f}_n$ when $d = 2$ and is 0 when $d > 2$.

As $n > 2d$, we may do this two more times on $\hat{h}$ and get $[\hat{f}_d, 0, \ldots, 0]$ of arity $k = n - 2d$. Now connect this signature back to $\hat{f}$ via $(\neq_2)$. We get a signature of arity $n - k = 2d$. This signature consists of the last $2d + 1$ signature entries of $\hat{f}$, up to a nonzero factor. We may repeat this operation zero or more times, which reduces the arity of $\hat{f}$ by k each time, until the arity k' of the resulting signature is less than or equal to k. We claim that this signature has the form $\hat{g} = [0, \ldots, 0, \hat{f}_n]$. In other words, the $k' + 1$ entries of $\hat{g}$ consist of the last $\hat{f}_n$ and k' many 0's in the signature $\hat{f}$, all appearing after $\hat{f}_d$. This is because there are $n - d - 1$ many 0 entries in the signature $\hat{f}$ after $\hat{f}_d$, and $n - d - 1 \geq k \geq k'$.

Hence, in the RHS of $\operatorname{Holant}(\neq_2 \mid \hat{f})$ we can get both $[\hat{f}_d, 0, \ldots, 0]$ of arity k and $[0, \ldots, 0, \hat{f}_n]$ of arity $k' \leq k$. If $k > k'$ we may repeat this process in the way of a subtractive Euclidean algorithm. The result is that we can get both $[1, 0]^{\otimes t}$ and $[0, 1]^{\otimes t}$, where $t = \gcd(k, k')$. Now putting k/t many copies of $[0, 1]^{\otimes t}$ together, we get $[0, 1]^{\otimes k}$.

Then we connect $[0, 1]^{\otimes k}$ back to $\hat{h} = [\hat{f}_{d-2}, \hat{f}_{d-1}, \hat{f}_d, 0, \ldots, 0, 0/\hat{f}_n]$ of arity $n - 2d + 4 = k + 4$ via $(\neq_2)$. Thus we get a signature of arity 4, which is $[\hat{f}_{d-2}, \hat{f}_{d-1}, \hat{f}_d, 0, 0]$. Note that the last entry is 0 (and not $\hat{f}_n$), because $k \geq 1$.

However, $\operatorname{Holant}(\neq_2 \mid [\hat{f}_{d-2}, \hat{f}_{d-1}, \hat{f}_d, 0, 0])$ is equivalent to $\operatorname{Holant}(\neq_2 \mid [0, 0, 1, 0, 0])$ when $\hat{f}_d \neq 0$. This is the Eulerian orientation problem #EO on 4-regular graphs and is #P-hard by Theorem 6.32. □

Finally we are ready to give the hardness proof in Theorem 7.19. We prove the contrapositive: If Holant($\mathcal{F}$) is not #P-hard then $\mathcal{F}$ must be in one of the five cases listed in Theorem 7.19. Combined with the tractability proof given in Section 7.2, this completes the proof of Theorem 7.19.

Proof of hardness in Theorem 7.19. Assume that Holant($\mathcal{F}$) is not #P-hard. If all of the nondegenerate signatures in $\mathcal{F}$ are of arity ≤ 2, then it is tractable case 1. Otherwise we have some nondegenerate signatures of arity ≥ 3. For any such f, by Theorem 7.52, $f \in \mathscr{P}_1 \cup \mathscr{P}_2 \cup \mathscr{A}_3$ or f is vanishing. If any of them is in $\mathscr{P}_1 \cup \mathscr{P}_2 \cup \mathscr{A}_3$, then by Lemma 7.49, Lemma 7.50, or Lemma 7.51, we have that $\mathcal{F}$ is $\mathscr{A}$- or $\mathscr{P}$-transformable, which are tractable cases 2 and 3.

Now we assume all nondegenerate signatures of arity ≥ 3 in $\mathcal{F}$ are vanishing, and there is a nonempty set of such signatures in $\mathcal{F}$. By Lemma 7.34, they must all be in $\mathscr{V}^\sigma$ for the same $\sigma \in \{+, -\}$. By Lemma 7.33, we know that any nondegenerate binary signature in $\mathcal{F}$ must belong to $\mathscr{R}_2^\sigma$. Furthermore, if there is an $f \in \mathscr{V}^\sigma$ from $\mathcal{F}$ such that $\text{rd}^\sigma(f) \geq 2$, then by Lemma 7.31, the only unary signature that is allowed in $\mathcal{F}$ is $[1, \sigma \mathfrak{i}]$, and all degenerate signatures in $\mathcal{F}$ are a tensor power of $[1, \sigma \mathfrak{i}]$, up to a scalar. Thus, all nondegenerate signatures of arity ≥ 3 as well as all degenerate signatures belong to $\mathscr{V}^\sigma$, and all nondegenerate binary signatures belong to $\mathscr{R}_2^\sigma$. This is tractable case 4.

Finally, we have the following: (i) all nondegenerate signatures of arity ≥ 3 in $\mathcal{F}$ belong to $\mathscr{V}^\sigma$; (ii) all signatures $f \in \mathcal{F} \cap \mathscr{V}^\sigma$ have $\text{rd}^\sigma(f) \leq 1$, which implies that $f \in \mathscr{R}_2^\sigma$; and (iii) all nondegenerate binary signatures in $\mathcal{F}$ belong to $\mathscr{R}_2^\sigma$. Hence all nondegenerate signatures in $\mathcal{F}$ belong to $\mathscr{R}_2^\sigma$. All unary signatures also belong to $\mathscr{R}_2^\sigma$ by definition. This is indeed tractable case 5. The proof is complete. □

7.9 Decidability of the Dichotomy

Theorem 7.19 is a full dichotomy for Holant($\mathcal{F}$) for a set of symmetric signatures on Boolean variables. It says that if $\mathcal{F}$ satisfies one of the tractability conditions then Holant($\mathcal{F}$) is computable in P; otherwise it is #P-hard. A natural question is given a finite set $\mathcal{F}$, how can one decide which case it is. This is the decidability question of the dichotomy theorem.

Clearly conditions 1, 4, and 5 are all easily decidable in polynomial time. Hence we may assume $\mathcal{F}$ contains at least one nondegenerate signature of arity ≥ 3, and the real question is how can one decide whether $\mathcal{F}$ is $\mathscr{A}$- or $\mathscr{P}$-transformable. This problem is nontrivial because it involves a potentially infinite set of transformational matrices.

Theorem 7.53. *Given any finite set $\mathcal{F}$ of symmetric signatures with complex weights in Boolean variables containing a nondegenerate signature f of arity ≥ 3, it is decidable in polynomial time in the input size of $\mathcal{F}$ whether $\mathcal{F}$ is $\mathscr{A}$- or $\mathscr{P}$-transformable. Thus the tractability criterion of Theorem 7.19 is decidable in polynomial time.*

In this book we will not give the full proof of this theorem; interested readers can find it in [CGW14]. In this section we discuss some necessary conditions of the tractability criterion based on some structural properties, that both shed new light on the dichotomy and also help in many practical cases to test the criterion on particular functions. The decision algorithm of Theorem 7.53 is built on top of these ideas with some additional technical extensions. We will omit some technical proofs in this section.

By Corollary 7.47, f is $\mathscr{A}$- or $\mathscr{P}$-transformable iff $f \in \mathscr{A}_1 \cup \mathscr{A}_2 \cup \mathscr{A}_3$, or $f \in \mathscr{P}_1 \cup \mathscr{P}_2$, where $\mathscr{A}_1 \subset \mathscr{P}_1$ and $\mathscr{A}_2 = \mathscr{P}_2$. We say a signature $f = [f_0, f_1, \ldots, f_n]$ satisfies a second order recurrence relation, if there exist $a, b, c \in \mathbb{C}$ not all zero, such that for all $0 \leq k \leq n-2$, $af_k - bf_{k+1} + cf_{k+2} = 0$. We say the recurrence has type $\langle a, b, c \rangle$. The characteristic equation of the recurrence relation is $a - bX + cX^2 = 0$. The discriminant is $b^2 - 4ac$.

Lemma 7.54. *Let f be a nondegenerate symmetric signature of arity $n \geq 3$. If f satisfies a second-order recurrence relation then the recurrence relation is unique up to a nonzero multiple.*

Under any holographic transformation, the property that a signature satisfies a second order recurrence relation, and if so, the condition $b^2 - 4ac \neq 0$ are both invariant.

Lemma 7.55. *Let f be a symmetric signature of arity n and $\hat{f} = T^{\otimes n} f$ for some $T \in \mathbf{GL}_2(\mathbb{C})$. Then f satisfies a second-order recurrence relation iff $\hat{f}$ does. Furthermore, in this case the discriminant for f is nonzero iff the one for $\hat{f}$ is nonzero.*

Proof. If we write the $2^2 \times 2^{n-2}$ signature matrix of f as M_f, with rows indexed by the first two input bits and the rest are the other $n-2$ input bits, as in Definition 1.4, then the corresponding signature matrix of $T^{\otimes n} f$ is $T^{\otimes 2} M_f (T^{\mathsf{T}})^{\otimes (n-2)}$. f satisfies a second-order recurrence relation of type $\langle a, b, c \rangle$ iff $(a, -b/2, -b/2, c)$ is a solution of $(x, y, y, z) M_f = 0$ iff $(a, -b/2, -b/2, c)(T^{-1})^{\otimes 2}$ is a solution of $(x', y', y', z') M_{T^{\otimes n} f} = 0$. In this case, the discriminant of f is $\det \begin{bmatrix} a & -b/2 \\ -b/2 & c \end{bmatrix}$, and that of $T^{\otimes n} f$ is multiplied by $(\det T)^2$. □

A simple but important observation is that if f is $\mathscr{A}$- or $\mathscr{P}$-transformable, then f satisfies a second-order recurrence relation with a nonzero discriminant. In fact, in their canonical forms, $f \in \mathscr{A}_1$, $\mathscr{A}_2$, and $\mathscr{A}_3$ have types $\langle -1, 0, 1\rangle$, $\langle 1, 0, 1\rangle$, and $\langle -\mathfrak{i}, 0, 1\rangle$, with characteristic polynomials $X^2 - 1$, $X^2 + 1$, and $X^2 - \mathfrak{i}$, respectively. For $f \in \mathscr{P}_1$ and $\mathscr{P}_2$, the types are $\langle -1, 0, 1\rangle$ and $\langle 1, 0, 1\rangle$. By Lemma 7.55, any $f \in \mathscr{P}_1 \cup \mathscr{A}_2 \cup \mathscr{A}_3$ satisfies a second-order recurrence relation with a nonzero discriminant. Hence,

Lemma 7.56. *Let $f \in \mathscr{P}_1 \cup \mathscr{A}_2 \cup \mathscr{A}_3$ be a nondegenerate symmetric signature of arity $n \geq 3$. Then f satisfies a second-order recurrence relation with nonzero discriminant. It has the form $f = \begin{bmatrix} \alpha \\ \gamma \end{bmatrix}^{\otimes n} + \begin{bmatrix} \beta \\ \delta \end{bmatrix}^{\otimes n}$, where $\det \begin{bmatrix} \alpha & \beta \\ \gamma & \delta \end{bmatrix} \neq 0$.*

The following definition of the θ function is crucial.

Definition 7.57. For a pair of linearly independent vectors $u = \begin{bmatrix} \alpha \\ \gamma \end{bmatrix}$ and $v = \begin{bmatrix} \beta \\ \delta \end{bmatrix}$, we define

$$\theta(u, v) = \left(\frac{\langle u, v\rangle}{\det[u, v]} \right)^2 = \left(\frac{\alpha\beta + \gamma\delta}{\alpha\delta - \beta\gamma} \right)^2,$$

where $\langle u, v\rangle$ is the dot product, and $\det[u, v]$ is the determinant with column vectors u and v. Furthermore, if $f = u^{\otimes n} + v^{\otimes n}$ for linearly independent u and v and $n \geq 3$, then we define $\theta(f) = \theta(u, v)$.

By linear independence, $\det[u, v] \neq 0$, so $\theta(u, v)$ is well defined.

Lemma 7.58. *For two linearly independent vectors $u, v \in \mathbb{C}^2$, and nonzero λ and $\mu \in \mathbb{C}$, $\theta(\lambda u, \mu v) = \theta(u, v)$. If $H \in \mathbf{O}_2(\mathbb{C})$, let $\widehat{u} = Hu$ and $\widehat{v} = Hv$. Then $\theta(\widehat{u}, \widehat{v}) = \theta(u, v)$. In particular $\theta(v, u) = \theta(u, v)$.*

Proof. The first claim is obvious. Because H is orthogonal, the dot product is preserved, $\langle \widehat{u}, \widehat{v}\rangle = \langle u, v\rangle$. Also $\det[\widehat{u}, \widehat{v}] = \pm \det[u, v]$. Hence $\theta(\widehat{u}, \widehat{v}) = \theta(u, v)$. □

Intuitively, $\theta(u, v)$ is the square of the cotangent of the angle from u to v. (The cotangent function is extended to the complex domain.) Let $f = u^{\otimes n} + v^{\otimes n}$ be a nondegenerate signature of arity $n \geq 3$. Since f is nondegenerate, u and v are linearly independent. The next proposition implies that this expression for f via u and v is unique up to a root of unity. Therefore, $\theta(f)$ from Definition 7.57 is well defined from f; it is independent of the particular expression of f.

Proposition 7.59. *Let $u, v, u', v' \in \mathbb{C}^2$. Suppose u and v are linearly independent. If for some $n \geq 3$, we have*

$$u^{\otimes n} + v^{\otimes n} = u'^{\otimes n} + v'^{\otimes n}, \tag{7.17}$$

then with possibly exchanging u and v, there exist ρ and $\omega \in \mathbb{C}$, such that $u' = \rho u$, $v' = \omega v$ where $\rho^n = \omega^n = 1$.

We omit the proof. Now we give some necessary conditions for membership in $\mathscr{A}_1 \cup \mathscr{A}_2 \cup \mathscr{A}_3$. Recall that $\mathscr{A}_1 \subseteq \mathscr{P}_1$.

Lemma 7.60. *Let f be a nondegenerate symmetric signature of arity ≥ 3. Then*

1. $f \in \mathscr{P}_1 \implies \theta(f) = 0$,
2. $f \in \mathscr{A}_2 \implies \theta(f) = -1$, *and*
3. $f \in \mathscr{A}_3 \implies \theta(f) = -\frac{1}{2}$.

Proof. The result clearly holds when f is in the canonical form of each set. This extends to the rest of each set by Lemma 7.58. □

These results imply the following lemma.

Lemma 7.61. *Let f be a nondegenerate symmetric signature f of arity $n \geq 3$. If f is $\mathscr{A}$-transformable, then f satisfies a second-order recurrence relation with nonzero discriminant, thus has the form $f = u^{\otimes n} + v^{\otimes n}$, where u and v are linearly independent, and $\theta(f) \in \{0, -1, -\frac{1}{2}\}$.*

Note that the condition in Lemma 7.61 on $\theta(f)$ is only a necessary condition for f being $\mathscr{A}$-transformable. This condition is not sufficient to determine if $f \in \mathscr{A}_1 \cup \mathscr{A}_2 \cup \mathscr{A}_3$. In fact, it can be directly verified that $\theta(u, v) = -1$ for linearly independent u and v if and only if at least one of u, v is $\begin{bmatrix} 1 \\ i \end{bmatrix}$ or $\begin{bmatrix} 1 \\ -i \end{bmatrix}$, up to a nonzero scalar. Thus, if $f = u^{\otimes n} + v^{\otimes n}$ with $u = \begin{bmatrix} 1 \\ i \end{bmatrix}$ and v is *not* a multiple of $\begin{bmatrix} 1 \\ \pm i \end{bmatrix}$, then $\theta(f) = -1$. But f is not in $\mathscr{A}_2$ by Proposition 7.59, and Lemma 7.41. A full characterization can be achieved with additional work; but we will not pursue it further here [CGW14].

To obtain the decision algorithm for Theorem 7.19 as stated in Theorem 7.53 will require more careful work concerning a common transformation for the whole set $\mathcal{F}$. We will not present all the details, but it starts with the following.

Lemma 7.38 showed that every $\mathscr{A}$-transformable set $\mathcal{F}$ is contained in either $H\mathscr{A}$ or $H\begin{bmatrix} 1 & 0 \\ 0 & \alpha \end{bmatrix}\mathscr{A}$ for some $H \in \mathbf{O}_2(\mathbb{C})$. For nondegenerate symmetric signatures, we can replace $\mathscr{A}$ by $\mathscr{F}_1 \cup \mathscr{F}_2 \cup \mathscr{F}_3$. We have the following classification.

$$\theta(f) = \begin{cases} 0 & \text{if } f \in \mathscr{F}_1 \cup \mathscr{F}_2 \cup \begin{bmatrix} 1 & 0 \\ 0 & \alpha \end{bmatrix}\mathscr{F}_1, \\ -1 & \text{if } f \in \mathscr{F}_3, \\ -\frac{1}{2} & \text{if } f \in \begin{bmatrix} 1 & 0 \\ 0 & \alpha \end{bmatrix}(\mathscr{F}_2 \cup \mathscr{F}_3). \end{cases} \tag{7.18}$$

Regarding $\mathscr{P}$-transformability, Lemma 7.60 also gives a useful necessary condition for f being $\mathscr{P}$-transformable. As $\mathscr{P}_2 = \mathscr{A}_2$, the same discussion for $\mathscr{A}_2$ applies to $\mathscr{P}_2$.

The next lemma tells us how to decide membership in $\mathscr{P}_1$ for signatures of arity ≥ 3.

Lemma 7.62. *Let $f = v_0^{\otimes n} + v_1^{\otimes n}$ be a symmetric signature of arity $n \geq 3$, where v_0 and v_1 are linearly independent. Then $f \in \mathscr{P}_1$ iff $\theta(f) = 0$.*

With some additional work one can obtain the decision algorithm stated in Theorem 7.53. Details can be found in [CGW14].

8
Planar #CSP for Symmetric Constraints

In Chapter 3 we proved a complexity dichotomy for #CSP($\mathcal{F}$), Theorem 3.7, where $\mathcal{F}$ is a set of constraint functions mapping Boolean inputs to complex numbers. The conclusion of Theorem 3.7 is that the problem #CSP($\mathcal{F}$) is #P-hard, unless $\mathscr{F} \subseteq \mathscr{A}$ or $\mathscr{F} \subseteq \mathscr{P}$, in which case the problem is computable in polynomial time. In this chapter, we would like to consider this problem restricted to planar input instances, i.e., the bipartite graph describing the connection between variables and constraints is a planar graph. Here we consider only symmetric constraint functions. We will show that in addition to $\mathscr{A}$ and $\mathscr{P}$, there is a third tractable class $\widehat{\mathscr{M}}$, which is the class of matchgate signatures $\mathscr{M}$ under a holographic transformation by the Hadamard matrix $H_2 = \begin{bmatrix} 1 & 1 \\ 1 & -1 \end{bmatrix}$. Everything else remains #P-hard.

8.1 Introduction

In terms of Holant problems, the problem #CSP($\mathcal{F}$) is equivalent to Holant($\mathcal{F} \cup \mathcal{EQ}$), where $\mathcal{EQ} = \{=_1, =_2, =_3, \dots\}$ and $(=_k) = [1, 0, \dots, 0, 1]$ is the equality signature of arity k. This equivalence is also valid between planar #CSP and planar Holant problems as well. Considering the edge–vertex incidence graph, one can see that Holant($\mathcal{F}$) is equivalent to Holant($=_2 | \mathcal{F}$). Then we can apply a holographic transformation, and see that for any nonsingular 2×2 matrix T, Holant($=_2 | \mathcal{F}$) is equivalent to Holant($(=_2)(T^{-1})^{\otimes 2} \mid T\mathcal{F}$). Here $T\mathcal{F} = \{T^{\otimes \text{arity}(f)} f | f \in \mathcal{F}\}$.

Using the operator Sym_n^{n-1} and the expression of a sum of tensor powers, Theorem 4.11 can be equivalently stated as

Theorem 8.1. *Let f be a symmetric signature in $\mathscr{M}$ of arity n. Then there exist $a, b, \lambda \in \mathbb{C}$ such that f takes one of the following forms:*

1. $[a, b]^{\otimes n} + [a, -b]^{\otimes n} = \begin{cases} 2[a^n, 0, a^{n-2}b^2, 0, \ldots, 0, b^n] & n \text{ is even}, \\ 2[a^n, 0, a^{n-2}b^2, 0, \ldots, 0, ab^{n-1}, 0] & n \text{ is odd}; \end{cases}$
2. $[a, b]^{\otimes n} - [a, -b]^{\otimes n} = \begin{cases} 2[0, a^{n-1}b, 0, a^{n-3}b^3, 0, \ldots, 0, ab^{n-1}, 0] & n \text{ is even}, \\ 2[0, a^{n-1}b, 0, a^{n-3}b^3, 0, \ldots, 0, b^n] & n \text{ is odd}; \end{cases}$
3. $\lambda \operatorname{Sym}_n^{n-1}([1, 0]; [0, 1]) = [0, \lambda, 0, \ldots, 0]$;
4. $\lambda \operatorname{Sym}_n^{n-1}([0, 1]; [1, 0]) = [0, \ldots, 0, \lambda, 0]$.

In (4.23) and (4.24) of Chapter 4 these expressions are proved to be equivalent to the list in Theorem 4.11.

From the Holant perspective, the signatures in $\mathcal{EQ}$ are always available in #CSP($\mathcal{F}$). By the signature theory of matchgates and their transformations, the Hadamard matrix H_2 is essentially the only holographic transformation under which all equalities in $\mathcal{EQ}$ become realizable as standard signatures of matchgates $\mathscr{M}$. To see this, recall the definition of the equivalence relation $\sim$ given in Proposition 4.24, which defines the basis manifold $\mathcal{B} = \mathrm{GL}_2(\mathbb{C})/\sim$. Then in Lemma 4.29, we take $\lambda = \mu = 1$ and $\alpha = 0$ to get the set of all holographic transformations T such that $(=_k)(T^{-1})^{\otimes k} \in \mathscr{M}$. As representatives in $\mathcal{B}$, these T take the form $T = \begin{bmatrix} 1 & \omega \\ 1 & -\omega \end{bmatrix}$, where $\omega^{2k} = 1$. Since $\mathcal{EQ}$ contains $(=_k)$ for all $k \geq 1$, we have $\omega = \pm 1$, i.e., the only common transformations T represented in $\mathcal{B}$ are $H_2 = \begin{bmatrix} 1 & 1 \\ 1 & -1 \end{bmatrix}$ or $\begin{bmatrix} 0 & 1 \\ 1 & 0 \end{bmatrix} H_2 = \begin{bmatrix} 1 & -1 \\ 1 & 1 \end{bmatrix}$. By the definition of $\mathcal{B}$, this means that T has the form $\begin{bmatrix} x & 0 \\ 0 & y \end{bmatrix} H_2$ or $\begin{bmatrix} 0 & x \\ y & 0 \end{bmatrix} H_2$, for $xy \neq 0$. However, for any recognizer signature Γ of arity n written as a row vector, $\Gamma \in \mathscr{M}$ iff $\Gamma \begin{bmatrix} x & 0 \\ 0 & y \end{bmatrix}^{\otimes n} \in \mathscr{M}$, and also iff $\Gamma \begin{bmatrix} 0 & x \\ y & 0 \end{bmatrix}^{\otimes n} \in \mathscr{M}$. Equivalently, if Γ is a generator signature of arity n written as a column vector, $\Gamma \in \mathscr{M}$ iff $\begin{bmatrix} 1/x & 0 \\ 0 & 1/y \end{bmatrix}^{\otimes n} \Gamma \in \mathscr{M}$, and also iff $\begin{bmatrix} 0 & 1/y \\ 1/x & 0 \end{bmatrix}^{\otimes n} \Gamma \in \mathscr{M}$. This can be seen by a combinatorial construction, as well as by MGI. For symmetric signatures this is also obvious by Theorem 8.1. Thus, given Holant($\mathcal{EQ} \mid \mathcal{F}$), to test whether there is any holographic transformation T such that $(\mathcal{EQ})T^{-1} \subseteq \mathscr{M}$ and $T\mathcal{F} \subseteq \mathscr{M}$, we may consider H_2 as the only possible transformation, namely whether $H_2\mathcal{F} \subseteq \mathscr{M}$.

Let $\widehat{\mathcal{F}}$ denote $H_2\mathcal{F}$ for any set $\mathcal{F}$ of signatures. Note that H_2^{-1} is H_2 up to a nonzero constant, hence $\widehat{\widehat{\mathcal{F}}} = \mathcal{F}$. It is easy to see that $\widehat{\mathcal{EQ}}$ is $\{[1, 0], [1, 0, 1], [1, 0, 1, 0], \ldots\}$. Therefore, #CSP($\mathcal{F}$) $\equiv_{\mathrm{T}}$ Holant($\mathcal{EQ} \mid \mathcal{F}$) $\equiv_{\mathrm{T}}$ Holant($\mathcal{F} \cup \mathcal{EQ}$), is equivalent to Holant($\widehat{\mathcal{F}} \cup \widehat{\mathcal{EQ}}$). The equivalence of #CSP($\mathcal{F}$), Holant($\mathcal{F} \cup \mathcal{EQ}$), and Holant($\widehat{\mathcal{F}} \cup \widehat{\mathcal{EQ}}$) is also valid over planar instances. We denote by Pl-#CSP($\mathcal{F}$), Pl-Holant($\mathcal{F} \cup \mathcal{EQ}$), and Pl-Holant($\widehat{\mathcal{F}} \cup \widehat{\mathcal{EQ}}$), the planar versions of the corresponding problems #CSP($\mathcal{F}$), Holant($\mathcal{F} \cup \mathcal{EQ}$), and Holant($\widehat{\mathcal{F}} \cup \widehat{\mathcal{EQ}}$) respectively.

The main dichotomy theorem of this chapter is due to Guo and Williams [GW13]. With their permission, we will closely follow their presentation. This dichotomy states that, for #CSP problems over any set of complex-valued *symmetric* constraint functions on Boolean variables, Valiant's holographic transformation followed by Kasteleyn's algorithm (Theorem 4.7) is a *universal* methodology to solve in polynomial time on planar graphs those counting problems that are #P-hard in general. This theorem generalizes a previous theorem with the same conclusion for real-valued symmetric signatures [CLX10].

Theorem 8.2 (Guo-Williams). *Let $\mathcal{F}$ be any set of symmetric, complex-valued signatures in Boolean variables. Then* Pl-#CSP($\mathcal{F}$) *is* #P*-hard unless $\mathcal{F}$ satisfies one of the following conditions, in which case it is tractable:*

1. #CSP($\mathcal{F}$) *is tractable, i.e., $\mathcal{F} \subseteq \mathscr{A}$ or $\mathscr{F} \subseteq \mathscr{P}$ (by Theorem 3.7).*
2. *$\widehat{\mathcal{F}}$ is realizable as standard matchgate signatures, i.e., $\mathcal{F} \subseteq \widehat{\mathscr{M}}$.*

Thus, for symmetric complex-valued signatures in Boolean variables, to be transformable to matchgates under the Hadamard transformation followed by Kasteleyn's algorithm (a.k.a. the FKT algorithm; see Chapter 4) is a *universal* methodology to solve Pl-#CSP problems in polynomial time over planar graphs that are #P-hard over general graphs.

We note that Theorem 8.2 is proved only for sets of *symmetric* signatures, whereas Theorem 3.7 does not have this restriction. In Chapter 10 we will state the dichotomy theorem for Pl-#CSP($\mathcal{F}$), Theorem 10.1, which is valid for general asymmetric signatures, i.e., without the restriction that the signatures be symmetric. We will not give the full proof of Theorem 10.1, but a main engine of the proof is to use Theorem 8.2. We also note that the tractability criteria of Theorem 8.2 are decidable in polynomial time in the length of the description of the symmetric signatures in $\mathcal{F}$. This follows from the forms of symmetric signatures $\mathscr{F}_{123}$ in $\mathscr{A}$, and those in $\mathscr{P}$ and in $\widehat{\mathscr{M}}$. Note that for $f = [a, b]^{\otimes n} \pm [a, -b]^{\otimes n} \in \mathscr{M}$, when written as a column vector, its transformation $H_2 f \in \widehat{\mathscr{M}}$ has the form $[x, y]^{\otimes n} \pm [y, x]^{\otimes n}$, where $x = a + b$ and $y = a - b$. (Recall that we write a row vector and its column vector transpose interchangeably, provided it is suitable for matrix-vector product.)

In proving dichotomy theorems for Boolean #CSP($\mathcal{F}$), pinning is an important tool. If one can realize the unary pinning signatures Is-Zero $\Delta_0 = [1, 0]$

and IS-ONE $\Delta_1 = [0, 1]$, then one can obtain all subsignatures of any $f \in \mathcal{F}$. However, the proof in Chapter 3 for #CSP($\mathcal{F}$) obtaining Δ_0 and Δ_1 is by a *nonplanar* reduction. More intrinsically, we note that in the nonplanar setting, Δ_0 and Δ_1 are contained in both maximal tractable sets $\mathscr{A}$ and $\mathscr{P}$. Therefore, it is plausible to prove the pinning reduction #CSP($\mathcal{F} \cup \{\Delta_0, \Delta_1\}$) $\leq_T$ #CSP($\mathcal{F}$), as was done in Lemma 3.13; pinning in this setting would not imply the collapse of #P to P. However, $\mathcal{EQ}$ and $\{[1,0], [0,1]\}$ are not simultaneously realizable as matchgates, under any holographic transformation. Suppose Theorem 8.2 is true. Then we should not expect to be able to pin for Pl-#CSP($\mathcal{F}$), i.e., to prove a general pinning reduction Pl-#CSP($\mathcal{F} \cup \{\Delta_0, \Delta_1\}$) $\leq_T$ Pl-#CSP($\mathcal{F}$), since otherwise #P collapses to P. The collapse is seen by taking $\mathcal{F} = \widehat{\mathscr{M}}$, which is neither a subset of $\mathscr{A}$ nor $\mathscr{P}$, and $\Delta_0 \notin \widehat{\mathscr{M}}$ since $H_2 \Delta_0 = [1, 1] \notin \mathscr{M}$.

So we cannot expect to pin in the *primal* setting Pl-#CSP($\mathcal{F}$) in general. Instead, we apply the Hadamard transformation and consider in the *dual* setting Pl-Holant($\widehat{\mathcal{F}} \cup \widehat{\mathcal{EQ}}$). In this Hadamard basis, it is at least possible again that we can prove a pinning reduction for planar Holant problems, since Δ_0 and Δ_1 are contained in $\mathscr{A} \cap \widehat{\mathscr{P}} \cap \mathscr{M}$. Note that $\widehat{\mathscr{A}} = \mathscr{A}$ and $\widehat{\mathscr{P}}$ still contains all unary signatures. If $\mathcal{F} \subseteq \mathscr{A}$, or $\mathscr{P}$, or $\widehat{\mathscr{M}}$, then $\widehat{\mathcal{F}} \cup \widehat{\mathcal{EQ}} \cup \{\Delta_0, \Delta_1\} \subseteq \mathscr{A}$, or $\widehat{\mathscr{P}}$ or $\mathscr{M}$ respectively. Indeed, we will prove our pinning result in this Hadamard basis in Section 8.6, see (8.5).

For convenience we will denote Pl-Holant($\mathcal{F} \cup \{\Delta_0, \Delta_1\}$) by Pl-Holant^c($\mathcal{F}$).

Our plan for the proof of Theorem 8.2 is as follows. Instead of proving in the primal setting, we will prove the theorem mainly in the dual setting Pl-Holant($\widehat{\mathcal{F}} \cup \widehat{\mathcal{EQ}}$). For Holant problems, as in #CSP problems, it is often important to understand the complexity of the small arity cases first. We already proved Theorem 6.3, a dichotomy for Pl-Holant(f), where f is a symmetric signature of arity 3. After some strengthening of unary interpolations, we will prove a dichotomy for Pl-Holant(f), where f is a symmetric signature of arity 4. This uses the idea of a *planar pairing*. If every signature in $\mathcal{F}$ has even arity, then by a simple parity argument, every $\mathcal{F}$-gate will also have even arity. Thus it is impossible to construct directly a gadget with an odd arity. After we prove some useful lemmas using a technique called *domain pairing* we will circumvent this difficulty and *functionally* realize an odd arity signature using only signatures of even arity. After that we show some no-mixing lemmas, which say that the known classes of tractable signatures for Pl-#CSP become #P-hard when mixed. Then we realize pinning in the Hadamard basis. After that we prove the Pl-#CSP dichotomy for a single signature. The main dichotomy for Pl-#CSP($\mathcal{F}$), Theorem 8.2, follows easily from this and the no-mixing lemmas.

For easy reference we list some results in a convenient form from previous chapters. Note that the problem Pl-Holant($[x, y, z] \mid \mathcal{EQ}$) is just Pl-#CSP($[x, y, z]$), which is also equivalent to a planar spin system with edge function $[x, y, z]$.

Theorem 8.3. *For any* $x, y, z \in \mathbb{C}$, *both* Pl-Holant($[x, y, z] \mid \mathcal{EQ}$) *and* Pl-Holant($[x, y, z] \mid \widehat{\mathcal{EQ}}$) *are* #P-*hard unless one of the following conditions holds, in which case both problems are computable in polynomial time:*

1. $xz = y^2$.
2. $y = 0$.
3. $xz = -y^2$ *and* $x = -z$.
4. $x = z$.

Proof. Theorem 5.1 (equivalently Theorem 5.3) was proved for the cases $k = 3$ and 4 in Chapter 5. For Pl-Holant($[x, y, z] \mid \mathcal{EQ}$) both $(=_3)$ and $(=_4)$ are present in $\mathcal{EQ}$, and thus Pl-Holant($[x, y, z] \mid \mathcal{EQ}$) is #P-hard unless both tractability criteria of Theorem 5.3 for $k = 3$ and $k = 4$ are satisfied. This gives the listed tractability criterion. Conversely, if one of the tractability conditions is satisfied then clearly Pl-Holant($[x, y, z] \mid \mathcal{EQ}$) is computable in P.

Under the transformation $H_2 = \begin{bmatrix} 1 & 1 \\ 1 & -1 \end{bmatrix}$, it is easy to see that the conditions $xz = y^2$ and $(xz = -y^2) \wedge (x = -z)$ are invariant while the conditions $y = 0$ and $x = z$ map to each other. Therefore, by an apparent coincidence, the tractability conditions remain the same for Pl-Holant($[x, y, z] \mid \widehat{\mathcal{EQ}}$). □

Theorem 8.4. *Let* $\mathcal{F}$ *be any set of symmetric, complex-valued signatures in Boolean variables. Then* Pl-Holant^c($\mathcal{F} \cup \widehat{\mathcal{EQ}}$) *is computable in polynomial time if* $\mathcal{F} \subseteq \mathscr{A}$, $\mathcal{F} \subseteq \widehat{\mathscr{P}}$, *or* $\mathcal{F} \subseteq \mathscr{M}$.

Proof. We only need to observe that $\widehat{\mathcal{EQ}} \cup \{[1, 0], [0, 1]\} \subset \mathscr{A} \cap \widehat{\mathscr{P}} \cap \mathscr{M}$. The case $\mathcal{F} \subseteq \mathscr{A}$ follows from Theorem 6.12. The case $\mathcal{F} \subseteq \widehat{\mathscr{P}}$ follows from Theorem 6.12 and Theorem 3.5. The case $\mathcal{F} \subseteq \mathscr{M}$ follows from Theorem 4.7. □

8.2 Unary Interpolation Revisited

In this section, we discuss *unary recursive construction* and give an exact condition for when it succeeds. The goal of this construction is to interpolate an arbitrary unary signature.

For a unary recursive construction in the Holant framework, there are two gadgets involved: a *starter* gadget of arity 1 and a *recursive* gadget of arity 2. In the Pl-Holant framework the gadgets must also be planar. The signature of the

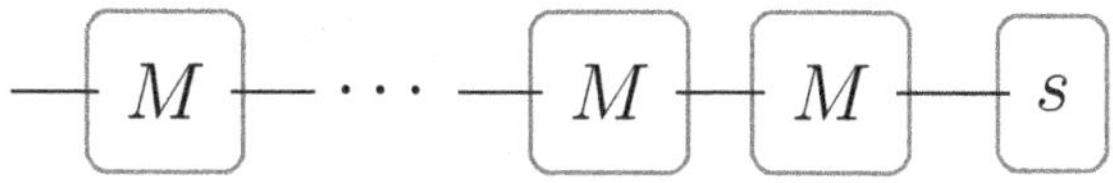

Figure 8.1. Unary recursive construction with starter gadget.

starter gadget is represented by a vector $s \in \mathbb{C}^2$. The signature of the recursive gadget is represented by a matrix $M \in \mathbb{C}^{2\times 2}$. The construction begins with the starter gadget and proceeds by connecting $k \geq 0$ recursive gadgets, one at a time, to the only available edge (see Figure 8.1). The signature matrix of the resulting gadget is $M^k s$. This construction is denoted by (M, s).

The most demanding part of the requirement in these constructions, for the purpose of polynomial interpolation, is to construct an infinite set of signatures that are pairwise linearly independent (see Chapter 5). The pairwise linear independence of signatures translates into distinct evaluation points for the polynomial being interpolated.

We say two vectors are orthogonal if their dot product is 0. We first prove a lemma that gives a determinantal condition for the statement that a vector s is not orthogonal to any row eigenvector of a matrix M.

Lemma 8.5. *Suppose $M \in \mathbb{C}^{n\times n}$ and $s \in \mathbb{C}^n$. Then s is not orthogonal to any row eigenvector of M iff* $\det([s\, Ms \ldots M^{n-1}s]) \neq 0$.

Proof. Suppose $\det([s\, Ms \ldots M^{n-1}s]) = 0$. Then there is a nonzero row vector v such that $v[s\, Ms \ldots M^{n-1}s] = 0$ is the zero vector. Consider the linear span S by row vectors in the set $\{v, vM, \ldots, vM^{n-1}\}$. By the Cayley–Hamilton theorem, M satisfies its own characteristic polynomial, which is a monic polynomial of degree n. Thus, M^n is a linear combination of $I_n, M, \ldots, M^{n-1}$. This shows that S is an invariant subspace of row vectors under the action of multiplication by M from the right. Being an invariant subspace over $\mathbb{C}$, there exists a row eigenvector $u \in S$ of M, and the dot product of s with u is 0.

Conversely, suppose $\det([s\, Ms \ldots M^{n-1}s]) \neq 0$ and assume for a contradiction that s is orthogonal to some row eigenvector v of M with eigenvalue λ. Then $v[s\, Ms \ldots M^{n-1}s] = 0$ is the zero vector because $vM^i s = \lambda^i vs = 0$. Since $v \neq 0$, this is a contradiction. □

To construct an infinite set of signatures that are pairwise linearly independent, a necessary condition is that M has infinite order modulo a scalar. Otherwise, $M^k = cI_n$ for some $k > 0$ and $c \neq 0$, then any vector of the form $M^\ell s$ for $\ell \geq k$ is a multiple of a vector in the set $\{M^i s\}_{0 \leq i < k}$.

Lemma 8.6. *Suppose $M \in \mathbb{C}^{n \times n}$ and $s \in \mathbb{C}^n$. If the following three conditions are satisfied,*

1. $\det(M) \neq 0$.
2. $\det([s\, Ms \ldots M^{n-1}s]) \neq 0$.
3. *M has infinite order modulo a scalar.*

then the vectors in the set $V = \{M^k s\}_{k \geq 0}$ are pairwise linearly independent.

Proof. Since $\det(M) \neq 0$, M is nonsingular and all eigenvalues are nonzero. Let $M = P^{-1}JP$ be the Jordan decomposition of M and let $v = Ps \in \mathbb{C}^n$. Clearly $v \neq 0$ since $s \neq 0$ by condition 2. Suppose for a contradiction that the vectors in V are not pairwise linearly independent. This means that there exist integers $k > \ell \geq 0$ such that $M^k s = cM^\ell s$ for some nonzero $c \in \mathbb{C}$. Let $m = k - \ell > 0$. Then we have $P^{-1}J^m Ps = M^m s = cs$ and thus $J^m v = cv$.

Suppose that J contains some nontrivial Jordan block and consider the 2×2 submatrix $\begin{bmatrix} \lambda & 1 \\ 0 & \lambda \end{bmatrix}$ in the bottom right corner of this block, where λ is an eigenvalue of M. Without loss of generality we may assume that the rows and colmuns of this 2×2 submatrix are indexed by $n-1$ and n. From this portion of J, the two equations given by $J^m v = cv$ are $\lambda^m v_{n-1} + m\lambda^{m-1} v_n = cv_{n-1}$ and $\lambda^m v_n = cv_n$. By Lemma 8.5, s is not orthogonal to any row eigenvector of M; therefore $v_n \neq 0$ since the last row of P is a row eigenvector of M. But then these equations imply that $m\lambda^{m-1} v_n = 0$, a contradiction.

Otherwise, J is a diagonal matrix $\operatorname{diag}\{\lambda_1, \ldots, \lambda_n\}$. From $J^m v = cv$, we get the equations $\lambda_i^m v_i = cv_i$ for $1 \leq i \leq n$. As $PM = JP$ the rows of P are row eigenvectors of M. By definition $v = Ps$, and since s is not orthogonal to any row eigenvector of M, $v_i \neq 0$ for $1 \leq i \leq n$. But then $M^m = cI_n$, which contradicts the assumption that M has infinite order modulo a scalar. □

These conditions guarantee success of interpolation by a unary recursive construction.

Lemma 8.7. *Let $\mathcal{F}$ be a set of signatures. If there exist a planar $\mathcal{F}$-gate with signature matrix $M \in \mathbb{C}^{2 \times 2}$ and a planar $\mathcal{F}$-gate with signature $s \in \mathbb{C}^2$ satisfying the following conditions:*

1. $\det(M) \neq 0$;
2. $\det([s\, Ms]) \neq 0$;
3. *M has infinite order modulo a scalar;*

then Pl-Holant$(\mathcal{F} \cup \{[a, b]\}) \leq_T$ Pl-Holant$(\mathcal{F})$ *for any $a, b \in \mathbb{C}$.*

Proof. Consider an instance Ω of Pl-Holant($\mathcal{F} \cup \{[a, b]\}$). Let V' be the subset of vertices assigned $[a, b]$ and suppose that $|V'| = n$. We construct from Ω a sequence of instances Ω_k of Pl-Holant($\mathcal{F}$) indexed by $k \geq 0$. We obtain Ω_k from Ω by replacing each occurrence of $[a, b]$ with the unary recursive construction (M, s) in Figure 8.1 containing k copies of the recursive gadget. This unary recursive construction has the signature $[x_k, y_k] = M^k s$.

By Lemma 8.6, the signatures in the set $V = \{[x_k, y_k] | 0 \leq k \leq n + 1\}$ are pairwise linearly independent. In particular, at most one y_k can be 0.

We stratify the assignments in Ω based on the assignment to $[a, b]$. Let c_ℓ be the sum over all assignments of products of evaluations at all vertices other than those from V' such that exactly ℓ occurrences of $[a, b]$ have their incident edge assigned 0 (and $n - \ell$ have their incident edge assigned 1). Then

$$\text{Holant}(\Omega) = \sum_{0 \leq \ell \leq n} a^\ell b^{n-\ell} c_\ell$$

and the value of the Holant on Ω_k, for $0 \leq k \leq n + 1$, is $\text{Holant}(\Omega_k) = \sum_{0 \leq \ell \leq n} x_k^\ell y_k^{n-\ell} c_\ell$, and for at least $n + 1$ of them where $y_k \neq 0$, we have

$$\frac{\text{Holant}(\Omega_k)}{y_k^n} = \sum_{0 \leq \ell \leq n} \left(\frac{x_k}{y_k} \right)^\ell c_\ell.$$

This is a linear system with unknowns c_ℓ and a Vandermonde coefficient matrix. Since the signatures in V are pairwise linearly independent, the ratios x_k/y_k are all distinct (and well-defined since $y_k \neq 0$), which implies that the Vandermonde matrix has full rank. Therefore, we can solve the linear system for the unknown c_ℓ's and obtain the value of Holant(Ω). □

With this lemma, we obtain a tight characterization for the success of interpolation by a unary recursive construction. For example, the construction using a recursive gadget with signature matrix $M = \begin{bmatrix} 1 & 1 \\ 0 & 1 \end{bmatrix}$ and a starter gadget with signature $s = \begin{bmatrix} 0 \\ 1 \end{bmatrix}$ is successful because M and s satisfy the conditions of Lemma 8.6 even though the eigenvalues of M are equal.

The first two conditions of Lemma 8.7 are easy to check. The third condition holds in one of these two cases: either the eigenvalues are the same but M is not a multiple of the identity matrix, or the eigenvalues are different but their ratio is not a root of unity.

These conditions work well with the anti-gadget technique (see Chapter 5). The power of this lemma is that when the third condition fails to hold, there exists an integer m such that $M^m = cI_2$, where $c \neq 0$ and I_2 is the 2×2 identity matrix. Therefore we can construct $M^{m-1} = cM^{-1}$, which is effectively the same as M^{-1}, and use this in other gadget constructions. We remark that if any such m exists then the minimum such m is bounded by a constant. Regardless of the number of internal nodes of the $\mathcal{F}$-gate, its signature entries in $M \in \mathbb{C}^{2\times 2}$ all belong to the fixed algebraic extension field $\mathbf{F}$ over $\mathbb{Q}$, from which functions in $\mathcal{F}$ take values. The ratio of the two eigenvalues λ and μ of M belongs to the splitting field which has a degree of extension at most $2[\mathbf{F} : \mathbb{Q}]$. If M has finite order modulo a scalar, then let m be the minimum exponent $\ell > 0$ such that $M^\ell = cI_2$ for some $c \neq 0$. Then $m = \min\{\ell > 0 \mid \left(\frac{\lambda}{\mu}\right)^\ell = 1\}$. Thus the mth primitive root of unity $e^{2\pi \mathrm{i}/m}$ belongs to the splitting field. However the mth cyclotomic field $\Phi_m = \mathbb{Q}(e^{2\pi \mathrm{i}/m})$ has degree of extension $[\Phi_m : \mathbb{Q}] = \phi(m)$ over the rational numbers $\mathbb{Q}$, where $\phi(m)$ is the Euler's totient function. By an elementary estimate $\phi(m)^2 \geq m/2$. Hence m is upper bounded by a constant only depending on $\mathbf{F}$.

8.3 Planar Pairing

In this section, we prove a dichotomy for Pl-Holant(f), where f is a symmetric signature of arity 4. The key part is to consider the problem Pl-Holant($[t, 1, 0, 0, 0]$) when t is different from 0. Over the next two lemmas, we prove that this problem is #P-hard by reducing from Pl-Holant($[t, 1, 0, 0]$). These problems are weighted versions of counting matchings over planar k-regular graphs for $k = 4$ and $k = 3$ respectively. We use the idea of a planar pairing to be defined in Definition 8.9, and apply Lemma 5.31.

In Lemma 8.8, we show how to use either the anti-gadget technique or the unary interpolation technique based on Lemma 8.7 from Section 8.2 to effectively obtain $[1, 0, 0]$. The construction in this proof is actually not a unary recursive construction, but a binary recursive construction. However, the particular forms in the starter and recursive gadgets permit analysis equivalent to that of a unary recursive construction.

Lemma 8.8. *For any* $t \in \mathbb{C}$ *and signature set* $\mathcal{F}$ *containing* $[t, 1, 0, 0, 0]$,

$$\text{Pl-Holant}(\mathcal{F} \cup \{[1, 0, 0]\}) \leq_T \text{Pl-Holant}(\mathcal{F}).$$

Proof. Consider the gadget construction in Figure 8.2. For $k \geq 0$, the signature of N_k is of the form $[a_k, b_k, 0]$, and $N_0 = [t, 1, 0]$. Since N_k is symmetric and always ends with 0, we can analyze this construction as though it were a unary

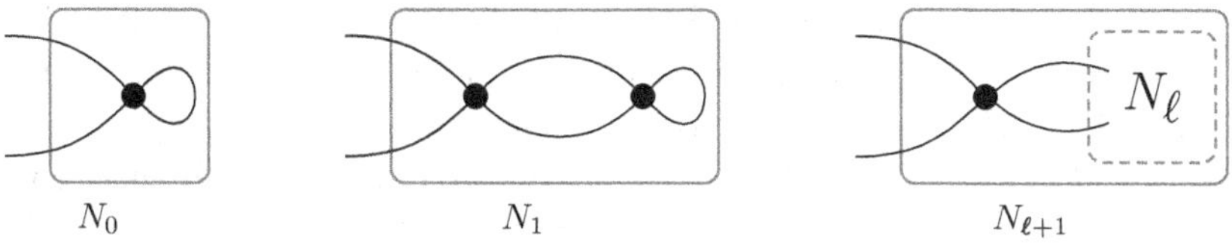

Figure 8.2. Binary recursive construction with starter gadget used to interpolate $[1, 0, 0]$. The vertices are assigned $[t, 1, 0, 0, 0]$.

recursive construction. Let $s_k = \begin{bmatrix} a_k \\ b_k \end{bmatrix}$, so $s_0 = \begin{bmatrix} t \\ 1 \end{bmatrix}$. It is clear that $s_k = M^k s_0$, where $M = \begin{bmatrix} t & 2 \\ 1 & 0 \end{bmatrix}$.

Now M is nonsingular since $\det(M) = -2$. If M has finite order modulo a scalar, then $M^m = cI_2$ for some integer $m \geq 1$ and some nonzero $c \in \mathbb{C}$. Then N_{m-1} is an anti-gadget of M with signature $M^{m-1}s_0 = cM^{-1}s_0 = c\begin{bmatrix} 1 \\ 0 \end{bmatrix}$. Note that $\begin{bmatrix} 1 \\ 0 \end{bmatrix}$ here denotes the signature $[1, 0, 0]$ as an abbreviated notation. Hence we have $[1, 0, 0]$ after normalizing.

Now assume that M has infinite order modulo a scalar. Since $\det([s_0 \, Ms_0]) = -2$, we can interpolate any signature of the form $[x, y, 0]$ by adapting Lemma 8.7, including $[1, 0, 0]$. □

Note that having $[1, 0, 0] = [1, 0]^{\otimes 2}$ is equivalent to be able to pin *two variables* to 0 simultaneously. To apply $[1, 0]^{\otimes 2}$ in a planar setting, one must be able to connect the two variables in a planar way.

Definition 8.9 (Planar pairing). A *planar pairing* M in a planar graph $G = (V, E)$ is a partition of V into pairs such that $(V, E \uplus M)$ is a planar multigraph, where $E \uplus M$ is the multiset formed by disjoint union.

Lemma 5.31 shows that for any 3-regular planar graph, a planar pairing exists and can be easily found in polynomial time. Now we use the planar pairing technique to show the following.

Lemma 8.10. *Let* $t \in \mathbb{C}$. *Then* Pl-Holant($[t, 1, 0, 0]$) $\leq_T$ Pl-Holant($[t, 1, 0, 0, 0]$).

Proof. An instance of Pl-Holant($[t, 1, 0, 0]$) is a signature grid Ω with underlying graph $G = (V, E)$ that is planar and 3-regular. By Lemma 5.31, there exists a planar pairing M in G and it can be found in polynomial time. Then $G' = (V, E \uplus M)$ is a planar and 4-regular multigraph. We assign $[t, 1, 0, 0, 0]$

to every vertex in G'. By Lemma 8.8, we can assume that we have $[1, 0, 0]$. We replace each edge in M with a path of length 2 to form a graph G'' and assign $[1, 0, 0] = [1, 0]^{\otimes 2}$ to the new vertices on the paths. Note that applying $[1, 0]$ to $[t, 1, 0, 0, 0]$ gives $[t, 1, 0, 0]$. Then the signature grid Ω'' with underlying graph G'' has the same Holant value as the original signature grid Ω. □

Corollary 8.11. *The problem* Pl-Holant($[t, 1, 0, 0, 0]$) *is* #P*-hard for all nonzero* $t \in \mathbb{C}$.

Proof. By Theorem 6.3, the planar Holant dichotomy for a single ternary signature, the problem Pl-Holant($[t, 1, 0, 0]$) is #P-hard. Then Corollary 8.11 follows from Lemma 8.10. □

One feature of this reduction is worth noticing. Note that from any set of signatures all with even arities, any gadget construction will produce a signature also of even arity. This is by a simple parity argument, because the sum of arities of all signatures used in the construction is twice the number of internal edges plus the arity of the signature produced. Thus, with $[t, 1, 0, 0, 0]$ alone one cannot construct any signature of arity 3 by gadget construction.

Now we are ready to prove our Pl-Holant dichotomy for a symmetric arity 4 signature.

Theorem 8.12. *If f is a nondegenerate, symmetric, complex-valued signature of arity* 4 *in Boolean variables, then* Pl-Holant(f) *is* #P*-hard unless f is $\mathscr{A}$-transformable or $\mathscr{P}$-transformable or vanishing or $\mathscr{M}$-transformable, in which case the problem is computable in polynomial time.*

Proof. Let $f = [f_0, f_1, f_2, f_3, f_4]$. Then its compressed matrix

$$\widetilde{M_f} = \begin{bmatrix} f_0 & 2f_1 & f_2 \\ f_1 & 2f_2 & f_3 \\ f_2 & 2f_3 & f_4 \end{bmatrix}.$$

If $\det(\widetilde{M_f}) \neq 0$, equivalently if there do not exist $a, b, c \in \mathbb{C}$, not all zero, such that for all $k \in \{0, 1, 2\}$, $af_k + bf_{k+1} + cf_{k+2} = 0$, then Pl-Holant($f$) is #P-hard by Corollary 7.25. Otherwise, there do exist such a, b, c. If $a = c = 0$, then $b \neq 0$, so $f_1 = f_2 = f_3 = 0$. In this case, $f \in \mathscr{P}$ is a generalized equality signature, so f is $\mathscr{P}$-transformable. Now suppose a and c are not both 0. If $b^2 - 4ac \neq 0$, then $f_k = \alpha^{4-k}\beta^k + \gamma^{4-k}\delta^k$, where $\det\begin{bmatrix} \alpha & \gamma \\ \beta & \delta \end{bmatrix} \neq 0$. We apply a holographic transformation by $M = \begin{bmatrix} \alpha & \gamma \\ \beta & \delta \end{bmatrix}$ to Pl-Holant(f) $\equiv_{\mathrm{T}}$ Pl-Holant($=_2 \mid f$). M transforms f to $(M^{-1})^{\otimes 4} f = (=_4)$, and the left-hand side ($=_2$) to a binary symmetric signature $g = (=_2)M^{\otimes 2}$. Then we apply Theorem 5.1 (the case $k = 4$ is proved in Chapter 5) to show that Pl-Holant(f) is #P-hard unless f is either $\mathscr{A}$-, $\mathscr{P}$-, or $\mathscr{M}$-transformable. Otherwise, $b^2 - 4ac = 0$ and there are two cases. In the first, for any $0 \le k \le 4$, $f_k = ck\lambda^{k-1} + d\lambda^k$, where $c \ne 0$. In the second, for any $0 \le k \le 4$, $f_k = c(4-k)\lambda^{3-k} + d\lambda^{4-k}$, where $c \ne 0$. These cases map between each other under a holographic transformation by $\begin{bmatrix} 0 & 1 \\ 1 & 0 \end{bmatrix}$, so assume that we are in the first case. If $\lambda = \pm i$, then f is vanishing. Otherwise, a further holographic transformation by $\frac{1}{\sqrt{1+\lambda^2}} \begin{bmatrix} 1 & \lambda \\ \lambda & -1 \end{bmatrix}$ transforms f to $\hat{f} = [t, 1, 0, 0, 0]$ for some $t \in \mathbb{C}$ after normalizing the second entry. If $t = 0$, then the problem is counting perfect matchings over planar 4-regular graphs, so $\hat{f} \in \mathscr{M}$ and f is $\mathscr{M}$-transformable. Otherwise, $t \ne 0$ and by Corollary 8.11, we are done. □

8.4 Domain Pairing

In this section, we discuss a technique called domain pairing, which pairs input variables to simulate a problem on a domain of size 4 and then reduces a problem in the Boolean domain to it. As explained in Section 8.1, we work in the Hadamard basis instead of the standard basis. The goal therefore is a dichotomy for Pl-Holant($\mathcal{F} \cup \widehat{\mathcal{EQ}}$).

By a simple parity argument, gadgets constructed with signatures of even arity can only realize other signatures of even arity. In particular, this means that $(=_4)$ cannot by itself be used to construct $(=_3)$. Nevertheless, there is an indirect method that can realize $(=_3)$ using $(=_4)$. The idea is to change the domain from individual elements to pairs of elements. Thus, we call this reduction technique *domain pairing*.

The following is a lemma in which the reduction is carried out by this method.

Lemma 8.13 (Domain pairing). *Let $a, b, x, y \in \mathbb{C}$. Suppose $f = [x, 0, y, 0]$ and $g = [a, 0, \ldots, 0, b]$ with arity ≥ 3. If $aby \ne 0$ and $x^2 \ne y^2$, then*

$$\text{Pl-Holant}([x, y, y] \mid \mathcal{EQ}) \le_T \text{Pl-Holant}(\{f, g\} \cup \widehat{\mathcal{EQ}}).$$

In particular, Pl-Holant($\{f, g\} \cup \widehat{\mathcal{EQ}}$) *is* #P-*hard.*

Proof. The problem Pl-Holant($[x, y, y] \mid \mathcal{EQ}$) is #P-hard by Theorem 8.3 since $y \ne 0$ and $x^2 \ne y^2$. Hence we only need to prove the reduction. Since we have $g = [a, 0, \ldots, 0, b]$ in Pl-Holant($\{f, g\} \cup \widehat{\mathcal{EQ}}$), by Lemma 7.48, we have $\mathcal{EQ}_2$, the set of all EQUALITIES of even arity.

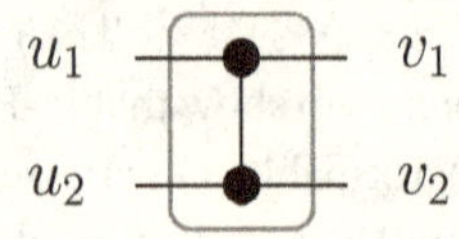

Figure 8.3. Gadget designed for the paired domain. One vertex is assigned $[1, 0, 1, 0]$ and the other is assigned $[x, 0, y, 0]$.

We remark that if one were able to obtain $(=_3)$ on the Boolean domain, then one could have constructed $[x, y, y]$ from $[x, 0, y, 0]$ easily. Simply taking a self-loop on $(=_3)$ gives $[1, 1]$, and connecting $[1, 1]$ to $[x, 0, y, 0]$ gives $[x, y, y]$. However, as noted earlier, there is an obstacle to obtaining $(=_3)$ due to parity. Instead we use the domain pairing technique.

An instance of Pl-Holant($[x, y, y] \mid \mathcal{EQ}$) is a planar bipartite graph $G = (U, V, E)$ in which every vertex in U has degree 2. We replace every vertex in V of degree k (which is assigned $(=_k) \in \mathcal{EQ}$) with a vertex of degree $2k$ and assign $(=_{2k}) \in \mathcal{EQ}_2$. Furthermore, we bundle two adjacent edges (representing variables) to form k bundles of two edges each. The k bundles correspond to the k incident edges of the original vertex with degree k.

If the inputs to these Equality signatures are restricted to $\{(0, 0), (1, 1)\}$ on each bundle, then these Equality signatures take value 1 on $((0, 0), \ldots, (0, 0))$ and $((1, 1), \ldots, (1, 1))$ and take value 0 elsewhere. Thus, if we restrict the domain to $\{(0, 0), (1, 1)\}$, it is the Equality signature $(=_k)$.

To simulate $[x, y, y]$, we connect $f = [x, 0, y, 0]$ to $e = [1, 0, 1, 0] \in \widehat{\mathcal{EQ}}$ by a single edge as shown in Figure 8.3 to form a gadget with signature

$$h(u_1, u_2, v_1, v_2) = \sum_{w=0,1} f(u_1, v_1, w) \cdot e(u_2, v_2, w).$$

We replace every (degree 2) vertex in U (which is assigned $[x, y, y]$) by a degree 4 vertex assigned h, where the variables of h are bundled as (u_1, u_2) and (v_1, v_2).

The vertices in this new graph G' are connected as in the original graph G, except that every original edge is replaced by two edges that connect to the same side of the gadget in Figure 8.3, either the left or the right side. Clearly G' is still a planar graph. Notice that h is only connected by (u_1, u_2) and (v_1, v_2) to some bundle of two incident edges of an Equality signature. Since this Equality signature enforces that the value on each bundle is either $(0, 0)$ or $(1, 1)$, we need only to consider the restriction of h to the domain $\{(0, 0), (1, 1)\}$. On this domain, h is exactly the *symmetric* binary signature $[x, y, y]$. Therefore, the Holant of G' has the same Holant value as the original graph G. □

Lemma 8.13 is mainly used to prove that generally speaking, GEN-EQ that do not belong to $\mathscr{A}$ cannot mix with $\widehat{\mathcal{EQ}}$. As $\widehat{\mathcal{EQ}} \subset \mathscr{A}$ and $\mathscr{A}$ is a tractable class, only having GEN-EQ that *do* belong to $\mathscr{A}$ together with $\widehat{\mathcal{EQ}}$ cannot lead to #P-hardness (if #P does not collapse to P). However, we show that having GEN-EQ of arity ≥ 3 that *do not* belong to $\mathscr{A}$ together with $\widehat{\mathcal{EQ}}$ will lead to #P-hardness. There are two scenarios, Corollary 8.14 and Lemma 8.16, that will let us apply Lemma 8.13. Lemma 8.16 says that a nondegenerate GEN-EQ of arity ≥ 3 that does not belong to $\mathscr{A}$ together with $\widehat{\mathcal{EQ}}$ will lead to #P-hardness. Corollary 8.14 says that even if this GEN-EQ of arity ≥ 3 may belong to $\mathscr{A}$, if there is another binary nondegenerate GEN-EQ that does not belong to $\mathscr{A}$, then together with $\widehat{\mathcal{EQ}}$ we still have #P-hardness.

Corollary 8.14 is an immediate corollary of Lemma 8.13, and is also used in Lemma 8.16, so we prove that first.

Corollary 8.14. *Let $a, b, x, y \in \mathbb{C}$. Suppose $f = [x, 0, y]$ and $g = [a, 0, \ldots, 0, b]$ with arity ≥ 3. If $abxy \neq 0$ and $x^4 \neq y^4$, then* Pl-Holant$(\{f, g\} \cup \widehat{\mathcal{EQ}})$ *is* #P-*hard.*

Proof. Connect three copies of $f = [x, 0, y]$ to $[1, 0, 1, 0]$, with one on each edge, to get $x[x^2, 0, y^2, 0]$ and we apply Lemma 8.13. □

The proof of Lemma 8.16 applies Corollary 8.14 and an interpolation of a unary signature. We prove this interpolation first; it will be used again later.

Lemma 8.15. *Let $x \in \mathbb{C}$. Suppose $\mathcal{F}$ is a set of signatures containing $f = [1, x, 1]$. If $x \notin \{0, \pm 1\}$ and the matrix form M_f of f has infinite order modulo a scalar, then*

$$\text{Pl-Holant}(\mathcal{F} \cup \{[a, b]\} \cup \widehat{\mathcal{EQ}}) \leq_{\mathrm{T}} \text{Pl-Holant}(\mathcal{F} \cup \widehat{\mathcal{EQ}})$$

for any $a, b \in \mathbb{C}$.

Proof. Consider the unary recursive construction (M_f, s) in Figure 8.1, where $s = \begin{bmatrix} 1 \\ 0 \end{bmatrix} \in \widehat{\mathcal{EQ}}$. The determinant of $M_f = \begin{bmatrix} 1 & x \\ x & 1 \end{bmatrix}$ is $1 - x^2 \neq 0$. The determinant of $[s\ M_f s] = \begin{bmatrix} 1 & 1 \\ 0 & x \end{bmatrix}$ is $x \neq 0$. By assumption, M_f has infinite order modulo a scalar. Therefore, we can interpolate any unary signature by Lemma 8.7. □

Lemma 8.16. *Let $a, b \in \mathbb{C}$. Suppose $f = [a, 0, \ldots, 0, b]$ with arity ≥ 3. If $ab \neq 0$ and $a^4 \neq b^4$, then* Pl-Holant$(\{f\} \cup \widehat{\mathcal{EQ}})$ *is* #P-*hard.*

Proof. Since $a \neq 0$, we normalize f to $[1, 0, \ldots, 0, x]$, where $x \neq 0$ and $x^4 \neq 1$. If the arity of f is even, then after some number of self-loops, we have $[1, 0, x]$ and are done by Corollary 8.14. Otherwise, the arity of f is odd. After zero or more self-loops, we have $g = [1, 0, 0, x]$. If we have the signature $[1, 1]$, then we can connect this to g to get $[1, 0, x]$ and be done by Corollary 8.14.

We use the signature $[1, x]$, which we obtain via a self-loop on g. Suppose $\Re(x)$, the real part of x, is nonzero. Connecting $[1, x]$ to $[1, 0, 1, 0] \in \widehat{\mathcal{EQ}}$ gives $h = [1, x, 1]$. The eigenvalues of M_h are $\lambda_\pm = 1 \pm x$. They are nonzero because $x \neq \pm 1$. Note that $|x - 1| = |x + 1|$ iff x lies on the pure imaginary line $\Re(x) = 0$, so we have $|\frac{\lambda_+}{\lambda_-}| \neq 1$. Hence the ratio of the eigenvalues is not a root of unity, and so M_h has infinite order modulo a scalar. Therefore, we can interpolate $[1, 1]$ by Lemma 8.15.

Otherwise, $\Re(x) = 0$ but x is not a root of unity since $x \neq \pm \mathfrak{i}$. Connecting $[1, x]$ to g gives $h = [1, 0, x^2]$. Clearly $\left(x^2\right)^4 \neq 1$. Hence we apply Corollary 8.14 on h and f, implying that Pl-Holant($\{f\} \cup \widehat{\mathcal{EQ}}$) is #P-hard. □

8.5 No-Mixing of Tractable Signatures

Perhaps a natural goal at this point would be to attempt a dichotomy Pl-#CSP(f) for a single signature in the form of Pl-Holant($\{\hat{f}\} \cup \widehat{\mathcal{EQ}}$), which should say that in order *not* to be #P-hard, $\hat{f}$ must belong to $\mathscr{A} \cup \widehat{\mathscr{P}} \cup \mathscr{M}$, the union of the three tractable classes. This will indeed be proved (Theorem 8.32) before we prove the main dichotomy (Theorem 8.34). However, seemingly taking a detour, we will actually prove a no-mixing theorem (Theorem 8.22) first. This no-mixing theorem assumes that a given signature set $\mathcal{F}$ already satisfies $\mathcal{F} \subseteq \mathscr{A} \cup \widehat{\mathscr{P}} \cup \mathscr{M}$ (a necessary condition that would be implied by the single signature dichotomy), and states that the problem is still #P-hard unless $\mathcal{F}$ is contained in a single tractable class. In other words, no mixing of distinct tractable classes can occur. The reason we organize the proof in this way is because the no-mixing theorem will be useful in the proof of the single signature dichotomy.

Thus in this section, we prove that various tractable signatures, if they do not belong to a single tractable class, will produce #P-hardness in combination. There are three tractable classes. To help understand these classes, Figure 8.4 contains a Venn diagram of symmetric signatures in $\mathscr{A}$, $\widehat{\mathscr{P}}$, and $\mathscr{M}$. Note that nontrivial intersections occur; but crucially $\mathscr{M} \cap \widehat{\mathscr{P}} \subset \mathscr{A}$, thus there is an empty cell in the Venn diagram

$$\mathscr{M} \cap \widehat{\mathscr{P}} - \mathscr{A} = \emptyset. \tag{8.1}$$

Exercise: Prove (8.1). Thus there is an empty cell in the Venn diagram (8.4).

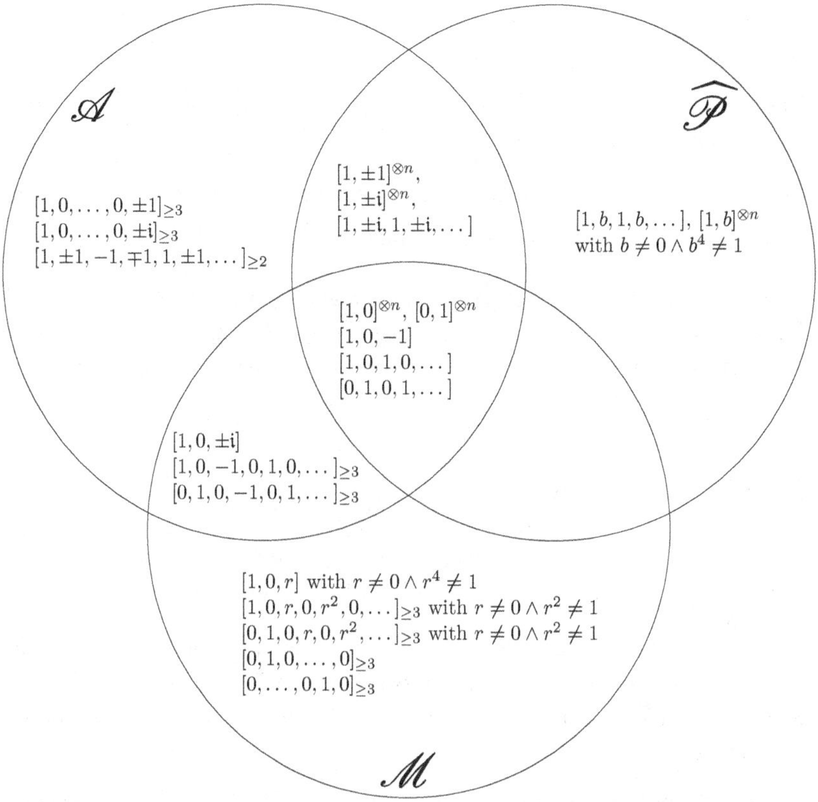

Figure 8.4. Venn diagram of the tractable Pl-#CSP signature sets in the Hadamard basis. Each signature has been normalized for simplicity of presentation. For a signature f, the notation "$f_{\geq k}$" is short for "arity$(f) \geq k$". Notice that $\mathscr{M} \cap \widehat{\mathscr{P}} - \mathscr{A}$ is empty.

Exercise: Verify that all nonzero symmetric signatures in $\mathscr{A}$, $\widehat{\mathscr{P}}$, and $\mathscr{M}$ up to a nonzero scalar are depicted correctly in the Venn diagram (8.4).

We remark that $\mathscr{P}$ (and hence $\widehat{\mathscr{P}}$ also) contains all degenerate signatures, in particular all unary signatures.

We will prove five lemmas giving pairwise no-mixing statements. In the first two lemmas one of the signatures is unary. The last three lemmas consider the general case of the three pairwise no-mixing statements between three tractable classes, respectively.

Lemma 8.17 is a no-mixing lemma of $\mathscr{A}$ versus $\widehat{\mathscr{P}}$, where the unary signature is not in $\mathscr{A}$. Note that for a unary signature $[a, b]$, the condition $ab \neq 0$ and

$a^4 \neq b^4$ is equivalent to $[a, b] \notin \mathscr{A}$. Also note that since $\mathscr{P}$ contains all unary signatures, so does $\widehat{\mathscr{P}}$.

Lemma 8.17. *Let $a, b \in \mathbb{C}$. Suppose $f \in \mathscr{A} - \widehat{\mathscr{P}}$. If $ab \neq 0$ and $a^4 \neq b^4$, then the problem* Pl-Holant($\{f, [a, b]\} \cup \widehat{\mathcal{EQ}}$) *is* #P-*hard.*

Proof. To understand what forms a symmetric signature $f \in \mathscr{A} - \widehat{\mathscr{P}}$ may take, we may consider a holographic transformation $H_2 = \begin{bmatrix} 1 & 1 \\ 1 & -1 \end{bmatrix}$. Note that f is nondegenerate since $f \notin \widehat{\mathscr{P}}$. As $H_2 \in \text{Stab}(\mathscr{A})$, we have $H_2 f \in \mathscr{A} - \mathscr{P}$. Since $\mathscr{F}_1 \subset \mathscr{P}$, we have $H_2 f \in \mathscr{F}_2 \cup \mathscr{F}_3 - \mathscr{P}$. By analyzing the forms in $\mathscr{F}_2 \cup \mathscr{F}_3 - \mathscr{P}$ we can obtain the following possibilities for f, up to a nonzero scalar.

- $[1, 0, \pm i]$.
- $[1, 0, \ldots, 0, x]$ of arity ≥ 3 with $x^4 = 1$.
- $[1, \pm 1, -1, \mp 1, 1, \pm 1, -1, \mp 1, \ldots]$ of arity ≥ 2.
- $[1, 0, -1, 0, 1, 0, -1, 0, \ldots, 0 \text{ or } 1 \text{ or } (-1)]$ of arity ≥ 3.
- $[0, 1, 0, -1, 0, 1, 0, -1, \ldots, 0 \text{ or } 1 \text{ or } (-1)]$ of arity ≥ 3.

We handle these cases below. Our general technique is to produce a binary signature that does not fit the tractability criterion of Theorem 8.3 for spin systems. In one case we also use Corollary 8.14 and Lemma 8.16 based on the domain pairing technique.

1. Suppose $f = [1, 0, \pm i]$. Connecting $[a, b]$ to $[1, 0, 1, 0]$ gives $[a, b, a]$, and connecting two copies of $[1, 0, \pm i]$ to $[a, b, a]$, one on each edge, gives $g = [a, \pm ib, -a]$. Then Pl-Holant($g \mid \widehat{\mathcal{EQ}}$) is #P-hard by Theorem 8.3.
2. Suppose $f = [1, 0, \ldots, 0, x]$ of arity ≥ 3 with $x^4 = 1$. Connecting $[a, b]$ to f gives $g = [a, 0, \ldots, 0, bx]$ of arity ≥ 2. Note that $(bx)^4 = b^4 \neq a^4$. If the arity of g is exactly 2, then Pl-Holant($\{f, g\} \cup \widehat{\mathcal{EQ}}$) is #P-hard by Corollary 8.14, so we are done. Otherwise, the arity of g is ≥ 3 and Pl-Holant($\{g\} \cup \widehat{\mathcal{EQ}}$) is #P-hard by Lemma 8.16.
3. Suppose $f = [1, \pm 1, -1, \ldots]$ of arity ≥ 2. Connecting some number of $[1, 0]$ gives $[1, \pm 1, -1]$ of arity exactly 2. Connecting $[a, b]$ to $[1, 0, 1, 0]$ gives $[a, b, a]$ and connecting two copies of $[a, b, a]$ to $[1, \pm 1, -1]$, one on each edge, gives $g = [a^2 \pm 2ab - b^2, \pm(a^2 + b^2), -a^2 \pm 2ab + b^2]$. This is easily verified by
$$\begin{bmatrix} a & b \\ b & a \end{bmatrix} \begin{bmatrix} 1 & \pm 1 \\ \pm 1 & -1 \end{bmatrix} \begin{bmatrix} a & b \\ b & a \end{bmatrix} = \begin{bmatrix} a^2 \pm 2ab - b^2 & \pm(a^2 + b^2) \\ \pm(a^2 + b^2) & -a^2 \pm 2ab + b^2 \end{bmatrix}.$$
Then Pl-Holant($g \mid \widehat{\mathcal{EQ}}$) is #P-hard by Theorem 8.3.

4. Suppose $f = [1, 0, -1, 0, \ldots]$ of arity ≥ 3. Connecting some number of $[1, 0]$ gives $g = [1, 0, -1, 0]$ of arity exactly 3. Connecting $[a, b]$ to g gives $h = [a, -b, -a]$. Then Pl-Holant($h \mid \widehat{\mathcal{EQ}}$) is #P-hard by Theorem 8.3.
5. The argument for $f = [0, 1, 0, -1, \ldots]$ is similar to the previous case. □

Before we move to the next lemma, we observe that in Lemma 8.17 we did not explicitly require that $\{f, [a, b]\} \not\subseteq \mathscr{M}$. In general, a pairwise no-mixing statement should say that for some f and g, if $\{f, g\}$ is not contained in a single tractable class, then their joint presence leads to #P-hardness. The condition in Lemma 8.17 ensures that $\{f, [a, b]\} \not\subseteq \mathscr{A}$ and $\{f, [a, b]\} \not\subseteq \widehat{\mathscr{P}}$. However, even though it was not explicitly stated that $\{f, [a, b]\} \not\subseteq \mathscr{M}$, this is implied since $ab \neq 0$ violates the parity requirement for $[a, b] \in \mathscr{M}$.

The next lemma is a no-mixing statement of $\mathscr{M}$ versus $\mathscr{A}$ involving a unary signature $[a, b]$. Note that for a unary signature $[a, b]$, the condition $ab \neq 0$ is equivalent to $[a, b] \notin \mathscr{M}$. We explicitly require that $f \in \mathscr{M} - \mathscr{A}$, but we do not require that $[a, b] \in \mathscr{A}$. This gives us a slightly wider applicability. Also note that we do not explicitly require that $\{f, [a, b]\} \not\subseteq \widehat{\mathscr{P}}$. This is because $f \in \mathscr{M} - \mathscr{A}$ implies that $f \notin \widehat{\mathscr{P}}$ by the observation (8.1).

Lemma 8.18. *Let $a, b \in \mathbb{C}$. If $f \in \mathscr{M} - \mathscr{A}$ and $ab \neq 0$, then* Pl-Holant($\{f, [a, b]\} \cup \widehat{\mathcal{EQ}}$) *is* #P-*hard.*

Proof. By considering the forms in Theorem 4.11 for symmetric matchgates $\mathscr{M}$ and the list of signatures in $\mathscr{F}_1 \cup \mathscr{F}_2 \cup \mathscr{F}_3$ in Section 3.2, it is easy to see that, up to a nonzero scalar, the possibilities for $f \in \mathscr{M} - \mathscr{A}$ are as follows:

- $[1, 0, r]$ with $r \neq 0$ and $r^4 \neq 1$.
- $[1, 0, r, 0, r^2, 0, \ldots]$ of arity ≥ 3 with $r \neq 0$ and $r^2 \neq 1$.
- $[0, 1, 0, r, 0, r^2, \ldots]$ of arity ≥ 3 with $r \neq 0$ and $r^2 \neq 1$.
- $[0, 1, 0, \ldots, 0]$ of arity ≥ 3.
- $[0, \ldots, 0, 1, 0]$ of arity ≥ 3.

We handle these cases below. Again our technique is to produce a binary signature that does not fit the tractability criterion of Theorem 8.3 for spin systems.

1. Suppose $f = [1, 0, r]$ with $r^4 \neq 1$ and $r \neq 0$. Connecting $[a, b]$ to $[1, 0, 1, 0]$ gives $[a, b, a]$ and connecting two copies of $[1, 0, r]$ to $[a, b, a]$, one on each edge, gives $g = [a, br, ar^2]$. If $a^2 \neq b^2$, then Pl-Holant($g \mid \widehat{\mathcal{EQ}}$) is #P-hard by Theorem 8.3.

   Otherwise, $a^2 = b^2$ and we first connect $[a, b]$ to $[1, 0, r]$ to get $[a, br]$. Then by the same construction, we have $g' = [a, br^2, ar^2]$ and Pl-Holant($g' \mid \widehat{\mathcal{EQ}}$) is #P-hard by Theorem 8.3.

2. Suppose $f = [1, 0, r, 0, \ldots]$ of arity ≥ 3 with $r^2 \neq 1$ and $r \neq 0$. Connecting some number of $[1, 0] \in \widehat{\mathcal{EQ}}$ gives $g = [1, 0, r, 0]$ of arity exactly 3. Connecting $[a, b]$ to g gives $h = [a, br, ar]$. If $a^2 \neq b^2 r$, then Pl-Holant$(h \mid \widehat{\mathcal{EQ}})$ is #P-hard by Theorem 8.3.

   Otherwise, $a^2 = b^2 r$ and we first connect $[1, 0]$ and $[a, b]$ to $[1, 0, r, 0]$ to get $[a, br]$. Then by the same construction, we have $g' = [a, br^2, ar]$ and Pl-Holant$(g' \mid \widehat{\mathcal{EQ}})$ is #P-hard by Theorem 8.3.
3. The argument for $f = [0, 1, 0, r, \ldots]$ is similar to the previous case.
4. Suppose $f = [0, 1, 0, \ldots, 0]$ of arity $k \geq 3$. Connecting $k - 2$ copies of $[a, b]$ to f gives $g = a^{k-3}[(k - 2)b, a, 0]$. Then Pl-Holant$(g \mid \widehat{\mathcal{EQ}})$ is #P-hard by Theorem 8.3.
5. The argument for $f = [0, \ldots, 0, 1, 0]$ is similar to the previous case. □

Now we consider the general case of two signatures from two different tractable sets. The three tractable classes $\mathscr{A}$, $\widehat{\mathscr{P}}$ and $\mathscr{M}$ give rise to three pairs of pairwise cases to consider, each of which is covered in one of the next three lemmas.

Lemma 8.19 **(($\mathscr{A}$, $\widehat{\mathscr{P}}$)-no-mixing).** *If $f \in \mathscr{A} - \widehat{\mathscr{P}}$ and $g \in \widehat{\mathscr{P}} - \mathscr{A}$, then* Pl-Holant$(\{f, g\} \cup \widehat{\mathcal{EQ}})$ *is* #P-*hard.*

Proof. The only possibility for g is $[a, b, a, b, \ldots]$, or $[a, b]^{\otimes n}$, where $n \geq 1$, $ab \neq 0$ and $a^4 \neq b^4$. Connecting some number of $[1, 0] \in \widehat{\mathcal{EQ}}$ to g gives $[a, b]$ up to a nonzero scalar, and we are done by Lemma 8.17. □

Note that implicitly the validity of Lemma 8.19 hinges on the fact that (8.1) holds, namely that empty cell in the Venn diagram (8.4). Were this not the case, that signature $g \in \widehat{\mathscr{P}} - \mathscr{A}$ could have been in $\mathscr{M}$, and that $f \in \mathscr{A} - \widehat{\mathscr{P}}$ may also belong to $\mathscr{M}$, then Pl-Holant$(\{f, g\} \cup \widehat{\mathcal{EQ}})$ would be tractable in P and Lemma 8.19 would collapse #P and P.

Similarly if (8.1) were not true, then the following Lemma 8.20 would also callapse #P and P, because that $f \in \mathscr{M} - \mathscr{A}$ could have belonged to $\widehat{\mathscr{P}}$, and it would be possible that $\{f, g\} \cup \widehat{\mathcal{EQ}} \subset \widehat{\mathscr{P}}$.

Lemma 8.20 **(($\mathscr{M}$, $\mathscr{A}$)-no-mixing).** *If $f \in \mathscr{M} - \mathscr{A}$ and $g \in \mathscr{A} - \mathscr{M}$, then* Pl-Holant$(\{f, g\} \cup \widehat{\mathcal{EQ}})$ *is* #P-*hard.*

Proof. If g is degenerate in $\mathscr{A} - \mathscr{M}$, then g does not contain any 0 entries. Suppose g does not contain a 0 entry. Then after connecting some number of $[1, 0]$ to g, we have a unary signature $[a, b]$ with $ab \neq 0$, and are done by Lemma 8.18.

Otherwise, g contains a 0 entry. In particular, g is nondegenerate. Then, by the forms of $\mathscr{F}_1 \cup \mathscr{F}_2 \cup \mathscr{F}_3$ and $\mathscr{M}$, it must be a GEN-EQ (see Figure 8.4),

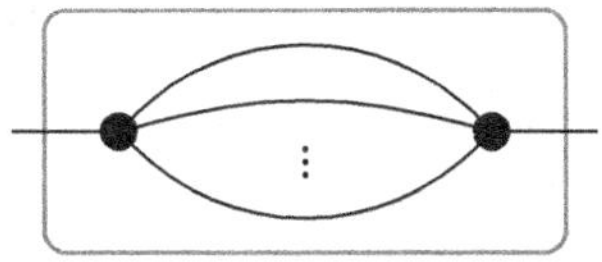

Figure 8.5. The vertices are assigned $f = [0, 1, 0, \ldots, 0]$.

$g = [x, 0, \ldots, 0, y]$ of arity ≥ 3 with $xy \neq 0$ (and $x^4 = y^4$). Up to a nonzero scalar, the possibilities for f are as follows:

- $[1, 0, r]$ with $r \neq 0$ and $r^4 \neq 1$.
- $[1, 0, r, 0, r^2, 0, \ldots]$ of arity ≥ 3 with $r \neq 0$ and $r^2 \neq 1$.
- $[0, 1, 0, r, 0, r^2, \ldots]$ of arity ≥ 3 with $r \neq 0$ and $r^2 \neq 1$.
- $[0, 1, 0, 0, \ldots, 0]$ of arity ≥ 3.
- $[0, \ldots, 0, 0, 1, 0]$ of arity ≥ 3.

We handle these cases below.

1. Suppose $f = [1, 0, r]$ with $r \neq 0$ and $r^4 \neq 1$. Then we are done by Corollary 8.14.
2. Suppose $f = [1, 0, r, 0, \ldots]$ of arity ≥ 3 with $r \neq 0$ and $r^2 \neq 1$. After connecting some number of $[1, 0]$ to f, we have $h = [1, 0, r, 0]$ of arity exactly 3. Then Pl-Holant($\{g, h\} \cup \widehat{\mathcal{EQ}}$) is #P-hard by Lemma 8.13.
3. Suppose $f = [0, 1, 0, r, \ldots]$ of arity ≥ 3 with $r \neq 0$ and $r^2 \neq 1$. After connecting some number of $[1, 0]$ to f, we have $h = [0, 1, 0, r]$ of arity exactly 3. Connecting two more copies of $[1, 0]$ to h gives $[0, 1]$. Then we apply a holographic transformation by $T = \begin{bmatrix} 0 & 1 \\ 1 & 0 \end{bmatrix}$, so g is transformed to $\hat{g} = [y, 0, \ldots, 0, x]$ and h is transformed to $\hat{h} = [r, 0, 1, 0]$. Since T is orthogonal, Pl-Holant($\{g, h\} \cup \widehat{\mathcal{EQ}}$) is transformed to Pl-Holant($\{\hat{g}, \hat{h}\} \cup T\widehat{\mathcal{EQ}}$). Every even arity signature in $\widehat{\mathcal{EQ}}$ remains unchanged after a holographic transformation by T. By attaching $[0, 1]T = [1, 0]$ to every even arity signature in $T\widehat{\mathcal{EQ}}$, we obtain all of the odd arity signatures in $\widehat{\mathcal{EQ}}$ again. Then Pl-Holant($\{\hat{g}, \hat{h}\} \cup \widehat{\mathcal{EQ}}$) is #P-hard by Lemma 8.13.
4. Suppose $f = [0, 1, 0, \ldots, 0]$ of arity $k \geq 3$. The gadget in Figure 8.5 with f assigned to both vertices has signature $h = [k - 1, 0, 1]$. Then Pl-Holant($\{g, h\} \cup \widehat{\mathcal{EQ}}$) is #P-hard by Corollary 8.14.
5. The argument for $f = [0, \ldots, 0, 1, 0]$ is similar to the previous case. □

Note that in Lemma 8.19 and 8.20 we do not explicitly require that $\{f, g\}$ is not contained in the third tractable class. But that is implied (as it should). In Lemma 8.19, $g \in \widehat{\mathscr{P}} - \mathscr{A}$ implies that $g \notin \mathscr{M}$. In Lemma 8.20, $f \in \mathscr{M} - \mathscr{A}$

implies that $f \notin \widehat{\mathscr{P}}$. Both these statements follow from the observation (8.1). However in Lemma 8.21 we do have to explicitly require that $\{f, g\} \not\subseteq \mathscr{A}$.

Lemma 8.21 **($(\widehat{\mathscr{P}}, \mathscr{M})$-no-mixing).** *If* $f \in \mathscr{M} - \widehat{\mathscr{P}}$ *and* $g \in \widehat{\mathscr{P}} - \mathscr{M}$ *and* $\{f, g\} \not\subseteq \mathscr{A}$, *then* Pl-Holant($\{f, g\} \cup \widehat{\mathcal{EQ}}$) *is* #P-*hard.*

Proof. The only possibilities for g are $[a, b, a, b, \ldots]$ or $[a, b]^{\otimes n}$ with $ab \neq 0$ (see Figure 8.4). Connecting some number of $[1, 0]$ to g gives $h = [a, b]$ up to a nonzero scalar. If $f \notin \mathscr{A}$, then Pl-Holant($\{f, h\} \cup \widehat{\mathcal{EQ}}$) is #P-hard by Lemma 8.18. Otherwise, $f \in \mathscr{A}$, so $g \notin \mathscr{A}$ and Pl-Holant($\{f, g\} \cup \widehat{\mathcal{EQ}}$) is #P-hard by Lemma 8.19. □

Observe that in the proof of Lemma 8.21 when we apply Lemma 8.18 to $\{f, h\}$, we have $f \in \mathscr{M} - \mathscr{A}$, but we do not necessarily have the condition $h \in \mathscr{A} - \mathscr{M}$. We only have $h \notin \mathscr{M}$. However, Lemma 8.18 does not require that $h \in \mathscr{A}$. Also note that we choose to prove the three lemmas, Lemma 8.19, 8.20, and 8.21, in that order so that the proof of Lemma 8.21 can apply Lemma 8.19.

We summarize this section with the following theorem, which says that the tractable signature sets cannot mix. Signatures from different tractable sets, when put together, lead to #P-hardness.

Theorem 8.22 (No-mixing theorem). *Let $\mathcal{F}$ be any set of symmetric, complex-valued signatures in Boolean variables. If $\mathcal{F} \subseteq \mathscr{A} \cup \widehat{\mathscr{P}} \cup \mathscr{M}$, then* Pl-Holant($\mathcal{F} \cup \widehat{\mathcal{EQ}}$) *is* #P-*hard unless $\mathcal{F} \subseteq \mathscr{A}$, or $\mathcal{F} \subseteq \widehat{\mathscr{P}}$, or $\mathcal{F} \subseteq \mathscr{M}$, in which case* Pl-Holant^c($\mathcal{F} \cup \widehat{\mathcal{EQ}}$) *is tractable.*

Proof. If $\mathcal{F}$ is a subset of $\mathscr{A}$, $\widehat{\mathscr{P}}$, or $\mathscr{M}$, then the tractability is given in Theorem 8.4. Suppose $\mathcal{F}$ is not a subset of $\mathscr{A}$, $\widehat{\mathscr{P}}$, or $\mathscr{M}$. Then $\mathcal{F}$ contains a signature $g \in (\widehat{\mathscr{P}} \cup \mathscr{M}) - \mathscr{A}$ since $\mathcal{F} \not\subseteq \mathscr{A}$. As $\mathcal{F} \subseteq \mathscr{A} \cup \widehat{\mathscr{P}} \cup \mathscr{M}$, either $\mathcal{F}$ contains a signature $f \in \mathscr{A} - (\widehat{\mathscr{P}} \cup \mathscr{M})$, or $\mathcal{F} \subseteq \widehat{\mathscr{P}} \cup \mathscr{M}$. Suppose it is the first case. If $g \in \widehat{\mathscr{P}} - \mathscr{A}$, then Pl-Holant($\mathcal{F} \cup \widehat{\mathcal{EQ}}$) is #P-hard by Lemma 8.19. Otherwise, $g \in \mathscr{M} - \mathscr{A}$ and Pl-Holant($\mathcal{F} \cup \widehat{\mathcal{EQ}}$) is #P-hard by Lemma 8.20.

Now suppose it is the second case that $\mathcal{F} \subseteq \widehat{\mathscr{P}} \cup \mathscr{M}$. Since $(\widehat{\mathscr{P}} \cap \mathscr{M}) - \mathscr{A}$ is empty by the observation (8.1) (see Figure 8.4), either $g \in (\widehat{\mathscr{P}} - \mathscr{M}) - \mathscr{A}$ or $g \in (\mathscr{M} - \widehat{\mathscr{P}}) - \mathscr{A}$. Suppose $g \in (\widehat{\mathscr{P}} - \mathscr{M}) - \mathscr{A}$. Since $\mathcal{F} \not\subseteq \widehat{\mathscr{P}}$, there exists a signature $f \in \mathscr{M} - \widehat{\mathscr{P}}$. In this case, Pl-Holant($\mathcal{F} \cup \widehat{\mathcal{EQ}}$) is #P-hard by Lemma 8.21. Finally suppose $g \in (\mathscr{M} - \widehat{\mathscr{P}}) - \mathscr{A}$. Since $\mathcal{F} \not\subseteq \mathscr{M}$, there exists a signature $f \in \widehat{\mathscr{P}} - \mathscr{M}$. In this case, Pl-Holant($\mathcal{F} \cup \widehat{\mathcal{EQ}}$) is also #P-hard again by Lemma 8.21. □

8.6 Pinning for Planar Graphs

The idea of *pinning* is a common reduction technique between counting problems. For the Boolean #CSP framework, pinning fixes some variables to constants 0 or 1, and is represented by the unary signatures $[1, 0]$ and $[0, 1]$ respectively.

Dyer, Goldberg, and Jerrum [DGJ09] showed that one can pin both to 0 and 1 in the #CSP framework; this is Lemma 3.13. However, this reduction is not planar. As it was explained in Section 8.1, it is not possible to prove a general pinning lemma as Lemma 3.13 for the Pl-#CSP framework, assuming #P does not collapse to P. However, in [CLX10] it is shown that for real-weighted Pl-#CSP this difficultly can be overcome if we first perform a holographic transformation by the Hadamard matrix $H_2 = \begin{bmatrix} 1 & 1 \\ 1 & -1 \end{bmatrix}$. Thus in this Hadamard basis, pinning can be done in a planar way, but this holographic transformation is necessary.

The expression of Pl-#CSP($\mathcal{F}$) in the Hadamard basis is Pl-Holant($\widehat{\mathcal{F}} \cup \widehat{\mathcal{EQ}}$). In $\widehat{\mathcal{EQ}} = \{[1, 0], [1, 0, 1], [1, 0, 1, 0], \ldots\}$ we already have "half" of the pinning pair $[1, 0]$. So pinning in Pl-Holant($\widehat{\mathcal{F}} \cup \widehat{\mathcal{EQ}}$) amounts to obtaining the missing signature $[0, 1]$. We remark that there is no intrinsic precedence of $[1, 0]$ over $[0, 1]$. This is merely an artifact of our choice of $H_2 = \begin{bmatrix} 1 & 1 \\ 1 & -1 \end{bmatrix}$. Recall the equivalence relation $\sim$ defining the basis manifold $\mathcal{B}$, we could have made an equally valid choice $H_2' = \begin{bmatrix} -1 & 1 \\ 1 & 1 \end{bmatrix} = \begin{bmatrix} 0 & -1 \\ 1 & 0 \end{bmatrix} H_2$. Then $[0, 1] \in H_2'\mathcal{EQ}$ but $[1, 0] \notin H_2'\mathcal{EQ}$. Choosing either H_2 or H_2' gets us one of the two pinning signatures; getting the other "half" of the pair $\{[1, 0], [0, 1]\}$ will be a significant step toward proving the dichotomy theorem Theorem 8.34.

The Road to Pinning

Our goal is to prove a dichotomy for Pl-Holant($\mathcal{C} \cup \widehat{\mathcal{EQ}}$), for $\mathcal{C} = \widehat{\mathcal{F}}$. An important step for the dichotomy of Pl-Holant($\mathcal{C} \cup \widehat{\mathcal{EQ}}$) is to prove that it is equivalent to Pl-Holant^c($\mathcal{C} \cup \widehat{\mathcal{EQ}}$).

We begin the road to pinning with a lemma that assumes the presence of $[0, 0, 1] = [0, 1]^{\otimes 2}$, which is the tensor product of two copies of $[0, 1]$. In our pursuit to realize $[0, 1]$, this may be as close as we can get to directly. For example, suppose every signature in the set has even parity. Recall that a signature has even parity if its support is on entries of even Hamming weight. By a simple parity argument, gadgets constructed with signatures of even parity can only realize signatures of even parity. But obviously $[0, 1]$ does not have even parity.

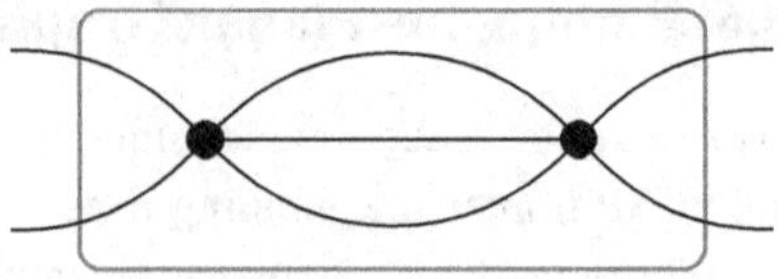

Figure 8.6. The circles are assigned $[a, 0, 0, 0, b, c]$.

However, if every signature has even parity and $[0, 0, 1]$ is present, then we can already prove a dichotomy. To prove this, we first prove the following simple lemma.

Lemma 8.23. *Let $a, b, c \in \mathbb{C}$. If $ab \neq 0$, then* Pl-Holant($[a, 0, 0, 0, b, c]$) *is* #P-*hard.*

Proof. Let f be the signature of the gadget in Figure 8.6 with $[a, 0, 0, 0, b, c]$ assigned to both vertices. The signature matrix of f is

$$\begin{bmatrix} a^2 & 0 & 0 & 0 \\ 0 & b^2 & b^2 & bc \\ 0 & b^2 & b^2 & bc \\ 0 & bc & bc & 3b^2 + c^2 \end{bmatrix},$$

which is redundant. Its compressed form is nonsingular since its determinant is $6a^2b^4 \neq 0$. Thus, we are done by Corollary 7.25. □

Lemma 8.24. *Suppose $\mathcal{F}$ is a set of symmetric signatures with complex weights containing* $[0, 0, 1]$. *If every signature in $\mathcal{F}$ has even parity, then either* Pl-Holant($\mathcal{F} \cup \widehat{\mathcal{EQ}}$) *is* #P-*hard or $\mathcal{F}$ is a subset of $\mathscr{A}$, $\widehat{\mathscr{P}}$, or $\mathscr{M}$, in which case* Pl-Holant^c($\mathcal{F} \cup \widehat{\mathcal{EQ}}$) *is tractable.*

Proof. The tractability is given in Theorem 8.4.

Any degenerate symmetric signature having even parity has the form $\lambda[1, 0]^{\otimes n}$, or $\lambda[0, 1]^{\otimes n}$ if the arity n is even. Both belong to $\mathscr{A} \cap \widehat{\mathscr{P}} \cap \mathscr{M}$. In the following we consider nondegenerate signatures of $\mathcal{F}$.

Every signature of arity ≤ 3 satisfying the parity condition is in $\mathscr{M}$. If $\mathcal{F}$ has no signature of arity > 3, then $\mathcal{F} \subseteq \mathscr{M}$ and we are done.

So we may suppose $\mathcal{F}$ contains some nondegenerate signature of arity ≥ 4. For every signature $f \in \mathcal{F}$ with $f = [f_0, f_1, \ldots, f_m]$ and $m \geq 4$, using $[0, 0, 1]$ and $[1, 0]$, we can obtain all subsignatures of the form $[f_{k-2}, 0, f_k, 0, f_{k+2}]$ for any even k such that $2 \leq k \leq m - 2$. If there is any subsignature g of this form that satisfies $f_{k-2}f_{k+2} \neq f_k^2$ and $f_k \neq 0$, then we are done by Corollary 7.25.

Otherwise all subsignatures of the above form satisfy

$$f_{k-2}f_{k+2} = f_k^2 \qquad \text{or} \qquad f_k = 0$$

for all even k such that $2 \le k \le m-2$. We classify all nondegenerate signatures $f \in \mathcal{F}$ with arity $m \ge 4$ into two types as follows. If there exists an even index $2 \le k \le m-2$ such that $f_k \ne 0$, then by the equation $f_{k-2}f_{k+2} = f_k^2 \ne 0$ we conclude that $f_\ell \ne 0$ for all even indices $0 \le \ell \le m$, and the signature entries of even Hamming weight form a geometric progression. These signatures have the form

$$[\alpha^n, 0, \alpha^{n-1}\beta, 0, \ldots, 0, \alpha\beta^{n-1}, 0, \beta^n]$$

or (8.2)

$$[\alpha^n, 0, \alpha^{n-1}\beta, 0, \ldots, 0, \alpha\beta^{n-1}, 0, \beta^n, 0]$$

for some $\alpha, \beta \in \mathbb{C}$, which are in $\mathscr{M}$. Otherwise, $f_k = 0$ for all even indices $2 \le k \le m-2$. Then the signature has the form

$$[x, 0, \ldots, 0, y] \tag{8.3}$$

of even arity $m \ge 4$, where $xy \ne 0$ by being nondegenerate, or

$$[x, 0, \ldots, 0, y, 0] \tag{8.4}$$

of odd arity $m \ge 5$, where $y \ne 0$ by being nondegenerate. We can group the subcase of (8.4) with $x = 0$ as part of the second case of (8.2) with $\alpha = 0$. Then we call signatures in (8.2) type 1 signatures. Type 2 signatures are those in (8.3) and (8.4), with both $x \ne 0$ and $y \ne 0$. Note that there are an odd number of 0's between x and y in type 2 signatures (since they have even parity). If all nondegenerate signatures of arity ≥ 4 in $\mathcal{F}$ are of type 1, then including all signatures of arity ≤ 3 in $\mathcal{F}$, we have $\mathcal{F} \subseteq \mathscr{M}$.

Otherwise $\mathcal{F}$ contains a signature of type 2. Suppose there is some signature $f = [x, 0, \ldots, 0, y, 0]$ of arity ≥ 5 with $xy \ne 0$ in (8.4). After zero or more self-loops, we have $g = [x, 0, 0, 0, y, 0]$ of arity exactly 5. Then we are done by Lemma 8.23.

Otherwise all type 2 signatures in $\mathcal{F}$ are from (8.3). If there is any such $f = [x, 0, \ldots, 0, y]$ of arity ≥ 4 with $xy \ne 0$ *and* $x^4 \ne y^4$, then we are done by Lemma 8.16. Otherwise all type 2 signatures in $\mathcal{F}$ are from (8.3) and satisfy $x^4 = y^4$. This puts every signature in $\mathcal{F}$ of type 2 in $\mathscr{A}$. Therefore $\mathcal{F} \subseteq \mathscr{A} \cup \mathscr{M}$, and by Theorem 8.22, we are done. □

The conclusion of every result in the remainder of Section 8.6 states that we are able to pin (under various assumptions on $\mathcal{F}$). Formally speaking, we repeatedly prove that Pl-Holant$^c(\mathcal{F} \cup \widehat{\mathcal{EQ}})$ is #P-hard (or in P) if and only if Pl-Holant$(\mathcal{F} \cup \widehat{\mathcal{EQ}})$ is #P-hard (or in P). The difference between these two counting problems is the presence of $[0, 1]$ in Pl-Holant$^c(\mathcal{F} \cup \widehat{\mathcal{EQ}})$. We always prove this statement in one of three ways:

1. We show that Pl-Holant$^c(\mathcal{F} \cup \widehat{\mathcal{EQ}})$ is tractable (so Pl-Holant$(\mathcal{F} \cup \widehat{\mathcal{EQ}})$ is as well).
2. Or we show that Pl-Holant$(\mathcal{F} \cup \widehat{\mathcal{EQ}})$ is #P-hard (so Pl-Holant$^c(\mathcal{F} \cup \widehat{\mathcal{EQ}})$ is as well).
3. Or we show how to reduce Pl-Holant$^c(\mathcal{F} \cup \widehat{\mathcal{EQ}})$ to Pl-Holant$(\mathcal{F} \cup \widehat{\mathcal{EQ}})$ by realizing $[0, 1]$ using signatures in $\mathcal{F} \cup \widehat{\mathcal{EQ}}$.

Note that these three cases together logically imply that

$$\text{Pl-Holant}^c(\mathcal{F} \cup \widehat{\mathcal{EQ}}) \leq_T \text{Pl-Holant}(\mathcal{F} \cup \widehat{\mathcal{EQ}}). \tag{8.5}$$

However, this reduction statement is not proved by directly obtaining the signature $[0, 1]$ with a gadget construction or interpolation in all cases.

Lemma 8.25. *Let $\mathcal{F}$ be any set of complex-weighted symmetric signatures containing* $[0, 0, 1]$. *Then* Pl-Holant$^c(\mathcal{F} \cup \widehat{\mathcal{EQ}})$ *is* #P-*hard (or in P) iff* Pl-Holant$(\mathcal{F} \cup \widehat{\mathcal{EQ}})$ *is* #P-*hard (or in P).*

Proof. If we had a unary signature $[a, b]$ where $b \neq 0$, then connecting $[a, b]$ to $[0, 0, 1]$ gives the signature $[0, b]$, which is $[0, 1]$ after normalizing. Thus, in order to reduce Pl-Holant$^c(\mathcal{F} \cup \widehat{\mathcal{EQ}})$ to Pl-Holant$(\mathcal{F} \cup \widehat{\mathcal{EQ}})$ by constructing $[0, 1]$, it suffices to construct a unary signature $[a, b]$ with $b \neq 0$.

For every signature $f \in \mathcal{F}$ with $f = [f_0, f_1, \ldots, f_m]$, using $[0, 0, 1]$ and $[1, 0]$, we can obtain all subsignatures of the form $[f_{k-1}, f_k]$ for any odd k such that $1 \leq k \leq m$. If any subsignature satisfies $f_k \neq 0$, where k is odd, then we can construct $[0, 1]$.

Otherwise all signatures in $\mathcal{F}$ have even parity and we are done by Lemma 8.24. □

There are two scenarios that lead to Lemma 8.25, which are the focus of the next two lemmas.

Lemma 8.26. *For $x \in \mathbb{C}$, let $\mathcal{F}$ be any set of complex-weighted symmetric signatures containing* $[1, 0, x]$ *such that* $x \notin \{0, \pm 1\}$. *Then* Pl-Holant$^c(\mathcal{F} \cup \widehat{\mathcal{EQ}})$ *is* #P-*hard (or in P) iff* Pl-Holant$(\mathcal{F} \cup \widehat{\mathcal{EQ}})$ *is* #P-*hard (or in P).*

Proof. There are two cases. In either case, we realize $[0, 0, 1]$ and finish by applying Lemma 8.25.

First we claim that the conclusion holds provided the norm $|x| \notin \{0, 1\}$. Combining k copies of $[1, 0, x]$ gives $[1, 0, x^k]$. Since $|x| \notin \{0, 1\}$, x is neither

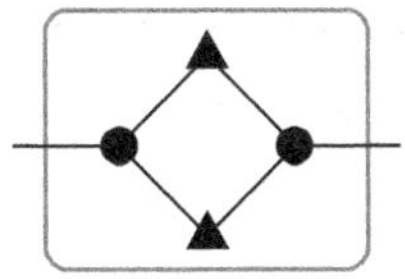

Figure 8.7. The circles are assigned $[1, 0, 1, 0]$ and the triangles are assigned $[1, 0, x]$.

0 nor a root of unity, so we can use polynomial interpolation to realize $[a, 0, b]$ for any $a, b \in \mathbb{C}$, including $[0, 0, 1]$.

Otherwise $|x| = 1$. The gadget in Figure 8.7 has signature $[f_0, f_1, f_2] = [1 + x^2, 0, 2x]$. If $x = \pm\mathrm{i}$, then we have $[0, 0, \pm 2\mathrm{i}]$, which is $[0, 0, 1]$ after normalizing.

Otherwise $x \neq \pm\mathrm{i}$, so $f_0 \neq 0$. Since $x \neq 0$, we have $f_2 \neq 0$. Since $x \neq \pm 1$ and the norm $|x| = 1$, we have $|f_0| < 2$. However, $|f_2| = 2$. Therefore, after normalizing, the signature $[1, 0, y]$ with $y = \frac{2x}{1+x^2}$ has $|y| > 1$, so it can interpolate $[0, 0, 1]$ by our initial claim since $|y| \notin \{0, 1\}$. □

Lemma 8.27. *Let $\mathcal{F}$ be any set of complex-weighted symmetric signatures containing a signature $[f_0, f_1, \ldots, f_n]$ that is not identically zero but has $f_0 = 0$. Then* Pl-Holant$^c(\mathcal{F} \cup \widehat{\mathcal{EQ}})$ *is* #P*-hard (or in P) iff* Pl-Holant$(\mathcal{F} \cup \widehat{\mathcal{EQ}})$ *is* #P*-hard (or in P).*

Proof. If $f_1 \neq 0$, then we connect $n - 1$ copies of $[1, 0]$ to f to get $[0, f_1]$, which is $[0, 1]$ after normalizing. If $f_1 = 0$, then $n \geq 2$. If $f_2 \neq 0$, then we connect $n - 2$ copies of $[1, 0]$ to f to get $[0, 0, f_2]$, which is $[0, 0, 1]$ after normalizing. Then we are done by Lemma 8.25. If $f_1 = f_2 = 0$, then $n \geq 3$, and $f = [0, 0, 0, f_3, \ldots, f_n]$. After one or more self-loops, we get a signature with exactly one or two initial 0's, which is one of the above scenarios. □

As a significant step toward pinning for any signature set $\mathcal{F}$, we show how to pin given any binary signature. Some cases resist pinning and are excluded.

Lemma 8.28. *Let $\mathcal{F}$ be any set of complex-weighted symmetric signatures containing a binary signature f. Then* Pl-Holant$^c(\mathcal{F} \cup \widehat{\mathcal{EQ}})$ *is* #P*-hard (or in P) iff* Pl-Holant$(\mathcal{F} \cup \widehat{\mathcal{EQ}})$ *is* #P*-hard (or in P) unless $f \in \{[0, 0, 0], [1, 0, -1], [1, r, r^2], [1, b, 1]\}$, up to a nonzero scalar, for any $b, r \in \mathbb{C}$.*

Proof. Let $f = [f_0, f_1, f_2] \neq [0, 0, 0]$. If $f_0 = 0$ then we are done by Lemma 8.27. Otherwise $f_0 \neq 0$, and we normalize f_0 to 1. If Pl-Holant$(f \mid \widehat{\mathcal{EQ}})$

is #P-hard by Theorem 8.3, then Pl-Holant($\mathcal{F} \cup \widehat{\mathcal{EQ}}$) is also #P-hard. Otherwise, f is one of the tractable cases, which implies that

$$f \in \{[1, r, r^2], [1, 0, x], [1, \pm 1, -1], [1, b, 1]\}.$$

If $f = [1, \pm 1, -1]$, then we connect f to $[1, 0, 1, 0] \in \widehat{\mathcal{EQ}}$ to get $[0, \pm 2]$, which is $[0, 1]$ after normalizing. If $f = [1, 0, x]$, then we are done by Lemma 8.26 unless $x \in \{0, \pm 1\}$. The remaining cases are all excluded by assumption, so we are done. □

Pinning in the Hadamard Basis

Before we show how to pin in the Hadamard basis, we handle two simple cases.

Lemma 8.29. *If $\mathcal{F}$ is a set of signatures containing* $[1, \mathrm{i}]$ *or* $[1, -\mathrm{i}]$, *then we have*

$$\text{Pl-Holant}^c(\mathcal{F} \cup \widehat{\mathcal{EQ}}) \leq_T \text{Pl-Holant}(\mathcal{F} \cup \widehat{\mathcal{EQ}}).$$

Proof. Connect two copies of $[1, \pm\mathrm{i}]$ to $[1, 0, 1, 0] \in \widehat{\mathcal{EQ}}$ to get $[0, \pm 2\mathrm{i}]$, which is $[0, 1]$ after normalizing. □

The next lemma considers the signature $[1, b, 1, b^{-1}]$, which we will also use in Theorem 8.32, the single signature dichotomy.

Lemma 8.30. *Let $b \in \mathbb{C}$. If $b \notin \{0, \pm 1\}$, then* Pl-Holant($\{[1, b, 1, b^{-1}]\} \cup \widehat{\mathcal{EQ}}$) *is* #P-*hard.*

Proof. Connect two copies of $[1, 0]$ to $f = [1, b, 1, b^{-1}]$ to get $[1, b]$. Connecting this back to f gives $g = [1 + b^2, 2b, 2]$. Then Pl-Holant($g \mid \widehat{\mathcal{EQ}}$) is #P-hard by Theorem 8.3. □

Now we are ready to prove our pinning result.

Theorem 8.31 (Pinning). *Let $\mathcal{F}$ be any set of complex-weighted symmetric signatures. Then* Pl-Holant^c($\mathcal{F} \cup \widehat{\mathcal{EQ}}$) *is* #P-*hard (or in P) iff* Pl-Holant($\mathcal{F} \cup \widehat{\mathcal{EQ}}$) *is* #P-*hard (or in P).*

This theorem does not exclude the possibility that either framework contains a problem of intermediate complexity, i.e., a problem that is neither in P nor #P-hard. It only says that if one framework does not contain a problem of intermediate complexity, then the other framework also does not. We will indeed prove neither happens. Our goal is to prove a dichotomy for Pl-Holant($\mathcal{F} \cup \widehat{\mathcal{EQ}}$). By Theorem 8.31, this is equivalent to proving a dichotomy for Pl-Holant^c($\mathcal{F} \cup \widehat{\mathcal{EQ}}$).

Proof of Theorem 8.31. For simplicity, we normalize the first nonzero entry of every nonzero signature in $\mathcal{F}$ to 1.

If $\mathcal{F}$ does not contain any nondegenerate signature, then $\mathcal{F} \subseteq \widehat{\mathscr{P}}$ and we are done. Otherwise, let $f \in \mathcal{F}$ be any nondegenerate signature. Being nondegenerate it has arity ≥ 2. We connect some number of $[1, 0]$ to f until we obtain a signature with arity exactly two. We call the resulting signature the binary prefix of f. If this binary prefix is not one of the exceptional forms in Lemma 8.28, then we are done, so assume that it is one of the exceptional forms.

Now we perform case analysis according to the exceptional forms in Lemma 8.28. There are five cases below because we consider $[1, r, r^2]$ as $[1, 0, 0]$ and $[1, r, r^2]$ with $r \neq 0$ as separate cases. In each case, we either show that the conclusion of the theorem holds or that $f \in \mathscr{A} \cup \widehat{\mathscr{P}} \cup \mathscr{M}$. After the case analysis, we then handle all of these tractable f together.

1. Suppose the binary prefix of f is $[0, 0, 0]$. Since f is not identically 0, and $f_0 = 0$, we are done by Lemma 8.27.
2. Suppose the binary prefix of f is $[1, 0, -1]$. If f is not of the form

$$[1, 0, -1, 0, 1, 0, -1, 0, \ldots, 0 \text{ or } 1 \text{ or } (-1)], \tag{8.6}$$

   then in particular f has arity ≥ 3 and after one self-loop, we have a signature of arity ≥ 1 with 0 as its first entry but is not identically 0, so by Lemma 8.27 we are done.

   Thus, in this case, we may assume f has the form given in (8.6).
3. Suppose the binary prefix of f is $[1, 0, 0]$. Since f is assumed to be nondegenerate, $f \neq [1, 0, \ldots, 0]$, then after connecting $[1, 0]$ for some $k \geq 0$ times, we have $[1, 0, \ldots, 0, x]$ of arity ≥ 3, where $x \neq 0$. If $x^4 \neq 1$, then Pl-Holant($\{f\} \cup \widehat{\mathcal{EQ}}$) is #P-hard by Lemma 8.16, so Pl-Holant($\mathcal{F} \cup \widehat{\mathcal{EQ}}$) is also #P-hard.

   Otherwise, $x^4 = 1$. Suppose that x is not the last entry in f. Then $k \geq 1$. By connecting one fewer $[1, 0]$ than before, we have $g = [1, 0, \ldots, 0, x, y]$ and there are two cases to consider. If the index of x in g is odd, then after zero or more self-loops, we have $h = [1, 0, 0, x, y]$. The determinant of the compressed signature matrix of h is $-2x^2 \neq 0$. Thus, Pl-Holant(h) is #P-hard by Corollary 7.25, so Pl-Holant($\mathcal{F} \cup \widehat{\mathcal{EQ}}$) is also #P-hard.

   Otherwise, the index of x in g is even. After zero or more self-loops, we have $h = [1, 0, 0, 0, x, y]$. Then by Lemma 8.23, Pl-Holant(h) is #P-hard, so Pl-Holant($\mathcal{F} \cup \widehat{\mathcal{EQ}}$) is also #P-hard.

   Thus, in this case, we may assume $f = [1, 0, \ldots, 0, x]$ with $x^4 = 1$.
4. Suppose the binary prefix of f is $[1, r, r^2]$, where $r \neq 0$. Since f is nondegenerate, $f \neq [1, r, \ldots, r^n]$, then after connecting some number of $[1, 0]$,

we have $[1, r, \ldots, r^m, y]$, where $y \neq r^{m+1}$ and $m \geq 2$. Using $[1, 0]$, we can get $[1, r]$. If $r = \pm \mathrm{i}$, then we are done by Lemma 8.29, so assume that $r \neq \pm \mathrm{i}$. Then we can attach $[1, r]$ back to the signature $[1, r, \ldots, r^m, y]$ some number of times to get $g = [1, r, r^2, x]$ after factoring out the normalizing constant $1 + r^2 \neq 0$, where $x \neq r^3$. We connect $[1, r]$ once more to get $h = [1 + r^2, r(1 + r^2), r^2 + rx]$. If h does not have one of the exceptional forms in Lemma 8.28, then we are done, so assume that it does.

Since the second entry of h is not 0 and $x \neq r^3$, the only possibility is that h has the form $[1, b, 1]$ up to a scalar. This gives $x = r^{-1}$. Note that $r \neq \pm 1$ since $x \neq r^3$. A self-loop on $g = [1, r, r^2, r^{-1}]$ gives $[1 + r^2, r + r^{-1}]$, which is $[1, r^{-1}]$ after normalizing. Connecting this back to g gives $h' = [2, 2r, r^2 + r^{-2}]$. We assume that h' has one of the exceptional forms in Lemma 8.28 since we are done otherwise. Since the middle entry $2r \neq 0$, the only possibilities for $h'/2 = [1, r, (r^2 + r^{-2})/2]$ are $[1, r, r^2]$ or $[1, b, 1]$. If it has the form $[1, r, r^2]$, then $(r^2 + r^{-2})/2 = r^2$ thus $r^4 = 1$, a contradiction. So it must be of the form $[1, b, 1]$. But then $(r^2 + r^{-2})/2 = 1$ thus $r^2 = 1$, which is also a contradiction.

5. Suppose the binary prefix of f is $[1, b, 1]$. If $b = \pm 1$, then this binary prefix is degenerate. This case has already been considered in the previous item. So we assume $b \neq \pm 1$. If f is not of the form $[1, b, 1, b, \ldots]$, then first suppose that the index of the first entry in f to break the pattern is even. Then after connecting some number of $[1, 0]$, we have $[1, b, 1, \ldots, b, y]$ of arity ≥ 4, where $y \neq 1$. Then after some number of self-loops and normalizing, we have $g = [1, b, 1, b, x]$, where $x \neq 1$. The determinant of its compressed signature matrix is $2(b^2 - 1)(1 - x) \neq 0$. Thus, Pl-Holant(g) is #P-hard by Corollary 7.25, so Pl-Holant$(\mathcal{F} \cup \widehat{\mathcal{EQ}})$ is also #P-hard.

   Otherwise, the index of the first entry in f to break the pattern is odd. Then after connecting some number of $[1, 0]$, we have $[1, b, 1, \ldots, 1, y]$ of arity ≥ 3, where $y \neq b$. Then after some number of self-loops and normalizing, we have $[1, b, 1, x]$, where $x \neq b$. We do a self-loop to get $g = [2, b + x]$. If $b = 0$, then connecting g to $[1, 0, 1, x]$ gives $h = [2, x, 2 + x^2]$. We assume that h has one of the exceptional forms in Lemma 8.28 since we are done otherwise. Because $x \neq 0$, the only possibility is that h has the form $[1, r, r^2]$ up to a scalar. Then we get $x^2 = -4$, so $g = [2, x] = 2[1, \pm \mathrm{i}]$ and we are done by Lemma 8.29. We use the signature g again below.

   Otherwise, $b \neq 0$. Using $[1, 0]$, we can get $h = [1, b, 1]$. As $b \neq \pm 1$, h is nondegenerate. If M_h has infinite order modulo a scalar, then we can interpolate $[0, 1]$ by Lemma 8.15 since $b \notin \{0, \pm 1\}$. If the signature matrix M_h of h has finite order modulo a scalar, then $M_h^\ell = \beta I_2$ for some positive integer ℓ and some nonzero complex value β. Thus after normalizing, we can construct the anti-gadget $[1, -b, 1]$ by connecting $\ell - 1$ copies of h together. Connecting $[1, 0]$ to $[1, -b, 1]$ gives $[1, -b]$ and connecting this to $[1, b, 1, x]$ gives $[1 - b^2, 0, 1 - bx]$. If $\frac{1-bx}{1-b^2} \notin \{0, \pm 1\}$, then by Lemma 8.26 we are done.

Otherwise, $y = \frac{1-bx}{1-b^2} \in \{0, \pm 1\}$. For $y = 0$, we get $x = b^{-1}$ and by Lemma 8.30 we are done since we have $[1, b, 1, b^{-1}]$ and $b \notin \{0, \pm 1\}$. For $y = 1$, we get $b = x$, a contradiction. For $y = -1$, we get $2 - b^2 - bx = 0$. Then connecting $g = [2, b + x]$ to $[1, -b, 1]$ gives $[2 - b^2 - bx, x - b] = [0, x - b]$, which is $[0, 1]$ after normalizing since $x \neq b$.

Thus, in this case, we may assume $f = [1, b, 1, b, \dots]$.

At this point, every nondegenerate signature in $\mathcal{F}$ must be of one of the following forms:

- $[1, 0, -1, 0, 1, 0, -1, 0, \dots, 0 \text{ or } 1 \text{ or } (-1)]$, which is in $\mathscr{A} \cap \mathscr{M}$.
- $[1, 0, \dots, 0, x]$, where $x^4 = 1$, which is in $\mathscr{A}$.
- $[1, b, 1, b, \dots, 1 \text{ or } b]$, which is in $\widehat{\mathscr{P}}$.

Also every degenerate signature is in $\widehat{\mathscr{P}}$. Therefore $\mathcal{F} \subseteq \mathscr{A} \cup \widehat{\mathscr{P}} \cup \mathscr{M}$ and we are done by Theorem 8.22, the no-mixing theorem. □

8.7 Planar #CSP Dichotomy

In this section, we prove our main dichotomy theorem of this chapter. We begin with a dichotomy for a single signature.

Recall that Pl-Holant($\{\hat{f}\} \cup \widehat{\mathcal{EQ}}$) is the equivalent expression of Pl-#CSP($\{f\}$), where $\hat{f} = H_2 f$. Theorem 8.32 is expressed in the Pl-Holant language.

Theorem 8.32. *If f is a nondegenerate symmetric signature of arity ≥ 2 with complex weights in Boolean variables, then* Pl-Holant($\{f\} \cup \widehat{\mathcal{EQ}}$) *is #P-hard unless $f \in \mathscr{A} \cup \widehat{\mathscr{P}} \cup \mathscr{M}$, in which case the problem is computable in polynomial time.*

Proof. When $f \in \mathscr{A} \cup \widehat{\mathscr{P}} \cup \mathscr{M}$, the problem is tractable by Theorem 8.4. When $f \notin \mathscr{A} \cup \widehat{\mathscr{P}} \cup \mathscr{M}$, we prove that Pl-Holant^c($\{f\} \cup \widehat{\mathcal{EQ}}$) is #P-hard, which is equivalent because of pinning (Theorem 8.31). The advantage of Pl-Holant^c is that we can use $[1, 0]$ and $[0, 1]$ to obtain any subsignature of f.

Notice that once we have $[0, 1]$ and $\widehat{\mathcal{EQ}}$, we can realize every signature in $T\widehat{\mathcal{EQ}}$, where $T = \begin{bmatrix} 0 & 1 \\ 1 & 0 \end{bmatrix}$. In fact, every even arity signature in $\widehat{\mathcal{EQ}}$ is also in

$T\widehat{\mathcal{EQ}}$, and we obtain all the odd arity signatures in $T\widehat{\mathcal{EQ}}$ by attaching $[0, 1]$ to all the even arity signatures in $\widehat{\mathcal{EQ}}$. Also note that $T[0, 1] = [1, 0]$ and $T[1, 0] = [0, 1]$. Therefore, Pl-Holant$^c(\{f\} \cup \widehat{\mathcal{EQ}}) \equiv_T$ Pl-Holant$^c(\{Tf\} \cup \widehat{\mathcal{EQ}})$. Furthermore, $\mathscr{A} \cup \widehat{\mathscr{P}} \cup \mathscr{M}$ is closed under T. We use these facts later.

The possibilities for f can be divided into three cases:

- f satisfies the parity condition.
- f does not satisfy the parity condition but does contain a 0 entry.
- f does not contain a 0 entry.

We handle these cases below.

1. Suppose that f satisfies the parity condition. If f has even parity, then by Lemma 8.24 we are done.

   Otherwise, f has odd parity. If f has odd arity, $f = [0, f_1, 0, \dots, 0, f_{2k+1}]$, then under a holographic transformation by $T = \begin{bmatrix} 0 & 1 \\ 1 & 0 \end{bmatrix}$, f is transformed to $f' = [f_{2k+1}, 0, \dots, 0, f_1, 0]$, which has even parity. Then either Pl-Holant$^c(\{f'\} \cup \widehat{\mathcal{EQ}})$ is #P-hard by Lemma 8.24 (and thus Pl-Holant$^c(\{f\} \cup \widehat{\mathcal{EQ}})$ is also #P-hard), or $f' \in \mathscr{A} \cup \widehat{\mathscr{P}} \cup \mathscr{M}$ (and thus $f \in \mathscr{A} \cup \widehat{\mathscr{P}} \cup \mathscr{M}$).

   Otherwise, f has odd parity and even arity, $f = [0, f_1, 0, \dots, 0, f_{2k-1}, 0]$. Connect $[0, 1]$ to f to get a signature $g = [f_1, 0, \dots, 0, f_{2k-1}, 0]$ with even parity and odd arity. Then either Pl-Holant$^c(\{g\} \cup \widehat{\mathcal{EQ}})$ is #P-hard by Lemma 8.24 (and thus Pl-Holant$^c(\{f\} \cup \widehat{\mathcal{EQ}})$ is also #P-hard), or $g \in \mathscr{A} \cup \widehat{\mathscr{P}} \cup \mathscr{M}$. In the latter case, the last entry of g is 0, since g has odd arity and even parity. Then inspection of Figure 8.4 shows that $g \in \mathscr{M}$. In particular, the even parity entries of g form a geometric progression, by Theorem 4.11 (case 2, this includes the form $[0, \dots, 0, *, 0]$). Therefore $f \in \mathscr{M}$ since f has odd parity and the same geometric progression among its odd parity entries.
2. Suppose that f does not satisfy the parity condition but does contain a 0 entry. Then there are two nonzero entries f_i and f_j, where $i < j$ and i and j are of the opposite parity. Take two such entries with minimum $j - i$, which is odd. If $j - i > 1$ then any entry f_ℓ with $i < \ell < j$ must be 0, since $j - i = (j - \ell) + (\ell - i)$ and one of them must be odd. Thus f_i and f_j are separated by an even number $j - i - 1 \geq 0$ of 0 entries. And so f contains a subsignature $g = [a, 0, \dots, 0, b]$ of odd arity $n = 2k + 1 \geq 1$, where $ab \neq 0$.

   If $k = 0$, then $n = 1$ and $g = [a, b]$. But since f does contain a 0 entry, f has arity greater than one. Then we can shift from the positions of g either to the right or to the left and find a 0 entry in f and obtain (by pinning) a binary subsignature h of the form $[c, d, 0]$ or $[0, c, d]$, where $cd \neq 0$. Then Pl-Holant($h \mid \widehat{\mathcal{EQ}}$) is #P-hard by Theorem 8.3, so Pl-Holant($\{f\} \cup \widehat{\mathcal{EQ}}$) is also #P-hard, by applying Theorem 8.31.

Otherwise $k \geq 1$, so $n \geq 3$. If $a^4 \neq b^4$, then Pl-Holant($\{g\} \cup \widehat{\mathcal{EQ}}$) is #P-hard by Lemma 8.16, so Pl-Holant($\{f\} \cup \widehat{\mathcal{EQ}}$) is also #P-hard, by applying Theorem 8.31.

Otherwise, $a^4 = b^4$, so $g \in \mathscr{A}$. If $f = g$, then we are done, so assume that $f \neq g$, which implies that there is another entry just before a or just after b. If this entry is nonzero, then f has a subsignature h of the form $[c, a, 0]$ ($ac \neq 0$) or $[0, b, d]$ ($bd \neq 0$). Then Pl-Holant($h \mid \widehat{\mathcal{EQ}}$) is #P-hard by Theorem 8.3, so Pl-Holant($\{f\} \cup \widehat{\mathcal{EQ}}$) is also #P-hard.

Otherwise, this entry is 0 and f has a subsignature h of the form

$$[0, a, 0, \dots, 0, b] \qquad \text{or} \qquad [a, 0, \dots, 0, b, 0]$$

of arity ≥ 4, whereas $g = [a, 0, \dots, 0, b]$ is a subsignature of h. Since g has odd arity, the arity of h is even. Then after some number of self-loops, we have a signature h' of the form $[0, a, 0, 0, b]$ or $[a, 0, 0, b, 0]$ of arity exactly 4. Then Pl-Holant(h') is #P-hard by Corollary 7.25 since $ab \neq 0$, so Pl-Holant($\{f\} \cup \widehat{\mathcal{EQ}}$) is also #P-hard.

3. Suppose f contains no 0 entry. If f has a binary subsignature g such that Pl-Holant($g \mid \widehat{\mathcal{EQ}}$) is #P-hard by Theorem 8.3, then Pl-Holant($\{f\} \cup \widehat{\mathcal{EQ}}$) is also #P-hard.

   Otherwise every binary subsignature $[a, b, c]$ of f satisfies the conditions of some tractable case in Theorem 8.3. Since there are no 0 entries, there are three possible tractable cases for $[a, b, c]$:

   (D) it is degenerate with condition $ac = b^2$ (case 1);
   (A) it has the affine type $\mathscr{A}$ with condition $[ac = -b^2] \wedge [a = -c]$ (case 3); or
   (P) it has a Hadamard-transformed product type $\widehat{\mathscr{P}}$ with condition $a = c$ (case 4).

Remark: Condition (D) refers to degenerate. Condition (A) refers to a subset of affine signatures. Some affine signatures are included in condition (P) for $\widehat{\mathscr{P}}$, e.g., $[1, i, 1]$. To group them this way is for the technical convenience in the following proof.

If every binary subsignature $[a, b, c]$ of f satisfies $ac = b^2$, then f is degenerate, a contradiction. If every binary subsignature $[a, b, c]$ of f satisfies $[ac = -b^2] \wedge [a = -c]$, then $f = [1, \pm 1, -1, \mp 1, 1, \pm 1, -1,$

$\mp 1, \ldots] \in \mathscr{A}$ (up to a nonzero scalar) and we are done. If every binary subsignature $[a, b, c]$ of f satisfies $a = c$, then $f \in \widehat{\mathscr{P}}$ and we are done.

Otherwise, no single one of the three tractable conditions (D), (A) or (P) is satisfied by all binary subsignatures of f. Hence f has arity $n \geq 3$. Let $h_i = [f_i, f_{i+1}, f_{i+2}]$ for all $0 \leq i \leq n-2$ be binary subsignatures of f. Suppose there exists an i such that h_i satisfies the condition (A) (case 3). We claim that there must exist two successive subsignatures $h = [a, b, c]$ and $h' = [b, c, d]$ such that h satisfies condition (A) and $h' = [b, c, d]$ satisfies either condition (D) or condition (P), up to a transformation of $\begin{bmatrix} 0 & 1 \\ 1 & 0 \end{bmatrix}$. This is because we can start from h_i and search in both directions h_{i-1} and h_{i+1} until we found such h and h'. It is always successful because not all h_j satisfies condition (A). Let $g = [a, b, c, d]$ be the ternary subsignature of f. Then for either case (D) or (P) of h', we have $g = [1, \varepsilon, -1, \varepsilon]$ after normalization, where $\varepsilon^2 = 1$. Connecting two copies of $[0, 1]$ to g gives $[-1, \varepsilon]$. Connecting this back to g gives $g' = [0, -2\varepsilon, 2]$. Then Pl-Holant($g' \mid \widehat{\mathcal{EQ}}$) is #P-hard by Theorem 8.3, so Pl-Holant($\{f\} \cup \widehat{\mathcal{EQ}}$) is also #P-hard.

Now we suppose no h_i satisfies condition (A). Then every h_i satisfies either (D) or (P), but no single condition (D) or (P) is satisfied by all. Therefore some h_i must not satisfy (D), which must satisfy (P), and some other h_j must satisfy (D); otherwise all would satisfy (P). Starting from h_i we can shift either to the right or to the left to find the first h_j satisfying (D). Up to a transformation $\begin{bmatrix} 0 & 1 \\ 1 & 0 \end{bmatrix}$, we may assume $j > i$. Then h_{j-1} does not satisfy (D), therefore h_{j-1} must satisfy (P). Thus there exists a ternary subsignature $g = [a, b, c, d]$ of f such that $h = [a, b, c]$ satisfies (P) but not (D), and $h' = [b, c, d]$ satisfies (D). Then $g = [1, b, 1, b^{-1}]$ after normalization, where $b^2 \neq 1$. Then Pl-Holant($\{g\} \cup \widehat{\mathcal{EQ}}$) is #P-hard by Lemma 8.30, so Pl-Holant($\{f\} \cup \widehat{\mathcal{EQ}}$) is also #P-hard. □

Now we are ready to prove our main dichotomy theorem of this chapter.

Theorem 8.33. *Let $\mathcal{F}$ be any set of symmetric, complex-valued signatures in Boolean variables. Then* Pl-Holant($\mathcal{F} \cup \widehat{\mathcal{EQ}}$) *is* #P-*hard unless $\mathcal{F} \subseteq \mathscr{A}$, $\mathcal{F} \subseteq \widehat{\mathscr{P}}$, or $\mathcal{F} \subseteq \widehat{\mathscr{M}}$, in which case the problem is computable in polynomial time.*

Proof. The tractability is given in Theorem 8.4. When $\mathcal{F}$ is not a subset of $\mathscr{A}$, $\widehat{\mathscr{P}}$, or $\widehat{\mathscr{M}}$, we prove that Pl-Holant^c($\mathcal{F} \cup \widehat{\mathcal{EQ}}$) is #P-hard, which is sufficient because of pinning (Theorem 8.31).

If $\mathcal{F} \not\subseteq \mathscr{A} \cup \widehat{\mathscr{P}} \cup \widehat{\mathscr{M}}$, since $\widehat{\mathscr{P}}$ contains all degenerate signatures (in particular all unary signatures), there exists some nondegenerate signature $f \in \mathcal{F} - (\mathscr{A} \cup \widehat{\mathscr{P}} \cup \widehat{\mathscr{M}})$ of arity ≥ 2. Then the problem is #P-hard by

Theorem 8.32. Otherwise, $\mathcal{F} \subseteq \mathscr{A} \cup \widehat{\mathscr{P}} \cup \mathscr{M}$ and we are done by Theorem 8.22. □

Since Pl-#CSP($\mathcal{F}$) $\equiv_T$ Pl-Holant($\widehat{\mathcal{F}} \cup \widehat{\mathcal{EQ}}$), where $\widehat{\mathcal{F}} = H_2\mathcal{F}$, and $\widehat{\mathcal{EQ}} = H_2\mathcal{EQ}$, we can translate Theorem 8.33 to obtain the main theorem of this chapter, Theorem 8.34, the Pl-#CSP dichotomy. Recall that $H_2\mathscr{A} = \mathscr{A}$, $H_2\mathscr{P} = \widehat{\mathscr{P}}$ and $H_2\mathscr{M} = \widehat{\mathscr{M}}$. Theorem 8.34 is Theorem 8.33 stated in the standard basis.

Theorem 8.34. *Let $\mathcal{F}$ be any set of symmetric, complex-valued signatures in Boolean variables. Then* Pl-#CSP($\mathcal{F}$) *is* #P-*hard unless $\mathcal{F} \subseteq \mathscr{A}$, $\mathcal{F} \subseteq \mathscr{P}$, or $\mathcal{F} \subseteq \widehat{\mathscr{M}}$, in which case the problem is computable in polynomial time.*

9

Planar Holant for Symmetric Constraints

In this chapter, we consider the complexity of planar Holant problems Pl-Holant($\mathcal{F}$), where $\mathcal{F}$ is any set of complex-valued, symmetric functions on Boolean variables.

9.1 Introduction

In Chapter 8 we presented a complexity dichotomy, Theorem 8.34, for planar #CSP problems Pl-#CSP($\mathcal{F}$), where $\mathcal{F}$ is any set of complex-valued symmetric constraint functions over the Boolean domain. This theorem states that Pl-#CSP($\mathcal{F}$) is either tractable in polynomial time or #P-hard. It is tractable precisely when #CSP($\mathcal{F}$) is tractable without the planarity restriction, and this happens precisely when $\mathcal{F} \subseteq \mathscr{A}$ or $\mathscr{F} \subseteq \mathscr{P}$ (by Theorem 3.7); or when $\mathcal{F}$ is transformable to a set of standard matchgate signatures, i.e., $\mathcal{F} \subseteq \widehat{\mathscr{M}} = H_2\mathscr{M}$, where the Hadamard matrix $H_2 = \begin{bmatrix} 1 & 1 \\ 1 & -1 \end{bmatrix}$. Thus, for those #CSP problems over the Boolean domain that are #P-hard in general but solvable in polynomial time over planar graphs, a *universal* methodology is a suitable holographic transformation followed by Kasteleyn's algorithm given in Chapter 4.

This has been a recurring theme. We have seen this phenomenon already several times in earlier chapters. For example, in Chapter 5, Theorems 5.1 and 5.3 are also examples of this. It can be verified that the complexity classification theorems for the Tutte polynomial $\mathrm{T}(G; x, y)$, Theorems 6.28 and 6.29, are also consistent with this theme. Indeed, the only additional tractable cases in Theorem 6.29 over Theorem 6.28 on planar graphs are all captured by $(x-1)(y-1) = 2$. By the correspondence with the Potts model (see Chapter 1, Section 1.2.1), this is precisely the case $q = 2$ (the Boolean domain; thus the Potts model becomes the Ising model, with a symmetric binary signature $[1+\gamma, 1, 1+\gamma]$ on each edge). Note that $[1+\gamma, 1, 1+\gamma] \in \widehat{\mathscr{M}}$.

In this chapter, we want to achieve a complexity classification of planar Holant problems Pl-Holant($\mathcal{F}$) for symmetric signatures $\mathcal{F}$. This would be a stronger theorem than Theorem 8.34, as Pl-#CSP($\mathcal{F}$) is just Pl-Holant($\mathcal{F} \cup \mathcal{EQ}$). All previously known classification theorems, such as those presented in earlier chapters, suggest the following putative classification of Pl-Holant($\mathcal{F}$) into *exactly* three categories, (1) those that are P-time solvable over general graphs; (2) those that are P-time solvable over planar graphs but #P-hard over general graphs; and (3) those that remain #P-hard over planar graphs. Moreover, category (2) consists of precisely those problems that are holographically reducible to Kasteleyn's algorithm. This theme is so strong that it has become an intuitive and trusty guide for us when we investigate unknown problems and plan proof strategies. In fact, many partial results leading to the main theorem in this chapter are proved in this way.

However, a surprising departure to this theme occurs with Pl-Holant($\mathcal{F}$). It is still true that all Pl-Holant($\mathcal{F}$) problems can be classified into exactly three categories; but there are more problems in category (2) than those that are holographically reducible to Kasteleyn's algorithm. We discover new planar tractable problems that are not expressible by a holographic reduction to matchgates. Thus a holographic reduction to Kasteleyn's algorithm is *not* a universal strategy for Holant problems that are #P-hard in general but solvable in polynomial time over planar graphs. So far, for any class of counting problems expressible in our frameworks for sum-of-product computations, this is the only known exception to the *universality* of holographic reductions to Kasteleyn's algorithm. Moreover, with the discovery of this new class of planar tractable problems the picture is complete for Pl-Holant($\mathcal{F}$): there are no more undiscovered planar tractable problems unless #P collapses to P.

This classification generalizes both the Holant dichotomy, Theorem 7.19, and the planar #CSP classification, Theorem 8.34. At a high level, we can state the main theorem of this chapter as follows.

Theorem 9.1. *Let $\mathcal{F}$ be a set of complex-valued, symmetric functions on Boolean variables. Then there is an effective classification for all possible $\mathcal{F}$, according to which,* Holant($\mathcal{F}$) *is either (1) P-time computable over general graphs, or (2) P-time computable over planar graphs but* #P*-hard over general graphs, or (3)* #P*-hard over planar graphs. Furthermore, category (2) consists precisely of those problems that are holographically reducible to Kasteleyn's algorithm, and those newly discovered problems.*

The complete statement is given in Theorem 9.29. The classification is explicit. The tractability criterion is decidable in polynomial time.

Let us meet some new tractable problems. A hypergraph $H = (V, E)$ consists of a finite vertex set V and a set E of hyperedges e, each of which is a subset of V. If every e has cardinality 2, then H is an ordinary graph. A perfect matching in H is a subset $M \subseteq E$ such that every $v \in V$ is contained in exactly one $e \in M$. Consider the problem of counting perfect matchings in a hypergraph H. In general, each hyperedge e has a cardinality k_e, and if $k_e = k$ for all e, then the hypergraph is called a k-uniform hypergraph. The number of hyperedges e containing a vertex v is called its degree; if every vertex has the same degree d, then it is called d-regular. Each hypergraph H can be represented by its bipartite edge–vertex incidence graph B_H with vertex bipartition (E, V) and an edge exists in B_H between $v \in V$ and $e \in E$ exactly when $v \in e$. Clearly H is a k-uniform d-regular hypergraph iff B_H is a (k, d)-biregular bipartite graph. We say a hypergraph H is planar when B_H is planar [Zyk74]. Some new tractable problems are counting perfect matchings in planar hypergraphs.

Suppose H is a k-uniform planar hypergraph. If $k = 2$, this is an ordinary planar graph, and Kasteleyn's algorithm can count the number of perfect matchings in polynomial time. We will show that for $k = 3$ and $k = 4$, the problem of counting perfect matchings in k-uniform planar hypergraphs is #P-hard. However, for $k \geq 5$, there is a big surprise that this problem is again computable in polynomial time. The formal statement is given in Theorem 9.28.

These hypergraph perfect matching problems are special cases of a class of graph orientation problems, and their complexity will be derived by looking at this broader class of problems. These can be described as follows. Given a planar (ordinary) graph G, we consider permissible orientations of all edges satisfying certain requirements locally at each incident vertex. We specify two kinds of vertices. The first kind can be either a sink or a source vertex in every permissible orientation. The second kind allows exactly one incoming edge. The goal is to count the number of permissible orientations, i.e., orientations that satisfy these constraints. Clearly this can be expressed as a bipartite Holant problem on the edge–vertex incidence graph of G, where we assign $(\neq_2)$ on each edge, and either the EQUALITY function or the EXACTONE function on each vertex. A vertex with an EQUALITY taking the value 0 or 1 corresponds to a source or a sink, respectively. A vertex with an EXACTONE is a vertex that allows exactly one incoming edge. Thus this problem can be expressed as Pl-Holant($\neq_2$| EQUALITY, EXACTONE). Under a complex-valued holographic transformation $Z = \begin{bmatrix} 1 & 1 \\ \mathfrak{i} & -\mathfrak{i} \end{bmatrix}$, we have $(\neq_2) = (=_2)Z^{\otimes 2}$. It follows that the problem can also be expressed in a nonbipartite Holant form Holant($\mathcal{F}$) under a Z-transformation, where $\mathcal{F}$ consists of the Z transformations of EQUALITY and EXACTONE. The problem of counting the number of perfect matchings in a hypergraph is the special case of Pl-Holant($\neq_2$| EQUALITY, EXACTONE), where the connections between EQUALITY and EXACTONE via edges labeled ($\neq_2$) form a bipartite graph. In this case, at every vertex with an EQUALITY we absorb all the incident edges with ($\neq_2$) into this EQUALITY, making it an EQUALITY again. Then we obtain the bipartite edge–vertex incidence graph B_H of a hypergraph H: Vertices labeled EXACTONE are vertices of H, and vertices labeled EQUALITY are hyperedges of H, and all connections are by ($=_2$). A valid configuration is to choose a subset S of hyperedges (set to 1), such that every vertex of H is incident to exactly one hyperedge in S.

Suppose vertices assigned EQUALITY functions all have degree k. If $k = 2$, then the bipartite problem Pl-Holant($\neq_2$| EXACTONE, ($=_k$)) can be solved by Kasteleyn's algorithm since every signature is realizable as a matchgate. We show that this problem is #P-hard if $k = 3$ or $k = 4$, but is tractable again if $k \geq 5$. The algorithm involves a recursive procedure that simplifies the instance until it can be solved by known algorithms, including Kasteleyn's algorithm. The algorithm crucially uses global topological properties of a planar graph, in particular Euler's characteristic formula. If the graph is not planar, then this algorithm does not work, and indeed the problem is #P-hard over general graphs.

More generally, we allow vertices of arbitrary degrees to be assigned EQUALITY, with at least one vertex of degree ≥ 3. It turns out that the complexity depends on the greatest common divisor (gcd) of the arities of the EQUALITY functions present. The problem is tractable if gcd ≥ 5 and #P-hard if gcd ≤ 4. Note that it is not the minimum arity k of the EQUALITY functions that is the determining factor, but rather it is the gcd of the arities. (We allow multiple edges in a planar signature grid for Pl-Holant($\neq_2$| EQUALITY, EXACTONE). If $k \geq 6$ and $d \geq 3$, a planar (k, d)-biregular bipartite graph must have multiple edges. This follows from Euler's formula; see Lemma 9.24.)

Most of the reductions in previous Holant dichotomy theorems [HL16, CGW16] do not hold for planar graphs, so new techniques must be developed. An important ingredient in the proof of the Holant dichotomy Theorem 7.19 is Theorem 6.16, the #CSP^d dichotomy by Huang and Lu. Here #CSP^d denotes #CSP where every variable appears a multiple of d times. However, the proof of Theorem 6.16 is not planar, because it uses a pinning technique, and multiple copies of an instance are created and vertices are connected across different copies, which violates planarity. Moreover, as a consequence of the new dichotomy in this chapter, this violation of planarity is unavoidable. Owing to our newly discovered tractable problems, the putative form of a planar #CSP^d dichotomy stating that a holographic reduction to Kasteleyn's algorithm is universal, is *false* when $d \geq 5$. Nevertheless, it turns out that a dichotomy for planar #CSP^2 in the putative form is true. Obtaining a dichotomy for planar #CSP^2 turns out to be an essential step because it captures a significant fraction of planar Holant problems either directly or through reductions. It turns out that one can prove the planar Holant dichotomy without having a planar #CSP^d dichotomy for $d \geq 3$, which is still open. An outline of the proof of the planar #CSP^2 dichotomy will be given.

Owing to the length of the proof of this planar Holant dichotomy, we will omit most of the #P-hardness proof of this dichotomy, as well as the proof of the planar #CSP^2 dichotomy. Our presentation will focus on the polynomial time algorithm for the new planar tractable problems and related issues.

9.2 Some Known Results

We consolidate some known results proved in previous chapters that will be useful for the discussion in this chapter.

Recall that a signature set $\mathcal{F}$ is $\mathscr{C}$-transformable if there exists a $T \in \mathbf{GL}_2(\mathbb{C})$ such that $[1, 0, 1]T^{\otimes 2} \in \mathscr{C}$ and $\mathcal{F} \subseteq T\mathscr{C}$. In this chapter the relevant classes $\mathscr{C}$ for planar Holant tractable problems will be $\mathscr{C} = \mathscr{A}$, $\mathscr{P}$ and $\mathscr{M}$. The putative form of the classification would state that these are the only tractable families for planar Holant problems. This turns out to be inadequate for the final classification; however, we will first need to get a good understanding of $\mathscr{C}$-transformable classes, especially the $\mathscr{M}$-transformable classes.

Here we list several known dichotomies. The first is the dichotomy for Holant problems, Theorem 7.19.

Theorem 9.2. *Let $\mathcal{F}$ be any set of symmetric, complex-valued signatures in Boolean variables. Then* Holant($\mathcal{F}$) *is* #P-*hard unless $\mathcal{F}$ satisfies one of the following conditions, in which case the problem is in* P*:*

1. *All nondegenerate signatures in $\mathcal{F}$ are of arity ≤ 2.*
2. *$\mathcal{F}$ is $\mathscr{A}$-transformable.*
3. *$\mathcal{F}$ is $\mathscr{P}$-transformable.*
4. *$\mathcal{F} \subseteq \mathscr{V}^\sigma \cup \{f \in \mathscr{R}_2^\sigma \mid \text{arity}(f) = 2\}$ for $\sigma \in \{+, -\}$.*
5. *All nondegenerate signatures in $\mathcal{F}$ are in $\mathscr{R}_2^\sigma$ for $\sigma \in \{+, -\}$.*

Next we restate a dichotomy for planar Holant problems with additional restrictions. This is a combination of Theorem 6.3 and Theorem 8.12 for arity 3 and 4, respectively. This theorem forms the base case of an inductive proof of our single signature dichotomy, Theorem 9.17.

Theorem 9.3. *If f is a nondegenerate, symmetric, complex-valued signature of arity 3 or 4 in Boolean variables, then* Pl-Holant(f) *is* #P*-hard unless f satisfies one of the following conditions, in which case the problem is computable in polynomial time:*

1. Holant(f) *is tractable (i.e., f is $\mathscr{A}$-transformable, $\mathscr{P}$-transformable, or vanishing).*
2. *f is $\mathscr{M}$-transformable.*

Next is a dichotomy theorem about spin systems with prescribed vertex degrees.

Theorem 9.4. *Let $S \subseteq \mathbb{Z}^+$ containing some $r \geq 3$, let $\mathcal{G} = \{=_k \mid k \in S\}$, and let $d = \gcd(S)$. Further suppose that $f_0, f_1, f_2 \in \mathbb{C}$. Then* Pl-Holant($[f_0, f_1, f_2] \mid \mathcal{G}$) *is* #P*-hard unless one of the following conditions holds:*

1. $f_0 f_2 = f_1^2$.
2. $f_0 = f_2 = 0$.
3. $f_1 = 0$.
4. $f_0 f_2 = -f_1^2$ *and* $f_0^d = -f_2^d \neq 0$.
5. $f_0^d = f_2^d \neq 0$.

In all exceptional cases, the problem is computable in polynomial time.

This theorem follows from Theorem 5.3. The tractability is straightforward after possibly a holographic transformation of the form $\begin{bmatrix} 1 & 0 \\ 0 & \omega \end{bmatrix}$ for some $\omega^{2d} = 1$ (for the matchgate class $\mathscr{M}$) or $\omega^{4d} = 1$ (for the affine class $\mathscr{A}$), where $d = \gcd(S)$. The #P-hardness follows from applying Theorem 5.3 individually to each single arity k for $(=_k) \in \mathcal{G}$.

Theorem 9.4 is explicit and easy to apply. Conceptually, it can be restated as Theorem 9.4′, which supports the putative form of the dichotomy that a holographic reduction to Kasteleyn's algorithm is universal. Note that this is for a single signature, and is consistent with the single signature dichotomy, Theorem 9.17. Denote by $\mathcal{T}_k = \left\{ \begin{bmatrix} 1 & 0 \\ 0 & \omega \end{bmatrix} \mid \omega^k = 1 \right\}$, a set of diagonal matrices of order dividing k.

Theorem 9.4′. *Let $S \subseteq \mathbb{Z}^+$ contain $k \geq 3$, let $\mathcal{G} = \{=_k \mid k \in S\}$, and let $d = \gcd(S)$. Further suppose that f is a nondegenerate, symmetric, complex-valued binary signature in Boolean variables. Then* Pl-Holant($f \mid \mathcal{G}$) *is* #P*-hard unless f satisfies one of the following conditions, in which case, the problem is computable in polynomial time:*

1. There exists $T \in \mathcal{T}_{4d}$ *such that* $T^{\otimes 2} f \in \mathscr{A}$.
2. $f \in \mathscr{P}$.
3. There exists $T \in \mathcal{T}_{2d}$ *such that* $T^{\otimes 2} f \in \widehat{\mathscr{M}}$.

Lastly, we quote the Pl-#CSP dichotomy, Theorem 8.34. It also supports the putative form of a dichotomy, which states that holographic algorithms using matchgates followed by the Kasteleyn's algorithm is a universal strategy.

Theorem 9.5. *Let* $\mathcal{F}$ *be any set of symmetric, complex-valued signatures in Boolean variables. Then* Pl-#CSP($\mathcal{F}$) *is* #P*-hard unless* $\mathcal{F} \subseteq \mathscr{A}$, $\mathcal{F} \subseteq \mathscr{P}$, *or* $\mathcal{F} \subseteq \widehat{\mathscr{M}}$, *in which case the problem is computable in polynomial time.*

9.3 $\mathscr{A}$-, $\mathscr{P}$-, and $\mathscr{M}$-transformable Signatures

In this section, we first restate the properties of $\mathscr{A}$- and $\mathscr{P}$-transformable signatures. Then we investigate the properties of $\mathscr{M}$-transformable signatures. Throughout, we define $\alpha = \frac{1+\mathrm{i}}{\sqrt{2}} = \sqrt{\mathrm{i}} = e^{\frac{\pi \mathrm{i}}{4}}$ and use $\mathbf{O}_2(\mathbb{C})$ to denote the group of 2×2 orthogonal matrices over $\mathbb{C}$.

9.3.1 Characterization of $\mathscr{A}$- and $\mathscr{P}$-transformable Signatures

We summarize some results from Section 7.7. The following definition summarizes Definitions 7.39, 7.40, and 7.42.

Definition 9.6. A symmetric signature f of arity n is in, respectively, $\mathscr{A}_1$, or $\mathscr{A}_2$, or $\mathscr{A}_3$ if there exist an $H \in \mathbf{O}_2(\mathbb{C})$ and a nonzero constant $c \in \mathbb{C}$ such that f has the form, respectively,

$$cH^{\otimes n} \left(\begin{bmatrix} 1 \\ 1 \end{bmatrix}^{\otimes n} + \beta \begin{bmatrix} 1 \\ -1 \end{bmatrix}^{\otimes n} \right),$$

or

$$cH^{\otimes n} \left(\begin{bmatrix} 1 \\ \mathrm{i} \end{bmatrix}^{\otimes n} + \begin{bmatrix} 1 \\ -\mathrm{i} \end{bmatrix}^{\otimes n} \right),$$

or

$$cH^{\otimes n} \left(\begin{bmatrix} 1 \\ \alpha \end{bmatrix}^{\otimes n} + \mathrm{i}^r \begin{bmatrix} 1 \\ -\alpha \end{bmatrix}^{\otimes n} \right),$$

where $\beta = \alpha^{tn+2r}$, $r \in \{0, 1, 2, 3\}$, and $t \in \{0, 1\}$.

For $k \in \{1, 2, 3\}$, when such an orthogonal H exists, we say that $f \in \mathscr{A}_k$ with transformation H. If $f \in \mathscr{A}_k$ with I_2, then we say f is in the canonical form of $\mathscr{A}_k$.

Lemma 7.41 characterizes the signatures in $\mathscr{A}_2$. We restate it as follows.

Lemma 9.7. *Let f be a symmetric signature of arity n. Then $f \in \mathscr{A}_2$ if and only if $f = c\begin{bmatrix}1\\ \mathfrak{i}\end{bmatrix}^{\otimes n} + d\begin{bmatrix}1\\ -\mathfrak{i}\end{bmatrix}^{\otimes n}$ for some nonzero $c, d \in \mathbb{C}$. In other words, $f \in \mathscr{A}_2$ iff $f = Z[c, 0, \ldots, 0, d]$.*

Membership in these three sets characterize the $\mathscr{A}$-transformable signatures. We restate Lemma 7.43.

Lemma 9.8. *Let f be a nondegenerate symmetric signature. Then f is $\mathscr{A}$-transformable if and only if $f \in \mathscr{A}_1 \cup \mathscr{A}_2 \cup \mathscr{A}_3$.*

There is a similar characterization for $\mathscr{P}$-transformable signatures. We restate the definitions for $\mathscr{P}_1$ and $\mathscr{P}_2$ (see Definition 7.45).

Definition 9.9. A symmetric signature f of arity n is in $\mathscr{P}_1$ if there exist an $H \in \mathbf{O}_2(\mathbb{C})$ and a nonzero $c \in \mathbb{C}$ such that

$$f = cH^{\otimes n}\left(\begin{bmatrix}1\\1\end{bmatrix}^{\otimes n} + \beta\begin{bmatrix}1\\-1\end{bmatrix}^{\otimes n}\right),$$

where $\beta \neq 0$.

We define $\mathscr{P}_2 = \mathscr{A}_2$.

For $k \in \{1, 2\}$, when such an H exists, we say that $f \in \mathscr{P}_k$ with transformation H. If $f \in \mathscr{P}_k$ with I_2, then we say f is in the canonical form of $\mathscr{P}_k$.

Membership in these two sets characterize the $\mathscr{P}$-transformable signatures. We restate Lemma 7.46.

Lemma 9.10. *Let f be a nondegenerate symmetric signature. Then f is $\mathscr{P}$-transformable if and only if $f \in \mathscr{P}_1 \cup \mathscr{P}_2$.*

9.3.2 Characterization of $\mathscr{M}$-transformable Signatures

Now we develop a similar theory for the $\mathscr{M}$-transformable signatures. Recall that a signature set $\mathcal{F}$ is $\mathscr{M}$-transformable, if there exists a $T \in \mathbf{GL}_2(\mathbb{C})$ such that $[1, 0, 1]T^{\otimes 2} \in \mathscr{M}$, and $\mathcal{F} \subseteq T\mathscr{M}$. Since $[1, 0, 1]$ is symmetric, $[1, 0, 1]T^{\otimes 2}$ is also symmetric. However, it is unnecessary to consider all binary signatures in $\mathscr{M}$. We can normalize via *right* multiplication by elements in

$$\text{Stab}(\mathscr{M}) = \{T \in \mathbf{GL}_2(\mathbb{C}) | \mathscr{M}T \subseteq \mathscr{M}\} = \{T \in \mathbf{GL}_2(\mathbb{C}) | T\mathscr{M} \subseteq \mathscr{M}\},$$

the stabilizer group of $\mathscr{M}$. The right and left stabilizer groups of $\mathscr{M}$ coincide, since they are both generated by nonzero scalar multiples of matrices of the form $\begin{bmatrix} 1 & 0 \\ 0 & \nu \end{bmatrix}$ for any nonzero $\nu \in \mathbb{C}$ and $X = \begin{bmatrix} 0 & 1 \\ 1 & 0 \end{bmatrix}$.

After this normalization, it is enough to consider only two cases (case 1 and case 3) in Proposition 7.35, which is restated here.

Proposition 9.11. *Let $T \in \mathbb{C}^{2\times 2}$ be a matrix. Then the following hold:*

1. $[1, 0, 1]T^{\otimes 2} = [1, 0, 1]$ *if and only if* $T \in \mathbf{O}_2(\mathbb{C})$.
2. $[1, 0, 1]T^{\otimes 2} = [0, 1, 0]$ *if and only if there exists an* $H \in \mathbf{O}_2(\mathbb{C})$ *such that* $T = \frac{1}{\sqrt{2}}H\begin{bmatrix} 1 & 1 \\ i & -i \end{bmatrix}$.

It follows that

Lemma 9.12. *Let $\mathcal{F}$ be a set of signatures. Then $\mathcal{F}$ is $\mathscr{M}$-transformable if and only if $\mathcal{F} \subseteq Z\mathscr{M}$ for $Z = \begin{bmatrix} 1 & 1 \\ i & -i \end{bmatrix}$, or there exists an $H \in \mathbf{SO}_2(\mathbb{C})$ such that $\mathcal{F} \subseteq H\mathscr{M}$.*

Proof. Sufficiency is easily verified by checking that $(=_2)$ is transformed into $\mathscr{M}$ in both cases. Indeed $(=_2)Z^{\otimes 2} = (\neq_2) \in \mathscr{M}$, and H leaves $(=_2) \in \mathscr{M}$ unchanged.

Suppose $\mathcal{F}$ is $\mathscr{M}$-transformable. The nondegenerate binary signatures in $\mathscr{M}$ are either $[0, 1, 0]$ or of the form $[1, 0, \nu]$, up to a scalar. However, notice that $[1, 0, 1] = [1, 0, \nu]\begin{bmatrix} 1 & 0 \\ 0 & \nu^{-\frac{1}{2}} \end{bmatrix}^{\otimes 2}$ and $\begin{bmatrix} 1 & 0 \\ 0 & \nu^{-\frac{1}{2}} \end{bmatrix} \in \mathrm{Stab}(\mathscr{M})$. Thus, we only need to consider $[1, 0, 1]$ and $[0, 1, 0]$. Now we apply Proposition 9.11.

1. If $(=_2)T^{\otimes 2} = [1, 0, 1]$, then by case 1 of Proposition 9.11, we have $T \in \mathbf{O}_2(\mathbb{C})$. If $T \in \mathbf{SO}_2(\mathbb{C})$, then we are done with $H = T$. Otherwise, $T \in \mathbf{O}_2(\mathbb{C}) - \mathbf{SO}_2(\mathbb{C})$. We want to find an $H \in \mathbf{SO}_2(\mathbb{C})$ such that $\mathcal{F} \subseteq H\mathscr{M}$. Let $H = T\begin{bmatrix} 1 & 0 \\ 0 & -1 \end{bmatrix} \in \mathbf{SO}_2(\mathbb{C})$. Then
$$\mathcal{F} \subseteq T\mathscr{M} = T\begin{bmatrix} 1 & 0 \\ 0 & -1 \end{bmatrix}\mathscr{M} = H\mathscr{M},$$
since $\begin{bmatrix} 1 & 0 \\ 0 & -1 \end{bmatrix} \in \mathrm{Stab}(\mathscr{M})$.
2. If $(=_2)T^{\otimes 2} = [0, 1, 0]$, then by case 2 of Proposition 9.11, there exists an $H \in \mathbf{O}_2(\mathbb{C})$ such that $T = \frac{1}{\sqrt{2}}H\begin{bmatrix} 1 & 1 \\ i & -i \end{bmatrix}$. Therefore $\mathcal{F} \subseteq H\begin{bmatrix} 1 & 1 \\ i & -i \end{bmatrix}\mathscr{M}$.

Furthermore, if $H = \begin{bmatrix} a & b \\ -b & a \end{bmatrix} \in \mathbf{SO}_2(\mathbb{C})$, then $a^2 + b^2 = 1$ and

$$\mathcal{F} \subseteq H \begin{bmatrix} 1 & 1 \\ \mathfrak{i} & -\mathfrak{i} \end{bmatrix} \mathscr{M} = \begin{bmatrix} 1 & 1 \\ \mathfrak{i} & -\mathfrak{i} \end{bmatrix} \begin{bmatrix} a + b\mathfrak{i} & 0 \\ 0 & a - b\mathfrak{i} \end{bmatrix} \mathscr{M} = \begin{bmatrix} 1 & 1 \\ \mathfrak{i} & -\mathfrak{i} \end{bmatrix} \mathscr{M},$$

since

$$H \begin{bmatrix} 1 & 1 \\ \mathfrak{i} & -\mathfrak{i} \end{bmatrix} = \begin{bmatrix} 1 & 1 \\ \mathfrak{i} & -\mathfrak{i} \end{bmatrix} \begin{bmatrix} a + b\mathfrak{i} & 0 \\ 0 & a - b\mathfrak{i} \end{bmatrix}$$

and

$$\begin{bmatrix} a + b\mathfrak{i} & 0 \\ 0 & a - b\mathfrak{i} \end{bmatrix} \in \mathrm{Stab}(\mathscr{M}).$$

Otherwise, $H = \begin{bmatrix} a & b \\ b & -a \end{bmatrix} \in \mathbf{O}_2(\mathbb{C}) - \mathbf{SO}_2(\mathbb{C})$, so $a^2 + b^2 = 1$ and

$$\mathcal{F} \subseteq H \begin{bmatrix} 1 & 1 \\ \mathfrak{i} & -\mathfrak{i} \end{bmatrix} \mathscr{M} = \begin{bmatrix} 1 & 1 \\ \mathfrak{i} & -\mathfrak{i} \end{bmatrix} \begin{bmatrix} 0 & a - b\mathfrak{i} \\ a + b\mathfrak{i} & 0 \end{bmatrix} \mathscr{M} = \begin{bmatrix} 1 & 1 \\ \mathfrak{i} & -\mathfrak{i} \end{bmatrix} \mathscr{M},$$

since

$$H \begin{bmatrix} 1 & 1 \\ \mathfrak{i} & -\mathfrak{i} \end{bmatrix} = \begin{bmatrix} 1 & 1 \\ \mathfrak{i} & -\mathfrak{i} \end{bmatrix} \begin{bmatrix} 0 & a - b\mathfrak{i} \\ a + b\mathfrak{i} & 0 \end{bmatrix}$$

and

$$\begin{bmatrix} 0 & a - b\mathfrak{i} \\ a + b\mathfrak{i} & 0 \end{bmatrix} \in \mathrm{Stab}(\mathscr{M}).$$ □

We use four sets to characterize the $\mathscr{M}$-transformable signatures. The notation Sym is from Definition 7.3.

Definition 9.13. A symmetric signature f of arity n is in, respectively, $\mathscr{M}_1$, or $\mathscr{M}_2$, or $\mathscr{M}_3$, or $\mathscr{M}_4$ if there exist an $H \in \mathbf{O}_2(\mathbb{C})$ and nonzero constants $c, \gamma \in \mathbb{C}$ such that f has the form, respectively,

$$cH^{\otimes n} \left(\begin{bmatrix} 1 \\ 1 \end{bmatrix}^{\otimes n} \pm i^n \begin{bmatrix} 1 \\ -1 \end{bmatrix}^{\otimes n} \right),$$

or

$$cH^{\otimes n} \left(\begin{bmatrix} 1 \\ \gamma \end{bmatrix}^{\otimes n} \pm \begin{bmatrix} 1 \\ -\gamma \end{bmatrix}^{\otimes n} \right),$$

or

$$cH^{\otimes n} \operatorname{Sym}_n^{n-1} \left(\begin{bmatrix} 1 \\ 0 \end{bmatrix}; \begin{bmatrix} 0 \\ 1 \end{bmatrix} \right),$$

or

$$cH^{\otimes n} \operatorname{Sym}_n^{n-1}\left(\begin{bmatrix}1\\ \mathfrak{i}\end{bmatrix}; \begin{bmatrix}1\\ -\mathfrak{i}\end{bmatrix}\right).$$

For $k \in \{1, 2, 3, 4\}$, when such an H exists, we say that $f \in \mathscr{M}_k$ with transformation H. If $f \in \mathscr{M}_k$ with I_2, then we say f is in the canonical form of $\mathscr{M}_k$.

Notice that $\left\{\begin{bmatrix}1\\ \mathfrak{i}\end{bmatrix}, \begin{bmatrix}1\\ -\mathfrak{i}\end{bmatrix}\right\}$ is set-wise invariant under any transformation in $\mathbf{O}_2(\mathbb{C})$ up to nonzero constants. Using this fact, the following lemma gives a characterization of $\mathscr{M}_4$. It says that any signature in $\mathscr{M}_4$ is essentially in canonical form.

Lemma 9.14. *Let f be a symmetric signature of arity n. Then $f \in \mathscr{M}_4$ if and only if*

$$f = c \operatorname{Sym}_n^{n-1}\left(\begin{bmatrix}1\\ \mathfrak{i}\end{bmatrix}; \begin{bmatrix}1\\ -\mathfrak{i}\end{bmatrix}\right) \text{ or } f = c \operatorname{Sym}_n^{n-1}\left(\begin{bmatrix}1\\ -\mathfrak{i}\end{bmatrix}; \begin{bmatrix}1\\ \mathfrak{i}\end{bmatrix}\right)$$

for some nonzero constant $c \in \mathbb{C}$.

Proof. Suppose $f \in \mathscr{M}_4$, so that $f = cH^{\otimes n} \operatorname{Sym}_n^{n-1}\left(\begin{bmatrix}1\\ \mathfrak{i}\end{bmatrix}; \begin{bmatrix}1\\ -\mathfrak{i}\end{bmatrix}\right)$. If $H \in \mathbf{SO}_2(\mathbb{C})$, then $H = \begin{bmatrix}a & b\\ -b & a\end{bmatrix}$ for some $a, b \in \mathbb{C}$ such that $a^2 + b^2 = 1$. Since $H\begin{bmatrix}1\\ \mathfrak{i}\end{bmatrix} = (a + b\mathfrak{i})\begin{bmatrix}1\\ \mathfrak{i}\end{bmatrix}$ and $H\begin{bmatrix}1\\ -\mathfrak{i}\end{bmatrix} = (a - b\mathfrak{i})\begin{bmatrix}1\\ -\mathfrak{i}\end{bmatrix}$, it follows that $f = c(a + b\mathfrak{i})^{n-1}(a - b\mathfrak{i}) \operatorname{Sym}_n^{n-1}\left(\begin{bmatrix}1\\ \mathfrak{i}\end{bmatrix}; \begin{bmatrix}1\\ -\mathfrak{i}\end{bmatrix}\right)$. Otherwise, $H \in \mathbf{O}_2(\mathbb{C}) - \mathbf{SO}_2(\mathbb{C})$, so $H = \begin{bmatrix}a & b\\ b & -a\end{bmatrix}$ for some $a, b \in \mathbb{C}$ such that $a^2 + b^2 = 1$. Then $f = c(a + b\mathfrak{i})(a - b\mathfrak{i})^{n-1} \operatorname{Sym}_n^{n-1}\left(\begin{bmatrix}1\\ -\mathfrak{i}\end{bmatrix}; \begin{bmatrix}1\\ \mathfrak{i}\end{bmatrix}\right)$.

Now suppose

$$f = c \operatorname{Sym}_n^{n-1}\left(\begin{bmatrix}1\\ \mathfrak{i}\end{bmatrix}; \begin{bmatrix}1\\ -\mathfrak{i}\end{bmatrix}\right)$$

or $f = c \operatorname{Sym}_n^{n-1}\left(\begin{bmatrix}1\\ -\mathfrak{i}\end{bmatrix}; \begin{bmatrix}1\\ \mathfrak{i}\end{bmatrix}\right)$. The first case is already in the standard form of $\mathscr{M}_4$. In the second case, we pick $H = \begin{bmatrix}1 & 0\\ 0 & -1\end{bmatrix} \in \mathbf{O}_2(\mathbb{C})$. Then $H^{\otimes n} f$ is in the standard form of $\mathscr{M}_4$. $\square$

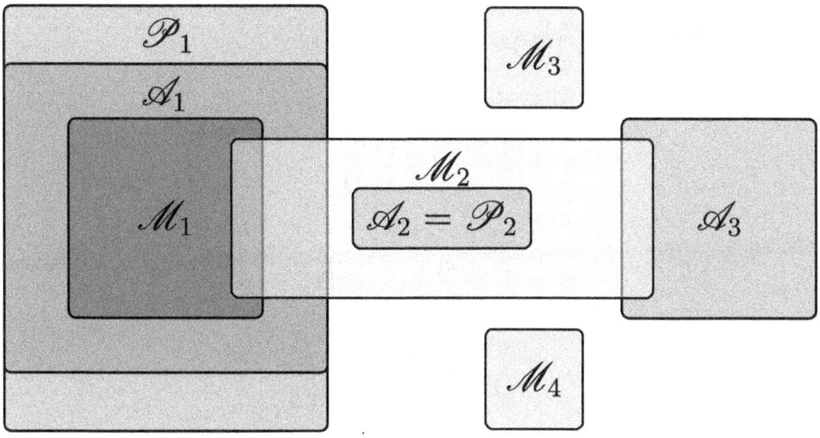

Figure 9.1. Relationships among $\mathscr{A}_1$, $\mathscr{A}_2$, $\mathscr{A}_3$, $\mathscr{P}_1$, $\mathscr{P}_2$, $\mathscr{M}_1$, $\mathscr{M}_2$, $\mathscr{M}_3$, and $\mathscr{M}_4$, for signatures of arity $n \geq 3$. Note that $\mathscr{P}_1 \cap \mathscr{M}_2 \subseteq \mathscr{A}_1$.

We can further split $\mathscr{M}_4$ into two classes $\mathscr{M}_4^{\pm}$. Define

$$\mathscr{M}_4^{\pm} = \left\{ f \,\middle|\, f = c\,\mathrm{Sym}_n^{n-1}\left(\begin{bmatrix}1\\ \pm\mathrm{i}\end{bmatrix}; \begin{bmatrix}1\\ \mp\mathrm{i}\end{bmatrix}\right)\right\}.$$

In other words, $\mathscr{M}_4^{+}$ contains signatures of the form $Z^{\otimes n}[0, 1, 0, \ldots, 0]$ and $\mathscr{M}_4^{-}$ contains signatures of the form $Z^{\otimes n}[0, \ldots, 0, 1, 0]$ up to a scalar, where $Z = \begin{bmatrix}1 & 1\\ \mathrm{i} & -\mathrm{i}\end{bmatrix}$. We will denote $[0, 1, 0, \ldots, 0]$ of arity k by $\textsc{ExactOne}_k$, and $[0, \ldots, 0, 1, 0]$ of arity k by $\textsc{AllButOne}_k$. Note that these are precisely the $\textsc{Perfect Matching}$ signatures and corresponding reversals.

Notice that $\mathscr{M}_1 \subset \mathscr{A}_1 \subset \mathscr{P}_1$ and $\mathscr{A}_2 = \mathscr{P}_2 \subset \mathscr{M}_2$. See Figure 9.1 for a visual description of the relationships among sets.

Next we show that $\mathscr{M}_k$ for $k = 1, 2, 3, 4$ captures all $\mathscr{M}$-transformable signatures.

Lemma 9.15. *Let f be a nondegenerate symmetric signature. Then f is $\mathscr{M}$-transformable if and only if $f \in \mathscr{M}_1 \cup \mathscr{M}_2 \cup \mathscr{M}_3 \cup \mathscr{M}_4$.*

Proof. Assume that f is $\mathscr{M}$-transformable of arity n. By applying Lemma 9.12 to $\{f\}$, we have $f \in \begin{bmatrix}1 & 1\\ \mathrm{i} & -\mathrm{i}\end{bmatrix}\mathscr{M}$ or there exists an $H \in \mathbf{SO}_2(\mathbb{C})$ such that $f \in H\mathscr{M}$. Proposition 8.1 lists all the symmetric signatures in $\mathscr{M}$. Since we are interested only in nondegenerate signatures, we consider only a, b, and λ that are nonzero. Now we consider the possible cases.

1. Suppose $f \in \begin{bmatrix} 1 & 1 \\ \mathfrak{i} & -\mathfrak{i} \end{bmatrix} \mathscr{M}$.
   - Further, suppose

$$f = \begin{bmatrix} 1 & 1 \\ \mathfrak{i} & -\mathfrak{i} \end{bmatrix}^{\otimes n} \left(\begin{bmatrix} a \\ b \end{bmatrix}^{\otimes n} \pm \begin{bmatrix} a \\ -b \end{bmatrix}^{\otimes n} \right)$$

   for some nonzero $a, b \in \mathbb{C}$. Let $T = \frac{1-\mathfrak{i}}{2} \begin{bmatrix} u & v \\ v & -u \end{bmatrix}$, where $u = a + b\mathfrak{i}$ and $v = \mathfrak{i}(a - b\mathfrak{i})$. Then

$$f = T^{\otimes n} \left(\begin{bmatrix} 1 \\ 1 \end{bmatrix}^{\otimes n} \pm \mathfrak{i}^n \begin{bmatrix} 1 \\ -1 \end{bmatrix}^{\otimes n} \right).$$

   Since $T \in \mathbf{O}_2(\mathbb{C})$ up to a nonzero factor of $\sqrt{2ab}$, we have $f \in \mathscr{M}_1$.
   - Further, suppose

$$f = \lambda \begin{bmatrix} 1 & 1 \\ \mathfrak{i} & -\mathfrak{i} \end{bmatrix}^{\otimes n} \operatorname{Sym}_n^{n-1} \left(\begin{bmatrix} 1 \\ 0 \end{bmatrix}; \begin{bmatrix} 0 \\ 1 \end{bmatrix} \right)$$

   for some nonzero $\lambda \in \mathbb{C}$. Then we have $f = \lambda \operatorname{Sym}_n^{n-1} \left(\begin{bmatrix} 1 \\ \mathfrak{i} \end{bmatrix}; \begin{bmatrix} 1 \\ -\mathfrak{i} \end{bmatrix} \right)$, so $f \in \mathscr{M}_4$.
   - Further, suppose

$$f = \lambda \begin{bmatrix} 1 & 1 \\ \mathfrak{i} & -\mathfrak{i} \end{bmatrix}^{\otimes n} \operatorname{Sym}_n^{n-1} \left(\begin{bmatrix} 0 \\ 1 \end{bmatrix}; \begin{bmatrix} 1 \\ 0 \end{bmatrix} \right)$$

   for some nonzero $\lambda \in \mathbb{C}$. Then we have $f = \lambda \operatorname{Sym}_n^{n-1} \left(\begin{bmatrix} 1 \\ -\mathfrak{i} \end{bmatrix}; \begin{bmatrix} 1 \\ \mathfrak{i} \end{bmatrix} \right)$, so $f \in \mathscr{M}_4$ by Lemma 9.14.
2. Suppose $f \in H\mathscr{M}$.
   - Further, suppose

$$f = H^{\otimes n} \left(\begin{bmatrix} a \\ b \end{bmatrix}^{\otimes n} \pm \begin{bmatrix} a \\ -b \end{bmatrix}^{\otimes n} \right)$$

   for some nonzero $a, b \in \mathbb{C}$. Then we have

$$f = a^n H^{\otimes n} \left(\begin{bmatrix} 1 \\ \gamma \end{bmatrix}^{\otimes n} \pm \begin{bmatrix} 1 \\ -\gamma \end{bmatrix}^{\otimes n} \right),$$

   where $\gamma = \frac{b}{a}$, so $f \in \mathscr{M}_2$.

- Further, suppose

$$f = \lambda H^{\otimes n} \operatorname{Sym}_n^{n-1}\left(\begin{bmatrix}1\\0\end{bmatrix};\begin{bmatrix}0\\1\end{bmatrix}\right)$$

for some nonzero $\lambda \in \mathbb{C}$. Then $f \in \mathscr{M}_3$.
- Further, suppose

$$f = \lambda H^{\otimes n} \operatorname{Sym}_n^{n-1}\left(\begin{bmatrix}0\\1\end{bmatrix};\begin{bmatrix}1\\0\end{bmatrix}\right)$$

for some nonzero $\lambda \in \mathbb{C}$. Let $H' = H\begin{bmatrix}0 & 1\\1 & 0\end{bmatrix} \in \mathbf{O}_2(\mathbb{C})$. Then we have

$$f = \lambda H'^{\otimes n} \operatorname{Sym}_n^{n-1}\left(\begin{bmatrix}1\\0\end{bmatrix};\begin{bmatrix}0\\1\end{bmatrix}\right),$$

so $f \in \mathscr{M}_3$.

Conversely, if there exists a matrix $H \in \mathbf{O}_2(\mathbb{C})$ such that $H^{\otimes n} f$ is in one of the canonical forms of $\mathscr{M}_1$, $\mathscr{M}_2$, $\mathscr{M}_3$, or $\mathscr{M}_4$, then one can directly check that f is $\mathscr{M}$-transformable. In fact, the transformations that we applied above are all invertible, except for $\mathscr{M}_1$, if the given orthogonal transformation is of the form $\begin{bmatrix}u & -v\\v & u\end{bmatrix}$, we do $\begin{bmatrix}1 & 0\\0 & -1\end{bmatrix}$ first followed by $\begin{bmatrix}u & v\\v & -u\end{bmatrix}$. □

Recall from Section 7.9 that a symmetric signature $f = [f_0, f_1, \ldots, f_n]$ satisfies a second-order recurrence relation, if there exist $a, b, c \in \mathbb{C}$ not all zero, such that for all $0 \le k \le n-2$,

$$af_k + bf_{k+1} + cf_{k+2} = 0.$$

Lemma 7.56 says that any $\mathscr{A}$- or $\mathscr{P}$-transformable signature $f \in \mathscr{P}_1 \cup \mathscr{A}_2 \cup \mathscr{A}_3$ of arity $n \ge 3$ satisfies a second-order recurrence relation with nonzero discriminant $b^2 - 4ac$.

This can be extended to $\mathscr{M}$-transformable signatures.

Exercise: Let f be a symmetric $\mathscr{M}$–transformable signature of arity $n \ge 3$. Then f satisfies a second-order recurrence relation. The discriminant is nonzero iff $f \in \mathscr{M}_1 \cup \mathscr{M}_2$. The recurrence relation for $f \in \mathscr{M}_4$ has a double root belonging to $\{\mathfrak{i}, -\mathfrak{i}\}$. The recurrence relation for $f \in \mathscr{M}_3$ has a double root not belonging to $\{\mathfrak{i}, -\mathfrak{i}\}$.

Hence $\mathscr{M}_1 \cup \mathscr{M}_2$, $\mathscr{M}_3$ and $\mathscr{M}_4$ for signatures of arity ≥ 3 are pairwise disjoint.

9.4 Dichotomy for Pl-#CSP^2

In this section, we state the dichotomy for Pl-#CSP^2. We provide only a sketch of the proof here; the interested reader can find the full proof in Part II of [CFGW15]. Let

$$\mathcal{T}_k = \left\{ \begin{bmatrix} 1 & 0 \\ 0 & \omega \end{bmatrix} \in \mathbb{C}^{2\times 2} \,\middle|\, \omega^k = 1 \right\}.$$

Theorem 9.16. *Let $\mathcal{F}$ be a set of symmetric signatures. Then* Pl-#CSP$^2(\mathcal{F})$ *is* #P-*hard unless $\mathcal{F}$ satisfies one of the following conditions:*

1. *There exists $T \in \mathcal{T}_8$ such that $\mathcal{F} \subseteq T\mathscr{A}$.*
2. *$\mathcal{F} \subseteq \mathscr{P}$.*
3. *There exists $T \in \mathcal{T}_4$ such that $\mathcal{F} \subseteq T\widehat{\mathscr{M}}$.*

In each exceptional case, Pl-#CSP$^2(\mathcal{F})$ *is computable in polynomial time.*

Proof Sketch. We first define some tractable families of signatures specific to the Pl-#CSP^2 framework. Let

$$\mathscr{T}_k = \mathcal{T}_{2k} \setminus \mathcal{T}_k = \left\{ \begin{bmatrix} 1 & 0 \\ 0 & \omega \end{bmatrix} \,\middle|\, \omega^k = -1 \right\}.$$

Let $\mathscr{A}^\dagger = \mathscr{T}_4\mathscr{A}$ and $\widehat{\mathscr{M}}^\dagger = \mathscr{T}_2\widehat{\mathscr{M}}$ be the sets of signatures transformed by $\mathscr{T}_4$ from the affine family $\mathscr{A}$ and transformed by $\mathscr{T}_2$ from $\widehat{\mathscr{M}}$, respectively. Let

$$\widetilde{\mathscr{A}} = \mathscr{A} \cup \mathscr{A}^\dagger \qquad \text{and} \qquad \widetilde{\mathscr{M}} = \widehat{\mathscr{M}} \cup \widehat{\mathscr{M}}^\dagger.$$

Thus $\widetilde{\mathscr{A}} = \mathscr{A} \cup \begin{bmatrix} 1 & 0 \\ 0 & e^{\pi i/4} \end{bmatrix}\mathscr{A}$ and $\widetilde{\mathscr{M}} = \widehat{\mathscr{M}} \cup \begin{bmatrix} 1 & 0 \\ 0 & i \end{bmatrix}\widehat{\mathscr{M}}$. These are the $\mathscr{A}$-transformable and $\mathscr{M}$-transformable signatures for Pl-#CSP^2. One can show that $\widetilde{\mathscr{A}}$ covers case 1 above, and $\widetilde{\mathscr{M}}$ covers case 3. The proof will revolve around these tractable classes.

The overall plan is to break the proof into two main steps.

The first step is to prove the dichotomy theorem for Pl-#CSP$^2(\mathcal{F})$ when there is at least one nonzero signature of *odd* arity in $\mathcal{F}$. In this case, we can simulate Pl-#CSP($\mathcal{F}$) by Pl-#CSP$^2(\mathcal{F})$ if $\mathcal{F}$ includes a unary signature $[a, b]$ with $ab \neq 0$. Then we can apply the known dichotomy Theorem 9.5 for Pl-#CSP. However, this strategy (provably) *cannot* work when every signature in $\mathcal{F}$ satisfies the *parity* constraint (in particular, $\mathcal{F}$ does not contain any unary signature $[a, b]$ with $ab \neq 0$). In that case we employ other means. This first step of the proof is relatively uncomplicated.

The second step is to deal with the case when all nonzero signatures in $\mathcal{F}$ have even arity. This is where the real difficulties lie. In this case it is impossible to directly construct *any* unary signature, which makes the proof more difficult. But we prove that there is a way to simulate Pl-#CSP($\mathcal{F}$) by Pl-#CSP^2($\mathcal{F}$) in a *global* fashion, *if* $\mathcal{F}$ includes some tensor power of the form $[a, b]^{\otimes 2}$, where $ab \neq 0$. Moreover, we have a lucky break (for the complexity of the proof) if $\mathcal{F}$ includes a signature that is in $\widehat{\mathscr{M}} \setminus (\mathscr{P} \cup \widetilde{\mathscr{A}})$. In this case, we can construct a special binary signature, and obtain $[1, 1]^{\otimes 2}$ by interpolation. This proof uses some properties of *cyclotomic fields*. This simplifies the proof greatly. For all other cases when $\mathcal{F}$ has only even-arity signatures, the proof gets going in earnest – we will attempt an induction on the arity of signatures.

The lowest arity of this induction will be 2. We will try to reduce the arity to 2 whenever possible; however, for many cases an arity reduction to 2 destroys the #P-hardness at hand. Therefore, the true base case of this induction proof of Pl-#CSP^2 starts with arity 4. Consequently, we will first prove a dichotomy theorem for Pl-#CSP^2(f), where f is a signature of arity 4. Several tools will be used. These include the rank criterion for redundant signatures Corollary 7.27, a dichotomy for planar spin systems Theorem 9.4, and a trick called the *Three Stooges* by domain pairing.

However, in the next step we do not attempt a general Pl-#CSP^2 dichotomy for a *single* signature of even arity. This would have been natural at this point, but it would have been too difficult. We will need some additional leverage by proving a conditional no-mixing lemma for pairs of signatures of even arity. So, seemingly taking a detour, we prove that for two signatures f and g both of even-arity, each individually belongs to some tractable class, but jointly do not belong to a single tractable class in the conjectured dichotomy, $\mathscr{P}$, or $\mathscr{A}$, or $\mathscr{A}^\dagger$, or $\widehat{\mathscr{M}}$, or $\widehat{\mathscr{M}}^\dagger$, the problem Pl-#CSP^2($f, g$) is #P-hard. We prove this no-mixing lemma for any pair of signatures f and g both of even arity, not restricted to arity 4. Even though at this point we only have a dichotomy for a single signature of arity 4, we prove this no-mixing lemma for higher even-arity pairs f and g by simulating two signatures f' and g' of arity 4 that belong to different tractable sets, from that of Pl-#CSP^2(f, g). After this arity reduction (within the no-mixing lemma), we prove that Pl-#CSP^2(f', g') is #P-hard by the dichotomy for a *single* signature of arity 4. After this, we prove a no-mixing lemma for a *set* of signatures $\mathcal{F}$ of even arities, which states that *if* $\mathcal{F}$ is contained in the union of all tractable classes, then it is still #P-hard unless it is *entirely* contained in one single tractable class. Note that at this point we still only have a *conditional* no-mixing lemma in the sense that we have to assume every signature in $\mathcal{F}$ belongs to some tractable set.

We then attempt the proof of a Pl-#CSP^2 dichotomy for a *single* signature of arbitrary even arity. This uses all the previous lemmas, in particular the (conditional) no-mixing lemma for a set of signatures. However, after completing the proof of this Pl-#CSP^2 dichotomy for a single signature of even arity, the no-mixing lemma becomes *unconditional.*

Finally the dichotomy for a single signature of even arity is logically extended to a dichotomy theorem for Pl-#CSP$^2(\mathcal{F})$ where all signatures in $\mathcal{F}$ have even arity. Together with the first main step when $\mathcal{F}$ contains some nonzero signature of odd arity, this completes the proof of Theorem 9.16. □

9.5 Single-Signature Dichotomy

By Lemmas 9.8, 9.10 and 9.15, f is $\mathscr{A}$-, or $\mathscr{P}$-, or $\mathscr{M}$-transformable, or vanishing if and only if

$$f \in \mathscr{A}_1 \cup \mathscr{A}_2 \cup \mathscr{A}_3 \cup \mathscr{P}_1 \cup \mathscr{P}_2 \cup \mathscr{M}_1 \cup \mathscr{M}_2 \cup \mathscr{M}_3 \cup \mathscr{M}_4 \cup \mathscr{V}.$$

However, since $\mathscr{M}_1 \subset \mathscr{A}_1 \subset \mathscr{P}_1$, $\mathscr{A}_2 = \mathscr{P}_2 \subset \mathscr{M}_2$, we have the equivalent expression

$$f \in \mathscr{P}_1 \cup \mathscr{M}_2 \cup \mathscr{A}_3 \cup \mathscr{M}_3 \cup \mathscr{M}_4 \cup \mathscr{V}.$$

In fact, for signatures f of arity $n \geq 3$, $f \in \mathscr{M}_4$ implies $f \in \mathscr{V}$. So for arity $n \geq 3$, we could further replace it by $f \in \mathscr{P}_1 \cup \mathscr{M}_2 \cup \mathscr{A}_3 \cup \mathscr{M}_3 \cup \mathscr{V}$. However, for the analysis of cases we will actually consider the equivalent expression

$$f \in \mathscr{P}_1 \cup \mathscr{P}_2 \cup (\mathscr{M}_2 \setminus \mathscr{P}_2) \cup \mathscr{A}_3 \cup \mathscr{M}_3 \cup \mathscr{M}_4 \cup (\mathscr{V} \setminus \mathscr{M}_4).$$

We state Theorem 9.17, the single-signature dichotomy for Pl-Holant problems. This theorem says that for a single signature the putative form of the Pl-Holant dichotomy holds. That is, for a nondegenerate symmetric signature f of arity ≥ 3, Pl-Holant(f) is either #P-hard, or f is $\mathscr{A}$-, or $\mathscr{P}$-, or $\mathscr{M}$-transformable, or vanishing.

Theorem 9.17. *If f is a nondegenerate symmetric signature of arity $n \geq 3$ with complex weights in Boolean variables, then* Pl-Holant(f) *is* #P-*hard unless* $f \in \mathscr{P}_1 \cup \mathscr{M}_2 \cup \mathscr{A}_3 \cup \mathscr{M}_3 \cup \mathscr{M}_4 \cup \mathscr{V}$, *in which case the problem is computable in polynomial time.*

The proof of Theorem 9.17 is by induction on the arity; but we will not give the details. In fact, the proof of Theorem 9.17 already has to investigate the mixing properties of various tractable signatures. Interested readers can find the proof in the paper [CFGW15].

Having established Theorem 9.17, the main question is for a *set* of tractable signatures what kinds of mixture are possible. A great deal of effort is spent trying to prove various tractable classes cannot mix, much in the same way as in the proof of the no-mixing lemma for Pl-#CSP^2 problems. Previously we had always found that two tractable signatures of different types cannot be mixed together; i.e., any signature set containing both signatures leads to #P-hardness. For Pl-Holant problems, it turns out that this is no longer true for all the tractable classes.

But the non-mixing property is still true for $f \in \mathscr{P}_1 \cup (\mathscr{M}_2 \setminus \mathscr{P}_2) \cup \mathscr{A}_3 \cup \mathscr{M}_3$. The technique is similar to that of Section 7.7.2, and we can reduce the case to Theorem 9.16, the dichotomy for Pl-#CSP^2 problems. For example, suppose $\mathcal{F}$ contains some $f \in \mathscr{P}_1$ of arity $n \geq 3$, we can write it as $f = H[a, 0, \ldots, 0, b]$, where $H \in \mathbf{O}_2(\mathbb{C})$ and $ab \neq 0$. After a holographic transformation by H in Pl-Holant($=_2| \mathcal{F}$), we have Pl-Holant($=_2| H^{-1}\mathcal{F}$), where the generalized equality $[a, 0, \ldots, 0, b] \in H^{-1}\mathcal{F}$. Then we can follow the proof of Lemma 7.48, which remains valid for planar Holant problems, and obtain all equalities of even arity $\mathcal{EQ}_2$. Basically, by repeated self-loops we can get a generalized equality of arity 3 or 4. If it is 3, connecting two copies with one edge gives a generalized equality of arity 4. From arity 4, a self-loop gives a generalized equality of arity 2. Then we can get generalized equalities of all even arities. Finally we get $\mathcal{EQ}_2$ by interpolation, or in case the ratio is a root of unity, by a direct construction. Once we get $\mathcal{EQ}_2$ we can apply Theorem 9.16, the dichotomy for Pl-#CSP^2 problems.

Some more effort is needed to prove non-mixing for $\mathscr{V} \setminus \mathscr{M}_4$. So the real difficulty is the mixing question of $\mathscr{P}_2$ and $\mathscr{M}_4$.

The following lemma deals with a case of mixing property for some signatures from $\mathscr{P}_2$ and $\mathscr{M}_4$. There are three signatures, the first two signatures $Z^{\otimes 4}[0, 1, 0, 0, 0]$ and $Z^{\otimes 4}[0, 0, 0, 1, 0]$ both belong to $\mathscr{M}_4$. The third signature $Z^{\otimes 4}g$ is the Z-transformation of the conjunction $g(x_1, x_2, x_3, x_4) = (x_1 = x_3) \wedge (x_2 = x_4) \wedge (x_1 \neq x_2)$. This is the conjunction of $Z^{\otimes 2}[1, 0, 1]_{x_1,x_3}$, $Z^{\otimes 2}[1, 0, 1]_{x_2,x_4}$ and $Z^{\otimes 2}[0, 1, 0]_{x_1,x_2}$; each belongs to $\mathscr{P}_2$. This lemma is already needed in the proof of Theorem 9.17, and is part of the proof of the no-mixing for Pl-Holant problems. We will also use it to investigate the complexity of hypergraph perfect matching problems in Section 9.6. The lemma is stated in the form after the Z-transformation.

Lemma 9.18. *Let g be the arity 4 signature whose matrix is*

$$M_g = \begin{bmatrix} g_{0000} & g_{0010} & g_{0001} & g_{0011} \\ g_{0100} & g_{0110} & g_{0101} & g_{0111} \\ g_{1000} & g_{1010} & g_{1001} & g_{1011} \\ g_{1100} & g_{1110} & g_{1101} & g_{1111} \end{bmatrix} = \begin{bmatrix} 0 & 0 & 0 & 0 \\ 0 & 0 & 1 & 0 \\ 0 & 1 & 0 & 0 \\ 0 & 0 & 0 & 0 \end{bmatrix}. \tag{9.1}$$

Then Pl-Holant($\neq_2 \mid [0, 1, 0, 0, 0], [0, 0, 0, 1, 0], g$) *is* #P-*hard.*

Proof. Recall that we list the columns of a signature matrix in reverse lexicographic order so that matrix product corresponds to linking the signature in a

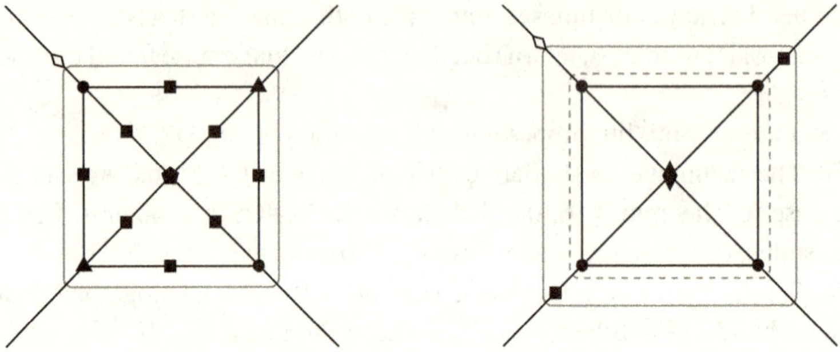

(a) $(\neq_2 \mid [0,1,0,0,0], [0,0,0,1,0], g)$-gate on right side (b) Simpler construction with the same signature

Figure 9.2. Two gadgets with the same signature used in Lemma 9.18.

chain. The only nonzero entries of M_g are $g_{0101} = g_{1010} = 1$. Thus g is a conditional crossover signature where in a planar cyclic order the two wires crossing each other must carry unequal values 0 and 1.

Consider the gadget in Figure 9.2a. We assign $[0, 0, 0, 1, 0]$ to the triangle vertices, $[0, 1, 0, 0, 0]$ to the circle vertices, g to the pentagon vertex, and $[0, 1, 0]$ to the square vertices. Let h be the signature of this gadget. If we append two binary DISEQUALITY signatures consecutively at the lower left and also at the upper right dangling edges, we do not change the signature h. Then we absorb four adjacent DISEQUALITY signatures surrounding each $[0, 0, 0, 1, 0]$ signature at the two triangle vertices, and we convert the ALLBUTONE$_4$ signature $[0, 0, 0, 1, 0]$ to the EXACTONE$_4$ signature $[0, 1, 0, 0, 0]$. If we also absorb the two DISEQUALITY signatures at the upper left and lower right positions of the pentagon vertex, the signature g there becomes $(=_4)$. This is now depicted as the diamond vertex in Figure 9.2b, which has the same signature h as in Figure 9.2a, where the circle vertices are still assigned $[0, 1, 0, 0, 0]$, the square vertices are still assigned $[0, 1, 0]$, and the diamond vertex is assigned $(=_4)$. To compute the signature h, first compute the signature h' of the inner gadget enclosed by the dashed line, which has signature matrix

$$M_{h'} = \begin{bmatrix} 3 & 0 & 0 & 1 \\ 0 & 1 & 0 & 0 \\ 0 & 0 & 1 & 0 \\ 1 & 0 & 0 & 1 \end{bmatrix}.$$

Then by Figure 9.3, the signature matrix of h is

$$M_h = \begin{bmatrix} 0 & 0 & 0 & 1 \\ 0 & 1 & 3 & 0 \\ 0 & 1 & 1 & 0 \\ 1 & 0 & 0 & 0 \end{bmatrix}.$$

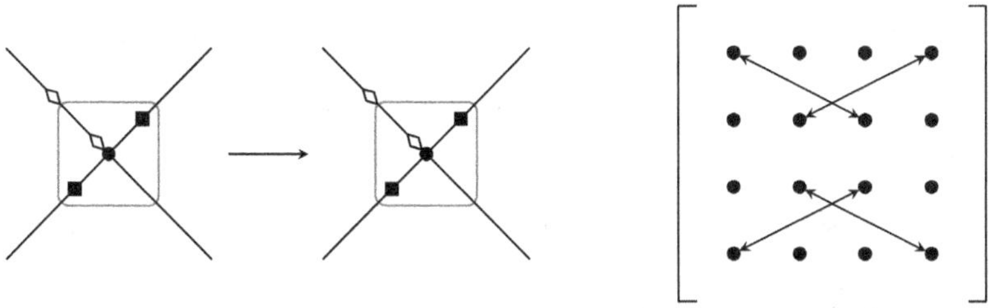

(a) Negating the second and fourth inputs (b) Movement of even Hamming weight entries

Figure 9.3. The movement of the even Hamming weight entries in the signature matrix of a quaternary signature under the negation of the second and fourth inputs (i.e., the square vertices are assigned [0, 1, 0]).

We need one more symmetrizing construction. Consider the gadget in Figure 9.4b. We assign h to the circle vertices and [0, 1, 0] to the square vertices. The signature of the resulting gadget is r with signature matrix M_r (see Figure 7.4 for the signature of a rotated copy of h that appears as the second circle vertex in Figure 9.4b), where

$$M_r = \begin{bmatrix} 0 & 0 & 0 & 1 \\ 0 & 1 & 3 & 0 \\ 0 & 1 & 1 & 0 \\ 1 & 0 & 0 & 0 \end{bmatrix} \left(\begin{bmatrix} 0 & 1 \\ 1 & 0 \end{bmatrix} \otimes \begin{bmatrix} 0 & 1 \\ 1 & 0 \end{bmatrix} \right) \begin{bmatrix} 0 & 0 & 0 & 1 \\ 0 & 1 & 1 & 0 \\ 0 & 3 & 1 & 0 \\ 1 & 0 & 0 & 0 \end{bmatrix} = \begin{bmatrix} 0 & 0 & 0 & 1 \\ 0 & 6 & 4 & 0 \\ 0 & 4 & 2 & 0 \\ 1 & 0 & 0 & 0 \end{bmatrix}.$$

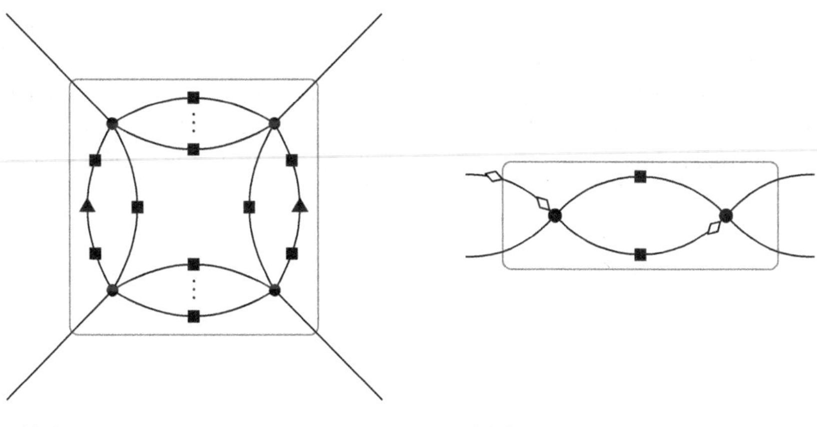

(a) Gadget that realizes a partial crossover (b) Gadget with a useful signature matrix

Figure 9.4. Two quaternary gadgets used in the proof of Lemma 9.18.

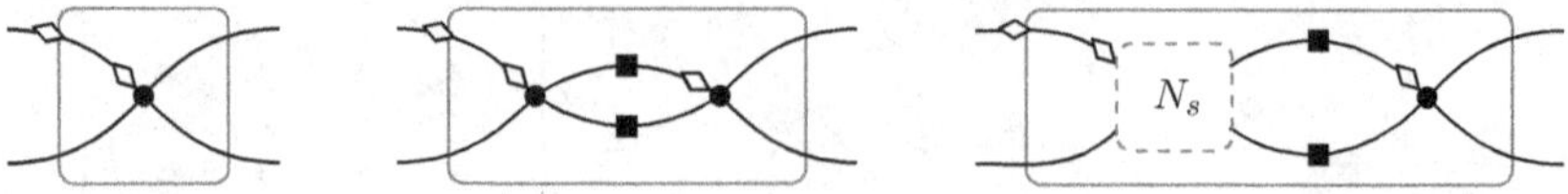

Figure 9.5. Linear recursive construction used for interpolation in a nonstandard basis.

Now we consider the signature

$$M_R = \begin{bmatrix} 0 & 0 & 0 & 1 \\ 0 & 3 & 1 & 0 \\ 0 & 1 & 1 & 0 \\ 1 & 0 & 0 & 0 \end{bmatrix}.$$

By a rotation (see Figure 7.4) R becomes R', with

$$M_{R'} = \begin{bmatrix} 0 & 0 & 0 & 3 \\ 0 & 1 & 1 & 0 \\ 0 & 1 & 1 & 0 \\ 1 & 0 & 0 & 0 \end{bmatrix}.$$

The signature matrix $M_{R'}$ is redundant and the compressed signature matrix is nonsingular. Hence by Corollary 7.27 using the Z transformation, Pl-Holant($\neq_2$| R') is #P-hard. Therefore so is Pl-Holant($\neq_2$| R).

Now we prove a reduction:

$$\text{Pl-Holant}(\neq_2 | \, R) \leq_T \text{Pl-Holant}(\neq_2 | \, r).$$

Consider an instance Ω of Pl-Holant($\neq_2$ | R). Suppose that R appears n times in Ω. We construct from Ω a sequence of instances Ω_s of Pl-Holant($\neq_2$ | r) indexed by $s \geq 1$. We obtain Ω_s from Ω by replacing each occurrence of R with the gadget N_s in Figure 9.5 with r assigned to the circle vertices and $[0, 1, 0]$ assigned to the square vertices. In Ω_s, the edge corresponding to the ith significant index bit of N_s connects to the same location as the edge corresponding to the ith significant index bit of R in Ω.

We can express the signature matrix of N_s as

$$M_{N_s} = X(XM_r)^s = XP \, \text{diag}(1, 4 + 2\sqrt{3}, 4 - 2\sqrt{3}, 1)^s P^{-1},$$

where

$$X = \begin{bmatrix} 0 & 0 & 0 & 1 \\ 0 & 0 & 1 & 0 \\ 0 & 1 & 0 & 0 \\ 1 & 0 & 0 & 0 \end{bmatrix} \quad \text{and} \quad P = \begin{bmatrix} 1 & 0 & 0 & 0 \\ 0 & 1 & 1 & 0 \\ 0 & \sqrt{3} & -\sqrt{3} & 0 \\ 0 & 0 & 0 & 1 \end{bmatrix}.$$

Since $M_R = XP\,\mathrm{diag}(1, 1+\sqrt{3}, 1-\sqrt{3}, 1)P^{-1}$, we can view our construction of Ω_s as first replacing M_R with $XP\,\mathrm{diag}(1, 1+\sqrt{3}, 1-\sqrt{3}, 1)P^{-1}$, which does not change the Holant value, and then replacing the diagonal matrix $\mathrm{diag}(1, 1+\sqrt{3}, 1-\sqrt{3}, 1)$ with the diagonal matrix $\mathrm{diag}(1, 4+2\sqrt{3}, 4-2\sqrt{3}, 1)^s$.

We stratify the assignments in Ω based on the assignments to the n occurrences of the signature whose signature matrix is the diagonal matrix

$$\begin{bmatrix} 1 & 0 & 0 & 0 \\ 0 & 1+\sqrt{3} & 0 & 0 \\ 0 & 0 & 1-\sqrt{3} & 0 \\ 0 & 0 & 0 & 1 \end{bmatrix}. \tag{9.2}$$

We only need to consider the assignments that assign

- i many times the bit patterns 0000 or 1111,
- j many times the bit pattern 0110, and
- k many times the bit pattern 1001,

since any other assignment contributes a factor 0. Let c_{ijk} be the sum over all such assignments of the products of evaluations of all signatures (including the signatures corresponding to the signature matrices X, P, and P^{-1}) in Ω except for the signature corresponding to the signature matrix in (9.2). Then

$$\mathrm{Holant}_\Omega = \sum_{i+j+k=n} (1+\sqrt{3})^j(1-\sqrt{3})^k c_{ijk}$$

and the value of the Holant on Ω_s, for $s \geq 1$, is

$$\begin{aligned} \mathrm{Holant}_{\Omega_s} &= \sum_{i+j+k=n} ((4+2\sqrt{3})^j(4-2\sqrt{3})^k)^s c_{ijk} \\ &= \sum_{i+j+k=n} ((4+2\sqrt{3})^{j-k}4^k)^s c_{ijk}. \end{aligned}$$

We argue that this Vandermonde system has full rank, which is to say that $(4+2\sqrt{3})^{j-k}4^k \neq (4+2\sqrt{3})^{j'-k'}4^{k'}$ unless $(j,k) = (j',k')$. If $(4+2\sqrt{3})^{j-k}4^k = (4+2\sqrt{3})^{j'-k'}4^{k'}$, then we have $(4+2\sqrt{3})^{j-k-(j'-k')}4^{k-k'} = 1$. Since any nonzero integer power of $4+2\sqrt{3}$ is not rational, we must have $j-k = j'-k'$. And in this case, $4^{k-k'} = 1$, and hence $k = k'$ and $j = j'$.

Therefore, we can solve for the unknown c_{ijk}'s and obtain the value of Holant_Ω. This completes the proof that Pl-Holant($\neq_2 | \, r$) is #P-hard by interpolation. Thus the problem Pl-Holant($\neq_2 \mid [0,1,0,0,0], [0,0,0,1,0], g$) is also #P-hard. □

9.6 Mixing $\mathscr{P}_2$ and $\mathscr{M}_4$ – Equalities and Matchgates in the Z Basis

Given a set $\mathcal{F}$ of symmetric signatures, by Theorem 9.17, Pl-Holant($\mathcal{F}$) is #P-hard unless every nondegenerate signature f of arity ≥ 3 in $\mathcal{F}$ is in $\mathscr{P}_1 \cup \mathscr{M}_2 \cup \mathscr{A}_3 \cup \mathscr{M}_3 \cup \mathscr{M}_4 \cup \mathscr{V}$. In this section we consider mixing signatures from $\mathscr{P}_2$ and $\mathscr{M}_4$. It is in this setting that the new planar tractable problems are discovered.

We first do a holographic transformation by Z. Then the problem becomes

$$\text{Pl-Holant}(\neq_2 \mid =_k, \textsc{ExactOne}_d), \tag{9.3}$$

with various arities k and d. Recall that $\textsc{ExactOne}_d$ denotes the exact one function $[0, 1, 0, \ldots, 0]$ of arity d. These are the signatures for Perfect Matching and they are the basic components of matchgates.

Surprisingly, contrary to the putative form of a complexity classification for planar counting problems, the complexity of Pl-Holant($\neq_2 \mid =_k, \textsc{ExactOne}_d$) depends nontrivially on the values of d and k, and the problem is tractable for all large k. Furthermore, the tractable cases of this problem *cannot* be captured by a holographic reduction to Kasteleyn's algorithm, or any other known algorithm. Thus for planar problems the paradigm of holographic algorithms using matchgates (i.e., being $\mathscr{M}$-transformable) is *not* universal.

The problem (9.3) is tractable if $d \leq 2$, since in this case all signatures are in $\mathscr{P}$. Hence we assume $d \geq 3$. Let $\mathcal{EO} = \{\textsc{ExactOne}_d \mid d \geq 3\}$.

9.6.1 Hardness When $k = 3$ or 4

We begin with some hardness results.

Lemma 9.19. Pl-Holant($\neq_2 \mid =_3, \textsc{ExactOne}_3$) *is* #P-*hard.*

Proof. By connecting two copies of $\textsc{ExactOne}_3 = [0, 1, 0, 0]$ together via $(\neq_2)$, we have $\textsc{ExactOne}_4 = [0, 1, 0, 0, 0]$ on the right. Consider the gadget in Figure 9.6a. We assign $(=_3)$ to the triangle vertices, $[0, 1, 0, 0]$ to the circle vertices, $(\neq_2)$ to the square vertices, and $[0, 1, 0, 0, 0]$ on the diamond vertex in the middle. Let f be the signature of this gadget.

We claim that this signature f is 0-1 valued, and its support is $\{0011, 0110, 1100, 1001\}$. To see this, for each $(=_3)$ at the four triangle vertices, we imagine all three adjacent Disequality signatures $(\neq_2)$ are absorbed to $(=_3)$, making it another $(=_3)$ signature. Thus the gadget in Figure 9.6a is functionally equivalent to another virtual gadget Figure 9.6a*, where all square

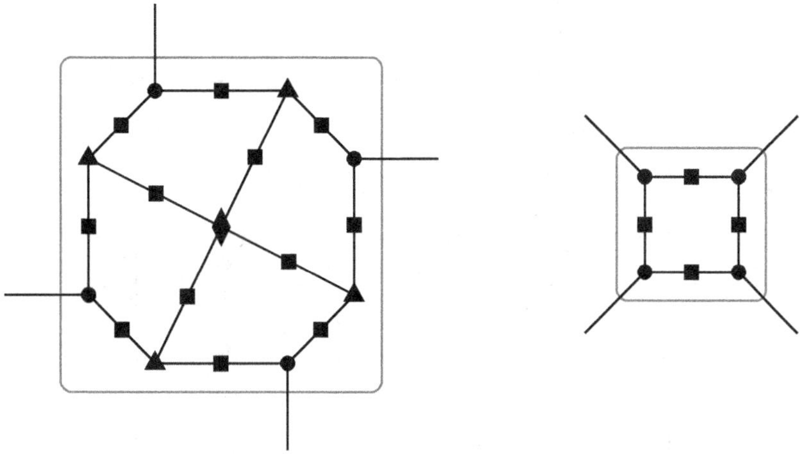

(a) Cycle-like gadget used more than once (b) Gadget to realize $\hat{g}$

Figure 9.6. Two gadgets used in the proof of Lemma 9.19.

vertices are given simply the EQUALITY signature $(=_2)$. Now in Figure 9.6a* the signature $[0, 1, 0, 0, 0]$ in the middle forces exactly one of the four equality signatures (triangles) to take value 1 and the other three equality signatures to take value 0. The two $[0, 1, 0, 0]$'s (circle vertices) adjacent to two equality signatures both assigned 0 must have 1 going out, and the other two $[0, 1, 0, 0]$'s having one adjacent equality assigned 0 and another adjacent equality assigned 1 must have 0 going out.

Now we consider the gadget in Figure 9.6a again. This time we place $[0, 1, 0, 0]$ on each triangle, $(=_3)$ on each circle, f on the middle diamond, and again $(\neq_2)$ on each square. In this gadget notice that each support vector of f makes two $[0, 1, 0, 0]$'s (triangles) that are cyclically adjacent on the outer cycle to become $[0, 1, 0]$ and the other two $[1, 0, 0]$. It is easy to see that the resulting signature is 0-1 valued, and has support $\{0111, 1011, 1101, 1110\}$. Therefore it is the reversed EXACTONE$_4$ signature $[0, 0, 0, 1, 0]$ (namely ALLBUTONE$_4$). The composite gadget as a construction for a signature on the RHS in Pl-Holant$(\neq_2 \mid =_3, \text{EXACTONE}_3)$ is illustrated in Figure 9.7, where each circle is assigned $[0, 1, 0, 0]$, triangle $(=_3)$, and square $(\neq_2)$.

Finally, we build the gadget in Figure 9.6b. We place $(=_3)$ on each circle and $(\neq_2)$ on each square. It is easy to see that there are only two support vectors of the resulting signature, which are 0101 and 1010. Recall the definition (9.1) of the conditional crossover g. This gadget realizes exactly g.

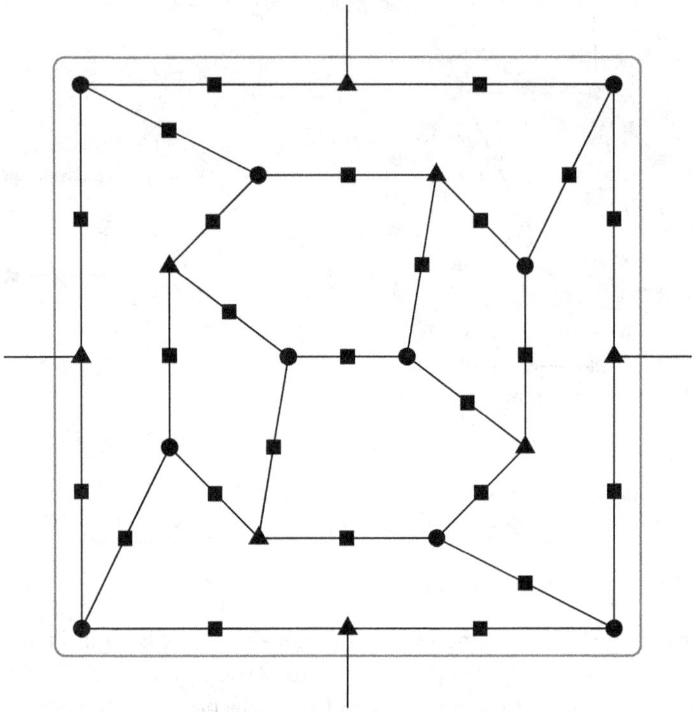

Figure 9.7. The whole gadget to realize [0, 0, 0, 1, 0].

By Lemma 9.18, Pl-Holant($\neq_2$ | [0, 1, 0, 0, 0], [0, 0, 0, 1, 0], g) is #P-hard. We have constructed [0, 1, 0, 0, 0], [0, 0, 0, 1, 0], and g on the right-hand side (RHS). Therefore Pl-Holant($\neq_2$ | $=_3$, [0, 1, 0, 0]) is #P-hard. □

For $k = 4$, we need the following lemma.

Lemma 9.20. *Let f be the arity 4 signature whose matrix is*

$$M_f = \begin{bmatrix} 2 & 0 & 0 & 0 \\ 0 & 1 & 0 & 0 \\ 0 & 0 & 1 & 0 \\ 0 & 0 & 0 & 1 \end{bmatrix}.$$

Then Pl-Holant(f) *is* #P-*hard.*

Proof. Let $h = [2, 1, 1]$. We show that Pl-#CSP(h) $\leq_T$ Pl-Holant(f) in two steps. In each step, we begin with a signature grid and end with a new signature grid such that the Holant values of both signature grids are the same.

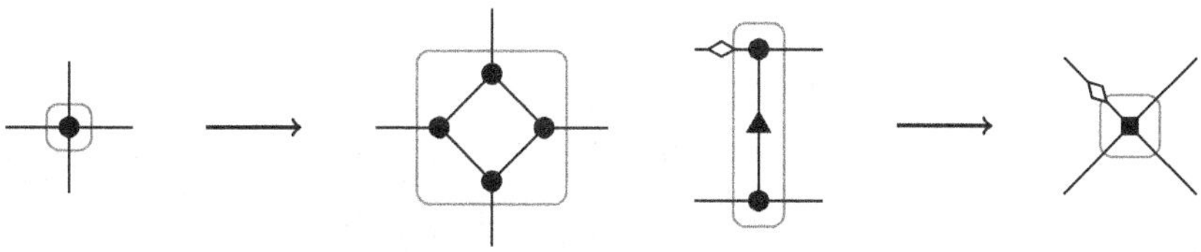

(a) Step one: Degree 4 vertex example (b) Step two: Contract edges

Figure 9.8. A reduction from Pl-Holant($\mathcal{EQ} \mid h$) to Pl-Holant(g) for any binary signature h and a quaternary signature g that depends on h. The circle vertices are assigned $(=_4)$ or $(=_3)$ respectively, the triangle vertex is assigned h, and the square vertex is assigned the signature of the gadget to its left.

Then by Theorem 9.5, we are done. Or more explicitly, since Pl-#CSP(h) $\equiv$ Pl-Holant($\mathcal{EQ} \mid h$), by Theorem 9.4, we are done.

For step 1, let $G = (U, V, E)$ be an instance of Pl-Holant($\mathcal{EQ} \mid h$). Fix an embedding of G in the plane. This defines a cyclic ordering of the edges incident to each vertex. Consider a vertex $u \in U$ of degree k. It is assigned the signature $(=_k)$. We split u into k vertices, and connect the k edges originally incident to u to these k new vertices so that each vertex is incident to exactly one edge. We also connect these k new vertices in a cycle according to the cyclic ordering induced on them by their incident edges. Each of these vertices has degree 3, and we assign them $(=_3)$. Clearly the Holant value is unchanged. This completes step 1. An example of this step applied to a vertex of degree 4 is given in Figure 9.8a. The resulting graph has the following properties: (1) it is planar; (2) every vertex has either degree 2 (in V and assigned h) or degree 3 (newly created and assigned $(=_3)$); (3) each degree 2 vertex is connected to two degree 3 vertices; and (4) each degree 3 vertex is connected to one degree 2 vertex and two other degree 3 vertices.

Now step 2. For every $v \in V$, v has degree 2. We contract the two edges incident to v, or equivalently, we replace the two circle vertices and one triangle vertex boxed in Figure 9.8b with a single (square) vertex of degree 4. The resulting graph $G' = (V', E')$ is planar and 4-regular.

Next we determine the signature on each $v' \in V'$ after this contraction. Note that the two inputs to each original circle have to be the same. Therefore the support of this signature is $\{0000, 0110, 1001, 1111\}$, listed starting from the diamond and going counterclockwise. Moreover, due to the triangle assigned h in the middle, the weight on 0000 is 2, and every other weight is 1. Hence it is exactly the signature f, with the diamond in Figure 9.8b marking the first input bit. This finishes the proof. $\square$

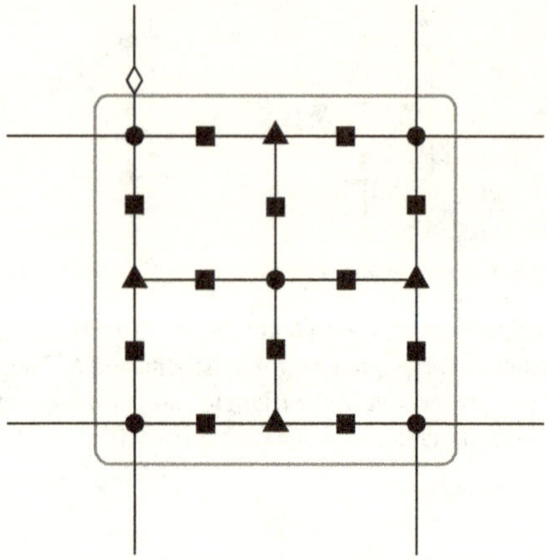

Figure 9.9. Grid-like gadget used in the proof of Lemma 9.21, whose support vectors are 00110011, 11001100, and 11111111. Each square is assigned a binary disequality ($\neq_2$), circle ($=_4$), and triangle [0, 1, 0, 0].

Remark: The two-step construction starting from a plane graph G produces what is known as the medial graph of G. See Definition 6.30 in Section 6.5, and Figure 6.9.

Our next goal is to prove the same result as Lemma 9.19, where ($=_3$) is replaced by ($=_4$). This proof is more challenging, because any gadget construction for the RHS of Pl-Holant($\neq_2 \mid =_4$, ExactOne$_3$) is very "fragile." It is not easy to produce gadgets with signatures that have nontrivial support with interesting nonzero values.

Lemma 9.21. Pl-Holant($\neq_2 \mid =_4$, ExactOne$_3$) *is* #P-*hard.*

Proof. Consider the gadget in Figure 9.9. We assign binary disequality ($\neq_2$) to the square vertices, ($=_4$) to the circle vertices, and ExactOne$_3$ = [0, 1, 0, 0] to the triangle vertices. We show that the resulting signature is 0-1 valued, and its support set is

$$\{00110011, 11001100, 11111111\},$$

where each vector is the assignment ordered counterclockwise starting from the diamond point.

We call the equality signature ($=_4$) in the middle the center vertex. There are two possible assignments at the center vertex. If it is assigned 0, then every adjacent perfect matching signature [0, 1, 0, 0] is matched to the half edge toward

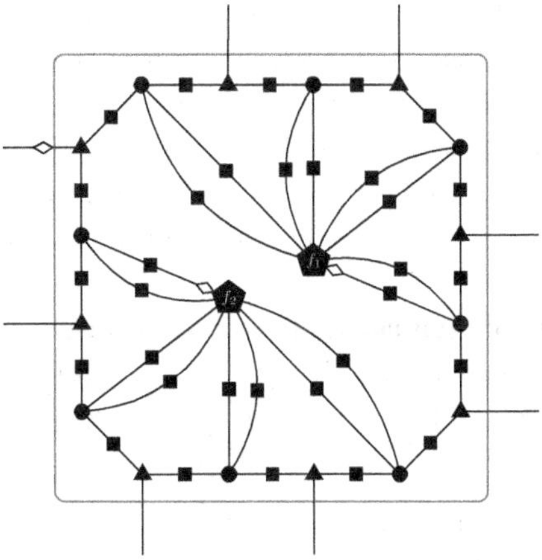

(a) Gadget with signature g. Each square is assigned a binary disequality ($\neq_2$), circle ($=_4$), triangle $[0,1,0,0]$, and pentagon f.

f_1	f_2	g
00000000	00000000	11111111
00110011	00000000	01111000
11001100	00000000	11110000
00000000	00110011	10000111
00110011	00110011	00000000
11001100	00110011	-
00000000	11001100	00001111
00110011	11001100	-
11001100	11001100	00000000

(b) Support of g. Each vector is an assignment ordered counterclockwise from the diamond.

Figure 9.10. Another gadget used in the proof of Lemma 9.21 and a table listing the support of its signature.

the center vertex, and every equality ($=_4$) is forced to be 1. This gives the support vector 11111111.

The other possibility is that the center vertex is 1. In this case, we can remove the center vertex and its four adjacent ($\neq_2$), changing the four $[0, 1, 0, 0]$ (triangles) into $(\neq_2) = [0, 1, 0]$. So the outer cycle effectively becomes a cycle of four equalities ($=_4$) (circle vertices) connected by ($\neq_2$), since a chain of three ($\neq_2$)'s is just a single ($\neq_2$). Hence in this case there are exactly two support vectors, which are 00110011 and 11001100.

Every pair of dangling edges at each corner vertex always takes the same value. We further connect each pair of these edges to a distinct copy of ($=_4$) via two copies of ($\neq_2$). This results in a gadget with a 0-1 valued signature f whose support is formed by taking complement bit patterns of the original support, namely, $\{11001100, 00110011, 00000000\}$.

Now consider the gadget in Figure 9.10a. We assign ($\neq_2$) to the square vertices, ($=_4$) to the circle vertices, $[0, 1, 0, 0]$ to the triangle vertices, and f to the pentagon vertex. In Figure 9.10a, two edges coming out of the pentagon vertex, which connect to the same circle vertex ($=_4$) on the outer cycle via ($\neq_2$), are from the same corner of the gadget in Figure 9.9 used to realize f, and thus must carry equal values 00 or 11. We now study the signature g of arity 8 defined by the gadget Figure 9.10a. The signature g is in the RHS of Pl-Holant($\neq_2 \mid =_4$, ExactOne$_3$).

Notice that if a $(=_4)$ on the outer cycle is assigned 0, then the two adjacent perfect matchings must match half edges toward that $(=_4)$, and their outgoing edges must be 0. Furthermore, the two $(=_4)$ one more step away must be 1. A further observation is that any pair of consecutive $(=_4)$'s on the outer cycle cannot be both 0, and if a pair of consecutive $(=_4)$'s are both 1, then the $[0, 1, 0, 0]$ in the middle must have a 1 going out. In Figure 9.10a, we call f_1 the pentagon connecting to four equalities $(=_4)$ on the upper right part, and call the other one f_2. We further order the external wires of f_1, f_2, and g counterclockwise, each starting from the edge marked with a diamond. With this notation and these observations, we get Table 9.10b listing the support of g. The support of g is

$$\{11111111, 01111000, 11110000, 10000111,$$

$$00000000, 00001111, 00000000\},$$

and 00000000 has multiplicity 2.

Next we use domain pairing. In Pl-Holant$(\neq_2 \mid =_4, \text{EXACTONE}_3, g)$, each $(=_4)$ is connected to four $(\neq_2)$'s on the left-hand side (LHS). We can move $(=_4)$ to the LHS by contracting four $(\neq_2)$'s into it. So we get a reduction Pl-Holant$(=_4 \mid g) \leq_T$ Pl-Holant$(\neq_2 \mid =_4, \text{EXACTONE}_3, g)$. We apply domain pairing on the problem Pl-Holant$(=_4 \mid g)$. Specifically, we can use $(=_4)$ as $(=_2)$ by pairing a pair of edges together, and also pair adjacent two outputs of g counterclockwise, starting from the diamond point. Each pair of output wires of g is connected to a pair of wires from $(=_4)$ on the left hand side. Note that $(=_4)$ enforces that each pair of edges always takes the same value. We reinterpret 00 or 11 as 0 or 1 in the Boolean domain. In this way, we can treat g as an arity 4 signature g' in the Boolean domain. So the reduction is

$$\text{Pl-Holant}(=_2 \mid g') \leq_T \text{Pl-Holant}(=_4 \mid g).$$

We get the expression of g' next. The two support bit strings 01111000 and 10000111 of g are eliminated as they do not agree on adjacent paired outputs. So in the paired (Boolean) domain, the support of g' becomes $\{1111, 1100, 0011, 0000\}$, where 0000 has multiplicity 2. We further rotate g' as a Boolean domain signature such that the support is $\{1111, 0110, 1001, 0000\}$. Now it is easy to see that the matrix of g', an arity 4 signature in the Boolean domain, is

$$M_{g'} = \begin{bmatrix} 2 & 0 & 0 & 0 \\ 0 & 1 & 0 & 0 \\ 0 & 0 & 1 & 0 \\ 0 & 0 & 0 & 1 \end{bmatrix}.$$

By Lemma 9.20 Pl-Holant(g') is #P-hard. Hence Pl-Holant($\neq_2 \mid =_4$, EXACTONE_3) is also #P-hard. □

To extend Lemma 9.19 and Lemma 9.21 to general EXACTONE_d functions, we show that we can always realize the pinning functions $\Delta_0 = [1, 0]$ and $\Delta_1 = [0, 1]$ in this setting.

Lemma 9.22. *For any integer $k \geq 3$ and $d \geq 3$,*

$$\text{Pl-Holant}(\neq_2 \mid =_k, \text{EXACTONE}_d, \Delta_0, \Delta_1)$$
$$\leq_T \text{Pl-Holant}(\neq_2 \mid =_k, \text{EXACTONE}_d).$$

Proof. Given an instance Ω of Pl-Holant($\neq_2 \mid =_k, \text{EXACTONE}_d, \Delta_0, \Delta_1$) with underlying planar bipartite graph G, if any Δ_0 from the RHS is used in Ω, then it is applied to a ($\neq_2$) of the LHS, and can be combined with ($\neq_2$) to become a Δ_1 on the LHS. Then it is applied to a signature of the RHS: If it's Δ_1 it becomes a constant 1 and they can be removed; if it's Δ_0 the Holant value is 0; if it's EXACTONE_d it becomes $\Delta_0^{\otimes(d-1)}$; if it's ($=_k$), it becomes $\Delta_1^{\otimes(k-1)}$. These newly created pinning signatures on the RHS can be applied recursively.

By a similar analysis, the only nontrivial occurrences of Δ_1's in Ω are those attached to EXACTONE_d via ($\neq_2$). We may therefore assume that the only appearances among Δ_0's and Δ_1's are possibly some Δ_1's applied to EXACTONE_d via ($\neq_2$).

We can construct ($=_{\ell k}$) for any integer $\ell \geq 1$, by ($\neq_2$) on the left and ($=_k$) on the right. In fact if we connect two copies of ($=_k$) via ($\neq_2$) on one wire as a planar gadget we get a signature of arity $2k - 2$ with $k - 1$ consecutive external wires labeled "+" and the others labeled "−," where wires of the same label take equal values, but oppositely labeled wires ("+−") take opposite values. As $k \geq 3$, we can take two wires of the $k - 1$ wires labeled "−" and attach to two copies of ($=_k$) via two ($\neq_2$). This creates a signature of arity $4k - 6$ with $3(k - 1)$ consecutive wires labeled "+" and the other $k - 3$ wires labeled "−". Finally connect $k - 3$ pairs of adjacent ("+−") labeled wires by ($\neq_2$) recursively. This creates a planar gadget with an equality signature of arity $2k$. This can be extended to any ($=_{\ell k}$) by applying the same process on any consecutive k wires.

Next we construct $\Delta_1^{\otimes r}$ in the RHS, for some integer $r \geq 1$. We get $\Delta_0^{\otimes(d-2)}$ by a self-loop of EXACTONE_d via ($\neq_2$), ignoring the factor 2. We pick an integer ℓ large enough so that $d - 2 < \ell k$. Then we connect $\Delta_0^{\otimes(d-2)}$ to ($=_{\ell k}$) via ($\neq_2$) to get $\Delta_1^{\otimes(\ell k-d+2)}$. Then we take $r = \ell k - d + 2$.

One more construction we will use is $\textsc{ExactOne}_{2+\ell(d-2)}$ for any integer $\ell \geq 1$. This is realizable by connecting ℓ many copies of $\textsc{ExactOne}_d$ sequentially via $(\neq_2)$.

Consider the dual graph G^* of G. Take a spanning tree T of G^*, with the external face as the root. In each face F, let c_F be the number of Δ_1's in the face. We start from the leaves to recursively move all the pinnings of Δ_1 to the external face. Suppose the face F is a current leaf of T. If $c_F = 0$ then we just remove the leaf from T and recurse on another leaf. Otherwise we temporarily remove all Δ_1's in F. Let s be the smallest integer such that $sr \geq c_F$. We replace the $(\neq_2)$ edge bordering between F and its parent F' by a path with three signatures: $(\neq_2)$, $\textsc{ExactOne}_{2+\ell(d-2)}$ and $(\neq_2)$, where ℓ is large such that $\ell(d-2) \geq sr - c_F$. The bipartiteness is preserved with the two $(\neq_2)$'s from the LHS and the $\textsc{ExactOne}_{2+\ell(d-2)}$ from the RHS. Two edges of $\textsc{ExactOne}_{2+\ell(d-2)}$ are connected to the two adjacent copies of $(\neq_2)$ on the path. Of the other $\ell(d-2)$ edges we will put $sr - c_F$ many dangling edges in F, and the remaining $\ell(d-2) - (sr - c_F)$ dangling edges in F'. Hence there are sr dangling edges in F, including those c_F many dangling edges that were connected to Δ_1's via $(\neq_2)$ before we removed them. We put s copies of $\Delta_1^{\otimes r}$ inside the face F to pin all of them in a planar way. Since this $\Delta_1^{\otimes r}$ is from the RHS, it is connected via $(\neq_2)$, and so the value being pinned are all 0, including its effect on $\textsc{ExactOne}_{2+\ell(d-2)}$. In the end all $\ell(d-2)$ dangling edges of $\textsc{ExactOne}_{2+\ell(d-2)}$ in both F and F' will be pinned to 0, thus its effect on the path $(\neq_2)$, $\textsc{ExactOne}_{2+\ell(d-2)}$ and $(\neq_2)$, is equivalent to a path of three $(\neq_2)$'s, thus equivalent to the original single $(\neq_2)$. To complete the description of the construction at face F, we add $\ell(d-2) - (sr - c_F)$ to $c_{F'}$, and they are all connected to Δ_1 via $(\neq_2)$. Remove the leaf F from T, and recurse.

After the process, all Δ_1's are in the external face of G. Suppose the number is p. We may use $\Delta_1^{\otimes r}$ to reduce this number to $0 \leq p < r$. We put r disjoint copies of G together to form a planar signature grid. Use p copies of $\Delta_1^{\otimes r}$ in the outer face in a planar way, making available a total of pr many Δ_1's. This is now an instance of Pl-Holant$(\neq_2 \mid =_k, \textsc{ExactOne}_d)$ and the Holant value is the rth power of that of Ω. Since the Holant value of Ω is a nonnegative integer, we can take the rth root and finish the reduction. □

Once we have the pinning functions, it is easy to construct $\textsc{ExactOne}_3$ from $\textsc{ExactOne}_d$. Therefore combining Lemmas 9.22 with 9.19 and 9.21 we get the following.

Corollary 9.23. *If $d \geq 3$ and $k \in \{3, 4\}$, then* Pl-Holant$(\neq_2 \mid =_k, \textsc{ExactOne}_d)$ *is #P-hard.*

The implication of Corollary 9.23 for counting planar hypergraph perfect matchings will be discussed in Theorem 9.28.

9.6.2 Tractability When $k \geq 5$

On the other hand, if the arity k of the equality signature is ≥ 5, then the problem Pl-Holant($\neq_2 \mid =_k, \mathcal{EO}$) is tractable. In this subsection we will first prove that the problem is tractable for $k \geq 6$. After that we will return to $(=_5)$.

In fact we consider the problem Pl-Holant($\neq_2 \mid =_k, \mathcal{EO}, \neq_2, \Delta_0, \Delta_1$). Note that $(\neq_2) = [0, 1, 0] = \text{EXACTONE}_2$ and $\Delta_1 = [0, 1] = \text{EXACTONE}_1$. Given an instance of the problem with underlying graph G, we can do some preprocessing as follows. Any occurrence of $(\neq_2)$ on the RHS is connected to two copies of $(\neq_2)$ on the LHS. We can replace the three $(\neq_2)$'s by a single $(\neq_2)$ on the LHS, thus eliminating the $(\neq_2)$ on the RHS. Any pinning signature Δ_0 or Δ_1 can be applied and possibly propagated. Any Δ_0 on the RHS is first transformed to Δ_1 by $(\neq_2)$ on the LHS and then applied to a signature on the RHS. If it is applied to Δ_1 then it produces a constant 1, and can be eliminated. If it is applied to Δ_0 then the Holant value is 0 and we are done. If it is applied to $(=_k)$ then this $(=_k)$ is changed to $\Delta_1^{\otimes(k-1)}$, which can be recursively applied. If it is applied to EXACTONE_d (for some $d \geq 3$) then this EXACTONE_d becomes $\Delta_0^{\otimes(d-1)}$, which can be recursively applied. The case with any occurrence of Δ_1 on the RHS is similar. It is first transformed to Δ_0 on the LHS. Then if it is applied to Δ_0 or Δ_1 on the RHS, the case is simple. Applying to $(=_k)$ produces $\Delta_0^{\otimes(k-1)}$, which can be recursively applied. Applying Δ_0 to EXACTONE_d (for some $d \geq 3$) produces EXACTONE_{d-1}, which is either in $\mathcal{EO}$ (if $d > 3$) or $(\neq_2)$ which can be eliminated.

Hence we may consider the instance is from Pl-Holant($\neq_2 \mid =_k, \mathcal{EO}$).

Any self-loop on $(=_k)$ by $(\neq_2)$ gives the Holant value 0. Any self-loop on an EXACTONE_d by $(\neq_2)$ changes it to a $\Delta_0^{\otimes(d-2)}$ with factor 2. These pinning signatures can be applied recursively. Whenever an EXACTONE_d and another EXACTONE_ℓ are connected by a $(\neq_2)$, we replace it by a single $\text{EXACTONE}_{d+\ell-2}$, shrinking the edge between (and remove the connecting $(\neq_2)$). (If there are $m > 1$ edges connecting them by $(\neq_2)$, then $m - 1$ self-loops result, which either lead to pinning signatures or Holant value 0 if $m > 2$.) On the other hand, consider a connected component made of $(=_k)$ and $(\neq_2)$. We call such a component an E_k-block. An E_k-block has either exactly two or zero support vectors. This depends on whether or not there exists a contradiction, which is formed by an odd cycle of $(=_k)$ connected by $(\neq_2)$. We say an E_k-block is trivial if it has no support. An E_k-block is nontrivial iff the graph with $(=_k)$ as vertices and $(\neq_2)$ as edges is bipartite. This is easy to check in

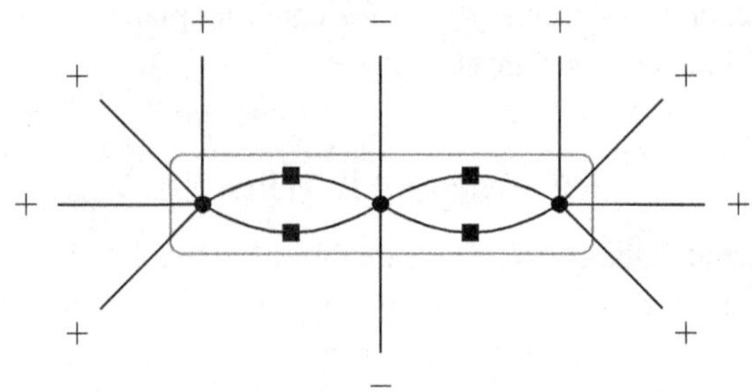

Figure 9.11. Example E_6-block. Circle vertices are assigned $(=_6)$ and square vertices are assigned $(\neq_2)$.

polynomial time, and if so we can find the two support assignments of a nontrivial E_k-block that are complements of each other. We mark dangling edges of a nontrivial E_k-block by "+" or "−" signs. Dangling edges marked with the same sign take the same value on both support assignments while dangling edges marked with different signs take opposite values on both support assignments. Let n_+ and n_- denote the number of dangling edges marked by $+$ and $-$ respectively. Then it is easy to show by induction that

$$n_+ \equiv n_- \pmod{k} \tag{9.4}$$

An example of an E_6-block is illustrated in Figure 9.11, with 8 "+" signs and 2 "−" signs.

Exercise: Prove (9.4).

Exercise: Prove that for any n_+ and $n_- \geq 0$ such that $n_+ + n_- = n$ and $n_+ \equiv n_- \mod k$, if $b_1 b_2 \ldots b_n \in \{+, -\}^n$ is a string with n_+ many $+$'s and n_- many $-$'s, then there is a linear sized E_k-block such that all its dangling edges are on the outer face, and has labels $b_1 b_2 \ldots b_n$ in cyclic order. (Since the string is arbitrary, the statements for clockwise cyclic order and counterclockwise cyclic order are equivalent.)

After contracting all edges between EXACTONE_d's and forming E_k-block's we obtain a bipartite graph G' connected between EXACTONE_d's and E_k-block's by edges labeled by $(=_2)$. We claim that G' is planar. Note that even though the given graph G is planar, and contracting edges is a planar operation, the planarity of the (abstract) graph G' where E_k-block's and (contracted) EXACTONE_d's are vertices requires a justification. In particular, an E_k-block B in G is a connected planar graph that may have dangling edges in several distinct faces, and the topological structure between the E_k-block's and EXACTONE_d's in G can be arbitrarily nested and complicated. Each E_k-block has one outer face and possibly several internal faces, each of which may have some dangling edges connected to EXACTONE_d's. To show planarity of G', we order the dangling edges on the outer face F of B counterclockwise (if any), and order the dangling edges on each internal face (if any) $F', F'', \ldots,$ clockwise, and then concatenate their labels into a single string σ of $+$'s and $-$'s. If an internal face F' with dangling edges exists, we will simply move the part within F' connected to these dangling edges to the outer face. Then we replace this E_k-block by a construction (see Exercise above) that has all dangling edges ordered cyclically according to σ and placed on the outer face. After the replacement we reconnect all dangling edges, now all on the outer face. This way we maintain planarity, and the replacement E_k-block can be contracted to a single point. After replacing every E_k-block by an equivalent E_k-block the planarity of the resulting graph shows that G' is also planar.

Now a key observation is that if a planar (bipartite) graph has moderately high degrees, then it cannot be a simple graph, i.e., it must have parallel edges.

Lemma 9.24. *Let $G = (L \cup R, E)$ be a planar bipartite graph with parts L and R. If every vertex in L has degree ≥ 6 and every vertex in R has degree ≥ 3, then G is not simple.*

Proof. Suppose G is simple. Let V, E, and F be the total number of vertices, edges, and faces, respectively. Let v_i are the number of vertices of degree i in L, where $i \geq 6$, and let u_j be the number of vertices of degree j in R, where $j \geq 3$. Since G is simple and bipartite, each face has at least four edges. Thus,

$$2\mathsf{E} \geq 4\mathsf{F}. \tag{9.5}$$

Furthermore, it is easy to see that

$$\mathsf{V} = \sum_{i\geq 6} v_i + \sum_{j\geq 3} u_j \quad \text{and} \quad \mathsf{E} = \sum_{i\geq 6} i v_i = \sum_{j\geq 3} j u_j. \tag{9.6}$$

Then starting from Euler's characteristic formula for planar graphs, we have

$$\begin{aligned} 2 &= \mathsf{V} - \mathsf{E} + \mathsf{F} \\ &\leq \mathsf{V} - \frac{\mathsf{E}}{2} && \text{(By (9.5))} \\ &= \sum_{i\geq 6} v_i + \sum_{j\geq 3} u_j - \frac{1}{6}\sum_{i\geq 6} i v_i - \frac{1}{3}\sum_{j\geq 3} j u_j && \text{(By (9.6))} \\ &= \sum_{i\geq 6} \frac{6-i}{6} v_i + \sum_{j\geq 3} \frac{3-j}{3} u_j \leq 0, \end{aligned}$$

a contradiction. □

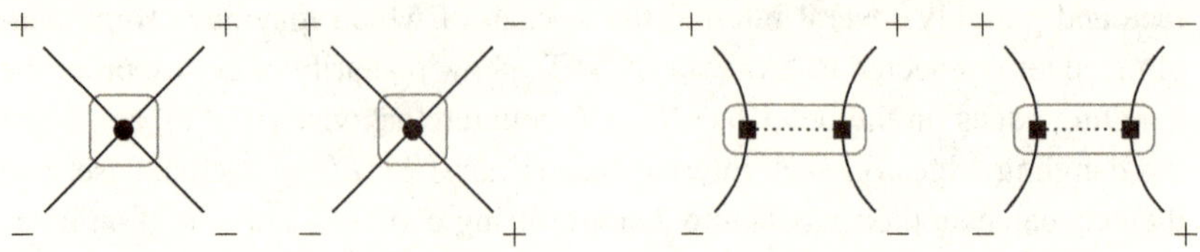

(a) Two different arity 4 E_k-blocks. (b) Replace them by parallel ($\neq_2$)'s.

Figure 9.12. Arity 4 E_k-blocks.

Lemma 9.24 does not give us tractability for the case of $k \geq 6$ yet. The reason is that, after preprocessing, given an instance of Pl-Holant($\neq_2 \mid =_k, \mathcal{EO}$), we form E_k-blocks to make the graph bipartite. It is possible to have E_k-blocks of arity < 6. If so Lemma 9.24 does not apply. However, for $k \geq 6$ and a nontrivial E_k-block of arity n where $n < 6$, by (9.4) and the fact that $0 \leq n_+, n_- \leq n < k$, we see that $n_+ = n_-$, and $n = n_+ + n_-$ must be even. Moreover, if $n = 2$, then this means that the E_k-block is just ($\neq_2$), in which case we can replace it by a single ($\neq_2$) connecting signatures from $\mathcal{EO}$. If this ($\neq_2$) connects two different signatures from $\mathcal{EO}$ then it produces a new EXACTONE signature in $\mathcal{EO}$. If it connects the same signature from $\mathcal{EO}$, this produces pinning Δ_0's and can be handled as before. So the only problematic case is when $n = 4$. We identify two possibilities of such E_k-blocks up to a rotation in Figure 9.12a.

We may replace any E_k-block of arity 4 by an equivalent E_k-block of arity 4 where all dangling edges are on the outer face, as depicted in Figure 9.12a. Functionally an E_k-block of arity 4 can be viewed as a pair of ($\neq_2$) in parallel, but there is a correlation between them, namely their support vectors are paired up in a unique way. If we replace the contracted E_k-block of arity 4 by two parallel edges with ($\neq_2$) as indicated in Fig. 9.12b, we maintain planarity (but we cannot claim it maintains functional equivalence due to the nontrivial correlation.)

We will show in the following lemma how to replace E_k-block of arity 4 by two parallel edges with ($\neq_2$) while keeping track of the Holant value.

Lemma 9.25. *For any integer* $k \geq 6$, Pl-Holant($\neq_2 \mid =_k, \mathcal{EO}, \neq_2, \Delta_0, \Delta_1$) *is tractable.*

Proof. Let Ω be an instance of Pl-Holant($\neq_2 \mid =_k, \mathcal{EO}, \neq_2, \Delta_0, \Delta_1$). Without loss of generality, we assume that Ω is connected. We already discussed the preprocessing so that we may assume Ω is an instance of Pl-Holant($\neq_2 \mid =_k, \mathcal{EO}$).

We call an edge pinned if it has the same value in all satisfying assignments, *if there is any*. When an edge is pinned to a known value, we can get a smaller instance of the problem Pl-Holant($\neq_2 \mid =_k, \mathcal{EO}$) without changing the number of satisfying assignments. In our algorithm we may also find a contradiction and simply return 0. If e is a pinned edge, then it is adjacent to another edge e' via $(\neq_2)$ on the LHS, and both e and e' are pinned. We remove e, e', and $(\neq_2)$, and perform the following on e (and on e' as well). If the other endpoint of e in the RHS is $(=_k)$, then all edges incident to this $(=_k)$ are pinned to the same value which we can recursively apply. If the other endpoint of e is $\text{EXACTONE}_d \in \mathcal{EO}$, then we replace this signature by EXACTONE_{d-1} if the pinned value is 0; or if the pinned value is 1 then the remaining $d-1$ edges of this EXACTONE_d are pinned to 0 which we recursively apply. Notice that we may create an EXACTONE_2 (i.e., $(\neq_2)$) on the RHS when we pin 0 on EXACTONE_3. Such $(\neq_2)$'s are replaced as described before. It is easy to see that all these procedures do not change the number of satisfying assignments, and work in polynomial time.

Claim: There always exists an edge in Ω that is pinned, unless Ω does not contain $(=_k)$, or does not contain EXACTONE_d functions (for some $d \geq 3$), or there is a contradiction. Furthermore, if there are $(=_k)$ and EXACTONE_d functions (for some $d \geq 3$), in polynomial time we can find a pinned edge with a known value, or return that there is a contradiction. (If there is a contradiction in Ω, we may still return a purported pinned edge with a known value, which we can apply and simplify Ω. The contradiction will eventually be found.)

Note that if Ω does not contain $(=_k)$, or does not contain EXACTONE_d functions (for some $d \geq 3$), then the problem is tractable, since Ω is an instance of $\mathscr{M}$, or an instance of $\mathscr{P}$, respectively. The lemma follows from the claim, for we either recurse on a smaller instance or have a tractable instance.

Suppose Ω is an instance where at least one $(=_k)$ and at least one $\text{EXACTONE}_d \in \mathcal{EO}$ appear. Let's examine how the signatures can be connected by $(\neq_2)$. Clearly we may assume there is no self-loop on any $(=_k)$; otherwise we return the Holant value 0. If a signature $\text{EXACTONE}_d \in \mathcal{EO}$ is connected to itself by a self-loop through a $(\neq_2)$, then there are two choices for the assignment on this pair of edges through the $(\neq_2)$, but the remaining $d-2 \geq 1$ edges are pinned to 0. We can keep track of the factor 2 and have found a pinned edge with a known value. Thus we may assume there are no self-loops via $(\neq_2)$ on EXACTONE signatures.

Next we consider the case that two separate signatures EXACTONE_d and EXACTONE_ℓ from $\mathcal{EO}$ are connected by some number of $(\neq_2)$'s. Depending on the number of connecting edges, there are three cases:

1. The connection is by a single $(\neq_2)$. We contract the connecting edge, maintaining planarity, and replace these three signatures by an $\text{EXACTONE}_{d+\ell-2}$

to get a new instance Ω'. If an edge is pinned in Ω' then it is also pinned in Ω to the same value. We continue with Ω'.

2. The connection is by two $(\neq_2)$'s. There are two choices for the assignment on these two pairs of edges through $(\neq_2)$, but the remaining $d + \ell - 4 \geq 2$ edges are pinned to 0.
3. The connection is by at least three $(\neq_2)$'s. The three $(\neq_2)$'s cannot be all satisfied, so there is no satisfying assignment, a contradiction. We return the value 0.

Hence, we may assume there is no connection via any number of $(\neq_2)$'s among ExactOne signatures in $\mathcal{EO}$.

Now we consider connections between various $(=_k)$'s by $(\neq_2)$. Define an E_k-block as a connected component composed of $(=_k)$ and $(\neq_2)$. All dangling edges of each E_k-block are marked with $+$ or $-$ and this can be found by testing bipartiteness of an E_k-block where we treat $(=_k)$'s as vertices and $(\neq_2)$'s as edges. If any E_k-block is not bipartite, we have found a trivial E_k-block, and we return 0. Thus we may assume all E_k-blocks are nontrivial. We replace all E_k-blocks by equivalent E_k-blocks with all dangling edges on the outer face, and maintain planarity. If there is a nontrivial E_k-block of arity 2, by (9.4) its signature is $(\neq_2)$. We replace it with an edge labeled by $(\neq_2)$ to form an instance Ω', maintaining planarity, such that any pinned edge in Ω' corresponds to a pinned edge in Ω. So we may assume that the arity of any E_k-block is ≥ 4. Since $k \geq 6$, the only possible E_k-blocks of arity 4 functionally are those in Figure 9.12a up to a rotation. Since there is at least one ExactOne$_d$ signature with $d \geq 3$, forming E_k-blocks does not consume all of Ω.

After these steps we may consider Ω a planar bipartite graph, with one side consisting of E_k-blocks of arity ≥ 4 and the other side ExactOne signatures in $\mathcal{EO}$. And they are now connected by edges labeled by $(=_2)$.

Suppose there are parallel edges between an E_k-block and an ExactOne$_d$ signature. We show that this always leads to some pinned edges. If two parallel edges are marked by the same sign in the E_k-block, then they must be pinned to 0. If they are marked by different signs, then the remaining $d - 2 \geq 1$ edges of the ExactOne$_d$ signature must be pinned to 0. Therefore, we may assume there are no parallel edges between any E_k-block and any ExactOne signature.

The next thing we do is to consider E_k-blocks of arity 4 with ExactOne signatures together. Call a connected component consisting of E_k-blocks of arity 4 and ExactOne signatures an *EO-Eq*-4-block. Figure 9.13a illustrates an example. As remarked earlier, the two possibilities of E_k-blocks of arity 4 can be viewed as two parallel $(\neq_2)$'s but with some correlation between them.

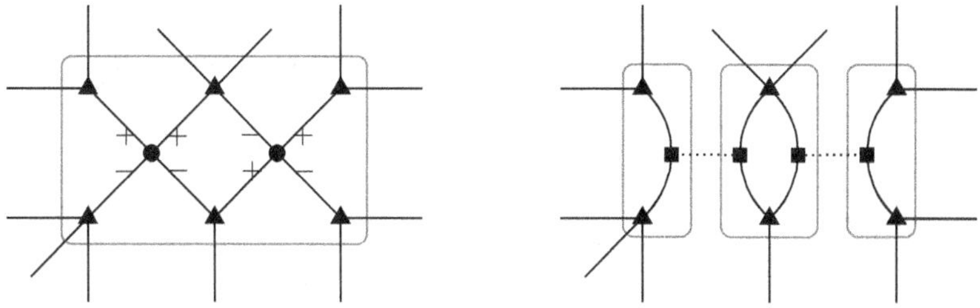

(a) An *EO-Eq*-4-block. Triangles are assigned ExactOne signatures and circles are E_k-blocks of arity 4.

(b) Break the *EO-Eq*-4-block into three components. Squares are assigned $(\neq_2)$. The component in the middle contains a cycle, and hence is degenerate. The other two are equivalent to ExactOne signatures.

Figure 9.13. *EO-Eq*-4-blocks.

This is illustrated in Figure 9.12b. Note that the two dotted lines in Figure 9.12b represent different correlations.

At this point we would like to replace every arity 4 E_k-block by two parallel $(\neq_2)$'s. However, this replacement destroys the equivalence of the Holant values, before and after.

The *surprising* move of this proof is that we shall do so anyway!

Suppose we ignore the correlation for the time being and replace every arity 4 E_k-block by two parallel $(\neq_2)$'s as in Figure 9.12b. This replacement produces a *planar* signature grid Ω_1. Every edge in Ω_1 corresponds to a unique edge in Ω. The set of satisfying assignments of Ω_1 is a superset of that of Ω. Moreover, if there is an edge pinned in Ω_1 to a known value, the corresponding edge is also pinned in Ω to the same value. Once we find that in Ω_1 we revert back to work in Ω and apply the pinning to the pinned edge.

All that remains to be shown is that pinning always happens in Ω_1. Each *EO-Eq*-4-block splits into some number of connected components in Ω_1. If any component contains a cycle, it must alternate between $(\neq_2)$, which are the newly created ones from the E_k-blocks of arity 4, and ExactOne$_d$ signatures for $d \geq 3$. Note that ExactOne$_d$ signatures on the cycle all have arity ≥ 3. The cycle has even length, and there are exactly two satisfying assignments, which assign exactly one 0 and one 1 to the two cycle edges incident to each ExactOne$_d$ signature. Hence any edge not on the cycle but incident to some vertex (with a signature from $\mathcal{EO}$) belonging to the cycle is pinned to 0. Moreover such edges must exist by the arity of ExactOne$_d$ signatures. This produces pinned edges.

Hence we may assume there are no cycles in these components. If we consider ExactOne functions as vertices, and $(\neq_2)$'s as edges, each component

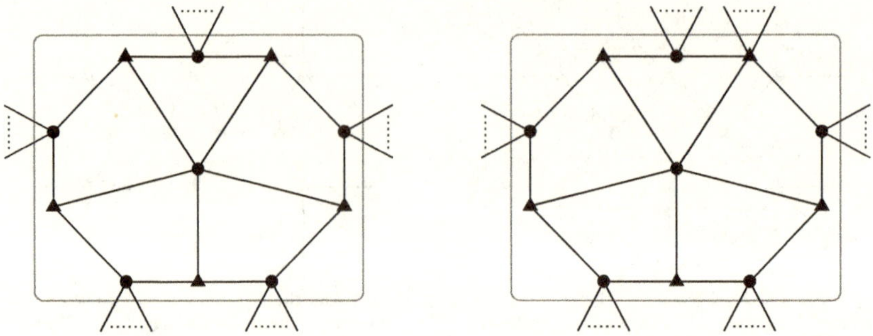

(a) Type 1 (b) Type 2

Figure 9.14. Two types of wheels. Each circle is an E_5-block and triangle an EXACTONE signature.

forms a tree. Suppose there are $n \geq 2$ vertices in such a tree. As discussed in item 1 above, the whole tree is an ExactOne_t function for some arity t. Since each vertex in the tree has degree ≥ 3, $t \geq 3n - 2(n-1) = n + 2 \geq 4$. We replace these components by ExactOne_t's. Since each connected component is a tree, we can contract it to a point, labeled by some $\text{ExactOne}_t \in \mathcal{EO}$, and maintain planarity. This defines a planar signature grid Ω_2.

Thus, each connected component in the graph underlying Ω_2 is a planar bipartite graph with E_k-blocks of arity ≥ 6 on one side and ExactOne_d signatures of arity ≥ 3 on the other. By Lemma 9.24, no component is simple, which means that there are parallel edges between some E_k-block and some ExactOne_d signature. As discussed earlier, there must exist some pinned edge, and we can find a pinned edge with a known value in polynomial time. This gives a pinned edge with a known value in Ω_1, and then in Ω. This finishes the proof. □

Now we go on to consider Pl-Holant($\neq_2 \mid =_5, \mathcal{EO}$). Unlike the situation in Lemma 9.24, a planar (5, 3)-regular bipartite graph *can* be simple. However, we show that any *simple* bipartite graph where the degrees are ≥ 5 from the left and ≥ 3 from the right must have a special induced subgraph. We call this structure a "wheel," which is pictured in Figure 9.14. There is a vertex v of degree 5 in the middle, and all faces adjacent to this vertex are 4-gons (i.e., quadrilaterals). Moreover, at least four neighbors of v have degree 3. Depending on the degree of the fifth neighbor (whether it is 3 or not), we have two types of wheel, which are pictured in Figures 9.14a and 9.14b.

Lemma 9.26. *Let $G = (L \cup R, E)$ be a planar bipartite graph with parts L and R. Every vertex in L has degree ≥ 5 and every vertex in R has degree ≥ 3.*

If G is simple, then there exists one of the two wheel structures in Figure 9.14 in G as an induced subgraph.

Proof. Let $V = L \cup R$ be the set of vertices and let F be the set of faces. We assign a *score* s_v to each vertex $v \in V$. We will define s_v so that $\sum_{v \in V} s_v = |V| - |E| + |F| = 2 > 0$. The score s_v will be the sum of values assigned to v locally. These values include a base score for the vertex v itself, and values from the $\deg(v)$ edges incident to v, and from the $\deg(v)$ faces containing v. The base score is $+1$ for each vertex, which accounts for $|V|$. For each k-gon face, we assign $\frac{1}{k}$ to each of its vertex. This accounts for $|F|$. As G is a simple bipartite graph, $k \geq 4$ and a score coming from a face to a vertex is at most $\frac{1}{4}$.

For $-|E|$, we separate two cases. For any edge if one of the two endpoints has degree 3, we give the degree 3 vertex a score of $-\frac{7}{12}$, and the other one $-\frac{5}{12}$. This is well defined because all degree 3 vertices are in R. If the endpoints are not of degree 3, we give each endpoint $-\frac{1}{2}$. This accounts for $-|E|$.

Claim 1: For any $v \in V$, $s_v \leq 0$ unless $v \in L$ and has degree 5.

To prove this claim, suppose $v \in L$ and has degree $d \geq 6$, then

$$s_v \leq 1 + \frac{d}{4} - \frac{5}{12}d = 1 - \frac{d}{6} \leq 0.$$

Now suppose $v \in R$ and v has degree $d \geq 4$. Then every edge adjacent to v gives a score $-\frac{1}{2}$. Hence,

$$s_v \leq 1 + \frac{d}{4} - \frac{1}{2}d = 1 - \frac{d}{4} \leq 0.$$

The remaining case is that $v \in R$ and v has degree 3. Then,

$$s_v \leq 1 + \frac{d}{4} - \frac{7}{12}d = 1 - \frac{d}{3} \leq 0.$$

Claim 1 is proved.

Since the total score is positive, there must exist $v \in L$, such that $\deg(v) = 5$ and $s_v > 0$.

Claim 2: There exists $v \in L$, such that $\deg(v) = 5$, $s_v > 0$, and all adjacent faces are 4-gons.

Suppose otherwise. Then any $v \in L$ with $\deg(v) = 5$ and $s_v > 0$ is adjacent to at least one k-gon for some $k \geq 6$. In this case,

$$s_v \leq 1 + \frac{1}{4} \cdot 4 + \frac{1}{6} - \frac{5}{12} \cdot 5 = \frac{1}{12}.$$

Moreover, if v is adjacent to more than one k-gon with $k \geq 6$, then

$$s_v \leq 1 + \frac{1}{4} \cdot 3 + \frac{1}{6} \cdot 2 - \frac{5}{12} \cdot 5 = 0,$$

contrary to the assumption that $s_v > 0$. Hence v is adjacent to exactly one k-gon with $k \geq 6$. Call this face F_v.

Along the boundary of F_v, v has two neighbors in R. We match each vertex v that has a positive score to the vertex that is on the boundary of F_v and next to v in clockwise order. By bipartiteness, v is matched to a vertex in R. We do this matching in all faces containing at least one positively scored vertex. It is possible that more than one such v are matched to the same $u \in R$. Suppose a vertex $u \in R$ is matched to from ℓ different such vertices of positive score. This means that u is adjacent to at least ℓ many k-gons with $k \geq 6$. The total number of faces that contain u is exactly $\deg(u)$ and therefore $\ell \leq \deg(u)$. Thus, if u has degree 3 then u has score

$$s_u \leq 1 + \frac{1}{4} \cdot (3 - \ell) + \frac{1}{6} \cdot \ell - \frac{7}{12} \cdot 3 = -\frac{\ell}{12}.$$

If u has degree $d \geq 4$ then u has score

$$s_u \leq 1 + \frac{1}{4} \cdot (d - \ell) + \frac{1}{6} \cdot \ell - \frac{1}{2} \cdot d \leq -\frac{\ell}{12}.$$

Hence in any case, we have $s_u \leq -\frac{\ell}{12}$. It implies that the sum of s_u and the scores of all positively scored vertices matched to u is ≤ 0. However each positively scored vertex is matched to a vertex in R. Hence the sum of all scores cannot be positive. This is a contradiction.

Therefore there exists $v \in L$ such that $s_v > 0$, $\deg(v) = 5$, and all adjacent faces are 4-gons. This proves claim 2.

We further note that at most one neighbor of v can have degree ≥ 4, for otherwise,

$$s_v \leq 1 + \frac{5}{4} - \frac{1}{2} \cdot 2 - \frac{5}{12} \cdot 3 = 0.$$

If all five neighbors of v have degree 3, that is a wheel of type 1 as in Figure 9.14a. If exactly one neighbor of v has degree ≥ 4, and the other four neighbors all have degree 3, that is a wheel of type 2 as in Figure 9.14b. □

As we shall see, either structure in Figure 9.14 leads to pinned edges.

Lemma 9.27. Pl-Holant$(\neq_2 \mid =_5, \mathcal{EO}, \neq_2, \Delta_0, \Delta_1)$ *is tractable.*

Proof. We proceed as in Lemma 9.25 up until the point of getting Ω_2. Note that due to (9.4) the only nontrivial E_5-blocks of arity ≤ 4 are $(\neq_2)$ and those

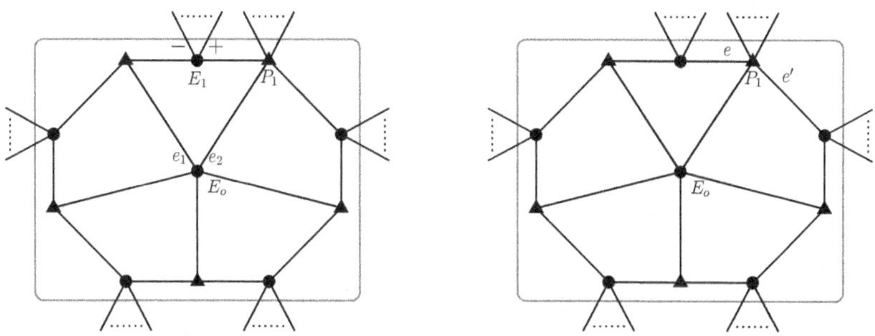

(a) Different signs of an E_5-block along the cycle lead to pinning

(b) Edges e and e' are pinned in wheels of type 2

Figure 9.15. Degeneracies in the wheel structure.

in Figure 9.12a, so that all the steps and transformations from Ω to Ω_1 and to Ω work exactly as in Lemma 9.25. Now each connected component of Ω_2 is planar and bipartite with vertices on one side having degree ≥ 5 and those on the other side ≥ 3. If it has some parallel edges, then we can find a pinned edge just as in Lemma 9.25. So we may assume it has no parallel edges. We only need to show that there are pinned edges in Ω_2.

Therefore each connected component of Ω_2 satisfies the condition of Lemma 9.26. By Lemma 9.26, the wheel structure in Figure 9.14 appears. All we need to show is that wheel structures of either type contain pinned edges.

First we claim that if a wheel of either type has a E_5-block, call it E_1, on the outer cycle which has different signs on the two edges incident to it along the cycle, then the middle $(=_5)$, denoted by E_o, is pinned. This is pictured in Figure 9.15a. It does not matter whether the wheel is type 1 or 2, or the position of E_1 relative to the special triangle P_1 in type 2. Because E_o is an equality, both e_1 and e_2, the two edges incident to E_o that are connected to the two ExactOne signatures flanking E_1, must take the same value. If both e_1 and e_2 are assigned 1, then the two incoming wires of E_1 along the cycle have to be both assigned 0, whereas they are marked by different signs. This is a contradiction. Hence both e_1 and e_2 are pinned to 0 as well as all edges of E_o.

We may therefore assume that each E_5-block has same signs along the outer cycle, either $++$ or $--$. If the wheel is of type 1, then there is no valid assignment such that E_o is assigned 0 because the cycle has odd length. In fact if E_o is assigned 0, then we can remove E_o and its incident edges, and effectively the five ExactOne signatures are now $(\neq_2)$'s forming a 5-cycle linked by binary equalities. Hence all incident edges of E_o are pinned to 1.

Otherwise the wheel is of type 2, and each E_5-block has signs $++$ or $--$ along the outer cycle. We denote by P_1 the special ExactOne$_d$ function that has

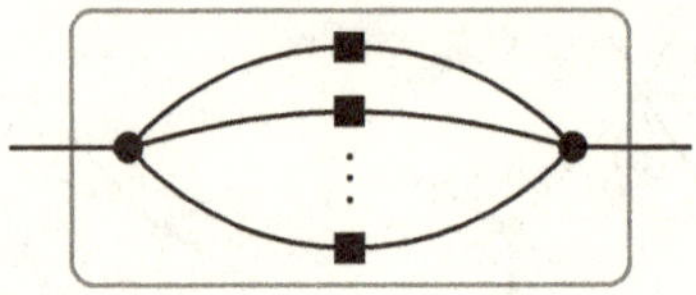

Figure 9.16. A binary gadget with $(=_r)$ assigned to circle vertices and $(\neq_2)$ assigned to square vertices.

arity $d > 3$. We claim that the two edges e and e' incident to P_1 along the cycle are both pinned to 0. This is illustrated in Figure 9.15b. As P_1 is EXACTONE_d, at most one of e and e' is 1. If one of e and e' is 1, the other is 0, and as P_1 is an EXACTONE_d function its edge to E_o is also 0, and thus all edges incident to E_o are 0. As all five neighbors of E_o are EXACTONE functions, the four EXACTONE_3 functions effectively become $(\neq_2)$ functions along the wheel, and we can remove E_o and its incident edges. This becomes the same situation as in the previous case of type 1, where effectively a cycle of five binary equalities are linked by five binary disequalities, which has no valid assignment. It implies that both e and e' are pinned to 0. This finishes the proof. □

Now we discuss the implications of Lemma 9.25 and Lemma 9.27 for planar hypergraph perfect matching.

Let $\mathcal{G} = \{=_k \mid k \in S\}$ be a set of EQUALITY signatures, where S is a set of positive integers containing at least one $r \geq 3$. Moreover let $\mathcal{EO}^+ := \{\text{EXACTONE}_d \mid d \in \mathbb{Z}^+\} = \mathcal{EO} \cup \{(\neq_2), [0, 1]\}$. Then Pl-Holant$(\mathcal{G} \mid \mathcal{EO}^+)$ is the problem of counting perfect matchings over hypergraphs with planar incidence graphs, where the hyperedge sizes are prescribed by S. In the incidence graph, vertices assigned signatures in $\mathcal{G}$ on the left represent hyperedges, and vertices assigned signatures in $\mathcal{EO}^+$ on the right represent vertices of the hypergraph. Let $t = \gcd(S)$. We show that this problem is tractable if $t \geq 5$ and #P-hard if $t \leq 4$. The tractability when $t \geq 5$ follows from Lemma 9.25 and 9.27, since we can reduce Pl-Holant$(\mathcal{G} \mid \mathcal{EO}^+)$ to Pl-Holant$(\neq_2 \mid =_t, \mathcal{EO}, (\neq_2), [0, 1])$. The reduction goes as follows. With $(\neq_2)$ on the LHS and $=_t$ on the RHS, we can construct all E_t-blocks and hence all of $\mathcal{EQ}_t$ on the right. Note that $\mathcal{G} \subseteq \mathcal{EQ}_t$. Then we move all signatures in $\mathcal{G}$ to the left via $(\neq_2)$.

The hardness of Pl-Holant$(\mathcal{G} \mid \mathcal{EO}^+)$ for $t \leq 4$ follows from Corollary 9.23. The reason is as follows. We construct $(\neq_2)$ on the left using the gadget pictured in Figure 9.16 with $(=_r) \in \mathcal{G}$ on the left side assigned to circle vertices and $(\neq_2)$ on the right side assigned to square vertices. Then we move $\mathcal{G}$ to the right side via $(\neq_2)$ on the right side. We construct $(=_t)$ on the right side in the same Euclidean process using $\mathcal{G}$ of the right side and $(\neq_2)$ of the left side. This gives us a reduction from Pl-Holant$(\neq_2 \mid =_t, \mathcal{EO})$, which is

#P-hard by Corollary 9.23 if $t = 3, 4$. Otherwise $t = 1, 2$. Recall that $(=_r) \in \mathcal{G}$ for some $r \geq 3$. We use $(=_t)$ to reduce the arity of $(=_r)$ to 3 or 4, if necessary. Again we are done by Corollary 9.23.

If there is no hyperedge of size ≥ 3 in Pl-Holant($\mathcal{G} \mid \mathcal{EO}^+$), then for $S = \{1\}$, the problem is tractable trivially, and for $S = \{2\}$, the problem is tractable by Kasteleyn's algorithm. The problem is #P-hard if $S = \{1, 2\}$, because we can realize nonperfect matching signatures $[1, 1, 0, \ldots, 0]$. In summary, we have the following theorem.

Theorem 9.28. *The problem* Pl-Holant($\mathcal{G} \mid \mathcal{EO}^+$) *counts perfect matchings over hypergraphs with planar incidence graphs, where the hyperedge sizes are prescribed by a set S of positive integers. Let $t = \gcd(S)$. If $t \geq 5$ or $S = \{1\}$ or $\{2\}$, then the problem is computable in polynomial time. Otherwise $S = \{1, 2\}$, or $t \leq 4$ and $S \not\subseteq \{1, 2\}$, then the problem is* #P*-hard.*

By looking at the proof of Lemma 9.19 and Lemma 9.21 closely, one can improve the #P-hardness part of Theorem 9.28 further.

Exercise: Prove that counting perfect matchings in 3-uniform 3-regular planar hypergraphs is also #P-hard.

Exercise: Prove the same result for 4-uniform 3-regular planar hypergraphs.

9.7 Dichotomy for Planar Holant

With the new tractable problems related to hypergraph perfect matchings found, together with $\mathscr{M}$-transformable signatures due to holographic reductions to Kasteleyn's algorithm, it turns out that there are no more planar tractable problems in the Pl-Holant framework for any set of symmetric signatures $\mathcal{F}$. The following is the main dichotomy theorem for planar Holant problems for complex valued symmetric signatures on Boolean variables. The proof of this theorem is omitted here; the interested readers can find it in [CFGW15].

For a set $\mathcal{F}$ of signatures, $\mathcal{F}_{nd}^{\geq 3}$ denotes the set of nondegenerate signatures in $\mathcal{F}$ of arity ≥ 3, and $\mathcal{F}^*$ denotes $\mathcal{F}$ with all degenerate signatures $[a, b]^{\otimes m}$ replaced by unary $[a, b]$.

Theorem 9.29. *Let $\mathcal{F}$ be any set of symmetric, complex-valued signatures in Boolean variables. Then* Pl-Holant($\mathcal{F}$) *is* #P*-hard unless $\mathcal{F}$ satisfies one of the following conditions:*

1. *All nondegenerate signatures in $\mathcal{F}$ are of arity ≤ 2;*
2. *$\mathcal{F}$ is $\mathscr{A}$-transformable.*

3. *$\mathcal{F}$ is $\mathscr{P}$-transformable.*
4. *$\mathcal{F} \subseteq \mathscr{V}^{\sigma} \cup \{f \in \mathscr{R}_2^{\sigma} \mid \text{arity}(f) = 2\}$ for some $\sigma \in \{+, -\}$.*
5. *All nondegenerate signatures in $\mathcal{F}$ are in $\mathscr{R}_2^{\sigma}$ for some $\sigma \in \{+, -\}$.*
6. *$\mathcal{F}$ is $\mathscr{M}$-transformable.*
7. *$\mathcal{F} \subseteq Z\mathscr{P} \cup \mathscr{M}_4^{\sigma}$ for some $\sigma \in \{+, -\}$, and the greatest common divisor of the arities of the signatures in $\mathcal{F}^* \cap \mathscr{P}_2$ is ≥ 5.*

In each exceptional case, Pl-Holant($\mathcal{F}$) *is computable in polynomial time. If $\mathcal{F}$ satisfies condition 1, 2, 3, 4, or 5, then* Holant($\mathcal{F}$) *is computable in polynomial time without planarity; otherwise* Holant($\mathcal{F}$) *is* #P*-hard.*

10

Dichotomies for Asymmetric Constraints

Most dichotomy theorems presented so far in Volume I are for counting problems specified by sets of *symmetric* constraint functions. This is true for the first nontrivial dichotomy for Holant* problems in Chapter 2, for spin systems on regular graphs in Chapter 5, for Holant^c problems and for CSP^d problems in Chapter 6, for the Holant dichotomy in Chapter 7, and for the planar Holant problems in Chapter 9. There is an exception for Boolean #CSP in Chapter 3, which is valid for all sets of complex-valued constraint functions that are not necessarily symmetric, over the Boolean domain. The planar version, Theorem 8.34 in Chapter 8, also applies only to sets of symmetric constraint functions.

In this final chapter of Volume I, we discuss the few results currently known regarding classification theorems specified by not necessarily symmetric constraint functions.

The first result we present is the generalization of Theorem 8.34 to the asymmetric setting. We prove that the classification in Theorem 8.34 still holds for sets of not necessarily symmetric constraint functions. Owing to its length, we will not give the full proof of this theorem, but only a proof outline, and illustrate some ideas in the proof of a few lemmas as an example. The full proof of this theorem can be found in [CF16].

The second result we present is a complexity dichotomy for Holant* problems specified by not necessarily symmetric constraint functions. This is a generalization of Theorem 2.12 from Chapter 2. The proof that will be presented also provides a better insight to the Holant* dichotomy, even for the symmetric case.

A common theme of the proofs of these theorems is a reduction to the dichotomies already proved for the symmetric case.

10.1 Planar #CSP with Asymmetric Constraints

In Chapter 8 we proved Theorem 8.34, which shows that the complexity dichotomy in Chapter 3 for #CSP problems on Boolean variables (Theorem 3.7) has an extension to take into account the planar case. Specifically, Theorem 8.34 shows that for #CSP problems over any set of complex-valued *symmetric* constraint functions on Boolean variables, Valiant's holographic transformation followed by Kasteleyn's algorithm (Theorem 4.7) is a *universal* methodology to solve in polynomial time on planar graphs those counting problems that are #P-hard in general.

The first main theorem of Chapter 10 states that this statement is also valid for complex-valued constraint functions on Boolean variables that are *not* necessarily symmetric.

Theorem 10.1. *For any finite set of constraint functions $\mathcal{F}$ over Boolean variables, each taking complex values and not necessarily symmetric, #CSP($\mathcal{F}$) belongs to exactly one of three types according to $\mathcal{F}$:*

1. *It is P-time solvable.*
2. *It is P-time solvable over planar graphs but #P-hard over general graphs.*
3. *It is #P-hard over planar graphs.*

Moreover, type (2) consists of precisely those problems that are holographically reducible to the Kasteleyn's algorithm, also known as the Fisher–Kasteleyn–Temperley (FKT) algorithm.

The theorem has an explicit form. The three fundamental tractable classes $\mathscr{A}$ (affine), $\mathscr{P}$ (product type), and $\mathscr{M}$ (matchgates) are all defined (Definition 3.6, Definition 3.2, and Definition 4.9 respectively) for complex-valued constraint functions that are not necessarily symmetric. Then #CSP($\mathcal{F}$) has type *(1)* if $\mathcal{F} \subseteq \mathscr{A}$ or $\mathcal{F} \subseteq \mathscr{P}$; otherwise it has type *(2)* if $\mathcal{F} \subseteq \widehat{\mathscr{M}}$, where $\widehat{\mathscr{M}}$ is $\mathscr{M}$ under the holographic transformation $H_2 = \frac{1}{\sqrt{2}}\begin{bmatrix} 1 & 1 \\ 1 & -1 \end{bmatrix}$; otherwise it has type *(3)*. This theorem finally settles the full reach of the power of Valiant's holographic algorithms in the #CSP framework over Boolean variables.

The theorem is also valid for infinite sets of constraint functions $\mathcal{F}$ in the following sense. If $\mathcal{F}$ satisfies one of the tractability criteria, then the problem is P-time solvable where the input size of an instance includes the description of the constraints used. On the other hand, if $\mathcal{F}$ does not satisfy the tractability criteria, then there is a finite subset $\mathcal{F}' \subset \mathcal{F}$ such that the problem defined by constraint functions $\mathcal{F}'$ is #P-hard.

From Chapter 9, we know that there are #P-hard Holant problems that become tractable in P-time on planar structures, but they are *not* holographically reducible to the FKT. In view of this, the validity of Theorem 10.1 should not be considered self-evident.

It turns out that the class of Holant problems is more than just a separate framework providing a cautionary reference to the validity of Theorem 10.1. In fact, they form the main arena where we carry out the proof of Theorem 10.1. A basic idea in this proof is a holographic transformation between the #CSP setting and the Holant setting via the Hadamard transformation H_2. As with classical Fourier transform, certain properties are easier to handle in one setting while others are easier after the transformation. We will go back and forth.

In Section 10.1.1 we give a proof outline of Theorem 10.1. In Section 10.1.2 we prove some results illustrating several technical ideas used. Among the new techniques, we introduce a tableau calculus, compressed signatures, integrality gap, and a reduction to symmetric planar $\#\text{CSP}^2$ dichotomy.

10.1.1 Outline of the Proof of Theorem 10.1

We now give a proof outline of Theorem 10.1 and also explain some overall vision that guided this proof. An important technique is to view our counting problems in the dual perspectives of planar #CSP *and* planar Holant problems, i.e., we make essential use of the equivalence Pl-#CSP($\mathcal{F}$) $\equiv_T$ Pl-Holant($\widehat{\mathcal{EQ}}$, $\widehat{\mathcal{F}}$), where $\widehat{\mathcal{F}} = H_2\mathcal{F}$. Some questions are easier to handle in one framework, while others are easier in the other.

We want to show that *either* $\mathcal{F} \subseteq \mathscr{A}$, or $\mathcal{F} \subseteq \mathscr{P}$, or $\mathcal{F} \subseteq \widehat{\mathscr{M}}$, in which case Pl-#CSP($\mathcal{F}$) is computable in P-time, *or else* Pl-#CSP($\mathcal{F}$) is #P-hard. In the Pl-Holant($\widehat{\mathcal{EQ}}$, $\widehat{\mathcal{F}}$) setting, the tractability condition is expressed as $\widehat{\mathcal{F}} \subseteq \widehat{\mathscr{A}}$, or $\widehat{\mathcal{F}} \subseteq \widehat{\mathscr{P}}$, or $\widehat{\mathcal{F}} \subseteq \mathscr{M}$.

Note that $\mathscr{A}$ is invariant under the transformation by H_2, i.e., $\widehat{\mathscr{A}} = \mathscr{A}$. However, $\widehat{\mathscr{P}}$ is more difficult to reason about than $\mathscr{P}$, while $\mathscr{M}$ is easier than $\widehat{\mathscr{M}}$ to handle. The former suggests that we carry out our proof in the Pl-#CSP setting, while the latter suggests the opposite, that we do so in the Pl-Holant setting instead.

One necessary condition for $\mathscr{M}$ is the parity condition for matchgates, i.e., all nonzero entries of a matchgate signature must have the same parity in Hamming weight. If some signature in $\widehat{\mathcal{F}}$ violates the parity condition, then we have eliminated one possibility $\widehat{\mathcal{F}} \subseteq \mathscr{M}$. In this case if we prove in the Pl-#CSP framework, we can avoid discussing the more difficult class $\widehat{\mathscr{M}}$. On the other hand, if $\widehat{\mathcal{F}}$ satisfies the parity condition, then we have a lucky situation, as follows.

Fact: For any signature $g \in \mathscr{P}$, if the transformation $\hat{g} = H_2^{\otimes \text{arity}(g)} g$ satisfies the parity condition, then $g \in \mathscr{A}$. Consequently $\hat{g} \in \mathscr{A}$ as well since $\widehat{\mathscr{A}} = \mathscr{A}$.

Exercise: Prove this fact for the special case in which g has at most two antipodal points in its support, and then conclude that the statement holds in general.

Hence if every function in $\widehat{\mathcal{F}}$ satisfies the parity condition, then $\widehat{\mathcal{F}} \subseteq \widehat{\mathscr{P}}$ already implies $\widehat{\mathcal{F}} \subseteq \mathscr{A}$, with the consequence that we do not need to specifically discuss the tractability condition $\widehat{\mathcal{F}} \subseteq \widehat{\mathscr{P}}$. Thus in this case we can avoid discussing the irksome class $\widehat{\mathscr{P}}$ altogether.

Therefore there are two main cases:

Case (1): $\widehat{\mathcal{F}}$ does not satisfy the parity condition. Then we want to show that Pl-#CSP($\mathcal{F}$) is #P-hard unless $\mathcal{F} \subseteq \mathscr{A}$ or $\mathcal{F} \subseteq \mathscr{P}$. From any signature in $\widehat{\mathcal{F}}$ violating the parity condition, we can construct a unary signature $[1, w]$ with $w \neq 0$, in the Holant framework Pl-Holant($\widehat{\mathcal{EQ}}, \widehat{\mathcal{F}}$). Any signature that violates the parity condition is a witness that $\widehat{\mathcal{F}} \not\subseteq \mathscr{M}$, or equivalently $\mathcal{F} \not\subseteq \widehat{\mathscr{M}}$. This unary signature $[1, w]$ with $w \neq 0$ is such a witness, and it is symmetric. If $\mathcal{F} \subseteq \mathscr{A}$ or $\mathcal{F} \subseteq \mathscr{P}$, then the problem Pl-#CSP($\mathcal{F}$) is tractable. Otherwise, there exist some signatures $f, g \in \mathcal{F}$ such that $f \notin \mathscr{A}$ and $g \notin \mathscr{P}$. We would like to construct some *symmetric* signatures from these that are also nonaffine and nonproduct type, respectively, and then apply the known dichotomy theorem for symmetric signatures. For the nonproduct type we will do so in the Pl-#CSP($\mathcal{F}$) setting, to avoid having to deal with $\widehat{\mathscr{P}}$. For the nonaffine signatures, we can do so in either the Pl-#CSP framework or the Pl-Holant framework as $\mathscr{A}$ is invariant $\widehat{\mathscr{A}} = \mathscr{A}$.

However, the obstacle in this plan is that it is generally difficult to construct *symmetric* signatures from *asymmetric* (i.e., not necessarily symmetric) signatures in a *planar* fashion, for arity > 3. Therefore, a main engine of the proof is *arity reduction*. Starting from a nonproduct type signature of arity $n > 3$, we construct in the Pl-#CSP setting a nonproduct type signature of arity $n - 1$. Then we show how to construct, from any nonproduct type signature of arity 3, a *symmetric* and nonproduct type signature of arity 3. These constructions will need suitable unary signatures which will be constructed, starting with that $[1, w]$ constructed in Pl-Holant($\widehat{\mathcal{EQ}}, \widehat{\mathcal{F}}$).

For the construction of nonaffine signatures, we will introduce a *tableau calculus*. Again we will carry out an arity reduction proof, this time all the way down to arity 1. We prove that with the help of three unary signatures $[1, 0], [0, 1], [1, x]$ with $x \neq 0$, we can get a unary nonaffine signature from any nonaffine signature of higher arity in the Pl-Holant setting. This proof heavily depends on the new tableau calculus. Then we construct $[1, 0], [0, 1], [1, x]$ by shuttling between Pl-#CSP($\mathcal{F}$) and Pl-Holant($\widehat{\mathcal{EQ}}, \widehat{\mathcal{F}}$). There is an exceptional case where all signatures in $\mathcal{F}$ are $\{0, 1\}$-valued in Pl-#CSP($\mathcal{F}$). In this case, we cannot construct $[1, 0]$, $[0, 1]$, $[1, x]$ simultaneously. We resolve this case separately. For $\{0, 1\}$-valued $\mathcal{F}$, we actually also cannot construct all the unary signatures in the arity reduction proof for nonproduct type, if we *only* assume the existence of some $g \in \mathcal{F} \setminus \mathscr{P}$. However, if we have both $g \in \mathcal{F} \setminus \mathscr{P}$ and some $f \in \mathcal{F} \setminus \mathscr{A}$, we can use f to produce the needed unary signatures to help the arity reduction on g. All these use tableau calculus.

Case (2): $\widehat{\mathcal{F}}$ satisfies the parity condition. In this case, if $\widehat{\mathcal{F}} \subseteq \mathscr{A}$, or $\widehat{\mathcal{F}} \subseteq \widehat{\mathscr{P}}$, or $\widehat{\mathcal{F}} \subseteq \mathscr{M}$, then the problem is tractable. However, due to the parity condition, there are only two consequential conditions here, $\widehat{\mathcal{F}} \subseteq \mathscr{A}$ or $\widehat{\mathcal{F}} \subseteq \mathscr{M}$; the containment $\widehat{\mathcal{F}} \subseteq \widehat{\mathscr{P}}$ is subsumed by $\widehat{\mathcal{F}} \subseteq \mathscr{A}$. Therefore we want to prove that if $\widehat{\mathcal{F}} \not\subseteq \mathscr{A}$ and $\widehat{\mathcal{F}} \not\subseteq \mathscr{M}$, then Pl-Holant($\widehat{\mathcal{EQ}}, \widehat{\mathcal{F}}$) is #P-hard.

Again a natural idea is to construct nonaffine and nonmatchgate *symmetric* signatures from any such asymmetric signatures, and then we can apply the known dichotomy theorem for symmetric signatures. The main difficulty of this approach lies in dealing with nonmatchgate signatures. Note that both $\widehat{\mathcal{F}}$ and $\widehat{\mathcal{EQ}}$ (being a subset of $\mathscr{M}$) satisfy the parity condition, and therefore the signature of any construction from $\widehat{\mathcal{EQ}} \cup \widehat{\mathcal{F}}$ must also satisfy the parity condition. On the other hand, it follows from matchgate identities (4.10) that any signature of arity ≤ 3 is a matchgate signature iff it satisfies the parity condition. Hence all constructible nonmatchgate signatures must have arity ≥ 4. But it is difficult to construct a symmetric signature from any asymmetric signature of arity ≥ 4 while preserving planarity.

So we take an alternative approach. For a given nonmatchgate signature, we first prove that we can get a nonmatchgate signature f of arity 4. Then we can construct from f either a special function called the crossover function $\mathfrak{X}$ or the EQUALITY function $(=_4)$ of arity 4. Here the crossover function $\mathfrak{X} = [(x_1 = x_3) \wedge (x_2 = x_4)]$ is a function on four variables x_1, x_2, x_3, x_4 in cyclic order, which ensures that pairs of diagonally crossed variables are equal.

If we have $\mathfrak{X}$, we can finish the proof by the nonplanar #CSP dichotomy theorem: To any nonplanar input signature grid, we just replace any crossing by $\mathfrak{X}$, and obtain an equivalent planar signature grid, and then we can apply Theorem 3.7.

If we have $(=_4)$, then we can get all $\mathcal{EQ}_2 = \{=_2, =_4, \ldots, =_{2k}, \ldots\}$ in Pl-Holant($\widehat{\mathcal{EQ}}, \widehat{\mathcal{F}}$). This implies that

$$\begin{aligned} \text{Pl-Holant}(\widehat{\mathcal{EQ}}, \widehat{\mathcal{F}}) &\equiv_{\mathrm{T}} \text{Pl-Holant}(\mathcal{EQ}_2, \widehat{\mathcal{EQ}}, \widehat{\mathcal{F}}) \\ &\equiv_{\mathrm{T}} \text{Pl-\#CSP}^2(\widehat{\mathcal{EQ}}, \widehat{\mathcal{F}}), \end{aligned} \tag{10.1}$$

where $\#\text{CSP}^2$ denotes #CSP in which every variable appears an even number of times.

Now comes a "cognitive dissonance." By (10.1), what used to be the "right-hand side" in the equivalence $\text{Pl-\#CSP}(\mathcal{F}) \equiv_T \text{Pl-Holant}(\widehat{\mathcal{EQ}}, \widehat{\mathcal{F}})$ will be treated as a $\text{Pl-\#CSP}^2$ problem:

$$\begin{array}{c}\text{Pl-\#CSP}(\mathcal{F}) \equiv_T \text{Pl-Holant}(\widehat{\mathcal{EQ}}, \widehat{\mathcal{F}}) \equiv_T \text{Pl-Holant}(\mathcal{EQ}_2, \widehat{\mathcal{EQ}}, \widehat{\mathcal{F}}) \\ \qquad\qquad\qquad\qquad\qquad\qquad\qquad \|\!|_T \\ \qquad\qquad\qquad\qquad\qquad\qquad\qquad \text{Pl-\#CSP}^2(\widehat{\mathcal{EQ}}, \widehat{\mathcal{F}}).\end{array}$$

A $\text{Pl-\#CSP}^2$ problem behaves more like a Pl-#CSP problem. For $\text{Pl-\#CSP}^2$ problems over *symmetric* signatures, the dichotomy theorem for $\text{Pl-\#CSP}^2$, Theorem 9.16 (the proof can be found in [CFGW15]), says that there are five tractable classes $\mathscr{P}$, $\mathscr{A}$, $\mathscr{A}^\dagger$, $\widehat{\mathscr{M}}$, and $\widehat{\mathscr{M}^\dagger}$. But now we will apply these on the "dual side" $\widehat{\mathcal{EQ}} \cup \widehat{\mathcal{F}}$, instead of the "primal side" $\mathcal{F}$. The "cognitive dissonance" is that the transformation from $(\mathcal{EQ}, \mathcal{F}) \mapsto (\widehat{\mathcal{EQ}}, \widehat{\mathcal{F}})$ is precisely for the purpose of transforming $\mathcal{EQ}$ to be $\widehat{\mathcal{EQ}}$, a subset of $\mathscr{M}$, but now we will subject $\widehat{\mathcal{EQ}}$ to tractability tests including $\widehat{\mathscr{M}}$ and $\widehat{\mathscr{M}^\dagger}$. But it is a simple fact that $\widehat{\mathcal{EQ}} \nsubseteq \mathscr{P} \cup \mathscr{A}^\dagger \cup \widehat{\mathscr{M}} \cup \widehat{\mathscr{M}^\dagger}$. Note that all signatures in $\widehat{\mathcal{EQ}}$ are symmetric. Therefore the only remaining possibility for tractability is $\mathscr{A}$. If we can construct a nonaffine symmetric signature in $\text{Pl-\#CSP}^2(\widehat{\mathcal{EQ}}, \widehat{\mathcal{F}})$, then the problem is #P-hard.

Of course if $\widehat{\mathcal{F}} \subseteq \mathscr{A}$, then $\text{Pl-Holant}(\widehat{\mathcal{EQ}}, \widehat{\mathcal{F}})$ is tractable. Suppose $\widehat{\mathcal{F}} \nsubseteq \mathscr{A}$, and we want to construct a *symmetric* nonaffine signature. We produce such a signature of arity 2 by arity reduction. From any $f \in \widehat{\mathcal{F}} \setminus \mathscr{A}$, which satisfies the parity condition, we can first get a nonaffine signature satisfying the even parity condition. Then every signature constructible from that has even parity, as $\widehat{\mathcal{EQ}}$ also has even parity. Any binary signature satisfying the even parity condition has the form $[x, 0, y]$ and it is automatically symmetric.

A technical difficulty is that when $\widehat{\mathcal{F}}$ satisfies the even parity condition, it is impossible to construct the pinning signature $[0, 1]$. Instead we can try to construct the double pinning signature $[0, 1]^{\otimes 2} = [0, 0, 1]$ and prove that $[0, 1]^{\otimes 2}$ is almost as good as $[0, 1]$ with help of the parity function $[1, 0, 1, 0] \in \widehat{\mathcal{EQ}}$. Then there are three cases.

1. If it is not the case that every function in $\widehat{\mathcal{F}}$ takes values in $\{0, 1, -1\}$ up to a constant, then we can construct $[0, 1]^{\otimes 2}$ and $[1, 0, -1]$ and complete the proof.

2. If every function in $\widehat{\mathcal{F}}$ takes values in $\{0, 1, -1\}$ up to a constant but not every function in $\widehat{\mathcal{F}}$ takes values in $\{0, 1\}$ up to a constant, then we can construct $[1, 0, -1]$ and complete the proof.
3. If every function in $\widehat{\mathcal{F}}$ takes values in $\{0, 1\}$ then we prove it separately.

In all cases, we use our tableau calculus. The full proof of Theorem 10.1 can be found in [CF16].

10.1.2 Arity Reduction for Nonaffine Asymmetric Signatures

In this subsection we prove a few specific lemmas to illustrate some new techniques. The lemmas are used for arity reduction of nonaffine asymmetric signatures in the dual setting Pl-Holant$(\widehat{\mathcal{EQ}}, \widehat{\mathcal{F}})$.

As described in the proof outline, for any signature $f \notin \mathscr{A}$ (resp. $\mathscr{P}, \widehat{\mathscr{M}}$), we want to construct a *symmetric* signature $g \notin \mathscr{A}$ (resp. $\mathscr{P}, \widehat{\mathscr{M}}$), and then apply known dichotomy theorems. If f has high arity, it is very difficult to construct a symmetric signature from a nonsymmetric f in a planar fashion. So our first step is to reduce the arity of f. This task can be delicate. One reason is that a minor modification can change a tractable asymmetric signature to a signature that defines a #P-hard problem. For example, let f be the signature with signature matrix

$$M_{x_1x_2,x_4x_3}(f) = \begin{bmatrix} f_{0000} & f_{0010} & f_{0001} & f_{0011} \\ f_{0100} & f_{0110} & f_{0101} & f_{0111} \\ f_{1000} & f_{1010} & f_{1001} & f_{1011} \\ f_{1100} & f_{1110} & f_{1101} & f_{1111} \end{bmatrix} = \begin{bmatrix} 1 & 0 & 0 & 1 \\ 0 & 1 & -1 & 0 \\ 0 & -1 & -1 & 0 \\ 1 & 0 & 0 & -1 \end{bmatrix},$$

listing the 16 values of $f_{x_1x_2x_3x_4}$. Then $f \in \mathscr{A} \cap \mathscr{M}$. But if we change the sign of any one nonzero entry of f, then it is not in *any* tractable class, and it defines a #P-hard problem.

We will prove two lemmas, Lemmas 10.4 and 10.10. They show, in the case when all signatures in $\widehat{\mathcal{F}}$ satisfy the parity condition, how to reduce the arity of a nonaffine asymmetric signature in Pl-Holant$(\widehat{\mathcal{EQ}}, \widehat{\mathcal{F}})$. In Lemma 10.4 all signatures in $\widehat{\mathcal{F}}$ take values in $\{0, 1\}$. In Lemma 10.10 all signatures in $\widehat{\mathcal{F}}$ take values in $\{0, 1, -1\}$ and we need an additional auxiliary function. These lemmas are used in the proof of Theorem 10.1 in the second main case according to the proof outline; they are used to handle the two special cases (2) and (3) listed toward the end of that proof outline, after we managed to get $(=_4)$ and hence a reduction to #CSP^2.

In order to prove Lemma 10.4 we first present Lemmas 10.2 and 10.3, which help regularize a signature. Lemma 10.2 shows how to use $[0, 1]^{\otimes 2}$ and

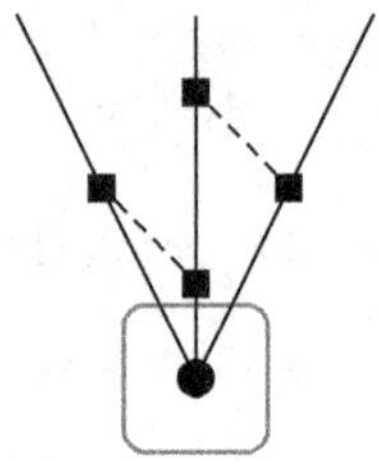

Figure 10.1. Flipping two variables of f that are not adjacent by $[0, 1, 0]^{\otimes 2}$ while preserving planarity. The circle vertex is labeled f and squares are $[0, 1, 0]$. A pair of squares connected by a dashed line forms $[0, 1, 0]^{\otimes 2}$.

$[1, 0, 1, 0] \in \widehat{\mathcal{EQ}}$ to get $[0, 1, 0]^{\otimes 2}$, then to flip any two variables that are not necessarily adjacent, while preserving planarity.

Lemma 10.2. *In Pl-Holant$(\widehat{\mathcal{EQ}}, [0, 1]^{\otimes 2}, f)$, if f has arity n, then for any $s \neq t \in [n]$, we can construct g such that $g(x'_1, x'_2, \ldots, x'_n) = f(x_1, x_2, \ldots, x_n)$, where $x'_k = \overline{x_k}$ for $k \in \{s, t\}$ and $x'_k = x_k$ otherwise. Moreover, $f \in \mathscr{A}$ iff $g \in \mathscr{A}$.*

Proof. That $f \in \mathscr{A}$ iff $g \in \mathscr{A}$ follows from the definition of $\mathscr{A}$.

By connecting $[0, 1]^{\otimes 2}$ to two disjoint copies of $[1, 0, 1, 0] \in \widehat{\mathcal{EQ}}$, we get $[0, 1, 0]^{\otimes 2}$. Note that this is a planar gadget where two adjacent pairs of variables are to be flipped. This function is $(x_1 \neq x_2) \wedge (x_3 \neq x_4)$.

Without loss of generality, we assume that $t > s$. If $t = s + 1$, then x_s and x_{s+1} are adjacent variables and we can directly flip both x_s and x_{s+1}. In general (see Figure 10.1 for an illustration), we can flip two adjacent variables sequentially as follows. Let $h^{(0)} = f$ and for $1 \leq j \leq t - s$, define $h^{(j)}$ to be the result of flipping both variables x_{s+j-1} and x_{s+j}, i.e., $h^{(j)}(x_1^{(j)}, x_2^{(j)}, \ldots, x_n^{(j)}) = h^{(j-1)}(x_1^{(j-1)}, x_2^{(j-1)}, \ldots, x_n^{(j-1)})$, where $x_i^{(j)} = \overline{x_i^{(j-1)}}$ for $i \in \{s + j - 1, s + j\}$ and $x_i^{(j)} = x_i^{(j-1)}$ for all others. Then we are done by letting $g = h^{(t-s)}$. In effect, all variables x_i with $s < i < t$ are flipped twice. □

Lemma 10.2 is also valid for $\mathscr{P}$ and $\mathscr{M}$ and is useful in the proof of Theorem 10.1, but we will not use this more generalized form in the exposition here.

Lemma 10.3 implies that in Pl-Holant$(\widehat{\mathcal{EQ}}, f)$, where $f \notin \mathscr{A}$, we can assume that $f_{00\cdots0} = 1$. Moreover, if f satisfies the parity condition, we can assume it satisfies the *even* parity condition.

Lemma 10.3. *If $\widehat{\mathcal{F}}$ contains a signature $f \notin \mathscr{A}$ of arity n, then we can construct a function $f' \notin \mathscr{A}$ of arity n with $f'_{00\cdots0} = 1$ such that*

$$\text{Pl-Holant}(\widehat{\mathcal{EQ}}, f', \widehat{\mathcal{F}}) \leq_T \text{Pl-Holant}(\widehat{\mathcal{EQ}}, \widehat{\mathcal{F}}).$$

Moreover, if f satisfies the parity condition, then f' satisfies the even *parity condition, and if f takes values in* $\{0, 1\}$ $(\{0, 1, -1\})$, *then f' also takes values in* $\{0, 1\}$ $(\{0, 1, -1\})$.

Proof. If $f_{00\cdots0} \neq 0$, then we simply normalize f by setting $f' = f/f_{00\cdots0}$. So we suppose $f_{00\cdots0} = 0$. By $f \notin \mathscr{A}$, clearly f is not identically 0. Let $\text{wt}(\alpha) = \min_{\beta\in\{0,1\}^n}\{\text{wt}(\beta) | f_\beta \neq 0\}$. Let $S = \{i \mid 1 \leq i \leq n,\ \text{the } i\text{th bit of } \alpha \text{ is } 0\}$. Since we have $[1, 0] \in \widehat{\mathcal{EQ}}$ we can get a nonzero multiple of $[0, 1]^{\otimes \text{wt}(\alpha)}$, by connecting a copy of $[1, 0]$ to each variable of f in S. Depending on whether $\text{wt}(\alpha)$ is odd or even, we can get either $[0, 1]$ or $[0, 1]^{\otimes 2}$ respectively, by taking zero or more self loops on $[0, 1]^{\otimes \text{wt}(\alpha)}$. Then we can get either $[0, 1, 0]$ or $[0, 1, 0]^{\otimes 2}$ respectively, by connecting $[0, 1]$ or $[0, 1]^{\otimes 2}$ to one or two copies of $[1, 0, 1, 0] \in \widehat{\mathcal{EQ}}$.

If $\text{wt}(\alpha)$ is odd, and this includes the case when f satisfies the odd parity condition, we have $[0, 1, 0]$ and can flip any variable of f individually. By flipping all variables in $[n] \setminus S$, and normalizing, we obtain f' with the required property. In particular, if f satisfies the odd parity condition, then f' satisfies the even parity condition.

If $\text{wt}(\alpha)$ is even, and this includes the case when f satisfies the even parity condition, we have $[0, 1]^{\otimes 2}$ and $[0, 1, 0]^{\otimes 2}$. By Lemma 10.2 we can flip any two variables of f. By applying the construction in Lemma 10.2 simultaneously on $\text{wt}(\alpha)/2$ pairs of variables of f, we can transform f to f' by a planar construction so that $f'_{00\cdots0} = f_\alpha \neq 0$. By normalizing, we obtain the required f' with $f'_{00\cdots0} = 1$. In particular, if f satisfies the even parity condition, then f' also satisfies the even parity condition. We get f' from f by flipping some variables in all cases. Thus if f takes values in $\{0, 1\}$ ($\{0, 1, -1\}$ respectively), then f' also takes values in $\{0, 1\}$ ($\{0, 1, -1\}$ respectively). □

Lemma 10.4. *Suppose all signatures in $\widehat{\mathcal{F}}$ take values in* $\{0, 1\}$ *and satisfy the parity condition. If $\widehat{\mathcal{F}}$ contains a signature $f \notin \mathscr{A}$ of arity $n \geq 3$, then there exists a signature $g \notin \mathscr{A}$ of arity $< n$, such that*

$$\text{Pl-Holant}(\widehat{\mathcal{EQ}}, g, \widehat{\mathcal{F}}) \leq_T \text{Pl-Holant}(\widehat{\mathcal{EQ}}, \widehat{\mathcal{F}}).$$

Furthermore, g satisfies the even *parity condition.*

Proof. The proof is by induction on n, and uses the tableau calculus.

If there exists $i \in [n]$ such that $f^{x_i=0} \notin \mathscr{A}$, then we are done since we have $[1, 0] \in \widehat{\mathcal{EQ}}$, and connecting $[1, 0]$ to x_i of f gives $f^{x_i=0}$. In the following, we assume that $f^{x_i=0} \in \mathscr{A}$ for $1 \leq i \leq n$. By Lemma 10.3, we can assume that $f_{00\cdots0} = 1$ and f satisfies the *even* parity condition.

As a $\{0, 1\}$-valued signature, $f \in \mathscr{A}$ iff $\text{supp}(f)$ is an affine subspace. So $\text{supp}(f)$ is not an affine subspace, in particular not a linear subspace. There exist $\mathbf{a} = a_1a_2\cdots a_n$, $\mathbf{b} = b_1b_2\cdots b_n$, such that $\mathbf{a}, \mathbf{b} \in \text{supp}(f)$, $\mathbf{c} = \mathbf{a} \oplus \mathbf{b} = c_1c_2\cdots c_n \notin \text{supp}(f)$. Since $00\ldots0 \in \text{supp}(f)$, $\mathbf{a} \neq \mathbf{b}$, so there exists $i \in [n]$ such that $a_i \neq b_i$. Without loss of generality, we assume that $a_1 = 0$, $b_1 = 1$. It follows that $c_1 = 1$. Let $\mathbf{a}' = a_3\cdots a_n$, $\mathbf{b}' = b_3\cdots b_n$, $\mathbf{c}' = c_3\cdots c_n$.

By connecting the variables x_2, x_1 of $[1, 0, 1, 0]$ to the variables x_1, x_2 of f respectively, the planar gadget gives the signature

$$g(x', x_3, x_4, \ldots, x_n) = \sum_{x_1,x_2\in\{0,1\}} [1, 0, 1, 0](x_2, x_1, x')f(x_1, x_2, \ldots, x_n).$$

If $x' = 0$ then the parity function $[1, 0, 1, 0](x_2, x_1, x')$ forces $x_1x_2 = 00$ or 11, and if $x' = 1$ then the parity function forces $x_1x_2 = 01$ or 10. So we have

$$g(x_1 \oplus x_2, x_3, \ldots, x_n) = f(x_1, x_2, x_3, \ldots, x_n) + f(\bar{x}_1, \bar{x}_2, x_3, \ldots, x_n).$$

We show that $g \notin \mathscr{A}$.

Now comes the tableau calculus. We will examine all four cases of membership of $\bar{a}_1\bar{a}_2\mathbf{a}'$ and $\bar{b}_1\bar{b}_2\mathbf{b}'$ in $\text{supp}(f)$.

- Suppose $\bar{a}_1\bar{a}_2\mathbf{a}' \in \text{supp}(f)$ and $\bar{b}_1\bar{b}_2\mathbf{b}' \in \text{supp}(f)$. By assumption $\mathbf{a} = 0a_2\mathbf{a}' \in \text{supp}(f)$ and $0\bar{b}_2\mathbf{b}' = \bar{b}_1\bar{b}_2\mathbf{b}' \in \text{supp}(f)$. Since $f^{x_1=0} \in \mathscr{A}$, $\text{supp}(f^{x_1=0})$ is an affine subspace. As $0\ldots0 \in \text{supp}(f^{x_1=0})$, $\text{supp}(f^{x_1=0})$ is a linear subspace. We have

$$\begin{array}{rll} & a_2\mathbf{a}' & \in \text{supp}(f^{x_1=0}) \\ \oplus & \bar{b}_2\mathbf{b}' & \in \text{supp}(f^{x_1=0}) \\ \hline & \bar{c}_2\mathbf{c}' & \end{array}$$

  Hence $\bar{c}_2\mathbf{c}' \in \text{supp}(f^{x_1=0})$. This means that $f_{\bar{c}_1\bar{c}_2\mathbf{c}'} \neq 0$ since $\bar{c}_1 = 0$. By definition

$$g_{(c_1\oplus c_2)\mathbf{c}'} = f_{c_1c_2\mathbf{c}'} + f_{\bar{c}_1\bar{c}_2\mathbf{c}'}.$$

  Thus $g_{(c_1\oplus c_2)\mathbf{c}'} = 1$ since $f_{c_1c_2\mathbf{c}'} = 0$ and $f_{\bar{c}_1\bar{c}_2\mathbf{c}'} = 1$ (as f is $\{0, 1\}$-valued). Moreover, by

$$g_{(a_1\oplus a_2)\mathbf{a}'} = f_{a_1a_2\mathbf{a}'} + f_{\bar{a}_1\bar{a}_2\mathbf{a}'},$$

  we have $g_{(a_1\oplus a_2)\mathbf{a}'} = 2$, since $f_{a_1a_2\mathbf{a}'} = f_{\bar{a}_1\bar{a}_2\mathbf{a}'} = 1$. This implies that g is not affine, as g has nonzero values of different norm, a necessary condition to be in $\mathscr{A}$.

- Suppose $\bar{a}_1\bar{a}_2\mathbf{a}' \notin \text{supp}(f)$ and $\bar{b}_1\bar{b}_2\mathbf{b}' \notin \text{supp}(f)$. By

$$g_{00\cdots0} = f_{000\cdots0} + f_{110\cdots0},$$

we have $g_{00\cdots0} \neq 0$ since $f_{000\cdots0} = 1$ and $f_{110\cdots0} \in \{0, 1\}$. Suppose g is affine, for otherwise we are done. Then $\text{supp}(g)$ is a linear subspace, since $00\cdots0 \in \text{supp}(g)$. By

$$g_{(a_1\oplus a_2)\mathbf{a}'} = f_{a_1a_2\mathbf{a}'} + f_{\bar{a}_1\bar{a}_2\mathbf{a}'},$$
$$g_{(b_1\oplus b_2)\mathbf{b}'} = f_{b_1b_2\mathbf{b}'} + f_{\bar{b}_1\bar{b}_2\mathbf{b}'},$$

we have $g_{(a_1\oplus a_2)\mathbf{a}'} = 1$, $g_{(b_1\oplus b_2)\mathbf{b}'} = 1$ since $f_{a_1a_2\mathbf{a}'} = f_{b_1b_2\mathbf{b}'} = 1$ and $f_{\bar{a}_1\bar{a}_2\mathbf{a}'} = f_{\bar{b}_1\bar{b}_2\mathbf{b}'} = 0$. This implies that $(a_1 \oplus a_2)\mathbf{a}'$, $(b_1 \oplus b_2)\mathbf{b}' \in \text{supp}(g)$. Then by

$$\begin{array}{rll} & (a_1 \oplus a_2)\mathbf{a}' & \in \text{supp}(g) \\ \oplus & (b_1 \oplus b_2)\mathbf{b}' & \in \text{supp}(g) \\ \hline & (c_1 \oplus c_2)\mathbf{c}' & \end{array}$$

we have $(c_1 \oplus c_2)\mathbf{c}' \in \text{supp}(g)$. This implies that

$$g_{(c_1\oplus c_2)\mathbf{c}'} = f_{c_1c_2\mathbf{c}'} + f_{\bar{c}_1\bar{c}_2\mathbf{c}'} \neq 0.$$

Thus $f_{\bar{c}_1\bar{c}_2\mathbf{c}'} \neq 0$ since $f_{c_1c_2\mathbf{c}'} = 0$. So $\bar{c}_1\bar{c}_2\mathbf{c}' \in \text{supp}(f)$. Hence, $\bar{c}_2\mathbf{c}' \in \text{supp}(f^{x_1=0})$ since $\bar{c}_1 = 0$. By

$$\begin{array}{rll} & a_2\mathbf{a}' & \in \text{supp}(f^{x_1=0}) \\ \oplus & \bar{c}_2\mathbf{c}' & \in \text{supp}(f^{x_1=0}) \\ \hline & \bar{b}_2\mathbf{b}' & \end{array}$$

we have $\bar{b}_2\mathbf{b}' \in \text{supp}(f^{x_1=0})$. Thus $\bar{b}_1\bar{b}_2\mathbf{b}' \in \text{supp}(f)$ as $\bar{b}_1 = 0$. This is a contradiction.
- Suppose $\bar{a}_1\bar{a}_2\mathbf{a}' \in \text{supp}(f)$ and $\bar{b}_1\bar{b}_2\mathbf{b}' \notin \text{supp}(f)$. Then we consider

$$\begin{aligned} g_{(a_1\oplus a_2)\mathbf{a}'} &= f_{a_1a_2\mathbf{a}'} + f_{\bar{a}_1\bar{a}_2\mathbf{a}'}, \\ g_{(b_1\oplus b_2)\mathbf{b}'} &= f_{b_1b_2\mathbf{b}'} + f_{\bar{b}_1\bar{b}_2\mathbf{b}'}. \end{aligned} \tag{10.2}$$

This implies that $g_{(a_1\oplus a_2)\mathbf{a}'} = 2$ and $g_{(b_1\oplus b_2)\mathbf{b}'} = 1$. Thus $g \notin \mathscr{A}$.
- Suppose $\bar{a}_1\bar{a}_2\mathbf{a}' \notin \text{supp}(f)$ and $\bar{b}_1\bar{b}_2\mathbf{b}' \in \text{supp}(f)$. Then (10.2) implies that $g_{(a_1\oplus a_2)\mathbf{a}'} = 1$ and $g_{(b_1\oplus b_2)\mathbf{b}'} = 2$. Thus $g \notin \mathscr{A}$. □

Looking at the proof of Lemma 10.4 closely, it becomes clear that to be able to move a support point to $00\cdots0$ using Lemma 10.3 simplifies greatly the argument in the tableau calculus. For affine spaces V, the closure property is that whenever $\mathbf{a}, \mathbf{b}, \mathbf{c} \in V$, we have $\mathbf{a} \oplus \mathbf{b} \oplus \mathbf{c} \in V$. But for linear spaces V, the closure property is simply whenever $\mathbf{a}, \mathbf{b} \in V$, we have $\mathbf{a} \oplus \mathbf{b} \in V$. Instead of having to deal with supp(f) as an affine space, by moving the support point to $00\cdots0$ we can deal with linear space, and as such we only need to have two known points in supp(f) to obtain another point in supp(f).

In order to prove Lemma 10.10, it is helpful to have some facts about multilinear representation of signatures. Lemma 10.5 shows that for a $\{1,-1\}$-valued signature of arity k, there exists a unique multilinear polynomial $P(x_1,\ldots,x_k) \in \mathbb{Z}_2[x_1,\ldots,x_k]$ such that

$$f(x_1,\ldots x_k) = (-1)^{P(x_1,\ldots,x_k)}. \tag{10.3}$$

Thus if (10.3) holds where P has total degree > 2, then $f \notin \mathscr{A}$.

Lemma 10.5. *If f is a signature of arity k taking values in $\{\pm1\}$, then there exists a unique multilinear polynomial $P(x_1,\ldots,x_k) \in \mathbb{Z}_2[x_1,\ldots,x_k]$ such that $f(x_1,\ldots x_k) = (-1)^{P(x_1,\ldots,x_k)}$.*

Proof. For any input $\alpha = a_1 \ldots a_k \in \{0,1\}^k$, there exists $r_\alpha \in \{0,1\}$ such that $f_\alpha = (-1)^{r_\alpha}$ since f takes values in $\{1,-1\}$. Let $P(x_1,\ldots,x_k) = \sum_{\alpha\in\{0,1\}^k} r_\alpha \prod_{i=1}^k \widetilde{x_i} \in \mathbb{Z}_2[x_1,\ldots,x_k]$, where $\alpha = a_1\ldots a_k$, $\widetilde{x_i} = x_i$ if $a_i = 1$ and $1-x_i$ if $a_i = 0$. Then $f(x_1,\ldots x_k) = (-1)^{P(x_1,\ldots,x_k)}$.

To prove uniqueness, it is equivalent to prove that if f is the constant 1 function then $P(x_1,\ldots,x_k) = 0$ in $\mathbb{Z}_2[x_1,\ldots,x_k]$. For a contradiction suppose $\prod_{i\in S} x_i$ is a nonzero term in $P(x_1,\ldots,x_k)$ with minimum cardinality $|S|$. Set $x_i = 1$ for all $i \in S$, and all other $x_i = 0$. Then P evaluates to 1 in $\mathbb{Z}_2$, and f evaluates to $-1 \neq 1$. This is a contradiction. □

Corollary 10.6. *Let $f \in \mathscr{A}$ be a $\{1,-1\}$-valued signature of arity n with support of dimension $k \geq 0$. Let $Y = \{x_{i_1},\ldots,x_{i_k}\}$ be a set of free variables. Then there exists a unique $Q(Y) \in \mathbb{Z}_2[Y]$ such that*

$$f(x_1,x_2,\ldots,x_n) = (-1)^{Q(Y)}$$

for $(x_1,x_2,\ldots,x_n) \in$ supp(f), where $Q(Y)$ is a multilinear polynomial of degree ≤ 2.

Corollary 10.7. *Let f be a binary signature and $M_{x_1,x_2}(f) = \begin{bmatrix} f_{00} & f_{01} \\ f_{10} & f_{11} \end{bmatrix} = \begin{bmatrix} 1 & b \\ c & d \end{bmatrix}$. If $b, c, d \in \{1,-1\}$ then $f \in \mathscr{A}$.*

Definition 10.8. If f has affine support of dimension k, and $Y = \{x_{i_1},\ldots,x_{i_k}\}$ is a set of free variables, then the *compressed signature* of f (with respect to Y)

is the unique function $\underline{f}$ defined on Y such that

$$\underline{f}(x_{i_1}, \ldots, x_{i_k}) = f(x_1, x_2, \ldots, x_n),$$

where $(x_1, x_2, \ldots, x_n) \in \text{supp}(f)$.

Corollary 10.9. *Let f be a signature of arity n having affine support of dimension k. Suppose f takes values in $\{0, \pm 1\}$, and $Y = \{x_{i_1}, \ldots, x_{i_k}\}$ is a set of free variables. Then $f \in \mathscr{A}$ iff $\underline{f} \in \mathscr{A}$.*

Proof. If $\underline{f}(x_{i_1}, \ldots, x_{i_k}) = (-1)^{Q(x_{i_1}, \ldots, x_{i_k})}$, then

$$f(x_1, \ldots, x_n) = \chi_{\text{supp}(f)}(x_1, \ldots, x_n) \cdot (-1)^{Q(x_{i_1}, \ldots, x_{i_k})}.$$

So $f \in \mathscr{A}$ iff $\underline{f} \in \mathscr{A}$. □

Now we are ready to prove Lemma 10.10. We will use the tableau calculus again, which is used numerous times in the proof of Theorem 10.1. Another technique is an integrality gap (0 mod 4 versus 2 mod 4) that guarantees that a function is not in $\mathscr{A}$. Here the arity reduction is more difficult, and we need the help of another auxiliary function $[1, 0, -1]$. When we use it in the proof of Theorem 10.1, this auxiliary function will have to be produced by other means.

Lemma 10.10. *Suppose all signatures in $\widehat{\mathcal{F}}$ take values in $\{0, 1, -1\}$ and satisfy the parity condition. If $\widehat{\mathcal{F}}$ contains the binary signature $[1, 0, -1]$ and a signature $f \notin \mathscr{A}$ of arity $n \geq 3$, then there exists a signature $h \notin \mathscr{A}$ of arity $< n$, such that*

$$\text{Pl-Holant}(\widehat{\mathcal{EQ}}, h, \widehat{\mathcal{F}}) \leq_{\text{T}} \text{Pl-Holant}(\widehat{\mathcal{EQ}}, \widehat{\mathcal{F}}).$$

Furthermore, h satisfies the even *parity condition.*

Proof. Again the proof is by induction on n. As in the proof of Lemma 10.4 we may assume that $f^{x_i=0} \in \mathscr{A}$ for $1 \leq i \leq n$, $f_{00\cdots0} = 1$ and f satisfies the *even* parity condition.

♣ Step 1: Suppose supp(f) is not an affine subspace.

Being not an affine subspace, supp(f) is certainly not a linear subspace. Thus there exist $\mathbf{a} = a_1a_2\cdots a_n$, $\mathbf{b} = b_1b_2\cdots b_n$, such that $\mathbf{a}, \mathbf{b} \in \text{supp}(f)$, $\mathbf{c} = \mathbf{a} \oplus \mathbf{b} = c_1c_2\cdots c_n \notin \text{supp}(f)$. Since $f_{00\cdots0} = 1$, i.e., $00\cdots0 \in \text{supp}(f)$, we have $\mathbf{a} \neq \mathbf{b}$. Without loss of generality, we assume that $a_1 = 0, b_1 = 1$. It follows that $c_1 = 1$. Let $\mathbf{a}' = a_3\cdots a_n$, $\mathbf{b}' = b_3\cdots b_n$, $\mathbf{c}' = c_3\cdots c_n$.

Connecting the first variable of $[1, 0, -1]$ to the first variable of f, this gives the signature

$$\widetilde{f}(x_1, x_2, \ldots, x_n) = \sum_{x_1' \in \{0,1\}} [1, 0, -1](x_1', x_1) f(x_1', x_2, \ldots, x_n).$$

Since $[1, 0, -1]$ is a weighted binary equality that takes value 1 on 00 and -1 on 11,

$$\widetilde{f}(x_1, x_2, \ldots, x_n) = (-1)^{x_1} f(x_1, x_2, \ldots, x_n).$$

Moreover, by connecting the variables x_2, x_1 of $[1, 0, 1, 0] \in \widehat{\mathcal{EQ}}$ to the variables x_1, x_2 of f respectively, the planar construction gives the signature

$$h(x', x_3, x_4, \ldots, x_n) = \sum_{x_1, x_2 \in \{0,1\}} [1, 0, 1, 0](x_2, x_1, x') f(x_1, x_2, \ldots, x_n).$$

As in the proof of Lemma 10.4, we have

$$h(x_1 \oplus x_2, x_3, \ldots, x_n) = f(x_1, x_2, x_3, \ldots, x_n) + f(\bar{x}_1, \bar{x}_2, x_3, \ldots, x_n).$$

Similarly, by connecting the variables x_2, x_1 of $[1, 0, 1, 0]$ to the variables x_1, x_2 of $\widetilde{f}$ respectively, the planar gadget gives the signature $\widetilde{h}(x', x_3, x_4, \ldots, x_n)$ with

$$\begin{aligned}\widetilde{h}(x_1 \oplus x_2, x_3, \ldots, x_n) &= (-1)^{x_1} f(x_1, x_2, x_3, \ldots, x_n) \\ &+ (-1)^{\bar{x}_1} f(\bar{x}_1, \bar{x}_2, x_3, \ldots, x_n).\end{aligned}$$

Since $f^{x_i=0} \in \mathscr{A}$ and affine functions have affine support, and furthermore $f^{x_i=0}_{0\cdots0} = 1$, we conclude that $\text{supp}(f^{x_i=0})$ is a linear subspace of $\mathbb{Z}_2^{n-1}$. Recall that all nonzero values of an affine signature have the same norm.

Next comes the tableau calculus. We will examine all four cases of membership of $\bar{a}_1\bar{a}_2\mathbf{a}'$ and $\bar{b}_1\bar{b}_2\mathbf{b}'$ in $\text{supp}(f)$. In each case we prove that either h or $\widetilde{h}$ is not affine with arity $n-1$.

- Suppose $\bar{a}_1\bar{a}_2\mathbf{a}' \in \text{supp}(f)$ and $\bar{b}_1\bar{b}_2\mathbf{b}' \in \text{supp}(f)$. By assumption $\mathbf{a} = 0a_2\mathbf{a}' \in \text{supp}(f)$ and $0\bar{b}_2\mathbf{b}' = \bar{b}_1\bar{b}_2\mathbf{b}' \in \text{supp}(f)$, and since $\text{supp}(f^{x_1=0})$ is a linear subspace, we have

$$\begin{array}{rcl} & a_2\mathbf{a}' & \in \text{supp}(f^{x_1=0}) \\ \oplus & \bar{b}_2\mathbf{b}' & \in \text{supp}(f^{x_1=0}). \\ \hline & \bar{c}_2\mathbf{c}' & \end{array}$$

Hence $\bar{c}_2\mathbf{c}' \in \text{supp}(f^{x_1=0})$. This means that $f_{\bar{c}_1\bar{c}_2\mathbf{c}'} \neq 0$ since $\bar{c}_1 = 0$. By definition of h and $\widetilde{h}$,

$$\begin{aligned} h_{(c_1\oplus c_2)\mathbf{c}'} &= \quad f_{c_1c_2\mathbf{c}'} + f_{\bar{c}_1\bar{c}_2\mathbf{c}'}, \\ \widetilde{h}_{(c_1\oplus c_2)\mathbf{c}'} &= -f_{c_1c_2\mathbf{c}'} + f_{\bar{c}_1\bar{c}_2\mathbf{c}'}. \end{aligned}$$

Thus $|h_{(c_1\oplus c_2)\mathbf{c}'}| = |\widetilde{h}_{(c_1\oplus c_2)\mathbf{c}'}| = 1$ since $f_{c_1c_2\mathbf{c}'} = 0$ and $f_{\bar{c}_1\bar{c}_2\mathbf{c}'} = \pm 1$. Moreover, by

$$\begin{aligned} h_{(a_1\oplus a_2)\mathbf{a}'} &= f_{a_1a_2\mathbf{a}'} + f_{\bar{a}_1\bar{a}_2\mathbf{a}'}, \\ \widetilde{h}_{(a_1\oplus a_2)\mathbf{a}'} &= f_{a_1a_2\mathbf{a}'} - f_{\bar{a}_1\bar{a}_2\mathbf{a}'}, \end{aligned} \tag{10.4}$$

if both $h_{(a_1\oplus a_2)\mathbf{a}'}$ and $\widetilde{h}_{(a_1\oplus a_2)\mathbf{a}'}$ are 0, then $f_{a_1a_2\mathbf{a}'} = 0$, by treating (10.4) as a linear equation system. This contradicts that $\mathbf{a} \in \text{supp}(f)$. Thus we have $h_{(a_1\oplus a_2)\mathbf{a}'} \neq 0$ or $\widetilde{h}_{(a_1\oplus a_2)\mathbf{a}'} \neq 0$. Both $f_{a_1a_2\mathbf{a}'} \neq 0$ and $f_{\bar{a}_1\bar{a}_2\mathbf{a}'} \neq 0$ by assumption, and since f takes values in $\{0, 1, -1\}$, we have $f_{a_1a_2\mathbf{a}'}, f_{\bar{a}_1\bar{a}_2\mathbf{a}'} \in \{1, -1\}$. This implies that one of $|h_{(a_1\oplus a_2)\mathbf{a}'}|$ or $|\widetilde{h}_{(a_1\oplus a_2)\mathbf{a}'}|$ is 2. So h or $\widetilde{h}$ is not affine, because at least one of them has nonzero values of unequal norm.

- Suppose $\bar{a}_1\bar{a}_2\mathbf{a}' \notin \text{supp}(f)$ and $\bar{b}_1\bar{b}_2\mathbf{b}' \notin \text{supp}(f)$. By treating the following as a linear system

$$\begin{aligned} h_{00\cdots0} &= f_{000\cdots0} + f_{110\cdots0}, \\ \widetilde{h}_{00\cdots0} &= f_{000\cdots0} - f_{110\cdots0}, \end{aligned} \tag{10.5}$$

we have $h_{00\cdots0} \neq 0$ or $\widetilde{h}_{00\cdots0} \neq 0$ since $f_{000\cdots0} \neq 0$. Without loss of generality, we assume that $h_{00\cdots0} \neq 0$. The same argument can be applied to $\widetilde{h}$ if $\widetilde{h}_{00\cdots0} \neq 0$. Suppose h is affine, for otherwise we are done. Since $00\cdots0 \in \text{supp}(h)$, $\text{supp}(h)$ is a linear subspace. By

$$\begin{aligned} h_{(a_1\oplus a_2)\mathbf{a}'} &= f_{a_1a_2\mathbf{a}'} + f_{\bar{a}_1\bar{a}_2\mathbf{a}'}, \\ h_{(b_1\oplus b_2)\mathbf{b}'} &= f_{b_1b_2\mathbf{b}'} + f_{\bar{b}_1\bar{b}_2\mathbf{b}'}, \end{aligned}$$

we have $|h_{(a_1\oplus a_2)\mathbf{a}'}| = 1$, $|h_{(b_1\oplus b_2)\mathbf{b}'}| = 1$ since $f_{\bar{a}_1\bar{a}_2\mathbf{a}'} = f_{\bar{b}_1\bar{b}_2\mathbf{b}'} = 0$ and $f_{a_1a_2\mathbf{a}'}, f_{b_1b_2\mathbf{b}'} \in \{1, -1\}$. This implies that $(a_1 \oplus a_2)\mathbf{a}', (b_1 \oplus b_2)\mathbf{b}' \in \text{supp}(h)$. Then by

$$\begin{array}{rll} & (a_1 \oplus a_2)\mathbf{a}' & \in \text{supp}(h) \\ \oplus & (b_1 \oplus b_2)\mathbf{b}' & \in \text{supp}(h) \\ \hline & (c_1 \oplus c_2)\mathbf{c}' & \end{array}$$

we have $(c_1 \oplus c_2)\mathbf{c}' \in \text{supp}(h)$. This implies that

$$h_{(c_1\oplus c_2)\mathbf{c}'} = f_{c_1c_2\mathbf{c}'} + f_{\bar{c}_1\bar{c}_2\mathbf{c}'} \neq 0.$$

Thus $f_{\bar{c}_1\bar{c}_2\mathbf{c}'} \neq 0$ since $f_{c_1c_2\mathbf{c}'} = 0$. So $\bar{c}_1\bar{c}_2\mathbf{c}' \in \text{supp}(f)$. Therefore, $\bar{c}_2\mathbf{c}' \in \text{supp}(f^{x_1=0})$ since $\bar{c}_1 = 0$. Recall that by assumption $\mathbf{a} = 0a_2\mathbf{a}' \in \text{supp}(f)$. By

$$\begin{array}{rll} & a_2\mathbf{a}' & \in \text{supp}(f^{x_1=0}) \\ \oplus & \bar{c}_2\mathbf{c}' & \in \text{supp}(f^{x_1=0}) \\ \hline & \bar{b}_2\mathbf{b}' & \end{array}$$

we have $\bar{b}_2\mathbf{b}' \in \text{supp}(f^{x_1=0})$. Thus $\bar{b}_1\bar{b}_2\mathbf{b}' \in \text{supp}(f)$ as $\bar{b}_1 = 0$. This is a contradiction.

- Suppose $\bar{a}_1\bar{a}_2\mathbf{a}' \in \text{supp}(f)$ and $\bar{b}_1\bar{b}_2\mathbf{b}' \notin \text{supp}(f)$. By treating the following as a linear system,

$$h_{(b_1 \oplus b_2)\mathbf{b}'} = \quad f_{b_1b_2\mathbf{b}'} + f_{\bar{b}_1\bar{b}_2\mathbf{b}'},$$
$$\widetilde{h}_{(b_1 \oplus b_2)\mathbf{b}'} = -f_{b_1b_2\mathbf{b}'} + f_{\bar{b}_1\bar{b}_2\mathbf{b}'},$$

we have $|h_{(b_1 \oplus b_2)\mathbf{b}'}| = |\widetilde{h}_{(b_1 \oplus b_2)\mathbf{b}'}| = 1$ since $f_{b_1b_2\mathbf{b}'} = \pm 1$ and $f_{\bar{b}_1\bar{b}_2\mathbf{b}'} = 0$. Then by

$$h_{(a_1 \oplus a_2)\mathbf{a}'} = f_{a_1a_2\mathbf{a}'} + f_{\bar{a}_1\bar{a}_2\mathbf{a}'},$$
$$\widetilde{h}_{(a_1 \oplus a_2)\mathbf{a}'} = f_{a_1a_2\mathbf{a}'} - f_{\bar{a}_1\bar{a}_2\mathbf{a}'},$$

and $f_{a_1a_2\mathbf{a}'}, f_{\bar{a}_1\bar{a}_2\mathbf{a}'} \in \{1, -1\}$, we have $|h_{(a_1 \oplus a_2)\mathbf{a}'}| = 2$ or $|\widetilde{h}_{(a_1 \oplus a_2)\mathbf{a}'}| = 2$. This implies that h or $\widetilde{h}$ is not affine, since one of them has nonzero values of unequal norm.
- The proof for the case $\bar{a}_1\bar{a}_2\mathbf{a}' \notin \text{supp}(f)$ and $\bar{b}_1\bar{b}_2\mathbf{b}' \in \text{supp}(f)$ is symmetric.

♣ Step 2: Now we may assume that supp(f) is an affine subspace with dimension $k \geq 0$.

Let Y be a set of k free variables, $Y = \{y_1, y_2, \ldots, y_k\} = \{x_{i_1}, x_{i_2} \ldots, x_{i_k}\}$. Let $\underline{f}(Y)$ be the compressed signature of f. By Corollary 10.9, $\underline{f} \notin \mathscr{A}$. However, for $\{1, -1\}$-valued $\underline{f}$, if $k = 2$, then $\underline{f} \in \mathscr{A}$ by Corollary 10.7, a contradiction. Also trivially $\underline{f} \in \mathscr{A}$ if $k < 2$. Hence $k \geq 3$.

By Lemma 10.5, and since $\underline{f}$ takes values in $\{1, -1\}$, there exists a unique multilinear polynomial $Q(y_1, y_2, \ldots, y_k) \in \mathbb{Z}_2[X]$ such that $\underline{f}(y_1, y_2, \ldots, y_k) = (-1)^{Q(y_1, y_2, \ldots, y_k)}$. $\underline{f}$ is affine iff $Q(y_1, y_2, \ldots, y_k)$ is a multilinear polynomial of total degree ≤ 2.

If $k \geq 4$ and there exists a term $y_{i_1}y_{i_2}\cdots y_{i_s}$ with nonzero coefficient, where $3 \leq s < k$, then there exists some $y_j \notin \{y_{i_1}, y_{i_2}, \ldots, y_{i_s}\}$, such that

$$\underline{f}^{y_j=0}(y_1, \ldots, \widehat{y_j}, \ldots, y_k) = (-1)^{Q'(y_1, \ldots, \widehat{y_j}, \ldots, y_k)},$$

where Q' is a polynomial on $k-1$ variables ($\widehat{y_j}$ denotes the absence of y_j), where the term $y_{i_1}y_{i_2}\cdots y_{i_s}$ of degree $s>2$ still appears. This implies that $f^{y_j=0}$ is not affine. This is a contradiction. Thus we may assume that $Q(y_1,y_2,\ldots,y_k)=P(y_1,y_2,\ldots,y_k)+ay_1y_2\cdots y_k$, where $P(y_1,y_2,\ldots,y_k)\in\mathbb{Z}_2[X]$ is a multilinear polynomial of total degree ≤ 2, and $a\in\mathbb{Z}_2$. Note that this statement is also vacuously true if $k=3$. If $a=0$, then f is affine. This is a contradiction. Otherwise, $Q(y_1,y_2,\ldots,y_k)=P(y_1,y_2,\ldots,y_k)+y_1y_2\cdots y_k$. Moreover, by connecting the first variable of $[1,0,-1]$ to y_i of f, the gadget gives the signature f' such that

$$f'(x_1,x_2,\ldots,x_n)=(-1)^{y_i}f(x_1,x_2,\ldots,x_n).$$

This implies that f' has the same support of f and

$$\underline{f'}(y_1,y_2,\ldots,y_k)=(-1)^{y_i+P(y_1,y_2,\ldots,y_k)+y_1y_2\cdots y_k},$$

where $\underline{f'}$ is the compressed signature of f' for Y. Thus $f'\notin\mathscr{A}$. This implies that we can add a linear term to $P(y_1,y_2,\ldots,y_k)$ in $\mathbb{Z}_2[X]$ freely.

We connect all variables of f except for y_1 to $n-1$ variables of the even parity function $\frac{1}{2}\{[1,1]^{\otimes n}+[1,-1]^{\otimes n}\}=[1,0,1,\ldots,0\text{ (or }1)]\in\widehat{\mathcal{EQ}}$. This planar construction gives a binary signature f^* that also has even parity, since both f and $[1,0,1,\ldots,0\text{ (or }1)]$ have even parity. Hence $f^*_{01}=f^*_{10}=0$, and the binary signature f^* has the form $[f^*_{00},0,f^*_{11}]$. In particular, f^* is symmetric. We have

$$\begin{aligned} f^*_{00} &= \sum_{y_2,y_3,\ldots,y_k\in\{0,1\}} \underline{f}^{y_1=0}(y_2,y_3,\ldots,y_k),\\ f^*_{11} &= \sum_{y_2,y_3,\ldots,y_k\in\{0,1\}} \underline{f}^{y_1=1}(y_2,y_3,\ldots,y_k). \end{aligned} \tag{10.6}$$

These values are as stated because for any $y_1=0,1$ as the first input bit to f^*, there is exactly one term $(x_1,x_2,\ldots x_n)\in\text{supp}(f)$ in the sum defining f^*, for every $(y_1,y_2,\ldots,y_k)$. Because f has even parity, $\sum_{i=1}^n x_i\equiv 0 \bmod 2$, thus the sum of the internal $n-1$ edges has the same parity as y_1. On the other hand, the even parity function $[1,0,1,\ldots,0\text{ (or }1)]$ ensures that this sum of internal edges has the same parity as the second input bit of f^*.

Now we take a closer look at f^*. First, we consider the special case that the coefficient of y_iy_j in $P(y_1,y_2,\ldots,y_k)$ is nonzero for all $1\le i<j\le k$.

- If $k=3$, we may assume that $P(y_1,y_2,y_3)=y_1+y_2+y_3+y_1y_2+y_1y_3+y_2y_3$ since we can add linear terms to $P(y_1,y_2,\ldots,y_k)$ at will. Then we have $\underline{f}(y_1,y_2,y_3)=(-1)^{P(y_1,y_2,y_3)+y_1y_2y_3}$. The polynomial $P(y_1,y_2,y_3)+y_1y_2y_3=1+(1+y_1)(1+y_2)(1+y_3)\in\mathbb{Z}_2[y_1,y_2,y_3]$ corresponds to the

OR function on three bits, $y_1 \vee y_2 \vee y_3$. Thus the signature matrix

$$M_{y_1,y_2y_3}(\underline{f}) = \begin{bmatrix} 1 & -1 & -1 & -1 \\ -1 & -1 & -1 & -1 \end{bmatrix}.$$

Thus $f^* = [-2, 0, -4]$, which has nonzero terms of unequal norm, thus not in $\mathscr{A}$ and we are done.

- For $k \geq 4$, we may assume that $P(y_1, y_2, \ldots, y_k)$ has no linear terms since we can add linear terms to $P(y_1, y_2, \ldots, y_k)$ freely. Since P has all terms y_iy_j, both $\underline{f}^{y_1=0}$ and $\underline{f}^{y_1=1}$ are symmetric signatures. For $\underline{f}^{y_1=0}$, the entry of Hamming weight ℓ is $(\underline{f}^{y_1=0})_\ell = (-1)^{\frac{\ell(\ell-1)}{2}}$ for $0 \leq \ell \leq k-1$. For $\underline{f}^{y_1=1}$, we have $(\underline{f}^{y_1=1})_\ell = (-1)^{\frac{\ell(\ell+1)}{2}}$ for $0 \leq \ell \leq k-2$ and $(\underline{f}^{y_1=1})_{k-1} = (-1)^{\frac{k(k-1)}{2}+1}$. This implies that

$$\begin{aligned}
\underline{f}^{y_1=0} &= [1, 1, -1, -1, \ldots, (-1)^{\frac{(k-1)(k-2)}{2}}] \\
&= \frac{1}{1+\mathfrak{i}}\left\{[1, \mathfrak{i}]^{\otimes k-1} + \mathfrak{i}[1, -\mathfrak{i}]^{\otimes k-1}\right\}, \\
\underline{f}^{y_1=1} &= [1, -1, -1, 1, \ldots, (-1)^{\frac{k(k-1)}{2}}] - 2(-1)^{\frac{k(k-1)}{2}}[0, 1]^{\otimes k-1} \\
&= \frac{1}{1-\mathfrak{i}}\left\{[1, \mathfrak{i}]^{\otimes k-1} - \mathfrak{i}[1, -\mathfrak{i}]^{\otimes k-1}\right\} - 2(-1)^{\frac{k(k-1)}{2}}[0, 1]^{\otimes k-1}.
\end{aligned}$$

Thus

$$\begin{aligned}
f_{00}^* = \sum_{\beta \in \{0,1\}^{k-1}} (\underline{f}^{y_1=0})_\beta &= \frac{1}{1+\mathfrak{i}}\sum_{w=0}^{k-1}\binom{k-1}{w}[\mathfrak{i}^w + \mathfrak{i}(-\mathfrak{i})^w] \\
&= \frac{1}{1+\mathfrak{i}}[(1+\mathfrak{i})^{k-1} + \mathfrak{i}(1-\mathfrak{i})^{k-1}] \\
&= (1+\mathfrak{i})^{k-2} + (1-\mathfrak{i})^{k-2} \\
&= 2^{\frac{k}{2}}\cos((k-2)\pi/4), \\
f_{11}^* = \sum_{\beta \in \{0,1\}^{k-1}} (\underline{f}^{y_1=1})_\beta &= \frac{1}{1-\mathfrak{i}}\sum_{w=0}^{k-1}\binom{k-1}{w}[\mathfrak{i}^w - \mathfrak{i}(-\mathfrak{i})^w] - 2(-1)^{\frac{k(k-1)}{2}} \\
&= \frac{1}{1-\mathfrak{i}}[(1+\mathfrak{i})^{k-1} - \mathfrak{i}(1-\mathfrak{i})^{k-1}] - 2(-1)^{\frac{k(k-1)}{2}} \\
&= -2^{\frac{k}{2}}\sin((k-2)\pi/4) - 2(-1)^{\frac{k(k-1)}{2}}.
\end{aligned}$$

For $k \equiv 1 \mod 2$, $|f_{00}^*| = 2^{\frac{k-1}{2}}$, and $|f_{11}^*| = 2^{\frac{k-1}{2}} \pm 2$ (since $k \geq 4$). Therefore $f_{11}^* f_{00}^* \neq 0$ and $|f_{11}^*| \neq |f_{00}^*|$ (again using $k \geq 4$). Thus $f^* \notin \mathscr{A}$ and we are done.

For $k \equiv 0 \mod 4$, $f^*_{00} = 0$, $|f^*_{11}| = 2^{\frac{k}{2}} \pm 2 \neq 0$. This implies that we have $f^* = f^*_{11}[0, 1]^{\otimes 2}$. Using $[1, 0, -1]$, $[0, 1]^{\otimes 2}$ and $f \notin \mathscr{A}$, by a separate lemma we can get a binary signature of the form $[x, 0, y] \notin \mathscr{A}$. (The proof of this separate lemma is nontrivial and is omitted here; see [CF16] for details.)

For $k \equiv 2 \mod 4$, $|f^*_{00}| = 2^{\frac{k}{2}} \geq 4$ since $k \geq 4$, and $|f^*_{11}| = 2$, so $f^* \notin \mathscr{A}$ and we are done.

Now we assume that there exist $i \neq j \in [k]$ such that the coefficient of $y_i y_j$ is 0 in $P(y_1, y_2, \ldots, y_k)$. For notational simplicity, without loss of generality we assume that $i = k-1$, $j = k$. Then we can assume that (with the linear term y_{k-1} and y_k removed if needed)

$$P(y_1, y_2, \ldots, y_k) = y_1(L_1 + \epsilon_1) + y_2(L_2 + \epsilon_2) + \cdots + y_{k-2}(L_{k-2} + \varepsilon_{k-2}),$$

where

$$L_1 = \sum_{i=2}^{k} a_{1i} y_1, \quad L_2 = \sum_{i=3}^{k} a_{2i} y_i, \quad \ldots, \quad L_{k-2} = \sum_{i=k-1}^{k} a_{(k-2)i} y_i$$

and $a_{ji} \in \mathbb{Z}_2$ are fixed, but we can choose $\epsilon_i \in \mathbb{Z}_2$ freely since we can add linear terms freely. Let $F_{(0)} = \underline{f}$, $F_{(i)} = \underline{f}^{y_1=0, y_2=0, \ldots, y_i=0}$ for $i \in [k-2]$. We claim that there exist $\epsilon_1, \epsilon_2, \ldots, \epsilon_{k-2} \in \mathbb{Z}_2$ such that

$$\sum_{y_{i+1}, y_{i+2}, \ldots, y_k \in \{0,1\}} F_{(i)}(y_{i+1}, y_{i+2}, \ldots, y_k) \geq 4,$$

for all $1 \leq i \leq k-2$. We prove this claim by induction. The base case is for $F_{(k-2)}$. Note that $P(0, \ldots, 0, y_{k-1}, y_k)$ is identically 0. Thus

$$\sum_{y_{k-1}, y_k \in \{0,1\}} F_{(k-2)}(y_{k-1}, y_k) = 4.$$

By induction, we may assume that

$$\sum_{y_{i+1}, y_{i+2}, \ldots, y_k \in \{0,1\}} F_{(i)}(y_{i+1}, y_{i+2}, \ldots, y_k) \geq 4$$

and prove that

$$\sum_{y_i, y_{i+1}, \ldots, y_k \in \{0,1\}} F_{(i-1)}(y_i, y_{i+1}, \ldots, y_k) \geq 4$$

for $i \leq k-2$. Note that

$$\begin{aligned} F_{(i-1)}(1, y_{i+1}, y_{i+2}, \ldots, y_k) &= (-1)^{(L_i + \epsilon_i) + y_{i+1}(L_{i+1} + \epsilon_{i+1}) + \cdots + y_{k-2}(L_{k-2} + \epsilon_{k-2})}, \\ F_{(i-1)}(0, y_{i+1}, y_{i+2}, \ldots, y_k) &= (-1)^{y_{i+1}(L_{i+1} + \epsilon_{i+1}) + \cdots + y_{k-2}(L_{k-2} + \epsilon_{k-2})}. \end{aligned} \tag{10.7}$$

By inductive hypothesis,

$$\sum_{y_{i+1},\ldots,y_k\in\{0,1\}} F_{(i-1)}(0,y_{i+1},\ldots,y_k) = \sum_{y_{i+1},\ldots,y_k\in\{0,1\}} F_{(i)}(y_{i+1},\ldots,y_k) \geq 4.$$

If L_i is identically 0, then we set $\epsilon_i = 0$. It follows that

$$\sum_{y_{i+1},\ldots,y_k\in\{0,1\}} F_{(i-1)}(1,y_{i+1},\ldots,y_k) = \sum_{y_{i+1},\ldots,y_k\in\{0,1\}} F_{(i-1)}(0,y_{i+1},\ldots,y_k).$$

Thus we have

$$\sum_{y_i,y_{i+1},\ldots,y_k\in\{0,1\}} F_{(i-1)}(y_i,y_{i+1},\ldots,y_k) = 2\sum_{y_{i+1},\ldots,y_k\in\{0,1\}} F_{(i-1)}(0,y_{i+1},\ldots,y_k) \geq 8$$

and we are done.

Otherwise, $L_i = 0$ defines a subspace V of $\mathbb{Z}_2^{k-i}$ that has dimension $k-i-1$. Let $V' = \mathbb{Z}_2^{k-i} \setminus V$; then V' is an affine space defined by $L_i = 1$ which has dimension $k-i-1$.

Let

$$a = \sum_{y_{i+1}y_{i+2}\cdots y_k\in V} F_{(i-1)}(0,y_{i+1},y_{i+2},\ldots,y_k),$$

$$b = \sum_{y_{i+1}y_{i+2}\cdots y_k\in V'} F_{(i-1)}(0,y_{i+1},y_{i+2},\ldots,y_k),$$

$$a' = \sum_{y_{i+1}y_{i+2}\cdots y_k\in V} F_{(i-1)}(1,y_{i+1},y_{i+2},\ldots,y_k),$$

and

$$b' = \sum_{y_{i+1}y_{i+2}\cdots y_k\in V'} F_{(i-1)}(1,y_{i+1},y_{i+2},\ldots,y_k).$$

Then

$$\sum_{y_i,y_{i+1},\ldots,y_k\in\{0,1\}} F_{(i-1)}(y_i,y_{i+1},\ldots,y_k) = a+b+a'+b'.$$

By induction, we have $a+b \geq 4$. Thus $a \geq 2$ or $b \geq 2$. If $a \geq 2$, we choose $\epsilon_i = 0$, then $a = a'$, $b = -b'$ by (10.7). Thus $a+a'+b+b' = 2a \geq 4$. If $b \geq 2$, we choose $\epsilon_i = 1$, then $a = -a'$, $b = b'$ by (10.7). Thus $a+a'+b+b' = 2b \geq 4$. This finishes the proof of the claim.

The claim shows that

$$f_{00}^* = \sum_{y_2,y_3,\ldots,y_k\in\{0,1\}} f^{y_1=0}(y_2,y_3,\ldots,y_k) = \sum_{y_2,y_3,\ldots,y_k\in\{0,1\}} F_{(1)}(y_2,y_3,\ldots,y_k) \geq 4.$$

Let g be the n-ary signature with the same support as f (thus satisfies the even parity condition) and on its support

$$g(x_1, x_2, \ldots, x_n) = (-1)^{P(y_1, y_2, \ldots, y_k)},$$

then $\underline{f}_\beta = \underline{g}_\beta$ for any $\beta \in \{0, 1\}^k$ other than $\beta = 11 \cdots 1$. For $\beta = 11 \cdots 1$,

$$\underline{f}_\beta = (-1)^{P(1,1,\ldots,1)+1\cdot 1 \cdots 1} = -(-1)^{P(1,1,\ldots,1)} = -\underline{g}_\beta.$$

This implies that

$$\underline{f} = \underline{g} \pm 2[0, 1]^{\otimes k}. \tag{10.8}$$

(Note that we do not really construct g in Pl-Holant$(\widehat{\mathcal{EQ}}, \widehat{\mathcal{F}})$. We just use g to argue that $f^* \notin \mathscr{A}$.) Since both $[1, 0, 1, \ldots, 0 \text{ (or } 1)]$ and g are affine signatures, the following construction would define an affine signature: Connect all variables of g other than y_1 to $n - 1$ variables of $[1, 0, 1, \ldots, 0 \text{ (or } 1)]$. This construction gives a binary signature $g^* = [g^*_{00}, 0, g^*_{11}]$. (By the even parity condition, the weight 1 entry must be 0.) Note that

$$g^*_{00} = \sum_{y_2, y_3, \ldots, y_k \in \{0,1\}} \underline{g}^{y_1=0}(y_2, y_3, \ldots, y_k),$$

$$g^*_{11} = \sum_{y_2, y_3, \ldots, y_k \in \{0,1\}} \underline{g}^{y_1=1}(y_2, y_3, \ldots, y_k).$$

Thus by (10.6) and (10.8), we have

$$f^*_{00} = g^*_{00}, \quad f^*_{11} = g^*_{11} \pm 2. \tag{10.9}$$

Since g^* is a binary affine signature, we must have

$$\text{either} \quad g^*_{00} = 0 \quad \text{or} \quad g^*_{11} = 0 \quad \text{or} \quad (g^*_{00})^4 = (g^*_{11})^4.$$

Since we have $g^*_{00} = f^*_{00} \geq 4$ and both g^*_{00} and g^*_{11} are real numbers, we must have

$$\text{either} \quad g^*_{11} = 0 \quad \text{or} \quad g^*_{11} = \pm g^*_{00}.$$

If $g^*_{11} = 0$ then by (10.9) $f^*_{11} = \pm 2$ has a different nonzero norm than f^*_{00}. If $g^*_{11} = \pm g^*_{00}$, then by (10.9) $g^*_{11} = \pm f^*_{00}$ has norm ≥ 4, and thus $f^*_{11} = g^*_{11} \pm 2$ has norm $|g^*_{11}| \pm 2 = |f^*_{00}| \pm 2$. And so in this case f^*_{11} also has a different nonzero norm than f^*_{00}. In each case, $|f^*_{00}| \neq |f^*_{11}|$ and $f^*_{00} f^*_{11} \neq 0$. This implies that $f^* \notin \mathscr{A}$ and we are done. □

10.2 Holant* Dichotomy

In this section we prove a complexity dichotomy theorem for Holant*($\mathcal{F}$), where $\mathcal{F}$ is a set of complex-valued, *not necessarily symmetric*, constraint functions on Boolean variables. In Theorem 10.14 we identify four classes of problems that are polynomial time computable; all other problems are proved to be #P-hard. The main proof technique and indeed the formulation of the theorem use holographic transformations.

Theorem 2.12 from Chapter 2 proved a complexity dichotomy theorem for Holant*($\mathcal{F}$), where $\mathcal{F}$ is a set of complex-valued *symmetric* constraint functions on Boolean variables. Theorem 3.5 from Chapter 3 is an equivalent statement of Theorem 2.12 stated in terms of holographic transformations (except for case 3, which has to do with vanishing signatures $\mathscr{V}$ introduced in Chapter 7). Theorem 10.14 is a generalization of Theorem 2.12 to asymmetric signatures; Theorem 2.12 can be viewed as a special case of Theorem 10.14 restricted to symmetric signatures. Furthermore, Theorem 10.14 also helps us gain a deeper and clearer understanding of the tractable cases for the symmetric ones.

10.2.1 Dichotomy Statement

For any (constraint) function F, or synonymously a signature, of arity n, we also use the same symbol F to denote the column vector indexed by $\{0, 1\}^n$ as an expression of F, listing all its values in lexicographic order. Sometimes we also partition the variable set into two parts $\{x_1, x_2, \ldots, x_n\} = I \cup J$, and write F as a matrix with rows indexed by $\{0, 1\}^{|I|}$ and columns indexed by $\{0, 1\}^{|J|}$. For example, for a ternary function F, we can write it as a 4×2 matrix $F = F_{x_1x_2,x_3}$. As another example, for a binary function $F(x, y)$ we can write it both as a vector in $\mathbb{C}^4$, or a 2×2 matrix $F = F_{x,y}$, with row index x and column index y ranging over $\{0, 1\}$. As constraint functions, F and cF, for any nonzero $c \in \mathbb{C}$, are equivalent in terms of the complexity of Holant problems they define, and we denote it by $F \cong cF$. $F \cong 0$ means that F is (identically) 0.

We use $F^{x_i=x_j}$ to denote the function of arity $n-2$ realized by a function F of arity $n \geq 2$ after taking a self-loop by connecting two edges corresponding to variables x_i and x_j. Another operation is called juxtaposition, or tensor product. Suppose F is a function of arity n and $\mathcal{I} = \{I_1, \ldots, I_k\}$ is a partition of $[n]$. If $F(X) = \prod_{j=1}^{k} F_j(X|_{I_j})$ for some functions $F_1, \ldots, F_k$, where $X = \{x_1, \ldots, x_n\}$ and $X|_{I_j} = \{x_s | s \in I_j\}$ (we also denote it by X_j), then we say F can be decomposed into type $\mathcal{I}$, or simply F has type $\mathcal{I}$. We say $\mathcal{I}$ is a *type specification*. We denote such an F by $F = \bigotimes_{\mathcal{I}}(F_1, \ldots, F_k)$. If each F_j is the function of some gadget, then $\bigotimes_{\mathcal{I}}(F_1, \ldots, F_k)$ is the function of the gadget which is the disjoint

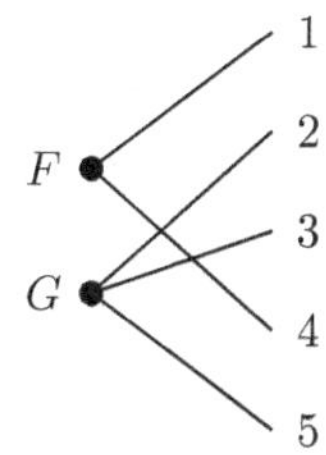

Figure 10.2. An example of juxtaposition $\bigotimes_{\mathcal{I}}(F, G)$, where $\mathcal{I} = \{\{1, 4\}, \{2, 3, 5\}\}$.

union of these gadgets for F_j, with variables arranged according to $\mathcal{I}$. When the indexing is clear, we also use notation $F_1 \otimes \cdots \otimes F_k$. Note that this tensor product notation $\otimes$ is consistent with tensor product of matrices and vectors (see Figure 10.2.) This definition of $F = \bigotimes_{\mathcal{I}}(F_1, \ldots, F_k)$ can be easily generalized to the case where $\mathcal{I} = \{I_1, \ldots, I_k\}$ is a partition of an arbitrary finite set of indices.

We use $F^{x_{j_1}=U_1,\ldots,x_{j_k}=U_k}$ to denote the function realized by connecting the unary function U_s to the input variable x_{j_s} of F, (for $s = 1, \ldots, k$). $F^{x_j=0}$ and $F^{x_j=1}$ denote respectively $F^{x_j=[1,0]}$ and $F^{x_j=[0,1]}$, and when x_j is clear we also use FU to denote $F^{x_j=U}$. The notation FU is also consistent with matrix-vector product. Matrix multiplication will be used often. For example, suppose $A = A_{x_1,x_2}$, $B = B_{x_3,x_4}$ and $C = C_{x_5,x_6}$ are three binary functions. Then the function $(A \otimes B \otimes C)^{x_2=x_3,x_4=x_5}$ is formed by linking the three functions A, B, C in a path, and its matrix form is just the matrix product ABC indexed by x_1 (for row) and x_6 (for column). Note that $A_{\emptyset,x_1}B_{x_2,\emptyset}$ or $A^{\mathrm{T}}B$ is the dot product of unary functions A and B. Similarly, $A_{x_1,\emptyset}B_{\emptyset,x_2}$ or AB^{T} is the matrix form of the tensor product function $\bigotimes_{\{\{1\},\{2\}\}}(A, B)$ (or just $A \otimes B$) of unary functions A and B.

We say a function set $\mathcal{F}$ is closed under tensor product if for any functions A and B on two disjoint sets of variables indexed by I and J respectively, $A, B \in \mathcal{F}$ implies that $\bigotimes_{\mathcal{I}}(A, B) \in \mathcal{F}$, where $\mathcal{I} = \{I, J\}$. Given $\mathcal{F}$, its tensor closure $\langle \mathcal{F} \rangle$ is the minimum set containing $\mathcal{F}$ and is closed under tensor product. This closure exists, being the set of all functions obtained by taking a finite sequence of tensor products from $\mathcal{F}$. Because tensor products are defined for functions on disjoint sets of variables, for a holographic transformation defined by any matrix M, $\langle M\mathcal{F} \rangle = M\langle \mathcal{F} \rangle$.

The following three classes of functions are important for our dichotomy.

Definition 10.11. A function $F : \{0, 1\}^n \to \mathbb{C}$ is in the class $\mathcal{E}$ iff F is zero except (possibly) on two inputs $(a_1, \ldots, a_n)$ and $(\bar{a}_1, \ldots, \bar{a}_n) = (1 - a_1, \ldots, 1 - a_n)$.

In other words, $F \in \mathcal{E}$ iff its support is contained in a pair of complementary points. We think of $\mathcal{E}$ as a generalization to the asymmetric setting of GENERALIZED EQUALITY $[a, 0, \ldots, 0, b]$. The tensor closure of $\mathcal{E}$ is a class we are already familiar with.

Exercise: Prove that the tensor closure $\langle \mathcal{E} \rangle$ is the class of signatures of product type $\mathscr{P}$.

Definition 10.12. A function $F : \{0, 1\}^n \to \mathbb{C}$ is in the class $\mathcal{M}$ iff F is 0 except (possibly) on $n + 1$ inputs whose Hamming weight is ≤ 1.

The name $\mathcal{M}$ is given for *matching*. But this is not to be confused with the matchgate signatures $\mathscr{M}$, which is defined by gadgets called matchgates using weighted *perfect* matching signatures.

Definition 10.13. $\mathcal{T}$ is the set of all functions of arity ≤ 2.

Note that $\mathcal{U}$ is a subset of all three classes, $\mathcal{E}$, $\mathcal{M}$, and $\mathcal{T}$. $\langle \mathcal{U} \rangle$ is the class of degenerate signatures. A binary function belongs to $\langle \mathcal{U} \rangle$ iff its matrix has determinant 0. A ternary function $F(x_1, x_2, x_3)$ belongs to $\langle \mathcal{T} \rangle$ iff $F^{x_j = U} \cong 0$ for some $1 \leq j \leq 3$ and some unary $U \ncong 0$. If furthermore the ternary function $F(x_1, x_2, x_3)$ is symmetric, then the following statements are all equivalent: (1) $F \in \langle \mathcal{T} \rangle$; (2) $F \in \langle \mathcal{U} \rangle$; (3) $F = [a, b]^{\otimes 3}$ for some unary $[a, b]$; and (4) $FU \cong 0$ for some unary $U \ncong 0$ (take $U = [b, -a]$ if $[a, b] \ncong 0$, or *any* unary $U \ncong 0$ if $[a, b] \cong 0$).

Denote

$$Z_1 = \begin{bmatrix} 1 & 1 \\ \mathrm{i} & -\mathrm{i} \end{bmatrix} \quad \text{and} \quad Z_2 = \begin{bmatrix} 1 & 1 \\ -\mathrm{i} & \mathrm{i} \end{bmatrix}.$$

Note that *symmetric* signatures f in $Z_1\mathcal{M}$ are the same as those in $\mathscr{R}_2^+$, and equivalently $\mathrm{rd}^+(f) \leq 1$. Similarly Z_2 corresponds to $\mathscr{R}_2^-$, and rd^-.

Theorem 10.14. *Let $\mathcal{F}$ be any set of complex-valued functions in Boolean variables. The problem Holant* $(\mathcal{F})$ is polynomial time computable, if*

1. *$\mathcal{F} \subseteq \langle \mathcal{T} \rangle$, or*
2. *there exists an orthogonal matrix H such that $\mathcal{F} \subseteq \langle H\mathcal{E} \rangle$, or*
3. *for $Z = Z_1$ or Z_2, $\mathcal{F} \subseteq \langle Z\mathcal{E} \rangle$, or*
4. *for $Z = Z_1$ or Z_2, $\mathcal{F} \subseteq \langle Z\mathcal{M} \rangle$.*

In all other cases, Holant $(\mathcal{F})$ is #P-hard. The dichotomy is still valid even if the inputs are restricted to planar graphs.*

Note that since $\langle M\mathcal{F}\rangle = M\langle\mathcal{F}\rangle$, and $\langle\mathcal{E}\rangle = \mathscr{P}$, the tractable classes can also be stated as $\mathcal{F} \subseteq \langle\mathcal{T}\rangle$, $\mathcal{F} \subseteq H\mathscr{P}$, $\mathcal{F} \subseteq Z\mathscr{P}$, or $\mathcal{F} \subseteq Z\langle\mathcal{M}\rangle$.

Compared to Theorems 2.12 and 3.5, the case $\mathcal{F} \subseteq \langle\mathcal{T}\rangle$ clearly corresponds to sets of functions of arity ≤ 2 (case 1 of Theorem 3.5, also Class $\mathfrak{A}$ of Theorem 2.12). The cases $\mathcal{F} \subseteq \langle H\mathcal{E}\rangle$ and $\mathcal{F} \subseteq \langle Z\mathcal{E}\rangle$ correspond to case 2 of Theorem 3.5, also to all of Class $\mathfrak{B}$ (except the case $\mathrm{I}(1, \pm 2\mathrm{i})$) as well as Class $\mathfrak{C}$ in Theorem 2.12. This is the same as $\mathscr{P}$-transformable for asymmetric signatures. The case $\mathcal{F} \subseteq \langle Z\mathcal{M}\rangle$ corresponds to case 3 of Theorem 3.5, also to the case $\mathrm{I}(1, \pm 2\mathrm{i})$ of Class $\mathfrak{B}$ in Theorem 2.12.

Exercise: Prove that, for symmetric signatures, Theorem 10.14 is equivalent to Theorems 3.5 and 2.12. Moreover, the tractable classes correspond to each other as described above.

Exercise: Explain how the tractable cases in Theorem 7.19 related to the vanishing signatures $\mathscr{V}$ present themselves in Theorem 10.14. Explain why the tractable classes $\mathscr{A}$-transformable signatures and $\mathscr{M}$-transformable signatures do not appear in Theorem 10.14.

10.2.2 Tractability

We first prove the four classes in Theorem 10.14 are polynomial time computable.

Theorem 10.15. *Holant*$^*(\langle\mathcal{T}\rangle)$, *Holant*$^*(\langle H\mathcal{E}\rangle)$, *Holant*$^*(\langle Z\mathcal{E}\rangle)$ *and Holant*$^*(\langle Z\mathcal{M}\rangle)$ *are polynomial time computable.*

Proof. By "decoupling" a vertex v into several vertices according to its tensor product factors of the function at v, Holant$^*(\langle\mathcal{F}\rangle)$ is trivially equivalent to Holant$^*(\mathcal{F})$, for any $\mathcal{F}$.

Firstly, to show the tractability of Holant$^*(\mathcal{T})$, we consider any input graph G. Each connected component of G is either a path or a cycle, since the maximum degree of G is 2. So we only need to compute some m steps of matrix multiplications and trace operations, where m is the number of edges in G. This is clearly polynomial time computable.

Next we prove the tractability of Holant$^*(H\mathcal{E})$ for any orthogonal matrix H, and of Holant$^*(Z\mathcal{E})$ for both $Z = Z_1$ and Z_2. Let $M = H$ or Z. We first reformulate it as a bipartite Holant problem Holant$(=_2 | M\mathcal{E})$ (since $\mathcal{U} = M\mathcal{U} \subset M\mathcal{E}$, we can drop the $*$ notation in Holant). Here the edges are replaced by the binary EQUALITY function $(=_2) = [1, 0, 1]$. Now we perform a holographic reduction by M, where the right-hand side (RHS) is transformed to $\mathcal{E}$ and the left-hand side (LHS) is transformed to $[1, 0, 1]H^{\otimes 2} = [1, 0, 1] \in \mathcal{E}$ or

$[1, 0, 1]Z^{\otimes 2} \cong [0, 1, 0] \in \mathcal{E}$. Hence after a holographic reduction, our problem becomes a case of Holant($\mathcal{E}$). This is clearly polynomial time computable: Any 0-1 assignment to a particular edge can be uniquely propagated to all edges in each connected component, unless we find a contradiction (in effect we discover a cycle, implied by the connections of the signatures from $\mathcal{E}$, that requires an odd number of switches of 0, 1 values. In that case we return 0.) So, Holant*($H\mathcal{E}$) is polynomial time computable.

Finally we prove the tractability of Holant*($Z\mathcal{M}$). After a holographic reduction by Z, it becomes Holant($\{\neq_2\}|\mathcal{M}$). (Again we can drop the $*$ notation in Holant, because $\mathcal{U} = Z\mathcal{U} \subset Z\mathcal{M}$.) If two unary functions from $\mathcal{M}$ are connected via ($\neq$), it produces a constant factor. If a unary function U in $\mathcal{M}$ is connected via ($\neq$) to any $F \in \mathcal{M}$ of arity $n > 1$, it produces FU' of arity $n - 1$ for another unary U'. Clearly $FU' \in \mathcal{M}$, since the function value of FU' on any input with Hamming weight ≥ 2 is certainly 0. Recursively apply these replacement steps until there are no more unary functions left. Hence, we only need to show that Holant($\{\neq_2\}|\mathcal{M} - \mathcal{U}$) is tractable. The input graph is a bipartite graph. Because all functions on the LHS vertex set are $\neq_2$, in order to have a nonzero evaluation, any assignment must have exactly half of all edges assigned 0 and the other half assigned 1. All functions on the RHS are from $\mathcal{M} - \mathcal{U}$, thus have arity ≥ 2. If there is a vertex of degree ≥ 3 in the RHS, then the RHS requires that strictly less than half of all edges are 1, so the value of this problem is 0. Thus we only need to calculate on graphs where all vertices have degree 2 (each connected component is a cycle), which is tractable by matrix multiplication and taking trace. □

10.2.3 Outline of the Hardness Proof

We outline the hardness proof of Theorem 10.14. The main idea is to reduce the general Holant* problem with asymmetric signatures to the symmetric case, and then apply Theorem 2.12. However, it is not easy to do that when functions have large arities. In Subsection 10.2.5, we first establish an arity reduction theorem. We show that for any of the four tractable families $\mathcal{F}'$, starting from any $F \in \mathcal{F} - \mathcal{F}'$, unless F already has some minimum arity, we can construct a function Q such that

1. Holant*($\mathcal{F} \cup \{Q\}$) $\equiv_T$ Holant*($\mathcal{F}$),
2. $Q \notin \mathcal{F}'$, and
3. Q has a reduced arity.

So assuming that the given set of functions is not a subset of any of the four tractable families, we can continue this arity reduction until reaching some minimum arity. More specifically this will end in a finite number of steps producing the following: In the case of $\langle \mathcal{T} \rangle$, we will end up with an arity 3 signature that is not in $\langle \mathcal{T} \rangle$. For any family of the type $\langle H\mathcal{E} \rangle$, $\langle Z\mathcal{E} \rangle$, $\langle Z\mathcal{M} \rangle$, we can get a signature of arity 2 that is not in the family.

Having these functions with small arities (2 or 3) in hand, we can construct some simple gadgets to get *symmetric* functions, which we do in Section 10.2.6. The hope is that these symmetric functions are also out of various tractable families. However, we come cross some difficulties by doing this. For example, using a single function of arity 3 alone that is not in $\langle \mathcal{T} \rangle$, we do not know how to construct a symmetric arity 3 function that is not in $\langle \mathcal{T} \rangle$ in all cases. In our proof, we get help from other signatures. Namely, we not only use a signature of arity 3 that is not in $\langle \mathcal{T} \rangle$, but also some binary signatures that are not in $\langle H\mathcal{E} \rangle$, $\langle Z\mathcal{E} \rangle$ or $\langle Z\mathcal{M} \rangle$, respectively, to construct a *symmetric* signature Q of arity 3 that is not in $\langle \mathcal{T} \rangle$. This is proved in Theorem 10.24. Similarly, in Theorem 10.25, we prove that we can also construct binary *symmetric* signatures Q' that are not in $\langle H\mathcal{E} \rangle$, $\langle Z\mathcal{E} \rangle$ or $\langle Z\mathcal{M} \rangle$, respectively, starting from a given binary signature not in the respective tractable class *together* with the ternary signature $Q \notin \langle \mathcal{T} \rangle$, *provided* Q happens to belong to the respective tractable class $\langle H\mathcal{E} \rangle$, $\langle Z\mathcal{E} \rangle$ or $\langle Z\mathcal{M} \rangle$.

Then we apply the dichotomy for symmetric signatures, Theorem 2.12, as follows. Either Holant*(Q) is #P-hard, or Q belongs to one of the tractable families, which are the restrictions of the listed four cases in Theorem 10.14 to symmetric signatures. If Holant*(Q) is already #P-hard then we are done. Otherwise, since $Q \notin \langle \mathcal{T} \rangle$, it must belong to $\langle H\mathcal{E} \rangle$, $\langle Z\mathcal{E} \rangle$ or $\langle Z\mathcal{M} \rangle$. So the additional condition on Q is fulfilled and by Theorem 10.25, we can construct binary symmetric signatures that are not in the same tractable family, and therefore we also get #P-hardness by Theorem 2.12.

We note that, all our starting problems for #P-hardness are already hard for planar graphs and all the gadgets we use in the reduction are planar. As a result, our final dichotomy also holds for planar graphs. In the proofs later, we will not explicitly state this.

One technical lemma is used extensively in both Sections 10.2.5 and 10.2.6, which substantially simplifies the proof. We call it the separation lemma, which is stated and proved in Section 10.2.4.

10.2.4 Separation Lemma

In this subsection, we introduce a simple lemma that is used frequently in the proof, and its main purpose is proof simplification. This lemma is applied in the following situation. We have identified a finite set of requirements, where the violation of each requirement can be expressed as a system of polynomial equations in the entries of some signatures. These entries can be ultimately expressed as polynomials in the free variables constituting the entries of arbitrary unary signatures we use. Then to show all these requirements can be simultaneously satisfied, we only need to prove each requirement can be individually satisfied, without regard to the consistency of the satisfying variable assignments.

We will state several versions of the separation lemma. The following lemma is well known. For completeness we give a proof.

Lemma 10.16. *Let $\mathbb{F}$ be an infinite field. Suppose $\{P_1, P_2, \ldots, P_m\}$ is a finite set of nonzero polynomials in $\mathbb{F}[x_1, x_2, \ldots, x_n]$. Then there exist values $a_1, a_2, \ldots, a_n \in \mathbb{F}$ such that $P_i(a_1, a_2, \ldots, a_n) \neq 0$ for all $1 \leq i \leq m$.*

Proof. For $n = 1$, the conclusion holds obviously.

Suppose the conclusion holds for $n-1$. Let $P_i = \sum_{j=0}^{m_i} p_{i,j}(x_1, \ldots, x_{n-1}) x_n^j$. Because P_i is not the zero polynomial, we may assume p_{i,m_i} is a nonzero polynomial in $\mathbb{F}[x_1, x_2, \ldots, x_{n-1}]$. By induction, there exist values $a_1, a_2, \ldots, a_{n-1} \in \mathbb{F}$ such that $p_{i,m_i}(a_1, a_2, \ldots, a_{n-1}) \neq 0$, and $P_i(a_1, a_2, \ldots, a_{n-1}, x_n) \in \mathbb{F}[x_n]$ is a nonzero polynomial in x_n, for all $1 \leq i \leq m$. It follows that there exists $a_n \in \mathbb{F}$ such that $P_i(a_1, a_2, \ldots, a_n) \neq 0$ for all $1 \leq i \leq m$. □

We will give various gadget constructions which use some unary functions $U_k = [x_k, y_k]$, $k = 1, 2, \ldots, m$. Technically the gadget is defined only when specific values for x_k, y_k have been chosen. A signature is expressed as an ordered set of values; this is true for the given constraint functions as well as the signature of the constructed gadget. The entries of the signature of the constructed gadget can be expressed as polynomials in x_k, y_k (the coefficients depend on the given constraint functions). Frequently we have a finite set of conditions, where the negation of each condition is expressible as polynomial equations on x_k, y_k. A construction succeeds if we satisfy all these conditions. The following lemma lets us deal with these conditions separately.

Lemma 10.17. *Let F be the signature of a gadget construction using unary functions $U_k = [x_k, y_k]$, $k = 1, 2, \ldots, m$.*

Suppose $S_1, S_2, \ldots, S_N$ are sets of functions, where a function $K \in S_i$ iff the signature entries of K satisfy a finite system of polynomial equations $\{P_{i,1}, P_{i,1}, \ldots, P_{i,m_i}\}$.

If for every $1 \leq i \leq N$, there exists an assignment σ of x_k and y_k ($k = 1, 2, \ldots, m$), such that $F \notin S_i$ under σ; then there exists an assignment σ of x_k and y_k ($k = 1, 2, \ldots, m$), such that for all $1 \leq i \leq N$, $F \notin S_i$ under σ.

Equivalently, suppose for every assignment σ *of* x_k *and* y_k $(k = 1, 2, \ldots, m)$*, there exists* $1 \le i \le N$*, such that* $F \in S_i$ *under* σ*, then there exists* $1 \le i \le N$*, such that for all assignments* σ *of* x_k *and* y_k $(k = 1, 2, \ldots, m)$*,* $F \in S_i$ *under* σ*.*

Proof. Every signature entry of F is expressible as a polynomial in x_k and y_k, $k = 1, 2, \ldots, m$. Substituting these expressions in $\{P_{i,1}, P_{i,1}, \ldots, P_{i,m_i}\}$ we can express the condition $F \in S_i$ by a finite set of polynomial equations $\{P'_{i,1}, P'_{i,1}, \ldots, P'_{i,m_i}\}$ on x_k and y_k. If for every $1 \le i \le N$, there exists an assignment σ of x_k and y_k $(k = 1, 2, \ldots, m)$, such that $F \notin S_i$ under σ; then for every $1 \le i \le N$, there exists some $1 \le j_i \le m_i$ such that P'_{i,j_i} is not identically 0 as a polynomial on x_k and y_k $(k = 1, 2, \ldots, m)$. Therefore there is an assignment σ on x_k and y_k $(k = 1, 2, \ldots, m)$, such that for all $1 \le i \le N$, $P'_{i,j_i}|_\sigma \ne 0$, hence $F \notin S_i$ under σ. □

The following lemma is another version of the same statement.

Lemma 10.18. *Suppose a gadget construction using unary functions* $U_k = [x_k, y_k]$ $(k = 1, 2, \ldots, m)$ *succeeds if it satisfies a finite set of properties* R_i*,* $i = 1, 2, \ldots, N$*. Suppose violation of each property* R_i *is specified by a finite set of polynomial equations. If for each* i *we can find unary functions* $U_k = [x_k, y_k]$ *to satisfy property* R_i*, then we can find unary functions* $U_k = [x_k, y_k]$ *so that the construction succeeds.*

10.2.5 Arity Reduction

In the next two subsections we prove the hardness part of Theorem 10.14, that is, if $\mathcal{F} \not\subseteq \langle \mathcal{T} \rangle$, $\mathcal{F} \not\subseteq \langle H\mathcal{E} \rangle$, $\mathcal{F} \not\subseteq \langle Z\mathcal{E} \rangle$, $\mathcal{F} \not\subseteq \langle Z\mathcal{M} \rangle$, then Holant*$(\mathcal{F})$ is #P-hard.

In this subsection, we show how to reduce the arity of a nontractable signature to a minimum.

Theorem 10.19. *Let* $\mathcal{F}'$ *be any one of* $\langle \mathcal{T} \rangle$*, or* $\langle H\mathcal{E} \rangle$ *where* H *is orthogonal, or* $\langle Z\mathcal{E} \rangle$*, or* $\langle Z\mathcal{M} \rangle$*, where* $Z = Z_1$ *or* Z_2*. Let* $r = 3$ *if* $\mathcal{F}' = \langle \mathcal{T} \rangle$*, and* $r = 2$ *in the other three cases.*

Suppose there is a function $F \in \mathcal{F} - \mathcal{F}'$*. If* $\text{arity}(F) > r$*, then we can realize a function* Q *by connecting* F *with some unary functions, such that:*

*(1) Holant**$(\mathcal{F} \cup \{Q\}) \equiv_{\mathrm{T}}$ *Holant**$(\mathcal{F})$*;*
(2) $Q \notin \mathcal{F}'$ *and*
(3) $r \le \text{arity}(Q) < \text{arity}(F)$*.*

In order to prove Theorem 10.19, we first prove two lemmas. First, we show that any *type specification* in a tensor product decomposition can be described by a finite system of polynomial equations. Note that the type a signature has is not unique; e.g., if $\mathcal{J}$ is a refinement of $\mathcal{I}$, then a signature having type $\mathcal{J}$ also has type $\mathcal{I}$.

Lemma 10.20. *For any type specification $\mathcal{I}$, there is a finite set of polynomial equations $E_{\mathcal{I}}$ in the entries of a signature F, such that F has type $\mathcal{I}$ iff F satisfies $E_{\mathcal{I}}$.*

Proof. If $\mathcal{I} = \{[n]\}$, there is no requirement on F to have type $\mathcal{I}$. We can use a trivial equation such as $0 = 0$.

Consider the case $\mathcal{I} = \{I_1, I_2\}$. Suppose F has type $\mathcal{I}$. Then obviously, for any two values a_1, b_1 of X_1 and any two values a_2, b_2 of X_2,

$$\det \begin{bmatrix} F(a_1, a_2) & F(a_1, b_2) \\ F(b_1, a_2) & F(b_1, b_2) \end{bmatrix} = 0. \tag{10.10}$$

Without loss of generality we list the indices in I_1 before those of I_2. Hence the collection of all these equations $E_{\{I_1, I_2\}}$ is a necessary condition that F has type $\mathcal{I}$. It is also a sufficient condition. Arrange the values of F into a matrix $F_{X_1, X_2} = F(X_1, X_2)$, where the row indices (respectively column indices) are all possible values of X_1 (respectively X_2). The conditions (10.10) for all a_1, b_1 and all a_2, b_2 imply that any 2×2 submatrix of F is singular, and so $\operatorname{rank}(F_{X_1, X_2}) \leq 1$. Hence, F_{X_1, X_2} is the product of a column vector and a row vector. It follows that F has type $\mathcal{I}$.

Now consider a general partition $\mathcal{I} = \{I_1, \ldots, I_k\}$, and again suppose F has type $\mathcal{I}$. It follows that for any $1 < i \leq k$, and for any fixed values $a_{i+1}, \ldots, a_k$ for $X_{i+1}, \ldots, X_k$ (this is vacuous if $i = k$), we get a function F' from F by setting variables in $X_{i+1}, \ldots, X_k$ to $a_{i+1}, \ldots, a_k$ respectively. F' has type $\{\bigcup_{j=1}^{i-1} I_j, I_i\}$. We define the following set of equations: For all $1 < i \leq k$, and for all assignment values $a_{i+1}, \ldots, a_k$ for $X_{i+1}, \ldots, X_k$, include all equations in $E_{\{\bigcup_{j=1}^{i-1} I_j, I_i\}}$ for the function F'. This is a finite set of polynomial equations. Obviously, this is a necessary condition for F to have type $\mathcal{I}$.

We prove that it is also a sufficient condition. If F is the zero function, then F has type $\mathcal{I}$ trivially. Assume F is not the zero function. Let $i = k$, by what has been proved when $k = 2$, $F = \bigotimes_{\{\bigcup_{j=1}^{k-1} I_j, I_k\}}(P_{k-1}, F_k)$ for some functions P_{k-1} and F_k. Because F is not the zero function, there exists a value a_k for X_k such that $F_k(a_k) \neq 0$. The remaining conditions, for $1 < i \leq k - 1$, yield a finite set of homogeneous equations for $F^{X_k = a_k}(X_1, \ldots, X_{k-1}) = P_{k-1}(X_1, \ldots, X_{k-1}) F_k(a_k)$. After canceling the nonzero factor $F_k(a_k)$, by induction, we obtain a sufficient condition that P_{k-1} has type $\{I_1, \ldots, I_{k-1}\}$. Hence F has type $\mathcal{I}$. □

Next, we prove a property of this decomposition. This property is used throughout in the proof of Theorem 10.19.

Lemma 10.21. *Suppose there exists some type $\mathcal{I} = \{I_1, \ldots, I_k\}$ over $[n-1]$, such that for all unary functions $U = [x, y]$, $F^{x_n=U} = FU = xF^{x_n=0} + yF^{x_n=1}$ has the same type $\mathcal{I}$. Furthermore, suppose $F^{x_n=0} = \bigotimes_{\mathcal{I}}(F_1, F_2, \ldots, F_k)$ and $F^{x_n=1} = \bigotimes_{\mathcal{I}}(K_1, K_2, \ldots, K_k)$ are linearly independent as two vectors. Then there exists* exactly one *index $i \in [k]$ such that F_i and K_i are linearly independent.*

Proof. If F_i and K_i are linearly dependent for all $i \in [k]$, then obviously $F^{x_n=0}$ and $F^{x_n=1}$ would be linearly dependent. Now for a contradiction, suppose there are two distinct indices $i \in [k]$, such that F_i and K_i are linearly independent. Without loss of generality, let $i = 1, 2$ respectively.

Because $F^{x_n=0}$, $F^{x_n=1}$ are linearly independent, F_j and K_j are not the zero function for any $j \in [k]$. For any $j \in [k] - \{1, 2\}$, by Lemma 10.16, there exist $|I_j|$ many unary functions such that both F_j and K_j become nonzero constants when combined with them. This is because we can write a finite system of polynomial equations Φ_j in a_k, b_k for the $|I_j|$ many unary functions $U_k = [a_k, b_k]$, that expresses the condition that F_j becomes 0 when connecting variables x_k to U_k, for all $k \in I_j$. Since F_j is not the zero function, some 0-1 assignment violates some equation in Φ_j. Hence at least one polynomial P_1 in Φ_j is not identically 0. Similarly there is a finite system of polynomial equations Ψ_j for K_j and some polynomial P_2 is not identically 0. Then by Lemma 10.16 there is a common assignment to a_k, b_k ($k \in I_j$) such that both F_j and K_j are nonzero. After combining $F^{x_n=0}$ and $F^{x_n=1}$ with these unary functions, for all $j \in [k] - \{1, 2\}$, we obtain respectively the functions $c_0 F_1 \otimes F_2$ and $c_1 K_1 \otimes K_2$ over the variables in $I_1 \cup I_2$, where $c_0, c_1 \neq 0$.

Suppose $U = [x, y]$ and $xy \neq 0$. If we combine $FU = xF^{x_n=0} + yF^{x_n=1}$ with the same set of $\left|\bigcup_{j=3}^{k} I_j\right|$ many unary functions, the resulting function is $c_0 x F_1 \otimes F_2 + c_1 y K_1 \otimes K_2$. By the assumption, FU has type $\mathcal{I}$, then this function has type $\{I_1, I_2\}$. However, we show that, for any $xy \neq 0$, this function does not have type $\{I_1, I_2\}$. The matrix form (row index is $X|_{I_1}$, column index is $X|_{I_2}$) of this function is the $2^{|I_1|} \times 2^{|I_2|}$ matrix

$$\begin{bmatrix} F_1 & K_1 \end{bmatrix} \begin{bmatrix} c_0 x & 0 \\ 0 & c_1 y \end{bmatrix} \begin{bmatrix} F_2^{\mathrm{T}} \\ K_2^{\mathrm{T}} \end{bmatrix},$$

where F_1, K_1, F_2, K_2 are column vectors. Since F_1 and K_1 are linearly independent, and F_2 and K_2 are linearly independent, this matrix has rank 2. If this function has type $\{I_1, I_2\}$, its matrix form would have rank ≤ 1. This contradiction proves the lemma. $\square$

Proof of Theorem 10.19: $\mathcal{F}' = \langle \mathcal{T} \rangle$

Suppose $F \in \mathcal{F} - \langle \mathcal{T} \rangle$, with arity$(F) > 3$. Being out of $\langle \mathcal{T} \rangle$, F is not the zero function. If for some unary function $U = [x, y]$, $FU \notin \langle \mathcal{T} \rangle$, then we are done by setting $Q = FU$. Hence we assume for any unary function $U = [x, y]$, $FU = xF^{x_n=0} + yF^{x_n=1}$ has some type $\mathcal{J}$, where each set $J_j \in \mathcal{J}$ has cardinality $|J_j| \leq 2$. For the fixed arity(F), there are only finitely many such types, which are specifiable by a finite set of polynomial equations in x, y. It follows from Lemma 10.17 that, there exists some type $\mathcal{I} = \{I_1, \ldots, I_k\}$, where each $|I_j| \leq 2$, such that for all x, y, FU has the same type $\mathcal{I}$. In particular, both $F^{x_n=0}$ and $F^{x_n=1}$ have type $\mathcal{I}$.

If $F^{x_n=0}$ and $F^{x_n=1}$ are linearly dependent, then $F \in \langle \mathcal{T} \rangle$, having type $\mathcal{I} \cup \{\{n\}\}$.

So we assume $F^{x_n=0} = \bigotimes_{\mathcal{I}}(F_1, F_2, \ldots, F_k)$ and $F^{x_n=1} = \bigotimes_{\mathcal{I}}(K_1, K_2, \ldots, K_k)$ are linearly independent. Being linearly independent, none of the tensor factors of $F^{x_n=0}$ and $F^{x_n=1}$ can be the zero function. By Lemma 10.21, there is exactly one pair of linearly independent tensor factors, without loss of generality, say F_1 and K_1. Expressing K_i in terms of F_i, for $i \geq 2$, there exists a nonzero constant c, such that $F^{x_n=1} = \bigotimes_{\mathcal{I}}(cK_1, F_2, \ldots, F_k)$. If $|I_1| = 1$, that is, F_1 and K_1 are unary functions, then $F \in \langle \mathcal{T} \rangle$, of type $\{I_1 \cup \{n\}, I_2, \ldots, I_k\}$. Thus, $|I_1| = 2$.

Without loss of generality, let $I_1 = \{1, 2\}$. We can fix the variables of F in $I_2, \ldots, I_k$ to some values, such that $F_2, \ldots, F_k$ each contributes a nonzero factor. By this we get a ternary function Q in variables x_1, x_2, x_n, and $F = Q \otimes F_2 \otimes \cdots \otimes F_k$. If $Q \in \langle \mathcal{T} \rangle$, then $F \in \langle \mathcal{T} \rangle$, contrary to assumption. Hence $Q \notin \langle \mathcal{T} \rangle$.

Proof of Theorem 10.19: $\mathcal{F}' = \langle H\mathcal{E} \rangle$ or $\langle Z\mathcal{E} \rangle$

For any function F of arity n and invertible matrix M, $F \in \langle M\mathcal{E} \rangle$ iff $(M^{-1})^{\otimes n} F \in \langle \mathcal{E} \rangle$. Since we will realize our function Q of a lower arity than F by connecting F with some unary functions, and since unary functions are transformed to other unary functions under any invertible holographic transformation, we only need to prove for $\langle \mathcal{E} \rangle$. Suppose $F \notin \langle \mathcal{E} \rangle$, and arity$(F) = n > 2$. F is not the zero function. If for some unary function $U = [x, y]$, $FU \notin \langle \mathcal{E} \rangle$, we are done with $Q = FU$. Hence we assume for any unary function $U = [x, y]$, $FU = xF^{x_n=0} + yF^{x_n=1} \in \langle \mathcal{E} \rangle$.

For any partition $\mathcal{I} = \{I_1, \ldots, I_k\}$ of $[n]$, and any $\mathcal{A} = \{A_1, \ldots, A_k\}$, such that $A_j \in \{0, 1\}^{|I_j|}$, we define a set of functions $(\mathcal{I}, \mathcal{A})$. Each A_j is a 0-1 string of length $|I_j|$. In the definition for $(\mathcal{I}, \mathcal{A})$ below we use the set $\{A_j, \bar{A}_j\}$ where A_j and its complement $\bar{A}_j$ play symmetric roles, and so we may normalize the first bit of A_j to be 0. We say a function P belongs to the set $(\mathcal{I}, \mathcal{A})$ iff P has type

$\mathcal{I}$, and $P = \bigotimes_{\mathcal{I}}(P_1, P_2, \ldots, P_k)$ for some functions $P_1, P_2, \ldots, P_k$ such that for any $j \in [k]$, $P_j(X|_{I_j})$ is zero if $X|_{I_j} \notin \{A_j, \bar{A}_j\}$. Thus, $P_j \in \mathcal{E}$ for each $j \in [k]$, and any function in $(\mathcal{I}, \mathcal{A})$ belongs to $\langle \mathcal{E} \rangle$.

Indeed functions in $\langle \mathcal{E} \rangle$ of arity n is the union of these finitely many function sets $(\mathcal{I}, \mathcal{A})$. Obviously, functions in $(\mathcal{I}, \mathcal{A})$ can be described by a finite system of polynomial equations. Since $FU \in \langle \mathcal{E} \rangle$ for all $U = [x, y]$, by Lemma 10.17, there must exist one $(\mathcal{I}, \mathcal{A})$, such that for any x, y, FU belongs to the same set $(\mathcal{I}, \mathcal{A})$. In particular, both $F^{x_n=0}$ and $F^{x_n=1}$ belong to $(\mathcal{I}, \mathcal{A})$.

If $F^{x_n=0}$ and $F^{x_n=1}$ are linearly dependent, then $F \in \langle \mathcal{E} \rangle$, because $F^{x_n=0}$, $F^{x_n=1} \in \langle \mathcal{E} \rangle$, and $F = F^{x_n=b} \otimes [\lambda, \mu]$ for some $b \in \{0, 1\}$, and $\lambda, \mu \in \mathbb{C}$, where $[\lambda, \mu] \in \mathcal{U} \subset \mathcal{E}$ is a unary function on x_n.

Let $F^{x_n=0} = \bigotimes_{\mathcal{I}}(F_1, F_2, \ldots, F_k)$ and $F^{x_n=1} = \bigotimes_{\mathcal{I}}(K_1, K_2, \ldots, K_k)$ be linearly independent. Being linearly independent, none of the tensor factors of $F^{x_n=0}$ and $F^{x_n=1}$ can be the zero function. By Lemma 10.21, there is exactly one pair of linearly independent tensor factors, say, F_1 and K_1. Expressing K_i in terms of F_i, for $i \geq 2$, there exists a nonzero constant c, such that $F^{x_n=1} = \bigotimes_{\mathcal{I}}(cK_1, F_2, \ldots, F_k)$.

We can fix the variables of F in $I_2, \ldots, I_k$ to some values, such that $F_2, \ldots, F_k$ each contributes a nonzero factor. We obtain a function K. After dividing out a nonzero constant, we have $K^{x_n=0} = F_1$ and $K^{x_n=1} = cK_1$, where $c \neq 0$. K evaluates to zero, except on possibly four inputs $\{A_1 0, \bar{A}_1 0, A_1 1, \bar{A}_1 1\}$. Combine the $|I_1| - 1$ variables of K other than x_n and the first variable in I_1 with the function $[1, 1]$, we get a binary function in matrix form

$$Q = \begin{bmatrix} K(A_1 0) & K(A_1 1) \\ K(\bar{A}_1 0) & K(\bar{A}_1 1) \end{bmatrix},$$

where we index the row by the first variable in I_1 and the column by x_n. Note that we have used the definition of $(\mathcal{I}, \mathcal{A})$. For example, in the sum defining $Q(0, 0)$, only one (possibly) nonzero term $K(A_1 0)$ is involved because the first and last bits are both set to 0. The first column in the 2×2 matrix form of Q is F_1 at entries $A_1, \bar{A}_1$, and similarly the second column is from cK_1. Because F_1 and K_1 are linearly independent, and these are the only possible nonzero entries of K, we have $\det Q \neq 0$. We claim that $Q \notin \langle \mathcal{E} \rangle$. For otherwise, being nondegenerate, $Q \in \mathcal{E}$, then the support of K is contained in either $\{A_1 0, \bar{A}_1 1\}$ or $\{A_1 1, \bar{A}_1 0\}$ This implies that $K \in \mathcal{E}$, and hence $F = K \otimes F_2 \otimes \cdots \otimes F_k \in \langle \mathcal{E} \rangle$.

Proof of Theorem 10.19: $\mathcal{F}' = \langle Z\mathcal{M} \rangle$

Again we only need to prove for $\langle \mathcal{M} \rangle$. Suppose $F \notin \langle \mathcal{M} \rangle$, and arity$(F) = n > 2$. Again we may assume for any unary function $U = [x, y]$, $FU = xF^{x_n=0} + yF^{x_n=1} \in \langle \mathcal{M} \rangle$; otherwise, we are done.

For any $U = [x, y]$, $FU \in \langle \mathcal{M} \rangle$ has some type $\mathcal{I}$, and each tensor factor belongs to $\mathcal{M}$, that is, each tensor factor is 0 except on inputs of Hamming weight ≤ 1. Each type $\mathcal{I}$ can be specified by a finite system of polynomial equations $E_{\mathcal{I}}$ by Lemma 10.20. Now we put $E_{\mathcal{I}}$ together with the requirement that for each tensor factor, all entries of Hamming weight > 1 are 0. This requirement can also be stated as a finite system of polynomial equations. (For example, if we require that $(a_{i,j})$ is a tensor product $(b_i) \otimes (c_j)$, where $1 \leq i \leq n, 1 \leq j \leq m$, and for some subsets $B \subseteq [n], C \subseteq [m]$, $[\forall i \in B, \forall j \in C, b_i = c_j = 0]$, then we include the equations for the type specification from Lemma 10.20 together with $[a_{i,j} = 0, \forall i \in B, \textit{or } \forall j \in C]$. The equations from Lemma 10.20 imply that a tensor factorization $(a_{i,j}) = (b_i) \otimes (c_j)$ exists. If for all $i \in [n]$, $b_i = 0$, or if for all $j \in [m]$, $c_j = 0$, then $a_{i,j}$ is identically 0. On the other hand, if for some $i_0 \in [n]$, and $j_0 \in [m]$, $b_{i_0} \neq 0$ and $c_{j_0} \neq 0$, Then $b_i = a_{i,j_0}/c_{j_0} = 0$ for all $i \in B$. Similarly $c_j = 0$ for all $j \in C$.)

Applying Lemma 10.17, we conclude that there is one type $\mathcal{I}$ such that for any $U = [x, y]$, FU has a decomposition having the same type specified by $\mathcal{I}$ with tensor factors from $\mathcal{M}$.

If $F^{x_n=0}$ and $F^{x_n=1}$ are linearly dependent, obviously, $F \in \langle \mathcal{M} \rangle$, as $U \in \mathcal{M}$, contrary to assumption.

Let $F^{x_n=0} = \bigotimes_{\mathcal{I}}(F_1, F_2, \ldots, F_k)$ and $F^{x_n=1} = \bigotimes_{\mathcal{I}}(K_1, K_2, \ldots, K_k)$ be linearly independent. As before, none of the tensor factors of $F^{x_n=0}$ and $F^{x_n=1}$ can be the zero function, and exactly one pair among F_i and K_i are linearly independent, say, F_1 and K_1. We can fix the variables of F in $I_2, \ldots, I_k$ to some values, such that $F_2, \ldots, F_k$ contribute a nonzero factor. We get a function in matrix form $K = \begin{bmatrix} F_1^{\mathrm{T}} \\ cK_1^{\mathrm{T}} \end{bmatrix}$, where the row index is $x_n = 0, 1$, columns are indexed by $\{0, 1\}^{|I_1|}$, and $c \neq 0$. Here the first row is $K^{x_n=0} = F_1^{\mathrm{T}}$. The second row is $K^{x_n=1} = cK_1^{\mathrm{T}}$. Columns are indexed by $A \in \{0, 1\}^{|I_1|}$. If the weight of A is greater than 1, then the column indexed by A is 0 because $F_1, K_1 \in \mathcal{M}$. Let S_0 denote the column indexed by $0 \cdots 0 \in \{0, 1\}^m$, and S_i denote the column indexed by the bit sequence $A \in \{0, 1\}^m$, where only the ith bit of A is 1.

For simplicity of notations, assume $I_1 = \{1, 2, \ldots, m\}$. There exists a 0-1 string $A \in \{0, 1\}^m$ of Hamming weight 1, such that $K_1(A) \neq 0$; otherwise, the only nonzero entry for K all have Hamming weight ≤ 1, and so $K \in \mathcal{M}$. This would imply $F \in \langle \mathcal{M} \rangle$, contrary to assumption. Hence there is a column S_i, $1 \leq i \leq m$, whose second entry is not zero. Without loss of generality we assume this S_i is S_m. Because F_1 and K_1 are linearly independent, There exists a column S_j linearly independent with the nonzero column S_m. If S_0 is such a column, then $Q = K^{x_1=0,\ldots,x_{m-1}=0}$ is a binary function on $\{x_n, x_m\}$, and has the matrix form $Q = [S_0, S_m]$. Note that the index for the column S_m is

$A = 0\cdots01 \in \{0,1\}^m$. If S_0 is linearly dependent with S_m. Then for some $1 \le j \le m-1$, S_j is linearly independent with S_m. Let $Q = K^{x_1=0,\ldots,x_{j-1}=0,x_j=[x,y],x_{j-1}=0,\ldots,x_{m-1}=0} = [xS_0 + yS_j, xS_m]$, where $x \neq 0$ and $y \neq 0$. (Here in Q, the row index is by x_n and column index is by x_m.) We have obtained our Q such that Q is not degenerate and $Q(1,1) \neq 0$, i.e., $Q \notin \langle \mathcal{M} \rangle$. □

10.2.6 From Asymmetric to Symmetric

In this subsection, we show how to get a symmetric function from some asymmetric functions, keeping the property of *not* belonging to any one of the four tractable classes, $\langle \mathcal{T} \rangle$, $\langle H\mathcal{E} \rangle$, $\langle Z\mathcal{E} \rangle$ or $\langle Z\mathcal{M} \rangle$.

Recall that $F \cong cF$ for $c \neq 0$. In the following lemma, when we count the number of solutions, we count in terms of equivalence classes under $\cong$.

Lemma 10.22. *Suppose F is ternary function. Then $F^{x_3=U} \cong 0$ for some $U \ncong 0$ iff F has type $\{\{1,2\},\{3\}\}$. Suppose $F \notin \langle \mathcal{T} \rangle$. Then $F^{x_3=U} \ncong 0$ for any nonzero unary function U, and there exist exactly one or two nonzero $U = [x,y]$ such that $F^{x_3=U}$ is degenerate.*

Proof. If F has type $\{\{1,2\},\{3\}\}$, then $F = T \otimes [a,b]$. If $a = b = 0$ then F is identically 0, and $F^{x_3=U} \cong 0$ for any unary U. If $[a,b] \ncong 0$, then $F^{x_3=U} \cong 0$ for $U = [b,-a] \ncong 0$. Conversely, if $F^{x_3=U} \cong 0$ for some $U \ncong 0$, then $F^{x_3=0}$ and $F^{x_3=1}$ are linearly dependent, and hence F has type $\{\{1,2\},\{3\}\}$. It follows that if $F \notin \langle \mathcal{T} \rangle$, then $F^{x_3=U} \ncong 0$ for any nonzero unary function U.

Let $F' = F^{x_3=U}$. Then F' is degenerate iff $\det \begin{bmatrix} F'(0,0) & F'(0,1) \\ F'(1,0) & F'(1,1) \end{bmatrix} = 0$. Let $U = [x,y]$, then the entries of F' are linear homogeneous polynomials of x and y, so the determinant equation is a quadratic homogeneous equation. It has either one or two solutions $U \ncong 0$, or it is identically zero. We only need to prove the latter case contradicts $F \notin \langle \mathcal{T} \rangle$.

Suppose F' is degenerate for all U, then we have in particular $F^{x_3=0} = F_1 \otimes F_2$ and $F^{x_3=1} = K_1 \otimes K_2$. If $F^{x_3=0}$ and $F^{x_3=1}$ are linearly dependent, then $F \in \langle \mathcal{T} \rangle$, and so $F^{x_3=0}$ and $F^{x_3=1}$ are linearly independent. Then by Lemma 10.21, exactly one of the two pairs of functions $\{F_1, K_1\}$ and $\{F_2, K_2\}$ is linearly independent, say, $\{F_1, K_1\}$ is linearly dependent. Then F is the tensor product of F_1 and one binary function on the remaining two variables $\{x_2, x_3\}$, and so $F \in \langle \mathcal{T} \rangle$. □

For any ternary function $F(x_1,x_2,x_3) \notin \langle \mathcal{T} \rangle$, the conclusion of Lemma 10.22 certainly applies to all three variables. There is a simple relationship, among $1 \le i \le 3$, between the nonzero unary functions U_i such that $F^{x_i=U_i}$ is degenerate. Suppose $F^{x_1=U_1}$ is degenerate, where $U_1 \not\cong 0$, then $F^{x_1=U_1} = L \otimes R$, where L and R are unary functions on x_2 and x_3 respectively. Since $F^{x_1=U_1} \not\cong 0$, both $L \not\cong 0$ and $R \not\cong 0$. The matrix form of $F^{x_1=U_1}$ has rank exactly 1. It follows that the decomposition $L \otimes R$ is unique under $\cong$. If we define $[x, y]^{\perp} = [y, -x]$, and let $U_2 = L^{\perp}$, then $F^{x_1=U_1,x_2=U_2}$ is identically 0. Given U_1 such that $F^{x_1=U_1}$ is degenerate but not identically 0, the property that $F^{x_1=U_1,x_2=U_2} \cong 0$ also uniquely determines a nonzero U_2 as $U_2 \cong L^{\perp}$, where L is uniquely determined by U_1. Because $(F^{x_2=U_2})^{x_1=U_1} \cong 0$, we have $F^{x_2=U_2}$ is degenerate. This mapping from $U_1 \mapsto U_2 = L^{\perp}$ is well defined under $\cong$. It is also 1-1: Suppose $F^{x_1=U_1,x_2=U_2}$ and $F^{x_1=U_1',x_2=U_2}$ are both identically 0 for $U_1 \not\cong 0$ and $U_1' \not\cong 0$. Then $F^{x_2=U_2}$ is degenerate, and expressible as $A(x_1) \otimes B(x_3)$, where A and B are nonzero unary functions. It follows that both $U_1 \cong A^{\perp}$ and $U_1' \cong A^{\perp}$. Thus $U_1 \cong U_1'$. By symmetry the inverse map is also well defined.

The same statement holds for x_3.

We summarize this in the following lemma. Suppose $F \notin \langle \mathcal{T} \rangle$ is a ternary function. Let

$$\mathcal{U}_i = \{U \not\cong 0 \mid F^{x_i=U} \text{ is degenerate}\}, \qquad 1 \le i \le 3.$$

Lemma 10.23. *There is a one-to-one correspondence between $\mathcal{U}_1, \mathcal{U}_2$, and $\mathcal{U}_3$, as follows. For $\{i, j, k\} = \{1, 2, 3\}$, each $U_i \in \mathcal{U}_i$ gives a unique factorization $F^{x_i=U_i} = V_j(x_j) \otimes V_k(x_k)$, where $V_j^{\perp} \in \mathcal{U}_j$ and $V_k^{\perp} \in \mathcal{U}_k$. In particular, $|\mathcal{U}_1| = |\mathcal{U}_2| = |\mathcal{U}_3| = 1$ or 2.*

Now we will prove a crucial theorem for the hardness part of Theorem 10.14.

Theorem 10.24. *Suppose in Holant*$(\mathcal{F})$, we can realize the following functions*

1. *$F \notin \langle \mathcal{T} \rangle$ of arity 3.*
2. *For any orthogonal matrix H, some $P_H \notin \langle H\mathcal{E} \rangle$ of arity 2.*
3. *For both $Z = Z_1$ or Z_2, some $P_Z \notin \langle Z\mathcal{E} \rangle$ of arity 2.*
4. *For both $Z = Z_1$ or Z_2, some $S_Z \notin \langle Z\mathcal{M} \rangle$ of arity 2.*

Then we can realize a symmetric *ternary function $Q \notin \langle \mathcal{T} \rangle$ in Holant*$(\mathcal{F})$.*

Proof. We use the gadget shown in Figure 10.3 to realize a symmetric ternary function Q. (In some cases we will need to modify it to define Q; this will be discussed later.) This gadget consists of nine copies of the function F, three copies of a unary function U_1, and three copies of a unary function U_2. The unary functions are to be determined later. Each shaded triangle labeled with

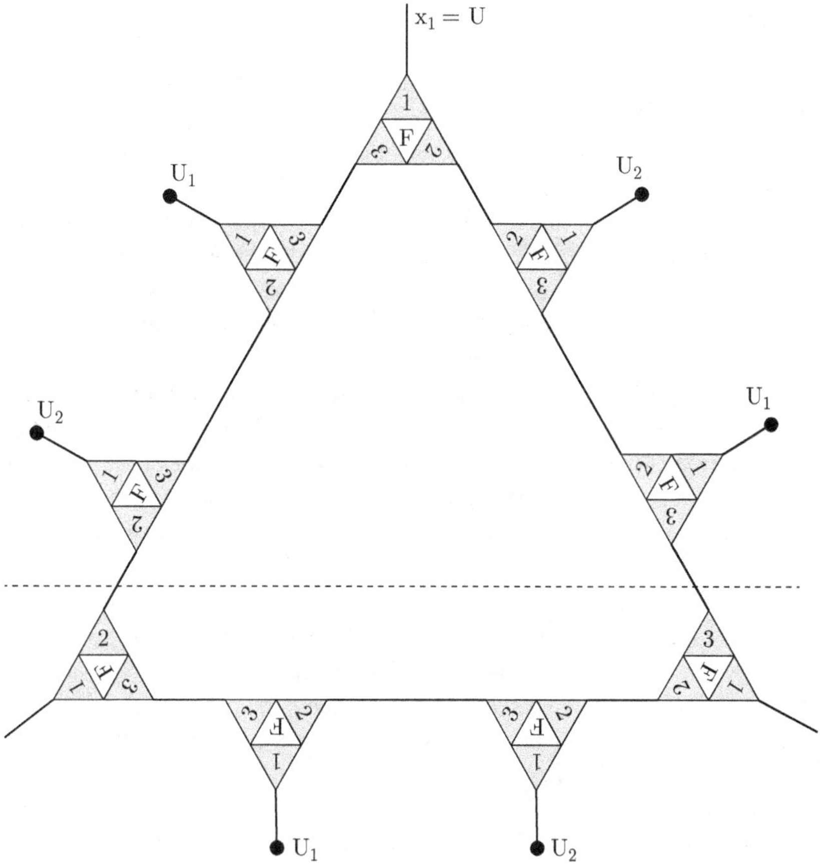

Figure 10.3. Gadget to realize a symmetric ternary function.

F in a central inner triangle represents the function $F(x_1, x_2, x_3) \notin \langle \mathcal{T} \rangle$. The labels $1, 2, 3$ inside the shaded triangle indicate which edge corresponds to variables x_1, x_2, x_3. This gadget remains unchanged if we rotate it $\frac{2\pi}{3}$. Hence, $Q(x_1, x_2, x_3) = Q(x_2, x_3, x_1) = Q(x_3, x_1, x_2)$. It follows that Q is symmetric (notice that this conclusion uses the fact that each variable x_i is a Boolean variable).

Our goal is to prove that there exist nonzero unary functions U_1 and U_2, such that $Q \notin \langle \mathcal{T} \rangle$. Since Q is symmetric, this is equivalent to: There exists no nonzero unary function U satisfying $Q^{x_1=U} \cong 0$, by Lemma 10.22.

To prove this, we divide the gadget into two parts, as shown by the dashed line in Figure 10.3. We establish two properties, one property for each part respectively. The upper part is a ternary function, denoted by S. The first property is that if $U \not\equiv 0$, then $S^{x_1=U} \not\equiv 0$. The matrix form of $S^{x_1=U}$ is the matrix product $F^{x_1=U_2}_{x_2,x_3} F^{x_1=U_1}_{x_2,x_3} F^{x_1=U}_{x_3,x_2} F^{x_1=U_2}_{x_2,x_3} F^{x_1=U_1}_{x_2,x_3}$, where $F^{x_1=U'}_{x_2,x_3}$ denotes the 2×2 matrix form of $F^{x_1=U'}$ with row index x_2 and column index x_3. Because $F \notin \langle \mathcal{T} \rangle$, if $U \not\equiv 0$, then $F^{x_1=U} \not\equiv 0$, by Lemma 10.22. To satisfy this property on S, we only need some U_1 and U_2 such that $F^{x_1=U_1}$ and $F^{x_1=U_2}$ are nondegenerate. By Lemma 10.22, there exist such U_1 and U_2.

The lower part is a function of arity 4, denoted by P. Two inputs of P are the original inputs x_2, x_3 of Q, corresponding to the lower left and lower right corners of the gadget respectively. The other two inputs correspond to edges connecting P with S, denoted by y_2, y_3 respectively. The second property is that the 4×4 matrix $P_{x_2x_3,y_2y_3}$ is nonsingular.

If there exist U_1 and U_2 such that both properties hold, then for any nonzero unary function U, the vector form of $Q^{x_1=U}$ is the matrix-vector product $P_{x_2x_3,y_2y_3} S^{x_1=U}$, where $S^{x_1=U}$ takes its vector form as a vector of dimension 4. Hence $Q^{x_1=U}$ is not the zero function, because $S^{x_1=U}$ is a nonzero column vector (the first property) and $P_{x_2x_3,y_2y_3}$ is a nonsingular matrix (the second property). This proves $Q \notin \langle \mathcal{T} \rangle$.

To establish the two properties, we can apply the separation lemma 10.18, and prove the two properties individually. We have proved the first one. Now we prove the second one. (The separation lemma allows us to choose unary functions U_1 and U_2 separately for the two parts in order to satisfy the two properties, even though in the actual gadget construction the three occurrences of U_1 must be the same, and similarly for U_2, in order to produce a symmetric Q.)

The idea for the proof of the second property on P will be counterintuitive. Our goal is to choose unary functions U_1 and U_2 such that the function P has a full-rank matrix. We will do this by a nonzero unary function U_1 such that $F^{x_1=U_1}$ has a *singular* matrix. (This should be surprising as we seem to go the opposite direction.) However, once $F^{x_1=U_1}$ is degenerate, this effectively severs the bottom path in this gadget P. (This entanglement on the path connecting the two copies of F on the lower two corners makes it difficult to analyze P.) Consequently the matrix $P_{x_2x_3,y_2y_3}$ become a tensor product of two matrices $A_{x_2,y_2} \otimes B_{x_3,y_3}$. We then aim to guarantee that both A_{x_2,y_2} and B_{x_3,y_3} are nonsingular 2×2 matrices.

Since $F \notin \langle \mathcal{T} \rangle$, by Lemma 10.22 there exists $U_1 \not\equiv 0$ such that $F^{x_1=U_1}$ is degenerate, and $F^{x_1=U_1} = L_L \otimes R_L$, or in more detail, $F^{x_1=U_1}(z_3, z_2) = L_L(z_3)R_L(z_2)$. L_L and R_L are not the zero function. We also want the matrix form A_{x_2,y_2} of $F^{x_3=L_L}$ to be nonsingular. In the notation of Lemma 10.23, by the 1-1 correspondence from $\mathcal{U}_1$ to $\mathcal{U}_3$, $U_1 \in \mathcal{U}_1$ gives L_L and then gives a corresponding $L_L^{\perp} \in \mathcal{U}_3$. Each one of (at most two) $U_1 \in \mathcal{U}_1$ gives a unique $L_L^{\perp} \in \mathcal{U}_3$. We want to choose a $U_1 \in \mathcal{U}_1$ such that its corresponding $L_L \notin \mathcal{U}_3$. By the 1-1

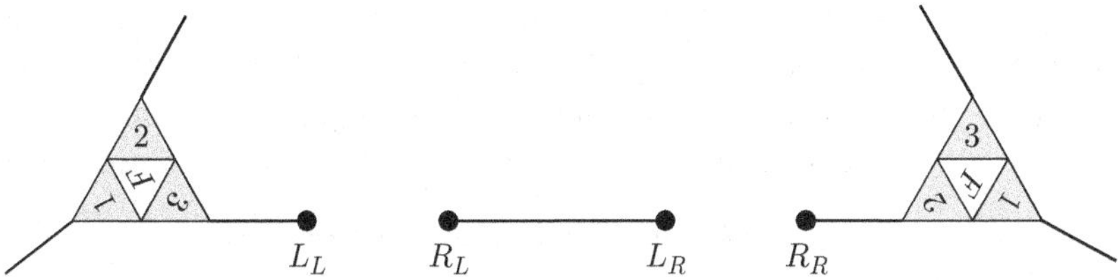

Figure 10.4. Replace $F^{x_1=U_1}$ by $L_L \otimes R_L$, and $F^{x_1=U_2}$ by $L_R \otimes R_R$.

correspondence, this is equivalent to choosing some unary $U \in \mathcal{U}_3$, such that $U^\perp \notin \mathcal{U}_3$. Such a $U \in \mathcal{U}_3$, by the inverse map of the 1-1 correspondence, gives us the desired $U_1 \in \mathcal{U}_1$.

We have a similar requirement for U_2 and $B = F^{x_2=R_R}$, on the right half of the gadget P: $U_2 \in \mathcal{U}_1$, $F^{x_1=U_2} = L_R \otimes R_R$, and $R_R \notin \mathcal{U}_2$. Suppose $F^{x_1=U_1} = L_L \otimes R_L$ and $F^{x_1=U_2} = L_R \otimes R_R$, then we can replace them by unary functions, and combine $R_L^{\mathrm{T}} L_R$ to get a scalar factor c. Then $P_{x_2x_3,y_2y_3}$ is $cA_{x_2,y_2} \otimes B_{x_3,y_3}$, where A_{x_2,y_2} and B_{x_3,y_3} are the matrix forms for $F^{x_3=L_L}$ and $F^{x_2=R_R}$ respectively (Figure 10.4). So we also want $c = R_L^{\mathrm{T}} L_R \neq 0$, in addition to A_{x_2,y_2} and B_{x_3,y_3} being nonsingular.

Note that if we write in matrix form for P from left to right (see Figures 10.3 and 10.4), we have another 4×4 matrix form for P as $P_{y_2x_2,y_3x_3} = F_{x_2x_1,x_3}(L_L R_L^{\mathrm{T}})(L_R R_R^{\mathrm{T}})F_{x_2,x_3x_1}$. Taking out the dot product value $c = R_L^{\mathrm{T}} L_R$ (a scalar), the remainder of the function is $(F_{x_2x_1,x_3} L_L)(R_R^{\mathrm{T}} F_{x_2,x_3x_1})$, a product of a 4×1 matrix with a 1×4 matrix. This matrix has rank 1. However, this matrix form for P is not the same as $P_{x_2x_3,y_2y_3}$, which is a rotated version. It is $P_{x_2x_3,y_2y_3}$ that we want to ensure that it has rank 4.

To summarize for P, for the second property, we identify three conditions whose conjunction is sufficient.

Condition (1): $F^{x_1=U_1} = L_L \otimes R_L$ is degenerate and $F^{x_3=L_L}$ is nondegenerate.
Condition (2): $F^{x_1=U_2} = L_R \otimes R_R$ is degenerate and $F^{x_2=R_R}$ is nondegenerate.
Condition (3): $R_L^{\mathrm{T}} L_R \neq 0$.

There are three cases, depending on $\mathcal{U}_3$, where one cannot pick U_1 to satisfy condition (1). (We will deal with conditions (2) and (3) separately.)

a. $|\mathcal{U}_3| = 1$ and for the unique $U \in \mathcal{U}_3$, it also holds that $U^\perp \in \mathcal{U}_3$.
b. $|\mathcal{U}_3| = 2$ and for both $U \in \mathcal{U}_3$, it also holds that $U^\perp \cong U \in \mathcal{U}_3$.
c. $|\mathcal{U}_3| = 2$ and $\mathcal{U}_3 = \{U, U^\perp\}$.

Now we will use the given binary signatures $P_H \notin \langle H\mathcal{E} \rangle$, $P_Z \notin \langle Z\mathcal{E} \rangle$, and $S_Z \notin \langle Z\mathcal{M} \rangle$.

In case (a.): $U^\perp \cong U$, and thus $U \cong [1, \mathfrak{i}]$ or $[1, -\mathfrak{i}]$.

If $U \cong [1, \mathfrak{i}]$ (resp. $[1, -\mathfrak{i}]$), we show that $S_{Z_1}U$ (resp. $S_{Z_2}U$) does not belong to $\mathcal{U}_3$. Because the binary $S_{Z_1} \notin \langle Z_1\mathcal{M}\rangle$, in matrix form $S_{Z_1} = Z_1 T Z_1^{\mathtt{T}}$ for some $T \notin \langle\mathcal{M}\rangle$. Note that $\mathcal{U} \subset \mathcal{M}$, so that if the binary signature T is degenerate, then $T \in \langle\mathcal{M}\rangle$, a contradiction. Hence not only $T \notin \langle\mathcal{M}\rangle$, but also $T \notin \mathcal{M}$, that is $T(1,1) \neq 0$. Note that the matrix-vector product $Z_1^{\mathtt{T}}U = \begin{bmatrix} 1 & \mathfrak{i} \\ 1 & -\mathfrak{i} \end{bmatrix}\begin{bmatrix} 1 \\ \mathfrak{i} \end{bmatrix} \cong \begin{bmatrix} 0 \\ 1 \end{bmatrix}$. Then we calculate $U^{\mathtt{T}}S_{Z_1}U = U^{\mathtt{T}}Z_1TZ_1^{\mathtt{T}}U \cong \begin{bmatrix} 0 & 1 \end{bmatrix} T \begin{bmatrix} 0 \\ 1 \end{bmatrix} = T(1,1) \neq 0$. Hence, $S_{Z_1}U \ncong [1, \mathfrak{i}]$, and $S_{Z_1}U \ncong [0, 0]$. In this case, $F^{x_3 = S_{Z_1}U}$ is nondegenerate, and we will modify the construction in Figure 10.3 by adding the binary gadget with signature S_{Z_1} to replace the three edges whose endpoints are small triangles both marked by 3 in the gadget. (This modification does change two edges in the construction of S; but it does not affect what has been proved for S, since S_{Z_1} is nondegenerate. The same is true for case (b.) and (c.) below.) The proof for $S_{Z_2}U$ is similar.

In case (b.), $\mathcal{U}_3 = \{[1, \mathfrak{i}], [1, -\mathfrak{i}]\}$.

We show in this case, one of $P_{Z_1}[1, \mathfrak{i}]$ or $P_{Z_1}[1, -\mathfrak{i}] \notin \mathcal{U}_3$. Because $P_{Z_1} \notin \langle Z_1\mathcal{E}\rangle$, in matrix form $P_{Z_1} = Z_1TZ_1^{\mathtt{T}}$ for some $T \notin \langle\mathcal{E}\rangle$. We claim that at least one of the two columns $\begin{bmatrix} e \\ f \end{bmatrix}$ of T have both entries nonzero. Otherwise, either T is degenerate or $T \in \mathcal{E}$, in either case $T \in \langle\mathcal{E}\rangle$, a contradictoion. If the first (resp. second) column has this property, $P_{Z_1}[1, -\mathfrak{i}]$ (resp. $P_{Z_1}[1, \mathfrak{i}]$) does not belong to $\mathcal{U}_3$. Indeed, if it is the first case,

$$P_{Z_1}\begin{bmatrix} 1 \\ -\mathfrak{i} \end{bmatrix} = Z_1TZ_1^{\mathtt{T}}\begin{bmatrix} 1 \\ -\mathfrak{i} \end{bmatrix} \cong Z_1T\begin{bmatrix} 1 \\ 0 \end{bmatrix} = \begin{bmatrix} 1 & 1 \\ \mathfrak{i} & -\mathfrak{i} \end{bmatrix}\begin{bmatrix} e \\ f \end{bmatrix} \ncong [1, \pm\mathfrak{i}].$$

In this case, $F^{x_3 = P_{Z_1}[1,-\mathfrak{i}]}$ (resp. $F^{x_3 = P_{Z_1}[1,\mathfrak{i}]}$) is nondegenerate, and we will modify the construction in Figure 10.3 by adding the binary gadget with signature P_{Z_1} to replace the three edges whose endpoints are small triangles both marked by 3 in the gadget.

In case (c.), we have $|\mathcal{U}_3| = 2$ and $\mathcal{U}_3 = \{U, U^{\perp}\}$. Hence $U \ncong U^{\perp}$, and U and $U^{\perp}$ are linearly independent. Then $U \not\perp U$, otherwise $U \cong 0$. Hence the dot product $U^{\mathtt{T}}U \neq 0$, and we may assume $U = [a, b]$ and $U^{\perp} = [b, -a]$ are unit vectors: $a^2 + b^2 = 1$. Let $H = \begin{bmatrix} a & b \\ b & -a \end{bmatrix}$, then H is an orthogonal matrix. Then it follows that one of $P_H[a, b]$ or $P_H[b, -a]$ does not belong to $\mathcal{U}_3$. The proof is similar with case (b). In case (c) we will modify the construction in Figure 10.3 by adding the binary gadget with signature P_H to replace the three edges whose endpoints are small triangles both marked by 3 in the gadget.

The proof for Condition (2) is similar to Condition (1). The replacement in the construction of Figure 10.3 happens at the three edges connecting the copy of F with U_2 and the corner F (the three edges whose endpoints are small triangles both marked by 2 in the gadget).

Now consider Condition (3). Suppose $R_L^{\mathrm{T}} L_R = 0$. We separate out the case $R_L \in \{[1, \mathfrak{i}], [1, -\mathfrak{i}]\}$, or not. If $R_L \in \{[1, \mathfrak{i}], [1, -\mathfrak{i}]\}$, and $R_L^{\mathrm{T}} L_R = 0$, then $R_L = L_R = [1, \mathfrak{i}]$ or $[1, -\mathfrak{i}]$. In this case $R_L^{\mathrm{T}} S_{Z_1} L_R \neq 0$ or $R_L^{\mathrm{T}} S_{Z_2} L_R \neq 0$ by a simple calculation as before. If $R_L \notin \{[1, \mathfrak{i}], [1, -\mathfrak{i}]\}$, and $R_L^{\mathrm{T}} L_R = 0$, then $L_R \notin \{[1, \mathfrak{i}], [1, -\mathfrak{i}]\}$ as well. We can assume $R_L = [a, b]$ and $L_R = [b, -a]$ and $a^2 + b^2 = 1$. Then one of $R_L^{\mathrm{T}} P_H L_R$ and $R_L^{\mathrm{T}} P_H^{\mathrm{T}} L_R$ is not zero, also by a simple calculation as before, where $H = \begin{bmatrix} a & b \\ b & -a \end{bmatrix}$. For condition (3), the replacement in the construction of Figure 10.3 happens at the three edges connecting the copy of F with U_1 with the copy of F with U_2 (the three edges whose endpoints are small triangles marked by 2 and 3 respectively in the gadget).

If conditions (1), (2), and (3) all hold, then the gadget satisfies the second property, and the theorem is proved. For each condition, if it does not hold, we have modified the gadget construction by adding some binary functions to rectify the construction, and these binary functions are available by the conditions of the theorem. With these modifications to the construction in Figure 10.3, the proof of the theorem is complete. □

Remark: In the proof of Theorem 10.24 we used the separation lemma to satisfy various conditions in isolation. In particular, for the second property (for the arity 4 signature P in the lower part of the gadget), we argued that we *could* satisfy the property by choosing unary functions U_1 and U_2 to make $F^{x_1=U_1}$ and $F^{x_1=U_2}$ degenerate. When the final gadget is produced by simultaneously satisfying both the first condition (for S) and the second condition (for P), there is no expectation that this degeneracy will persist. In fact, the we proved that the first condition (for S) *could* be satisfied by choosing U_1 and U_2 to make $F^{x_1=U_1}$ and $F^{x_1=U_2}$ nondegenerate. There is some nonconstructiveness by using the separation lemma to argue for the success of such constructions. If necessary, one can make it constructive (in the sense of Turing computability), by looking more closely at the sets of polynomial equations (which could be large, but specifiable by at most polynomially many bits). The reduction proved to exist is a polynomial time reduction, but the proof does not give it explicitly.

We will prove the #P-hardness part of Theorem 10.14 by appealing to Theorem 2.12 for symmetric Holant* problems. For that purpose we need to construct appropriate *symmetric* binary signatures as well.

Theorem 10.25. *Let $\mathcal{F}$ denote any one of the function sets $\langle H\mathcal{E}\rangle$ (for an orthogonal matrix H), $\langle Z\mathcal{E}\rangle$ or $\langle Z\mathcal{M}\rangle$ (for the matrix $Z = Z_1$ or Z_2). Suppose we can realize a symmetric ternary function $F \in \mathcal{F} - \langle\mathcal{T}\rangle$ and a binary function $P \notin \mathcal{F}$. Then we can realize a* symmetric *binary function $Q \notin \mathcal{F}$.*

Proof. We define the *symmetric* binary function $Q = PF^{x_1=U}P^{\mathrm{T}}$ in matrix form, for some unary function U. This Q is realizable by a gadget consisting of a path of three signatures P, $F^{x_1=U}$ and P^{T}. At both ends we have a copy of P and we connect the second variable of each copy of P to the two remaining variables of the symmetric $F^{x_1=U}$.

The essence of the proof is an appropriate holographic transformation. We are given $F \in \langle H\mathcal{E}\rangle$ (for an orthogonal matrix H), or $\langle Z\mathcal{E}\rangle$ or $\langle Z\mathcal{M}\rangle$. Let $M = H$ or Z depending on the cases. Let $F = M^{\otimes 3}F_1$ and $P = MP_1M^{\mathrm{T}}$ in matrix form. $F \notin \langle\mathcal{T}\rangle$ implies that $F_1 \notin \langle\mathcal{T}\rangle$. We also have $F^{x_1=U} = MF_1^{x_1=MU}M^{\mathrm{T}}$ in matrix form for the binary $F^{x_1=U}$. Let $Q_1 = P_1M^{\mathrm{T}}MF_1^{x_1=MU}M^{\mathrm{T}}MP_1^{\mathrm{T}}$; then

$$Q = MQ_1M^{\mathrm{T}} = MP_1M^{\mathrm{T}}MF_1^{x_1=MU}M^{\mathrm{T}}MP_1^{\mathrm{T}}M^{\mathrm{T}}.$$

Case (1) $\mathcal{F}$ is $\langle H\mathcal{E}\rangle$.

We take $M = H$. Since H is orthogonal, $Q = HP_1F_1^{x_1=HU}P_1^{\mathrm{T}}H^{\mathrm{T}}$. We have $F_1 \notin \langle\mathcal{T}\rangle$. We also have $P_1 \notin \langle\mathcal{E}\rangle$, since $P \notin \langle H\mathcal{E}\rangle$. But by $F \in \langle H\mathcal{E}\rangle$, we have $F_1 \in \langle\mathcal{E}\rangle$. By the condition $F_1 \notin \langle\mathcal{T}\rangle$, we must actually have $F_1 \in \mathcal{E}$, because F_1 only has arity 3 and therefore any tensor factorization will put F_1 in $\langle\mathcal{T}\rangle$. Being symmetric and nondegenerate, $F_1 = [u, 0, 0, v]$, where $u \neq 0$ and $v \neq 0$. We only need to prove $Q_1 = P_1F_1^{x_1=HU}P_1^{\mathrm{T}} \notin \langle\mathcal{E}\rangle$, which is the same as $Q \notin \langle H\mathcal{E}\rangle$. Because we can pick any $U' = HU$, for any x, y, we can realize $F_1^{x_1=HU} = [x, 0, y]$. Suppose $P_1 = \begin{bmatrix} a & b \\ c & d \end{bmatrix}$. Then $Q_1 = \begin{bmatrix} a^2x + b^2y & acx + bdy \\ acx + bdy & c^2x + d^2y \end{bmatrix}$.

We need $Q_1 \notin \langle\mathcal{E}\rangle$. This is translated into three conditions: (1) Q_1 is nondegenerate, (2) Q_1 is not of the form $[*, 0, *]$, and (3) Q_1 is not of the form $[0, *, 0]$. By the separation lemma, we only need to prove that there is some $[x, y]$ to satisfy each condition individually. If $x \neq 0$ and $y \neq 0$, then $F_1^{x_1=HU} = [x, 0, y]$ in nondegenerate. Also P_1 is nondegenerate because $P_1 \notin \langle\mathcal{E}\rangle$. Thus Q_1 is nondegenerate. Again because $P_1 \notin \langle\mathcal{E}\rangle$, either $ac \neq 0$ or $bd \neq 0$. There exists some $[x, y]$ such that $acx + bdy \neq 0$; thus Q_1 is not of the form $[*, 0, *]$. Similarly, it is easy find $[x, y]$ such that Q_1 is not of the form $[0, *, 0]$.

Case (2) $\mathcal{F}$ is $\langle Z\mathcal{E}\rangle$.

Take $M = Z$. Note that $Z^{\mathrm{T}}Z \cong (\neq_2)$. We have $Q \cong ZP_1(\neq_2)F_1^{x_1=ZU}(\neq_2)P_1^{\mathrm{T}}Z^{\mathrm{T}} = ZQ_1Z^{\mathrm{T}}$. We have $F_1 \notin \langle\mathcal{T}\rangle$, $F_1 \in \mathcal{E}$, and $P_1 \notin \langle\mathcal{E}\rangle$. We only need to prove $Q_1 \notin \langle \mathcal{E} \rangle$. For any x, y, we can pick U to realize $(\neq_2)F_1^{x_1=ZU}(\neq_2) = [x, 0, y]$. This is seen by the fact that $(\neq_2)[x, 0, y](\neq_2) = [y, 0, x]$. The remaining proof is the same as case (1) for $\langle H\mathcal{E} \rangle$.

Case (3) $\mathcal{F}$ is $\langle Z\mathcal{M} \rangle$.

Take $M = Z$. Since $Z^{\mathrm{T}}Z \cong (\neq_2)$, we have $Q_1 \cong P_1(\neq_2)F_1^{x_1=ZU}(\neq_2)P_1^{\mathrm{T}}$.

We have $F_1 \notin \langle \mathcal{T} \rangle$, and $F_1 \in \langle \mathcal{M} \rangle$. Again because F_1 has arity 3, from these two conditions we conclude $F_1 \in \mathcal{M}$, since any tensor factorization for F_1 would place it in $\langle \mathcal{T} \rangle$. Being symmetric and nondegenerate, F_1 has the form $F_1 \cong [f, 1, 0, 0]$. Let $P_1 = \begin{bmatrix} a & b \\ c & d \end{bmatrix}$, and we have $P_1 \notin \langle \mathcal{M} \rangle$, since $P \notin \langle Z\mathcal{M} \rangle$. For any x, there is some U, such that $F_1^{x_1=ZU} = [x, 1, 0]$. We only need to prove that $Q_1 = P_1(\neq_2)F_1^{x_1=ZU}(\neq_2)P_1^{\mathrm{T}} \notin \langle \mathcal{M} \rangle$.

Because $P_1 \notin \langle \mathcal{M} \rangle$, certainly P_1 is nondegenerate. Therefore $\det Q_1 \neq 0$, and Q_1 is nondegenerate. Also, the entry $P_1(1, 1) = d \neq 0$, because $P_1 \notin \mathcal{M}$. It follows that there exists x such that $Q_1(1, 1) = 2cd + d^2x \neq 0$. Hence, $Q_1 \notin \langle \mathcal{M} \rangle$. □

Now we are ready to finish the proof Theorem 10.14.

Proof of Theorem 10.14 (#P-hardness part)

Suppose $\mathcal{F} \nsubseteq \langle \mathcal{T} \rangle$, $\mathcal{F} \nsubseteq \langle H\mathcal{E} \rangle$ for all orthogonal H, $\mathcal{F} \nsubseteq \langle Z\mathcal{E} \rangle$ and $\mathcal{F} \nsubseteq \langle Z\mathcal{M} \rangle$ for $Z = Z_1$ and Z_2. By Lemma 10.19, we can realize functions of arity of 2 or 3 *not* belonging to these function sets respectively.

The conditions in Theorem 10.24 are satisfied, so we can realize a symmetric ternary function $Q_3 \notin \langle \mathcal{T} \rangle$ (with the help of those binary functions). Certainly Q_3 is nondegenerate. If Holant*($\{Q_3\}$) is #P-hard, then the theorem is proved. Otherwise, by Theorem 2.12 for symmetric Holant* problems, Q_3 belongs to one of the special function families listed in the theorem. It can be shown that these are precisely restrictions of $\langle H\mathcal{E} \rangle$, $\langle Z\mathcal{E} \rangle$ or $\langle Z\mathcal{M} \rangle$ to symmetric signatures. By Theorem 10.25, we can realize a symmetric binary function Q_2 not in this family. By Theorem 2.12 again, Holant*($\{Q_3, Q_2\}$) is #P-hard, and therefore Holant*($\mathcal{F}$) is also #P-hard. □

Bibliography

[AB09] Sanjeev Arora and Boaz Barak. *Computational complexity: A modern approach.* Cambridge University Press, 2009.

[ALM+98] Sanjeev Arora, Carsten Lund, Rajeev Motwani, Madhu Sudan, and Mario Szegedy. Proof verification and the hardness of approximation problems. *Journal of the ACM* **45** (1998), no. 3, 501–555.

[Bar83] Francisco Barahona. *Balancing signed toroidal graphs in polynomial time.* Preprint, University of Chile, 1983.

[Bar04] Régis Barbanchon. On unique graph 3-colorability and parsimonious reductions in the plane. *Theoretical Computer Science* **319** (2004), no. 1–3, 455–482.

[Bax82] Rodney J. Baxter. *Exactly solved models in statistical mechanics.* Academic press London, 1982.

[BCSS98] Lenore Blum, Felipe Cucker, Michael Shub, and Steve Smale. *Complexity and real computation*, Springer, 1998.

[BD07] Andrei A. Bulatov and Victor Dalmau. Towards a dichotomy theorem for the counting constraint satisfaction problem. *Information and Computation* **205** (2007), no. 5, 651–678.

[BG05] Andrei A. Bulatov and Martin Grohe. The complexity of partition functions. *Theoretical Computer Science* **348** (2005), no. 2, 148–186.

[BOKR86] Michael Ben-Or, Dexter Kozen, and John Reif. The complexity of elementary algebra and geometry. *Journal of Computer and Systems Sciences* **32** (1986), no. 2, 251–264.

[Bol98] Béla Bollobás. *Modern graph theory*. Springer-Verlag, 1998.

[BPR06] Saugata Basu, Richard Pollack, and Marie-Françoise Roy. *Algorithms in real algebraic geometry*, Springer-Verlag, 2006.

[BS94] Robert Burton and Jeffrey E. Steif. Non-uniqueness of measures of maximal entropy for subshifts of finite type. *Ergodic Theory and Dynamical Systems* **14** (1994), no. 2, 213–235.

[BS95] ______. New results on measures of maximal entropy. *Israel Journal of Mathematics* **89** (1995), 275–300.

[BSS89] Lenore Blum, Mike Shub, and Steve Smale. On a theory of computation and complexity over the real numbers: NP-completeness, recursive functions and universal machines. *Bulletin of the American Mathematical Society* **21** (1989), 1–46.

[Bul13] Andrei A. Bulatov. The complexity of the counting constraint satisfaction problem. *Journal of the ACM* **60** (2013), no. 5, 34:1–34:41.

[BW05] Graham R. Brightwell and Peter Winkler. Counting Eulerian circuits is #P-complete, in *Proceedings of the 2nd Workshop on Analytic Algorithmics and Combinatorics*, 2005.

[CC06] Jin-Yi Cai and Vinay Choudhary. Some results on matchgates and holographic algorithms, in *Proceedings of the 33rd International Colloquium on Automata, Languages, and Programming*, 2006, pp. 703–714.

[CC07] ______. Valiant's Holant theorem and matchgate tensors. *Theoretical Computer Science* **384** (2007), no. 1, 22–32.

[CC12] Jin-Yi Cai and Xi Chen. Complexity of counting CSP with complex weights, in *Proceedings of the 44th Annual ACM Symposium on Theory of Computing*, 2012, pp. 909–920.

[CCL09] Jin-Yi Cai, Vinay Choudhary, and Pinyan Lu. On the theory of matchgate computations. *Theory of Computing Systems* **45** (2009), no. 1, 108–132.

[CCL13] Jin-Yi Cai, Xi Chen, and Pinyan Lu. Graph homomorphisms with complex values: A dichotomy theorem. *SIAM Journal on Computing* **42** (2013), no. 3, 924–1029.

[CCL16] ______. Nonnegative weighted #CSP: An effective complexity dichotomy. *SIAM Journal on Computing* **45** (2016), no. 6, 2177–2198.

[CF16] Jin-Yi Cai and Zhiguo Fu. Holographic algorithm with matchgates is universal for planar #CSP over boolean domain, CoRR **abs/1603.07046** (2016).

[CFGW15] Jin-Yi Cai, Zhiguo Fu, Heng Guo, and Tyson Williams. A Holant dichotomy: Is the FKT algorithm universal?, in *Proceedings of the IEEE 56th Annual Symposium on Foundations of Computer Science*, 2015, pp. 1259–1276.

[CG14] Jin-Yi Cai and Aaron Gorenstein. Matchgates revisited. *Theory of Computing* **10** (2014), no. 7, 167–197.

[CGG+16] Jin-Yi Cai, Andreas Galanis, Leslie Ann Goldberg, Heng Guo, Mark Jerrum, Daniel Štefankovič, and Eric Vigoda. #BIS-hardness for 2-spin systems on bipartite bounded degree graphs in the tree non-uniqueness region. *Journal of Computer and System Sciences* **82** (2016), no. 5, 690–711.

[CGW14] Jin-Yi Cai, Heng Guo, and Tyson Williams. Holographic algorithms beyond matchgates. *Proceedings of the 41st International Colloquium on Automata, Languages, and Programming*, 2014, pp. 271–282.

[CGW16] Jin-Yi Cai, Heng Guo, and Tyson Williams. A complete dichotomy rises from the capture of vanishing signatures. *SIAM Journal on Computing* **45** (2016), no. 5, 1671–1728.

[CK12] Jin-Yi Cai and Michael Kowalczyk. Spin systems on k-regular graphs with complex edge functions. *Theoretical Computer Science* **461** (2012), 2–16.

[CK13] ______. Partition functions on k-regular graphs with 0, 1-vertex assignments and real edge functions. *Theoretical Computer Science* **494** (2013), 63–74.

[CKS01] Nadia Creignou, Sanjeev Khanna, and Madhu Sudan. Complexity classifications of boolean constraint satisfaction problems. *Society for Industrial and Applied Mathematics*, 2001.

[CLX10] Jin-Yi Cai, Pinyan Lu, and Mingji Xia. Holographic algorithms with matchgates capture precisely tractable planar #CSP. *Proceedings of the IEEE 51st Annual Symposium on Foundations of Computer Science*, 2010, pp. 427–436.

[CLX11] Jin-Yi Cai, Pinyan Lu, and Mingji Xia. A computational proof of complexity of some restricted counting problems. *Theoretical Computer Science* **412** (2011), no. 23, 2468–2485.

[CNR89] Hyeong-Ah Choi, Kazuo Nakajima, and Chong S. Rim. Graph bipartization and via minimization. *SIAM Journal on Discrete Mathematics* **2** (1989), no. 1, 38–47.

[DFJ02] Martin Dyer, Alan Frieze, and Mark Jerrum. On counting independent sets in sparse graphs. *SIAM Journal on Computing* **31** (2002), no. 5, 1527–1541.

[DG00] Martin E. Dyer and Catherine Greenhill. The complexity of counting graph homomorphisms. *Random Structures and Algorithms* **17** (2000), 260–289.

[DGJ09] Martin Dyer, Leslie Ann Goldberg, and Mark Jerrum. The complexity of weighted boolean CSP. *SIAM Journal on Computing* **38** (2009), no. 5, 1970–1986.

[DGJ10] ______. An approximation trichotomy for boolean #CSP. *Journal of Computer and System Sciences* **76** (2010), no. 3, 267–277.

[DGJR12] Martin Dyer, Leslie Ann Goldberg, Markus Jalsenius, and David Richerby. The complexity of approximating bounded-degree boolean #CSP. *Information and Computation* **220–221** (2012), 1–14.

[DL92] Paul Dagum and Michael Luby. Approximating the permanent of graphs with large factors. *Theoretical Computer Science* **102** (1992), no. 2, 283–305.

[DP91] Christopher T.J. Dodson and Timothy Poston. Tensor geometry: The geometric viewpoint and its uses, Graduate Texts in Mathematics 130, Springer-Verlag, 1991.

[DR13] Martin E. Dyer and David Richerby. An effective dichotomy for the counting constraint satisfaction problem. *SIAM Journal on Computing* **42** (2013), no. 3, 1245–1274.

[Eul36] Leonhard Euler. Solutio problematis ad geometriam situs pertinentis. *Commentarii Academiae Scientiarum Petropolitanae* **8** (1736), 128–140.

[GG16] Andreas Galanis and Leslie Ann Goldberg. The complexity of approximately counting in 2-spin systems on k-uniform bounded-degree hypergraphs. *Information and Computation* **251** (2016), 36–66.

[GGJ16] Andreas Galanis, Leslie Ann Goldberg, and Mark Jerrum. A complexity trichotomy for approximately counting list H-colourings, in *Proceedings of the 43rd International Colloquium on Automata, Languages, and Programming*, 2016, pp. 46:1–46:13.

[GGJT10] Leslie Ann Goldberg, Martin Grohe, Mark Jerrum, and Marc Thurley. A complexity dichotomy for partition functions with mixed signs. *SIAM Journal on Computing* **39** (2010), no. 7, 3336–3402.

[GGv$^+$11] Andreas Galanis, Qi Ge, Daniel Štefankovič, Eric Vigoda, and Linji Yang. Improved inapproximability results for counting independent sets in the hard-core model, in *Proceedings of the 15th International Workshop on Randomization and Computation*, 2011.

[GGY16] Andreas Galanis, Leslie Ann Goldberg, and Kuan Yang. Approximating partition functions of bounded-degree boolean counting constraint satisfaction problems. CoRR **abs/1610.04055** (2016).

[GJP03] Leslie Ann Goldberg, Mark Jerrum, and Mike Paterson. The computational complexity of two-state spin systems. *Random Structures and Algorithms* **23** (2003), no. 2, 133–154.

[GL16] Heng Guo and Pinyan Lu. Uniqueness, spatial mixing, and approximation for ferromagnetic 2-spin systems, in *Proceedings of the 20th International Workshop on Randomization and Computation*, 2016, pp. 31:1–31:26.

[GM94] G. Galbiati and F. Maffioli. On the computation of Pfaffians. *Discrete Applied Mathematics* **51** (1994), no. 3, 269–275.

[Gov15] Artem Govorov, private communication, 2015.

[GŠ12] Qi Ge and Daniel Štefankovič. The complexity of counting Eulerian tours in 4-regular graphs. *Algorithmica* **63** (2012), no. 3, 588–601.

[GT11] Martin Grohe and Marc Thurley. Counting homomorphisms and partition functions. *Model Theoretic Methods in Finite Combinatorics*. Contemporary Mathematics 558 (Martin Grohe and Johann Makowsky, eds.), American Mathematical Society, 2011.

[GW95] Michel X. Goemans and David P. Williamson. Improved approximation algorithms for maximum cut and satisfiability problems using semidefinite programming. *Journal of the ACM* **42** (1995), no. 6, 1115–1145.

[GW13] Heng Guo and Tyson Williams. The complexity of planar boolean #CSP with complex weights, in *Proceedings of the 40th International Colloquium on Automata, Languages, and Programming*, 2013, pp. 516–527.

[Had75] F. Hadlock. Finding a maximum cut of a planar graph in polynomial time, *SIAM Journal on Computing* **4** (1975), no. 3, 221–225.

[Hal66] John H. Halton. A combinatorial proof of Cayley's theorem on Pfaffians. *Journal of Combinatorial Theory* **1** (1966), no. 2, 224–232.

[Hås01] Johan Håstad. Some optimal inapproximability results. *Journal of the ACM* **48** (2001), no. 4, 798–859.

[Hie73] Carl Hierholzer. Ueber die möglichkeit, einen linienzug ohne wiederholung und ohne unterbrechung zu umfahren. *Mathematische Annalen* **6** (1873), no. 1, 30–32.

[HL16] Sangxia Huang and Pinyan Lu. A dichotomy for real weighted holant problems. *Computational Complexity* **25** (2016), no. 1, 255–304.

[HT74] John Hopcroft and Robert E. Tarjan. Efficient planarity testing. *Journal of the Association for Computing Machinery* **21** (1974), no. 4, 549–568.

[IMRS98] Harry B. Hunt III, Madhav V. Marathe, Venkatesh Radhakrishnan, and Richard E. Stearns. The complexity of planar counting problems. *SIAM Journal on Computing* **27** (1998), no. 4, 1142–1167.

[Isi25] Ernst Ising. *Beitrag zur theorie des* ferromagnetismus. *Zeitschrift für Physik* **31** (1925), no. 1, 253–258.

[Jac85a] Nathan Jacobson. *Basic algebra*, volume 1, W.H. Freeman, 1985.

[Jac85b] ______. *Basic algebra*, volume 2, W.H. Freeman, 1985.

[Jer87] Mark Jerrum. Two-dimensional monomer-dimer systems are computationally intractable. *Journal of Statistical Physics* **48** (1987), no. 1, 121–134.

[Jer15] Mark Jerrum, private communication, 2015.

[JS93] Mark Jerrum and Alistair Sinclair. Polynomial-time approximation algorithms for the Ising model. *SIAM Journal on Computing* **22** (1993), no. 5, 1087–1116.

[JVW90] F. Jaeger, D.L. Vertigan, and D.J.A. Welsh. On the computational complexity of the Jones and Tutte polynomials. *Mathematical Proceedings of the Cambridge Philosophical Society* **108** (1990), no. 1, 35–53.

[KC10] Michael Kowalczyk and Jin-Yi Cai. Holant problems for regular graphs with complex edge functions, in *Proceedings of the 27th International Symposium on Theoretical Aspects of Computer Science*, 2010, pp. 525–536.

[KD79] M.S. Kirshnamoorthy and Narsingh Deo. Node-deletion NP-complete problems. *SIAM Journal on Computing* **8** (1979), no. 4, 619–625.

[Kir47] G. Kirchhoff. Ueber die auflösung der gleichungen, auf welche man bei der Untersuchung der linearen Verteilung galvanischer ströme gefuhrt wird. *Annalen der Physik und Chemie* **12** (1847), 497–508.

[Kow10] Michael Kowalczyk. *Dichotomy theorems for Holant problems*, Ph.D. thesis, University of Wisconsin–Madison, 2010.

[Lic82] David Lichtenstein. Planar formulae and their uses. *SIAM Journal on Computing* **11** (1982), no. 2, 329–343.

[Lie67a] Elliott H. Lieb. Exact solution of the F model of an antiferroelectric. *Physical Review Letters* **18** (1967), no. 24, 1046–1048.

[Lie67b] ______. Exact solution of the problem of the entropy of two-dimensional ice. *Physical Review Letters* **18** (1967), no. 17, 692–694.

[Lie67c] ______. Exact solution of the two-dimensional Slater KDP model of a ferroelectric. *Physical Review Letters* **19** (1967), no. 3, 108–110.

[Lie67d] ______. Residual entropy of square ice. *Physical Review* **162** (1967), no. 1, 162–172.

[Lin86] Nathan Linial. Hard enumeration problems in geometry and combinatorics. *SIAM Journal on Algebraic and Discrete Methods* **7** (1986), no. 2, 331–335.

[LL15] Jingcheng Liu and Pinyan Lu. FPTAS for counting monotone CNF, in *Proceedings of the 26th Annual ACM-SIAM Symposium on Discrete Algorithms*, 2015, pp. 1531–1548.

[LLY12] Liang Li, Pinyan Lu, and Yitong Yin. Approximate counting via correlation decay in spin systems, in *Proceedings of the 23rd ACM-SIAM Symposium on Discrete Algorithms*, 2012.

[LLZ14] Jingcheng Liu, Pinyan Lu, and Chihao Zhang. The complexity of ferromagnetic two-spin systems with external fields, in *Proceedings of the 18th International Workshop on Randomization and Computation*, 2014, pp. 843–856.

[LOT03] Maciej Liśkiewicz, Mitsunori Ogihara, and Seinosuke Toda. The complexity of counting self-avoiding walks in subgraphs of two-dimensional grids and hypercubes. *Theoretical Computer Science* **304** (2003), no. 1, 129–156.

[LW16] Jiabao Lin and Hanpin Wang. The complexity of Holant problems over boolean domain with non-negative weights. CoRR **abs/1611.00975** (2016).

[LY80] John M. Lewis and Mihalis Yannakakis. The node-deletion problem for hereditary properties is NP-complete. *Journal of Computer and System Sciences* **20** (1980), no. 2, 219–230.

[MSV04] Meena Mahajan, P.R. Subramanya, and V. Vinay. The combinatorial approach yields an NC algorithm for computing Pfaffians. *Discrete Applied Mathematics* **143** (2004), no. 1–3, 1–16.

[Mur00] Kazuo Murota. *Matrices and matroids for systems analysis.* Springer, Berlin, 2000.

[MW96] Milena Mihail and Peter Winkler. On the number of eulerian orientations of a graph. *Algorithmica* **16** (1996), no. 4–5, 402–414.

[MWW09] Elchanan Mossel, Dror Weitz, and Nicholas Wormald. On the hardness of sampling independent sets beyond the tree threshold. *Probability Theory and Related Fields* **143** (2009), no. 3, 401–439.

[Nes96a] Yu V. Nesterenko. Modular functions and transcendence problems. *Comptes Rendus de l'Académie des Sciences – Series I – Mathematics* **322** (1996), 909–914.

[Nes96b] ______. Modular functions and transcendence questions. *Matematicheskii Sbornik* **187** (1996), no. 9, 65–96.

[Oht92] Yasuhiro Ohta. *Bilinear theory of soliton*, Ph.D. thesis, University of Tokyo, 1992.

[Pap94] Christos H. Papadimitriou. *Computational complexity*, Addison-Wesley, 1994.

[Pau35] Linus Pauling. The structure and entropy of ice and of other crystals with some randomness of atomic arrangement. *Journal of the American Chemical Society* **57** (1935), no. 12, 2680–2684.

[PB83] J. Scott Provan and Michael O. Ball. The complexity of counting cuts and of computing the probability that a graph is connected. *SIAM Journal on Computing* **12** (1983), no. 4, 777–788.

[Pot52] Renfrey B. Potts. Some generalized order-disorder transformations. *Mathematical Proceedings of the Cambridge Philosophical Society* **48** (1952), no. 1, 106–109.

[Rom12] Antonio Romano. *Classical mechanics with mathematica.* Birkhäuser, 2012.

[Sip96] Michael Sipser. *Introduction to the theory of computation.* PWS Publishing Company, 1996.

[Sly10] Allan Sly. Computational transition at the uniqueness threshold, in *Proceedings of the 51st IEEE Annual Symposium on Foundations of Computer Science*, 2010, pp. 287–296.

[SS12] Allan. Sly and Nike Sun. The computational hardness of counting in two-spin models on d-regular graphs, in *Proceedings of the IEEE 53rd Annual Symposium on Foundations of Computer Science*, 2012, pp. 361–369.

[SST12] Alistair Sinclair, Piyush Srivastava, and Marc Thurley. Approximation algorithms for two-state anti-ferromagnetic spin systems on bounded degree graphs, in *Proceedings of the 23rd ACM-SIAM Symposium on Discrete Algorithms*, 2012.

[Tar51] Alfred Tarski. *A decision method for elementary algebra and geometry*. Berkeley and Los Angeles, University of California Press, 1951.

[Thu09] Marc Thurley. *The complexity of partition functions*, Ph.D. thesis, Humboldt-Universität zu Berlin, 2009.

[Tod91] S. Toda. PP is as hard as the polynomial hierarchy. *SIAM Journal on Computing* **20** (1991), no. 5, 865–877.

[TS41] W.T. Tutte and C.A.B. Smith. On unicursal paths in a network of degree 4. *American Mathematical Monthly* **48** (1941), 233–237.

[Tut48] W.T. Tutte. The disection of equilateral triangles into equilateral triangles. *Mathematical Proceedings of the Cambridge Philosophical Society* **44** (1948), 463–482.

[Vad01] Salil P. Vadhan. The complexity of counting in sparse, regular and planar graphs. *SIAM Journal on Computing* **8** (2001), no. 1, 398–427.

[vAEdB51] T. van Aardenne-Ehrenfest and N.G. de Bruijn. Circuits and trees in oriented linear graphs. *Simon Stevin* **28** (1951), 203–217.

[Val79a] Leslie G. Valiant. The complexity of computing the permanent. *Theoretical Computer Science* **8** (1979), no. 2, 189–201.

[Val79b] ______. The complexity of enumeration and reliability problems. *SIAM Journal on Computing* **8** (1979), no. 3, 410–421.

[Val02a] ______. Expressiveness of matchgates. *Theoretical Computer Science* **289** (2002), no. 1, 457–471.

[Val02b] ______. Quantum circuits that can be simulated classically in polynomial time. *SIAM Journal on Computing* **31** (2002), no. 4, 1229–1254.

[Val05] Leslie G. Valiant. Completeness for parity problems, in *Proceedings of the 11th International Computing and Combinatorics Conference*, 2005, pp. 1–8.

[Val06] Leslie G. Valiant. Accidental algorithms, in *Proceedings of the IEEE 54th Annual Symposium on Foundations of Computer Science*, 2006, pp. 509–517.

[Val08] ______. Holographic algorithms. *SIAM Journal on Computing* **37** (2008), no. 5, 1565–1594.

[Val10] Leslie G. Valiant. Some observations on holographic algorithms, in *Proceedings of the 9th Latin American Theoretical Informatics Symposium*, 2010, pp. 577–590.

[Ver88] Michel Las Vergnas. On the evaluation at $(3,3)$ of the Tutte polynomial of a graph. *Journal of Combinatorial Theory, Series B* **45** (1988), no. 3, 367–372.

[Ver05] Dirk Vertigan. The computational complexity of Tutte invariants for planar graphs. *SIAM Journal on Computing* **35** (2005), no. 3, 690–712.

[VV86] Leslie G. Valiant and Vijay V. Vazirani. NP is as easy as detecting unique solutions. *Theoretical Computer Science* **47** (1986), 85–93.

[Wei06] Dror Weitz. Counting independent sets up to the tree threshold, in *Proceedings of the 38th Annual ACM Symposium on Theory of Computing*, 2006, pp. 140–149.

[Wel93] Dominic Welsh. *Complexity: Knots, colourings and countings*. London Mathematical Society Lecture Note Series. Cambridge University Press, 1993.

[WR70] B. Widom and J.S. Rowlinson. New model for the study of liquid-vapor phase transitions. *Journal of Chemical Physics* **52** (1970), no. 4.

[XZZ07] Mingji Xia, Peng Zhang, and Wenbo Zhao. Computational complexity of counting problems on 3-regular planar graphs. *Theoretical Computer Science* **384** (2007), no. 1, 111–125.

[Yan81a] Mihalis Yannakakis. Edge-deletion problems. *SIAM Journal on Computing* **10** (1981), no. 2, 297–309.

[Yan81b] ———. Node-deletion problems on bipartite graphs. *SIAM Journal on Computing* **10** (1981), no. 2, 310–327.

[Zyk74] A.A. Zykov. Hypergraphs. *Uspekhi Matematicheskikh Nauk* **6** (1974), no. 180, 89–154.

Index